T0183095

Lecture Notes in Computer Science 9243

More information about this series at http://www.springer.com/series/7412

Xiaofei He · Xinbo Gao
Yanning Zhang · Zhi-Hua Zhou
Zhi-Yong Liu · Baochuan Fu
Fuyuan Hu · Zhancheng Zhang (Eds.)

Intelligence Science and Big Data Engineering

Big Data and Machine Learning Techniques

5th International Conference, IScIDE 2015
Suzhou, China, June 14–16, 2015
Revised Selected Papers, Part II

Editors
Xiaofei He
Zhejiang University
Hangzhou
China

Xinbo Gao
Xidian University
Xi'an
China

Yanning Zhang
Northwestern Polytechnical University
Xi'an
China

Zhi-Hua Zhou
Nanjing University
Nanjing
China

Zhi-Yong Liu
Chinese Academy of Sciences
Beijing
China

Baochuan Fu
Suzhou University of Science and Technology
Suzhou
China

Fuyuan Hu
Suzhou University of Science and Technology
Suzhou
China

Zhancheng Zhang
Suzhou University of Science and Technology
Suzhou
China

ISSN 0302-9743 ISSN 1611-3349 (electronic)
Lecture Notes in Computer Science
ISBN 978-3-319-23861-6 ISBN 978-3-319-23862-3 (eBook)
DOI 10.1007/978-3-319-23862-3

Library of Congress Control Number: 2015948168

LNCS Sublibrary: SL6 – Image Processing, Computer Vision, Pattern Recognition, and Graphics

Springer Cham Heidelberg New York Dordrecht London

Printed on acid-free paper

Springer International Publishing AG Switzerland is part of Springer Science+Business Media
(www.springer.com)

Preface

ISciDE 2015, the International Conference on Intelligence Science and Big Data Engineering, took place in Suzhou, China, June 14–16, 2015. As one of the annual events organized by the Chinese Golden Triangle ISIS (Information Science and Intelligence Science) Forum, this meeting was scheduled as the fifth in a series of annual meetings promoting the academic exchange of research on various areas of intelligence science and big data engineering in China and abroad. In response to the call for papers, a total of 416 papers were submitted from 14 countries and regions. Among them, 18 papers were selected for oral presentation, 32 for spotlight presentation, and 76 for poster presentation, yielding an acceptance rate of 30.3 % and an oral presentation rate of about 4.3 %. We would like to thank all the reviewers for spending their precious time on reviewing papers and for providing valuable comments that helped significantly in the paper selection process.

We would like to express special thanks to the Conference General Co-chairs, Yanning Zhang, Zhi-Hua Zhou, and Zhigang Chen, for their leadership, advice, and help on crucial matters concerning the conference. We would like to thank all Steering Committee members, Program Committee members, Invited Speakers' Committee members, Organizing Committee members, and Publication Committee members for their hard work. We would like to thank Prof. Lionel M. Ni, Prof. Deyi Li, and Prof. Lei Xu for delivering the keynote speeches, and Tony Jebara, Houqiang Li, Keqiu Li, Shutao Li, Bin Cui, and Xiaowu Chen for delivering the invited talks and sharing their insightful views on ISIS research issues. Finally, we would like to thank all the authors of the submitted papers, whether accepted or not, for their contribution to the high quality of this conference. We count on your continued support of the ISIS community in the future.

June 2015

Xiaofei He
Xinbo Gao

Organization

General Chairs

Yanning Zhang	Northwestern Polytechnical University, Xi'an, China
Zhi-Hua Zhou	Nanjing University, Nanjing, China
Zhigang Chen	Suzhou University of Science and Technology, Suzhou, China

Technical Program Committee Chairs

Xiaofei He	Zhejiang University, Hangzhou, China
Xinbo Gao	Xidian University, Xi'an, China

Local Arrangements Chair

Baochuan Fu	Suzhou University of Science and Technology, Suzhou, China

Local Arrangements Members

Fuyuan Hu	Suzhou University of Science and Technology, Suzhou, China
Xue-Feng Xi	Suzhou University of Science and Technology, Suzhou, China
Ze Li	Suzhou University of Science and Technology, Suzhou, China
Hongjie Wu	Suzhou University of Science and Technology, Suzhou, China

Publicity Chair

Zhi-Yong Liu	Chinese Academy of Sciences, Beijing, China

Program Committee Members

Deng Cai	Zhejiang University, Hangzhou, China
Fang Fang	Peking University, Beijing, China
Jufu Feng	Peking University, Beijing, China
Xinbo Gao	Xidian University, Xi'an, China
Xin Geng	Southeast University, Nanjing, China
Ziyu Guan	Northwest University, Xi'an, China
Xiaofei He	Zhejiang University, Hangzhou, China
Kalviainen Heikki	Lappeenranta University of Technology, Finland
Akira Hirose	The University of Tokyo, Japan
Dewen Hu	National University of Defense Technology, Changsha, China
Hiroyuki Iida	Japan Advanced Institute of Science and Technology, Japan
Zhong Jin	Nanjing University of Science and Technology, Nanjing

Ikeda Kazushi	Nara Advanced Institute of Science and Technology, Japan
Andrey S. Krylov	Lomonosov Moscow State University, Russia
James Kwok	Hong Kong University of Science and Technology, Hong Kong, China
Jian-huang Lai	Sun Yat-sen University, Zhongshan, China
Shutao Li	Hunan University, Changsha, China
Xuelong Li	Xi'an Optics and Fine Mechanics, Chinese Academy of Sciences, Xi'an, China
Yi Li	Australian National University, Australia
Binbin Lin	University of Michigan, USA
Zhouchen Lin	Peking University, Beijing, China
Cheng Yuan Liou	National Taiwan University, Taiwan, China
Qingshan Liu	Nanjing University of Information Science and Technology, Nanjing, China
Yiguang Liu	Sichuang University, Chengdu, China
Bao-Liang Lu	Shanghai Jiao Tong University, Shanghai
Seiichi Ozawa	Kobe University, Japan
Yuhua Qian	Shanxi University, Taiyuan
Karl Ricanek	University of North Carolina Wilmington, USA
Shiguang Shan	Institute of Comp. Tec. Chinese Academy of Sciences, Beijing, China
Chunhua Shen	University of Adelaide, Australia
Changyin Sun	Southeast University, Nanjing, China
Dacheng Tao	University of Technology, Sydney, Australia
Vincent S. Tseng	National Cheng Kung University, Taiwan, China
Liang Wang	Institute of Automation, Chinese Academy of Sciences, Beijing, China
Liwei Wang	Peking University, Beijing, China
Yishi Wang	University of North Carolina, Wilmington, USA
Jian Yang	Nanjing University of Science and Technology, Nanjing, China
Changshui Zhang	Tsinghua University, Beijing, China
Daoqiang Zhang	Nanjing University of Aeronautics and Astronautics, Nanjing, China
Lei Zhang	Hong Kong Polytechnic University, Hong Kong, China
Lijun Zhang	Nanjing University, Nanjing, China
Yanning Zhang	Northwestern Polytechnical University, Xi'an, China

Contents – Part II

Contents – Part I

Exhaustive Hybrid Posting Lists Traversing Technique

Kun Jiang(✉) and Yuexiang Yang

College of Computer, National University of Defense Technology,
Changsha 410073, China
{jiangkun,yyx}@nudt.edu.cn

Abstract. A large amount of optimization techniques have been studied in addressing the performance challenges of web search engines, but still leave much room for further improvement. In this paper, we focus on the inverted index traversal techniques, which make directly scans of the posting lists with different loop schemes, providing preliminary results for a complicated ranking procedure. We propose a novel exhaustive index traversal technique called hybrid-scoring at a time (HAAT) on document-ordered indexes, which can reduce memory consumption and candidate selection cost of existing document at a time (DAAT) and term at a time (TAAT) at the expense of revisiting the posting lists of the remaining query terms. Preliminary analysis show comparable computational complexity between HAAT and existing methods. Experimental results with the TREC GOV2 collection show that our approach is comparable with the existing DAAT baseline and considerable performance gains compared to TAAT baseline.

Keywords: Inverted index · Query processing · Index traversal · Hybrid-scoring at a time

1 Introduction

Large-scale search engines process thousands of queries per second over billions of documents. One major problem in query processing is that the length of the inverted list can easily grow to hundreds of MBs or even GBs for common terms [1]. Given that search engines need to answer queries within fractions of a second, naively traversing the basic index structure, which could take hundreds of milliseconds, is not acceptable. Efficient index traversal techniques can quicken the query processing procedure and reduce large amounts of hardware cost of search engines.

The schemes for traversing posting lists for a query fall into two main classes: term at a time (TAAT) and document at a time (DAAT) [2]. In TAAT technique, the contributions of each term to the score of each document are completely processed before moving to the next term. In DAAT technique, the score of each document is completely computed with all the posting lists before advancing to the next document. The smallest current docid of all the posting lists is always selected as the candidate for scoring. As the main disadvantage, TAAT technique requires large amounts of memory to store partial scores when traversing each posting list. The iterative candidate

X. He et al. (Eds.): IScIDE 2015, part II, LNCS 9243, pp. 1–11, 2015.
DOI: 10.1007/978-3-319-23862-3_1

selection of DAAT technique has high costs associated with the random access of all of the posting lists, which requires jumping back and forth between different posting lists [3, 4]. In this paper, we propose a novel exhaustive index traversal technique called hybrid-scoring at a time (HAAT), which can reduce the memory consumption of TAAT and avoid the costly candidate selection of DAAT, at the expense of revisiting some of the left posting lists.

The rest of this paper is organized as follows. We provide background and related works on inverted index and matching techniques in Sect. 2. In Sect. 3, we present our HAAT index traversal. Experimental comparisons of HAAT and existing exhaustive index traversal techniques are demonstrated in Sect. 4. Finally, Sect. 5 concludes this paper and presents a prospective of future work.

2 Background and Related Work

2.1 Inverted Index

The inverted index plays a vital role in the efficient processing of ranked and Boolean queries [4, 5]. Given a collection of N documents, we assume that each document is identified by a unique document identifier (*docid*) between 0 and $N - 1$. An inverted index can be seen as an array of lists or postings, where each entry of the array corresponds to a different term or word in the collection. For each term t, the index contains an inverted list I_t consisting of a number of index postings describing all places where term t occurs in the collection. We assume postings have *docids* and *frequencies* but do not consider other data such as positions or contexts; thus, each posting is of the form (d_i, f_i). The set of terms is called the lexicon of the collection, which is comparatively small in most cases. However, the inverted lists may consist of many millions of postings which could be roughly linear with the size of the collection.

To allow faster access and limit the amount of memory needed, search engines use various compression techniques that significantly reduce the size of the inverted lists, see [6] for some recent work. A data structure named skiplists, which allows fast jumps in the compressed inverted lists, is adopted to accelerate the query processing [3, 7, 8]. Skiplists divide the posting lists into blocks of entries and provide pointers for fast access to the compressed blocks. Thus, additional skipping pointers have to be embedded in the posting list or a skipping table describing the front of a few selected blocks have to be stored together with the lexicon, both describing a potential skip when decoding. This allows entire blocks that would be evaluated by an exhaustive evaluation to be skipped searching for a given docid.

2.2 Index Traversal

As mentioned above, the schemes to iterate through posting lists for a query fall into two main classes: term-centric TAAT and document-centric DAAT [2, 4]. With TAAT technique, the query results are obtained by sequentially traversing one inverted list at a time. For this reason, it requires large amounts of memory to store partial scores when traversing each posting list. With DAAT technique, the posting lists are traversed in

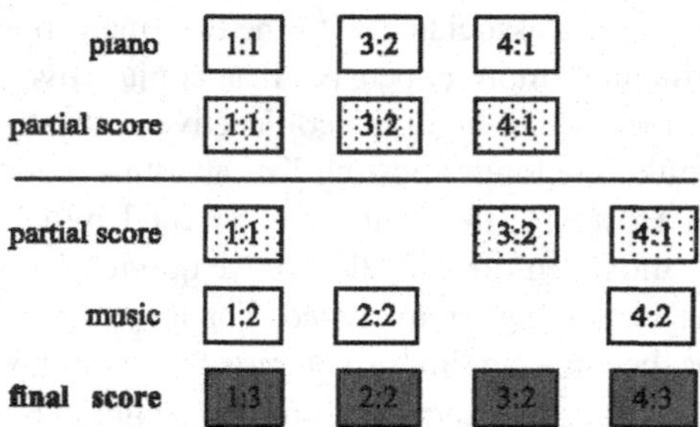

Fig. 1. An example shows the index traversing procedure of TAAT with query terms 'piano' and 'music'. The tuple $(d : f_{t,d})$ represents docid d and score $f_{t,d}$ (term frequency) in document d. The solid line separates different iteration of the procedure, and the scoring operation scans from up to down in each iteration to summing up all the term frequencies $f_{t,d}$.

parallel. Each iteration selects a candidate docid with the least of the current docids of all posting lists. The iterative candidate selection of DAAT technique costs much in random accessing all the posting lists, which need to jump back and forth between different posting lists. See [9] for an experimental comparison of the two exhaustive traversal techniques.

Furthermore, search engines use a simple ranking function to compute a score for each document that passes a simple Boolean query filter [10]. The most commonly used Boolean query filters are conjunctive (AND) and disjunctive (OR) query filters. Conjunctive queries make a compulsive requirement that documents matching all of the query terms can be returned as a result. In contrast, disjunctive queries relax the requirement such that documents matched by any of the query terms may be reported as a result. Thus, the DAAT and TAAT techniques can be enhanced into four algorithms with different query filters, i.e., AND_DAAT, OR_DAAT, AND_TAAT and OR_TAAT. The two disjunctive filters with TAAT and DAAT are shown in Figs. 1 and 2. For more details of the implementation of DAAT and TAAT with different query filters can be referred to [4, 10].

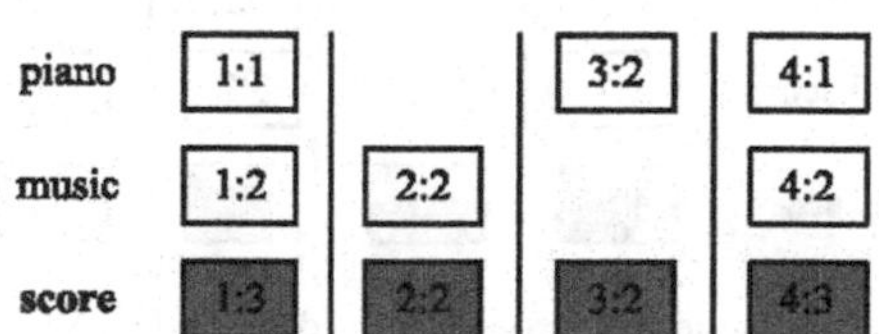

Fig. 2. An example shows the index traversing procedure of DAAT with query terms 'piano' and 'music'. The tuple $(d : f_{t,d})$ represents docid d and score $f_{t,d}$ (term frequency) in document d. The solid line separates different iteration of the procedure, and the scoring operation scans from left to right in each iteration to accumulate the partial scores.

One of the main challenges associated with the two query filters is that disjunctive queries tend to be significantly more expensive than conjunctive queries, as they have to evaluate many more documents. In all the conjunctive cases, the shortest posting list can be used to skip through the longer posting lists and thus reduce the amount of data to be processed [3, 4]. However, the result sets returned by conjunctive queries are potentially smaller than those returned by disjunctive queries. This might be a problem when one of the terms is missing or mistyped. For this reason, the optimization of disjunctive queries has become an important research in index traversal, including skipping in disjunctive queries and dynamic pruning techniques [11, 12].

3 Hybrid Exhaustive Index Traversal

In this section, we present hybrid scoring at a time (HAAT) index traversal technique and give a detailed implementation description of its four main methods.

3.1 Hybrid Scoring at a Time

We describe a new hybrid scoring index traversal technique, in which postings are accessed in an order that is neither strictly term-centric nor strictly document-centric. In HAAT, we separate query terms into candidate selection term and scoring adding term. Every time we select candidate document in candidate selection term sequentially and score the postings of the candidate selection term for all terms when it appears in its posting list. Thus, we access all posting lists of the query terms simultaneously. Finally, we insert the result document and its score into the result heap. One simple example

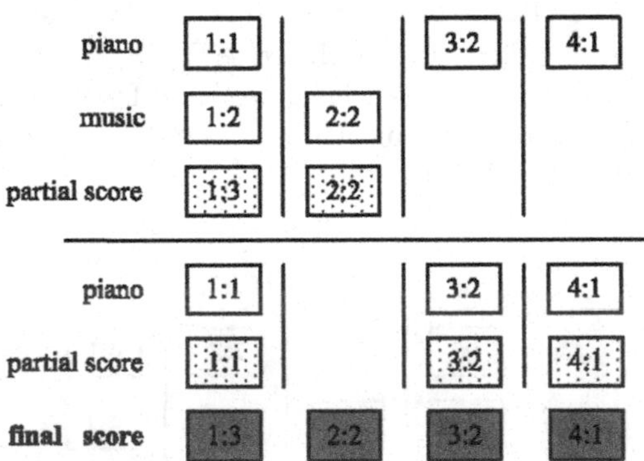

Fig. 3. An example shows the index traversing procedure of HAAT with query terms 'piano' and 'music'. The tuple $(d : f_{t,d})$ represents docid d and score $f_{t,d}$ (term frequency) in document d. The dashed line separates different candidate document scorings of a given posting list from left to right to get partial scores. The solid line separates different query term iteration to scan another posting list for candidate documents. The final score of a candidate document can be achieved by comparing different partial scores of different phrased separated by solid lines.

showing the procedure of HAAT is depicted in Fig. 3. Given the inverted index and the result heap, HAAT returns the k documents that have the highest score according to the scoring function $S(q, d), d \in D$. The detail of the technique is depicted in Algorithm 1.

As shown in Algorithm 1, HAAT iterates the posting list for every query term by first sorting the posting lists by ascending order of its frequency so that we evaluate the shortest list first. Not surprisingly, the posting list with the highest document frequency is almost always the one with the smallest term importance in common ranking functions such as BM25. This means that the usual estimates of term importance (e.g. *idf*) will decrease for lists encountered later in the evaluation. The iteration from most significant term to the less ones makes the most promising documents appear early for scoring. Thus, the threshold of the result heap will rise up quickly, which can avoid lower scored documents from being inserted into the heap. For a given posting list, it selects a candidate document d and checks for the scoring posting lists. Specifically, every posting list should first skip to the candidate document d, and if the candidate does not exist in the list, it will point to the first document larger than d. If it points to the candidate, just add its contribution to the accumulated score using a simple ranking function. When all the posting lists are considered for d, we insert the candidate d into the result heap if its score is larger than the threshold. When the posting list of candidate selection term reaches the end, the scoring posting lists should be reset to make the cursor points to its first document. Finally, we return the top k score document of the results in the heap.

Algorithm 1. Exhaustive Disjunctive HAAT Traversal

Input: index iterators $\{I_{t_1},\cdots,I_{t_q}\}$ sorted by ascending lists length

Output: top k query results sorted by descending score

```
resHeap(k) ← ∅
for all inverted list I_{t_i} (1 ≤ i ≤ q) do
while I_{t_i} is not finished do
   d ← I_{t_i}.getdid(), score ← s(I_{t_i});
   for all inverted lists I_{t_j} (i < j ≤ q)
      I_{t_j}.skipto(d);
      if I_{t_j} points to d then
         score ← score + s(I_{t_j});
      end if
   end for
   resHeap.insert(d, score);
   I_{t_i}.next();
end while
I_{t_j}.reset();
end for
return the top k results.
```

For q query terms and n matching documents (containing at least one query term), the first pass of the query costs $\Theta(N_{t_1} \times n + N_{t_1} \times \log_2 k)$. Here, N_t is the number of documents containing t, $N_{t_1} \times n$ corresponds to the first loop starting in line 8 of scoring and $N_{t_1} \times \log_2 k$ corresponds to the sorting of search results in line 11. For the *ith* query pass corresponding to the *ith* loop in line 8, the time is $\Theta(N_{t_i} \times (n - (i - 1)) + N_{t_i} \times \log_2 k)$. Depending on how often the query terms co-occur in a document, the posting can be scored $n - (i - 1)$ times for *ith* loop (line 8) in the worst case. Thus, for all n loop iterations, the overall time complexity of the algorithm is

$$\Theta\left(\sum (N_{t_i} \times (n - (i - 1))) + N_q \times \log_2 k\right) \tag{1}$$

where $N_q = N_{t_1} + N_{t_2} + \ldots + N_{t_n}$. It is reported in [10] that the time complexities of DAAT and TAAT with normal implementation are both $\Theta\left(N_q \times n + N_q \times \log_2 k\right)$. We also find that

$$\begin{aligned}\Theta\left(\sum (N_{t_i} \times (n - (i - 1))) + N_q \times \log_2 k\right) &< \Theta\left(\sum (N_{t_i} \times n) + N_q \times \log_2 k\right) \\ &= \Theta\left(N_q \times n + N_q \times \log_2 k\right)\end{aligned} \tag{2}$$

Although the complexity of the three techniques is almost the same, the run time cost is quite different. The bottleneck of DAAT lies in the generation of the smallest docid of all the posting lists' current docid. In DAAT, the implementation can be achieved by using a min-heap structure or a comparison of the current docids. Also, the TAAT technique needs large amount of accumulators stored in memory for partial scoring. The superiority of HAAT is that it avoids having a large memory footprint and the generation of the smallest docid in candidate selection. The disadvantage of HAAT processing is that the postings can be rescored whenever a term is considered. However, the reconsidered posting will just score the remaining terms and obtain a partial score because it has already been considered in the past terms (line 5).

3.2 Posting List Iterator

The operation of posting lists of a fixed index organization is the basis of query processing. For a given posting, it contains the two important attributes: *currentptr* is the address of a pointer of the posting list, and *currentdid* is the document identifier of a given posting. There are also other attributes that can be used to score the document, but we do not need them when traversing the posting list. During index traversal, a cursor is created for each term t in the query, and is used to traverse t's posting list. Regardless of the underlying index structure, we define four basic methods on the posting list iterator as an abstract data type.

- **getdid**() returns the docid of the current position in the posting list where the pointer address is *currentptr*.

- **reset**() sets the current address of the posting list to zero and initializes its current posting to be the first posting element in the posting list, i.e., *currentptr* $\leftarrow$ 0, currentposting $\leftarrow$ t.iterator[*currentptr*].
- **next**() increments the current address to return the next posting of term t's posting list, i.e., *currentptr* ++, currentposting $\leftarrow$ t.iterator[*currentptr*].
- **skipto**(*d*) randomly accesses to the position with docid equal to docid d or the one next to docid *d* in the posting list. The detail of the skipping procedure depends on the implementation of the skipping structure.

The skipping structure we used is similar to [3, 12], in which the postings are separated into data chunks and the skipping pointers are built upon to generate hierarchical skipping chunks. More precisely, we start with a basic partitioning into chunks containing 128 postings. We build a skipping trunk structure on top of the 128 chunks, where each higher-up chunk stores the last docid and the end-offset from each chunk below, respectively. The offsets stored in the chunks represent the length of a corresponding lower level chunk. These result in the lowest level of skipping chunks. Similar to the posting list itself, we divide skipping chunks into groups of 128 elements. The last docid and the end-offset of each skip-chunk are recursively stored in a skip-chunk in the level above. Finally, the root level contains a group with no more than 128 elements indicating equal number of chunks below the root level.

The skipto(*d*) operation based on the above skipping structure can be performed by comparing *d* to the last docid of the active chunk. When *d* is greater, the higher level chunk is considered as the active chunk. Conversely, the chunk referred by the offset of the docid is decoded. The search goes down iteratively until the data-chunk is reached. The binary search technique is used within each chunk for the element indicating the lower level chunk or the exact target docid. The search stops when the current docid is equal to the target docid or the upper bound on the bottom level.

4 Experiments

In our experiments, we use the TREC GOV2 collection containing about 25.2 million documents and about 32.8 million terms in its vocabulary with an uncompressed size of 426 GB. We build inverted index structures with 128 docids per block, using the PForDelta compression algorithm [6], removing the standard English stopwords, and applying Porter's English stemmer. The final compressed index size is 7.57 GB. We use the first 10000 queries selected from the TREC2005 Efficiency track queries using distinct amounts of terms with $|q| \geq 2$. We use the standard Okapi BM25 bag-of-words similarity scoring function. Our experiments were performed on an Intel(r) Xeon(r) E5620 processor running at 2.40 GHz with 8 GB of RAM and 12,288 KB of cache. The default physical block size is 16 KB, unless stated otherwise. The index was preloaded into memory for warm-up. All solutions were implemented in JAVA on the well-known Terrier information retrieval platform [13]. All the codes is openly available by contacting the authors.

4.1 Query Latency

As mentioned above, one of the main challenges associated with the processing of disjunctive (OR) queries is that they tends to be much slower than conjunctive (AND) queries, as they have to evaluate more documents. The two types of queries can be applied to each of the index traversal strategies mentioned above. We implement six types of exhaustive query processing algorithms, namely, disjunctive DAAT, disjunctive TAAT, disjunctive HAAT, conjunctive DAAT conjunctive TAAT and conjunctive HAAT. To verify the effectiveness of our proposed HAAT scheme, we perform a explicit comparison of the exhaustive index traversal techniques, depicted in Table 1 with different number of results k = 10, 100, 1000. The performance is always measured by average time per query in microseconds, and the running time is averaged over 5 independent runs.

Table 1. Average processing times in *ms* of the exhaustive index traversal techniques

	$k = 10$	$k = 100$	$k = 1000$
OR_TAAT	2732.0	2735.7	2736.8
OR_DAAT	231.6	232.6	237.9
OR_HAAT	279.8	284.4	285.7
AND_TAAT	314.5	316.0	318.3
AND_DAAT	24.1	24.1	24.3
AND_HAAT	34.2	34.3	34.5

From Table 1, we can see that disjunctive queries tend to be significantly more expensive (by about an order of magnitude) than conjunctive queries with all the index traversal techniques. This finding suggests the necessity of reducing the huge performance gap between disjunctive queries and conjunctive queries. Thus, we focus on disjunctive queries of HAAT in this paper, which do not disqualify one document when it misses some of the results. In addition, we can also find that DAAT runs much faster than TAAT in both disjunctive and conjunctive queries. The HAAT traversal performs a little slower than DAAT but much better than TAAT in both disjunctive and conjunctive queries. This is mainly due to the shortage of the main memory used for storing all document information during TAAT processing with increasing data set, and the cache misses caused by random access in accumulators used for storing intermediate results. Overall, the HAAT exhaustive index traversal technique gives comparable performance results to those of the DAAT exhaustive index traversal technique.

4.2 Processed Elements

We also measure the performance by other criteria, including a number of processed elements, i.e., the total number of candidates inserted into the result heap, calls to the scoring function, docids evaluated and decompressed chunks. These criteria provide in-depth descriptions of the above query latency measurement and are also used in previous work [9, 11, 12]. Table 2 shows other criteria for the six index traversal techniques with the number of results $k = 10$.

Table 2. Number of processed elements for different index traversal techniques with the number of results $k = 10$.

	Heap inserts	Scorings called	Docids evaluated	Chunks decoded
OR_TAAT	120794	2733003465	2733003465	21696082
OR_DAAT	120447	2733003465	2733003465	21696082
OR_HAAT	84234	3094181690	3485686272	56815118
AND_TAAT	79918	219938564	2733003465	21696082
AND_DAAT	50459	76794374	227718630	8775132
AND_HAAT	50459	198111226	266950244	8784707

As can be seen in Table 2, the disjunctive queries processed considerably more elements than did the conjunctive queries in all cases with different exhaustive traversing techniques. This is because skipping can be used in conjunctive queries to avoid processing a large number of postings by allowing the shortest posting list to skip through other posting lists to the pointer address with the same docid. With disjunctive queries, the scorings function called, docid evaluated and chunks decoded are the same for both the DAAT and TAAT schema. This is because all the postings or chunks are considered just at different periods in the schema. However, the three criteria of HAAT increase significantly due to the revisiting of chunks and postings. With conjunctive queries, the three criteria of DAAT are significantly reduced compared to TAAT and HAAT. This is because the largest docid of the current position can be select as the candidate document in DAAT, and lots of postings that only appear in part of the lists can be avoided for processing.

Here, we mainly look at the heap inserts of the three exhaustive traversal techniques. It is worth noting that the heap inserts of HAAT are considerably reduced compared to DAAT and TAAT for both disjunctive and conjunctive queries. With conjunctive queries, the criteria is the same for HAAT and DAAT; this is because the posting lists of the two are read in sequential order which get the same candidate and the same threshold changes. However, TAAT calls the heap inserts when all results are ready in the accumulators and makes the threshold of the result heap grow much slower when inserting. This causes lots of heap mis-inserts of the schema in both disjunctive and conjunctive queries. With disjunctive queries, the surprising result is that the heap inserts of HAAT are almost 30 % less than the other two schemes. In essence, this is mainly because HAAT sorts the terms in descending frequency order and processes important postings in important terms first. This makes the threshold get very good promotion and greatly reduces the number of candidates that are inserted into the result heap. This allows us to consider whether the advantage of HAAT can be used to solve the slow startup problem in dynamic pruning techniques [11, 12]. Overall, we find that the HAAT index traversal achieves much performance gains than the existing traversing techniques, but still leaves much room for further query processing improvements.

5 Conclusions

In this paper we have presented a novel exhaustive index traversal technique called hybrid scoring at a time (HAAT), which can reduce the memory consumption and candidate selection cost of existing methods, at the expense of revisiting the posting lists of the left query terms. Analysis and experiments show less computational complexity and comparable performance between HAAT and existing exhaustive index traversal baselines. Another conclusion is that our HAAT index traversal is more suitable for the dynamic pruning due to its consideration of term importance. Further investigations will extend to the reducing of postings revisitings in HAAT and dynamic pruning techniques [11, 12] for further efficiency improvements.

References

1. Dean, J.: Challenges in building large-scale information retrieval systems: invited talk. In: Proceedings of the Second ACM International Conference on Web Search and Data Mining, pp. 1–1. ACM (2009)
2. Turtle, H., Flood, J.: Query evaluation: strategies and optimizations. Inf. Process. Manage. **31**(6), 831–850 (1995)
3. Moffat, A., Zobel, J.: Self-indexing inverted files for fast text retrieval. ACM Trans. Inf. Syst. (TOIS) **14**(4), 349–379 (1996)
4. Croft, W.B., Metzler, D., Strohman, T.: Search Engines: Information Retrieval in Practice. Addison-Wesley Reading, Boston (2010)
5. Zobel, J., Moffat, A.: Inverted files for text search engines. ACM Comput. Surv. (CSUR) **38** (2), 6 (2006)
6. Zukowski, M., Heman, S., Nes, N., Boncz, P.: Super-scalar RAM-CPU cache compression. In: Proceedings of the 22nd International Conference on Data Engineering, ICDE 2006, pp. 59–59. IEEE (2006)
7. Chierichetti, F., Lattanzi, S., Mari, F., Panconesi, A.: On placing skips optimally in expectation. In: Proceedings of the 2008 International Conference on Web Search and Data Mining, pp. 15–24. ACM (2008)
8. Boldi, P., Vigna, S.: Compressed perfect embedded skip lists for quick inverted-index lookups. In: Consens, M.P., Navarro, G. (eds.) SPIRE 2005. LNCS, vol. 3772, pp. 25–28. Springer, Heidelberg (2005)
9. Lacour, P., Macdonald, C., Ounis, I.: Efficiency comparison of document matching techniques. In: Proceedings of ECIR (2008)
10. Büttcher, S., Clarke, C., Cormack, G.V.: Information Retrieval: Implementing and Evaluating Search Engines. The MIT Press, Boston (2010)

11. Ding, S., Suel, T.: Faster top-k document retrieval using block-max indexes. In: Proceedings of the 34th international ACM SIGIR conference on Research and development in Information Retrieval, pp. 993–1002. ACM (2011)
12. Jonassen, S., Bratsberg, S.E.: Efficient compressed inverted index skipping for disjunctive text-queries. In: Clough, P., Foley, C., Gurrin, C., Jones, G.J., Kraaij, W., Lee, H., Mudoch, V. (eds.) ECIR 2011. LNCS, vol. 6611, pp. 530–542. Springer, Heidelberg (2011)
13. Ounis, I., Amati, G., Plachouras, V., He, B., Macdonald, C., Lioma, C.: Terrier: a high performance and scalable information retrieval platform. In: Proceedings of SIGIR OSIR Workshop (2006)

Control Parameters Optimization for Spacecraft Large Angle Attitude Based on Multi-PSO

Wenbo Zhao, Jianxin Zhang, Qiang Zhang(✉), and Xiaopeng Wei

Key Lab of Advanced Design and Intelligent Computing, Dalian University, Ministry of Education, Dalian 116622, People's Republic of China
Zhangq26@126.com

Abstract. In order to meet the requirement of spacecraft large angle attitude maneuver, the paper presents a spacecraft parameters design and optimization methods for large angle attitude and rapid maneuver. Firstly, Lyapunov controller model is described by quaternion with the output torque of the actuator as constraint condition. Then, the optimal time and power consumption are regarded as the optimized objectives. Finally, a multi-objective particle swarm optimization (MOPSO) algorithm is proposed to solve above control parameter optimization problem. Simulations show that the optimized control parameters which satisfy the constraint of the output torque can make the control system with lower consumption, higher stability and better convergence.

Keywords: Multi-objective particle swarm optimization algorithms · Weighting coefficient · Large angle attitude · Parameter optimization

1 Introduction

As we all known, wide variation of attitude angle and dramatic changes in angular rate are required in many space missions [1–3], so with the completion of space missions such as rendezvous and docking of spacecraft, initial orientation, gesture capture, the spacecraft is demanded to have the capability of rapid mobility and less power consumption. In addition, uncertain factors such as the gravity gradient parameters, disturbance torques and atmosphere icpneumatic parameters are existed in those space tasks, so the robustness of the controller is also becoming a focus issue to be considered [4].

Particle swarm optimization (PSO) is a comparatively recent heuristic inspired by the choreography of a bird flock. PSO emulates the swarm behavior of birds flocking, where these swarm individuals search for food in a collaborative way. Each member in the swarm adapts its search patterns by learning from its own experience and other members' experiences simultaneously. These phenomena are studied and corresponding mathematical models are constructed. In PSO, a member in the swarm, called a particle, represents a potential solution which is a point in the search space. The global optimum is regarded as the location of food. Each particle has a fitness value and a velocity to adjust its flying direction according to the best experiences of the swarm to search for the global optimum in the D-dimensional solution space [5].

X. He et al. (Eds.): IScIDE 2015, Part II, LNCS 9243, pp. 12–19, 2015.
DOI: 10.1007/978-3-319-23862-3_2

In the paper, spacecraft attitude tracking control model based on the method of Lyapunov was designed, and we devise the index which evaluate the ability of attitude tracking control for the parameters that impact the performance. Under the corresponding constraint conditions, a set of the optimal parameters combination of better convergence and lower energy consumption based on the attitude tracking control model were achieved by adopted MOPSO with two optimized goals of time and power consumption.

2 Mathematical Model and Control Law Design

2.1 The Attitude Dynamics and Kinematics Model

A rigid spacecraft rotating under the influence of body-fixed devices is considered. The attitude dynamics of spacecraft are a set of uncertainty nonlinear differential equations. In this paper, the attitude dynamics of a rigid spacecraft can be expressed as follows:

$$\left.\begin{aligned} M_{l1} + d_1 &= I_1\dot{\omega}_1 + (I_3 - I_2)\omega_3\omega_2 \\ M_{l2} + d_2 &= I_2\dot{\omega}_2 + (I_1 - I_3)\omega_1\omega_3 \\ M_{l3} + d_3 &= I_3\dot{\omega}_3 + (I_2 - I_1)\omega_2\omega_1 \end{aligned}\right\} \tag{1}$$

Where ω_i $(i = 1, 2, 3)$ is the vector of spacecraft angular rate, expressed in body frame, I_i is the inertia matrix, M_{li} $(i = 1, 2, 3)$ is the vector of magnetic control torquesand $d_i(i = 1, 2, 3)$ is the vector of external disturbance torques.

The most common parameterization for spacecraft attitude is given by quaternion elements, which lead to the following representation for the attitude kinematics:

$$\dot{Q}_d = \frac{1}{2}G(\omega)Q_d \tag{2}$$

Where $Q_d = \begin{bmatrix} q_0 \\ q_1 \\ q_2 \\ q_3 \end{bmatrix}$ is the quaternion elements, and $G(\omega) = \begin{bmatrix} 0 & -\omega_1 & -\omega_2 & -\omega_3 \\ \omega_1 & 0 & \omega_3 & -\omega_2 \\ \omega_2 & -\omega_3 & 0 & \omega_1 \\ \omega_3 & \omega_2 & -\omega_1 & 0 \end{bmatrix}$

2.2 Control Law Design

The process of spacecraft tracking attitude control can be summarized as: The initial position of the spacecraft as Q_0 and the desired position as Q_e are given. The controller which make the spacecraft attitude from Q_0 to Q_e by continuous tracking and hold the spacecraft at the equilibrium position is designed.

Suppose the desired attitude $Q_e = [\,q_{0e} \quad q_{1e} \quad q_{2e} \quad q_{3e}\,]^{\mathrm{T}}$. And the control strategy of large angle attitude maneuver is given by the following theorem.

Theorem: Using control law [1, 2]

$$\left.\begin{aligned} M_{l1} &= -[\rho_1 sign(\omega_1) + kI(\varepsilon + 1 - e^{-\omega_0 t})(q_{0e}q_1 - q_{1e}q_0 - q_{2e}q_3 - q_{3e}q_2 + k_1\omega_1)] \\ M_{l2} &= -[\rho_2 sign(\omega_2) + kI(\varepsilon + 1 - e^{-\omega_0 t})(q_{0e}q_2 - q_{1e}q_3 - q_{2e}q_0 - q_{3e}q_1 + k_2\omega_2)] \\ M_{l3} &= -[\rho_3 sign(\omega_3) + kI(\varepsilon + 1 - e^{-\omega_0 t})(q_{0e}q_3 - q_{1e}q_2 - q_{2e}q_1 - q_{3e}q_0 + k_3\omega_3)] \end{aligned}\right\} \tag{3}$$

Where *sign* is sign function, and k, ε, ω_0, k_1, k_2, k_3 are positive design parameters of the actuator. Besides $I = (I_1 + I_2 + I_3)/3$,and $\rho_i = \max(d_i)$. Assuming $K = k(\varepsilon + 1 - e^{-\omega_0 t})$, so K is the variable exponential gain, which to be selected can reduce the initial control torque.

Proof: the candidate Lyapunov is defined as

$$V = \frac{1}{2}[I_1\omega_1^2 + I_2\omega_2^2 + I_3\omega_3^2] + K[(q_0 - q_{0e})^2 + (q_1 - q_{1e})^2 + (q_2 - q_{2e})^2 + (q_3 - q_{3e})^2] \tag{4}$$

As quaternion theorem shows $Q_d^T Q_d = 1$, namely $q_0^2 + q_1^2 + q_2^2 + q_3^2 = 1$
With the substitution of $Q_d^T Q_d = 1$ we obtain

$$V = \frac{1}{2}[I_1\omega_1^2 + I_2\omega_2^2 + I_3\omega_3^2] + 2K(1 - q_{0e}q_0 - q_{1e}q_1 - q_{2e}q_2 - q_{3e}q_3) \tag{5}$$

And the time derivative of V is

$$\dot{V} = I_1\omega_1\dot{\omega}_1 + I_2\omega_2\dot{\omega}_2 + I_3\omega_3\dot{\omega}_3 - 2K(q_{0e}\dot{q}_0 + q_{1e}\dot{q}_1 + q_{2e}\dot{q}_2 + q_{3e}\dot{q}_3) \tag{6}$$

Substituting Eqs. (1), (2) and (3) into Eq. (6) the time derivative of V can be written as

$$\begin{aligned} \dot{V} = -(\rho_1|\omega_1| - d_1\omega_1) - (\rho_2|\omega_2| - d_2\omega_2) - -(\rho_3|\omega_3| - d_3\omega_3) - k_1\omega_1^2 - k_2\omega_2^2 - k_3\omega_3^2 \\ \leq k_1\omega_1^2 - k_2\omega_2^2 - k_3\omega_3^2 < 0 \end{aligned}$$

Select appropriate nonnegative combination of $[k, k_1, k_2, k_3]$, which make the controller has good control performance, besides that $\dot{V}$ is the semi-negative definite matrix can been sured.

3 The Controller Parameter Optimization

3.1 Attitude Maneuver Controller Parameter Optimization Model

The parameters of system control model are optimized and rational objective optimization functions are designed, then the relationship of quantitative evaluation system index and parameter optimization is built, which mathematical nature is the optimal control function approximation problems. In the essay, the definition of the error angular velocity of spacecraft systems and error status of quaternion respectively are

$\omega_{er} = \omega - \omega_e, q_{er} = q - q_e. \omega_{er}$ and q_{er} as optimal design parameters are quickly made close to zero.

In the paper, the system consists of six independent state variables as follows:

$$x = [\omega_1 \ \omega_2 \ \omega_3 \ q_1 \ q_2 \ q_3] \tag{7}$$

When using quaternion to describe the attitude, since q_0 and $[q_1 q_2 q_3]$satisfy the constraint equation of quaternion, and their changes cannot be independent, so with satisfied the quaternion constraint, the convergence of $[q_1 q_2 q_3]$ can uniquely determine the value of q_0. According to formula (3), we can see that the dynamic change process of six independent state variables are closely related to the selection of $[kk_1 k_2 k_3]$, so the different options of $[kk_1 k_2 k_3]$ greatly affect the control performance of controller.

The two system optimization indexes are considered in the paper. First, to realize rapid mobility, as the time when the spacecraft achieve the desired attitude is shorter. As we all known, T is the time that ω_{er} and q_{er} spent on approaching zero. In order to quantitatively evaluate of the performance of the controller,t_i is denoted as the required time that the error state variables x_{ei} drop from the initial value to the maximum value of 2 % during the adjustment process, and then the adjustment ability of the controller is defined by the form of the time index as follows [6, 7]:

$$T = \begin{cases} \sum_{i=1}^{6} a_i t_i & (T_{\max} < T_p) \\ \sum_{i=1}^{6} a_i t_i + \lambda (T_{\max} - T_p) & (T_{\max} > T_p) \end{cases} \tag{8}$$

Where a_i is weight coefficient, and $\sum_{i=1}^{6} a_i = 1$. Besides $T_{\max}$ is maximum torque of control and T_p is the control torque limit of the actuator. Denoting λ as the penalty parameter.

Secondly, to achieve less power consumption of attitude tracking control. In this essay the work of torque is merely considered. And the corresponding power consumption is defined as follows:

$$W = M\theta = \sum_{i=1}^{3} \left(\int |M_i(t)| |\omega_i(t)| dt \right) \tag{9}$$

Where θ is the total amount change in attitude angle, $\omega_i(t)$ is the time-varying angular rate based on the i-spindle of the body coordinate system, and $M_i(t)$ is the torque with time-varying around the body of the i-spindle. During the optimization process, penalty function is written in the paper by the form of system performance index, and the maximum control torque is acted as the constraint [3].

In summary, the best optimal parameters of the spacecraft control model are $[kk_1 k_2 k_3]$, which makes the performance index T and W simultaneously obtain the smallest value.

3.2 Multi-objective PSO Algorithm

The PSO algorithm is apt to accomplish and has been empirically shown to carry out well on many optimization problems. Whereas it have not been extended to handle multiple objectives problems until recently. Since the PSO algorithm presents the high speed of convergence on single-objective optimization, the algorithm seems to particularly be fit for multi-objective optimization. In this paper, we put forward a proposal, called "multi-objective particle swarm optimization" (MOPSO) [8], which allows the PSO algorithm to be able to dispose of multi-objective optimization problems. The current proposal is an improved version of the algorithm, in which we have added mutation operator and constraint-handling mechanism that relatively improves the exploratory capabilities of our original algorithm. MOPSO is validated using several standard test functions reported in the specialized literature as a result of the ability to quickly converge to a reasonably good solution and its simplicity of implementation.

In the essay, MOPSO algorithm based on the variable weight coefficient is adopted. By turning Multi-objective optimization problem into single objective optimization problem, and taking different weighting factor for each single objective optimization problem, you can use the algorithm to work out a Pareto optimal solution of the multi-objective optimization problem. In this paper, combined with the optimization problem, the corresponding MOPSO algorithm as follows:

a. Initialize all the particle individuals of the population.
b. Initialize the speed of each particle.
c. Calculate individual fitness value: According to formulas (8) and (9), denote the fitness function as $\varphi(k) = \sigma_1 T(k_1 k_2 k_3 k) + \sigma_2 W(k_1 k_2 k_3 k k_3$, Where σ_1 and σ_2 is the variable weight coefficient, $\sigma_1 + \sigma_2 = 1$.
d. For each particle, its fitness value is compared with its experience fitness values with the best position where it has ever experienced. If better, then replace the previous best position with the current best position. Then the best individual fitness value is compared with swarm experienced fitness values, if better, then correspondingly denote it as the current global best position.
e. Based on the velocity and position update equation, update the velocity and position of all the particles.
f. If the preset maximum number of iterations has not reached, or has failed to reach the end condition, then return b. After the occurrence of multiple iterations the change of the global optimal position is less than the threshold value,it's considered to find the global best solution.

4 Simulation

In the paper physical parameters and requirements of the spacecraft are adopted in literature [1, 2]. The main parameters of Spacecraft are as follows: $I_1 = 720\ kg.m^2$, $I_2 = 800\ kg.m^2$, $I_3 = 550\ kg.m^2$. Supposing spacecraft the initial state of attitude is $[\omega_{10}\ \omega_{20}\ \omega_{30}\ q_{00}\ q_{10}\ q_{20}\ q_{30}]$ = [00 0.2362 -0.8128 0.3430-0.4073], which is equivalent to $\theta(t_0) = 210°$, $\phi(t_0) = 90°$, $\gamma(t_0) = 50°$ in the resting position. Presuming spacecraft attitude state of termination is $[\omega_{1e}\ \omega_{2e}\ \omega_{3e}\ q_{0e}\ q_{1e}\ q_{2e}\ q_{3e}] = [0\ 0\ 0\ 1\ 0\ 0\ 0]$, which is equivalent to attitude angular and angle velocity are 0.And disturbance torque is taken as $d_1 = d_2 = d_3 = \pm\ 0.005\ \sin(t)$ (N/m), so $\rho_1 = \rho_2 = \rho_3 = 0.005$.

MOPSO algorithm simulation with weighting coefficients is adopted, population size is set to 100 and the number of iteration is set to 200. The simulation results are shown in the following Table 1.

Based on the actual needs of spacecraft attitude control, the best solution should meet constraint condition of the output torque, regard the convergence of various parameters as primary goal, and take power consumption and time of convergence into a count. From Table 1 we can see that the proper power consumption of the controller is between 8 and 9, and the appropriate value of the convergence time in the vicinity of 40. In order to satisfy the control requirements of time and power consumption, and

Table 1. The simulation results

σ_1	σ_2	k_1	k_2	k_3	k	T(s)	W (J)
0	1.0	6.5374	13.6851	9.1775	0.0311	99.99	4.2113
0.1	0.9	7.6839	9.3189	8.9724	0.0211	63.93	3.4255
0.2	0.8	7.7459	6.8767	2.9492	0.0814	59.08	7.4173
0.3	0.7	5.0734	5.9838	5.7039	0.0872	40.84	8.6158
0.4	0.6	4.4697	5.5043	5.3917	0.0805	39.11	9.4132
0.5	0.5	4.1447	5.2304	5.2301	0.0771	38.49	9.9149
0.6	0.4	3.7805	4.9034	5.0258	0.0733	37.97	10.5676
0.7	0.3	7.2956	6.2393	2.4552	0.1001	53.97	8.9470
0.8	0.2	6.8438	3.2165	4.5645	0.1075	62.61	9.4526
0.9	0.1	6.9961	6.1605	2.5836	0.1092	53.04	9.2126
1.0	0	6.9856	6.1494	2.5874	0.1095	53.01	9.2264

meet the convergence of the parameter, in this case the selection that weight coefficient are $\sigma_1 = 0.3$ and $\sigma_2 = 0.7$ can ensure the system having less convergence time and lower power consumption, correspondingly optimal parameters are $k = 0.0872$, $k_1 = 5.0734$, $k_2 = 5.9838$, $k_3 = 5.7039$, then the system power consumption and convergence time are respectively 8.6158 J and 40.84S.

Compared with the literature [3] the convergence time of system is reduced, while the power consumption increased slightly, obviously the overall performance of the controller has been significantly improved. Furthermore, under the same initial conditions, by using different optimization algorithms, to compare the number of iterations required for the best solution, the number of iterations required for the best solution in the literature [3] with GA is 82, however, the MOPSO algorithm in the paper only needs 56, so the faster convergence speed and the higher optimization efficiency are shown. With the optimal combination of parameters, the system simulation is shown in Fig. 1.

Seen from the Fig. 1(d), in the whole process of system control, the control torque is always less than the maximum torque of 4 N.m, which satisfies the constraint. As been worth mentioning, the method of parameter optimization based on MOPSO change the fact that the selection of the design parameters mostly depended on the experience of designers. Compared with simulation results of the reference [6, 7], the optimization of the multi-objective optimization problem is not a single objective

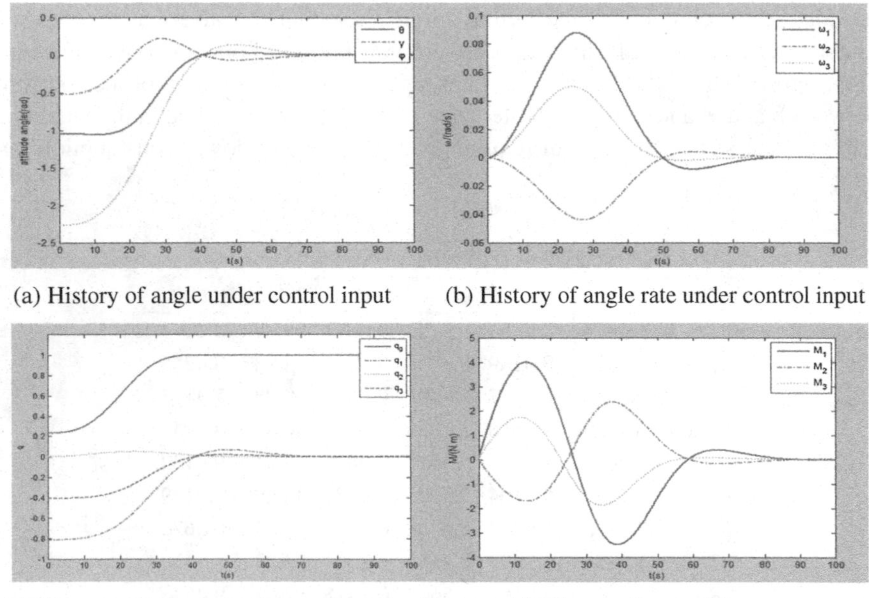

(a) History of angle under control input (b) History of angle rate under control input

(c) History attitude quaternion under control input (d) History of control torque

Fig. 1. The simulation results by using optimal parameters

optimization problem to obtain an optimal solution, but the search for group solutions with highest priority level. The solution may be optimal for this goal, to other target can only non-deteriorate, because usually there is conflict and constraint between the objectives of the optimization problem,therefore, the balance between the goals should been kept. As a consequence, contrasted with the reference [6, 7], the overall performance of the controller has been significantly improved.

5 Conclusion

In the essay the nonlinear controller parameters design problem of spacecraft large angle maneuver was solved, and the specific method of parameter optimization was gotten. Based on the spacecraft nonlinear dynamics model and Lyapunov method, a large angle attitude maneuver controller was proposed. And the effect of controller performance resulting from the control parameters was discussed in the paper, and two optimization indexes of the controller were defined. With the indexes optimized by MOPSO algorithm, the optimal parameter combination has been obtained under the actual requirements and constraints. It was verified that the control system with low energy consumption and system stability had significant performance under the premise of applying a large disturbance. The simulation results showed that the proposed method was superior to the traditional single-objective optimization design method. What is more, the parameter selection method take precedence over the traditional design approach depended on the designer experience, which has important practical significance. All in all, the method was highly expandability, and had better application prospect.

6 Acknowledgements

This work is supported by the National Natural Science Foundation of China (No. 61202251, 60875046), the Program for Liaoning Excellent Talents in University (No. LJQ2013133), and the Program for Liaoning Innovative Research Team in University (No. LT2010005), Natural Science Foundation of Liaoning Province (No. 201102008).

References

1. Wie, B., Barba, P.: Quaternion feedback for large angle maneuvers. J. Guid. Control Dyn. **8** (3), 360–365 (1985)
2. Pukdeboon, C., Zinober, A.S.I.: Control Lyapunov function optimal sliding mode controllers for attitude tracking of spacecraft. J. Franklin Inst. **349**, 456–475 (2012)
3. Huang, Q., Zhang, J.X., Zhang, Q., et al.: Optimization of control parameters based on multi-objective genetic algorithms for spacecraft large angle attitude. Appl. Mech. Mater. **538**, 470–475 (2014)
4. Lu, K.F., Xia, Y.Q., Zhu, Z., et al.: Sliding mode attitude tracking of rigid spacecraft with disturbances. J. Franklin Inst. **349**, 413–440 (2012)
5. Liang, J.J., Qin, A.K.: Comprehensive learning particle swarm optimizer for global optimization of multimodal functions. IEEE Trans. Evol. Comput. **10**(3), 281–295 (2006)
6. Jin, J., Zhang, J.R., Liu, Z.Z.: Optimized design of controller parameters for large angle spacecraft attitude maneuver. J. Tsinghua Univ. **49**(2), 289–292 (2009)
7. Zhang, J.R., Wang, Y.: PSO-based controller parameters design for spacecraft. Trans. Beijing Instit. Technol. **30**(4), 425–428 (2010)
8. Coello, C.A.C., Pulido, G.T., Lechuga, M.S.: Handling multiple objectives with particle swarm optimization. IEEE Trans. Evol. Comput. **8**(3), 256–279 (2004)

Analysis of the Time Characteristics of Network Water Army Based on BBS Information

Chunlong Fan, Chang Liu(✉), Chi Zhang, and Hengchao Wu

School of Computer Science,
Shenyang Aerospace University, Shenyang 110136, China
liuchang1125@163.com

Abstract. Water army is the network groups which engaged in the topic of speculation, marketing events, and the others to interfere normal information transmission, the information they released includes its behavioral characteristics. Therefore there is a need for in-depth discussion of their behavior and characteristics. In this paper, we adopt the opinion information extraction algorithm, and collect data released on the forum, including the posts and comments information of the Water army and the internet users. The background data is the time distribution of the numbers of QQ online in this paper, and analyzes the special characters of the collected information release time, finding the time features of the Water army is consistent with the general life cycle, to prove the Water army is the full-time; then make the accounts and posting comments constitute a network, analyzing its community structure using community division algorithm, and the results was compared with the water army organization behavior, to make sure the behavior of Water army has the organization.

Keywords: Water army · Spread of public opinion · Information extraction · Community divided

1 Introduction

Water army is a network organization with the forum, blog, micro-blogging and other forms of social media development and rapid rise, engaged in Gailou, irrigation, the topic of speculation, marketing events and other activities, to mislead the audience, interference spontaneous spread of public opinion; from "defamation event of Mengniu" in April 2010 to "present Xi take a taxi" in 2013, all are under the Water army manipulation [1]; In 2013, the highest law clearly statement that rumor of network will be held criminally responsible, from the network pushing Hands "Qin Huohuo" was sentenced in 2014, indicating that it's harmfulness has risen to aspects of Laws by country. Since 2009, Afer the CCTV [16] exposures the Water army's backroom operations, literature studies of the Water army appeared in the field of journalism or other fields in 2011 [2, 3, 6–8, 14, 18]. Literature [4, 5] that the Water army be divided into two categories of full-time and part-time, and considered part-time workers are the main army of the water; Which part-time workers are mainly students, the unemployed,

X. He et al. (Eds.): IScIDE 2015, part II, LNCS 9243, pp. 20–28, 2015.
DOI: 10.1007/978-3-319-23862-3_3

and so on, and full-time staff are usually experienced users, and has a much higher grade or high degree of concern account. Literature [6–10] from hype to analyze the process of Water army, the main interest of the proposed network marketing needs, to benefit the network of public relations company to receive orders, geographically dispersed Water army accept the offer, then all kinds of the delete, top, turn, posting, and the other actions, to meet the demands of the business results of their own interests. [18] From the Water army to manipulate public opinion to analyze the impact of the activities mainly Water army hide the true opinion, increased mental confusion and affect users of Internet users in the purchase of merchandise choices. Studies of abroad of the Water army focused on work patterns and geographic distribution, as in [11], from the geographical distribution of the water forces analysis, Water army organization focused on the central and southern China, and showing characteristics which is higher in east and lower in west, the time consistent with the Beijing time zone. According to foreign media reports [12, 15], the US Department of Defense and private companies are developing a software, using the false identity to distribute information on social media networks, the purpose is to accelerate pro-American propaganda. The above are the research perspectives of the Water army that are the operating target, staff composition, communication methods and the social influence. But the Water army is an autonomous group, we couldn't find the studies about the part-time Water army and organizational characteristics in time; The literature [19, 20] use the readers' borrow time as the readers' interests to the book, and using it to recommend books of library, and found that users in different time of the day and at different times of the week, will show obvious regularity query behavior. Therefore, in this paper we will make some research by the special distribution of the time, and using based semantic information extraction technology with a wrapper function and related algorithms [16], analysis of the collected text, comments data and numbers of 24 h online forums, in 24 h and a week. Get the typical U-shaped curve and the typical three peak distribution mode in 24 h. We find the working time is consistent with the normal schedule by analyzing, providing a valid evidence that the Water army is organizational. After using the complex network algorithm to divide the constructed network, then can get a clear community structure of the Water army, the Water army is network group with the clear community structure from the perspective of the structure characteristic.

2 Data Acquisition and Processing

2.1 Data Acquisition

By hiring the Water army to promote virtual goods information in Sina post bar of IT [16], and record the online numbers in per hour on the Forum. Then use JAVA to make a crawler to collect data samples appeared on the post bar before 30 pages of IT from 2013-6-6 to 2013-6-20. Collected 894 posts, 220 733 comments, including the Water army forces 10 posts, 2870 comments.

2.2 Data Processing

To use the tool of Html-Parser and pattern matching to extracted some accounts, posts and comments from the collected data. Then to edit number, remove duplicate data and establish the correspondence relationship on the extracted data, then save the accounts, posts and comments in the database. Then make the saved data match with the Water army data by SQL statements in the database, and identify the data of the Water army, and save the results in the database.

2.3 Network Building

By extracting information of Posts and comments, using the relationship between the use and the account, We construct the water army of the network. so-called relationship of the same posts means that If both accounts simultaneously been commented on a post, there is an edge between the two account in the actual constructed network. The constructed networks to have 1734110 and 1396944 edges and, diameter of the network are 8 and 2, the average path length are 3.0145 and 1.5202. The network is in line with the characteristics of small-world network, distribution of degree to meet power-law distribution.

3 Time Features of the Water Army

3.1 Data Analysis

Water army has always been believed that part-time as main, full-time is complementary, that means Water army is fewer groups. Therefore, we extracted comments of 24 h and a week, to study if the Water army consistent with the lifestyle and the working time of normal internet users. The basic experimental data extracts from the previous experiment posts' commentary time, processing the comments' time shown as following steps:

1. Every piece of information recorded as a post people. Extracting every comments' specific time, which day of a week and which hour of 24 h in the day,
2. Extracting the information of all the data and Published time of the feedback data from the Water army, showing the statistical data of 24 h, the numbers vector PA = < A1, A2..A24 > , the feedback data from the Water army vector PS = < S1, S2..S24 > ; Extracting published every piece of information is which day, statistics numbers of posted PW = < W1, W2..W7 > from Monday to Sunday; for forum of IT accounts and QQ accounts bounded, statistics the numbers of online forums of QQ at the same time, recording the online numbers in per hour PB = < B1, B2.. B24 > and the statistics online numbers of QQ from Monday to Sunday online PQ = < Q1, Q2..Q7 >.

3.2 Analysis the Water Army Behavior by the Hour Cycle

To compare the data of the above groups, to use the first-order deviation is very suitable, it can make all vectors are standard, so that data are comparable between the groups. Statistics of hours of vector PA, PN and PB changes as follows: First, suppose $\bar{x} = \frac{1}{24}\sum_{i=1}^{24} x_i$ as the average of a set of data, and then make the following changes for each set of data $y_i = \frac{x_i - \bar{x}}{\bar{x}} = \frac{x_i}{\bar{x}} - 1$ (value $y_i = x_i / \bar{x}$).

Figure 1 shows the first-order deviation of numbers of comments in the forum, the online numbers of QQ and comments of Water army; From this figure, we can obtain two conclusions:

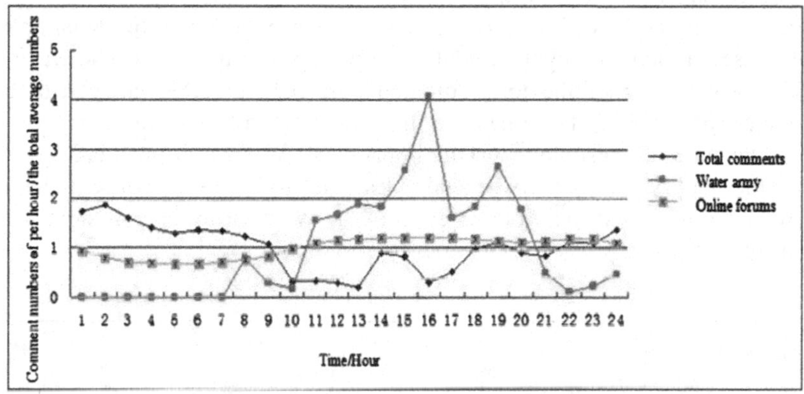

Fig. 1. The numbers of all comments, water army comments and online forums

(1) Calculate the case of deviation of average numbers of posts is much larger than deviation of the average numbers of onlinc in thc two data sets The reason is that this behavior is not consistent with online numbers' distribution of the forum, but affected by the number of the Water army and these posting behavior, making the whole comments increases with increases of the Water army comments.
(2) Information dissemination behavior has obvious characteristics of the work cycle. forum Online numbers of forum is lower in the morning, comment numbers is very high, but during working time with a higher online numbers of working hours forum, posts is lower, in line with modern life and work patterns. There are three peaks and three troughs on the distribution of the curve of posts,

This results is consistent with the rules in the paper [13]. The online numbers is lower on the total comment curve from 10:00 to 21:00, indicating during the time, the number of online and posting numbers and online numbers are disproportionate, but the distribution of, these peaks and troughs is consistent with people's work hours: 12:00 pm, 18:00 pm are the typical mealtime, 24:00 am, 8:00 am are the time for sleep and breakfast. Additionally the peak of 19:00 pm is significantly lower than the peak of 16:00 pm. We can know that the water army posting links submission time is usually around 15:00, 18:00 or 20:00 pm, a lot of water army will be after the break to find a new job or task after finishing that by the actual water army participate in the posting. For this reason, the trough significantly reduced, and increased at 15:00 pm.

3.3 The Water Army Behavior Analysis by a Day

The numbers changes have shown as the following from Monday to Sunday: set $\bar{x} = \frac{1}{7}\sum_{i=1}^{24} x_i$ as the average of a set of data, and then make the following changes for each set of data $y_i = \frac{x_i - \bar{x}}{\bar{x}} = \frac{x_i}{\bar{x}} - 1$ (value $y_i = x_i / \bar{x}$). Thus we can observe the total numbers of 24 h, QQ online numbers of 24 h and comments of the Water army of 24 h. Calculate the deviation of the numbers in a week and the QQ online numbers in a week to the average, shown as in Fig. 2.

From Fig. 2, it can be concluded: you can clearly see the extent of deviation of the three sets in seven days as a cycle, calculate the variance of the data of the water army is 1.2677, the variance of the total comments are 0.30529, the variance of online forums data is 0.00054, then we can see, the comments of the Water army are more volatile on the time. Because the working hours of the Water army will changes as the request of the employers, but the weekend is less, the working time concentrated in the normal working time, and the working style of the Water army is consistent with the normal internet users.

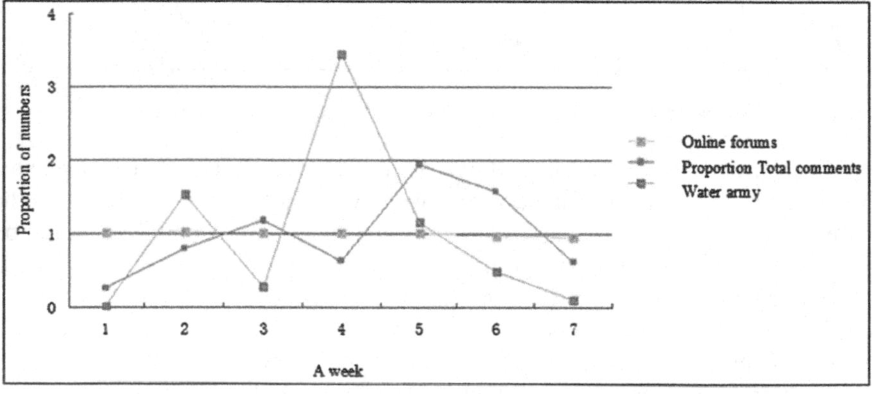

Fig. 2. posts and online numbers in a week

3.4 Absolute and Relative Time Analysis

We can get data in relative time from the data in the absolute time. Vector described in the second point reflects characters of the information publish in the absolute time. After extracting posting numbers of 24 h in a day, we set the time of the post publish as the basic time, and statist the comments in 24 h, the results shown in the Fig. 3.

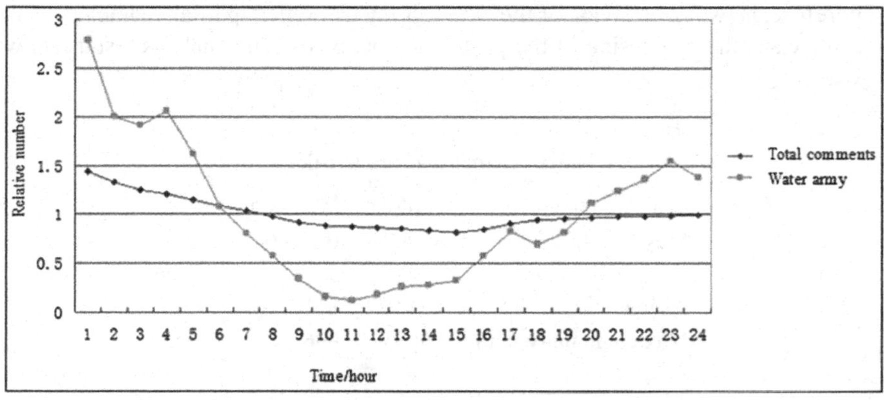

Fig. 3. Ideally with 24 points relative to the number of people

From the analysis of Fig. 3, we can see that curve reflect the activities of the Water army on the forums meets the normal daily physiological activities of the people. And compared with the posting numbers in the 24 h, there is a big difference between the two curves. The curve obviously showing an irregular shape of the U-shaped, there is the peak on the ends of the curve in 24 h, and the trough on the middle of the curve. People do a similar or the same job at the same time every day for the regular work habits and work patterns. After receiving the task, the water army will show the peak of working time. But because of the water army repeatedly engaged in a lot of boring activities of repetition posting, commenting, deleting posts, and the time and the amount of activities are strictly limited to the top organization, the Water army formed a regular work habits in this mode of operation required. Therefore, the operating model of the Water army in line with the U-shaped curve, is a process through peaks and valleys, in line with people's normal work habits.

By comparison with the ideal curve, we can see that the peaks of ideal curve is also slightly higher, trough of it curve also declined slightly in 24 h. But there is subtle similarities between the two curves, that means the numbers of pots is similar with the ideal numbers of posts. This explain such a special Water army group is consistent with ordinary normal operating mode and work habits. The groupment and the synergy of the Water army never change the fact that they meets the normal work of a person physical and psychological cycle.

4 Analysis the Organizational Structure of the Water Army

Most posting behavior of the Water army is organized and coordinated through an intermediary. the intermediary decide the content, way, quantity of the posts and way of the verification and payment after finishing it, forming the Water army formed a organization structure of network or the network organization in actual [17].

Therefore, network behavior of the Water army can reflect partial characters of the Water army,and the processing of the posts and comments. The analysis results shown in Table 1.

Table 1. Statistics characteristics

Statistics content	All data	Water army
Max Q value	0.4238	0.3369
Community num	52	2
Averge community num	81.87	811.5
Max community num	3685	996

In Table 1, Q values shows that community structure of the existing network is very clear. Q value of the total data is greater than the Q value of the Water army, which shows the organization of the Water army on the forums is more clear. The largest nodes of community of the former is big, indicating there is a large close-knit Water army community, shown in Fig. 4, which have five communities with the clear community structure, the numbers is 996 peoples, 166 peoples, 133 peoples, 125 peoples and 113 peoples. There are two water army from the feedback of the real water army in the five communities. And the other three are made by the posts and comments from other water army by artificial identification. You can see that there is a obvious central node in the Fig. 4, the community has the more clear hierarchy, which is in line with the fact that a water army only belong to the one or several specific groups, and there are the characters of the complex process of publishing, organizing and implementing by few organizers.

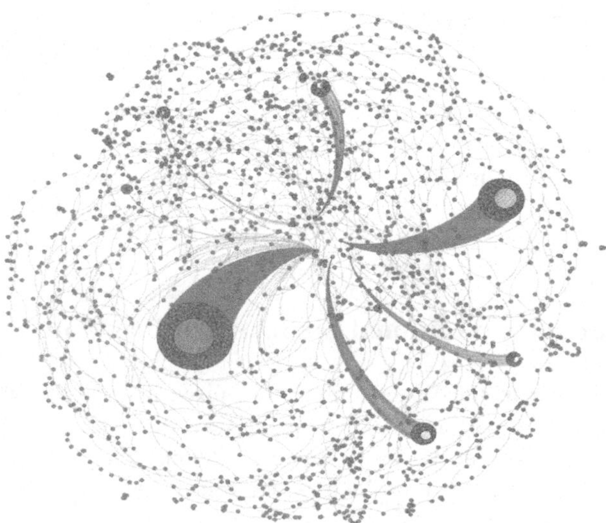

Fig. 4. All data network organization

5 Conclusion

Through practical manual analysis and statistics, we can draw the following conclusions:

(1): Water army is a full-time profession. By analyzing three curves of the reviews of the Water army, forums comments and online forums numbers. Get the lifestyle and working style of the water army, is consistent with the normal internet users, thus the (2): Water army is a full-time profession, part-time supplement

The Water army organizations has significant community structure. After dividing the established military network by the Fast algorithms, the division results contrast with the feedback of the real Water army, there are two communities are the Water army putted in this experiement. And large-scale Water army community networks is larger than the size of other network communities.

We get the charactersitics of the behavior and time of the Water army combined with the existing research. We will study the community discovery algorithm, the rapid detection the hot issues made by the Water army, the rumors spread mechanism of the Water army and dynamics analysis in the future.

References

1. Zhu, X.: The legal thinking of water army phenomenon of Mengniu and Yili PR. Lanzhou University, Gansu (2013)
2. Han, Z., Xu, F., Duan, D.: Equal probability map navy recognition model for microblogging. Comput. Res. Dev. **S2**, 180–186 (2013)
3. Secret network of water army survival: 50 % of the network security report posted by PR China, China Anti-Counterfeiting Report, 4, 31–33(2011)
4. Huang, X.: Water army: film vicious internet marketing lurker. Media Obs. **4**, 20–21 (2014)
5. Zhang, Y.: The intervention strategies of the network water army and the impact on the public opinion. Friends Ed. **1**, 68–70 (2014)
6. http://media.people.com.cn/n/2014/0818/c120837-25484101-2.html
7. Fan, X.: Information dissemination model of the network of water army. Media **9**, 54–56 (2011)
8. Lou, X., Liu, P.: Propagation analysis of network of water army. Contemp. Commun. **4**, 76–77 (2011)
9. Zhihuang, Y.: Network water army types, multiple credit and governance. Guangdong Inst. Public Adm. **4**, 16–22 (2011)
10. Lingfei, W.: Network of water army is what the military. Inf. Shanghai **04**, 82–83 (2011)
11. Zhijun, Y., Gensheng, G.: The formation mechanism and propagation efficacy study of network of water army. New Media Res. **05**, 53–56 (2011)
12. Linghung, C.: The study of effect network of water army of online public opinion. Anhui University, Hefei (2012)
13. Fan, X.: The propagation mechanisms and countermeasures of network of water army. News Enthus. **14**, 56–57 (2011)
14. Chen, C., Wu, K., Srinivasan, V.: Battling the internet water army: detection of hidden paid posters (2011). arXiv:1111.429v1 [cs.SI]

15. Fielding, N., Cobain, I.:Revealed: US spy operation that manipulates social media. http://www.theguardian.com/technology/2011/mar/17/us-spy-operation-social-networks
16. http://club.tech.sina.com.cn/thread-2115773-3-1.html
17. Fan, C.L., Xiao, X., Yu, L., Xu, L.: Behavior analysis of network navy organization based on web forums. Shenyang Aerosp. Univ. **29**(5), 64–67 (2012)
18. Newman, M.E.J.: Fast algorithm for detecting community structure in very large networks. Phys. Rev. E **69**(6), 0633 (2004)
19. Guo, C.H., Wang, X., Sun, L.: simulation research network culture of internet and market of the water army. Dalian Univ. Technol. Nat. Sci. **35**(3), 28–32 (2014)
20. Li, S., Wang, J., et al.: borrow time one binding characteristics of readers interested in the visual recognition method of library and information technology. Mod. Inf. Technol. Lib. **5**, 46–53 (2013)

Aspect and Sentiment Unification Model for Twitter Analysis

Hui Zhang[1,2](✉), Tong-xin Wang[3], Yi-qun Liu[1], and Shao-ping Ma[1]

[1] State Key Laboratory of Intelligent Technology and Systems, Tsinghua National Laboratory of Information Science and Technology, Department of Computer Science and Technology, Tsinghua University, Beijing, China
zhanghui_china@yeah.net, {yiqunliu,msp}@tsinghua.edu.cn

[2] Operation Experiment Center, Nanjing Army Command College, Nanjing, China

[3] Department of Physics, Fudan University, Shanghai, China
txwang11@fudan.edu.cn

Abstract. With the special "@, #, //" symbols, which include a lot of emotional symbols and pictures etc., tweets are different with other user-generated general texts, such as blogs, forums, reviews. Considering structural features and content of tweets, we present a semi-supervised Aspect and Sentiment Unification Model(PL-SASU). Using more information rather than solo texts, this model can model tweets better. The experiments of sentiment classification and aspect identification on real twitter data show that PL-SASU outperforms JTS, ASUM and UTSU model.

Keywords: Tweet · Sentiment classification · Aspect identification · Aspect and sentiment unification model · Emotional symbol

1 Introduction

With the development of Tweet platforms and the rapid increasing number of users in recent years, the sentiment analysis and aspect extraction of twitter becomes an important research direction, which has been widely applied in product satisfaction assessment [1], incident detection [2], pandemic forecasting [3], stock price prediction [4], personal health monitoring [5], etc. It is very difficult to label huge amount of tweets, because they contain big data, colloquial and non-standard language. Then unsupervised and semi-supervised learning algorithms become important and challenging research direction of twitter analysis.

Tweets are user-generated contents(UGC). Although a tweet is limited to no more than 140 characters, it contains not only text, but also pictures, videos and other content. Compared with general text, tweets contain a large number of emoticons (such as :), :(,🙂,😩, etc.) as well as special semi-structured information (such as"//, #, @", etc.). In previous academic researches, emoticons have been proved to be obviously related with author's emotion; "#hashtag", "//forward" and "@contact" features are

X. He et al. (Eds.): IScIDE 2015, Part II, LNCS 9243, pp. 29–42, 2015.
DOI: 10.1007/978-3-319-23862-3_4

closely related with aspects of a tweet. Aspect and sentiment models already exist, according to LDA [6], LSM [7], JTS [8], ASUM [9], UTSU [10] and so on, in which models are built based on pure text without considering above features of tweets, and the effect is not good when directly using those models.

In this paper, in order to improve the accuracy of sentiment and aspect analysis, we present Partly Labelled and semi-supervised Aspect and Sentiment Unification Model for Twitter analysis (PL-SASU), combining both sentiment and aspect analysis of tweets. This model has the following characteristics: (1) The semi-structured information affects the aspects and sentiment of tweets as an important factor. #hashtag labels are prior knowledge of aspects, and emoticons are prior knowledge of sentiments. A communicating tweet is relevant with "@contact", and a//forward tweet is relevant with a original tweet. (2) We assume that the generating process of a tweet is document-sentiment-sentence and document-aspect-words (including emoticons and labels), and the words in the same sentence have the same sentiment and different aspect. The rest of this paper is organized as follows: we introduce related work in Sect. 2. In Sect. 3, we present the graph model representation and inference of the PL-SASU model. The experimental setup and prior knowledge is discussed in Sect. 4. The experimental results and analysis are shown in Sect. 5, followed by the conclusion in Sect. 6.

2 Related Work

The words with the same semantics or emotion have the feature of co-occurrence. For example, if the document's aspect is "mobile phone" and sentiment is "positive", "fast" and "clarity" are likely to appear, and if the document's aspect is "Hotel" and sentiment is "positive", the words "comfort" and "clean" are likely to occur. Considering this feature, Chenghua Lin proposed JTS model [8] in 2011. In this JTS model, the generation process of a document is document-sentiment-aspect-words; the document is represented by the words bag model and the interrelation between words is not considered. But in fact, words are not completely independent. By adding different constraints to words, Li Fangtao et al. proposed the Dependency-Sentiment-LDA model [11]; Yohan Jo et al. proposed ASUM model [9]; SUN Yan et al. proposed UTSU model [10]. Dependency-Sentiment-LDA model considers the relationship between two words, which are connected by "and", "but" or other conjunctions. ASUM model assumes that the words in the same sentence have the same aspect and sentiment. UTSU model assumes that the words in the same sentence have the same sentiments and deferent aspects, which is consistent with the natural language expressions. The sentiment classification and aspect identification experiments on real product reviews data show that UTSU outperforms JTS and ASUM model.

As discussed earlier, a tweet has a lot of information: the author's user name and avatar, text, content, picture, video, publishing time, publishing location, publishing equipment, the number of praise, comment, forwarding and collection and so on. This paper only focuses on the own feature of tweet text content, including emotions and "//, @, #"symbols, which are stored in the form of text. Emotions have also been widely used in sentiment classification of tweets. Some supervised methods use emotional symbols which represent strong emotional feelings to supplement deficiency of the training set [12, 13]; some unsupervised methods added emotional symbols to sentiment dictionary as prior knowledge [14]. "#hashtag" is widely used to identify aspects of tweet, discover key events, etc. [15, 16]; "@Contacts" and "//forward" are used in tweet aspects recognition[17, 18].

In summary, the existing sentiment and aspect models consider only plain text features, without considering some special symbols of tweets. Similarly, although there are some sentiment classification and aspect identification models which consider not only features used by tweet language, but also special symbols, but aspect identification and sentiment classification are studied separately, and the common impact of these features on aspects and sentiments in text generation process has not been taken into account. This paper aims to develop a sentiment-aspect model, combining both aspect identification and sentiment classification, while considering the own feature of tweet.

3 PL-SASU Model

In this section, we will introduce a unification model for sentiment classification and aspect identification: PL-SASU model, which extend UTSU model [10], one of the most effectively used probabilistic aspect models. UTSU model and PL-SASU model are shown in Fig. 1(a) and (b). The symbols are described in Table 1.

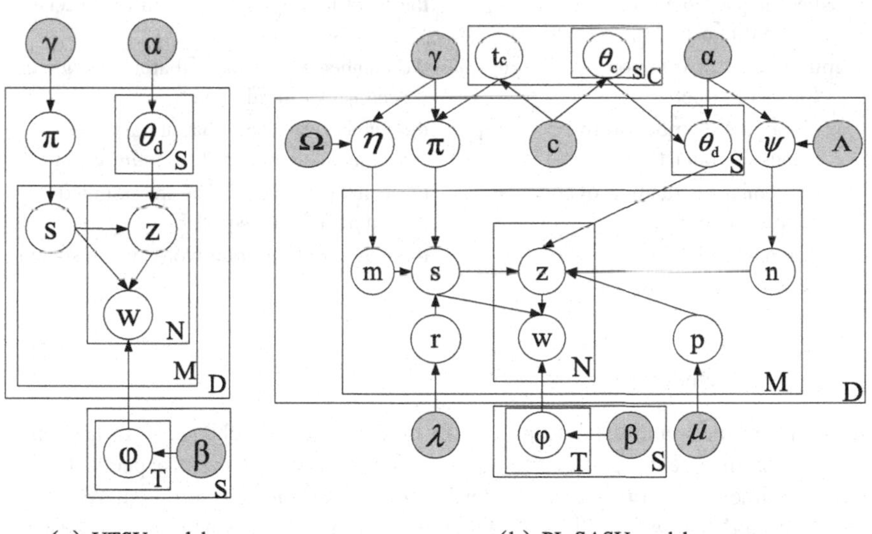

Fig. 1. Graphical representation of the Aspect and Sentiment Unification Model. (Nodes are random variables, edges are dependencies, and plates are replications. Only shaded nodes are observable)

Table 1. Meanings of notations

D	The number of tweets in a corpus	φ	multinomial distribution over words of topic
M	The number of sentence in a tweet	Ψ	multinomial distribution over aspects of aspect label
N	The number of words in a sentence	η	multinomial distribution over sentiments of sentiment label
T	The number of Aspect categories	α	Dirichlet prior vector for θ_d
S	The number of sentiment categories	γ	Dirichlet prior vector for π
V	the vocabulary size	β	Dirichlet prior vector for φ
w	word	S_l	subspace of S
z	Aspect	λ	Dirichlet prior vector for r
s	Sentiment	μ	Dirichlet prior vector for p
m	sentiment label	c	Dirichlet prior vector for t and θc
n	topic sentiment	s_i	the sentiment of *i*th sentence
Ω	sentiment label space	z_i	the aspect of *i*th word
Λ	aspect label space	s_{-i}	the sentiment assignments for all sentences except *i*th sentence
r	sentiment impact factor of forward tweet	z_{-i}	the aspect assignments for all words except *i*th word
p	aspect impact factor of forward tweet	w_i	the word list representation of *i*th tweet
θ_d	multinomial distribution over topics of tweet	n_{dj}	the number of sentences that are assigned sentiment *j* in *d*th tweet
θ_c	multinomial distribution over topics of contact	n_{djk}	the number of words that are assigned sentiment *j* and aspect *k* in *d*th tweet
π	multinomial distribution over sentiments of tweet	n_{jkw}	the number of words that are assigned sentiment *j* and aspect *k*
t	multinomial distribution over sentiments of contact	m_{iw}	the number of the total words in *i*th sentence

3.1 PL-SASU Model

UTSU is a sentiment and aspect unification model. This model makes an assumption that the words in a sentence have the same sentiment and deferent aspects. Both the sentence sentiment categories and the words aspect categories are sampled. We propose a generative model based on UTSU. Our goal is to improve the accuracy of sentiment classification and aspect identification of tweet.

PL-SASU adds #hashtag labels and emotions feature to UTSU model. This model has the following four assumptions:

- If tweet d has some #hashtag labels, then the document-aspect distribution(θ_d) will be affected by the #hashtag. A #hashtag label is added by the user or twitter platform, which represents aspect(key word) of a tweet and can be considered as a "display" aspect of tweet. Not every tweet has #hashtag label, the "latent" aspects of tweet without #hashtag label could be from any aspect. We proposed the transition variable x which determines whether the tweet d has #hashtag or not. If $x = 0$, the tweet has no "#hashtag", else $x = 1$.

- If tweet d has emotions, the document-sentiment distribution(π_d) will be affected. One emotional symbol often represents some certain sentiments. For example, the sentiment of tweet with may be "happiness" or "like", unlikely to be "sad" or "anger". An emotional symbol represents possible sentiment of a tweet and can be considered as a "display" sentiment of tweet. Not every tweet has emotions, the "latent" sentiments of tweet without emotional symbols might be any possible sentiment. We proposed the transition variable y which determines whether the tweet d has emotions or not. If $y = 0$, the tweet has no emotions, else $y = 1$.
- If tweet d starts at "@Contact", which usually implies an answer, the document-aspect distribution(θ_d) is affected by the contact-aspect distribution(θ_c) and the document-sentiment distribution(π_d) is affected by the contact-sentiment distribution(t_c) with some probability.
- If tweet d is a forwarded tweet, the document-aspect distribution(θ_d) affected by the document-aspect distribution(θ_{RT}) and the document-sentiment distribution(π_d) is affected by the contact-sentiment distribution(π_{RT}) with some probability.

As shown in Fig. 1(b), PL-SASU is a four-layer aspect model. The generative process of PL-SASU model is shown as below:

1. For every pair of sentiment s and aspect z, draw a word distribution $\varphi_{sz} \propto Dirichlet(\beta_s)$.
2. For each tweet d, if x=0 and y=0, the generative process of PL-STSU is the same with UTSU.
   - If y=0, Draw the document's sentiment distribution $\pi_d \propto Dirichlet(\gamma)$、the original document's sentiment distribution $\pi_{RT} \propto Dirichlet(\gamma)$,the contact-sentiment distribution $t_c \propto Dirichlet(c)$, weighted sum π_d.else if y=1, π_d is replaced by $\eta_{ls}, l \in [1, \ldots L], s \in [1, \ldots S_l]$.
   - If x=0, for each sentiment s, draw an topic distribution $\theta_{ds} \propto Dirichlet(\alpha)$, the original document's topic distribution $\theta_{RTs} \propto Dirichlet(\alpha)$,the contact-topic distribution $\theta_c \propto Dirichlet(\alpha)$,else if y=1,θ_{ds} is replaced by $\Psi_{hz}, h \in [1, \ldots H], z \in [1, \ldots T_h]$.
3. For each sentence, choose a sentiment $s \propto Multinomial(\pi_d)$.
4. For each word, firstly choose a topic $z \propto Multinomial(\theta_{ds})$, then generate words w $Multinomial(\varphi_{sz})$.

In summary, as shown in Fig. 1(b), the joint probability distribution of w, s, z in a tweet is:

$$\begin{aligned}&\mathrm{p}(w,s,z|\alpha,\beta,\gamma,\lambda,\mu,c,\Lambda,\Omega)\\&=p(w|s,z,\phi)p(z|\alpha,\mu)p(s|\gamma,\lambda)p(\phi|\beta)\end{aligned} \tag{1}$$

The sentiment variable s is mainly affected by π_d , π_{RT}, t_c and η_{ls}.

$$\begin{aligned}&\mathrm{p}(s|\gamma,\lambda)=(p(\mathrm{s}|\mathrm{m})\mathrm{p}(m|\eta_m)p(\eta_m|\gamma,\Lambda))^{\mathrm{x}}\times\\&[\alpha_{\mathrm{t}}\mathrm{p}(\mathrm{s}|\pi_{\mathrm{d}})\mathrm{p}(\pi_d|\gamma)^{\mathrm{r}}(\mathrm{p}(\mathrm{s}|\pi_{\mathrm{RT}})\mathrm{p}(\pi_{\mathrm{RT}}|\gamma)\mathrm{p}(\mathrm{r}|\lambda))^{1-r})\\&+(1-\alpha_{\mathrm{t}})\mathrm{p}(\mathrm{s}|\mathrm{t}_c)p(t_c|c)]^{1-x}\end{aligned} \tag{2}$$

The aspect variable z is mainly affected by θ_d, θ_c, θ_{RT} and ψ_n.

$$\begin{aligned}&\mathrm{p}(\mathrm{z}|\alpha,\mu)=p(z|n)p(n|\psi_n)p(\psi_n|\alpha,\Omega)^{\mathrm{y}}\\&\times[\tau(p(z|\theta_{RT})p(\theta_{RT}|\alpha))^{p}(p(z|\theta_d)p(\theta_d|\alpha))^{1-p}p(p|\mu)\\&+(1-\tau)p(z|\theta_c)p(\theta_c|c)]^{1-y}\end{aligned} \tag{3}$$

3.2 Model Inference

In this section, we describe the inference algorithms for PL-SASU.As shown in Fig. 1, original data of a tweet such as "#hashtag", "@Contacts" and "//forward" are directly obtained; aspects and sentiments of a tweet are latent variables. Generative model calculates latent variables by maximizing the joint distribution probability. The accurate calculation of PL-SASU model is very difficult, so the Gibbs Sampling is adopted here for the approximate estimation of parameters θ_d, θ_c, π, η, φ, ϕ.

In order to perform Gibbs sampling with PL-SASU, we need to compute the conditional probability$p(s_m = j,\ z_i = k|s_{-m},\ z_{-i}, w)$, where "$i$" represents index number of word token, $i = (d,m,n)$ represents nth word in mth sentence of dth tweet. s_m represents the sentiment of mth sentence, s_{-m} represents sentiments for all the sentences in the collection except for the sentence m. z_i represents aspect of ith word, z_{-i} is vectors of assignments of aspects for all the words in the collection except for the word i. Due to space limit, we show the sampling formulas without derivation:

$$\begin{aligned}&\mathrm{p}(s_{\mathrm{m}}=j,z_i=k|s_{-\mathrm{m}},z_{-i},w)\\&=\frac{p(s,z,w)}{p(s_{-m},z_{-i},w)}\\&=\frac{p(w|s,z)p(s,z)}{p(w_{-i}|s_{-m},z_{-i})p(w_i)p(s_{-m},z_{-i})}\\&\propto\frac{n_{dj}-1+\gamma}{\sum_{j'=1}^{S}(n_{dj'}-1+\gamma)}\times\frac{n_{djk}-1+\alpha}{\sum_{k'=1}^{T}(n_{djk'}-1+\alpha)}\times\frac{n_{jkw}-1+\beta_j}{\sum_{w'=1}^{V}(n_{jkw'}-1+\beta_j)}\end{aligned} \tag{4}$$

The notations are described in Table 1. After repeated iteration of formula (4), combined with sampling in all sentiments and aspects until the sampling result is stable, we can get the approximate probability of sentiment j in tweet d:

$$\pi_{\mathrm{dj}} = \frac{\mathrm{n}_{\mathrm{dj}} + \gamma}{\sum_{j'=1}^{S} n_{dj} + S\gamma} \tag{5}$$

The approximate probability of aspect k of sentiment j in tweet d is:

$$\theta_{\mathrm{djk}} = \frac{\mathrm{n}_{\mathrm{djk}} + \alpha}{\sum_{k'=1}^{T} n_{djk'} + T\alpha} \tag{6}$$

Using formulas (5) and (6) above, the approximate probability of aspect k in tweet d is:

$$\theta_{\mathrm{dk}} = \sum_{\mathrm{j}=1}^{S} (\theta_{djk} \times \pi_{dj}) \tag{7}$$

The approximate probability of word w in sentiment-aspect$\{j,k\}$ is:

$$\varphi_{\mathrm{jkw}} = \frac{\mathrm{n}_{\mathrm{jkw}} + \beta_{\mathrm{j}}}{\sum_{\mathrm{w}'=1}^{V} n_{jkw'} + V\beta_j} \tag{8}$$

Similarly, the approximate probability of aspect k and sentiment j in contact c is:

$$\theta_{\mathrm{ck}} = \frac{\mathrm{n}_{\mathrm{ck}} + \mathrm{c}}{\sum_{\mathrm{k}'=1}^{T} n_{ck} + Tc} \tag{9}$$

$$\mathrm{t}_{\mathrm{cjk}} = \frac{\mathrm{n}_{\mathrm{cjk}} + \mathrm{c}}{\sum_{\mathrm{c}'=1}^{\mathrm{C}} n_{c'\mathrm{jk}} + Tc} \tag{10}$$

The approximate probability of sentiment j in sentiment label m is:

$$\eta_{\mathrm{mj}} = \frac{\mathrm{n}_{\mathrm{mj}} + \gamma}{\sum_{\mathrm{j}'=1}^{S} n_{mj'} + S\gamma} \tag{11}$$

The approximate probability of aspect k in aspect label h is:

$$\psi_{hk} = \frac{n_{hk} + \alpha}{\sum_{k'=1}^{T} n_{hk'} + T\alpha} \tag{12}$$

3.3 Sentiment and Aspect Classification

With the above formulas, the model parameters θ_d, ϑ_c, π_d, t_c, φ and η can be estimated. Based on these parameters, we can know the sentiments and aspects of each tweet, the most likely sentiment of each emotion, the most likely aspect of each #hashtag label and the most interested aspects and emotional tendencies of each contact(author). The sentiment of Tweet d in training set can be achieved by calculating $s_d = argmax\{\pi_{dj}|j \in \{1,\ldots,S\}\}$. The aspect of Tweet d in training set can be achieved by calculating $\theta_d = argmax\{\theta_{dk}|k \in \{1,\ldots,T\}\}$.

4 Experimental Setup

In this section, we describe our data sets, the sentiment seed words, the priors knowledge and hyper parameter Settings.

4.1 Datasets

The data set used in this paper consists of two parts. One part comes from Chinese Sina Weibo(NLP&CC 2013), including one training set and one testing set, and there are totally 10000 tweets[1] which are taken randomly. Each tweet is labeled as one of the 8 sentiments (neutral, like, happiness, sadness, disgust, anger, surprise, fear). The other part consists of 10000 tweets which are random collected from tweets released between November 1, 2013 to November 30, 2013, which contain the keywords "transgenic" or "#transgenic#" or "@ Cui yongyuan", and these 10000 tweets are not labeled. The tweets used are pretreated: word segmentation and POS tagging are completed by ICTCLAS50 which was developed by institute of Computing Technology Chinese Academy of Sciences, deactivated words are deleted, transitional complex sentence is transferred to two sentences, repeated modal and modal at end of sentence are reserved, repeated punctuation are reserved. Repeating twice serves as one lexical item, repeating three times serves as another lexical item.

4.2 Sentiment Seed Words

This paper uses the affective lexicon Ontology [18] proposed by Information Retrieval Laboratory of Dalian University of Technology as prior knowledge. It contains a total of 20000 words, and emotions are split into 7 main categories and 21 sub categories.

[1] http://tcci.ccf.org.cn/conference/2013/dldoc/evdata02.zip.

4.3 Sentiment Labels

In this paper, we take into account 102 symbols[2] frequently used by Sina Weibo as sentiment labels, which are classified by its polarity and emotion categories, as shown in Table 2. The sentiment polarity classification includes positive, negative and neutral; if emotional symbol m is positive, distribution of symbol-sentiment is [0.8, 0.1, 0.1]. Considering that there is negative sentence and noise, positive emotional symbols don't necessarily appear in positive Tweets. If the emotional symbol m is negative, distribution of symbol-sentiment is [0.1, 0.8, 0.1]; and if the emotional symbol m is neutral, distribution of symbol-sentiment is [0.2, 0.2, 0.6].When sentiment classification is a multi-emotional classification problem, the distance among sentiments are different. For example, like and happiness is closer than like and sadness. That means, the possibility that two sentiments appear at the same time will be defined as the distance between them. Assumption: the distance between sentiments with same polarity is closer. Then like is close to happiness; sadness and disgust are close to anger. If emotional symbol is like, the symbol-sentiment distribution is $\eta_m = [0.5, 0.2, 0.05, 0.05, 0.05, 0.05, 0.05, 0.05]$, and it's the same for the emotional symbol of happiness. If emotional symbol is sadness, the symbol distribution is $\eta_m = [0.05, 0.05, 0.45, 0.15, 0.15, 0.05, 0.05, 0.05]$, and it's the same for the emotional symbol of disgust and anger. If emotional symbol is surprise, the symbol distribution is $\eta_m \in [0.05, 0.05, 0.05, 0.05, 0.05, 0.65, 0.05, 0.05]$, and it's the same for the emotional symbol of fear and neutral.

Table 2. Sentiment classification assignment of emotional symbols

Polarity	Sentiment	Emotional symbols	num.
neutral	neutral		23
positive	happiness	:)	20
	like	]	19
negative	sadness	:(	17
	disgust		11
	anger		7
*	surprise		5
*	fear	none	0

[2] Sina Weibo reserve emotional symbol picture in the form of "[characters]", for example is reserved as "[hehe]".

4.4 Aspect Labels

"#hashtag" can directly act as an aspect label of tweet, each aspect label corresponds to a distribution of aspects, but each aspect only takes part in exactly one aspect label. The "#apple" may come from the topic "computer", "plant" or "company", and the aspect "computer" can only belong to "#computer" label.

4.5 Hyperparameter Settings

Setting for other parameters refer to Reference[10, 11], $\alpha = c = \gamma = 1$, $\beta = 0.01$, $T = 40$, $\lambda = 1$, $\alpha_t = \tau = 0.5$, indicating that the forwarded tweet and original tweet have perfect correlation, the tweet and "@contact" have some relevance, with iterations $N = 2000$.

5 Experimental Results and Analysis

We performed some experiments to evaluate o PL-SASU. In one experiment, we test the sentiment classification performance of PL-SASU, and in another experiment, we evaluate the aspects discovered by PL-SASU.

5.1 Sentiment Classification

Comparison between PL-SASU, JTS, ASUM,UTSU and Best result of NLP&CC2013 has been proposed. PL-SASU, JTS, ASUM and UTSU adopt seeds emotional word as prior knowledge [18]. Precision of polarity identifying, value of recall rate and F-measure are evaluation indexes for polarity identification; precision of Micro average, value of recall rate and F-measure are evaluation indexes for sentiment classification (Table 3).

Table 3. Comparison effect of sentiment classification with other models

Model	*Polarity identification*			*Sentiment classification*		
	accuracy	*Recall rate*	*F-measure*	*accuracy*	*Recall rate*	*F-measure*
JTS	60 %	58 %	59 %	19 %	24 %	21 %
ASUM	70 %	67 %	68 %	24 %	29 %	26 %
UTSU	71 %	65 %	68 %	33 %	30 %	31 %
Best*	64 %	84 %	73 %	32 %	39 %	35 %
PL-SASU	73 %	70 %	71 %	42 %	34 %	38 %

No matter in polarity identification or sentiment classification, PL-SASU is better than JTS, ASUM and UTSU. In polarity identification, PL-SASU is 12 % higher than JTS, 3 % higher than ASUM and UTSU, but is 2 % lower than the best evaluating result of NLP&CC2013(Best*). In sentiment classification, PL-SASU is 17 % better than JTS, 12 % better than ASUM, 7 % better than UTSU, and worse 2 % than the best evaluating result of NLP&CC2013(Best*). In sentiment identification, PL-SASU is17 % better JTS, 12 % better than ASUM,7 % better than UTSU, and 3 % better than Best*. These results are statistically significant, and PL-SASU performs better than JTS,ASUM and UTSU, which indicate that this model is effective for Polarity identification and sentiment classification. Let the #hashtag or emotion knowledge merged into the course of modeling of tweet, so the generation process of tweet could be described more accurate. Only in Polarity identification is this model inferior to the Best*, but the Best* method is a supervised machine learning method which require a large number of labeled data. PL-SASU is semi-supervised method with no need for manually labeling; what's more, it is a full-probability Bayesian model, so it's easy to add prior knowledge for example the time or location of when a tweet is published.

5.2 Results with Different Aspect Numbers

The impacts on sentiment classification accuracy of different aspect numbers have been tested in this paper. As shown in Fig. 2, when the number of topic is 1, the model degenerates to traditional LDA model, in which Emotional categories act as topics. The document's aspect feature is neglected and it's considered that the generation of words is only affected by sentiment; and in this condition, the model has low accuracy. When the number of topic increases, the accuracy of the model also increases, which indicates that words are not only related with topics, but also sentiments. With further increase of the number of aspects, the performance of the model decreases. The number of topics is an important parameter influencing the performance of sentiment classification, and it's necessary to find the best parameter by unceasing experiments based on data sets.

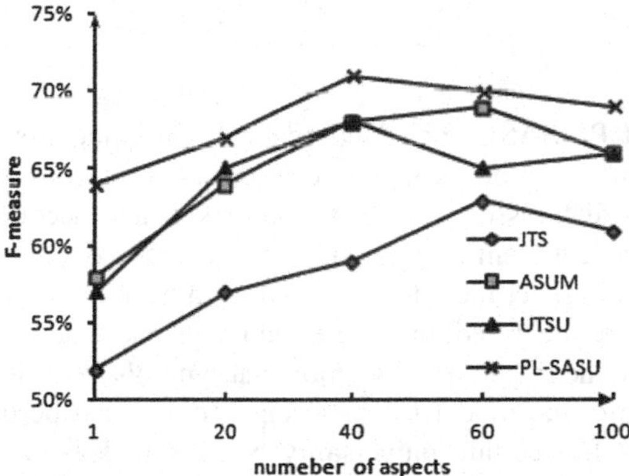

Fig. 2. Sentiment classification by the four sentiment and topic models, JTS, ASUM, UTSU, and PL-SASU model. The F measure represents the average value from 5 samples.

5.3 Aspect Identification

The second goal of PL-SASU is to extract topics from the tweet dataset and evaluate the effectiveness of aspect sentiment captured by the model. This paper takes perplexity [17, 19] as an index for evaluating Aspect Identification. The perplexity of the joint topic and sentiment model M is defined as:

$$\text{Perplexcity(M)} = \exp\{-\frac{\ln(\sum_{m} p(w_n))}{\sum_{m} N_n}\} \quad (13)$$

M represents the corpus word number (the number of words), the number of Nn represents the nth word, p(wn) represents the probability of the nth word. Better models M of the unknown distribution p will tend to assign higher probabilities p(wn) to the test events. Thus, they have lower perplexity (Table 4).

Table 4. Perplexity of the four sentiment and topic models

Iterations	JTS	ASUM	UTSU	PL-SASU
100	6341	6300	6320	6290
500	6137	5987	5534	5395
1000	5964	5835	5314	5279
2000	5930	5780	5278	5250

Under the same setting of parameters, the perplexity of PL-SASU is always lower than JTS,ASUM and UTSU in different iterations. It is confirmed that PL-SASU model is better than other three topic and sentiment model, because this model considers the "@, #, //" and other semi-structured information, which does improve the performance of aspect identification.

6 Conclusions

This paper proposed PL-SASU model based on UTSU, and it's a semi-supervised generative model which can be used in tweet sentiment analysis and aspect identification, with the own features of the tweet being taken into account. Compared with other three generative sentiment-topic models, the accuracy of sentiment classification and the performance of aspect identification have both been improved a lot; aspect and sentiment words can be identified, the interested topics and emotional tendency of the contacts can be found, the role played by emotional symbols in expressing mood can be analyzed. What's more, the model is cross-language and can be directly applied for English tweet; and it has certain universality because it doesn't emphasize domain knowledge. Certainly, this model is also worth studying further. First, tweets have extremely abundant information like authors, time and location of posting, etc. Due to restrictions of the data set, this paper studies tweet without considering these useful features. Secondly, due to the habits of Chinese Weibo users, #hashtag labels in data set is relatively less. Thus, the impact of aspect labels on aspect expression can't be verified very well. Further experimental research will be carried out in English tweets and super-popular tweet event.

Acknowledgments. This work was supported by National Key Basic Research Program (2015CB358700) and Natural Science Foundation (61472206, 61073071) of China.

References

1. Jmal, J., Faiz, R.: Customer review summarization approach using Twitter and SentiWordNet. In: Proceedings of the 3rd International Conference on Web Intelligence, Mining and Semantics. ACM (2013)
2. Abel, F., et al.: Twitcident: fighting fire with information from social web streams. In: Proceedings of the 21st International Conference Companion on World Wide Web. ACM (2012)
3. Signorini, A., Segre, A.M., Polgreen, P.M.: The use of Twitter to track levels of disease activity and public concern in the US during the influenza A H1N1 pandemic. PLoS One **6** (5), e19467 (2011)
4. Bollen, J., Mao, H., Zeng, X.: Twitter mood predicts the stock market. J. Comput. Sci. **2**(1), 1–8 (2011)
5. Triantafyllidis, A.K., et al.: A pervasive health system integrating patient monitoring, status logging, and social sharing. IEEE J Biomed. Health Inform. **17**(1), 30–37 (2013)
6. Blei, D.M., Ng, A.Y., Jordan, M.I.: Latent dirichlet allocation. J. Mach. Learn. Res. **3**, 993–1022 (2003)
7. He, Y.: Latent sentiment model for weakly-supervised cross-lingual sentiment classification. In: Clough, P., Foley, C., Gurrin, C., Jones, G.J., Kraaij, W., Lee, H., Mudoch, V. (eds.) ECIR 2011. LNCS, vol. 6611, pp. 214–225. Springer, Heidelberg (2011)
8. Lin, C., He, Y.: Joint sentiment/topic model for sentiment analysis. In: Proceedings of the 18th ACM Conference on Information and Knowledge Management. ACM (2009)
9. Jo, Y., Oh, A.H.: Aspect and sentiment unification model for online review analysis. In: Proceedings of the Fourth ACM International Conference on Web Search and Data Mining. ACM (2011)
10. Sun, Y., Zhou, X.G., Fu, W.: Unsupervised text sentiment analysis based on Topic-Sentiment mixed model. Acta Scientiarum Naturalium Universitatis Pekinensis **1**, 017 (2013)
11. Li, F., Huang, M., Zhu, X.: Sentiment analysis with global topics and local dependency. In: AAAI (2010)
12. Liu, K.L., Li, W.J., Guo, M.: Emoticon smoothed language models for twitter sentiment analysis. In: AAAI (2012)
13. Quercia, D., Capra, L., Crowcroft, J.: The social world of twitter: topics, geography, and emotions. In: ICWSM (2012)
14. Zhao, J., et al. Moodlens: an emoticon-based sentiment analysis system for chinese tweets. In: Proceedings of the 18th ACM SIGKDD International Conference on Knowledge Discovery and Data Mining. ACM (2012)
15. Boyd, D., Golder, S., Lotan, G.: Tweet, tweet, retweet: conversational aspects of retweeting on twitter. In: 2010 43rd Hawaii International Conference on System Sciences (HICSS). IEEE (2010)
16. Becker, H., Naaman, M., Gravano, L.: Beyond trending topics: real-world event identification on twitter. ICWSM **11**, 438–441 (2011)
17. Wang, Y., Agichtein, E., Benzi, M.: Tm-lda: efficient online modeling of latent topic transitions in social media. In: Proceedings of the 18th ACM SIGKDD International Conference on Knowledge Discovery and Data Mining. ACM (2012)

18. Linhong, X., Hongfei, L., Pan, Yu., Hui, R., Hianmei, C.: Constructing the affective lexicon ontology. J. China Soc. Sci. Tech. Inf. **27**(2), 180–185 (2008)
19. Newman, D., et al.: Distributed inference for latent dirichlet allocation. In: Advances in Neural Information Processing Systems (2007)

A Balanced Vertex Cut Partition Method in Distributed Graph Computing

Rujun Sun(✉), Lufei Zhang, Zuoning Chen, and Ziyu Hao

State Key Laboratory of Mathematical Engineering and Advanced Computing, Wuxi, China
rujuns@gmail.com

Abstract. Graph computing plays an important role in mining data at large scale. Partition is the primary step when we process large graph in a distributed system. A good partition has less communication and memory cost as well as more balanced load to take advantage of the whole system. Traditional edge cut methods introduce large communication cost for realistic power law graphs. Current vertex cut methods perform poorly with little consideration on load balance especially for online streaming vertex cut partition. In this paper, we formulate the total cost (partition cost, communication cost and computing cost) of graph computing especially that in iterating algorithms and analyze the cost of current partitioning methods. In addition, we explore a novel vertex cut method to ensure lower total cost. It has more balanced load with fewer communications. Experiments show that our method outperforms in state of the art graph computing frameworks at an average of 10 percent.

Keywords: Graph partition · Vertex cut · Streaming heuristics · Load balance

1 Introduction

The demand for large scale data analysis has driven distributed graph computing frameworks. Graph partition is of great significance as primary step in graph parallel computing. The goal of graph partition is to limit communication cost and balance computing load between machines. As we are more interested in iterative graph computing, these two aspects can make iterations more efficient with fewer synchronization cost.

Graph partition is an NP complete problem [1,2], traditional partition methods assign vertices to machines while edges got cut when adjacent vertices assigned to different machines. It finally contributes to the communication cost. It is called edge cut with the goal to make all machines have nearly the same number of vertices. Edge cut has been fully researched during the last twenty years with a number of high qualified algorithms by spectral methods, geometric methods or multilevel partitioning methods. The best known approximation

X. He et al. (Eds.): IScIDE 2015, Part II, LNCS 9243, pp. 43–54, 2015.
DOI: 10.1007/978-3-319-23862-3_5

guarantee [4] is $O(\sqrt{\log k \log n})$ where k is the cluster number and n is the size of the graph. Many applications get benefit from the development of edge cut such as biological networks, scientific computing and VLSI design.

However, vertices have not only data but also computations in graph parallel abstraction [5,7,8]. Computation cost has a positive relationship to the number of the outgoing or all edges of a vertex. For instance, when we calculate PageRank in a graph, a vertex needs to sum all information from its neighboring vertices which is its in-degree or sometimes its degree. Balanced vertex numbers do not always lead to balanced computing. Moreover, once a higher degree vertex gets separated from some of its neighbors, the communication cost will probably be high. The situation occurs in the scope of realistic graphs especially power law graphs from web or social networks where high degree vertices have so many neighbors that can't be placed in a single machine.

A new approach of graph partition was recently developed by cutting vertices but not edges of a graph [7,8]. It can avoid communication bottleneck in power law graphs, but theoretical computational complexity and approximation guarantee have not been extensively studied yet. Current vertex cut methods mainly focus on lowering the communication cost and care little about load balance. Sometimes the communication cost can be reduced while the assignment is highly imbalanced. But it has a serious impact on parallelization especially when iterations are needed. This situation occurs because that load balance is not a main factor in the evaluation of these graph partition methods, but it is only considered while comparison can not be made by other factors. In addition, the quality of streaming vertex cut methods relies on the order of input [9], but current vertex cut methods can't perform well in common graph input orders [10].

In this paper, we analyze communication, computing and partition cost of both edge cut and vertex cut in graph partition and develop a new vertex cut method which ensures load balance with little extra cost. It outperforms in state-of-the-art distributed graph computing frameworks and shows good scalability. Our contributions lie in following aspects:

- Provide a theoretical analysis for distributed graph computing cost in different partitions.
- Analyze current partition methods especially vertex cut methods.
- Propose a novel vertex cut partition method which can balance the load while not introduce extra communication cost.

The structure of the paper is as follows: Sect. 2 formulates the problem of graph partition and analyze communication and computing cost in graph partition, Sect. 3 provides our new vertex cut method and its comparison to previous methods, Sect. 4 gives the experiment result of our new method, related work is discussed in Sect. 5 and we get the conclusion in Sect. 6.

2 Graph Partition Formulation

We discuss a directed graph $G = (V, E)$ in this paper, V is the vertex set which has n members and E is the edge set which has m members and $E \subset V \times V$. Graph G needs to be divided into k partitions.

2.1 Partition Methods

Partition methods can be divided into two categories: edge cut and vertex cut.

Edge Cut: In an edge cut partition method, graph G is partitioned into k parts $S_1, S_2, \cdots, S_i, \cdots, S_k$. $S_i = G(V_i, E_i)$. Partition parameter $y_{v,i}$ is defined by $y_{v,i} = \begin{cases} 1, & \text{if vertex } v \text{ is assigned to machine } P_i \\ 0, & \text{otherwise.} \end{cases}$ $V_i = \{v \in V | y_{v,i} = 1\}$ Vertices are assigned to machines without replication. $\forall i \neq j$, we have $V_i \cap V_j = \emptyset$ And each vertex is assigned to a particular machine. $\cup_{i=0}^{k} V_i = V$ But edges whose adjacent vertices are on different machines will have replicas on all of these machines. $E_i \cap E_j = \{e(u, v) | u \in V_i, v \in V_j)\}$

Vertex Cut: In a vertex cut partition method, graph G partition can also be notated as k parts $S_1, S_2, \cdots, S_i, \cdots, S_k$. $S_i = G(V_i, E_i)$. Partition parameter $y_{e,i}$ is defined by $y_{e,i} = \begin{cases} 1, & \text{if edge } e \text{ is assigned to machine } P_i \\ 0, & \text{otherwise.} \end{cases}$ $E_i = \{v \in V | y_{v,i} = 1\}$ Edges are assigned to machines according to partition parameter $y_{e,i}$, $\forall i \neq j$, there is $E_i \cap E_j = \emptyset$ and all edges are assigned. $\cup_{i=0}^{k} E_i = E$ Thus a vertex will have replicas in different machines when its adjacent edges are assigned to more than one machine. $V_i \cap V_j = \{v | e_k \in E_i \text{ and } e_l \in E_j, \text{ where } e_k, e_l \in \Psi(v)\}$

2.2 Communication Cost

The communication cost can be calculated once we obtain a partition. In this section, we discuss the communication cost in a single iteration step where all vertices have done their defined computations.

In an edge cut partition, let $C_{comm}(P)$ be the communication cost of partition P. Communication occurs when an edge is across different machines. If each message is sent separately, the final communication cost can be represented by $C_{comm}(P) = \sum_{i=1}^{k} \sum_{j=1}^{k} \sum_{e \in E_i \cap E_j} c$ where c is the communication cost for each message. The above equation is for undirected graphs. For directed graphs, we should only change E_i, E_j to certain direction of edges such as $E_{in}(V_i) \cap E_{out}(V_j)$ or $E_{out}(V_i) \cap E_{in}(V_j)$ to describe the message passed along all out-edges or in-edges.

If messages from the same source machine to the same destination machine can be aggregated, communication cost will be reduced and it is usually used in graph computing frameworks. The communication cost will thus be $C_{comm}(P) = \sum_{i=1}^{k} \sum_{j=1}^{k} x_{i,j} c$ where $x_{i,j} = \begin{cases} 1, & \text{if } E_i \cap E_j \neq \emptyset \\ 0, & \text{otherwise.} \end{cases}$ It should be noticed that if the aggregated massage is too large, the cost will be larger since aggregation needs to be performed in more than one single message in realistic. c should be upper bound of the communication cost of an aggregation. The above equation is correct in theoretic analysis.

In a vertex cut partition, communication occurs when replicas of a vertex synchronize. $C_{comm}(P) = \sum_{i=1}^{k}\sum_{j=1}^{k}\bar{x}_{i,j}c$ where $\bar{x}_{i,j} = \begin{cases} 1, & \text{if } V_i \cap V_j \neq \emptyset \\ 0, & \text{otherwise.} \end{cases}$ As we have mentioned previously, vertex cut method can reduce communication cost especially in realistic power law graphs.

2.3 Computing Cost

In both vertex cut and edge cut graph partition methods, the computations we care are the same. Computing cost here is also limited to one iteration which is similar to that of communication cost.

For a vertex v, the computing cost may (1) be a constant value or (2) be related to its degree.

In case (1), the computing cost of a single machine is $C_{cp}(P_i) = |V_i|c_{cp}$, where c_{cp} is unit computing cost. And the total computing cost is $C_{cp} = \sum_{i=1}^{k}|V_i|c_{cp}$ In traditional edge cut partition, it is proportional to the number of vertices in the graph $C_{cp} = |V|c_{cp} = nc_{cp}$. In vertex cut partition, it is proportional to the total vertex replicas numbers.

In case (2), the computing cost in a single machine which has a partition of the whole graph is $C_{cp}(P_i) = |E_i|c_{cp}$. And the total computing cost is $C_{cp} = \sum_{i=1}^{k}|E_i|c_{cp}$ If the graph is partitioned by edge cut, edge will have replicas on different machines. The total computing cost C_{cp} will be the number of all edges with replicas. However, in a vertex cut partition, edges have no replicas which ensures a constant computing cost $C_{cp} = |E|c_{cp} = mc_{cp}$.

Iterative algorithms in graph computing are usually related to the degree of a vertex. We need to compute all its adjacent edges or incoming/outgoing edges. Fewer replicas of edges means less computing cost and less communication cost as a company. It proves that vertex cut partition is more suitable for iterative algorithms which needs calculation of all adjacent edges of vertex in data mining and machine learning.

However, minimum computing cost is not always the goal, since the computing cost we have calculated is within one iteration but communication happens between iterations. We should make a tradeoff between them to take full advantage of the distributed system and get optimal performance. The relationship of computing cost and communication cost will be discussed later in this section.

2.4 Partition Cost

Partition is the method to assign a graph to a distributed system. No matter how perfect the partition is, we care more about total processing cost of the graph but not the preprocessing of it. A good partition helps to reduce the computing cost and communication cost but doesn't need too much computing resources for itself.

However, there is a lack of research in theoretical guarantee of partition cost for graph partition algorithms as it is an NP complete problem. With a variety of heuristic methods to reach a good approximation, research goal for a new algorithm is to prove its efficiency in practice.

In offline partition, graph loading is separated from partitioning. We can have a full view of the whole graph while assigning a vertex or an edge. Nevertheless, offline partition can not always meet our needs since graphs are getting larger and it would be impossible to have a full view of the graph while assigning a single edge or vertex. Online streaming partition is essential.

In streaming partition, graph is partitioned while loading. The minimum assigning unit of a graph is edge. Vertex will be assigned with its adjacent edges. The cost of partition depends on the number of edges and the computing cost for assigning an edge. $C_{par} = \sum_{e_i \in E} c(e_i)$ In a simple partition method, an edge is randomly assigned to a processor by a hash function, so the unit cost is $c(e_i) = O(1)$. Total partition cost is $C_{par} = O(m)$.

In a method considering assigning history, the unit partition cost can be $c(e_i) = O(k \log |E_{\hat{j}}|)$ or $c(e_i) = O(k|E_{\hat{j}}|)$ etc. by different ways dealing with history information, where $E_{\hat{j}}$ is the number of edges having assigned to a machine until e_i comes. For example, if an edge will be assigned to the machine which has its adjacent vertex or adjacent edges and which stores edges in a vector with index, the unit cost will be the partition number k times search cost which is at least $O(\log |E_j|)$. Hash methods can also be used here to reduce search cost.

2.5 Total Cost

The total cost of graph computing includes that in partition, computing and communication periods.$C = C_{par} + t \cdot (C_{comm} + C_{cp})$, where t is iteration times of graph computing in most algorithms. If an algorithm needs no iteration, t will be 1. Most graph algorithms are practiced to be able to converge within hundreds iteration. It should be taken into consideration when designing a partition method.

For realistic distributed graph computing systems, what we care about is streaming vertex cut partition method for iterating algorithms where vertex calculate its value according to the degree of itself. The total cost can be expressed by $C = \sum_{e_i \in E} c_{par}(e_i) + t \cdot (\sum_{i=1}^{k} \sum_{j=1}^{k} \bar{x}_{i,j} c_{comm} + \sum_{i=1}^{k} |E_i| c_{cp})$.

In distributed systems, we need to not only minimize C but also balance the cost between machines to reach minimum processing time. Thus, the equation can be expressed as $\bar{C} = \bar{C}_{par} + t \cdot (\bar{C}_{comm} + \bar{C}_{cp})$, where $\bar{C}$ is the cost for parallel processing. In synchronous iterating model, a.k.a BSP model, $\bar{C}_{par,comm,cp}$ is the max processing cost among all machines.

3 Load Balance Vertex Cut

In this section, we will first analyze two current vertex cut partition methods implemented in GraphLab/PowerGraph. The framework is among only a few frameworks supporting vertex cut and is proved to be more efficient. Later we will show the shortcomings of its original partition methods. Then the relationship between edge arrival order and cost in different partitions will be discussed. Finally, a new balanced vertex cut considering the total cost will be introduced.

3.1 Random Assignment

Random assignment is the simplest edge assignment with $O(1)$ cost for each edge $\bar{C}_{par} = mc_{par}$. If edges are assigned uniformly random to machines, load balance will be achieved naturally. The addend $\bar{C}_{cp}$ of the total parallel processing cost $\bar{C} = \bar{C}_{par} + t \cdot (\bar{C}_{comm} + \bar{C}_{cp})$ will be less. But communication cost $\bar{C}_{comm}$ would be large.

The expected computing cost for one machine is $\bar{C}_{cp} = \frac{m}{k} c_{cp}$ where c_{cp} is the same as previously defined.

If messages can be aggregated, the expected communication cost for a vertex is $C_{comm}(v) = \sum_{i=1}^{k}(1 - \frac{1}{k})^{|E_{in}(v)|} - 1$ and the average communication cost for a single processing unit (a machine) is $\bar{C}_{comm} = C_{comm}(P_i) = \sum_{v \in V_i} \sum_{i=1}^{k}$ $(1 - \frac{1}{k})^{|E_{in}(v)|} - \frac{n}{k} r = n[1 - (1 - \frac{1}{k})^{d_{in}}] - \frac{n}{k} r$where r is the replication factor of vertices, d_{in} is the average in-degree of a vertex.

3.2 Streaming Heuristic

Greedy assignment is to heuristically minimize communication cost. Its algorithm can be described as follows for the assignment of edge $e(u, v)$.

1. Let $P(u)$ be the set of machines where each member contains vertex u, $P(v)$ be the set of machines where each member contains vertex v.
2. If $P(u) \cap P(v) \neq \emptyset$, assign $e(u, v)$ to $P_i \in P(u) \cap P(v)$ where P_i has minimum load.
3. If $P(u) \cap P(v) = \emptyset$ and $P(u) \cup P(v) \neq \emptyset$, assign $e(u, v)$ to $P_i \in P(u) \cup P(v)$ where P_i has minimum load.
4. If $P(u) \cup P(v) = \emptyset$, assign $e(u, v)$ to $P_i \in \cup_{j=1}^{k}\{P_j\}$ where P_i has minimum load.

Greedy assignment may work well when edges arrive in random order, but may fail when graph load in structured order such as BFS or DFS. However, most realistic graphs are obtained by grabbing links such as web graph and social graphs. For instance, if edges come in BFS order, previous greedy method may assign all edges to a single machine which definitely has minimum $C_{comm} = 0$ but it will lose the benefit of distributed computing and result in high computing cost. It becomes worse for large graphs. In addition, partition cost is high since global search is performed at each assignment of an edge.

Practical partition methods should support common graph search orders as well as random orders.

3.3 Load Balance in Specific Edge Arrival Orders

The arrival order of edges can be divided into BFS order, random order and other orders. Total vertex number increases in constant speed when edges arrive in BFS order. However in random order, it increases sharply at first and then reaches a stable situation, which matches the relationship between number of edges and vertices.

Other orders fall between random line and BFS line. The more random the arrival order is, the more balanced the final load will be.

We can use a parameter to express the order of edge arrival. Let $\mu(\xi)$ be the parameter of graph G's edge arrival order. It is defined by: $\mu(\xi) = \frac{|V(\xi)|}{|V|}$, $V(\xi)$ denotes the number of vertices occurred after $\xi|E|$ edges arrives.

We introduce a novel vertex cut partition method to obtain balanced load with low total cost to solve previous problems in graph partition. Upon the arrival of an edge, a serial of scores will be calculated to choose the best processor, which can stand for total cost C. $Score_i = f(\Delta\bar{C}_{comm}) + g(\Delta\bar{C}_{cp})$, where $\Delta\bar{C}_{comm}$ is the additional parallel communication cost after assigning edge e to machine i, and $\Delta\bar{C}_{cp}$ is the additional parallel computing cost.

Algorithm can be described as follows for the assignment of edge e(u,v).

1. Let $P(u)$ be the set of machines where each member contains vertex u, $P(v)$ be the set of machines where each member contains vertex v.
2. Calculate the score if edge $e(u, v)$ is assigned to machine P_i for all is, assign $e(u, v)$ to P_i which has minimum $Score_i$.

4 Evaluation

In this section, we will evaluate our vertex cut partition method and make a comparison to current methods.

The experiments were conducted on a cluster of 8 machines. Each one has 8 GB DDR2 RAM, eight 2.50 GHz Intel(R) Xeon(R) E5420 CPUs. The intra-connection network is 1 GB Ethernet. Unfortunately, the hardware of our cluster is old fashioned, which may cause longer execution time. But we mainly need to compare our method to that in GraphLab. Same platform with both Graphlab and ours installed can help to make the comparison. All execution time is normalized. Experiments of large cluster scale are conducted by virtual machine environment.

4.1 Data Preparation

The graphs we used is listed by Table 1. Twitter(TW), Google(GL), Epinions(EP) and RoadNet-CA(RN) datasets are from Stanford Data Collection [12], Kronecker generated (KG) dataset is generated by Graph 500 program [13]. As we need to value the performance of partition in different edge arrival orders, we preprocessed these graph to get several typical orders as listed. Parameter $\mu(\xi)$

Table 1. Datasets used in our experiments

Graph	Edges	Nodes	$\mu(0.5)$ BFS-like	$\mu(0.5)$ randomized	$\mu(0.5)$ id sorted
TW	1768149	81306	0.56	0.95	0.876
GL	5105039	875713	0.604	0.877	0.780
EP	508837	75879	0.340	0.740	0.364
KG	130636	3351	0.891	0.953	0.942
RN	5522314	1965206	0.510	0.949	0.519

for BFS order increases nearly at a constant speed, but that for random order increases sharply at first and then slower in Google and Twitter graph. For generated graph, $\mu(\xi)$ plots a more random curve, all orders of $\mu(\xi)$ grow rapidly then slower, but that of random order is more obvious.

4.2 Case Study

In this section, we will analyze all aspects of cost we have proposed. Kronecker generated (KG) graph will be our use case.

Partition Time: Partition time is too less comparing to total processing time, which is at most 10 percent for most of the algorithms. For iterating algorithms such as pagerank and label propagation, partition time is negligible.

Figure 1(a) shows the partition cost of different partition methods(LB-load balance/R-random/G-greedy). It agrees with previous analysis that partition cost for random partition method is least. Partition cost of our load balance method is comparable to that of greedy one, and less than 2x of that of random one, which is still negligible.

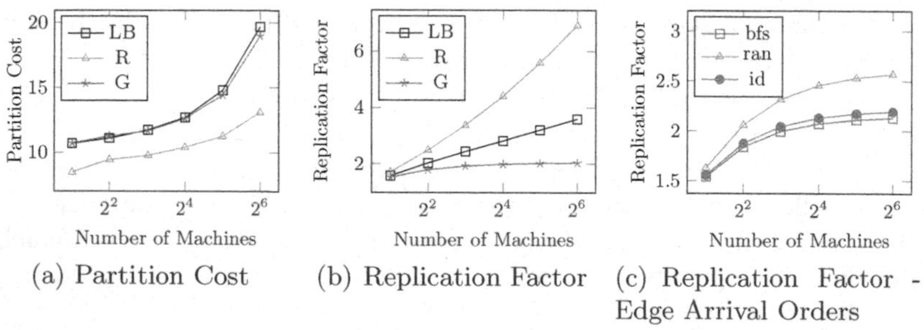

(a) Partition Cost (b) Replication Factor (c) Replication Factor - Edge Arrival Orders

Fig. 1. Partition cost and replication factor

Replication Factor: To prove the effectiveness of a partition method we should not only compare its balance of load to other method but also value its communication cost. We have analyzed in previous section that communication cost can be reflected by replication factor in vertex cut for graph partition.

Figure 1(b) shows replication factor of different partitions(LB/R/G) at different scale of system. As we can see from the figure, random method has the highest replication factor. Situation becomes worse when number of machine scales. For as many as 128 machines, a vertex will need an average of 25 replicas in random method, and only 8 in greedy or load balance one. It shows that random methods have more than 3 times communication cost to historical information based ones.

Our load balance method has comparable replication factor as greedy one, which means it wouldn't cause more communications. Later experiments of total cost will prove that.

When the replication factor is large, the computing cost is also increased since computations are abstracted in a vertex and the total computation unit is represented by total vertex replica number. Although main computations is proportional to the number of edges, the additional cost to implement computations on edges such as gathering the value from all adjacent edges is introduced.

Figure 1(c) shows the relationship between replication factor and edge arriving orders(BFS-like, random, id-sorted). When edges arrive in randomized order in which its adjacent vertices have the same probability to be occurred or not previously, they have the same chance to be assigned to a previous machine (which contains its vertex) or a new one. The imbalance is no longer existed. But we must ensure that the possibility that an edge's vertex has been occurred is not too small, otherwise, the slight change that it assigned to a new machine is negligible.

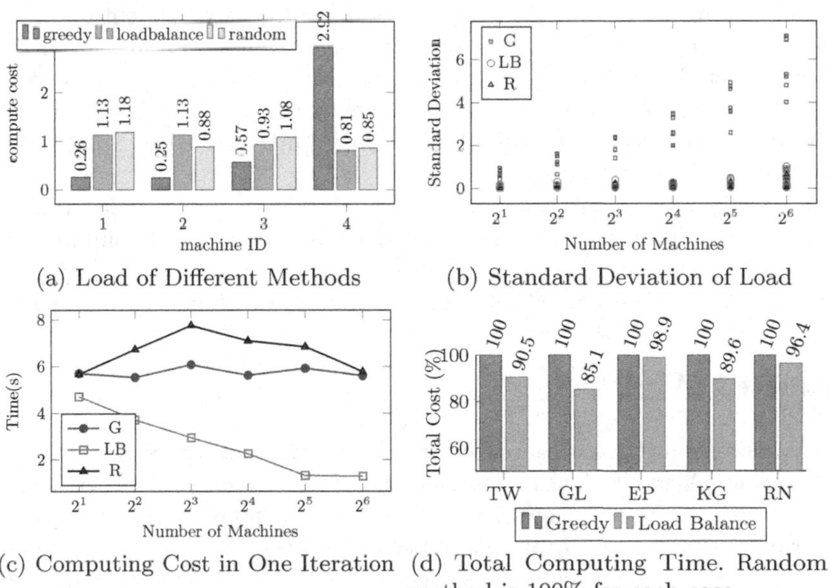

(a) Load of Different Methods (b) Standard Deviation of Load

(c) Computing Cost in One Iteration (d) Total Computing Time. Random method is 100% for each case

Fig. 2. Load balance factors(G-greedy, LB-load balance, R-random)

Load Balance: We compare our load balance method to greedy and random methods in different edge arrival orders. Greedy and random methods are originally supported by GraphLab. To make a clear figure, result of four-machine distributed graph computing is plotted in Fig. 2(a). Result for load balance at different scale of machines is similar and described by Fig. 2.

As is shown in Fig. 2(a), average computing cost of all machines is normalized to 1. Random method can naturally assign edges equally and is not subject to edge arrival order. Load of greedy method is highly imbalanced between machines. In BFS edge order, all edges are occurred after its neighbor except for a disconnected component. As a result, nearly all edges will be assigned to the machine which has the first edge. Just as the load of Machine 4, which is nearly 73 % percent of all load. Our load balance method successfully partitioned the graph to nearly the same subgraphs no matter in which order do edges arrive. Thus, total computing cost is reduced from imbalanced partition.

Figure 2 shows some factors (normalized max load, standard deviation) of load balance at different scale of machines. Greedy method has the worst factors (both high max load and standard deviation). Imbalanced load would lead to large computing cost at each iteration. Our load balance method has comparable max load and standard deviation to random method, which is much smaller than greedy one. Figure 2(c) shows the computing cost of pagerank in each iteration of these three methods. Cost of other algorithms such as SSSP, K-core decomposition are similar. Load balance method has smaller cost than greedy one and random one, and have good scalability.

Overall Performance: As we can see from Fig. 2(d), the total cost of iterating graph computing at different scales varies. We have experiment on different graphs as listed by Table 1. Typical iterating algorithm PageRank is run in different graphs at the scale of 8 to 64 machines which is selected to be suit for the scale of the graph. KG graphs are run on 8 machines and TW(original order – BFS like), GL(original order – BFS like) are on 64 machines. The result is shown as Fig. 2(d).

In Previous subsections we have shown that our new load balance method doesn't increase communication cost but decrease the parallel computing cost as well as communication cost. The total computing time of our load balance method is 10 % less than greedy method in average let alone random method.

5 Related Work

Graph Partition has been researched for more than twenty years as an NP complete problem. A great deal of methods has been developed. Previous graph partitions mainly focus on specific graphs such as planar graphs, meshes or VLSI graphs. Power law graphs have been researched only recently. Different from meshes or other simple graphs, power law graphs widely exist in modern relationships such as web graphs [14] and social network [15], which reflect the imbalance of nature.

Graph computing frameworks first came to public in this decade with Google's Pregel [5], where graph is still partitioned by edge cut. Such frameworks have by Bulk Synchronous Parallel(BSP) [16] computing abstraction. Thus, iterating algorithms are well supported and elegantly described. It greatly improve graph computing abstraction from MapReduce [17] by a higher level for programmers. But it suffers from large communication cost [18] because each vertex needs to send message to its neighbors, and the more edges it has cut, the more messages it will send. Graph computing frameworks have been booming since then [19].

Multilevel graph partitioning [11] has been proved to be the most state-of-the-art edge cut method by its effectiveness and high quality. But it is offline, which need the whole view of the graph while partitioning.

It was not until 2010 that vertex cut for graph partition has been proposed. [7] GraphLab (later PowerGraph) by CMU first support vertex cut for graph partition. Later, an improved branch of PowerGraph, PowerLyra [20] has been proposed to support power law graphs more efficiently. But it loses the benefit of distributed system which introduces all edges of a high degree vertex to the same machine while streaming partition and doesn't change it until whole graph is read. Specific design [6] for bipartite graphs was also made, but it is still limited to offline partition which need to reassign edges while finalizing the graph. More research for vertex cut partition is done by [3,10].

Large scale graphs are becoming more important and valuable which emphasize on distributed computing. We can hardly have a whole view of the graph while partitioning. Streaming graph partition turns out an effective solution. Stanton et al. [9] analyzed different streaming methods for distributed graph partition.

6 Conclusion

In this paper, we analyzed graph partition cost by edge cut and vertex cut and proved that vertex cut is less consuming for power law graphs. We also analyzed offline and streaming partition method, and reached a conclusion that streaming partition is subject to edge arrival orders if edges are assigned by historical information. But assigning edges by historical information is essential for reducing communication cost. To improve current streaming vertex cut partition method which suffers from imbalanced load when edges arrive by a graph-search order, we introduced a new vertex cut partition method. Our method balances the load while doesn't introduce additional communication cost. Experiments show that our method reduces computing cost in each iteration from greedy one and has good scalability. Total performance has been improved by at 10 percent in average.

Acknowledgment. This work has been supported by National High Technology Research and Development 863 Program of China under Grant No.2013AA013205 and Program of State Key Laboratory of High-end Server & Storage Technology under Grant No.2014HSSA16.

References

1. Holyer, I.: The np-completeness of some edge-partition problems. SIAM J. Comput. **10**(4), 713–717 (1981)
2. Finding good approximate vertex and edge partitions is np-hard. Inf. Process. Lett. **42**(3), 153–159 (1992)
3. Zhou, J., Bruno, N., Lin, W.: Advanced partitioning techniques for massively distributed computation. In: Proceedings of the 2012 ACM SIGMOD International Conference on Management of Data, pp. 13–24. ACM (2012)
4. Andreev, K., Rcke, H.: Balanced graph partitioning. In: Proceedings of the Sixteenth Annual ACM Symposium on Parallelism in Algorithms and Architectures, SPAA 04, pp. 120–124 (2004)
5. Malewicz, G., Austern, M.H., Bik, A.J., Dehnert, J.C., Horn, I., Leiser, N., Czajkowski, G.: Pregel: a system for large-scale graph processing. In: Proceedings of the 2010 ACM SIGMOD International Conference on Management of data, pp. 135–146. ACM (2010)
6. R. Chen et al.: Bigraph: Bipartite-aware distributed graph partition for big learning. Institute of Parallel and Distributed Systems Technical report, Number: IPADSTR-2013-002 (2013)
7. Low, Y., Gonzalez, J., Kyrola, A., Bickson, D., Guestrin, C., Hellerstein, J.M.: Graphlab: A new framework for parallel machine learning. CoRR, vol. abs/1006.4990 (2010)
8. Low, Y., Bickson, D., Gonzalez, J., Guestrin, C., Kyrola, A., Hellerstein, J.M.: Distributed graphlab: a framework for machine learning and data mining in the cloud. Proc. VLDB Endowment **5**(8), 716–727 (2012)
9. Stanton, I., Kliot, G.: Streaming graph partitioning for large distributed graphs. In: Proceedings of the 18th ACM SIGKDD International Conference on Knowledge Discovery and Data Mining, pp. 1222–1230. ACM (2012)
10. Bourse, F., Lelarge, M., Vojnovic, M.: Balanced graph edge partition in MSR Technical report, MSR-TR-2014-20, February 2014
11. Karypis, G., Kumar, V.: A fast and high quality multilevel scheme for partitioning irregular graphs. SIAM J. Sci. Comput. **20**(1), 359–392 (1998)
12. Stanford large network dataset collection. http://snap.stanford.edu/data/
13. Graph 500. http://www.graph500.org/
14. Faloutsos, M., Faloutsos, P., Faloutsos, C.: On power-law relationships of the internet topology. In: ACM SIGCOMM Computer Communication Review, vol. 29, no. 4, pp. 251–262. ACM (1999)
15. Mislove, A., Marcon, M., Gummadi, K.P., Druschel, P., Bhattacharjee, B.: Measurement and analysis of online social networks. In: Proceedings of the 7th ACM SIGCOMM Conference on Internet Measurement, pp. 29–42. ACM (2007)
16. Isard, M., Budiu, M., Yu, Y., Birrell, A., Fetterly, D.: Dryad: distributed data-parallel programs from sequential building blocks. ACM SIGOPS Operating Syst. Rev. **41**(3), 59–72 (2007)
17. Dean, J., Ghemawat, S.: Mapreduce: simplified data processing on large clusters. Commun. ACM **51**(1), 107–113 (2008)
18. Salihoglu, S., Widom, J.: Optimizing graph algorithms on pregel-like systems (2014)
19. Angles, R., Gutierrez, C.: Survey of graph database models. ACM Comput. Surv. (CSUR) **40**(1), 1 (2008)
20. Chen, R. et al.: Powerlyra: Differentiated graph computation and partitioning on skewed graphs. Institute of Parallel and Distributed Systems Technical report, Number:IPADSTR-2013-001 (2013)

Non-convex Regularized Self-representation for Unsupervised Feature Selection

Weizhi Wang[1], Hongzhi Zhang[1], Pengfei Zhu[2], David Zhang[1,2], and Wangmeng Zuo[1(✉)]

[1] Computational Perception and Cognition Centre, School of Computer Science and Technology, Harbin Institute of Technology, Harbin 150001, China
{weizhiwanghit, zhanghz0451, cswmzuo}@gmail.com, csdzhang@comp.polyu.edu.hk

[2] Biometrics Research Centre, Department of Computing, Hong Kong Polytechnic University, Hung Hom, Kowloon, Hong Kong
zhupengfeifly@gmail.com

Abstract. Feature selection aims to select a subset of features to decrease time complexity, reduce storage burden and improve the generalization ability of classification or clustering. For the countless unlabeled high dimensional data, unsupervised feature selection is effective in alleviating the curse of dimension-ality and can find applications in various fields. In this paper, we propose a non-convex regularized self-representation (RSR) model where features can be represented by a linear combination of other features, and propose to impose $L_{2,p}$ norm ($0 < p < 1$) regularization on self-representation coefficients for unsupervised feature selection. Compared with the conventional $L_{2,1}$ norm regularization, when $p < 1$, much sparser solution is obtained on the self-representation coefficients, and it is also more effective in selecting salient features. To solve the non-convex RSR model, we further propose an efficient iterative reweighted least squares (IRLS) algorithm with guaranteed convergence to fixed point. Extensive experimental results on nine datasets show that our feature selection method with small p is more effective. It mostly outperforms features selected at $p = 1$ and other state-of-the-art unsupervised feature selection methods in terms of classification accuracy and clustering result.

Keywords: Unsupervised feature selection · Sparse representation · L_{2p} norm · Self-representation

1 Introduction

Large amounts of high-dimensional data appear with the explosive use of electronic sensors and social media [1]. The high dimension will result in higher time and space complexity, and features which are irrelevant and redundant may also greatly affect the performance of classification and clustering [2]. Feature selection, a method of selecting a subset of features, is an important means in building machine learning models for classification, clustering, and other tasks. Feature selection is promising in reducing the computational and memory complexity of learning methods, enhancing

X. He et al. (Eds.): IScIDE 2015, part II, LNCS 9243, pp. 55–65, 2015.
DOI: 10.1007/978-3-319-23862-3_6

the model generalization capability, and alleviating the curse of dimensionality [3]. In recent years, continuous efforts have been made to develop feature selection algorithms.

In general, feature selection methods can be categorized into three groups: filter, wrapper, and embedded methods. Filter methods [4, 5] select a subset of features by using some feature evaluation indices, namely some statistical properties of data. For wrapper methods [6, 7], the space of feature subset is searched in and then the classification performance on selected features is taken as the evaluation criterion. Wrapper methods usually require the training of models for many times, significantly increasing the computational burden and limiting their applications. For enhancing computational efficiency, embedded approaches [8, 9] incorporate the selection process in the learning model to simultaneously learn the optimal classifier while finding salient features.

From the perspective of label availability, feature selection models can also be grouped into two categories: supervised and unsupervised ones. In this paper, we focus on unsupervised feature selection. For early studies on unsupervised feature selection, some evaluation indices were introduced to evaluate the importance of each individual feature or feature subset and features were selected one by one [10]. These important indices can be designed based on clustering performance, redundancy, sample similarity, manifold structure, etc., where some representative indices are Laplacian score [10], variance [11], and trace ratio [12]. However, the dependence on searching makes these methods computationally expensive. Naturally, the methods without searching have been investigated. Based on feature similarity, a feature clustering method is proposed to find the representative features in [13]. To best preserve sample similarity, a series of spectral clustering based feature selection methods have been developed [10, 14, 15]. Zhu et al. [16] proposed a regularized self-representation method for unsupervised feature selection, In their method, one feature is represented as a linear combination of other features, which is called self-representation property of features. By minimizing the self-representation error, a feature weight matrix is learned and a feature subset can be selected.

Recently, sparsity regularization has been widely used in dimensionality reduction. Researchers also started to use it to select features and have obtained some favorable results. By using the L_1 norm regularization to acquire sparse solution, L_1-SVM was proposed to perform feature selection [6]. The $L_{2,1}$ norm group sparsity has also been introduced to feature selection [5, 17, 18]. In [16], $L_{2,1}$ norm was used to regularize the feature weights matrix and has achieved state-of-the-art results.

In this paper, we propose to use $L_{2,p}$ norm regularization to select feature with emphasis on small p ($0 < p < 1$). Compared with the standard $L_{2,1}$ norm, when $0 < p < 1$, the nonzero rows of the resolved representation coefficient matrix will become sparser, the corresponding features are more effective. At the same time, to eliminate the adverse effect of outliers, we use the standard $L_{2,1}$ norm to regularize the loss term. An improved Iterative Reweighted Least-Squares (IRLS) algorithm is proposed to solve the model and its convergence is proved as well. Experiments on real world datasets validate that features selected with small p outperform features selected by standard $L_{2,1}$ norm regularization and other popular feature selection methods in terms of classification and clustering measurements.

The rest of this paper is organized as follows: Sect. 2 introduces the unsupervised feature selection; Sect. 3 presents the optimization and algorithms; Sect. 4 conducts experiments and Sect. 5 concludes this paper.

2 Non-convex Regularized Self-representation Model

2.1 Problem Statement

For real-world datasets without label information, which are often redundant in features and may contain outliers, unsupervised feature selection is an important manner to select a feature subset.

Let $\mathbf{X} \in \mathbb{R}^{m \times n}$. be the original data space, where m is the sample number and n is the feature dimension. We use $\mathbf{f}_i \in \mathbb{R}^m$ to represent the i-th feature vector of data matrix $\mathbf{X}$, then $\mathbf{X} = [\mathbf{f}_1, \ldots, \mathbf{f}_i, \ldots, \mathbf{f}_n]$. The purpose of feature selection is to select k features, and use them for classification, clustering or other tasks. By using the sample similarity or sample manifold structure, we construct a response matrix $\mathbf{Y}$. Feature selection can then be formulated as a multi-output regression problem:

$$J_0(\mathbf{X}_k) = \min_{k \subset D, \mathbf{w}} l(\mathbf{Y} - \mathbf{X}_k\mathbf{W}). \tag{1}$$

where $D = \{1, 2, \ldots, n\}$ is the dimension, and k is the selected subsets, $\mathbf{X}_k$ is the corresponding k columns of $\mathbf{X}$, $\mathbf{W}$ is the corresponding features weight matrix, and $l(\mathbf{Y} - \mathbf{X}_k\mathbf{W})$ is the loss term.

Obviously, this is a discrete optimization problem which needs to search $C_n^k = n!/k!(n-k)!$ feature subsets. It is a NP-hard problem. Rather than directly solve the challenging discrete optimization problem, we incorporate some regularization on $\mathbf{W}$, resulting in the following formulation,

$$\min_{\mathbf{W}} l(\mathbf{Y} - \mathbf{X}\mathbf{W}) + \lambda R(\mathbf{W}). \tag{2}$$

where $l(\mathbf{Y} - \mathbf{X}\mathbf{W})$ is the loss term, $R(\mathbf{W})$ is the regularization term imposed on $\mathbf{W}$ and λ is a positive constant.

2.2 Loss Term and Regularization Term

Inspired by the sample representation models [19–21], considering the purpose of feature selection, we utilize the property of feature self-representation. Like RSR [16], we use the original space data matrix $\mathbf{X}$ as the response matrix, namely $\mathbf{Y} = \mathbf{X}$, then each feature can be linearly represented by other features (include itself), i.e., for each feature vector $\mathbf{f}_i$ in $\mathbf{X}$, it can be represented as follows:

$$\mathbf{f}_i = \sum\nolimits_{j=1}^{n} \mathbf{f}_j w_{ji} + \mathbf{b}_i. \tag{3}$$

where w_{ji} is the represent coefficient in $\mathbf{W}$ and $\mathbf{b}_i \in \mathbb{R}^m$ is the bias vector. Then for all the features, we have

$$\mathbf{X} = \mathbf{XW} + \mathbf{B}. \tag{4}$$

where $\mathbf{X} = [\mathbf{f}_1, \mathbf{f}_2, \ldots, \mathbf{f}_n] \in \mathbb{R}^{m\times n}$, $\mathbf{W} = [w_{ji}] \in \mathbb{R}^{n\times n}$, $\mathbf{B} = [\mathbf{b}_1, \mathbf{b}_2, \ldots, \mathbf{b}_n] \in \mathbb{R}^{m\times n}$.

In this model, we will use the learned matrix $\mathbf{W}$ to reflect the importance of different features while the bias is small. Let $\mathbf{W} = [\mathbf{w}_1;\ldots,\mathbf{w}_i;\ldots;\mathbf{w}_n]$, where $\mathbf{w}_i$ is the i-th row of $\mathbf{W}$. Because $\|\mathbf{w}_i\|_2$ can reflect the importance of the i-th feature in the representation, it is used as the feature weight. e.g., if the i-th feature contributes nothing to other features' representation, then $\|\mathbf{w}_i\|_2$ will be 0. If the i-th feature is frequently used to represent most of other features, then $\|\mathbf{w}_i\|_2$ must be significant. Thus, the row-sparsity is expected for regularizing the coefficients matrix $\mathbf{W}$.

For the regularization term of $\mathbf{W}$, we select the $L_{2,p}$ norm regularizer, which satisfies the constraint that the representation coefficients matrix $\mathbf{W}$ is row sparse. When $0 < p < 1$, the solution is sparse, and $\mathbf{W}$ will be even sparser in rows. So we let $R(\cdot) = \|\cdot\|_{2,p}^{p}$. To reduce the sensitivity to outliers, we use $L_{2,1}$ norm to replace Frobenius norm to constrain the residual, namely $l(\cdot) = \|\cdot\|_{2,1}$, the nonzero rows correspond to the outlier samples. Upon this we can get the following objective function:

$$J(\mathbf{W}) = \arg\min_{\mathbf{w}} \|\mathrm{X} - \mathrm{XW}\|_{2,1} + \lambda \|\mathrm{W}\|_{2,p}^{p}. \tag{5}$$

where the $L_{2,p}$ norm on $\mathbf{W}$ is defined as:

$$\|\mathbf{W}\|_{2,p}^{p} = \sum\nolimits_{i=1}^{n} \left(\sum\nolimits_{j=1}^{n} w_{ij}^{2}\right)^{p} = \sum\nolimits_{j=1}^{n} \|\mathbf{w}_i\|^{p}. \tag{6}$$

where $\mathbf{w}_i$ is the i-th row of $\mathbf{W}$, λ is a positive tradeoff parameter. In this paper, we investigate feature selection performance with three typical p values, i.e., 0.4, 0.6, 0.8, to verify whether small p value ($p < 1$) would be more effective in selecting important features.

3 Iterative Reweighted Least-Squares Algorithm

The loss term is non-smooth and the regularization term is even non-convex, making the objective function in Eq. (5) non-convex and difficult to solve. In this section, we improve the Iterative Reweighted Least-Squares (IRLS) [16, 22] algorithm to solve the problem and then prove its convergence to a fixed point.

In the improved IRLS algorithm, given the current $\mathbf{W}^t$, the diagonal weighting matrices G_B^t and G_W^t are defined by,

$$g_{B,i}^{t} = \frac{1}{2\|\mathbf{x}_i - \mathbf{x}_i\mathbf{W}^t\|}. \tag{7}$$

$$g_{w,j}^{t} = \frac{p}{2}\left\|\mathbf{w}_j^{t}\right\|_2^{p-2}. \tag{8}$$

where $g^t_{B,i}$ and $g^t_{w,j}$ are the diagonal elements of G^t_B and G^t_W. $\mathbf{x}_i$ and $\mathbf{w}_j$ are the i-th and j-th rows of $\mathbf{X}$ and $\mathbf{W}$. Then $\mathbf{W}^{t+1}$ is updated by solving the following weighted least squares problem,

$$\mathbf{W}^{t+1} = \arg\min_{\mathbf{W}} Q(\mathbf{W}|\mathbf{W}^t) = \arg\min_{\mathbf{W}} \left\{ \begin{array}{l} tr((\mathbf{X} - \mathbf{XW})^T \mathbf{G}^t_B (\mathbf{X} - \mathbf{XW})) \\ + \lambda tr(\mathbf{W}^T \mathbf{G}^t_W \mathbf{W}) \end{array} \right\}. \tag{9}$$

Let $\frac{\partial}{\partial \mathbf{W}} Q(\mathbf{W}|\mathbf{W}^t) = 0$, we can get

$$(\mathbf{X}^T \mathbf{G}^t_B \mathbf{X} + \lambda \mathbf{G}^t_W)\mathbf{W} - \mathbf{X}^T \mathbf{G}^t_B \mathbf{X} = 0. \tag{10}$$

then we will get the closed form solution of $\mathbf{W}^{t+1}$:

$$\mathbf{W}^{t+1} = ((\mathbf{G}^t_W)^{-1} \mathbf{X}^T \mathbf{G}^t_B \mathbf{X} + \lambda \mathbf{I})^{-1} (\mathbf{G}^t_W)^{-1} \mathbf{X}^T \mathbf{G}^t_B \mathbf{X}. \tag{11}$$

To improve the stability of the solution, a sufficiently small tolerance ε is introduced by defining

$$g^t_{B,i} = \frac{1}{\max(2\|\mathbf{x}_i - \mathbf{x}_i \mathbf{W}^t\|, \varepsilon)}. \tag{12}$$

After obtaining the final $\mathbf{W}$, we can select the top k features within the non-zeros subsets by ranking the features according to the values of $\|\mathbf{w}_i\|_2$.

The algorithm for solving the $L_{2,p}$ norm regularization is described in Algorithm 1. In each iteration, the computing scale in Steps 3 and 4 are $O(n^2 m)$ and $O(n^2)$ respectively. When updating $\mathbf{W}$, it needs a computational complexity of $O(n^3 + n^2 m)$. In the above expression, m and n are the numbers of features and samples, respectively. In summary, the time complexity of our method is $O(T(n^3 + n^2 m))$, where T is the total number of iterations.

The algorithm monotonically decreases the objective value of Eq. (5) in each iteration, and guarantees to converge to a fixed point. To prove its convergence, we introduce a surrogate function $Q(\mathbf{W}|\mathbf{W}^t)$ in Eq. (9), and define $F(\mathbf{W}) = J(\mathbf{W}) - J(\mathbf{W}|\mathbf{W}^t)$. By solving Eq. (9), we can get an optimal solution sequence $\mathbf{W}_{optimal} = [\mathbf{W}^1, \mathbf{W}^2, \ldots, \mathbf{W}^T]$. First, we prove $F(\mathbf{W})$ is monotone descending on $\mathbf{W}_{optimal}$. Then, we prove that for any t = 1, 2, …, T-1, $J(\mathbf{W}^{t+1}) \leq J(\mathbf{W}^t)$ is permanently established, which means the objective function is monotone descending on $\mathbf{W}_{optimal}$. For that, we should prove the following two lemmas.

Algorithm 1: IRLS for solving the non-convex RSR model

Input: Data matrix $\mathbf{X} \in \mathbb{R}^{m \times n}$, and $\lambda > 0$
Output: Feature weights matrix $\mathbf{W}$
Procedure:

```
Set t = 1. Initialize W
Repeat
    Calculate G_B^t using Eq. (12)
    Calculate G_W^t using Eq. (8)
    Update: W^{t+1} = ((G_W^t)^{-1} X^T G_B^t X + λI)^{-1} (G_W^t)^{-1} X^T G_B^t X
    t = t+1
Until convergence
```

Lemma 1. $Q(\mathbf{W}|\mathbf{W}^t)$ is a surrogate function as above, we define $F(\mathbf{W}) = J(\mathbf{W}) - Q(\mathbf{W}|\mathbf{W}^t)$, then $F(\mathbf{W})$ get its maximum when $\mathbf{W} = \mathbf{W}^t$.

Proof. In the following, we will prove that for any $\mathbf{W}$, there is $F(\mathbf{W}^t) - F(\mathbf{W}) \geq 0$

$$F(\mathbf{W}^t) - F(\mathbf{W}) = J(\mathbf{W}^t) - Q(\mathbf{W}^t|\mathbf{W}^t) - J(\mathbf{W}) + Q(\mathbf{W}|\mathbf{W}^t). \tag{13}$$

Bringing $J(\mathbf{W})$ and $Q(\mathbf{W}|\mathbf{W}^t)$ into Eq. (10), then it can be rewritten as,

$$\sum\nolimits_i \frac{\left(\|\mathbf{X}_i - \mathbf{X}_i\mathbf{W}^t\|_2 - \|\mathbf{X}_i - \mathbf{X}_i\mathbf{W}\|_2^2\right)^2}{2\|\mathbf{X}_i - \mathbf{X}_i\mathbf{W}^t\|_2} + \sum\nolimits_j \left(\left(1-\frac{p}{2}\right)\left\|\mathbf{W}_j^t\right\|_2^p - \|\mathbf{W}_j\|_2^p + \frac{p\|\mathbf{W}_j\|_2^2}{2\left\|\mathbf{W}_j^t\right\|_2^{2-p}}\right). \tag{14}$$

Due to $\sum_i \frac{1}{2\|\mathbf{X}_i - \mathbf{X}_i\mathbf{W}^t\|_2}\left(\|\mathbf{X}_i - \mathbf{X}_i\mathbf{W}^t\|_2 - \|\mathbf{X}_i - \mathbf{X}_i\mathbf{W}\|_2^2\right)^2 \geq 0$, we further need to prove

$$\sum\nolimits_j \left(\left(1-\frac{p}{2}\right)\left\|\mathbf{W}_j^t\right\|_2^p - \|\mathbf{W}_j\|_2^p + \frac{p\|\mathbf{W}_j\|_2^2}{2\left\|\mathbf{W}_j^t\right\|_2^{2-p}}\right) \geq 0. \tag{15}$$

By defining $a_j = \left\|\mathbf{W}_j^t\right\|_2$, $y_j = \|\mathbf{W}_j\|_2$ and $h(y_j) = (1-\frac{p}{2})\left\|\mathbf{W}_j^t\right\|_2^p - \|\mathbf{W}_j\|_2^p + \frac{p\|\mathbf{W}_j\|_2^2}{2\left\|\mathbf{W}_j^t\right\|_2^{2-p}}$, then we have $a_j > 0$, $y_j \geq 0$, $0 < p < 1$, and:

$$h(y_j) = (1 - \frac{p}{2})a_j^p - y_j^p + \frac{p}{2}y_j^2 a_j^{p-2}. \tag{16}$$

Equation (16) is a polynomial about y_j. If $y_j = 0$, then $h(y_j) > 0$. When $a_j > 0$, $y_j > 0$, taking the first- and second-order derivatives of $h(y_j)$ w.r.t y_j are:

$$h'(y_j) = \frac{\partial h}{\partial y_j} = p y_j (a_j^{p-2} - y_j^{p-2}). \tag{17}$$

$$h''(y_j) = \frac{\partial^2 h}{\partial y_j^2} = pa_j^{p-2} - (p-1)y_j^{p-2}. \quad (18)$$

Because of $a_j > 0$, $y_j > 0$ and $0 < p < 1$, we get $h''(y_j) > 0$, $h'(a_j) = 0$ and $h(a_j) = 0$. Thus according to the theory of convex optimization,

$$h(y_j) \geq h(a_j) = 0. \quad (19)$$

then $\sum_j h(y_j) \geq 0$, namely Eq. (15) is permanently established. So does $F(\mathbf{W}^t) - F(\mathbf{W}) \geq 0$. Obviously, when $p = 1$, the conclusion is also satisfied. □

Lemma 2. Let $\mathbf{W}^{t+1} = \arg\min_{\mathbf{W}} Q(\mathbf{W}|\mathbf{W}^t)$, we will get $J(\mathbf{W}^{t+1}) \leq J(\mathbf{W}^t)$.

Proof. We can see that,

$$\begin{aligned} J(\mathbf{W}^{t+1}) \quad &= J(\mathbf{W}^{t+1}) - Q(\mathbf{W}^{t+1}|\mathbf{W}^t) + Q(\mathbf{W}^{t+1}|\mathbf{W}^t) \\ &= F(\mathbf{W}^{t+1}) + Q(\mathbf{W}^{t+1}|\mathbf{W}^t) \\ &\overset{F(\mathbf{W}^{t+1}) \leq F(\mathbf{W}^t)}{\Rightarrow} \leq F(\mathbf{W}^t) + Q(\mathbf{W}^{t+1}|\mathbf{W}^t) \\ &- J(\mathbf{W}^t) - Q(\mathbf{W}^t|\mathbf{W}^t) + Q(\mathbf{W}^{t+1}|\mathbf{W}^t) \\ &\overset{\mathbf{W}^{t+1} = \arg\min_{\mathbf{W}} Q(\mathbf{W}|\mathbf{W}^t)}{\Rightarrow} \leq J(\mathbf{W}^t) - Q(\mathbf{W}^t|\mathbf{W}^t) + Q(\mathbf{W}^t|\mathbf{W}^t) \end{aligned}$$

that is to say,

$$J(\mathbf{W}^{t+1}) \leq J(\mathbf{W}^t). \quad (20)$$

So we can update $\mathbf{W}^{t+1}$ by minimizing the surrogate function: $\mathbf{W}^{t+1} = \arg\min_{\mathbf{W}} Q(\mathbf{W}|\mathbf{W}^t)$. Therefore, our algorithm will converge to a fixed point. □

Thus, the objective function is decreasing in each iteration by the proposed improved IRLS algorithm, and $J(\mathbf{W})$ can be minimized by iteratively minimizing $Q(\mathbf{W}|\mathbf{W}^t)$. Based on the proof above and noting that $\mathbf{W}^{t+1}$ in Eq. (9) has a closed form solution in each iteration, IRLS can converge to a fixed point efficiently.

4 Experiments

In order to validate the performance of our $L_{2,p}$ regularization feature selection method, we apply it on two categories of public datasets: four human face datasets (orlraws10P[1], pixraw10P[2], warpAR10P[3], warPIE10P[4]) and five microarray datasets

[1] http://www.cl.cam.ac.uk/research/dtg/attarchive/facedatabase.html.

[2] http://peipa.essex.ac.uk/ipa/pix/faces/.

[3] http://featureselection.asu.edu/datasets.php.

[4] http://www.ri.cmu.edu/research_project_detail.html?project_id=418&menu_id=261.

(TOX-171[5], Prostate-GE[6], Carcinoma [23], LUNG [24], GLIOMA [25]). The number of features in all these datasets varies from 2400 to 11340. We summarize the detailed information of the nine datasets in Table 1. We compare the proposed $L_{2,p}$ regularization method with the standard $L_{2,1}$ RSR method and three other state-of-the-art unsupervised feature selection methods: Laplacian Score [10], UDFS [5], SPEC [14].

Table 1. Summary of the benchmark datasets

Data	Instances	Feature	Class	Keywords
orlraws10P	100	10304	10	Image,Face
pixraw10P	100	10000	10	Image,Face
warpAR10P	130	2400	10	Image,Face
warPIE10P	210	2420	10	Image,Face
TOX-171	171	5748	4	Microarray,Bio
Carcinoma	174	9182	11	Microarray,Bio
LUNG	203	3312	5	Microarray,Bio
Prostate-GE	102	5966	2	Microarray,Bio
GLIOMA	50	4434	4	Microarray,Bio

4.1 Classification Accuracy Comparison

All the data sets are normalized to have mean of 0 and standard deviation of 1. Nearest neighbor classifier (NNC) has been individually performed on all data sets using 10-fold cross validation. We evaluate the proposed feature selection method at three typical p values, i.e., 0.4, 0.6, and 0.8, and compare with the four competing methods mentioned above. For Laplacian Score, and UDFS, following the previous works, the size of neighborhood k is fixed at 5 on all the datasets. For UDFS and standard $L_{2,1}$ RSR and our proposed method, the regularization parameter needs to be tuned. The Gaussian kernel bandwidth parameters in Laplacian Score and SPEC should also be chosen. To fairly compare these different methods, the regularization and bandwidth parameters are tuned from {0.001, 0.005, 0.01, 0.05, 0.1, 0.5, 1, 5, 10, 100} and the best result is recorded. For feature dimension, because of the sparse properties of $L_{2,p}$ norm, we consider that the advantages of the proposed method would be prominent when the number of selected features is small. Hence, we set the number of features as {10, 15, 20, 25, 30}.

Figure 1 shows the classification accuracy of all feature selection methods on nine datasets with different feature numbers. Table 2 shows the average results with different dimensions. As shown in Table 2, our proposed methods ($0 < p < 1$) outperform the standard $L_{2,1}$ RSR and other popular feature selection methods in most cases.

[5] http://www.ncbi.nlm.nih.gov/sites/GDSbrowser.

[6] http://www.ncbi.nlm.nih.gov/pubmed/12381711.

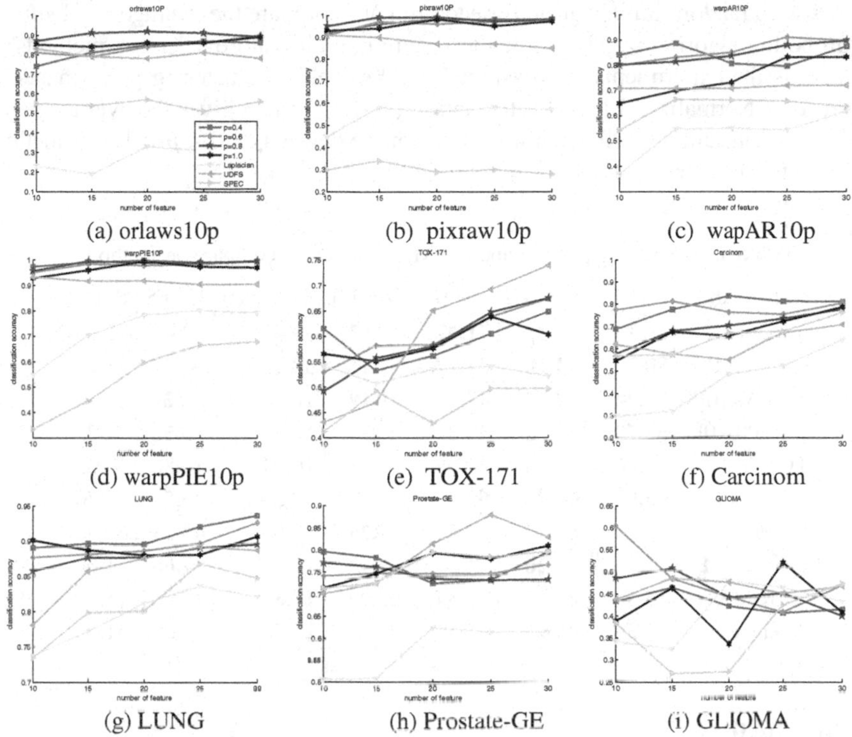

Fig. 1. Classification accuracy with different feature numbers on the nine datasets

Table 2. Classification accuracy of different comparison methods

Data\Method	p = 0.4	p = 0.6	p = 0.8	p = 1.0	Laplacian	UDFS	SPEC
orlraws10P	81.40	80.20	**88.80**	**86.00**	27.40	79.80	54.60
pixraw10P	**95.80**	95.40	**97.40**	95.40	54.80	88.00	30.20
warpAR10P	**84.20**	**85.20**	83.80	74.10	52.60	71.40	63.30
warpPIE10P	**98.60**	97.83	**98.30**	96.23	72.67	91.63	54.50
TOX-171	57.31	57.03	**57.34**	55.08	52.98	**59.72**	46.62
Carcinom	**71.48**	**73.36**	69.68	66.09	45.49	62.75	65.28
LUNG	**89.80**	88.98	85.57	**89.02**	79.60	85.88	81.01
Prostate-GE	75.80	74.47	71.10	74.53	57.40	**79.03**	**76.33**
GLIOMA	42.87	**44.83**	44.14	42.37	40.89	**46.39**	36.55
Average	**77.47**	**77.48**	77.35	75.43	53.76	73.85	56.49

4.2 Clustering Effectiveness Comparison

We compare the proposed method with other feature selection methods in terms of clustering result. We perform K-means clustering algorithm on the selected features. Since the K-Means depends on initialization, as in [16], we run the experiment for

20 times with random initialization for each run and compare the average results for all the competing algorithms. The parameters setting in all feature selection methods are the same as the experiments in classification. We evaluate clustering performance by the metric of Normalized Mutual Information (NMI). Table 3 gives the average values of the NMI with different dimensions. One can see that, when $0 < p < 1$, our methods can get better clustering results in most cases.

Table 3. Clustering performance (NMI) of different comparison methods

Data\Method	p = 0.4	p = 0.6	p = 0.8	p = 1.0	Laplacian	UDFS	SPEC
orlraws10P	69.14	64.15	**71.47**	**74.87**	39.46	62.40	42.95
pixraw10P	80.73	**84.10**	**84.17**	81.12	58.26	64.70	36.97
warpAR10P	42.74	42.43	47.63	**50.19**	16.90	**48.21**	44.38
warpPIE10P	**52.79**	51.14	48.81	43.14	18.31	**53.52**	24.23
TOX-171	21.06	**22.73**	**24.05**	16.73	10.14	10.86	10.03
Carcinom	**65.94**	**66.84**	63.96	57.58	42.37	46.51	57.76
LUNG	53.98	**55.89**	48.59	**57.26**	40.57	43.24	47.94
Prostate-GE	5.57	**6.26**	4.68	5.41	3.69	**7.08**	1.64
GLIOMA	17.59	**20.19**	12.54	13.90	**17.82**	17.06	15.36
Average	**45.50**	**45.97**	45.10	44.46	27.50	39.29	31.25

5 Conclusions

A novel non-convex regularized self-representation model is developed for unsupervised feature selection. Based on the self-representation property of features and the property that L_p norm regularizer generally favors sparser solution than the L_1 norm regularizer, we propose to use $L_{2,p}$ norm regularizer for unsupervised feature selection. When $0 < p < 1$, the problem is non-convex but can be efficiently solved using our improved Iterative Reweighted Least-Squares (IRLS) algorithm. The convergence of the IRLS algorithm is also analyzed. Experimental results on real-world datasets show that our method is more effective in selecting less and salient features compared with RSR and other state-of-the-art unsupervised feature selection algorithms. In our future work, we will investigate other forms of non-convex regularizer and extend the proposed model for supervised feature selection.

References

1. Tang, J., Liu, H.: Unsupervised feature selection for linked social media data. In: Proceedings of the 18th ACM SIGKDD International Conference on Knowledge Discovery and Data Mining, pp. 904–912 (2012)
2. Liu, H., Wu, X., Zhang, S.: Feature selection using hierarchical feature clustering. In: Proceedings of the 20th ACM International Conference on Information and Knowledge Management, pp. 979–984 (2011)

3. Liu, H., Motoda, H.: Feature Selection for Knowledge Discovery and Data Mining. Springer, New York (1998)
4. Langley, P.: Selection of Relevant Features in Machine Learning. Defense Technical Information Center, New Orleans (1994)
5. Yang, Y., Shen, H.T., Ma, Z., Huang, Z., Zhou, X.: $L_{2,1}$-norm regularized discriminative feature selection for unsupervised learning. Proc. Int. Joint Conf. Artif. Intell. **22**(1), 1589–1594 (2011)
6. Kohavi, R., John, G.H.: Wrappers for feature subset selection. Artif. Intell. **97**(1), 273–324 (1997)
7. Guyon, I., Elisseeff, A.: An introduction to variable and feature selection. J. Mach. Learn. Res. **3**, 1157–1182 (2003)
8. Vapnik, V.: The Nature of Statistical Learning Theory. Springer, New York (2000)
9. Hou, C., Nie, F., Yi, D., Wu, Y.: Feature selection via joint embedding learning and sparse regression. Proc. Int. Joint Conf. Artif. Intell. **22**(1), 1324–1329 (2011)
10. He, X., Cai, D., Niyogi, P.: Laplacian score for feature selection. In: Advances in Neural Information Processing Systems, pp. 507–514 (2005)
11. Dy, J.G., Brodley, C.E.: Feature selection for unsupervised learning. J. Mach. Learn. Res. **5**, 845–889 (2004)
12. Nie, F., Xiang, S., Jia, Y., Zhang, C., Yan, S.: Trace ratio criterion for feature selection. AAAI **2**, 671–676 (2008)
13. Mitra, P., Murthy, C.A., Pal, S.K.: Unsupervised feature selection using feature similarity. IEEE Trans. Pattern Anal. Mach. Intell. **24**(3), 301–312 (2002)
14. Zhao, Z., Liu, H.: Spectral feature selection for supervised and unsupervised learning. In: Proceedings of the 24th International Conference on Machine Learning, pp. 1151-1157 (2007)
15. Li, Z., Yang, Y., Liu, J., Zhou, X., Lu, H.: Unsupervised feature selection using nonnegative spectral analysis. In: AAAI, pp. 1026-1032 (2012)
16. Zhu, P., Zuo, W., Zhang, L., Hu, Q., Shiu, S.C.: Unsupervised feature selection by regularized self-representation. Pattern Recogn. **48**(2), 438–446 (2014)
17. Hou, C., Nie, F., Yi, D., Wu, Y.: Feature selection via joint embedding learning and sparse regression. Proc. Int. Joint Conf. Artif. Intell. **22**(1), 1324–1329 (2011)
18. Zhao, Z., Wang, L., Liu, H.: Efficient spectral feature selection with minimum redundancy. In: AAAI, pp. 673-678 (2010)
19. Elhamifar, E., Vidal, R.: Sparse subspace clustering. In: Computer Vision and Pattern Recognition, pp. 2790–2797 (2009)
20. Wright, J., Yang, A.Y., Ganesh, A., Sastry, S.S., Ma, Y.: Robust face recognition via sparse representation. IEEE Trans. Pattern Anal. Mach. Intell. **31**(2), 210–227 (2009)
21. Liu, G., Lin, Z., Yu, Y.: Robust subspace segmentation by low-rank representation. In: Proceedings of the 27th International Conference on Machine Learning, pp. 663–670 (2010)
22. El-Shaarawi, A.H., Piegorsch, W.W. (eds.): Encyclopedia of Environ Metrics, vol. 1. John Wiley and Sons, New York (2001)
23. Su, A.I., Welsh, J.B., Sapinoso, L.M., et al.: Molecular classification of human carcinomas by use of gene expression signatures. Cancer Res. **61**(20), 7388–7393 (2001)
24. Bhattacharjee, A., Richards, W.G., Staunton, J., et al.: Classification of human lung carcinomas by mRNA expression profiling reveals distinct adenocarcinoma subclasses. Proc. Natl. Acad. Sci. **98**(24), 13790–13795 (2001)
25. Nutt, C.L., Mani, D.R., Betensky, R.A., et al.: Gene expression-based classification of malignant gliomas correlates better with survival than histological classification. Cancer Res. **63**(7), 1602–1607 (2003)

Ranking Web Page with Path Trust Knowledge Graph

YaJun Du(✉), Qiang Hu, XiaoLei Li, XiaoLiang Chen, and ChenXing Li

School of Computer and Soft Engineering, Xihua University, Chengdu 610039, China
dyjdoc2003@aliyun.com

Abstract. How to find and discover useful information from Internet is a real challenge in information retrieval (IR) and search engines (SE). In this paper, we propose and construct Path Trust Knowledge Graph $PTKG$ model for assigning priority values to the unvisited web pages. For a given user specific topic t, its $PTKG$ contains five parts: (1) The context graph $G(t) = (V, E)$, where V is the crawled history web page set and E includes the hyper link set among the history web pages; (2) Retrieving knowledge implied in the paths among these web pages and finding their lengths; (3) Building the trust degrees among the web pages; (4) Constructing topic specific language model and general language model by using the trust degrees; (5) Assigning the priority values of web pages for ranking them. Finally, we perform an experimental comparison among our proposed $PTKG$ approach with the classic LCG and RCG. As a result, our method outperforms LCG and RCG.

Keywords: Knowledge graph · Path trust degree · Topic-specific crawlers

1 Introduction

Up to now, there are the larger amount web pages on the Web. The more and more web pages are related to user topics, before submitting a user topic, it is difficult that search engines as an information retrieving tool return the high precision web pages to the user. The famous search engines, such as Google, Bing, etc. aim at retrieving the most suitable web pages for user topics from the Web. However, with the increasing number of personalized requirements and the number of web pages, these search engines continuously face some challenges: No matter how the scientists improve search algorithms, the obtained search results still exist a lot of irrelevant web pages. Generally, a user only browse $TopN$ $(50 \geq N \geq 10)$ web pages within the several million returned web pages. More than ninety percent of the returned web pages are not browsed [1,2].

Y. Du—Project supported by the National Nature Science Foundation of China (No. 61271413, 61472329).

X. He et al. (Eds.): IScIDE 2015, Part II, LNCS 9243, pp. 66–75, 2015.
DOI: 10.1007/978-3-319-23862-3_7

In the last 20 years, the research of topic-specific web crawlers which are considered as an important component of a search engine became more and more hot spot. The topic specific web crawlers spend a less expensive hardware resource to get the more accurate search result. As much as possible, the general web crawlers download all web pages from the Web. However, a topic specific web crawler only downloads web pages related to user specific topics from the Web. Newly, the web site in [3] reports some open source of the topic specific web crawler, such as tkWWW Robot, Googlebot [4], SEO Spider [5], etc. Some specific topic web crawlers are widely applied to business search engines. For example, shoppers make ideally buying decisions [6] with the help of Become.com (http://www.become.com/) search engine. They provide the product price comparison service which prices are retrieved from the Web by the specific topic web crawler. Hotjobs.com (http://help.yahoo.com//l/us/yahoo/hotjobs/) provides an online jobs for job seekers and recruiters. The job web crawler gets jobs from the Web. To acquire specific topic-directed web pages, the ranking web page is usually adopted to rank the unvisited URLs (web pages) on the crawling queue of the specific topic web crawler and the web pages of search result. For the topic specific web crawlers, the key research is how to assign a proper order to unvisited web pages. There are three ranking approaches which are adopted to different web crawlers: (1) Hyperlink-based ranking approaches [7], (2) Content-based ranking approaches [8], (3) Hyperlink-content-based ranking approaches [9].

2 Prior Work of Context Graph

This section, to understand the our proposed the path trust knowledge graph, recalls some knowledge and references about the context graph (Link Context Graph [10], Relevancy Context Graph [11], Concept Context Graph [12]) in which they are adopted to the specific topic web crawlers.

2.1 Link Context Graph-*LCG*

For a user given topic, *LCG* collects some web pages from the famous search engine like Alta Vista or Google by feeding the core topic, and divides them into the target web pages and the related web pages, and clarifies the hyperlinks among these web pages. The hierarchical graph can be formalized with web pages as nodes and hyper links as arcs, such as Fig. 1. Every seed web page can construct a context graph, where the seed (target) web page [13] as node is located in 0th-layer. And so on, the related web pages as the nodes locating in ith-layer must point to the web pages as nodes locating in $(i-1)$th-layer with hyper links. These hyper links are considered as the arcs linking the nodes in ith-layer to the nodes in the $(i-1)$th-layer. The process is iterative until Nth-layer ($7 \geq N \geq 4$) is constructed. TF-IDF procedure (function) describes a web page into a vector, the Naive Bayes Classifier for each layer mapped an unvisited web page to i ($0 \leq i \leq$N) layer. The layer number i is the ranking of the unvisited web page.

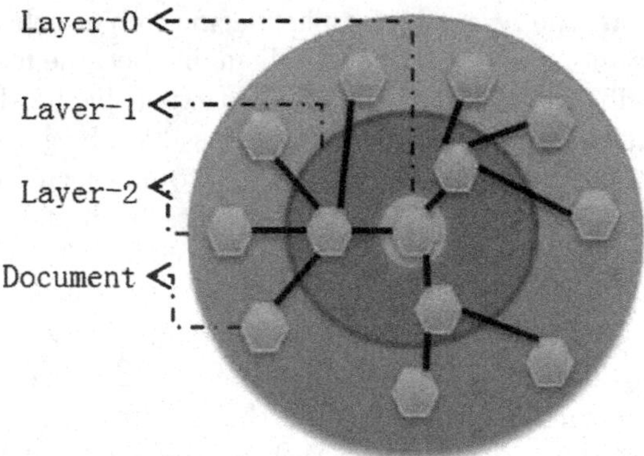

Fig. 1. Structure of Link Context Graph (*LCG*): hierarchical graph with web pages as nodes and hyper links as arcs.

2.2 Relevant Context Graph-*RCG*

On the foundation of *LCG*, C. C. Hsu's [11] research finds some deficiencies with respect to the link context graph, and proposes the relevant context graph (*RCG*). They consider that the topic spatial distributions of web pages are coincident to the spatial distributions and the similar web pages have the similar semantic to the target web pages. And so, they assign the priority α to web pages in the same layer, but not priority level of web pages, and the priority are decreased by their semantic meanings from the 0th-layer to Nth-layer. The web pages located in ith-layer have a relevant degree α^i with respect to the specific topic. For a specific topic t, when the specific topic web crawler based *RCG* approach meets an unvisited web page D, the *RCG* can assign a priority value to the web page D according to the bayes probability $P(t|D)$ that can be compute as follows:

$$P(t|D) = \frac{P(D|t)P(t)}{P(D)} \propto \frac{P(D|t)}{P(D)} = \frac{\sum_{w_i \in D} P(w_i|t)}{\sum_{w_i \in D} P(w_i)} \quad (1)$$

where $P(w_i|t)$ stands for the distribution of specific topic feature word w_i for user topic t, $P(w_i)$ stands for the distribution of the word w_i in our language. The $P(w_i|t)$ can be computed as follows,

$$P(w_i|t) = \frac{\sum_{\forall p \in G(t) \bigwedge w_i \in \pi(p)} R_p \cdot C_{ip}}{\sum_{w_j} R_p \cdot C_{jp}} \quad (2)$$

where $G(t)$ is the context graph with user topic t; $\pi(p)$ is a key word set for the web page p; R_p is the relevant degree between web page p that is contained in $G(t)$ and the target web page, $R_p = \alpha^m$, m is the layer number of web page p located in $G(t)$; C_{kp} is the number of occurrence of word $w_i (k = i \; or \; j)$ in web page p.

2.3 Concept Context Graph-CCG

Inspired by the LCG and RCG, Y. J. Du, etc. [12,14,15] propose Concept Context Graph (CCG) to express the context knowledge of the topic-specific crawler. The CCG consider the web pages in $G(t)$ as the objects and their feature words as the attributes. By using formal concept analysis [16], all objects O and attributes A in $G(t)$ constructed a formal context. Let $x \in O$, $a \in A$, xIa be the object x owns the attribute a, the object set $X \subseteq O$ and attribute $Y \subseteq A$ on the formal context can take respectively their common attributes and objects by Eq. (3). A pair (X, Y) satisfies $X^{\uparrow} = Y$, and $Y^{\downarrow} = X$, then (X, Y) is called as a concept. All concepts can be extracted from these web pages in context graph $G(t)$. For any two concepts (X_1, Y_1) and (X_2, Y_2), put $(X_1, Y_1) \leq (X_2, Y_2)$, if and only if $X_1 \subseteq X_2$, (X_1, Y_1) is called the sub-concept of (X_2, Y_2), and (X_2, Y_2) is the super-concept of (X_1, Y_1). We called relation $\leq$ as the hierarchical order of concepts. Let T be user query, a core concept (X, Y) satisfied $|\frac{Y \bigcap T}{Y \bigcup T}| \geq k$, where k is a given value.

$$X^{\uparrow} = \{a | a \in A, x \in X, xIa\}, Y^{\downarrow} = \{x | x \in O, y \in Y, xIa\}. \quad (3)$$

The core concepts locate in 0th-layer, it is a specific crawling topic. A number of sub-concepts or super-concepts of the core concept can be first retrieved. They are inserted into 1th-layer. The process is iterative until Nth-layer is constructed. It can form the concept context graph for the context web pages of the topic-specific crawling. For a waiting-crawling web page, we can consider itself as an object set X_w and its feature words as an attribute set Y_w, by using Finding appropriate position of virtual concept algorithm [12] of the virtual concept (X_w, Y_w) on the concept context graph, we assign the page ranking to the matched the waiting-crawling web page.

3 Knowledge Graph with Path Trust-$PTKG$

3.1 Path Analysis

Let $G(t) = (V, E)$ denote a context graph, where V is the set of crawled history web pages in which each one is divided into the different layers of $G(t)$[17]. The most related (target) web pages are located in inner-layer and the less related web pages are located in outer-layer; E includes the hyperlink set among the history web pages, from the inner-layer to outer-layer, in turn, any hyperlink in E is a back-link. Apparently, the context graph $G(t)$ of our proposed $PTKG$ is a directed graph. We can use an asymmetric adjacency matrix $M(x, y)$ (Eq. (4)) to express a context graph. If web page x has a hyperlink pointing to web page y.

$$M(x, y) = \begin{cases} 1, if \quad exist\ x \rightarrow y, \\ 0, otherwise. \end{cases} \quad (4)$$

By using the asymmetric adjacency matrix M, we can solve all paths and their lengths from web page x to y.

Definition 1. *For a given context graph of PTKG, let M^k $(k \geq 1)$ be the power (Eq. (5)) of asymmetric adjacency matrix M. M^k is the product of the k numbers' M. Generally,*

$$M^k = M \cdot M \cdots M. \tag{5}$$

where every elements of $M^k(x, y)$ in M^k is the number of paths with length k from web page x to y.

$$M^k(x, y) = \sum_{h=1}^{|V|} M_{xh}^{(k-1)} a_{hy}. \tag{6}$$

We notice that there is no path with length k from x to y. then, $M^k(x, y) = 0$. The algorithm of link prediction [18] is appropriate for social network that can be considered as a small-world network. World Wide Web is a social network constructed by web pages and hyperlinks. Two experiments have proved that WWW is a small-world network. The one is that two randomly chosen web pages on the web are on average 19 clicks away from each other [19]. Another experiment shows that the average shortest path of WWW is about 15 hops by analyzing the network link data [20]. In order to simplify problem, we assume that each link in a context graph has the same trust attenuation for the relevance between two web pages. Based on the above explanation and assumption, we give the definition of relevance trust degree.

3.2 Trust Degree

Definition 2. ***Trust Degree of Web Page****, the trust degree $TD(d_k, d_t)$ of web page d_k to the target web page d_t is defined based on the count of paths and length from d_k to d_t:*

$$TD(d_k, d_t) = \begin{cases} \frac{\sum_{i=\xi_k}^{\tau_k} (1-\beta)^i \times M^i(d_k, d_t)}{\sum_{p \in G(T)-\{d_t\}} \sum_{j=\xi_p}^{\tau_p} (1-\beta)^j \times M^j(d_p, d_t)} & k \neq t, \\ max_{d_\lambda \in G(T)-\{d_t\}} TD(d_\lambda, d_t) & k = t. \end{cases} \tag{7}$$

Where, $G(T)$ is the context graph with topic T; ξ_k (ξ_p) is the minimum length of a path taken into consideration between web pages d_k (d_p) and d_t, τ_k (τ_p) is the maximum length of a path taken into consideration between web pages d_k (d_p) and d_t, all paths are simple path (no circles), and $\tau_k \geq \xi \geq 1$; d_t is the target web page, which is located in 0th layer, d_p is the web page, which is located in ith-layer($i \geq 1$), d_k is a given web page, which is located in ξ_kth layer. $(1 - \beta)^i$ is a trust factor with weights paths according to their length i, β is trust attenuation factor, $M^i(d_k, d_t)$ or $M^j(d_k, d_t)$ is the number of directed paths with length-i (length-j) from d_k (d_p) to d_t.

In Eq. (7), the trust degree of web page d_k is impacted by three factors: (1) the count of paths of web page d_k to the target web page d_t; (2) the length of the path from d_k to d_t; (3) the trust degree of all web pages in $G(T) - \{d_t\}$. Obviously, $1 \geq TD(d_k, d_t) > 0$ and $\sum_{p \in G(T)-\{d_t\}} TD(d_k, d_t) = 1$ are hold.

3.3 Topic Specific Language Model-*TSLM*

In *PTKG* model, topic-specific feature words are the words appeared in its context graph. For specific topic T, the distribution of topic-specific feature word is a probability $P(w_i|T)$ of a word w_i related to T. The distribution is related to three factors: (1) the number of the web pages which includes the word w_i in the context graph of *PTKG* model; (2) The trust degree *TD* of these web pages; (3) The frequency which the word w_i appears in these web page.

Definition 3. ***Word Trust Frequency with*** $G(T)$*, in PTKG model, the trust frequency* $Trust_F$ *(Eq. (8)) of the word* w_i *is related to the web page* d_k *including the word* w_i*, the* w_i*'s frequency appearing the web page* d_k*, and the* d_k*'s trust degree. It is the product of the* w_i*'s frequency appearing* d_k *and trust degree* $TD(d_k, d_t)$ *of the web page* d_k*.*

$$Trust_F(w_i, d_k) = TD(d_k, d_t) \times f_{w_i,d_k}. \tag{8}$$

Where, $d_k \in G(T)$*;* d_t *is the target web page located in 0th-layer of* $G(T)$*;* $TD(d_k, d_t)$ *is the trust degree of the web page* d_k*;* f_{w_i,d_k} *is the occurrence number of word* w_i *in web page* d_k*.*

The distributions of all words in context graph $G(T)$ construct the Topic Specific Language Model (*TSLM*) of *PTKG* model. According to Eq. (2), Topic Specific Language Model (*TSLM*) of *PTKG* model is defined as follows:

Definition 4. ***Topic Specific Language Model****, in PTKG model, the words appeared in* $G(T)$ *and their trust distributions (probabilities) in* $G(T)$ *construct the topic specific language model of the PTKG. The probability* $P_{PTKG}(w_i|t)$ *of the word* w_i *is the proportion of the word* w_i*'s trust frequency to the trust frequencies of all words appearing the G(T).*

$$P_{PTKG}(w_i|t) = \frac{\sum_{\forall d_k \in G(T) \bigwedge w_i \in \pi(d_k)} Trust_F(w_i, d_k)}{\sum_{\forall d_\lambda \in G(T)} \sum_{w_j}^{\pi(d_\lambda)} Trust_F(w_j, d_\lambda)} \tag{9}$$

Where, $d_k \in G(T)$*,* $d_\lambda \in G(T)$*,* $G(T)$ *is a context graph of topic* T*;* $\pi(d_k)$ *and* $\pi(d_\lambda)$ *are the word sets of the web pages* d_k *and* d_λ*, respectively;* $Trust_F(w_i, d_k)$ *and* $Trust_F(w_j, d_\lambda)$ *are trust frequency of the word* w_i *in web page* d_k *and word* w_j *in web page* d_λ*, respectively.*

3.4 General Language Model-*GLM*

In our *PTKG* model, the topic-specific language model expresses the topic-specific feature words and their frequencies. The other words besides topic-specific feature words are necessary for describing *PTKG*. These words, which stem from the web pages out of $G(T)$, are not topic-specific feature words, but related to the topic. In *LCG* and *RCG*, when submitting the topic T to a famous search engine, such as google, bing, etc., they return TopN web pages. A web page set ζ filtering the web pages of $G(T)$ from the TopN web pages is used to construct the general language model. The general language model is made up of the words from ζ and their distributions (probabilities) in ζ. For given w_i, its' probability can be computed by Eq. (10). The probability of w_i is the proportion of the sum of frequencies, in which the word w_i appears in web pages of ζ, to the sum of the frequencies of all words appeared in web pages of ζ.

$$p_{w_i} = \frac{\sum_{d_k \in \zeta \bigwedge w_i \in \pi(d_k)} f_{w_i,d_k}}{\sum_{d_\lambda \in \zeta} \sum_{w_j \in \pi(d_\lambda)} f_{w_j,d_\lambda}}. \tag{10}$$

Where, $d_k \in \zeta$, $d_\lambda \in \zeta$; $\pi(d_k)$ and $\pi(d_\lambda)$ are the word sets of the web pages d_k and d_λ, respectively; f_{w_i,d_k} and f_{w_j,d_λ} are the occurrence numbers of words w_i in web page d_k and w_j in web page d_λ, respectively.

On the other hand, the words and their frequencies can stem from the knowledge database, such as HowNet, Encyclopaedia, etc. We can compute their distributions (probabilities) by using Eq. (10).

3.5 Ranking of Web Page

In *LCG* and *RCG*, the priority value of a given web page on the Web can be predicted by Eq. (1). The priority value is similarity degree of the web page with the topic T. We clearly consider that the words of the web page include the topic specific feature words (denoted by W_T) and the other words related to the topic T. In our proposed *PTKG* model, there are two factors determining the priority value for the given web page, (1) The occurrence probabilities of the topic specific feature words of the web page in *TSLM* and their frequencies on the web page; (2) The probabilities of the non topic specific feature words of the web page in *GLM* and their frequencies on the web page. For the words of the web page without including the *TSLM* and *GLM*, their probabilities take zero. The priority value of web page x can be computed as follows:

$$P_r(t|x) = \frac{P(x|t)P(t)}{P(x)} \propto \frac{P(x|t)}{P(x)} = \frac{\sum_{w_i \in \pi(x) \bigwedge w_i \in W_T} p_{PTKG}(w_i|t) \times f_{w_i,x}}{\sum_{w_j \in \pi(x)} p(w_j) \times f_{w_j,x}}. \tag{11}$$

Where, $p_{PTKG}(w_i|t)$ is the probabilities of the topic specific feature word w_i of the web page x in *TSLM*; $p(w_j)$ is the probabilities of the word w_j of the web page x in *GLM*; $\pi(x)$ is the word set of the web pages x; $f_{w_i,x}$ and $f_{w_j,x}$ are the occurrence numbers of words w_i in web page x and w_j in web page x, respectively.

4 Experimental Analysis

To evaluate the performance of our proposed *PTKG* model, the ranking of web pages will be applied to our designed specific topic web crawler. On the other hand, we given some specific topics, such as College Entrance Examination, shorten for CEE and cookbook, etc. and obtain their relevant web pages and average relevancy under three approaches *LCG*, *RCG*, *PTKG*. At last, we evaluate our proposed *PTKG* with other two approaches. The seed URLs are retrieved through www.baidu.com which is the most powerful Chinese search site. But the top 30 web pages returned by Baidu are not ideal. We summarize three kinds of interferential web pages: (1) Some web pages have few words and the rest of the pages are pictures. (2) Some web pages may have no backlink, because they located in deep layer. (3) The websites of some web pages are invalid. Therefore, we cannot get the content to extract topic words.

Figure 2(a), (b) show the number of relevant web pages contrasting with the number of downloaded web pages for the two topics (Cookbook, CEE). When the crawled web pages increase at 2000 step, we retrieve the number of relevant web pages by three web crawlers with *PTKG*, *RCG* and *LCG*. In Fig. 2(a), for the topic 'Cookbook', three web crawlers have the different increasing rates from 0 to 1750 crawled web pages. The web crawlers with *PTKG* and *RCG* keep almost the same increasing rate from 1750 to 18000 crawled web pages, and the web crawler with *LCG* has a low increasing rate. However, it still prove that the web crawler with *PTKG* can retrieve the more relevant web pages than the web crawlers with *RCG* and *LCG*. In Fig. 2(b), for the topics 'CEE', we obtain the same conclusions: (1) When three web crawlers with *PTKG*, *RCG* and *LCG* start to crawl the Web in primary stage, their performances have the fewer diversity for retrieving the web pages from the Web. (2) Totally, the web crawler with *PTKG* can retrieve the more relevant web pages than the web crawlers with *RCG* and *LCG*. The performance of our proposed approach outperforms the other two web crawlers. (3) Along with further crawling, the web crawlers with *PTKG* keep the more increasing rate than the web crawlers with *RCG* and *LCG*.

Figure 3(a), (b) show the average relevancy contrasting with the number of downloaded web pages for the two topics (Cookbook, CEE). When the crawled web pages increase at 2000 step, we retrieve their average relevancy by three web crawlers with *PTKG*, *RCG* and *LCG*, respectively. In Fig. 3(a), for the

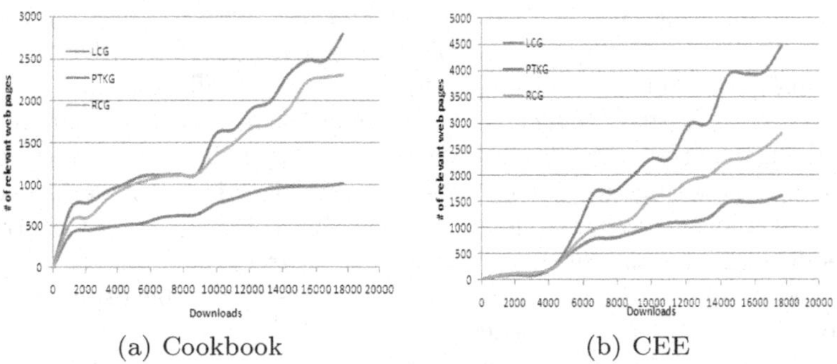

(a) Cookbook (b) CEE

Fig. 2. The number of the relevant documents contrast with the number of downloaded documents.

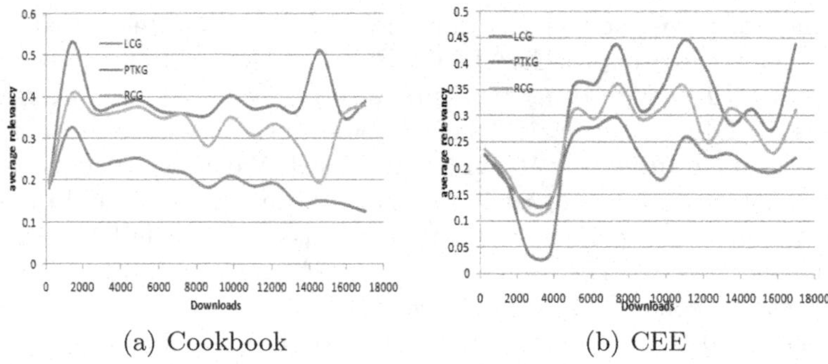

(a) Cookbook (b) CEE

Fig. 3. The average relevance of each 2000 neighboring downloaded web pages.

topic 'Cookbook', three web crawlers have the different increase rates of average relevancy from 0 to 1750 crawled web pages. The web crawlers with *PTKG* and *RCG* keep the higher relevancy than the web crawlers with *RCG* and *LCG*. In Fig. 3(b), for the topic 'CEE', the three web crawlers keep almost the same increasing rates at the initial stage (from 0 to 1750) of crawled web pages. Their performance keeps the fewer difference. At later stage, the web crawler with *PTKG* can retrieve the higher average relevancy than the web crawlers with *RCG* and *LCG*, the lower average relevancy occasionally take place at small interval. It reveals that the performance of our proposed approach outperforms the other two web crawlers, and shows that our proposed approach-*PTKG* can guide the web crawler to search in a good direction and to find more relevant web pages than *RCG* and *LCG*.

5 Conclusion and Future Work

In this paper, we propose *PTKG* to represent the context knowledge of web crawlers. Based on the foundation of the Link Context Graph (*LCG*), Relevancy Context Graph (*RCG*) and Concept Context Graph (*CCG*), we construct the Path Trust Knowledge Graph *PTKG* model for assigning the priority values to web pages. The *PTKG* model is applied to guide a fused crawler to retrieve the more and more related web pages for the user specific topics. For a given user specific topic t, its *PTKG* contains serve parts: (1) The context graph $G(t) = (V, E)$, V is the crawled history web pages, E includes the hyperlinks set among the history web pages; (2) Retrieving knowledge about the paths and their lengths; (3) Building the trust degree among the web pages; (4) Constructing topic specific language model by using the trust degree; (5) Establishing general language model by using the trust degree; (6) Assigning the priority value of web pages for ranking the web pages.

In the future, we plan to take semantic relationships of words into consideration to optimize the performance of *PTKG*. Another research is how to evaluate good backlinks for the basic context graph.

References

1. Liu, W.J., Du, Y.J.: A novel focused crawler based on cell-like membrane computing optimization algorithm. Neurocomputing **123**, 266–280 (2014)
2. The size of the World Wide Web (2014). http://www.worldwidewebsize.com/
3. Web crawler (2014). http://en.wikipedia.org/wiki/Web_crawler
4. Brin, S., Page, L.: The anatomy of a large-scale hypertextual Web search engine. Comput. Netw. ISDN Syst. **30**(1–7), 107–117 (1998)
5. Developed by WebBee Team, WebBee SEO Spider. Java based Desktop (SEO Spider) application (2014)
6. Chris, S.: Become.com Launches Shopping Search Engine. In: SES Conference and Expo (2005)
7. Kleinberg, J.M.: Authoritative sources in a hyperlinked environment. J. ACM **46**(5), 604–632 (1999)
8. Wu, B., Yang, J., He, L.: Chinese hownet-based multi-factor word similarity algorithm integrated of result modification. In: Huang, T., Zeng, Z., Li, C., Leung, C.S. (eds.) ICONIP 2012, Part V. LNCS, vol. 7667, pp. 256–266. Springer, Heidelberg (2012)
9. Du, Y.J., Hai, Y.F.: Semantic ranking of web pages based on formal concept analysis. J. Syst. Softw. **86**, 187–197 (2013)
10. Diligenti, M., Coetzee, F.M., Lawrence, S., Giles, C.L.: Focused crawling using context graphs. In: The 26th International Conference on Very Large Database (VLDB), pp. 527–534 (2000)
11. Hsu, C.C., Wu, F.: Topic-specific crawling on the web with the measurements of the relevancy context graph. Inf. Syst. **31**, 232–246 (2006)
12. Du, Y.J., Peng, Q.Q., Gao, Z.Q.: A topic-specific crawling strategy based on semantics similarity. Data Knowl. Eng. **88**, 75–93 (2013)
13. Du, Y.J., Hai, Y.F., Xie, C.Z.: An approach for selecting seed urls of focused crawler based on user-interest ontology. Appl. Soft Comput. **14**(C), 663–676 (2014)
14. Liu, Z.J., Du, Y.J., Zhao, Y.: Focused crawler based on domain ontology and FCA. J. Inf. Comput. Sci. **8**(10), 1909–1917 (2011)
15. Du, Y.J., Dong, Z.B.: Focused web crawling strategy based on concept context graph. J. Inf. Comput. Sci. **5**(3), 1097–1106 (2009)
16. Wille, R.: An approach based restructuring lattice theory hierarchies of concepts. In: Rival, I. (ed.) Ordered Sets, vol. 83, pp. 445–470. Springer, The Netherlands (1982)
17. Liu, Q., Tu, Z.P., Lin, S.X.: A novel graph-based compact representation of word alignment. In: Proceedings of Annual Meeting of the Association for Computational Linguistics ACL 2013 (2013)
18. Alexis, P., Panagiotis, S., Yannis, M.: Fast and accurate link prediction in social networking systems. J. Syst. Softw. **85**, 2119–2132 (2012)
19. Albert, R., Jeong, H., Barabasi, A.: Internet:diameter of the world-wide web. Nature **401**(6749), 130–131 (1999)
20. Guo, Y., Liu, Z.W., Zhao, Z.X.: Complexity analysis on link structure of world wide web. Comput. Eng. **37**(23), 105–106, 109 (2011)

Locality Preserving One-Class Support Vector Machine

Xiaoming Wang(✉), Yong Tian, Xiaohuan Yang, and Yajun Du

School of Computer and Software Engineering, Xihua University, Chengdu 610039, China
wangxmwm@gmail.com

Abstract. One-class support vector machine (OCSVM) tries to find a hyperplane to distinguish normal data from all other possible outliers or abnormal data. However, only support vectors determine the hyperplane. In this paper, we propose a novel data description method called locality preserving one-class support vector machine (LPOCSVM). It takes the intrinsic manifold structure of data into full consideration. In the paper, we discuss the linear and nonlinear case of LPOCSVM, and detail how to tackle the singularity of the locality preserving scatter matrix. Experimental results on several toy and benchmark datasets indicate the effectiveness and advantage of LPOCSVM by comparing it with OCSVM.

Keywords: Kernel methods · Unsupervised learning · Data description · One-class SVM

1 Introduction

During the past decade, kernel methods [1] are widely studied and applied. In supervised learning, the representative method is support vector machine (SVM), which is a powerful machine learning method based on Vapnik's Statistical Learning Theory [2]. On the other hand, one-class support vector machine (OCSVM) [3] is a powerful method in unsupervised learning scenarios and is developed to descript normal data. To be specific, OCSVM uses a decision hyperplane to distinguish a set of normal data from all other possible abnormal data. In order to build the decision hyperplane, OCSVM makes use of the strategy that the origin has maximal distance to the hyperplane. The final decision hyperplane of OCSVM is constructed by some of the data points called support vector. It is worthwhile to note that OCSVM is equivalent to support vector data description (SVDD) [4] in a special case where the RBF kernel is used. OCSVM has drawn many attention of the researchers in pattern recognition [5–10].

However, The key limitation in OCSVM is that only support vectors determine the hyperplane whereas all other data points have no influence on the hyperplane. This means that the characteristic of data is irrelevant to the decision hyperplane that is found to descript normal data. This drawback of OCSVM may result in a non-robust solution. The reason is that OCSVM ignores the data characteristic when trying to find the decision hyperplane.

X. He et al. (Eds.): IScIDE 2015, part II, LNCS 9243, pp. 76–85, 2015.
DOI: 10.1007/978-3-319-23862-3_8

Actually, in some algorithms the intrinsic characteristic of data has been explicitly considered in a direct way. A representative algorithm is locality preserving projections (LPP) [11], which is a linear dimensionality reduction method. It overcomes the shortcoming of the traditional linear methods that are unable to maintain the nonlinear manifold of original data.

Inspired by LPP and aiming at the drawback of OCSVM, in this paper we propose a novel data description method called locality preserving one-class support vector machine(LPOCSVM). First, by using the basic idea of the LPP, we define the locality preserving scatter matrix. Then, the optimization problem of LPOCSVM is formulated by using the defined locality preserving scatter matrix. Actually, both OCSVM and LPOCSVM attempts to find the hyperplane that has maximal distance to the origin. However, a key difference between OCSVM and LPOCSVM is that the latter employs the Mahalanobis distance metric rather than the Euclidean distance metric in OCSVM. The Mahalanobis distance metric used in LPOCSVM is induced by the defined locality preserving scatter matrix and explicitly reflects the data manifold structure. Therefore, LPOCSVM also takes the intrinsic manifold structure of data into full consideration.

The rest of this paper is organized as follows. The related works will be reviewed in Sect. 2. In Sect. 3, the locality preserving scatter matrix is first defined by using the basic idea of LPP, and then the optimization problem of LPOCSVM is formulated and solved. In Sect. 4, we discuss the nonlinear case. The experimental results are reported in Sect. 5. Finally, conclusions are drawn in Sect. 6.

2 Related Work

2.1 One-Class SVM

Given a dataset $\mathbf{X} = \{\mathbf{x}_i | \mathbf{x}_i \in \mathcal{R}^d, i = 1, \cdots, N\}$, in the linear case, the primal optimization problem of OCSVM is defined as [3]

$$\begin{aligned} &\min \ \frac{1}{2}\mathbf{w}^T\mathbf{w} - \rho + C\sum_{i=1}^{N}\xi_i \\ &s.t. \ \rho - \mathbf{w}^T\mathbf{x}_i - \xi_i \leq 0, \ \ \xi_i \geq 0, \ \ i = 1, \cdots, N \end{aligned} \tag{1}$$

Here, ξ_i's are slack variables, the parameter C needs to be predefined and controls the fraction of outliers in $\mathbf{X}$. Intuitively, OCSVM is to try to find the hyperplane that has the maximal margin distance from the origin to the hyperplane. Note, here the margin distance is $\frac{\rho}{\sqrt{\mathbf{w}^T\mathbf{w}}}$.

Assuming that $\{\mathbf{w}^*, \rho^*, \boldsymbol{\xi}^*\}$ is the solution to the optimization problem (1), the decision function of OCSVM can be formulated as

$$f(\mathbf{x}) = sign((\mathbf{w}^*)^T\mathbf{x} - \rho^*) \tag{2}$$

For one test example $\mathbf{x}$,if $f(\mathbf{x}) = -1$ then it is viewed as an outlier.

2.2 Locality Preserving Project

LPP is a linear dimensionality reduction method by feature projection. Let $N_k(\mathbf{x}_i)$ denotes k nearest neighbors of node i. Here the ith node corresponds to the data point $\mathbf{x}_i$. Nodes i and j are connected by an edge if i is among k nearest neighbors of j or j is among k nearest neighbors of i, i.e. $\mathbf{x}_j \in N_k(\mathbf{x}_i)$ or $\mathbf{x}_i \in N_k(\mathbf{x}_j)$. In order to weigh the edges of the adjacency graph G, ones generally choose the heat kernel to calculate the weight matrix $\mathbf{W}$ as follows

$$W_{ij} = \begin{cases} \exp(-\frac{||\mathbf{x}_i - \mathbf{x}_j||^2}{t}), & if\ \mathbf{x}_i \in N_k(\mathbf{x}_j)\ \ or\ \mathbf{x}_j \in N_k(\mathbf{x}_i) \\ 0, & other. \end{cases} \tag{3}$$

where $||\mathbf{x}|| = \left(\sum_{i=1}^{M} \mathbf{x}_i^2\right)^{\frac{1}{2}}$ is the usual Euclidean (L_2) norm in $\mathcal{R}^d$ and $t > 0$ is the heat kernel parameter and can be empirically determined. LPP tries to find the linear transformation vector $\mathbf{w} = \mathcal{R}^d$ by minimizing the following objective function

$$\begin{aligned} &\min_{\mathbf{w}} \mathbf{w}^T \mathbf{X} \mathbf{L} \mathbf{X}^T \mathbf{w} \\ &s.t.\ \ \mathbf{w}^T \mathbf{X} \mathbf{D} \mathbf{X}^T \mathbf{w} = 1 \end{aligned} \tag{4}$$

Where $\mathbf{D}$ is a diagonal matrix and its entries are column (or row, since $\mathbf{W}$ is symmetric) sum of $\mathbf{W}$, i.e. $D_{ii} = \sum_j W_{ij}$, and $\mathbf{L} = \mathbf{D} - \mathbf{W}$ is the Laplacian matrix. The transformation vector $\mathbf{w}$ that minimizes the objective function is given by the minimum eigenvalue solution to the generalized eigenvalue problem [11].

3 Locality Preserving One-Class Support Vector Machine

3.1 Locality Preserving Scatter Matrix

In order to facilitate our discussion, here we would like to give the following definitions by employing the basic idea of LPP.

Definition 1(Locality Preserving Scatter Matrix). Let $\mathbf{L} = \mathbf{D} - \mathbf{W}$ be the Laplacian matrix of the dataset $\mathbf{X}$, the matrix $\mathbf{Z} = \mathbf{X}\mathbf{L}\mathbf{X}^T = \mathbf{X}(\mathbf{D} - \mathbf{W})\mathbf{X}^T$ is called the locality preserving scatter matrix.

Here, the metrics $\mathbf{W}$ and $\mathbf{D}$ are the same in Subsect. 2.2.

Obviously, the locality preserving scatter matrix $\mathbf{Z}$ stems from the objective function of LPP. It is worthwhile to note that $\mathbf{Z}$ is formally similar to the scatter matrix. However, the locality preserving scatter matrix $\mathbf{Z}$ incorporates the intrinsic geometry and local structure of data.

3.2 Locality Preserving One-Class Support Vector Machine

First, in the case of no noise, we define the margin distance from the origin to the hyperplane $\rho - \mathbf{w}^T\mathbf{x} = 0$ as follows

$$\frac{\rho}{\sqrt{\mathbf{w}^T\mathbf{Z}\mathbf{w}}} \tag{5}$$

In order to find the hyperplane $\rho - \mathbf{w}^T\mathbf{x} = 0$, as OCSVM, LPOCSVM employs the maximal margin principle, i.e., the hyperplane $\rho - \mathbf{w}^T\mathbf{x} = 0$ has maximal distance to the origin. Therefore, we define the following problem

$$\begin{aligned} &\max \ \frac{\rho}{\sqrt{\mathbf{w}^T\mathbf{Z}\mathbf{w}}} \\ &s.t. \ \ \rho - \mathbf{w}^T\mathbf{x}_i \leq 0, \ \ i = 1, \cdots, N \end{aligned} \tag{6}$$

Thus, by transforming the above problem (6), we formulate the optimization problem of LPOCSVM as

$$\begin{aligned} &\min \ \frac{1}{2}\mathbf{w}^T\mathbf{Z}\mathbf{w} - \rho \\ &s.t. \ \ \rho - \mathbf{w}^T\mathbf{x}_i \leq 0, \ \ i = 1, \cdots, N \end{aligned} \tag{7}$$

Obviously, the key insight of our method is that it defines the margin distance from the origin to the hyperplane $\mathbf{w}^T\mathbf{x} - \rho = 0$ as $\frac{\rho}{\sqrt{\mathbf{w}^T\mathbf{Z}\mathbf{w}}}$ by using the Mahalanobis distance metric which is induced by the locality preserving covariance matrix $\mathbf{Z}$. However, in OCSVM the margin distance is defined as $\frac{\rho}{\sqrt{\mathbf{w}^T\mathbf{w}}}$ by directly using the Euclidean distance metric. Therefore, LPOCSVM incorporates the intrinsic geometry and local structure of data.

Further, considering the presence of possible outliers or noise, similarly to OCSVM, LPOCSVM introduces slack variables and defines the following optimization problem

$$\begin{aligned} &\min \ \frac{1}{2}\mathbf{w}^T\mathbf{Z}\mathbf{w} - \rho + C\sum_{i=1}^{N}\xi_i \\ &s.t. \ \ \rho - \mathbf{w}^T\mathbf{x}_i - \xi_i \leq 0, \ \ \xi_i \geq 0, \ \ i = 1, \cdots, N \end{aligned} \tag{8}$$

As in OCSVM, we can solve the above problem (8) by transforming it into its corresponding dual problem. By using Wolfe duality [12], the Wolf dual problem of (8) can be formulated as follows

$$\min \ \frac{1}{2}\sum_{i=1}^{N}\sum_{j=1}^{N}\alpha_i\alpha_j H_{ij}$$
$$s.t. \ \sum_{i=1}^{N}\alpha_i = 1, \ \ 0 \le \alpha_i \le C, \ \ i = 1,\cdots,N \tag{9}$$

where $H_{ij} = \mathbf{x}_i^T\mathbf{Z}^{-1}\mathbf{x}_j$. Suppose $\boldsymbol{\alpha}^* = [\alpha_1^*,\cdots,\alpha_N^*]^T$ can be used to solve the above optimization problem, then the optimal weight vector

$$\mathbf{w}^* = \mathbf{Z}^{-1}\sum_{i=1}^{N}\alpha_i^*\mathbf{x}_i \tag{10}$$

If $i : 0 < \alpha_i < C$, similarly to OCSVM, the corresponding data point $\mathbf{x}_i$ can be called a support vector. Here, let D_{SV} consist of all support vectors, $\mathbf{x}_k$ be any data point with $0 < \alpha_k^* < C \ |\forall k \in \{1,\cdots,N\}$ and N_{SV} refer to the number of support vectors, then we can calculate the optimal threshold ρ^* as follows

$$\rho^* = \frac{1}{N_{SV}}\sum_{i=1}^{N_{SV}}\sum_{j=1}^{N}\alpha_j^* H_{ij}, \quad \mathbf{x}_i \in D_{SV} \tag{11}$$

Thus, the decision function of LPOCSVM can be written as

$$f(\mathbf{x}) = sign((\mathbf{w}^*)^T\mathbf{x} - \rho^*) \tag{12}$$

3.3 The Singularity Problem of the Locality Preserving Scatter Matrix

In the above discussion, we suppose that the locality preserving scatter matrix $\mathbf{Z}$ is nonsingular. However, in practical application, $\mathbf{Z}$ may be singular. Next, we will discuss how to deal with the issue.

Let $\boldsymbol{\Psi}$ and $\boldsymbol{\Pi}$ be the complementary dimensional spaces spanned by the orthogonal eigenvectors of $\mathbf{Z}$ that correspond to nonzero eigenvalues and to zero eigenvalues, respectively. Thus, we rewrite the optimization problem (8) of LPOCSVM as

$$\min \ \frac{1}{2}\mathbf{v}^T\mathbf{Z}\mathbf{v} - \rho + C\sum_{i=1}^{N}\xi_i$$
$$s.t. \ \rho - \mathbf{v}^T\mathbf{x}_i - \xi_i \le 0, \ \ \xi_i \ge 0, \ i = 1,\cdots,N \tag{13}$$

where $\mathbf{v} \in \boldsymbol{\Psi}$. Actually, we transform the optimization problem of LPOCSVM in $\mathcal{R}^d$ into one in $\boldsymbol{\Psi}$. Suppose $\mathbf{Z}$ has m nonzero eigenvectors and let the column vectors of $\mathbf{P}$ are eigenvectors corresponding to nonzero eigenvectors of $\mathbf{Z}$, by linear algebra theory, $\boldsymbol{\Psi}$ is isomorphic to m-dimensional Euclidean space $\mathcal{R}^m$ [13]. And the corresponding isomorphic mapping is

$$\mathbf{v} = \mathbf{P}\eta,\ \mathbf{v} \in \boldsymbol{\Psi},\ \boldsymbol{\eta} \in \mathcal{R}^m \tag{14}$$

Thus, in $\mathcal{R}^m$ the optimization problem of LPOCSVM can be written as

$$\begin{aligned} \min\ & \frac{1}{2}\boldsymbol{\eta}^T\widehat{\mathbf{Z}}\eta - \rho + C\sum_{i=1}^{N}\xi_i \\ s.t.\ & \rho - \mathbf{v}^T\mathbf{y}_i - \xi_i \leq 0,\ \ \xi_i \geq 0,\ i = 1, \cdots, N \end{aligned} \tag{15}$$

where $\boldsymbol{\eta} \in \mathcal{R}^m$, $\mathbf{y}_i = \mathbf{P}^T\mathbf{x}_i$ and $\tilde{\mathbf{Z}} = \mathbf{P}^T\mathbf{Z}\mathbf{P}$. However, in $\mathcal{R}^m$ the locality preserving scatter matrix $\tilde{\mathbf{Z}}$ will be nonsingular, and we can solve (18) according to the discussion in Subsect.3.2.

Suppose $\{\boldsymbol{\eta}^*, \rho^*, \boldsymbol{\xi}^*\}$ solves the above optimization problem, the decision function is

$$f(\mathbf{X}) = \mathrm{sgn}(\boldsymbol{\eta}^{*T}\mathbf{y} - \rho^*) = \mathrm{sgn}(\boldsymbol{\eta}^{*T}\mathbf{P}^T\mathbf{X} - \rho^*) \tag{16}$$

4 The Nonlinear Case

In order to handle with nonlinear problems, we can seek to use the kernelization trick [1], i.e., define a mapping $\varphi(\mathbf{x}) : \mathcal{R}^d \rightarrow \mathcal{H}$ to map the d-dimensional data points into a high-dimensional feature space $\mathcal{H}$. In the feature space $\mathcal{H}$ the optimization problem of LPOCSVM is defined as

$$\begin{aligned} \min\ & \frac{1}{2}\mathbf{w}^T\mathbf{Z}_\phi\mathbf{w} - \rho + C\sum_{i=1}^{N}\xi_i \\ s.t.\ & \rho - \mathbf{w}^T\phi(\mathbf{x}_i) - \xi_i \leq 0,\ \ \xi_i \geq 0,\ i = 1, \cdots, N \end{aligned} \tag{17}$$

where $\phi(\mathbf{x}_i)$ $(i = 1, \cdots, N)$ denotes data in feature space, $\mathbf{Z}_\phi$ is the corresponding locality preserving scatter matrix and defined as

$$\mathbf{Z}_\phi = \mathbf{X}_\phi\mathbf{L}\mathbf{X}_\phi^T = \mathbf{X}_\phi(\mathbf{D} - \mathbf{W})\mathbf{X}_\phi^T \tag{18}$$

where $\mathbf{X}_\phi = [\phi(\mathbf{x}_1), \cdots, \phi(\mathbf{x}_N)]$.

However, in the high-dimensional feature space LPOCSVM could not directly get the solution because the locality preserving scatter matrix $\mathbf{Z}_\phi$ is generally singular. According to the representation theorem for Reproducing Kernel Hilbert Spaces [1], of which the vector $\mathbf{w}$ can be formulated as

$$\mathbf{w} = \sum_{i=1}^{N} a_i\phi(\mathbf{x}_i) \tag{19}$$

Thus, the optimization problem (22) can be reformulated as

$$\begin{aligned} &\min \ \frac{1}{2}\mathbf{a}^{\mathrm{T}}\mathbf{KLKa} - \rho + C\sum_{i=1}^{N}\xi_i \\ &s.t. \ \ \rho - \mathbf{a}^{\mathrm{T}}\mathbf{k}_i - \xi_i \le 0, \ \ \xi_i \ge 0, \ i = 1, \cdots, N \end{aligned} \tag{20}$$

where $\mathbf{K} = \left(k(\mathbf{x}_i, \mathbf{x}_j)\right)_{N\times N}$ is the kernel matrix, the vectors $\mathbf{k}_i$ and $\mathbf{a}$ are respectively defined as $\mathbf{k}_i = [k(\mathbf{x}_i, \mathbf{x}_1), k(\mathbf{x}_i, \mathbf{x}_2), \cdots, k(\mathbf{x}_i, \mathbf{x}_N)]^T$ and $\mathbf{a} = [\alpha_1, \cdots, \alpha_N]^T$. Here $k(\mathbf{x}_i, \mathbf{x}_j) = \phi^T(\mathbf{x}_i)\phi(\mathbf{x}_j)$ is a predefined kernel function. Let $\widehat{\mathbf{H}} = \mathbf{KLK}$, the above optimization problem (20) can be further written as

$$\begin{aligned} &\min \ \frac{1}{2}\mathbf{a}^{\mathrm{T}}\widehat{\mathbf{H}}\mathbf{a} - \rho + C\sum_{i=1}^{N}\xi_i \\ &s.t. \ \ \rho - \mathbf{a}^{\mathrm{T}}\mathbf{k}_i - \xi_i \le 0, \ \ \xi_i \ge 0, \ i = 1, \cdots, N \end{aligned} \tag{21}$$

It should be noted that the above optimization problem (21) is a optimization problem defined by the linear LPOCSVM algorithm since $\widehat{\mathbf{H}} = \mathbf{KLK}$ is the locality preserving scatter matrix of the dataset which consists of $\mathbf{k}_i(i = 1, \cdots, N)$. So, it can be efficiently solved according to the previous discussion. Suppose $\{\boldsymbol{\alpha}^*, \rho^*\}$ solves the above optimization problem, we can obtain the corresponding decision function as follows

$$f(\mathbf{x}) = sign((\boldsymbol{\alpha}^*)^T\mathbf{k} - \rho^*) \tag{22}$$

Here $\mathbf{k} = [k(\mathbf{x}_1, \mathbf{x}), k(\mathbf{x}_2, \mathbf{x}), \cdots, k(\mathbf{x}_N, \mathbf{x})]^T$.

5 Experiments

In this section, the experimental results will be reported. Generally, outlier detection is an important application of data description method. In order to evaluate the performance of outlier detection of our method, we choose several datasets from the University of California at Irvine (UCI) [14]. Table 1 depicts the characteristics of the selected datasets. In our method, we introduce two parameters including the heat parameter t and the neighborhood parameter k when calculating the weigh matrix according to (3). In [15], the authors proposed a method to compute the heat parameter t. In our work, we adopt the method to calculate the heat parameter t.

To compare the performance of the related methods, we adopt the F-measure as the performance metric, which is used in [16] and defined as

$$F - measure = \frac{2 \times precision \times recall}{precision + recall} \tag{23}$$

Table 1. Characteristics of the selected datasets

Dataset		No. of patterns	No. of features
Breast	Class1	241	8
	Class2	458	8
Heart	Class1	150	13
	Class2	120	13
Sonar	Class1	97	60
	Class2	111	60
Ionosphere	Class1	225	34
	Class2	126	34
Diabetes	Class1	268	8
	Class2	500	8
Wdbc	Class1	212	30
	Class2	357	30
Iris	Class1	50	4
	Class2	50	4
	Class3	50	4
wine	Class1	59	13
	Class2	71	13
	Class3	48	13

where $precision = \frac{TP}{TP+FP}$ and $recall = \frac{TP}{TP+FN}$. Here *TP* (True Positives) is the number of items correctly labeled as belonging to the positive class, *FP* (False Positive) the number of items incorrectly labeled as belonging to the positive class, and *FN* (False Negative) is the number of items incorrectly labeled as belonging to the Negative class. A good novelty detector should attain high values on F-measure.

For each dataset, we take a class as normal data and the other as outliers. We randomly sample 80 % of the normal data points for training. The remaining 20 % of the normal patterns and all the outliers are used for testing. Here, the RBF kernel is used and the results are based on averages over 50 random repetitions.

In [17], a method was proposed to select parameters for OCSVM. Here we adapted the same strategy. In the experiments, the RBF kernel parameter γ is selected in the set $\{2^{-5},\ldots, 2^{5}\}$ and the regularization parameter C is chosen from the set $\{2^{-5},\ldots, 2^{0}\}$. The compared methods include Parzen [18], OCSVM and LPOCSVM. Parzen is a density estimator and also can be used as a outlier detection method.

The experimental results are reported in Table 1. It can be found that OCSVM performs better in contrast with Parzen since it obtains better F-measure in most cases. This shows that OCSVM is a good method for data description. However, on the whole, LPOCSVM achieves an improvement in F-measure in contrast with OCSVM. This means that it is helpful to incorporate the intrinsic manifold structure of the data space in OCSVM. The proposed method embodies the idea and so outperforms Parzen and OCSVM.

Table 2. The values on F-measure of Parzen, OCSVM, LPOCSVM on the selected datasets

Dataset		Parzen	OCSVM	LPOCSVM
Breast	Class1	92.9	95.1	**96.7**
	Class2	93.1	96.2	**98.3**
Heart	Class1	69.6	71.4	**72.2**
	Class2	65.9	69.3	**69.6**
Sonar	Class1	**67.1**	66.5	66.5
	Class2	**64.9**	64.6	64.6
Ionosphere	Class1	79.8	81.0	**83.5**
	Class2	76.1	76.9	**77.6**
Diabetes	Class1	68.2	66.9	**70.4**
	Class2	69.8	70.1	**73.9**
Wdbc	Class1	73.1	75.5	**77.1**
	Class2	70.7	72.2	**75.2**
Iris	Class1	94.3	94.5	**95.1**
	Class2	89.2	88.3	**89.9**
	Class3	90.6	91.7	**93.2**
Wine	Class1()	80.9	82.3	**84.7**
	Class2()	62.7	65.5	**66.1**
	Class3()	79.8	78.6	**80.2**

6 Conclusions

In the paper, we propose a novel data description method called LPOCSVM. The key idea of the proposed method is that when finding the hyperplane that has the maximal distance to the origin it uses the Mahalanobis distance, which is induced by the defined locality preserving scatter matrix and explicitly reflects the data manifold structure. The experimental results show that LPOCSVM is effective and outperforms OCSVM.

Acknowledgements. This work is supported in part by the Key Scientific Research Foundation of Sichuan Provincial Department of Education (Grant No.11ZA004), the National Science Foundation of China (Grant No. 61103168, 61271413, 61472329).

References

1. Scholkopf, B., Smola, A.: Learning With Kernels. MIT Press, Cambridge, MA (2002)
2. Vapnik, V.: The Nature of Statistical Learning Theory. Springer, New York (1995)
3. Scholkopf, B., Platt, J., Shawe-Taylor, J., Smola, A., Williamson, R.: Estimating the support of a high dimensional distribution. Neural Comput. **13**, 1443–1471 (2001)
4. Tax, D.M.J., Duin, R.P.W.: Support vector data description. Mach. Learn. **54**, 45–66 (2004)

5. Tran, Q., Li, X., Duan, H.: Efficient performance estimate for one-class support vector machine. Pattern Recogn. Lett. **26**, 1174–1182 (2005)
6. Choi, Y.S.: Least squares one-class support vector machine. Pattern Recogn. Lett. **30**, 1236–1240 (2009)
7. Bilgin, G., Erturk, S., Yildirim, T.: Segmentation of hyperspectral images via subtractive clustering and cluster validation using one-class support vector machines. IEEE Trans. Geosci. Remote Sens. **49**, 2936–2944 (2011)
8. Gomez-Verdejo, V., Arenas-Garcia, J., Lazaro-Gredilla, M., Navia-Vazquez, A.: Adaptive one-class support vector machine. IEEE Trans. Signal Process. **59**, 2975–2981 (2011)
9. Li, Y.H.: Selecting training points for one-class support vector machines. Pattern Recogn. Lett. **32**, 1517–1522 (2011)
10. Yin, S., Zhu, X., Jing, C.: Fault detection based on a robust one-class support vector machines. Neurocomputing **145**, 263–268 (2014)
11. He, X., Niyogi, P.: Locality preserving projections. In: Proceedings of the Conference on Advances in Neural Information Processing Systems, pp. 585–591 (2003)
12. Fletcher, R.: Practical Methods of Optimization, 2nd edn. Wiley, New York (1987)
13. Yang, J., Yang, J.Y.: Why can LDA be performed in PCA transformed space? Pattern Recogn. **36**, 563–566 (2003)
14. Blake, C., Merz, C.: UCI Repository of machine learning databases. http://www.ics.uci.edu/_mlearn/MLRepository.html.
15. Manor, L.Z., Perona, P.: Self-tuning spectral clustering. In: Advances in Neural Information Processing Systems, pp:1601–1608 (2005)
16. Kwok, J.T., Tsang, I.W., Zurada, J.M.: A class of single-class minimax probability machines for novelty detection. IEEE Trans. Neural Netw. **18**, 778–785 (2007)
17. Zhang, L., Zhou, W.D.: 1-norm support vector novelty detection and its sparseness. Neural Netw. **48**, 125–132 (2013)
18. Hodge, V.J., Austin, J.: A survey of outlier detection methodologies. Artif. Intell. Rev. **22**, 85–126 (2004)

Locality Preserving Based K-Means Clustering

Xiaohuan Yang(✉), Xiaoming Wang, Yong Tian, and Yajun Du

School of Computer and Software Engineering, Xihua University, Chengdu 610039, China
yangxihu_2009@sina.com

Abstract. K-Means is a powerful clustering method, in which the Euclidean distance is usually employed. In the paper, by following the basic idea of locality preserving projection (LPP), we first define locality preserving scatter matrix, then introduce a new Mahalanobis distance by using the defined matrix, finally propose a novel K-Means clustering algorithm based on the given Mahalanobis distance. Different from the traditional K-Means algorithm, the proposed method considers fully the intrinsic manifold structure of data. Experimental results show that the proposed method can achieve better clustering accuracy in contrast with the traditional K-Means algorithm.

Keywords: Clustering · K-Means · Locality preserving projection · Mahalanobis distance

1 Introduction

Cluster analysis is one of the important tools of scientific research and has been extensively used [1, 2]. The aim of clustering is to group all data points into several clusters according to the similarity measure. It maximizes the same class with high similarity and minimizes inhomogeneity with low similarity.

K-Means is a well-known clustering algorithm and distance metric plays a very important role in it. Traditional K-Means generally employs Euclidean distance. Yet, Euclidean metric is not a good measure in that it only considers the geometric distance of each data point to the cluster center and assumes each feature of the data space is equally important. Aiming at this drawback of K-Means, many of the related research literatures attempted to find a suitable distance measure for it. In [3], Song et al. proposed weighted Euclidean distance to get effective clustering method by using the multiple correlation coefficients. In [4], by introducing variation coefficient weight vector, Fan et al. presented a fusion of variation coefficient of clustering method to reduce the influence of irrelevant attributes. Wu et al. utilized the weight of variables in distance measure to improve clustering accuracy [5]. In [6], Zhang et al. applied Mahalanobis distance in K-Means. Their method showed promising performance because Mahalanobis distance is independent of dimension and can remove the correlations among variables. However, Mahalanobis distance only considers data distribution and ignores the intrinsic manifold structure of data.

On the other hand, the intrinsic manifold structure of data has been explicitly considered in the locality preserving projection (LPP) algorithm which was proposed

X. He et al. (Eds.): IScIDE 2015, Part II, LNCS 9243, pp. 86–95, 2015.
DOI: 10.1007/978-3-319-23862-3_9

by He et al. in [7, 8]. LPP is an unsupervised linear dimensionality reduction method and it overcomes the shortcoming of the traditional linear methods that are unable to maintain the nonlinear manifold of original data. The basic idea of LPP is that it maps the data points to a subspace according to a transformation matrix which is computed by using the notion of the Laplacian of the graph. LPP builds a graph incorporating neighborhood information of data and optimally preserves local neighborhood information in a certain sense. LPP has been successfully applied in many domains [8, 9] and its basic idea has been extended to other methods [10, 11].

In this paper, we propose a novel clustering algorithm called locality preserving based K-Means (LPK-Means), in which the intrinsic manifold structure of data is explicitly considered. First, by using the basic idea of LPP, we define locality preserving scatter matrix, which reflects the intrinsic manifold structure of data. Then, we introduce a new Mahalanobis distance metric according the defined locality preserving scatter matrix. Finally, the LPK-Means clustering algorithm is proposed based on the given Mahalanobis distance. Different from traditional K-Means in which Euclidean distance and traditional Mahalanobis distance are usually used, the proposed method shares the basic characteristic of LPP, i.e., the local manifold structure of data is fully considered. Experimental results indicate that the proposed method can achieve better clustering performance.

The paper is organized as follows: The related works will be reviewed in Sect. 2. In Sect. 3, the locality preserving scatter matrix is defined and a novel Mahalanobis distance is suggested; after that, the optimization problem formulation of the proposed algorithm is proposed; finally, a discussion is carried out. Experimental results are reported in Sect. 4. Finally, conclusions are drawn in Sect. 5.

2 Related Work

In the paper, we suppose a dataset containing N data points, represented by $X = \{x_i | i = 1, \cdots, N\}$, where $x_i \in R^M$, M denotes the dimension of the data space.

2.1 The K-Means Algorithm

The K-Means algorithm groups N data points into k disjoint clusters, where the parameter k is predefined. The objective function J_c of K-Means can be expressed as follows

$$J_c = \sum_{j=1}^{k} \sum_{i=1}^{n_j} d(x_i^{(j)}, C_j) \tag{1}$$

where n_j is the sample number of the jth class, $C_j = 1/n_j \sum_{i=1}^{n_j} x_i^{(j)}$ is the cluster center of the jth class, $d(x_i^{(j)}, C_j)$ represents the distance between data point $x_i^{(j)}$ and the cluster center C_j.

The commonly used distance metric in the K-Means algorithm is Euclidean distance, which is defined as $d(x_i^{(j)}, C_j) = ||x_i^{(j)} - C_j||^2$. The drawback of Euclidean distance is that it only considers the geometric distance of each point to the cluster center and assuming each point's attribute is equally important and independent from others. The Mahalanobis metric is also used in the K-Means algorithm [12]. Mahalanobis distance attempts to calculate the correlations among variables and assigns different weights to the data points using the covariance matrix. It can be calculated as follows

$$d(x_i^{(j)}, C_j) = (x_i^{(j)} - C_j)S^{-1}(x_i^{(j)} - C_j)^T \tag{2}$$

where S is the covariance matrix and defined as follows:

$$S = E((X - C)(X - C)^T) \tag{3}$$

Usually, the Mahalanobis distance used in K-Means is induced by the covariance matrix. Obviously, it can incorporate the geometrical distribution of data. However, the calculation of covariance matrix is based on overall data points which may lead to exaggerate the small change of variables and ignore the intrinsic manifold structure of data.

2.2 Locality Preserving Projections (LPP)

Locality Preserving Projection (LPP) is a linear dimensionality reduction algorithm by feature projection. When the high-dimensional data points near a low dimensional manifold, LPP seeks a linear approximation of the nonlinear Laplacian Eigenmap, then maps the data points into a subspace and optimally preserves the local neighborhood structure of dataset. LPP finds the transformation matrix by minimizing the following objective function:

$$\arg\min_{w} w^T XLX^T w \tag{4}$$

where $L = D - W$ is the Laplacian matrix. Here W is a weight matrix incorporating the local neighborhood information of data, and D is a diagonal matrix and its entries are column (or row) sum of W, i.e. $D_{ii} = \sum_{j=1}^{N} W_{ij}$.

3 Locality Preserving Based K-Means Clustering

In this section, we first define the locality preserving scatter matrix and introduce a novel Mahalanobis distance based on the defined matrix. Then, we present the optimization problem formulation of the proposed algorithm and discuss how to solve it. Finally, a discussion is carried out about the proposed algorithm.

3.1 Locality Preserving Scatter Matrix and Distance Measure

As we know, LPP keeps the embedded structure of local information by constructing adjacency graph. Here by using the basic idea of LPP, we define locality preserving scatter matrix as follows.

Definition 1 (Locality Preserving Scatter Matrix). *Let L be the Laplacian matrix of data X, the matrix S_p is called the locality preserving scatter matrix and formulated as follows.*

$$S_P = X(D - W)X^T = XLX^T \tag{5}$$

here, D is a sparse matrix, and $D_{ii} = \sum_{j=1}^{N} W_{ij}$, the rest is 0. W is the weight matrix of the adjacency graph of dataset X. Let $N_r(x_i)$ denote the r nearest neighbors of node i. Node i and j are connected by an edge if i is among r nearest neighbors of j or j is among r nearest neighbors of i. A common choice for the weight matrix is the Gaussian kernel [13] as follows:

$$W_{ij} = \begin{cases} \exp(-\frac{\|x_i - x_j\|^2}{t}), & \text{if } x_i \in N_r(x_j) \text{ or } x_j \in N_r(x_i) \\ 0 & , \text{other} \end{cases} \tag{6}$$

where $\|x\| = (\sum_{i=1}^{M} x_i^2)^{1/2}$ is the usual Euclidean (L_2) norm in R^M. $t > 0$ is the Gaussian kernel parameter and controls the width of the neighborhoods. It is worthwhile to note that the locality preserving scatter matrix S_p effectively reveals the internal geometry and local structure of points.

By using the above defined locality preserving scatter matrix S_p, we introduce a new Mahalanobis distance which can be formulated as

$$(X - C)S_p^{-1}(X - C)^T = d(X, C) \tag{7}$$

Obviously, the above distance measure is also a kind of Mahalanobis distance. However, different from the traditional Mahalanobis distance that only considers data distribution, the proposed distance measure stems from the locality preserving scatter matrix and incorporates the intrinsic manifold structure of data.

3.2 The Proposed K-Means Clustering Algorithm

By using the introduced Mahalanobis distance, we define the optimization problem of the proposed method LPK-Means as follows

$$J_c = \sum_{j=1}^{k} \sum_{i=1}^{n_j} (x_i^{(j)} - C_j)S_p^{-1}(x_i^{(j)} - C_j)^T \tag{8}$$

where $x_i^{(j)}$, C_j and n_j are described in (1), the locality preserving scatter matrix S_p is defined as in (5). Obviously, the essential difference between our method and traditional K-Means is the distance metric or similarity metric used in K-Means procedure. Our method uses a novel distance metric which is introduced in (7) when measuring the similarity between data point and the cluster center of each class, and so takes the intrinsic structure characteristic of data into full consideration. Nevertheless, the K-Means that uses Euclidean distance only considers the geometric distance of each point to the cluster center and one ignores the intrinsic manifold structure of data although it takes data distribution into consideration when the traditional Mahalanobis distance metric induced by the covariance matrix.

The proposed method has similar solving procedure with traditional K-Means. The only difference is the way to compute the distance from data point to cluster center. The main procedure is outlined in Algorithm 1.

Algorithm 1: The LPK-Means algorithm

Step 1:	Set $I = 1$, $\varepsilon = 10^{-6}$, select the initial cluster centers $C_j \in X$;
Step 2:	Choose the nearest neighbors r and the Gaussian kernel parameter t, calculate the weight matrix $W = (W_{ij})$, then compute the locality preserving scatter matrix S_p;
Step 3:	Compute distance $d(x_i^{(j)}, C_j)$ based on S_p, choose the point x_i which corresponds $d(x_i^{(j)}, C_j) = \min\{D(x_i, C_j)\}$, then make $x_i \in \omega_j$;
Step 4:	Compute optimization function J_c by formula(8);
Step 5:	While $\lvert J_c(I) - J_c(I-1) \rvert < \varepsilon$, the procedure stops; or else $I = I + 1$, calculate new cluster center C_j, and return Step 3.

3.3 Discussion

3.3.1 Connection to the K-Means Algorithm Based on Mahalanobis Distance

In LPK-Means, if the weight matrix W is defined as $W_{ij} = 1/n^2$, according to [14], $D_{ii} = \sum_j W_{ij} = 1/n$, and $L = D - W = (1/n)I - (1/n^2)ee^T$, the locality preserving scatter matrix S_p is equivalent to the covariance matrix S, i.e.

$$S = E((X - C)(X - C)^T) = XLX^T = S_p \tag{9}$$

This suggests that the proposed algorithm can be also seen as a generalized version since its adjacency matrix is more general.

3.3.2 The Singularity of the Locality Preserving Scatter Matrix

In the previous section, we assumed that the locality preserving scatter matrix S_p is nonsingular. Actually, similarly to the covariance matrix S, S_p may encounter the singularity, especially when the dimension of features exceeds the sample size. In order to deal with the singularity problem of matrix, generally, we can employ the regularization method [15], i.e.

$$S_{P'} = S_P + \mu I \tag{10}$$

here, μ is the regularization parameter and usually is $\mu = 10^{-8}$.

3.3.3 Time Complexity

The computation of K-Means is mainly concentrated on the iterative process of calculating the optimization function, so the time complexity of K-Means is $O(kNI)$. In general, $k < < N, I < < N$. The proposed method not only need to calculate J_c, but also must compute the locality preserving scatter matrix S_p, it incurs an $O(rN^2)$ computational time. So the proposed method suffers higher time complexity than traditional K-Means.

4 Experimental Results

In this section, experimental results will be reported. In the first experiment, we illustrate the difference between the proposed method and other methods on an artificial dataset. In the second experiment, we evaluate the clustering performance of the proposed method (LPKM) by comparing it with K-Means based on Euclidean distance (KM-E) and the traditional Mahalanobis distance (KM-M) on several benchmarking datasets from the UCI repository [16]. In the third experiment, we demonstrate the capabilities of our proposed method in the recognition of the handwritten digits.

4.1 Artificial Dataset

In order to evaluate the effectiveness of K-Means based on different distance, we use an artificial dataset $X = \{x_1, x_2, \cdots, x_{200}\}$ which consists of 200 two-dimensional samples and is divided into two classes.

First, we randomly select initial centers from the artificial dataset. In all experiments, the parameter r and t are set as: $r = 5$, $t = 0.05$. The clustering results of K-Means using different distance are illustrated in Fig. 1. Different symbols and colors represent the cluster assignments and green boxes represent the initial centers. Figure 1(a) shows that the dataset can be obviously divided into two classes. However, the clustering result is poor in Fig. 1(b). This is because the Euclidean distance only considers the geometric distance of each point to the cluster center. Figure 1(c) is the result of KM-E which employs traditional Mahalanobis distance inducing by the covariance matrix. The result of the proposed method is reported in Fig. 1(d). Obviously, it is the best and achieving the clustering accuracy more than 90 %. The reason is that the proposed method considers the intrinsic manifold structure of data.

We further investigate influence of the initial centers on the clustering performance. Table 1 reports the clustering accuracy of KM-E, KM-M, and the proposed method when using different initial cluster centers. Obviously, KM-E performs the worst.

In some cases KM-M can obtain good clustering accuracy, but is sensitive to the initial cluster centers. On the contrary, the proposed method achieves the best result and outperforms KM-E and KM-M.

4.2 UCI Datasets

In order to further evaluate the performance of the proposed method, we selected eight datasets from the UCI repository which are widely used to evaluate the performance of

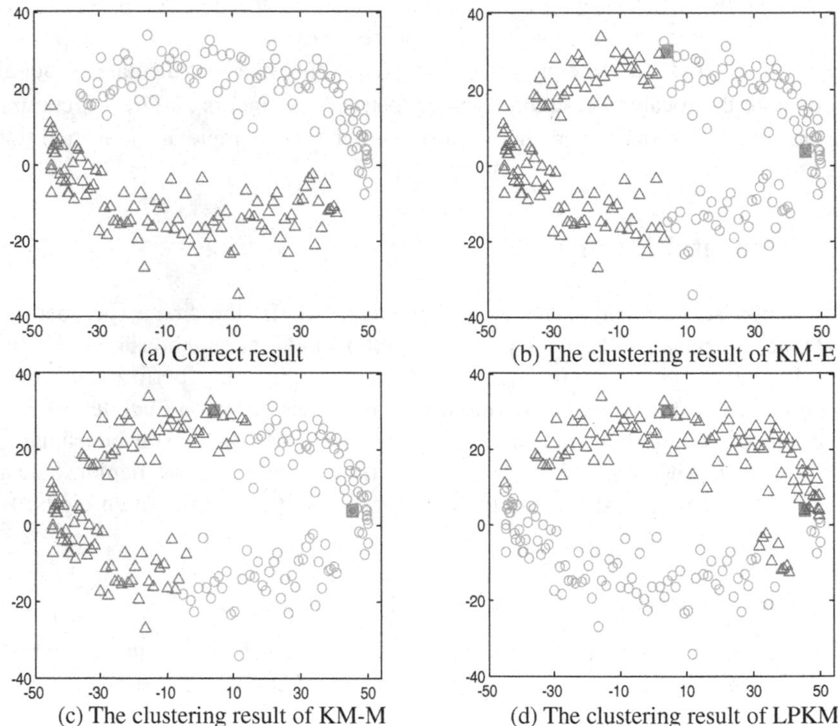

(a) Correct result
(b) The clustering result of KM-E
(c) The clustering result of KM-M
(d) The clustering result of LPKM

Fig. 1. Different clustering result when the initial cluster centers are $\{x_{19}, x_{63}\}$

Table 1. Performance comparison in the case of using different initial cluster centers

Initial cluster center	Clustering accuracy		
	KM-E	KM-M	LPKM
$x_{19} = (45.43, 3.81), x_{63} = (4.32, 30.12)$	0.6450	0.5650	0.9150
$x_{42} = (27.79, 15.94), x_{145} = (-19.73, -10.93)$	0.6500	0.8000	0.9150
$x_{168} = (6.65, -16.60), x_{196} = (37.34, -5.02)$	0.6450	0.8000	0.9150
$x_{125} = (-36.97, 4.80), x_{196} = (13.79, -6.67)$	0.6450	0.6250	0.9150
$x_5 = (49.77, 7.64), x_{18} = (45.91, 11.63)$	0.6450	0.5650	0.9150
$x_{97} = (-33.24, 16.06), x_{101} = (-45.00, 10.91)$	0.6450	0.8000	0.9150
$x_{162} = (-0.51, -16.56), x_{127} = (-35.62, -2.27)$	0.6500	0.8000	0.9150

clustering algorithm. At present, how to choose the parameters for Gaussian kernel t and the nearest neighbors r is an open problem. In particular we set as $r = 5$ and use a strategy described $t = (\log(n)/n)^M$ in [17] to choose the parameter t. A summary of the characteristics of the selected datasets are shown in Table 2. All data sets have been normalized to have zero mean and unit variance on each feature. Each experiment is repeated 50 times.

Table 3 shows the mean clustering accuracy of KM-E, KM-M, and LPKM on the selected datasets. From Table 3, on the whole, it can be found that LPKM has an improvement in the clustering accuracy in contrast with KM-E and KM-M. These experimental results indicate that the clustering performance can be improved when the intrinsic manifold structure of data is considered in K-Means.

Table 2. Characteristics of datasets

Dataset	No. of patterns	No. of features	No. of classes
Breast	699	9	2
Glass	214	9	6
Heart	270	13	2
Iris	150	4	3
Seeds	210	7	3
Wdbc	569	30	2
Wine	178	13	3
Landsat	2000	36	6

Table 3. Mean clustering accuracy on the eight datasets

Dataset	Mean accuracy		
	KM-E	KM-M	LPKM
Breast	0.6623	0.7167	**0.7286**
Glass	0.5836	0.5614	**0.6352**
Heart	0.7556	0.6978	**0.7852**
Iris	0.8192	0.8551	**0.8725**
Seeds	0.8211	0.8068	**0.8469**
Wdbc	0.8805	0.8615	**0.8825**
Wine	0.6892	0.6839	**0.7288**
Landsat	0.6771	0.6750	**0.7199**

4.3 USPS Dataset

To demonstrate the capabilities of the proposed method in the recognition of the handwritten digits, we conducted experiments on the popular USPS dataset [18]. The USPS dataset is the well-known United States Postal Service handwritten digits recognition corpus. It includes 10 classes from ‘0’ to ‘9’. Each class has 1100 samples and the size of each sample is 16×16 pixels with 256 gray levels. The dataset has traditionally been used in a splitting of 7291 samples for training and 2007 samples for testing. In this subsection, we randomly select some subsets from the testing samples as clustering datasets and each experiment is repeated 50 times. The mean clustering accuracy of KM-E, KM-M, and LPKM on the selected datasets are shown in Table 4. From the experimental results, we can see again that the proposed method better than

Table 4. Mean clustering accuracy on the USPS dataset datasets

Subsets of digits	No. of classes	No. of cases	Mean clustering accuracy		
			KM-E	KM-M	LPKM
{‘0’, ‘8’}	2	525	0.8166	0.7540	**0.9150**
{‘3’, ‘6’, ‘8’}	3	502	0.8511	0.8227	**0.8781**
{‘3’, ‘4’, ‘5’, ‘6’}	4	696	0.7394	0.7155	**0.7629**
{‘6’, ‘7’, ‘8’, ‘9’, ‘0’}	5	1019	**0.7028**	0.6932	0.7007

the KM-E and KM-M in terms of clustering accuracy on the USPS dataset.

5 Conclusion

In this paper, we propose a novel K-Means algorithm named LPK-Means. Different from traditional K-Means, the method stems from on the basic idea of LPP and utilizes the inherent structure characteristic of data. Experimental results indicate that the proposed algorithm is effective and can achieve better clustering accuracy by comparing with traditional K-Means. However, the Gaussian kernel parameter and neighborhood number are still the directions of future research.

Acknowledgments. This work is supported by the Graduate Innovation Fund of Xihua University (Grant No.ycjj2014032), the Key Scientific Research Foundation of Sichuan Provincial Department of Education (Grant No.11ZA004) and the National Science Foundation of China (Grant No. 61103168, 61271413, 61472329).

References

1. Rokach, L.: A survey of clustering algorithms. In: Maimon, O., Rokach, L. (eds.) Data Mining and Knowledge Discovery Handbook, pp. 269–298. Springer, Heidelberg (2010)
2. Xu, R., Wunsch II, D.: Survey of clustering algorithms. IEEE Trans. Neural Netw. **16**(3), 645–678 (2005)
3. Yu-chen, S., Yu-ying, Z., Hai-dong, M.: Research based on Euclid distance with weights of clustering method. Comput. Eng. Appl. **43**(4), 179–180 (2007)

4. Alin, F., Shuhua, R.: K-Means clustering algorithm based on coefficient of variation. Comput. Eng. Appl. **48**(35), 114–117 (2012)
5. Xianghus, W., Niu Shengjie, W., Chengou, W.: An improvement on estimating covariance matrix during cluster analysis using Mahalanobis distance. Appl. Stat. Manage. **30**(2), 240–245 (2011)
6. Xiang, Z., Shitong, W.: Mahalanbbis distance based possibilistic clustering algorithm and its analysis. Data Acquisit. Process. **23**(8), 86–88 (2011)
7. He, X.F., Niyogi, P.: Locality preserving projections. In: Proceedings of the Conference on Advances in Neural Information Processing Systems, pp. 585–591 (2003)
8. He, X.F., Yan, S.C., Hu, Y.X., Niyogi, P., Zhang, H.J.: Face recognition using Laplacian faces. IEEE Trans. Pattern Anal. Mach. Intell. **27**(3), 328–340 (2005)
9. Zhonghua, S., Yonghui, P., Shitong, W.: A supervised locality preserving projection algorithm for dimensionality reduction. Pattern Recog. Artif. Intell. **21**(2), 232–239 (2008)
10. Chuanliang, C., Rongfang, B., Ping, G.: Combining LPP with PCA for Microarray Data Clustering. IEEE Congress on Evolutionary Computation, pp. 2081–2086 (2008)
11. Sun, X., Zhang, Q., Wang, Z.: Using LPP and LS-SVM for spam filtering. In: 2009 ISECS International Colloquium on Computing, Communication, Control, and Management, vol. 2, pp. 451–454, 8–9 August 2009
12. Melnykov, I., Melnykov, V.: On K-means algorithm with the use of Mahalanobis distances. Stat. Probab. Lett. **84**(2014), 88–95 (2014)
13. Kokiopoulou, E., Saad, Y.: Orthogonal neighborhood preserving projections a projectionbased dimensionality reduction technique. IEEE Trans. Pattern Anal. Mach. Intell. **29**(12), 2143–2156 (2007)
14. Cai, D., He, X., Han, J., Zhang, H.J.: Orthogonal laplacianfaces for face recognition. IEEE Trans. Image Process. **15**(11), 3608–3614 (2006)
15. Kakade, S.M., Shalev-Shwartz, S., Tewari, A.: Regularization Techniques for Learning with Matrices. Mach. Learn. Res. **13**(1), 1865–1890 (2012)
16. Blake, C., Merz, C.: UCI Repository of machine learning databases. http://www.ics.uci.edu/~mlearn/MLRepository.html
17. von Luxburg, U.: A tutorial on spectral clustering. Stat. Comput. **17**(4), 395–416 (2007)
18. Keysers, D.: Usps dataset (1994). http://wwwi6.informatik.rwth-aachen.de/keysers/usps.html

Auroral Oval Boundary Modeling Based on Deep Learning Method

Bing Han[1,2(✉)], Xinbo Gao[1], Hui Liu[1], and Ping Wang[1]

[1] School of Electronic Engineering, Xidian University, Xi'an 710071, China
bhan@xidian.edu.cn, xbgao@mail.xidian.edu.cn,
liuhuilucky@yeah.net, winnie_wangping@163.com

[2] State Key Laboratory of Remote Sensing Science, Beijing 100101, China

Abstract. Research on the location of the auroral oval is important to understand the coupling processes of the Sun-Earth system. The equatorward boundary and poleward boundary of the auroral oval are significant parameters of the auroral oval location. Thus auroral oval boundary modeling is an efficient way to study the location of auroral oval. As the location of the auroral oval boundary is subject to a variety of geomagnetic factors, there are some limitations on traditional methods, which express the auroral oval boundary as a function of only one or several geomagnetic activity index. Deep learning method is used in this paper to learn the essential features of the inputs, which are a large number of geomagnetic parameters and the former locations of aurora boundary. Furthermore, a model is established to forecast the location of the auroral oval boundary. The experiment results show that our method can model and forecast the boundary of aurora oval efficiently on the data set obtained from Ultraviolet Imager (UVI) on Polar satellite and OMNI database on NASA.

Keywords: Auroral oval boundary modeling · Restricted bolzmann machine · Deep learning

1 Introduction

The light ring region of energetic particle precipitation caused by solar wind in the sky of north and south poles above the Earth is called auroraloval [1]. Auroral oval is regarded as a representative of space weather, thus it is significant to establish a complete model of the auroral oval boundary for predicting space weather [2].

The position of the auroral oval is expressed as a function of single variable in traditional studies. Feldstein propos a model based on the original data in 1967, he depicts the possibility of auroral oval occurrence as a function of Q index [3]. Starkov express the polewardand equatorward and diffuse of the auroral oval boundary as a function of AL index [4]. Zhang and Paxton use Epstein equation to calculate the electron flux and average energy flux, and express the location of the auroral oval as a function of Kp index [5]. Carbary also uses Kp index to investigate the auroral oval [2]. Milan's research shows that the radius of the auroral oval increases with the ring current during a magnetic storm [6]. Lukianova investigates the relationship between the polar cap regions of the border with the interplanetary magnetic field, and points out

X. He et al. (Eds.): IScIDE 2015, part II, LNCS 9243, pp. 96–106, 2015.
DOI: 10.1007/978-3-319-23862-3_10

that the polar cap boundary region at midday-midnight direction is controlled by the z-direction component of the interplanetary magnetic field, in terminator line direction by the y-direction [7]. Since the location of the auroral oval boundary cannot be determined by one single geomagnetic activity index, the effectiveness of traditional studies is limited.

The current trend is to study the position and size of the auroral oval using a variety of geomagnetic physical parameters. Yang uses multiple regression method and characterizes the auroral oval location as a function of several physical parameters [8], which is affected by pre-hypothetical regression equation, while the physical principles cannot be accurately reflected by this pre-hypothetical regression equation.

In this paper, the deep learning method is introduced to overcome the drawbacks of existing methods and to investigate how the location of the auroral oval boundary changes with a variety of geomagnetic physical parameters. Yang's method extract only 9 physical parameters according to the human experience and literature to constract the prediction model, which has the limitation. The deep learning method we used in this paper don't extract feature parameters firstly and all physical parameters obtained from NASA can be used as the whole input. Hence, the proposed deep learning based auroral oval boundary prediction method can better reflect law of the interaction of solar wind and the Earth's magnetic field. Furthermore, this paper gives the new way to study space weather forecasting.

2 Restricted Bolzmann Machine

RBM is an effective feature extraction method, which can significantly improve the generalization ability [9]. Furthermore, the depth of the belief network stacked by multiple RBM can extract more abstract features. Therefore, we choose the RBM model to learn the essential features of the geomagnetic parameters.

RBM consists of visual layer v and hidden layer h. Full probability distribution between visual layer and hidden layer is $p(v,h|\theta)$ which meets the Boltzmann distribution. When input v, we can get the hidden layer by $p(h|v)$, then we can get the visual layer v_1 by $p(v|h)$ similarly. By adjusting the network parameters, making the visible layer obtained from the hidden layer v_1 as consistent as possible with the original visible layer v. Thus the obtained hidden layer is an expression of visible layer, which can be used as input data features [10, 11].

RBM contains a set of visible units v and a set of hidden units h. There are no edges going form the neuron to itself. The energy of the state $\{v,h\}$ is defined as:

$$E(v,h|\theta) = -\sum_{i=1}^{n} a_i v_i - \sum_{j=1}^{m} b_j h_j - \sum_{i=1}^{n}\sum_{j=1}^{m} v_i W_{ij} h_j \tag{1}$$

where $\theta = \{W_{ij}, a_i, b_j\}$ are the model parameters. W_{ij} denotes the weight associated with the connection between unit i and unit j, a_i is the bias associated with visible unit i, b_j is the bias associated with visible unit j.

The joint probability distribution over the visual layer and hidden layer is:

$$p(v, h|\theta) = \frac{e^{-E(v,h|\theta)}}{Z(\theta)} \tag{2}$$

where $Z(\theta) = \sum_{v,h} e^{-E(v,h|\theta)}$ is the partition function, which is used to normalize the probability distribution.

The edge of the visible layers probability distribution, also called the likelihood function is:

$$p(v|\theta) = \frac{1}{Z(\theta)} \sum_{h} e^{-E(v,h|\theta)} \tag{3}$$

RBM learning task is to determine the value of the parameter, which can be obtained by maximizing RBM likelihood function on the training set, namely

$$\theta^* = \arg\max_{\theta} L(\theta) = \arg\max_{\theta} p(v|\theta) \tag{4}$$

The depth belief network stacked by multiple RBM, which is a deep learning network, is able to extract more abstract features.

3 Boundary Modeling Method for Auroral Oval

Modeling process of the auroral oval boundary is discussed in this section. Firstly, auroral oval is segmented from the image in order to extract boundary points. Then coordinate of aurora boundary is transformed into MLT-MLAT coordinate system and the auroral oval boundary points are extracted. Finally, auroral oval boundary model is established by the RBM model. The details for auroral oval boundary segmentation are referred in [12].

After the aurora boundary are segmented by [12], we extract the 24 points uniformly on the equatorward boundary and 24 points uniformly on poleward boundary of

Fig. 1. Points extracted on equatorward and poleward boundary in MLT-MLAT coordinate system.

aurora oval respectively at 1 MLT interval in MLT-MLAT coordinate system, shown in Fig. 1. In this figure, triangles and circles represent poleward boundary points and equatorward boundary points respectively.

The overview of the auroral oval boundary location forecast model is given in Fig. 2, which consists of RBM with two layers and RBF (Radial Basis Function) network with one layer [13]. The model will be used to learn how the geomagnetic physical parameters, obtained from the OMNI database, influence the auroral oval boundary position on the 24 magnetic local times. RBM network is used to get the features of geomagnetic parameters, while RBF network is used to simulate the mapping function which these features affect on the location of the auroral oval.

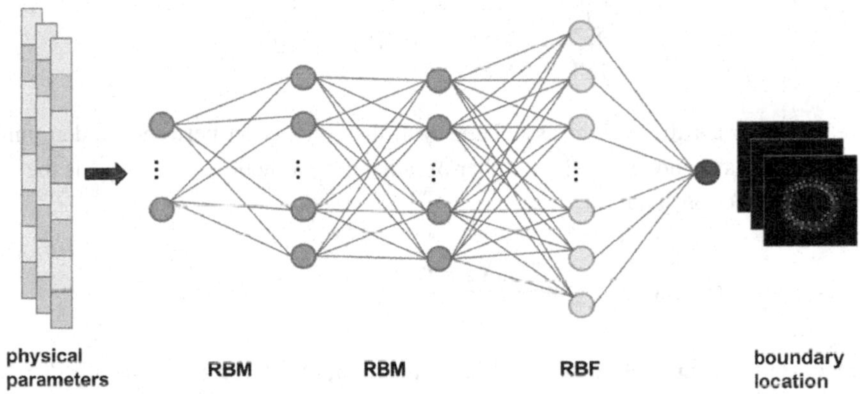

Fig. 2. The overview of the auroral oval boundary location forecast model.

The model input parameter from the OMNI database is donated as $\mathbf{X} = [x_1, x_2, \ldots, x_m]^T$, where m is the number of input parameters. Parameters of the first layer RBM network are $\theta_1 = \{w_{i_1 j_1}, a_{i_1}, b_{j_1}\}$, where $w_{i_1 j_1}$ denotes the weight associated with the connection between visible unit i_1 and hidden unit j_1, a_{i_1} is the bias associated with visible unit i_1, b_{j_1} is the bias associated with hidden unit j_1. The hidden layer of the first RBM network is visible layer of the second layer RBM network, and parameters of the second layer RBM network are $\theta_2 = \{w_{j_1 j_2}, a_{j_1}, b_{j_2}\}$, where $w_{j_1 j_2}$ denotes the weight associated with the connection between visible unit j_1 and hidden unit j_2, a_{j_1} is the bias associated with visible unit j_1, b_{j_2} is the bias associated with visible unit j_2. Output of the first layer RBM network is $\mathbf{Y_1} = [y_{11}, y_{12}, \ldots, y_{1n}]^T$, where n is the number of hidden layer node in the first RBM network. Therefore,

$$y_{1j_1} = \sum_{i_1=1}^{m} x_{i_1} w_{i_1 j_1} + b_{j_1} \qquad j_1 = 1, 2, \ldots, n \tag{5}$$

The output of the second layer RBM network is $\mathbf{Y_2} = [y_{21}, y_{22}, \ldots, y_{2c}]^T$, where c is the number of hidden layer node in the first RBM network. Therefore,

$$y_{2j_2} = \sum_{j_1=1}^{n} x_{j_1} w_{j_1 j_2} + b_{j_2} \qquad j_2 = 1, 2, \ldots, c \tag{6}$$

Contrastive Divergence (CD) is an approximation of the log-likelihood gradient that has been found to be a successful update rule for training RBM. We use the fast training algorithm of RBM based on CD [14] in this paper.

The input of RBF network is the output of the second layer RBM network, and the output of the RBF is $\mathbf{Y} = [y_1, y_2, \ldots, y_d]^T$, which is the auroral oval boundary position on the 24 magnetic local times, where d is the number of the output node. Therefore,

$$y_o = \sum_{j_3=1}^{l} \omega_{j_3 o} \varphi(\mathbf{Y_2}, \mathbf{c}_{j_3}) \qquad o = 1, 2, \ldots, d \tag{7}$$

where $\omega_{j_3 o}$ denotes the weight associated with the connection between hidden unit j_3 and the output layer node. l is the number of radial basis function, $\varphi(\mathbf{Y_2}, \mathbf{c}_{j_3})$ is the j_3-th radial basis function, and

$$\varphi(\mathbf{Y_2}, \mathbf{c}_{j_3}) = \exp(-\frac{\|\mathbf{Y_2} - \mathbf{c}_{j_3}\|^2}{\sigma_{j_3}^2}) \qquad j_3 = 1, 2, \ldots, l \tag{8}$$

where $\mathbf{c}_{j_3}$ is the center of j_3-th radial basis function, σ_{j_3} is the center width of radial basis function.

4 Experiments

In this section, the database for auroraboundary prediction is constructed firstly and then several experiments are design to verify the validity of our model.

4.1 Database Construction

In order to train and verify the model, a database with magnetic latitude information of the auroral oval boundary and geomagnetic parameters information should be constructed. Images captured by ultraviolet aurora imager on the Polar satellite and geomagnetic parameters in NASA OMNI database from December 1996 to February 1997 are used in this paper. The details of these parameters are found in the website: http://omniweb.gsfc.nasa.gov/html/HROdocum.html. Due to the change of the view of satellite photography, some of the auroral ovals in the image are full ovals and some are gap ovals. As there are whole magnetic latitude information of auroral oval boundary in full oval images because full oval boundary cover all magnetic local time, images with full oval are selected in our experiments.

Because of the time resolutions of physical parameters and UVI images are not the same, we need check all the dataset and select UVI images with physical parameters available as our experimental data. Finally, 4432 images with 32 geomagnetic parameters [15] are selected in our dataset in chronological order.

4.2 Experimental Results and Analysis

Several experiments are designed to verify the validity of the model. We use the mean absolute error (MAE) to evaluate the effectiveness of the model prediction results.

The MAE of network on the test set can be calculated by the following equation

$$\mathrm{MAE} = \frac{1}{24}\sum_{i=1}^{24}\left(\frac{1}{k}\sum_{j=1}^{k}\left|F_{MLAT}^{ij} - S_{MLAT}^{ij}\right|\right) \tag{9}$$

where S_{MLAT}^{ij} is magnetic latitude of the j-th test sample at i-thMLT taken by segmentation, F_{MLAT}^{ij} is magnetic latitude of the j-th test sample at i-th MLT taken by the model. k means the number of the test samples.

4.2.1 Experiment on Different Numbers of Hidden Layer Node in RBM

For RBM network, the effectiveness of feature extraction is influenced by the number of hidden layer node. This experiment attempts to find a suitable number of hidden nodes for our model.

In this experiment, 3500 samples are selected randomly as training data and the samples left in dataset as test data. The inputs of net are 32 physical parameters and the outputs of net are MLATs value of equatorward and poleward auroral boundaries at 24 MLTs, and the training error of RBF network are set tobe 4 geomagnetic latitude. In general, the number of neurons in hide layers is more than the numbers of inputs. So, we set 56, 64, 96, 128 and 160 neurons in hide layers respectively, prediction result of a single experiment is given in Fig. 3 (a), (b), (c) and (d), we call them the condition A, B, C, D, E. Each experiment in this part has been run 100 times. In Fig. 3, circles and

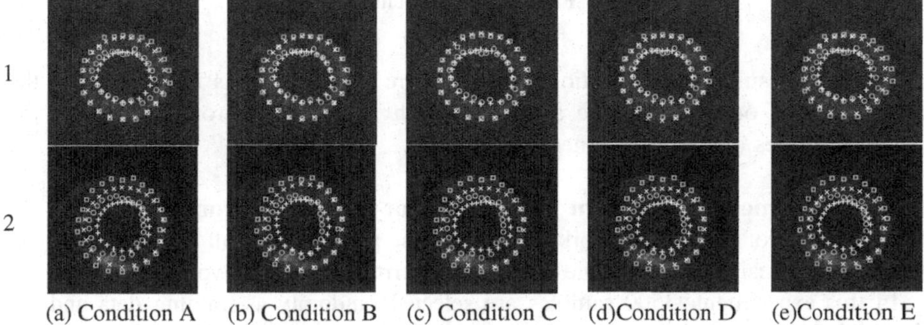

(a) Condition A (b) Condition B (c) Condition C (d)Condition D (e)Condition E

Fig. 3. Prediction results.

squares stand for the poleward boundary points and the equatorward boundary points respectively, crosses are poleward boundary points gotten by our model and x marks are equatorward boundary points taken by our method.

Table 1 gives the ratio of the points whose error is within three magnetic degrees of a single experiment. In Table 1, the ratio is highest when experiment condition is set the Condition A, which there are 64 neurons in hide layers.

Table 1. The ratio of the points whose error is within three magnetic degrees.

	Condition A	Condition B	Condition C	Condition D	Condition E
poleward	82.08 %	84.72 %	79.30 %	78.76 %	78.00 %
equatorward	83.77 %	87.67 %	82.7 %	82.78 %	81.02 %

The metric, MAE, are used to measure our model and results are shown in Fig. 4, which is the average absolute error of 100 experiments. Small value of MAE means the better performance of the method. As shown in Fig. 4, when there are 64 nodes in the hidden layer of RBM, mean absolute error of the network has the smallest value, which is no more than 2 in not only poleward boundary prediction but also equatorward boundary prediction.

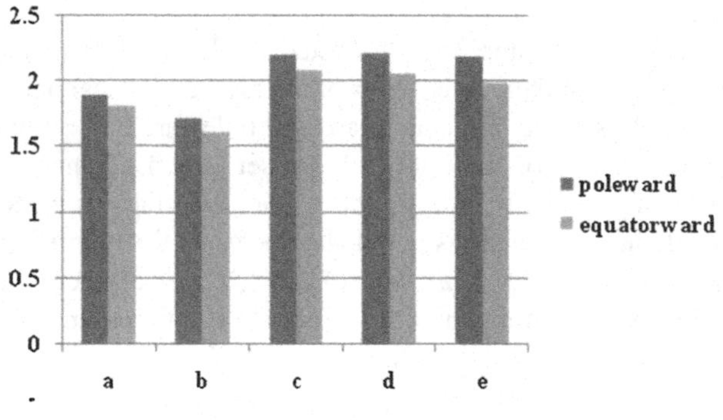

Fig. 4. MAE results.

From the results above mentioned, when there are 64 neurons in hide layers, the performance of our network are satisfied, which means 64 neurons can reflect the essential features of the geomagnetic parameters.

4.2.2 Experiment on Different Training Error of RBF Network

The training error of RBF network can impact the prediction result of the model, this experiment is designed to find the best training error of RBF network.

In this experiment, 3500 samples are selected randomly as training data and the samples left in dataset as test data. The inputs of the net are 32 physical parameters and the outputs of the net are MLATs of equatorward and poleward boundaries at 24 MLTs and there are 64 neurons in RBM hide layers. We set the training error of RBF network is 2, 4, 6, 8 geomagnetic latitude respectively, prediction result of a single experiment is given in Fig. 5 (a), (b), (c) and (d), we call them Condition A, B, C and D. Each experiment in this part has been run 100 times. Table 2 gives the ratio of the points whose error is within three magnetic degrees of a single experiment. In Table 3, the ratio is highest when the training error of RBF network is 4 geomagnetic latitude.

MAE result is shown in Fig. 6, which is the average absolute error of 100 experiments. As shown in Fig. 6, the mean absolute error of the network a minimum number when the training error of RBF network is 4 geomagnetic latitude. When the training error of RBF network is set as 2 geomagnetic latitude, the MAE is highest. This is an over-fitting phenomenon for the reason that the training sample distribution

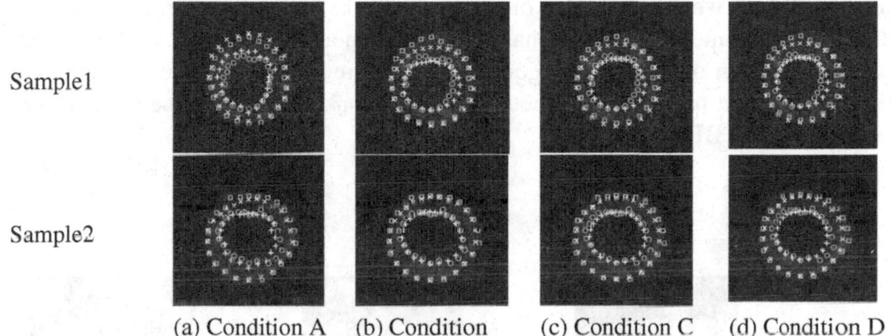

Fig. 5. Prediction results

Table 2. The ratio of the points whose error is within three magnetic degrees.

	a	b	c	d
poleward	79.20 %	84.72 %	81.60 %	80.97 %
equatorward	82.63 %	87.67 %	84.08 %	81.5 %

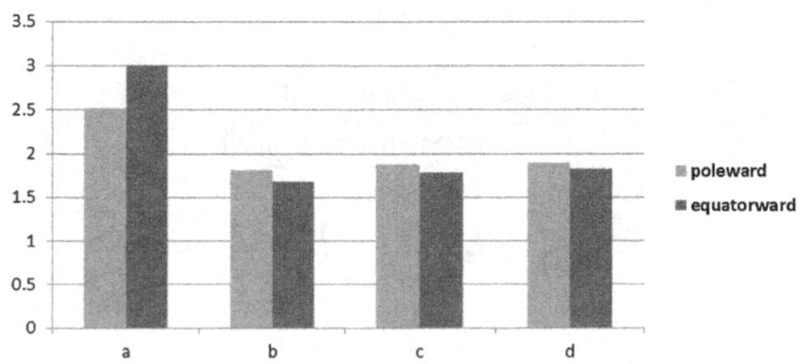

Fig. 6. MAE result.

and test sample distribution are not the same. The network on the training samples fits very well but fits bad on the test samples.

Thus, the result is best when the training error of RBF network is set as 4 geomagnetic latitude.

4.2.3 Comparison with BP Network

In this section, we compare the proposed method with the Back Propagation method in predicting the boundary of aurora. The experiment results of the two methods are shown in Fig. 7. The inputs of our network and BP network are the 32 physical parameters. The outputs of the two networks are MLATs of equator-ward and pole-ward auroral boundaries at 24 MLTs. 64 neurons are set in hide layers of the two RBM structure respectively and the training error of RBF network is set as 4 geomagnetic latitude in our network. There are two hidden layers in the BP network, the number of the node in each hidden layer is 64.

From Table 3, we conclude that our network works better than the BP network. As we know, the parameters of traditional network, such as BP network, are initialized at randomly. While in our model, the good data structure obtained by the first network is input of the second network, RBF network, which makes our model be more accurate and effective than BP network.

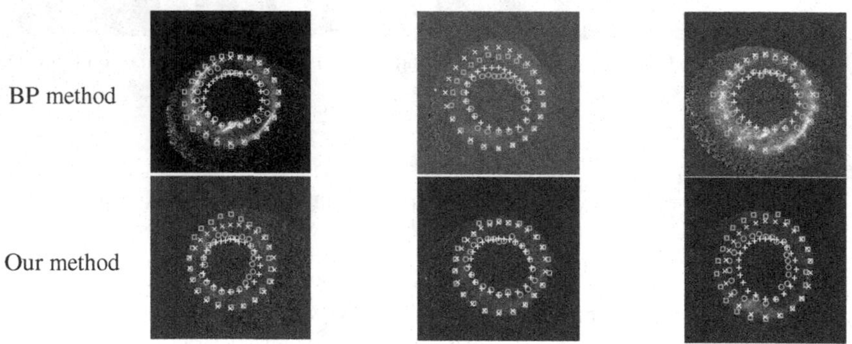

Fig. 7. Prediction results between BP method and our method.

Table 3. MAE result.

	BPMethod	Our method
poleward	2.21	1.82
equatorward	2.20	1.68

5 Conclusion

In this paper, aurora ovals boundary location model is established by deep learning. The prediction error of the model on the poleward aurora oval boundary is only 1.8 geomagnetic latitude and the prediction error of the model on the equatorward aurora oval boundary is only 1.7 geomagnetic latitude, so the model has good prediction effect.

Now, this is the first attempt to use the technique of Machine Learning and Artificial Intelligence to solve the problem about space physics. In the future, we can apply a nonlinear (nonparametric) multitask method such as kernel multitask learning to the problem.

At present, there is no clear relationship between physical parameters and location of aurora oval boundary, most of which are obtained from the observation of amounts of data. So, this is only an attempt to use deep learning method to model this phenomenon of space physics. In the future, we can do some experiments without some certain physical parameters and compare that is there relationship between boundary location and certain physical parameters.

Acknowledgments. This research is supported by the Special Scientific Research of Marine Public Welfare Industry (201005017), the Basic Foundation for Scientific Research, the Fundamental Research Funds for the Central Universities (K5051302008), the Project Sponsored by the Scientific Research Foundation for the Returned Overseas Chinese Scholars, State Education Ministry and the Open Funding of State Key Laboratory of Remote Sensing Science (OF-SLRSS201415), the Project Funded by China Postdoctoral Science Foundation.

References

1. Akasofu, S.I.: The development of the auroralsubstorm. Planet. Space Sci. **12**(4), 273–282 (1964)
2. Carbary, J.F.: A Kp-based model of auroral boundaries. Space Weather, 3(10) (2005)
3. Feldstein, Y.I., Starkov, G.V.: Dynamics of auroral belt and polar geomagnetic disturbances. Planet. Space Sci. **15**(2), 209–229 (1967)
4. Starkov, G.V.: Mathematical model of the auroral boundaries. Geomag. Aeron. **34**, 331–336 (1994)
5. Zhang, Y., Paxton, L.J.: An empirical Kp-dependent global auroral model based on TIMED/GUVI FUV data. J. Atmos. Solar Terr. Phys. **70**(8), 1231–1242 (2008)
6. Milan, S.E., Hutchinson, J., Boakes, P.D., Hubert, B.: Influences on the radius of the auroral oval. Annales Geophysicae **27**(7), 2913–2924 (2009). Copernicus GmbH
7. Lukianova, R., Kozlovsky, A.: Dynamics of polar boundary of the auroral oval derived from the IMAGE satellite data. Cosm. Res. **51**(1), 46–53 (2013)
8. Yang, Q.J.: Auroral Events Detection and Analysis Based on ASI and UVI Images. Ph.D. thesis (2013)
9. Hinton, G.E., Osindero, S., Teh, Y.: A fast learning algorithm for deep belief nets. Neural Comput. **18**, 1527–1554 (2006)
10. Dong, Y., Li, D.: Deep learning and its applications to signal and information processing. IEEE Signal Process. Mag. **28**, 145–154 (2011)

11. Bengio, Y.: Learning deep architectures for AI. Mach. Learn. **2**(1), 1–127 (2009)
12. Liu, H., Gao, X., Han, B., Yang, X.: An automatic MSRM method with a feedback based on shape information for auroral oval segmentation. In: Sun, C., Fang, F., Zhou, Z.-H., Yang, W., Liu, Z.-Y. (eds.) IScIDE 2013. LNCS, vol. 8261, pp. 748–755. Springer, Heidelberg (2013)
13. Lukaszyk, S.: A new concept of probability metric and its applications in approximation of scattered data sets. Comput. Mech. **33**, 299–304 (2004)
14. Hinton, G.: A Practical Guide to Training products of experts by minimizing contrastive divergence. Neural Comput. **14**(8), 1771–1800 (2002)
15. OMNI dataset description: http://omniweb.gsfc.nasa.gov/html/HROdocum.html.

Study on Prediction Method of Quality Uncertainty in the Textile Processing Based on Data

Jingfeng Shao[1(✉)], Jinfu Wang[1], Xiaobo Bai[1], Yong Liu[1], Congying Liu[2], and Xiaohong Ma[1]

[1] School of Management, Xi'an Polytechnic University, Xi'an 710048, China
shaojingfeng1980@aliyun.com

[2] Department of Power and Energy, Xianyang Huarun Textile Co., Ltd., Xianyang 712000, China

Abstract. To further predict textile quality fluctuation from the perspective of uncertainty factors, first, the reasons and regularities of quality fluctuation in the industrial textile processing were analyzed, and knowledge representation of textile quality attributes was studied. Second, through man-machine-environment system engineering (HMESE) theory, the uncertainty factors that affect textile quality were extracted, and its generation mechanism, interaction relationship and behavioral characteristics was explored. Then, an improved man-machine-environment brittle model oriented to the textile processing was built. As verified by the experiment, the results have shown that the improved brittle model has achieved a full range analysis of quality uncertainty of the textile, which are from the reason and regularity of quality fluctuation to generation mechanism, mutual relations, and behavior identification of the uncertainty factors.

Keywords: Textile processing · Prediction · Quality · Uncertainty

The textile quality prediction theories and methods were studied in the early 1970 s [1]. For example, Selvanayaki M., et al. studied the SVM-based yarn strength prediction method [2], Fattahi S., et al. proposed a fuzzy least square regression method of cotton yarn production process control [3], Mokhtar S., et al. put forward the fabric weaving process uncertainty test method [4], and so on. With the extension of the industrial textile processing theories, the involved variables are more and more [5]. For instance, Mohamed Naglaa, et al. [6] used regression model to predict the mixing properties of cotton fiber, Mwasiagi Josphat Igadwa, et al. [7] made use of hybrid algorithm to construct a yarn parameters to improve the performance of predictive model, and Mardani Mehrabad N, et al. [8] used finite element and multivariate spinning tension to analyze the uncertainty factors.

Many Chinese textile scholars also studied and proposed some prediction models and methods for the textile quality prediction and control [9], such as Yang Jianguo, who proposed yarn quality prediction model based on support vector machine (SVM) via statistical theory [10], Lv Zhijun, et al. used search optimization technique

X. He et al. (Eds.): IScIDE 2015, Part II, LNCS 9243, pp. 107–115, 2015.
DOI: 10.1007/978-3-319-23862-3_11

based on genetic algorithm to optimize yarn quality prediction model parameters with SVM [11], Li Beizhi, et al. took advantage of global search capability of genetic algorithm to solve the time-consuming problem with the selection parameters of SVM [12]. Hereafter, some scholars analyzed the reason why quality fluctuation of the textile from textile production processing theory [13, 14]. i.e., Zhao Bo, et al. predicted the yarn quality through neural network [15].

With previous studies in different ways, we start from uncertainty factors that affect quality uncertainty of industrial textiles. The main goal is to explore generation mechanism and behavior characteristic of the uncertainty factor. Accordingly, we construct a prediction model for quality uncertainty to analyze its fluctuation regular, study its generation mechanism, and dig its behavior characteristics.

1 Objective

To further explore the reason and regular of quality fluctuation in the textile processing, and forecast and indentify the relationship and behavior characteristic of quality uncertainty of industrial textiles, we began from the above four problems, and studied the following four aspects via the theories of HMESE and textile quality management.

(1) Explore fluctuation reason of quality characteristic value of industrial textiles.
(2) Study fluctuation regular of textile quality characteristic values, and define random error and systematic error.
(3) Dialysis generation and interaction mechanism of uncertainty factors caused textile quality fluctuation.
(4) Identify behavior characteristics of quality uncertainty caused systematic errors.

2 The Relationships Between Quality Fluctuation and Uncertainty Factor

2.1 Fluctuation Mechanism of Textile Quality

The textile processing is a process that multi-factor (e.g., temperature, humidity, materials, human, equipment and environment, etc.) and multi-process are interaction, and this interaction process relationship can make multiply factors be divided into the external factors and the internal factors.

Now, assumed that x represents the different combinations of process parameters, and y denotes the output characteristic values of the textile quality, then, according to textile quality management theory, the relationships between x and y would become a nonlinear function relationship, was shown in Fig. 1. When $x = x_1$, a corresponding quality output characteristic value is y_1. That is, when $x = x_1$, the fluctuation error of x is Δx_1, and the corresponding fluctuation error of y is Δy_1. While $x = x_2$, a combination of process parameters is x_2, the fluctuation error of x is Δx_2, its quality output characteristic is y_2, and the fluctuation error of y_2 is Δy_2. At this time, when $\Delta x_1 = \Delta x_2$, there exists $\Delta y_1 > \Delta y_2$.

To solve this problem, according to textile process design characteristics and specific configuration cotton conditions, the quality output characteristic value y_2 need to be corrected. Furthermore, through the function changing relationship $y = f(x)$ of textile technology parameter x, we defined two elements k and b, and k, b is constant and greater than zero, respectively. Then, we made k and x exist a linear function dependency relationship $x = f(k) + b$, where b is the offsetting variable of x in the function $y = f(x)$. Thus, the relationships between k and y can also be expressed a linear relationship $y = f(k) = ak + b$.

Where a is an unknown constant. There is a slope relationship between k and y, and named $\frac{dy}{dk} = \frac{df(k)}{dk} = a$.

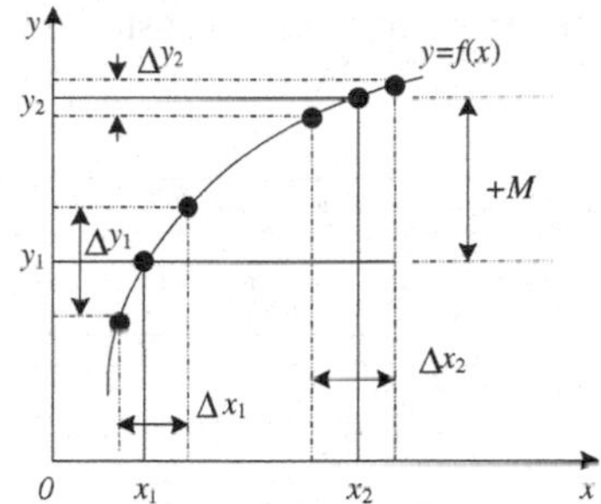

Fig. 1. Relation schema of technology parameter and quality output characteristic value

2.2 Fluctuation Regular of the Textile Quality

According to statistical theory, any one of random variable exist a corresponding probability distribution. Then, for Δy, it also should follow a certain probability distribution [16]. However, under the role of different nature properties of the uncertainty, the probability distribution of Δy is exactly inconsistent. Meanwhile, the overall distribution (i.e., a procedure) of n independent variables and the sum of identically distributed random variables is approximate to normal distribution.

Duo to the fluctuation of textile quality characteristic value obeying normal distribution, its fluctuation property constitutes a certain probability distribution. According to statistical probability theory, the determined probability distribution can be described by measuring numerical centralization and numerical characteristic value of decentralization degree (i.e., μ and δ), the meaning is that the size of μ and δ is decided by the operating state of the machine. Namely, the fluctuation error Δy follows $\Delta y \sim N(\mu,\delta^2)$, the result reflects how the defect formed and grade level in each fabric processing generated.

2.3 Interaction Mechanism of Quality Uncertainty

To further explore the relationships among all kinds of uncertainty factors, and analyze the reason of the fluctuation of Δy, we assumed that u represents an uncertainty factor set, its size is n, and the element number is m, which comes from abnormal events of the textile processing. Meanwhile, y denotes a quality fluctuation set. Then, u and y is expressed as $u = (u_1, u_2, u_3, \ldots, u_n)$ and $y = (y_1, y_2, y_3, \ldots, y_m)$, respectively.

Through analyzing generation mechanism of uncertainty factors, we found that both uncertainty factor u and quality fluctuation y is not a linear relationship, but a mutual influence and intersection tree relationship, because y has a root node and multiple child nodes. Furthermore, multiple u_i can simultaneously act on a y, and even a u_i can divide the entire y.

In order to correctly represent the above tree diagram relationship between u and y, and to explore generation mechanism of quality uncertainty factor, we studied from six aspects for generation mechanism of uncertainty factor, and then an improved man-machine-environment brittle model was constructed for quality fluctuation of industrial textiles via the brittle theory of complex system, which was shown in Fig. 2.

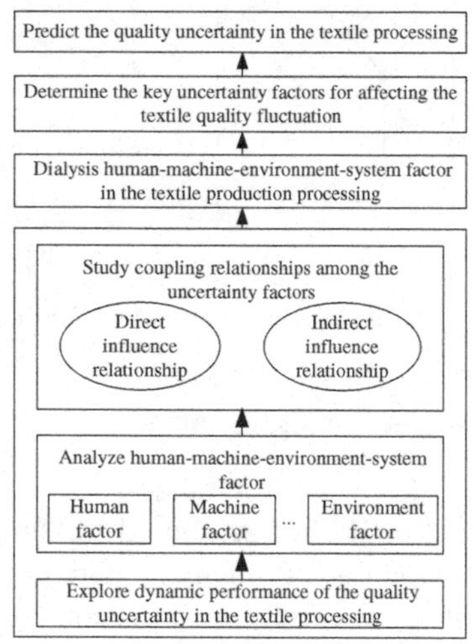

Fig. 2. The improved man-machine-environment brittle model

2.4 Behavior Identification of Quality Uncertainty

Now, for the textile quality fluctuation, the vector of uncertainty factors was denoted as $v = (v_1, v_2, \cdots, v_r)^T$, where v_k $(k = 1,2,\cdots,r)$ represents that the kth uncertainty factor is a function of q quality characteristic values. There exists the following function relationship.

$$u_k = g_k(y_1, y_2, \cdots, y_q) \tag{1}$$

We launched the Eq. (1) with Taylor series, and eliminated more than the second-order and higher-order terms, then u_k can be expressed as follows.

$$u_k = m_k + \sum_{j=1}^{q} \left(\frac{\partial_{g_k}}{\partial_{y_j}}|_{\bar{y}} \Delta y_j\right) \tag{2}$$

On this basis, we assumed that

$$d_k = y_k - m_k = \sum_{j=1}^{q} \left(\frac{\partial_{g_k}}{\partial_{y_j}}|_{\bar{y}} \Delta y_j\right) \tag{3}$$

In the above equation, Δy_j represents the deviation between quality characteristic value y_j and the average value $\bar{y}_j$, that is, $\Delta y_j = y_j\text{-}\bar{y}_j$ and $m_k = g_k(\bar{y})$.

According to the literature [6], for the textile processing with multiple associated quality characteristic values, the expected quality loss was expressed as follows.

$$E(L(d)) = tr(AV(d)) \tag{4}$$

Wherein, $V(d)$ was obtained from the following matrix.

$$V(d) = \begin{bmatrix} \operatorname{var}(d_1) & \operatorname{cov}(d_1, d_2) & \cdots & \operatorname{cov}(d_1, d_r) \\ \operatorname{cov}(d_1, d_2) & \operatorname{var}(d_2) & \cdots & \operatorname{cov}(d_2, d_r) \\ \vdots & \vdots & & \vdots \\ \operatorname{cov}(d_1, d_r) & \cdots & \cdots & \operatorname{var}(d_r) \end{bmatrix} \tag{5}$$

Where d is a textile quality uncertainty deviation vector, A is the quality loss coefficient matrix and $V(d)$ is the variance-covariance matrix of the vector d.

Assumed that the fluctuation regular of the textile quality follows normal distribution, and the quality characteristic value d_k also follows normal distribution. Then, according to statistical theory, the variance of d_k has the following relationship.

$$\operatorname{var}(d_k) = \sum_{j=1}^{q} \left(\left(\frac{\partial_{g_k}}{\partial_{y_j}}|_{\bar{y}}\right)^2 \operatorname{var}(\Delta y_j)\right) \tag{6}$$

Where $k = 1, 2, \cdots, r$. If there are common features among quality characteristic values of textiles, and they are associated with each other. Then, the covariance of the mth and nth correlation characteristic values can be expressed the following.

$$cov(d_m, d_n) = \sum_{j=1}^{q} \left(\left(\frac{\partial_{g_m}}{\partial_{y_j}}|_{\bar{y}}\right)\left(\frac{\partial_{g_n}}{\partial_{y_j}}|_{\bar{y}}\right) \operatorname{var}(\Delta y_j)\right) \tag{7}$$

Where $m, n = 1, 2, \ldots, r,$ and $m \neq n$

3 Experiment and Comparative Analysis

The experimental design solution was the following. First, we accessed to the same data from the same equipment via two methods, one is uncertainty factor prediction (UFP), and the other is the artificial neural network (ANN) method. The specific experimental process was shown as follows.

The experimental environment: temperature is 20°C and relative humidity is 65 %.

The purpose of the experiment: in the same order, we selected 45 batches yarn from one factory, used the same process and specifications, and did the experiment for two groups of data results.

The experimental scheme was shown as follows.

(1) All of yarn data were measured by CSIRO, Uster Tester III, Sirolan-tensor, and Tensorapid.
(2) The feature values were collected by computer monitoring system, which includes yarn fineness unevenness (%), details (*/km*), slub (*/km*), strength (*cN/tex*), and breaking elongation (%).

The experiment results were shown in Table 1.

Table 1. The prediction results comparison with two methods

Sample	Fineness unevenness (%)			Nip(*/km*)			Nub(*/km*)			Strength(*cN/tex*)			Breaking elongation (%)		
	UFP result	ANN result	Error	UFP result	ANN result	Error	UFP result	ANN result	Error	UFP Result	ANN result	Error	UFP result	ANN result	Error
1	21.59	21.79	−0.20	654	656	−2	246	249	−3	5.98	5.91	0.07	8.27	7.77	0.5
2	21.66	21.74	−0.08	719	722	−3	218	219	−1	5.91	6.03	−0.12	8.88	8.32	0.56
3	21.98	22.19	0.79	894	893	1	268	269	−1	5.70	5.72	−0.02	8.86	7.35	1.51
4	18.59	18.87	−0.28	201	203	−2	88	90	−2	6.26	5.94	0.32	6.88	7.21	−0.33
5	21.84	21.97	−0.13	766	767	−1	247	248	−1	6.36	6.50	−0.14	10.16	10.01	0.15
6	21.84	22.08	0.76	783	784	−1	234	237	−3	6.12	6.20	−0.08	10.26	9.67	0.59
7	18.40	17.65	0.75	196	193	3	52	53	−1	7.27	7.50	−0.23	14.46	16.32	−1.86
8	21.95	21.64	0.31	808	806	2	217	219	−2	6.03	6.36	−0.33	9.23	9.61	-0.38
9	20.82	20.97	−0.15	644	645	−1	186	189	−3	5.46	5.58	−0.12	6.35	7.45	−1.1
10	17.06	17.31	−0.25	124	125	−1	31	32	−1	6.86	6.77	0.09	15.08	14.02	1.06

Can be seen from Table 1, under the experimental condition of the same equipment and assortment, we obtained textile quality data by above two different methods, the error comparison results were shown that: in the time interval Δt, the errors of yarn quality characteristic values for one process and specification are different, and the variation tendency of the errors does not follow the normal distribution $N(\mu,\delta^2)$.

To this end, the main question now is how to make linear relationship in the time interval Δt more explicit. The corresponding solution is to solve the value of b according to the distribution regulars of Δx and Δy. In reality, this involves behavior identification problem of uncertainty factors. According to the formula $p_t = y_t$-$\mu = \frac{1}{(1-\phi_1 B)}\varepsilon_1$, we can consider b as the delay factor B, and then obtain the value of b.

Under the same condition, aiming at the acquired direct source of quality fluctuation of industrial textiles, we simulated two kinds of the textile quality methods, and the result was shown in Fig. 3.

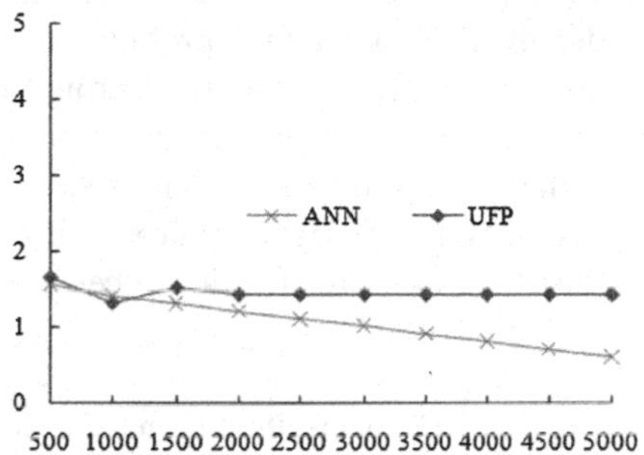

Fig. 3. The simulation results about two methods

We knew from Fig. 3, the larger systematic errors came from ANN method, and its variation trend did not follow a normal distribution. Furthermore, with the continuous advancement of the textile processing, the change trend of systematic errors would gradually approach zero, but not equal to zero. Meanwhile, we can be clearly seen from Fig. 4(*a*) and (*b*), the prediction results were more accurate and intuitive, and the main reasons that cause the fluctuation of the textile quality characteristic values were provoked by the wrong data source.

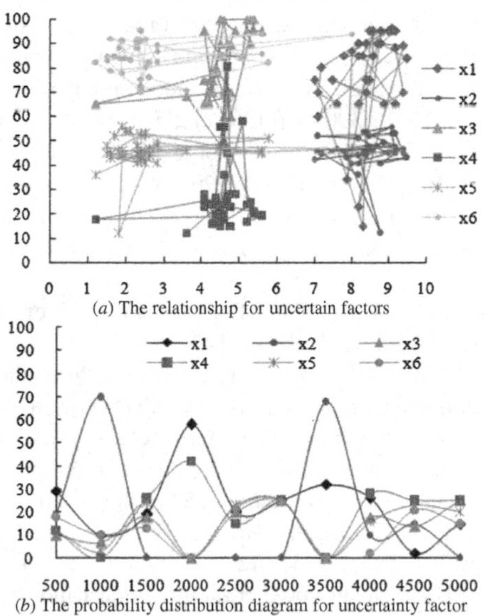

Fig. 4. The prediction simulation results

4 Conclusions

In the textile processing, the relationships among a number of factors were alternation and comprehension role, and the behavior characteristics caused systematic errors (abnormal fluctuation) were difficult to be expressed with mathematical relationship.

Therefore, to analyze deeply the uncertainty problem of the textile production processing, it not only can enhance effectively the monitoring in the textile processing procedure of ‘abnormal event’, but also can ensure the processing continuity of the enterprises or workshop. Furthermore, the proposed four-step prediction method can improve the management standard for production processing control of the textile enterprise, and provide a theoretical basis for textile processing to solve the ‘unusual event’.

Acknowledgments. The authors gratefully thank the Shaanxi Science and Technology Plan Project (Granted No. 2013KRM07), Textile Industry Association Science and Technology Guidance Plan Project in China (Granted No. 2014076, 2013068, 2011081), and Shaanxi Province Education Department (Granted No. 2013JK0742 and 11JK1055) and Beilin District Applied Technology Research and Development Project By Shaanxi Province in China (Granted No. GX1510).

References

1. Mattes, A., Pusch, T., Cherif, C.: Numerical simulation of yarn tensile force for dynamic yarn supply systems of textile machines. J. Text. Inst. **103**(1), 70–79 (2012)
2. Selvanayaki, M., Vijava, M.S., Jamuna, K.S., et al.: An interactive tool for yarn strength prediction using support vector regression. In: Proceedings of the 2nd International Conference on Machine Learning and Computing (ICMLC 2010), pp. 335–339 (2010)
3. Fattahi, S., Taheri, S.M., Ravandi, H.: Cotton yarn engineering via fuzzy least squares regression. Fibers Polym. **13**(3), 390–396 (2012)
4. Mokhtar, S., Ben, A.S., Sakli, F.: Optimization of textile parameters of plain woven vascular prostheses. J. Text. Inst. **101**(12), 1095–1105 (2010)
5. Fallahpour, A.R., Moghassem, A.R.: Spinning preparation parameters selection for rotor spun knitted fabric using VIKOR method of multicriteria decision-making. J. Text. Inst. **104** (1), 7–17 (2013)
6. Mohamed, N., Samar, A.E.: Prediction of some cotton fiber blends properties using regression models. Alexandria Eng. J. **47**(2), 147–153 (2008)
7. Mwasiagi, J.I., Huang, X.B., Wang, X.H.: The use of hybrid algorithms to improve the performance of yarn parameters prediction models. Fibers Polym. **13**(9), 1201–1208 (2012)
8. Mardani, M.N., Safar, J.M., Aghdam, M.M.: Finite-element and multivariate analyses of tension distribution and spinning parameter effects on a ring-spinning balloon. Proc. Inst. Mech. Eng. Part C: J. Mech. Eng. Sci. **224**(2), 253–258 (2010)
9. Kelly, C.M., Hequet, E.F., Devera, J.K.: Breeding for improved yarn quality: modifying fiber length distribution. Ind. Crops Prod. **42**(1), 386–396 (2013)
10. Yang, J.G., Lv, Z.J., Li, B.Z.: Quality prediction in complex industrial process with support vector machine and genetic algorithm optimization: A case study. Appl. Mech. Mater. **232**, 603–608 (2012)

11. Lv, Z., Yang, J., Xiang, Q.: GA Based parameters optimization on prediction method of yarn quality. J. Donghua Univ. (Natural Science) **38**(5), 519–523 (2012). (in Chinese)
12. Li, B., Li, L., Yang, J., et al.: Design and implementation of quality prediction system based on GA-SVM. Comput. Eng. **37**(1), 167–169 (2011). (in Chinese)
13. Pei, Z.G., Chen, G., Liu, C., et al.: Experimental study on the fiber motion in the nozzle of vortex spinning via high-speed photography. J. Nat. Fibers **9**(2), 117–135 (2012)
14. Lv, J., Cao, C.H.: Prediction of yarn quality based on differential evolutionary BP neural network. In: Proceedings of the 4th International Conference on Computational and Information Sciences (ICCIS 2012), pp. 1232–1235 (2012)
15. Zhao, B.: Prediction of thin place of polyester/cotton ring yarn properties from process parameters by using neural network and regression analysis. In: Proceedings of the 7th International Conference on System of Systems Engineering (SoSE 2012), pp. 18–20 (2012)
16. Duan, S., Fan, S., Huang, T., et al.: The research of incremental model based on multivariate quality loss function. Mach. Des. Manuf. **51**(7), 253–256 (2013)

Verification of Hibernate Query Language by Abstract Interpretation

Angshuman Jana[1], Raju Halder[1(✉)], and Agostino Cortesi[2]

[1] Indian Institute of Technology Patna, Patna, India
{ajana.pcs13,halder}@iitp.ac.in
[2] Università Ca′ Foscari Venezia, Venezia, Italy
cortesi@unive.it

Abstract. In this paper, we propose an abstract interpretation framework of Hibernate Query Language (HQL), aiming at automatically and formally verifying enterprise policy specifications on persistent objects which have permanent representation in the underlying database. To this aim, we extend the abstract interpretation approach for object-oriented languages, combined with an abstract semantics of structured query languages.

Keywords: Hibernate Query Language · Static analysis and verification · Abstract interpretation

1 Introduction

Hibernate Query Language (HQL) provides a unified platform for the programmers to develop object-oriented applications to interact with databases, without knowing much details about the underlying databases [1,2,8,13]. HQL is treated as an object-oriented variant of SQL, which allows to represent SQL queries in object-oriented terms and mitigates the paradigm mismatch between object modeling and relational modeling. Hibernate is basically an object-relational mapping tool that simplifies the data creation, data manipulation and data access. Various methods in "`Session`" are used to propagate object's states from memory to the database (or vice versa) and to synchronize both states when a change is made to persistent objects [12]. A HQL query is translated by Hibernate into a set of conventional SQL queries during run time which in turn performs actions on the database.

It is particularly important, in this context, to provide formal verification methods for behavioral properties like absence of run-time errors, absence of confidential information leakage, etc. Abstract Interpretation [7] is a well-established semantics-based static analysis framework which provides a sound approximation of program semantics focussing on a particular property. The intuition of Abstract Interpretation is to lift the concrete semantics to an abstract domain, by replacing concrete values by suitable properties of interest and simulating the

X. He et al. (Eds.): ISciDE 2015, Part II, LNCS 9243, pp. 116–128, 2015.
DOI: 10.1007/978-3-319-23862-3_12

operations in the abstract domain *w.r.t.* their concrete counterparts, in order to ensure the soundness.

F. Logozzo [10] introduced an Abstract Interpretation-based framework of Object-Oriented Programming (OOP) languages, aiming at verifying whether the programs respect the specifications correctly. The framework is used to ensure the class invariant, a property which is valid for all the instances of the class, before and after the execution of any method. Moreover, it can also be used for optimization of the code at class-level. For instance, if a class invariant states that the class will never throw a given exception, then in the corresponding code for throwing/handling the exception can be dropped.

Unfortunately, the usual framework of Abstract Interpretation of OOP languages [3,10,11] can not be directly applied to Hibernate Query Language (HQL) if one wants to verify the properties of persistent objects only, rather than transient objects, which have permanent representation in the underlying database. On the other side, the existing work on abstract interpretation of query languages [9] did not consider an access to the database operations through a high-level object-oriented language. The aim of this paper is to fill the gap between these two theories.

As an example, consider the HQL program depicted in Fig. 1. The `Session` methods of this program allow to update information (like age and salary) of the employees and to make simple queries to that database[1]. Consider the following enterprise policies given by the following three constraints:

Policy 1: *Employees age should be greater than or equal to 18 and less than or equal to 62.*

Policy 2: *The salary of employees with age greater than 30 should be at least 1500 euro.*

Policy 3: *Employees salary should not be more than three times of the lowest salary.*

Figure 2 depicts different states of the underlying database table when various `Session` methods of the program are executed. For instance, after executing statement 13, a tuple corresponding to the object 'obj' of class `emp` is inserted into the corresponding database state t_1, resulting in a new state t_2. Similarly, update and delete operations on the objects at 14–16 and 19–20, by the corresponding `Session` methods yield the states t_3 and t_5 respectively. Observe that, selection of objects at 17–18 produces the result shown in table t_{sel}, and of course, it does not change the database state (*i.e.* $t_3 = t_4$). The code satisfies policy 1, whereas it does not satisfy policies 2 and 3.

This can be formally and automatically verified by extending the Abstract Interpretation theory to the case of HQL: in general, it can be applied to formally verify some properties of persistent objects which have permanent representation in the underlying relational databases (or to find possible violation of the policy rules), by analyzing the HQL code on non-relational or relational abstract

[1] Observe at program points 13, 14–16, 17–18 that the basic differences between HQL and SQL.

```
class emp {
    private int id, age, dno, sal;
    emp(){ }
    public int getId() { return id;}
    public void setId(int id) { this.id = id;}
    public int getage() {return age;}
    public void setage(int age) { this.age = age;}
    public int getdno() { return dno;}
    public void setdno(int dno) { this.dno = dno;}
    public int getsalary() { return sal;}
    public void setsalary(int sal) { this.sal = sal;}
}
```

(a) POJO Class emp

```
public class ExClass{
  public static void main(String[] args) {

    Configuration cfg=new Configuration();
    cfg.configure("hibernate.cfg.xml");
    SessionFactory sf=cfg.buildSessionFactory();
    Session ses=sf.openSession();
    Transaction tr=ses.beginTransaction();

  % Creating emp object and stores into database %
    emp obj=new emp( );
    obj.setId(4);
    obj.setage(32);
    obj.setdno(1);
    obj.setsalary(1000);
    ses.save( obj );

  % Updating persistent emp objects %
    Query q1 = ses.createQuery("UPDATE emp e SET e.age= e.age+1, e.sal= e.sal + :inc×2 WHERE
                                                                              e.sal > 1600");
    q1.setParameter("inc",100);
    int r1 = q1.executeUpdate();

  % Selecting from persistent emp objects %
    Query q2 = ses.createQuery("SELECT e.dno, MAX(e.sal), AVG(DISTINCT e.age) FROM emp e
                  WHERE e.sal ≥ 1000 GROUP BY e.dno HAVING MAX(e.sal) < 4000 ORDER BY e.dno");
    List r2 = q2.list();

  % Deleting persistent emp objects %
    Query q3 = ses.createQuery("DELETE FROM emp e WHERE e.age > 50");
    int r3 = q3.executeUpdate();

    tr.commit();
    ses.close();}}
```

(b) Service class

Fig. 1. A HQL program P

domains [4,5]. The key point is the formalization of the abstract semantics of `Session` methods relating persistent objects to the database [9].

The structure of the paper is as follows: Sects. 2 and 3 recall the basics on the concrete/abstract semantics of query languages and object-oriented languages respectively. We formalize the concrete and abstract semantics of HQL in Sects. 4 and 5 respectively, by showing its applications in a simple yet general example. Section 6 concludes.

tid	tage	tdno	tsal
1	35	3	1600
2	19	2	900
3	50	3	2550

(a) Orginal Table t_1

tid	tage	tdno	tsal
1	35	3	1600
2	19	2	900
3	50	3	2550
4	32	1	1000

(b) Table t_2: After executing statement 13

tid	tage	tdno	tsal
1	35	3	1600
2	19	2	900
3	51	3	2750
4	32	1	1000

(c) Table t_3: After executing statements 14-16

tid	tage	tdno	tsal
1	35	3	1600
2	19	2	900
3	51	3	2750
4	32	1	1000

(d) Table t_4: After executing statement 17-18 (no change in database)

tdno	MAX(tsal)	AVG(tage)
1	1000	32
3	2750	43

(e) Table t_{sel}: Result of Selection at 17-18

tid	tage	tdno	tsal
1	35	3	1600
2	19	2	900
4	32	1	1000

(f) Table t_5: After executing statements 19-20

Fig. 2. Snapshot of database states after executing various `Session` methods

2 Semantics of Query Languages

Halder and Cortesi [6,9] formalized the semantics of query languages. The basic functionality of SQL statements can be stated as "Any SQL statement Q first identifies an active data set from the database using a pre-condition ϕ that follows first-order logic, and then performs the appropriate operations A on the selected data set". Therefore, the abstract syntax of SQL statements is denoted by a tuple $\langle A, \phi\rangle$. For instance, the query "SELECT a_1, a_2 FROM t WHERE $a_3 \leq 30$" is denoted by $\langle A, \phi\rangle$ where A represents the action-part "SELECT a_1, a_2 FROM t" and ϕ represents the conditional-part "$a_3 \leq 30$".

Table 1 depicts the syntactic sets, the abstract syntax of SQL, the environments and states associated with the SQL programs, and the semantics of SQL statements. Observe that all the syntactic elements in SQL statements (for example, `GROUP BY`, `ORDER BY`, `DISTINCT` clauses, etc.) are represented as functions and the semantics are described as a partial functions on the states which specify how expressions are evaluated and instructions are executed. A state in the program is represented by the tuple $(\ell,\ \rho_d)$ where $\ell \in \mathbb{L}$ is a program label and $\rho_d \in \mathfrak{E}_d$ is a database environment. Interested readers may refer to [9] for more details on the semantics of SQL statements.

3 Semantics of Object-Oriented Programming (OOP)

F. Logozzo in [10] formalized the concrete and abstract semantics of object-oriented programming languages as follows. Object-oriented programming languages consist of a set of classes including a main class from where execution starts. Each class contains a set of attributes and a set of methods - called members of the class. Therefore, a program P in OOP is defined as $P = \langle c_{main}, \mathtt{L}\rangle$ where `Class` denotes the set of classes, $c_{main} \in$ `Class` is the main class, $\mathtt{L} \subset$ `Class` are the other classes present in P.

Table 1. Syntax and semantics of programs embedding SQL statements

Syntactic Sets
$n : \mathbb{Z}$ (Integer)
$s : \mathbb{S}$ (String)
$k : \mathbb{C}$ (Constants)
$v_d : \mathbb{V}_d$ (Database Variables)
$e : \mathbb{E}$ (Arithmetic Expressions)
$b : \mathbb{B}$ (Boolean Expressions)
$A : \mathbb{A}$ (Action)
$\tau : \mathbb{T}$ (Terms)
$a_f : \mathbb{A}_f$ (Atomic Formulas)
$\phi : \mathbb{W}$ (Pre-condition)
$Q : \mathbb{Q}$ (SQL statements)
$\ell : \mathbb{L}$ (Set of Program Labels)

Abstract Syntax
$k ::= n \mid s$
$e ::= k \mid v_d \mid op_u\, e \mid e_1\, op_b\, e_2$, where $op_u \in \{+,-\}$ and $op_b \in \{+,-,*,/,\ldots\ldots\}$
$b ::= e_1\, op_r\, e_2 \mid \neg b \mid b_1 \vee b_2 \mid b_1 \wedge b_2 \mid true \mid false$, where $op_r \in \{=,\geq,\leq,<,\ldots\}$
$\tau ::= k \mid v_d \mid f_n(\tau_1,\tau_2,\ldots,\tau_n)$, where f_n is an n-ary function.
$a_f ::= R_n(\tau_1,\tau_2,\ldots,\tau_n) \mid \tau_1 = \tau_2$, where $R_n(\tau_1,\tau_2,\ldots,\tau_n) \in \{true, false\}$
$\phi ::= a_f \mid \neg\phi_1 \mid \phi_1 \vee \phi_2 \mid \phi_1 \wedge \phi_2 \mid \forall x_i\, \phi \mid \exists x_i\, \phi$
$g(\vec{e}) ::= \texttt{GROUP BY}(\vec{e}) \mid id$
$r ::= \texttt{DISTINCT} \mid \texttt{ALL}$
$s ::= \texttt{AVG} \mid \texttt{SUM} \mid \texttt{MAX} \mid \texttt{MIN} \mid \texttt{COUNT}$
$h(e) ::= s \circ r(e) \mid \texttt{DISTINCT}(e) \mid id$
$h(*) ::= \texttt{COUNT(*)}$
$\vec{h}(\vec{x}) ::= \langle h_1(x_1),\ldots,h_n(x_n)\rangle$, where $\vec{h} = \langle h_1,\ldots,h_n\rangle$ and $\vec{x} = \langle x_1,\ldots,x_n\rangle$
$f(\vec{e}) ::= \texttt{ORDER BY ASC}(\vec{e}) \mid \texttt{ORDER BY DESC}(\vec{e}) \mid id$
$A ::= \texttt{SELECT}(f(\vec{e'}),\ r(\vec{h}(\vec{x})),\ \phi,\ g(\vec{e})) \mid \texttt{UPDATE}(\vec{v_d},\ \vec{e}) \mid \texttt{INSERT}(\vec{v_d},\ \vec{e}) \mid \texttt{DELETE}(\vec{v_d})$
$Q ::= \langle A,\phi\rangle \mid Q'\ \texttt{UNION}\ Q'' \mid Q'\ \texttt{INTERSECT}\ Q'' \mid Q'\ \texttt{MINUS}\ Q'' \mid Q';\ Q''$

Database Environment	A database is a set of tables $\{t_i \mid i \in I_x\}$ for a given set of indexes I_x. A database environment is defined as a function ρ_d whose domain is I_x, such that for $i \in I_x$, $\rho_d(i) = t_i$.
Table Environment	A table environment ρ_t for a table t is defined as a function such that $\forall a_i \in attr(t)$, $\rho_t(a_i) = \langle \pi_i(l_j) \mid l_j \in t\rangle$ where π is the projection operator, i.e., $\pi_i(l_j)$ is the ith element of the l_j-th row.
State	The set of states is defined as $\Sigma_d = \mathbb{L} \times \mathfrak{E}_d$ where $\mathfrak{E}_d$ is the set of all database environments.
Semantics	Given a state $(\ell, \rho_d) \in \Sigma_d$, the semantics of SQL statement Q on (ℓ, ρ_d) is defined as $\mathbf{S}_{sql}[\![Q]\!](\ell,\rho_d) = \mathbf{S}_{sql}[\![Q]\!](\ell,\rho_t) = (\ell',\rho_{t'})$ where $\mathbf{S}_{sql}[\![.]\!]$ is the semantic function, $target(Q) = t \in d$, and $\ell' \in \mathbb{L}$ is the label of the successor statement in the program.

A class $c \in$`Class` is defined as a triplet $c = \langle$`init, F, M`$\rangle$ where `init` is the constructor, `F` is the set of fields, and `M` is the set of member methods in c.

Let `Var`, `Val` and `Loc` be the set of variables, the domain of values and the set of memory locations respectively. The set of environments, stores and states are defined below:

- The set of environments is defined as `Env: Var`$\longrightarrow$`Loc`
- The set of stores is defined as `Store: Loc`$\longrightarrow$`Val`
- A state is denoted by a tuple $\langle e, s\rangle$ where $e \in$`Env` and $s \in$`Store`.

It is assumed that a state contains some special variables $\{pc, V_{in}, V_{out}\} \subseteq$ `Var`, where pc denotes the current program counter, V_{in} denotes the method input variable (if any), and V_{out} denotes the method output variable (if any).

3.1 Constructor and Method Semantics

During object creation, the class constructor is invoked and object fields are instantiated by input values. Given a store s, the constructor maps its fields to fresh locations and then assigns values into those locations. The constructor never returns any output.

```
1. class Sample {
2. int a;
3. Sample(int i) {
4.   a = i;
5. }
6.  int parity() {
7.    if(a % 2 == 0)
8.      return 1;
9.    else return 0;
10. }
11. int * incr( int j ) {
12.   a = a + j;
13.   return &a;
14. }
15. }
```

Fig. 3. An example class

Definition 1 (Constructor Semantics). *Given a store s. Let* $\{a_{in}, a_{pc}\} \subseteq$ $\mathtt{Loc}$ *be the free locations,* $\mathtt{Val}_{in} \subseteq \mathtt{Val}$ *be the semantic domain for input values. Let* $v_{in} \in \mathtt{Val}_{in}$ *and* pc_{exit} *be the input value and the exit point of the constructor. The semantic of the class constructor* $\mathtt{init}$, $S[\![\mathtt{init}]\!] \in (\mathtt{Store} \times \mathtt{Val} \rightarrow \wp(\mathtt{Env} \times \mathtt{Store}))$, *is defined by:*

$$S[\![\mathtt{init}]\!](s, v_{in}) = \{(e_0, s_0) \mid (e_0 \triangleq V_{in} \rightarrow a_{in}, pc \rightarrow a_{pc}) \wedge (s_0 \triangleq s[a_{in} \rightarrow v_{in}, a_{pc} \rightarrow pc_{exit}])\}$$

Definition 2 (Method Semantics). *Let* $\mathtt{Val}_{in} \subset \mathtt{Val}$ *and* $\mathtt{Val}_{out} \subseteq \mathtt{Val}$ *be the semantic domains for the input values and the output values respectively. Let* $v_{in} \in \mathtt{Val}_{in}$ *be the input values,* a_{in} *and* a_{pc} *be the fresh memory locations, and* pc_{exit} *be the exit point of the method m. The semantic of a method m,* $S[\![m]\!] \in (\mathtt{Env} \times \mathtt{Store} \times \mathtt{Val}_{in} \rightarrow \wp(\mathtt{Env} \times \mathtt{Store} \times \mathtt{Val}_{out}))$, *is defined as:*

$$S[\![m]\!](e, s, v_{in}) = \{(e', s', v_{out}) \mid (e' \triangleq e[V_{in} \rightarrow a_{in}, pc \rightarrow a_{pc}]) \wedge (s' \triangleq s[a_{in} \rightarrow v_{in}, a_{pc} \rightarrow pc_{exit}]) \wedge v_{out} \in \mathtt{Val}_{out}\}$$

Example 1. Consider the example of Fig. 3. The class constructor `Sample()` creates a new environment consisting of field a. The semantics of constructor `Sample()`, semantics of the methods `parity()` and `incr()` are defined below:

$$S[\![\mathtt{Sample()}]\!](s, i) = \{(e_0, s_0) \mid (e_0 \triangleq a \rightarrow a_{in}, pc \rightarrow a_{pc}) \wedge (s_0 \triangleq s[a_{in} \rightarrow i, a_{pc} \rightarrow 5])\}$$

$$S[\![\mathtt{parity()}]\!](e, s, \varnothing) = \{(e, s', v_{out}) \mid (s' \triangleq s[e(pc) \rightarrow 10]) \wedge (v_{out} = \text{if}(s(e(a))\%2)\ ?1 : 0)\}$$

$$S[\![\mathtt{incr()}]\!](e, s, j) = \{(e, s', v_{out}) \mid (s' \triangleq s[e(a) \rightarrow s(e(a)) + j, e(pc) \rightarrow 14]) \wedge v_{out} = e(a)\}$$

Observe that `parity()` takes no input and returns an integer value as output, whereas `incr()` takes an integer value as input and returns an address as output.

3.2 Object and Class Semantics

The set of interaction states is defined by Σ=$\texttt{Env} \times \texttt{Store} \times \texttt{Val}_{out} \times \wp(\texttt{Loc})$ where Env, Store, $\texttt{Val}_{out}$, and Loc are the set of environments, the set of stores, the set of output values, and the set of addresses respectively.

Object semantics is defined in terms of interaction history between the program-context and the object. A direct interaction takes place when the program-context calls any member-method of the object, whereas an indirect interaction occurs when the program-context updates any address escaped from the object's scope. However, both direct or indirect interaction can cause a change in an interaction state.

The transition relation $\mathscr{T}$ includes both direct and indirect interactions.

Objects Fix-Point Semantics Given a store $s \in \texttt{Store}$, the set of initial interaction states is defined as $\mathcal{I}_0 = \big\{\langle e_0, s_0, \phi, \emptyset\rangle \mid S[\![\texttt{init}]\!](v_{in}, s) \ni \langle e_0, s_0\rangle, v_{in} \in \texttt{Val}_{in}\big\}$. The fix-point trace semantics of obj, is defined as: $\mathrm{T}[\![\texttt{obj}]\!](\mathcal{I}_0) = \mathrm{lfp}_{\emptyset}^{\subseteq} \mathcal{F}(\mathcal{I}_0) = \bigcup_{i \leq \omega} \mathcal{F}^i(\mathcal{I}_0)$ where

$$\mathcal{F}(\mathcal{I}) = \lambda \mathcal{T}.\, \mathcal{I} \cup \big\{\sigma_0 \xrightarrow{\ell_0} \dots \xrightarrow{\ell_{n-1}} \sigma_n \xrightarrow{\ell_n} \sigma_{n+1} \mid \sigma_0 \xrightarrow{\ell_0} \dots \xrightarrow{\ell_{n-1}} \sigma_n \in \mathcal{T} \wedge (\sigma_{n+1}, \ell_n) \in \mathscr{T}(\sigma_n)\big\}$$

4 Concrete Semantics of Hibernate Query Language

We are now in position to formalize the concrete and abstract semantics of HQL. We obtain it by (*i*) extending the OOP semantics and (*ii*) defining the semantics of Session methods combining with the abstract interpretation of query languages.

4.1 Syntax

Like OOP, a program P in HQL is also defined as $P = \langle c_{main}, \texttt{L}\rangle$ where $c_{main} \in \texttt{Class}$ is the main class, $\texttt{L} \subset \texttt{Class}$ are the other classes present in P. Similarly, a class $c \in \texttt{Class}$ is defined as a triplet $c = \langle \texttt{init}, \texttt{F}, \texttt{M}\rangle$ where init is the constructor, F is the set of fields, and M is the set of member methods in c.

An additional and attractive feature of Hibernate is the presence of Hibernate Session which provides a central interface between the application and database and acts as a persistence manager. In HQL, an object is transient if it has just been instantiated using the new operator. Transient instances will be destroyed by the garbage collector if the application does not hold a reference anymore. A persistent instance, on the other hand, has a representation in the database and an identifier value assigned to it. Given an object, the Hibernate Session is used to make the object persistent. Various methods in Hibernate Session are used to propagate object's states from memory to the database

Set of Classes

$c \in$ `Class`

$c ::= \langle$`init,F,M`$\rangle$

where `init` is the constructor, `F` $\subseteq$ `Var` is the set of fields, and `M` is the set of methods.

Session methods

$m_{ses} \in \mathtt{M}_{ses}$

$m_{ses} ::= \langle \mathtt{C}, \phi, \mathtt{OP} \rangle$ where `C` $\subseteq$ `Class` and ϕ represents 'WHERE' clause.

`OP` ::= `SAVE(obj)`

| $\mathtt{UPD}(\vec{v}, e\vec{x}p)$

| `DEL()`

| $\mathtt{SEL}\big(f(e\vec{x}p'),\ r(\vec{h}(\vec{x})),\ \phi,\ g(e\vec{x}p)\big)$

where ϕ represents 'HAVING' clause and `obj` denotes a class-instance.

HQL Programs

$p \in \mathbb{P}$

$p ::= \langle c_{main}, \mathtt{L} \rangle$ where $c_{main} \in$ `Class` is the main class and `L` $\subset$ `Class`.

Fig. 4. Abstract Syntax of `Session` methods and HQL programs

(or vice versa) and to synchronize both states when a change is made to persistent objects.

Abstract Syntax of `Session` *Methods* In abstract syntax, we denote a `Session` method by a triplet $\langle \mathtt{C}, \phi, \mathtt{OP} \rangle$ where `OP` is the operation to be performed on the tuples satisfying ϕ in the database tables corresponding to the set of POJO classes `C`. Four basic `OP` that cover a wide range of operations are `SAVE`, `UPD`, `DEL`, and `SEL`.

- $\big\langle \mathtt{C}, \phi, \mathtt{SAVE}(\mathtt{obj}) \big\rangle = \big\langle \{c\}, false, \mathtt{SAVE}(\mathtt{obj}) \big\rangle$: Stores the state of the object `obj` in the database table t, where t corresponds to the POJO class c and `obj` is the instance of c. The pre-condition ϕ is $false$ as the method does not identify any existing tuples in the database.
- $\big\langle \mathtt{C}, \phi, \mathtt{UPD}(\vec{v}, e\vec{x}p) \big\rangle = \big\langle \{c\}, \phi, \mathtt{UPD}(\vec{v}, e\vec{x}p) \big\rangle$: Updates the attributes corresponding to the class fields $\vec{v}$ by $e\vec{x}p$ in the database table t for the tuples satisfying ϕ, where t corresponds to the POJO class c.
- $\big\langle \mathtt{C}, \phi, \mathtt{DEL}() \big\rangle = \big\langle \{c\}, \phi, \mathtt{DEL}() \big\rangle$: Deletes the tuples satisfying ϕ in t, where t is the database table corresponding to the POJO class c.
- $\big\langle \mathtt{C}, \phi', \mathtt{SEL}\big(f(e\vec{x}p'),\ r(\vec{h}(\vec{x})),\ \phi,\ g(e\vec{x}p)\big) \big\rangle$: Selects information from the database tables corresponding to the set of POJO classes `C`, and returns the equivalent representations in the form of objects. This is done only for the tuples satifying ϕ'. The descriptions of f, r, h, g, ϕ, etc. are already mentioned in Table 1.

Observe that as `SAVE()`, `UPD()` and `DEL()` always target single class, the set `C` is a singleton $\{c\}$. However, `C` may not be singleton in case of `SEL()`. The syntax is defined in Fig. 4.

4.2 Semantics

The semantics of conventional constructors, methods, objects, classes in HQL are defined in the same way as in the case of OOP.

The `Session` methods require an 'ad-hoc' treatment. We define their concrete semantics by specifying how the methods are executed on (e, s, ρ_d) where $e \in \texttt{Env}$ is an environment, $s \in$`Store` is a store, and $\rho_d \in \mathfrak{E}_d$ is a database environment, resulting in new state $(e', s', \rho_{d'})$. The semantic definitions are expressed in terms of the semantics of database statements `SELECT, INSERT, UPDATE, DELETE` [9].

We use the following functions in the subsequent part: *map(v)* maps v to the underlying database object; *var(exp)* returns the variables appearing in *exp*; *attr(t)* returns the attributes associated with table t; *dom(f)* returns the domain of f.

The semantic function $\mathbf{S}_{hql}$, $\mathbf{S}_{hql} \in ((\texttt{Env} \times \texttt{Store} \times \mathfrak{E}_d) \rightarrow \wp(\texttt{Env} \times \texttt{Store} \times \mathfrak{E}_d))$, for a given session method $m_{ses} = \langle \texttt{C}, \phi, \texttt{OP}\rangle$ is defined as:

$$\mathbf{S}_{hql}[\![m_{ses}]\!](e,s,\rho_d) = \begin{cases} \mathbf{S}_{hql}[\![m_{ses}]\!](e,s,\rho_{t'}) \text{ if } \exists t_1,\dots,t_n \in dom(\rho_d): \texttt{C} = \{c_1,\dots,c_n\} \\ \qquad \wedge(\forall i \in 1\dots n.\ t_i = map(c_i)) \wedge t' = t_1 \times t_2 \times \dots \times t_n. \\ \\ \bot \quad otherwise. \end{cases}$$

Semantics of Session Method $\langle\{c\}, \phi, \texttt{UPD}(\vec{v}, \vec{exp})\rangle$. The semantics of $\langle\{c\}, \phi$, $\texttt{UPD}(\vec{v}, \vec{exp})\rangle$ is defined as[2]:

$$\begin{aligned}\mathbf{S}_{hql}[\![\langle\{c\},\phi,\texttt{UPD}(\vec{v},\vec{exp})\rangle]\!] &= \lambda(e,s,\rho_t).\text{let } c = \langle\texttt{init},\texttt{F},\texttt{M}\rangle \text{ such that } map(\texttt{F}) = attr(t) \\ &\text{and } map(\vec{v}) = \vec{a} \subseteq attr(t) \text{ where } \vec{v} \subseteq \texttt{F}, \text{ and let } \phi_d = \texttt{PE}[\![\phi]\!](e,s,\texttt{F}) \text{ and} \\ &\vec{exp}_d = \texttt{PE}[\![\vec{exp}]\!](e,s,\texttt{F}) \text{ in } \big\{\langle e,s,\rho_{t'}\rangle \mid \rho_{t'} \in \mathbf{S}_{sql}[\![\langle\texttt{UPDATE}(\vec{a},\vec{exp}_d),\phi_d\rangle]\!](\rho_t)\big\}.\end{aligned}$$

The auxiliary function $\texttt{PE}[\![X]\!]$ (which stands for partial evaluation) is used in the definition above to convert variables in X into the corresponding database objects. This is defined by $\texttt{PE}[\![X]\!](e,s,\texttt{F}) = X'$, where

$$X' = X[x_i/v_i] \text{ for all } v_i \in var(X) \text{ and } x_i = \begin{cases} map(v_i) & \text{if } v_i \in \texttt{F} \\ E[\![v_i]\!](e,s) & \text{otherwise} \end{cases}$$

Example 2. Consider the HQL example in Fig. 1. The abstract syntax of the `Session` method corresponding to the statements 14–16 is $\big\langle\{c\}, \phi, \texttt{UPD}(\vec{v}, \vec{exp})\big\rangle$, where

- {c}= {`emp`},
- ϕ= "`emp`.*sal* > 1600",
- $\texttt{UPD}(\vec{v}, \vec{exp})$=$\texttt{UPD}\big(\langle age, sal\rangle,\ \langle age+1, sal+:inc \times 2\rangle\big)$

[2] Observe that, for the sake of simplicity, we do not consider here the method `REFRESH`() which synchronize the in-memory objects state with that of the underlying database.

Given the table environment ρ_{t_2} in Fig. 2(b), the semantics is:

$$\begin{aligned}
&\mathbb{S}_{hql}[\![\langle\{\texttt{emp}\},\ (\texttt{emp}.sal > 1600),\ \texttt{UPD}(\langle age, sal\rangle,\ \langle age+1, sal+ : inc \times 2\rangle)\rangle]\!] = \\
&\quad \lambda(e, s, \rho_{t_2}).\ \text{let } \texttt{emp} = \langle \texttt{emp()}, \texttt{F}, \texttt{M}\rangle \text{ such that } \texttt{F} = \langle id, age, dno, sal\rangle \text{ and} \\
&\quad map(\texttt{F}) = attr(t) = \langle tid, tage, tdno, tsal\rangle \text{ and } map(\vec{v}) = map(\langle age, sal\rangle) = \langle tage, tsal\rangle \subseteq attr(t), \\
&\quad \text{and let } \phi_d = (tsal > 1600) = \texttt{PE}[\![(\texttt{emp}.sal > 1600)]\!](e, s, \texttt{F}) \text{ and} \\
&\quad e\vec{x}p_d = \langle tage+1, tsal + 100 \times 2\rangle = \texttt{PE}[\![\langle age+1, sal+ : inc \times 2\rangle]\!](e, s, \texttt{F}) \text{ in} \\
&\quad \{\langle e, s, \rho_{t_3}\rangle \mid \rho_{t_3} \in \mathbb{S}_{sql}[\![\langle \texttt{UPDATE}(\langle tage, tsal\rangle, e\vec{x}p_d),\ \phi_d\rangle]\!](\rho_{t_2})\}.
\end{aligned}$$

Semantics of $\langle \texttt{C}, \phi, \texttt{SEL}(f(e\vec{x}p'),\ r(\vec{h}(\vec{x})),\ \phi',\ g(e\vec{x}p))\rangle$. The semantics of `Session` method $\langle \texttt{C}, \phi, \texttt{SEL}(f(e\vec{x}p'),\ r(\vec{h}(\vec{x})),\ \phi',\ g(e\vec{x}p))\rangle$ is defined as:

$$\begin{aligned}
&\mathbb{S}_{hql}[\![\langle \texttt{C},\ \phi,\ \texttt{SEL}(f(e\vec{x}p'),\ r(\vec{h}(\vec{x})),\ \phi',\ g(e\vec{x}p))\rangle]\!] = \lambda(e, s, \rho_t).\ \text{let } \texttt{C} = \{\langle \texttt{init}_i, \texttt{F}_i, \texttt{M}_i\rangle \mid i = 1, \ldots, n\}, \\
&\quad \text{and } \texttt{F} = \bigcup_{i=1,\ldots,n} \texttt{F}_i, \text{ and } \langle e\vec{x}p'_d,\ \vec{x_d},\ \phi'_d,\ e\vec{x}p_d,\ \phi_d\rangle = \texttt{PE}[\![\langle e\vec{x}p',\ \vec{x},\ \phi',\ e\vec{x}p,\ \phi\rangle]\!](e, s, \texttt{F}), \\
&\quad \text{and let } \rho_{t'} = \mathbb{S}_{sql}[\![\langle \texttt{SELECT}(f(e\vec{x}p'_d),\ r(\vec{h}(\vec{x_d})),\ \phi'_d,\ g(e\vec{x}p_d)), \phi_d\rangle]\!](\rho_t) \\
&\quad \text{and } (e', s') = \bigsqcup_{\forall l_i \in t'} \mathbb{S}_{hql}[\![\texttt{Object()}]\!](s, \texttt{val}(l_i)) \text{ in } \{\langle e', s', \rho_t\rangle\}.
\end{aligned}$$

Observe that $\texttt{val}(l_i)$ converts each tuple $l_i \in t'$ into input values, and $\mathbb{S}_{hql}[\![\texttt{Object()}]\!]\ (s, \texttt{val}(l_i))$ invokes the object constructor `Object()` which creates an object by initializing the fields with $\texttt{val}(l_i)$. This is done for all tuples $l_i \in t'$, resulting in new (e', s').

We skip the semantic definition of Session Methods $\langle\{c\}, false, \texttt{SAVE}(\texttt{obj})\rangle$ and $\langle\{c\}, \phi, \texttt{DEL}()\rangle$ for the sake of space.

Fix-Point Semantics of `Session` Objects. Let `Env` and `Store` be the set of HQL environments and stores respectively. Let $\mathfrak{E}_d$ be the set of database environments. The set of interaction states of `Session` objects is defined below:

Definition 3 (Interaction States of Session Objects). *The set of interaction states of* `Session` *objects is defined by* $\Sigma = \texttt{Env} \times \texttt{Store} \times \mathfrak{E}_d$. *Therefore, an interaction state of a* `Session` *object is a triplet* $\langle e, s, \rho_d\rangle$ *where* $e \in \texttt{Env}$, $s \in \texttt{Store}$ *and* $\rho_d \in \mathfrak{E}_d$.

Because of nondeterministic executions, the transition relation is defined as $\mathscr{T} : \texttt{M}_{ses} \times \Sigma \rightarrow \wp(\Sigma)$ specifying which successor interaction states $\sigma' = \langle e', s', \rho_{d'}\rangle \in \Sigma$ can follow when a `Session` method $m_{ses} = \langle \texttt{C}, \phi, \texttt{op}\rangle \in \texttt{M}_{ses}$ is invoked on an interaction state $\sigma = \langle e, s, \rho_d\rangle$. That is,

$$\mathscr{T}[\![m_{ses}]\!](\langle e, s, \rho_d\rangle) = \{\langle e', s', \rho_{d'}\rangle \mid S[\![m_{ses}]\!](\langle e, s, \rho_d\rangle) \ni \langle e', s', \rho_{d'}\rangle \wedge m_{ses} \in \texttt{M}_{ses}\}$$

We denote a transition by $\sigma \xrightarrow{m_{ses}} \sigma'$ when application of a `Session` method m_{ses} on interaction state σ results in a new state σ'.

Let $\mathcal{I}_0$ be the set of initial interaction states. The semantics of `Session` object $\texttt{obj}_{ses}$ is defined as $\mathbb{T}[\![\texttt{obj}_{ses}]\!](\mathcal{I}_0) = \text{lfp}^{\subseteq}_{\emptyset} \mathcal{F}(\mathcal{I}_0) = \bigcup_{i \leq \omega} \mathcal{F}^i(\mathcal{I}_0)$, where

$$\mathcal{F}(\mathcal{I}) = \lambda\mathcal{T}.\mathcal{I} \cup \{\sigma_0 \xrightarrow{m_0} \ldots \xrightarrow{m_{n-1}} \sigma_n \xrightarrow{m_n} \sigma_{n+1} \mid \sigma_0 \xrightarrow{m_0} \ldots \xrightarrow{m_{n-1}} \sigma_n \in \mathcal{T} \wedge \sigma_n \xrightarrow{m_n} \sigma_{n+1} \in \mathscr{T}\}$$

Method Projected Collecting Semantics. Given a Session object trace $\tau = \sigma_0 \xrightarrow{m_1} \sigma_1 \xrightarrow{m_2} \dots \xrightarrow{m_n} \sigma_n$, where labels m_i $(i = 1, \dots, n)$ denotes Session method $\langle \mathtt{C}_i, \phi_i, \mathtt{op}_i \rangle$. Let $\mathtt{lab}(\tau[i])$ and $\mathtt{State}(\tau[i])$ denote the ith label m_i and the ith state σ_i respectively in a given trace τ. We define the following function which collects all states obtained after performing a specific Session method m_i with an operation op:

$$g[\![\tau]\!](\mathtt{op}) = \{\sigma_i \mid \exists i.\ \mathtt{lab}(\tau[i]) = m_i \text{ with operation } \mathtt{op} \text{ and } \mathtt{State}(\tau[i]) = \sigma_i\}$$

Given a set of traces of Session objects $\mathcal{T}$. The method projection function over $\mathcal{T}$ is defined as:

$$\mathtt{Projection}[\![\mathcal{T}]\!](\mathtt{op}) = \bigcup_{\tau \in \mathcal{T}} g[\![\tau]\!](\mathtt{op})$$

5 Verifying HQL Programs by Lifting Semantics from Concrete to Abstract Domains

As it is usual, in the Abstract Interpretation framework, once the concrete semantics is formulated, it can be lifted to an abstract semantics by simply making correspondence of concrete objects (variables values, object instances, stores, states, traces, etc.) into abstract ones representing partial information on them.

Given the set of concrete interaction states Σ. Let $\mathtt{D}^\sharp$ be an abstract domain representing properties of objects fields and database attributes. The concrete powerset domain $\wp(\Sigma)$ can be over-approximated by the abstract domain $\mathtt{D}^\sharp$ following a Galois connection $\langle \wp(\Sigma), \alpha, \gamma, \mathtt{D}^\sharp \rangle$, where α and γ represent abstraction and concretization function respectively. We denote the abstract version[3] session methods as $\mathtt{m}^\sharp_{ses} ::= \langle \mathtt{C}^\sharp, \phi^\sharp, \mathtt{OP}^\sharp \rangle$, where

$$\mathtt{OP}^\sharp ::= \mathtt{SEL}^\sharp\big(f^\sharp(\vec{exp'}^\sharp),\ r^\sharp(\vec{h}^\sharp(\vec{x}^\sharp)),\ \phi^\sharp,\ g^\sharp(\vec{exp}^\sharp)\big)\ |\mathtt{UPD}^\sharp(\vec{v}^\sharp, \vec{exp}^\sharp)\ |\mathtt{SAVE}^\sharp(\mathtt{obj}^\sharp)\ |\mathtt{DEL}^\sharp()$$

The abstract semantics of $\mathtt{m}^\sharp_{ses}$ is defined in terms of the abstract semantic of $\mathtt{INSERT}^\sharp$, $\mathtt{UPDATE}^\sharp$, $\mathtt{DELETE}^\sharp$, $\mathtt{SELECT}^\sharp$ [9].

Given two abstract states $\sigma^\sharp_1, \sigma^\sharp_2 \in \mathtt{D}^\sharp$, the transition relation in the abstract domain is denoted by $\sigma^\sharp_1 \xrightarrow{\mathtt{m}^\sharp_{ses}} \sigma^\sharp_2$, where the application of $\mathtt{m}^\sharp_{ses}$ on $\sigma^\sharp_1$ results in $\sigma^\sharp_2$. The computation of sound abstract fixed-point trace semantics of session objects in the abstract domain $\mathtt{D}^\sharp$ is straightforward.

A sound abstract projection function "$\mathtt{projection}^\sharp$" on a given set of abstract traces $\mathcal{T}^\sharp$ of session object, similarly, collects all the abstract states obtained after performing session methods $\mathtt{m}^\sharp_{ses}$.

In the following example, we show how this can be applied when considering simple abstract domains like intervals, reduced cardinal product, etc. [5] to over-approximate numerical values.

[3] The apex $\sharp$ represents an abstract version of the elements in the abstract domain.

Example 3. Recall the HQL code (Fig. 1) and the policies from Sect. 1.

Policy 1: *Employees age should be greater than or equal to 18 and less than or equal to 62.*
Policy 2: *The salary of employees with age greater than 30 should be at least 1500 euro.*
Policy 3: *Employees salary should not be more than three times of the lowest salary.*

Verifying Policy 1. Let us consider the domain of intervals INT representing properties of numerical values. Since we are interested only on numerical attribute 'age' in the policy, considering objects state and database state, we choose the abstract domain $\mathtt{D}^\sharp = \mathtt{INT} \times \mathtt{INT}$. Intuitively, an element in $\mathtt{D}^\sharp$ is a tuple $\langle [l_1, h_1], [l_2, h_2] \rangle$ where the first component upper-approximates the values taken by the field 'age' in `emp` class and the second component upper-approximates the values taken by the database attribute 'tage'.

The abstract initial interaction state in the example program is $\sigma^\sharp = \langle \bot, [19, 50] \rangle$ where $\bot$ represents the bottom element in the abstract domain INT. The set of abstract traces of `Session` object 'ses' in the program is $\mathcal{T}^\sharp = \{\ \tau^\sharp\ \}$ $= \{\sigma_0^\sharp \xrightarrow{ses.\mathtt{SAVE}^\sharp()} \sigma_1^\sharp \xrightarrow{ses.\mathtt{UPD}^\sharp()} \sigma_2^\sharp \xrightarrow{ses.\mathtt{SEL}^\sharp()} \sigma_3^\sharp \xrightarrow{ses.\mathtt{DEL}^\sharp()} \sigma_4^\sharp\}$ where $\sigma_1^\sharp = \langle [32, 32], [19, 50] \rangle$ and $\sigma_2^\sharp = \sigma_3^\sharp = \langle [32, 32], [19, 51] \rangle$ and $\sigma_4^\sharp = \langle [32, 32], [19, 50] \rangle$.

According to the abstract projected collecting semantics, we get

$$\begin{aligned}
\mathtt{Projection}^\sharp[\![\mathcal{T}^\sharp]\!](\mathtt{SAVE}^\sharp()) &= \{\sigma_1^\sharp\} = \{\langle [32, 32], [19, 50] \rangle\} \\
\mathtt{Projection}^\sharp[\![\mathcal{T}^\sharp]\!](\mathtt{UPD}^\sharp()) &= \{\sigma_2^\sharp\} = \{\langle [32, 32], [19, 51] \rangle\} \\
\mathtt{Projection}^\sharp[\![\mathcal{T}^\sharp]\!](\mathtt{SEL}^\sharp()) &= \{\sigma_3^\sharp\} = \{\langle [32, 32], [19, 51] \rangle\} \\
\mathtt{Projection}^\sharp[\![\mathcal{T}^\sharp]\!](\mathtt{DEL}^\sharp()) &= \{\sigma_4^\sharp\} = \{\langle [32, 32], [19, 50] \rangle\}
\end{aligned}$$

It is evident that the collecting projected semantics satisfy Policy 1.

Verifying Policy 2. Consider the relational abstract domain of reduced cardinal power $\mathtt{INT}^{\mathtt{INT}}$ where the base INT represents abstract *salary* values in the domain of intervals and the exponent INT represents the abstract *age* values in the domain of intervals [5]. We choose the abstract domain $\mathtt{D}^\sharp = \mathtt{INT}^{\mathtt{INT}} \times \mathtt{INT}^{\mathtt{INT}}$ where the first component in an element of $\mathtt{D}^\sharp$ upper-approximates the values taken by fields 'sal' and 'age' in `emp` class, whereas the second component upper-approximates the values taken by the database attributes 'tsal' and 'tage'. Following the similar method as in the case of Policy 1, it is immediate to say that the abstract projected collecting states on $\mathtt{SAVE}^\sharp()$ may not satisfy the Policy 2 because the base of the second component in $\sigma_1^\sharp$ has lower limit of salary below 1500 euro with valid age interval in the exponent. Hence, a possible policy violation is detected.

Verifying Policy 3. In this policy, since we are interested only on 'salary', we choose the abstract domain $\mathtt{D}^\sharp = \mathtt{INT} \times \mathtt{INT}$. According to the policy, an abstract state $\langle [h_1, k_1], [h_2, k_2] \rangle$ respects the policy if $k_2 < 3 * h_2$. The analysis says that the projected abstract collecting state $\sigma_2^\sharp$ on $\mathtt{UPD}^\sharp()$ may not satisfy Policy 3. Therefore in this case also a possible policy violation is detected by the analysis.

6 Conclusions

The contribution in this paper is not only the extension of Abstract Interpretation of object-oriented languages to the case of HQL, but also an interesting example of combination of concrete/abstract semantics of different languages for verification purposes. This generic framework can have many applications, *e.g.* formal verification of security issues like database access control, specification-based slicing of HQL programs, language-based information-flow analysis, *etc.*

References

1. Bauer, C., King, G.: Hibernate in Action. Manning Publications Co., Greenwich (2004)
2. Bauer, C., King, G.: Java Persistence with Hibernate. Manning Publications Co., Greenwhich (2006)
3. Bouaziz, M., Logozzo, F., Fähndrich, M.: Inference of necessary field conditions with abstract interpretation. In: Jhala, R., Igarashi, A. (eds.) APLAS 2012. LNCS, vol. 7705, pp. 173–189. Springer, Heidelberg (2012)
4. Chen, L., Miné, A., Cousot, P.: A sound floating-point polyhedra abstract domain. In: Ramalingam, G. (ed.) APLAS 2008. LNCS, vol. 5356, pp. 3–18. Springer, Heidelberg (2008)
5. Cortesi, A., Costantini, G., Ferrara, P.: A survey on product operators in abstract interpretation. EPTCS **129**, 325–336 (2013)
6. Cortesi, A., Halder, R.: Abstract interpretation of recursive queries. In: Hota, C., Srimani, P.K. (eds.) ICDCIT 2013. LNCS, vol. 7753, pp. 157–170. Springer, Heidelberg (2013)
7. Cousot, P., Cousot, R.: Abstract interpretation: a unified lattice model for static analysis of programs by construction or approximation of fixpoints. In: Proceedings of the POPL 1977. pp. 238–252. ACM Press, Los Angeles, CA, USA (1977)
8. Elliott, J., O'Brien, T., Fowler, R.: Harnessing Hibernate, 1st edn. O'Reilly Media, Sebastopol (2008)
9. Halder, R., Cortesi, A.: Abstract interpretation of database query languages. Comput. Lang. Syst. & Struct. **38**, 123–157 (2012)
10. Logozzo, F.: Class invariants as abstract interpretation of trace semantics. Comput. Lang. Syst. & Struct. **35**, 100–142 (2009)
11. Logozzo, F.: Practical verification for the working programmer with codecontracts and abstract interpretation. In: Jhala, R., Schmidt, D. (eds.) VMCAI 2011. LNCS, vol. 6538, pp. 19–22. Springer, Heidelberg (2011)
12. O'Neil, E.J.: Object/relational mapping 2008: hibernate and the entity data model (edm). In: Proceedings of the 2008 ACM SIGMOD International Conference on Management of Data (SIGMOD 2008). pp. 1351–1356. ACM, New York, USA (2008)
13. Wiśniewski, P., Stencel, K.: Universal query language for unified state model. Fundam. Inform. **129**(1–2), 177–192 (2014)

An Efficient MapReduce Framework for Intel MIC Cluster

Wenzhu Wang(✉), Qingbo Wu, Yusong Tan, and Yaoxue Zhang

College of Computer, National University of Defense Technology, Changsha, China
{wenzhuw,wu.qingbo2008,yusong.tan}@gmail.com,
cszyx@tsinghua.edu.cn

Abstract. MapReduce is a distributed programming framework to process large scale data set by employing clusters in scale-out ways. However, scaling-up the single node is better than scale-out solution because of less communication overhead. As Intel MIC has a higher performance than ordinary CPU, we propose an efficient MapReduce framework for Intel MIC cluster. Our framework provides several new features, such as fault tolerant mechanism for MIC management, efficient buffer management in MIC memory, and asynchronous task transfer between CPU and MIC. It could manage a large scale MIC cluster and exploit applications in MapReduce like ways. The experimental results show that our system is up to 1.35x and 6.8x faster than Hadoop on ordinary CPU cluster.

Keywords: MapReduce · Hadoop · Many integrated core · Big data

1 Introduction

With the explosion of abundant data in many areas, Big Data has become a hot topic in recent years. However, a challenge on both hardware infrastructure and software is to dig the huge value from large scale data set in an acceptable time. To deal with this urge, MapReduce framework [1] is proposed to process huge data set in parallel by utilizing large scale clusters. Moreover, the open-source implementations, such as Hadoop and Spark, have been widely used in internet enterprises and research communities.

The original goal of MapReduce is to employ common computer clusters in scale-out ways. However, study [2] has claimed that scale-up way is better than scale-out in terms of performance, cost, power and server density. However, scale-up a node by general purpose processor has encountered an obstacle: the frequency of a single processor has been almost impossible to improve due to the power consumption and transistor density, and increasing the number of cores also incurs the system complexity and processor cost.

As a result, we have witnessed a high speed development of coprocessors, such as GPU, FPGA, and coupled CPU-GPU, of which the developing speed are even faster than Moore's law. Moreover, several studies have implemented MapReduce framework on single coprocessor [3] or coprocessor clusters [4]. Many Integrated

X. He et al. (Eds.): IScIDE 2015, Part II, LNCS 9243, pp. 129–139, 2015.
DOI: 10.1007/978-3-319-23862-3_13

Core (MIC) is a new kind of coprocessor architecture designed by Intel Corporation, which has great computing power and friendly programming models [5,6]. There is also study about MapReduce framework on MIC architecture [7].

In this paper, we try to utilize the advantages of both MapReduce and MIC cluster to process large scale data set. As far as we know, this is the first study about MapReduce framework for Intel MIC cluster. Different from the ordinary CPUs or GPUs, MIC has many new characteristics, such as wider Vector Process Units (VPU), hundreds of threads, and PCIe data transfer. Therefore, we find several challenges during the system design and implementation, such as using MIC efficiently, communication overhead, fault tolerance, and system scalability, etc. We try to address these challenges and mainly make the following contributions.

- For using MIC efficiently, we provide several user APIs, such as buffer allocation, task transfer, and hash combine. Moreover, we relax the constraint of map/reduce function, therefore users could optimize the MIC codes further according to the specific application.
- To reduce the size of data storage and transfer, we design a Hash Combine stage on MIC to combine the <key, value> pairs with the same key between all threads. The hash combine operation uses low cost synchronous operations and efficient SIMD hash algorithm to make the key indexing O(1) time complexity.
- We design a token mechanism for cluster scalability and fault tolerance, which could avoid multiple tasks conflicting on using MICs within the same node, handle MIC failure, and adapt MIC dynamical changing in the cluster.
- An efficient memory management scheme is designed for MIC memory allocation, in which multiple types of buffer can be used by users. The Independent Buffer has no synchronization overhead but a large memory space, the Global Buffer has a low synchronization overhead but a small memory space, and the task buffer is used for storing tasks. All buffers are scalable for multi-threads and reused during the job running.
- Finally, we develop a asynchronous task transfer method for hiding parts of the communication overhead between CPU and MIC, which could improve the system overall efficiency further.

A prototype system is implemented on a MIC cluster with 4 computing nodes and 6 MIC coprocessors. To compare the performance of our system, we also build up a CPU cluster with 8 CPUs, and run three common MapReduce test programs on MIC cluster and CPU cluster respectively. The experimental results show that our system is up to 1.35x and 6.8x faster than Hadoop running on the ordinary CPU cluster.

The rest of this paper is organized as follows: The MIC architecture and MapReduce framework are introduced in Sect. 2. Section 3 gives the details of system design and implementation. Section 4 evaluates the performance of our system compared with Hadoop running on a CPU cluster. Finally, we conclude this paper in Sect. 5.

2 Background

2.1 Intel MIC Architecture

Many Integrated Core (MIC) architecture has 50+ x86 cores, and each core supports four hardware threads running on a dual in-order execution pipeline. The multiple threads running in round-robin manner could hide the latencies of in-order instruction execution. There are 512-bit width Vector Processing Units (VPUs) for SIMD (Single Instruction Multi Data) processing on each core. In addition, each core has 32K L1 data cache, 32K L1 instruction cache, and 512 KB L2 cache. The high speed bi-directional ring connecting all L2 caches could improve the cache efficiency by trying to access data from other core's L2 cache directly. Because of no L3 cache, the penalty of cache miss is as high as about 200 cycles. However, it has high memory bandwidth and memory speed, which are about 350 GB/s and 5 GT/s.

MIC mainly supports two programming models: native and offload. In native model, MIC can be seen as an independent SMP computing system. The software stack provides many interface supports standard APIs, such as TCP/IP and MPI communication interface. In offload model, MIC works as a coprocessor and connects with CPU via PCIe bus. The data and program transferred to MIC are initiated and controlled by host CPU. Because CPU and MIC share different memory address space, programmers need to manage two copies of data on both sides. Both of the programming models need to utilize the multithread program models, such as OpenMP, TBB, Cilk, and OpenCL.

2.2 MapReduce

MapReduce is a distributed programming framework for processing large scale data set in ordinary CPU clusters. In this programming model, developers mainly need to write two serial functions: map and reduce. Other works, such as data distribution, parallel processing, fault tolerance, and load balance, are handled by MapReduce runtime in transparent manners.

The map and reduce functions are both defined with restricted data structure of <key, value> pairs. Firstly, the Map function maps a pair of input data to another data domain:

Map: <k1, v1> → <k2, v2>

After map stage, the runtime shuffles all the intermediate results and groups the <key, value> pairs according to the value of key. Then, the Reduce function is executed for each group and produces a collection of values in the same key:

Reduce: <k2, list(v2)> → list(v3)

With the rapid development of coprocessors, many studies for adapting MapReduce framework on CPU-coprocessor architecture have emerged. Researches [3,8–10] optimize the MapReduce framework on single GPU coprocessor, and [4,11–13] leverage GPU clusters to run MapReduce for large scale data processing. There are also similar studies on other processors, like multi-core system [14,15], IBM Cell [16], and Intel MIC [7].

3 Design and Implementation

In this section, we introduce the design and implementation of our system. The goal of our system is to utilize the computing power of multiple MICs in a cluster to process large scale data set in efficient, fault tolerant, and easy programming ways.

3.1 The Overall Workflow

Like other MapReduce frameworks, the overall workflow of our system includes task distribution, map, shuffle, and reduce, etc. However, the most difference is that each task is processed by MIC instead of CPU or GPU. Figure 1 shows the overall workflow of map stage. Firstly, a new task is distributed to a computing node by Job Manager based on the distributed file system. Then, the computing node transfers the task to an idle MIC coprocessor via PCIe bus. The MIC processes the new task by multi-threads and SIMD in parallel, stores the results in MIC memory, and transfers the results back to computing node finally. As MIC has an outstanding computing power, a task should be large enough to fill up the computing units in MIC for getting desirable speedups. Therefore, different from the ordinary MapReduce tasks, our tasks usually contain one or more input files. The computing node reads all data from the input file set and stores them to the task buffer in host memory, then transfers all data from task buffer to MIC memory.

To use MIC efficiently and easily, we provide several user APIs, such as buffer allocation, task transfer, and hash combine, which will be introduced in details in the following subsections. Furthermore, we relax the constraint of map/reduce function, so users can optimize the MIC codes further according to the specific application. For example, users can set the thread affinity, leverage high-level MIC instructions, and optimize the SIMD operations.

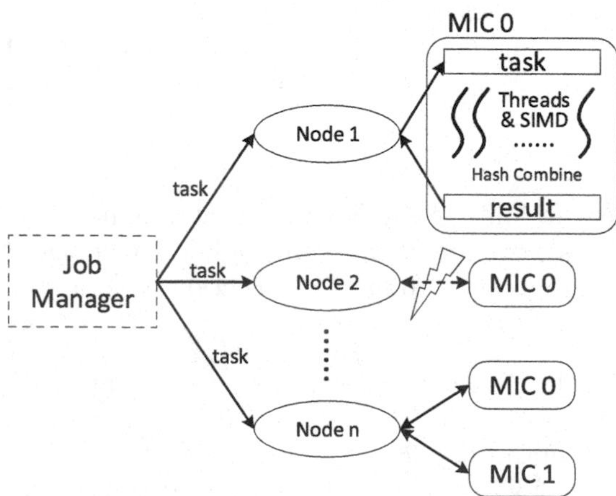

Fig. 1. The workflow of map stage

During the map stage, the intermediate results are stored in MIC memory temporarily. In order to reduce the memory usage, we design a Hash Combine stage on MIC to combine the <key, value> pairs with the same key between all threads. When a <key, value> pair is produced, Hash Combine calculates the hash value of the key by an efficient SIMD hash algorithm, which could utilize the VPUs on each core efficiently. The hash value can locate the storage location with an O(1) time complexity. Then, the values with the same key are combined by combine function.

Another challenge is the system scalability and fault tolerance. As Fig. 1 illustrated, there are different number of MICs on each node (node 1 and node 2 has one MIC, and node 3 has two MICs), and some MICs may be failed by some reasons, such as PCIe bus failed or offload service stopped (node 2). To handle this situation, we design a token mechanism. Each node uses the heartbeat-based communication mechanism to probe the usability of MICs, and give each MIC a unique token. A task must get a MIC token before running to avoid conflicting on the same MIC. Moreover, when a MIC is failed, the according token is also deleted. Furthermore, the token could be changed dynamically with the MIC coprocessors changing in cluster.

After map stage, the system runtime collects all intermediate results and stores them in host memory or hard disk. If there is a reduce stage, the system distributes all reduce tasks among computing nodes evenly for avoiding data skew, and collects all reduce results as the final output results. The workflow of reduce stage is similar to the map stage as Fig. 1 illustrated.

3.2 Memory Management Scheme

In the map stage, all threads running on MIC store their intermediate <key, value> pairs in MIC memory simultaneously. Therefore, the synchronization overhead is a potential bottleneck in this process. To handle this challenge, we design two types of result buffer for storing results efficiently: Independent Buffer and Global Buffer.

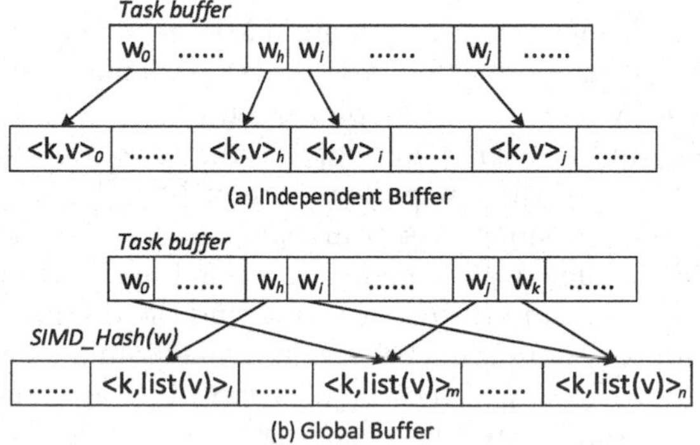

Fig. 2. MIC memory management

In the Independent Buffer, each thread has an independent memory space to store their results as Fig. 2 (a) illustrated. Each element in task w_i has a unique position to store its result, and the according result is stored as <key, value>$_i$. Although Independent Buffer occupies a large space, it has no locks or atomic operations for threads synchronization. Independent Buffer is applicable when there is no combining operations between all intermediate results, or combining operation can't save much memory space, like K-Means and Matrix Multiplication.

In the Global Buffer, all threads on MIC share the same memory space for results storing as Fig. 2 (b) illustrated. When using this buffer, multiple threads may access the same position at the same time. Therefore, we use a low cost synchronous operation on the Global Buffer, which is well supported on MIC architecture and doesn't have significant overhead. Furthermore, the probability of accessing the same position at the same time is very low in most applications, such as language and logs processing, as the range of hash value is very large. In Fig. 2 (b), w_i and w_k have the same hash value, and their values are combined into the <k,list(v)>$_n$. In brief, Global Buffer is applicable when combining can save much memory space and have a low probability of conflict on hash value.

Users could choose the type of result buffer in terms of the specific application. In addition, MIC also has a task buffer to store the tasks transferred from computing node. The task buffer is 64-bit aligned, which could get a high performance based on MIC architecture. To avoid massive requests for memory allocation, all buffers are page-locked and reused during job running. All tasks in a job are processed iteratively, and the intermediate/final results are stored in host memory or hard disk, so the data set processed by our system is not limited by the size of MIC memory or host memory.

3.3 Asynchronous Task Transfer

During the job running, tasks and results are transferred between CPU and MIC iteratively. The common method in a task processing cycle can be described as four parts as Fig. 3 (a) illustrated. (1) CPU processes the results produced by MIC and prepares a new task; (2) CPU transfers the task to MIC memory; (3) MIC processes the task and stores the results in MIC memory; (4) MIC transfers the results back to host memory. This process is synchronous, which means each part is executed one by one. In a task processing cycle, task transfer takes up half of the cycle: (2) and (4). Therefore, optimizing task transfer is meaningful for system efficiency due to the low performance of PCIe bus.

We develop an asynchronous task transfer method as Fig. 3 (b) illustrated. There are two results buffer in host memory: one is used for storing the results transferred from MIC side, and the other one storing the results of former task is used for CPU Compute. Therefore, the Results Transfer and CPU Compute can be executed at the same time, which could hide part of the Results Transfer time. Similarly, we could also allocate two task buffers in MIC memory and overlap the Task Transfer with MIC Compute. However, we found that the former one

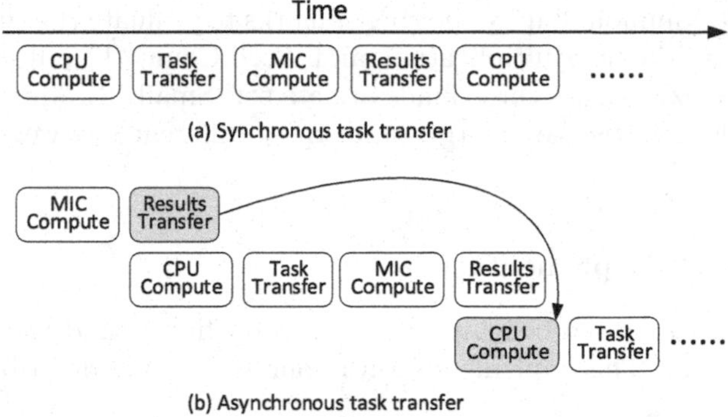

Fig. 3. Task transfer between CPU and MIC

has a better performance due to more appropriate overlap, and allocating buffers in host memory is more efficient than in MIC memory.

4 Experimental Evaluation

In this section, we evaluate the performance of our system in a MIC cluster. Firstly, we introduce the experimental setup, including hardware platform configuration and testing programs. Then, we show the comparing results of our system with Hadoop running on an ordinary CPU cluster.

4.1 Experimental Setup

We deploy a MIC cluster and an ordinary CPU cluster respectively. The MIC cluster includes 4 computing nodes and 6 MICs in total. The MIC coprocessor is Intel Xeon Phi 3120P, which has 6 GB memory, 57 cores working at 1.1 GHz, and 512-bit VPUs on each core. In contrast, the CPU cluster includes 4 computing nodes and 8 CPUs. The CPU is Intel Xeon E5-2670, which has 8 cores working at 2.6 GHz.

Others configuration: The host memory is 136G, and the hard drive is a 500G SATA3 magnetic hard disk. The 64 bit RedHat 6.2 is used as the operating system, and Intel composer_xe_2013 is used as the compile environment.

Table 1. The data size used for applications.

Application	Data sets		
K-Means	S:240M	M:2.4G	L:24G
Matrix Multiplication	S:240M	M:2.4G	L:24G
Word Count	S:680M	M:6.8G	L:68G

We use three common MapReduce benchmarks to evaluate the system performance: K-Means, Matrix Multiplication, and Word Count. To reflect the impact of different data size, we use three kinds of data set: Small (S), Medium (M) and Large (L). Table 1 shows the description of these different size we used for each application.

4.2 Benchmark Implementation

We implement these three benchmarks in Hadoop firstly, and then implement them in our system. The Implementation in our system are described briefly as follows.

K-Means (KM): K-Means clustering is a method of grouping thousands of items into k groups. Each map task contains n items. The map task calculates the distance from each item to all central nodes of the k groups, then finds the group with the minimum distance to the given item. The reduce task sums all items within the same group and recalculates the central node. The map and reduce stage won't stop until all the central nodes aren't changed anymore or reach the iteration number predefined. Combining the items within the same group after map stage can't save much memory space, therefore we use the Independent Buffer for intermediate results storing.

Matrix Multiplication (MM): Matrix Multiplication is a binary operation that a pair of matrices produces another matrix, which has been widely used in eigenvalue analysis. In a map task, m rows from the first matrix and n columns from the second matrix are multiplied, then all the outputs are summed up as the final result. The reduce stage isn't required in this application. There is no combine stage, so we use the Independent Buffer for results storing.

Word Count (WC): Word Count counts the number of each word in a collection of document files. A map task contains a set of data read from the input files, and gives each key a value of 1. The same keys in the intermediate results could be combined after map stage. The reduce task counts the number of each word collected from the map stage. Because there is a combining stage, we use the Global Buffer and Hash Combine to store the intermediate results.

4.3 Experimental Results

Firstly, we compare our system running on MIC cluster with Hadoop running on CPU cluster. The experimental results are illustrated in Fig. 4. Overall, our system could achieve up to 1.35x and 6.8x speedups over Hadoop based on different applications and data sizes. Specifically, KM and MM have higher speedups than WC. The main reason is that KM and MM have more SIMD operations to utilize MIC better. However, WC is still able to take advantage of multiple threads on MIC and get a desirable speedup. Moreover, the speedups are increased with the increasing data size due to the more utilization of MIC.

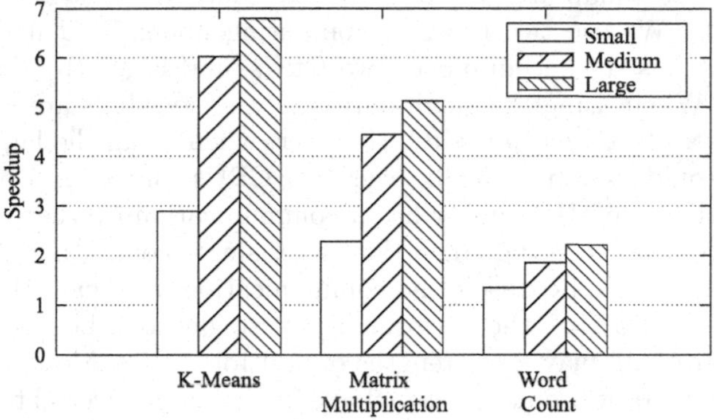

Fig. 4. Speedup over CPU cluster

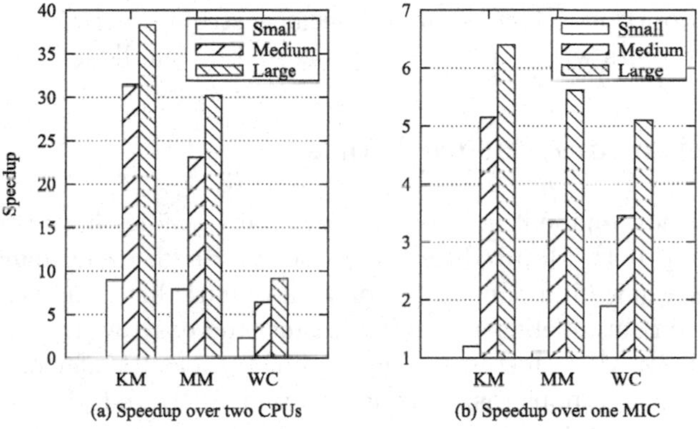

Fig. 5. Speedup over CPU and MIC

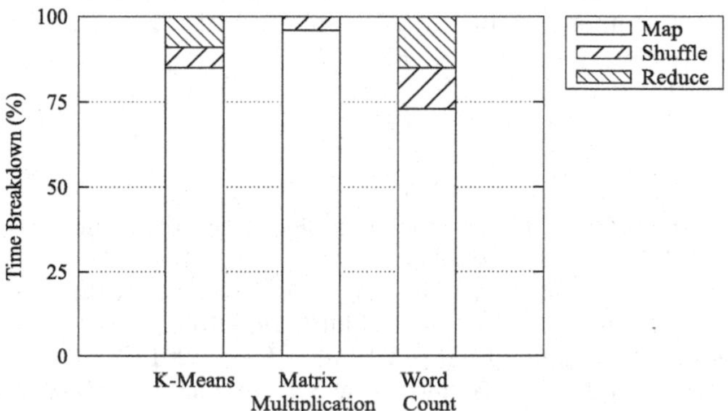

Fig. 6. Time breakdown

To evaluate the scalability of our system, we compare our system with single computing node. We use two types of computing node: CPU node and MIC node. Figure 5 (a) is the speedup over two CPUs, and Fig. 5 (b) is the speedup over one MIC. When using the small data size, the speedups are less than 10x over two CPUs and 2x over one MIC. However, when using the large data size, the speedups could be up to 38.3x over two CPUs and 6.5x over one MIC. The reason is that the MIC cluster has a communication overhead between all computing nodes. When using small data size, single computing node can get a desirable performance because of no communication overhead. However, with the increasing of data size, the performance of single node begins to decrease, and MIC cluster could play the strengths of multiple MICs. Moreover, from the results in (a) and (b), we could also see that MIC is more powerful than ordinary CPUs, as our system gets a higher speedup over CPUs than MIC.

Finally, we illustrate the time breakdown of these benchmarks in Fig. 6. For KM and MM, the main time is in Map stage, as these two applications have much computing in map stage. In contrast, for WC, the shuffle and reduce also occupies a large time proportion, which is mainly because string processing produces lots of <key, value> pairs which need to be processed by shuffle and reduce stage.

5 Conclusion and Future Work

With the rapid development of coprocessors, many studies have tried to utilize the advantages of both MapReduce framework and high-performance coprocessors. As MIC is a new kind of coprocessor with remarkable computing power, we design and implement an efficient, fault tolerant, and easy programming MapReduce framework for MIC cluster. Our system has efficient buffer management in MIC memory, asynchronous task transfer between CPU and MIC, and fault tolerant mechanism for MIC management. The experimental results have shown that our system has an exciting performance on MIC cluster compared with Hadoop on ordinary CPU cluster.

The future work includes improving the system programmability, optimizing MIC memory management, reducing the communication overhead further, and scaling our system on larger MIC clusters.

References

1. Dean, J., Ghemawat, S.: MapReduce: simplified data processing on large clusters. Commun. ACM **51**(1), 107–113 (2008)
2. Appuswamy, R., Gkantsidis, C., Narayanan, D., Hodson, O., Rowstron, A.: Scale-up vs scale-out for hadoop: time to rethink?. In: Proceedings of the 4th Annual Symposium on Cloud Computing, p. 20. ACM Press (2013)
3. He, B., Fang, W., Luo, Q., Govindaraju, N.K., Wang, T.: Mars: a MapReduce framework on graphics processors. In: Proceedings of the 17th International Conference on Parallel Architectures and Compilation Techniques, pp. 260–269. ACM Press, Toronto (2008)

4. Stuart, J.A., Owens, J.D.: Multi-GPU MapReduce on GPU clusters. In: 25th IEEE International Parallel & Distributed Processing Symposium, pp. 1068–1079. IEEE Press, Anchorage, Alaska (2011)
5. Heinecke, A., Klemm, M., Pflger, D., Bode, A., Bungartz, H.J.: Extending a highly parallel data mining algorithm to the intel many integrated core architecture. In: Alexander, M., et al. (eds.) Euro-Par 2011: Parallel Processing Workshops. LNCS, vol. 7156, pp. 375–384. Springer, Heidelberg (2012)
6. Schulz, K.W., Ulerich, R., Malaya, N., Bauman, P.T., Stogner, R., Simmons, C.: Early experiences porting scientific applications to the Many Integrated Core (MIC) platform. In: TACC-Intel Highly Parallel Computing Symposium. Austin, Texas (2012)
7. Lu, M., Zhang, L., Huynh, H. P., Ong, Z., Liang, Y., He, B., Huynh, R.: Optimizing the mapreduce framework on intel xeon phi coprocessor. In: International Conference on Big Data, pp. 125–130. IEEE Press, Santa Clara, California (2013)
8. Basaran, C., Kang, K.D.: Grex: an efficient MapReduce framework for graphics processing units. J. Parallel Distrib. Comput. **73**(4), 522–533 (2013)
9. Hong, C., Chen, D., Chen, W., Zheng, W., Lin, H.: MapCG: writing parallel program portable between CPU and GPU. In: Proceedings of the 19th International Conference on Parallel Architectures and Compilation Techniques, pp. 217–226. ACM Press, Vienna (2010)
10. Chen, L., Huo, X., Agrawal, G.: Accelerating mapreduce on a coupled cpu-gpu architecture. In: International Conference for High Performance Computing, Networking, Storage and Analysis, p. 25. IEEE Press, Salt Lake, Utah (2012)
11. Farivar, R., Verma, A., Chan, E.M., Campbell, R.H.: Mithra: Multiple data independent tasks on a heterogeneous resource architecture. In: IEEE International Conference on Cluster Computing, pp. 1–10. IEEE Press, New Orleans, Louisiana (2009)
12. Chen, Y., Qiao, Z., Jiang, H., Li, K.-C., Ro, W.W.: MGMR: Multi-GPU based MapReduce. In: Park, J.J.J.H., Arabnia, H.R., Kim, C., Shi, W., Gil, J.-M. (eds.) GPC 2013. LNCS, vol. 7861, pp. 433–442. Springer, Heidelberg (2013)
13. Fang, W., He, B., Luo, Q., Govindaraju, N.K.: Mars: accelerating mapreduce with graphics processors. IEEE Trans. Parallel Distrib. Syst. **22**(4), 608–620 (2011)
14. Ranger, C., Raghuraman, R., Penmetsa, A., Bradski, G., Kozyrakis, C.: Evaluating mapreduce for multi-core and multiprocessor systems. In: IEEE 13th International Symposium on High Performance Computer Architecture, pp. 13–24. IEEE Press, Phoenix, Arizona (2007)
15. Talbot, J., Yoo, R.M., Kozyrakis, C.: Phoenix++: modular MapReduce for shared-memory systems. In: Proceedings of the Second International Workshop on MapReduce and its Applications, pp. 9–16. ACM Press, San Jose, California (2011)
16. de Kruijf, M., Sankaralingam, K.: MapReduce for the Cell BE architecture. University of Wisconsin Computer Sciences Technical report CS-TR-2007-1625 (2007)

A Novel K-harmonic Means Clustering Based on Enhanced Firefly Algorithm

Zhiping Zhou(✉), Shuwei Zhu, and Daowen Zhang

School of Internet of Things Engineering, Jiangnan University, Wuxi, China
zzp@jiangnan.edu.cn, {zhushuwei32, zdwjndx}@gmail.com

Abstract. The K-harmonic means is a center based clustering algorithm, which is suffering from falling into local optima easily. To solve this problem, a hybrid K-harmonic means algorithm using enhanced firefly algorithm is proposed. Combining with parallel chaotic optimization, a novel chaotic local search method which has the capability of full dimensional and part dimensional searching is used to enhance the original firefly algorithm. Then this enhanced version of firefly algorithm is integrated into K-harmonic means algorithm to take full advantage of merits of both algorithms. The proposed method is compared with KHM and other two hybrid algorithms on four data sets, and the experiment results show the superiority of it that can escape from local optima effectively.

Keywords: K-harmonic means (KHM) · Clustering · Firefly algorithm (FA) · Parallel chaotic optimization · Chaotic local search

1 Introduction

Cluster analysis is a method of creating groups of objects based on similarity degrees of relevant features, which has been used in many areas such as data mining, pattern recognition and machine learning. K-means is the most popular partitioned clustering algorithm, and it has a rich and diverse history in the literature. Ease of implementation, efficiency and simplicity are the main reasons of its popularity, but it also has two main shortages including be sensitive to the selection of the initial cluster centers and easily run into local optima [1]. As an improved version of K-means, the K-harmonic means (KHM) algorithm is proposed by Zhang [2] in 1999, and modified by Hamerly [3] to define a general model. The KHM algorithm is not sensitive to the initial centers, but also suffers from the problem of running into local optima.

Recently many researchers have proposed different types of improved algorithms based on KHM, aiming at enhancing its clustering performance. Among them, the most popular methods are the combination of KHM with other heuristic algorithms, which can take full advantage of global search ability of heuristic algorithm and local search ability of KHM. Earlier Güngör and Ünler respectively combined KHM with simulated annealing (SA) [4] and Tabu-search (TS) [5] to build two efficient clustering algorithms. Afterwards, some other more efficient hybrid clustering algorithms were proposed which can get better performance. Such as combining KHM with Particle Swarm Optimization (PSO) [6], Ant Swarm Optimization (ACO) [7], Variable neighborhood

X. He et al. (Eds.): IScIDE 2015, Part II, LNCS 9243, pp. 140–149, 2015.
DOI: 10.1007/978-3-319-23862-3_14

search (VNS) [8], gravitational search algorithm (GSA) [9], Candidate groups search (CGS) [10] and modifier imperialist competitive algorithm (ICA) [11]. However, there is no comprehensive comparison among them, and it is known that there is no algorithm which can achieve better results than others for all data sets. Firefly algorithm (FA) is a novel swarm-intelligence method developed by Yang [12] in 2008, and it is a kind of heuristic algorithm that can be applied to solve many hard optimization problems. Many researchers have adopted FA in practical use, and it has already been proved to be useful in cluster analysis [13]. Adaniya also proposed a hybrid data clustering algorithm based on KHM and FA [14], called firefly harmonic clustering algorithm (FHCA), which was used to cluster the Digital Signature of Network Segment data and the network traffic samples. However, the concrete experiment analysis of FHCA was not given, and the superiority of it to KHM cannot be visually understood.

According to the above analysis, a novel hybrid K-harmonic means algorithm based on enhanced firefly algorithm is proposed in this paper. An efficient chaotic local search(CLS) method using parallel chaotic optimization(PCO) [15] is build to enhance the ability of FA to jump out of local optimal and find out global optimal. Then, the enhanced FA is integrated into KHM and experiment results for all the four data sets indicate that the proposed algorithm can produce accurate and stable clusters, and also escape from local optima effectively.

2 K-Harmonic Means Clustering

The KHM algorithm, similar to K-means, is a center based partitioned clustering algorithm, but the most obvious difference is that it adopt harmonic means(HM) instead of arithmetic means to calculate the objective function. Because of HM's characteristics of minimum deviation within groups and maximum deviation between groups, KHM algorithm is not sensitive to the initial centers and has better clustering performance than K-means. Assuming that a data set $\boldsymbol{X} = (\boldsymbol{x}_1, \boldsymbol{x}_2, ..., \boldsymbol{x}_N)$, $\boldsymbol{x}_i = (x_{i1},..., x_{im})$, which has N objects and each object contains m attributes, is partitioned into K clusters and the result can be defined as a set $P_K = \{C_1, C_2, ..., C_K\}$ in which C_j denotes jth cluster with the center $\boldsymbol{c}_j$. The objective function of the KHM is shown as Eq. (1):

$$KHM(X, C) = \sum_{i=1}^{N} \frac{K}{\sum_{j=1}^{K} 1/\|\boldsymbol{x}_i - \boldsymbol{c}_j\|^p}, \forall i = 1, \cdots, N. \tag{1}$$

where p is an important input parameter for the algorithm and typically $p \geq 2$ [2].

During clustering process, the objective function KHM(X,C) is minimized and keep steady until the end of the iteration, for each data point $\boldsymbol{x}_i$, its membership $m_{KHM}(\boldsymbol{c}_j/\boldsymbol{x}_i)$ and weight $w_{KHM}(\boldsymbol{x}_i)$ can be calculated according to Eq. (2).

$$m_{KHM}(\boldsymbol{c}_j/\boldsymbol{x}_i) = \frac{\|\boldsymbol{x}_i - \boldsymbol{c}_j\|^{-p-2}}{\sum_{j=1}^{K} \|\boldsymbol{x}_i - \boldsymbol{c}_j\|^{-p-2}}, w_{KHM}(\boldsymbol{x}_i) = \frac{\sum_{j=1}^{K} \|\boldsymbol{x}_i - \boldsymbol{c}_j\|^{-p-2}}{(\sum_{j=1}^{K} \|\boldsymbol{x}_i - \boldsymbol{c}_j\|^{-p})^2} \tag{2}$$

Then the center $c_j(j = 1,...,K)$ can be updated according to Eq. (3), and data point x_i is assigned to cluster j with the biggest $m_{KHM}(c_j/x_i)$.

$$c_j^{new} = \frac{\sum_{i=1}^{N} m_{KHM}(c_j/x_i) \times w_{KHM}(x_i) \times x_i}{\sum_{i=1}^{N} m_{KHM}(c_j/x_i) \times w_{KHM}(x_i)} \tag{3}$$

3 The Enhanced Firefly Algorithm

3.1 Original Firefly Algorithm

In the FA, the population individuals attract each other that primarily depend on two factors: the brightness and the attractiveness, the former determines the quality of each individual's position and moving direction, and the latter determines the moving distance. During the search process, brightness and attractiveness will be constantly updated. Usually the brightness of firefly i can be denoted by the objective function value I_i directly, as $I_i = f(x_i)$, $x_i = (x_{i1},...,x_{id})$. The attractiveness β between individual i and j which mainly depends on their space distance r_{ij} can be calculated according to the following generalized equation:

$$\beta = \beta_0 \times e^{-\gamma r_{ij}^2} \tag{4}$$

where β_0 denotes the attractiveness at $r = 0$ which can be set as 1 for most problems, γ is the light absorption coefficient which can be taken as a constant, and r_{ij} is generally the Euclidean distance.

The movement of a firefly i at location x_i that is attracted to another more brighter firefly j at location x_j is determined by Eq. (5).

$$x_i^{new} = x_i + \beta \times (x_j - x_i) + \alpha \times \varepsilon_i \tag{5}$$

where α is a randomization parameter in the interval (0,1) and ε_i is a vector of random numbers which are drawn from a Gaussian distribution.

3.2 Parallel Chaotic Local Search Firefly Algorithm

Even though FA has been proved to be an efficient optimization algorithm, the ability of it for both local and global exploration can be improved further, as well as the possibility of trapping into local optima can be reduced, hence a new modified version of FA is proposed. Chaos optimization algorithm (COA) which adopts chaotic sequences instead of random sequences is an efficient method to provide the search diversity in the optimization procedure, and it has the features of easy implementation, short execution time and robust mechanisms of escaping from the local optima. The research about chaotic sequences in the existing literature is very extensive, such as logistic map, sinusoidal map, gauss map and so on. An improved logistic map was adopted for chaotic local search(CLS) and incorporated into PSO and difference evolution(DE) respectively to solve the short-term scheduling problem of cascade hydropower system successfully in [16, 17]. It is formulated as:

$$z(l+1) = 1 - 2(z(l))^2,\ z(l) \in (-1,1) \tag{6}$$

where l is the iteration number, and it deserves to notice that $z(0) \notin \{0.25, 0.5, 0.75\}$. The probability density distribution of improved logistic map is shown as follows:

$$f(z) = \begin{cases} 1/\pi\sqrt{1-z^2}, & z \in (-1,1) \\ 0, & z \notin (-1,1) \end{cases} \tag{7}$$

It can be seen from Eq. (7) that the search space of chaotic variables is extended to symmetrical region (-1,1) by improved logistic map, and larger values of probability density around the border -1 and 1 indicate that larger chaotic variables with high probability can be generated in (-1,1). As a result of increasing chaotic local search region, it is suitably utilized to enhance the search ability of FA. Firstly, FA is executed to obtain the current optimal solution $\boldsymbol{x}_{pg}$, as it is obvious that the region around $\boldsymbol{x}_{pg}$ is most favorable to find the global optima. Thus the modified version of FA utilizes an extra procedure to search around $\boldsymbol{x}_{pg}$ directly, and this method is in nature the chaotic disturbance approach which generates many neighborhood points of local optimal solution in the process of optimization to increase the probability of finding the global optimal solution. Meanwhile, N different chaotic variables nearby $\boldsymbol{x}_{pg}$ are generated in each iteration based on the principle of PCO, in order to overcome the disadvantages of traditional COA with serial mechanism that is not good at searching for precise solution and converging stably.

Among the existing CLS methods, the vast majority are searching for all dimensions of variable space, letting every dimension to change. However, some dimensions of $\boldsymbol{x}_{pg}$ have already been in relatively good position during certain stages and there is no need to change their location. So far, no good method is built to determine whichever specific dimension that is located close to the global optima. In this paper, a part dimensional CLS strategy of selecting each dimension with an equal probability is proposed, and it combines with full dimensional CLS strategy to establish a highly efficient parallel chaotic local search (PCLS) method. During each iteration of PCLS, the optimal solution in parallel variables is compared to $\boldsymbol{x}_{pg}$ to judge whether it should be replaced or not, which improves the ability of local exploration and avoids the premature convergence effectively. The detailed procedure of the modified firefly algorithm can be summarized as follows.

Step 1: Initialize the position of firefly populations $\boldsymbol{x}_i (i = 1,\ldots,P_{size})$, and calculate objective function $f(\boldsymbol{x}_i)$ as the brightness I_i of the ith individual, set the initial iteration number $t = 0$ and the max total iteration number T_{max}.

Step 2: Execute FA to search for better solutions which update attractiveness and location respectively according to Eqs. (4) and (5), then find out the current optimal solution $\boldsymbol{x}_{pg}$.

Step 3: Execute PCLS based on $\boldsymbol{x}_{pg}$, set the initial chaotic iteration number $l = 0$ and the max chaotic iteration number C_{max} each time, initialize the chaotic variables $-1 < z_{ij}^{(0)} < 1$ ($i = 1,\ldots,N$; $j = 1,\ldots,n$) except for {0.25,0.5,0.75}, where N is the size of parallel chaotic variables, n is the size of variable's dimension, z_{ij} denotes the jth dimension of the ith parallel variable. The intermediate variable $\boldsymbol{y}$ is used below, and the size of part dimension is set as $n/2$.

Step 3.1: Let N intermediate variables $\boldsymbol{y}_i^{(l)} = \boldsymbol{x}_{pg}(i = 1,\ldots,N)$, and generate every chaotic variable $z_{ij}^{(l+1)}$ according to Eq. (6) to produce new parallel variable.

Step 3.2: Randomly generate a variable r_i in the interval (-1,1) for ith parallel variable to chose the strategy of CLS. If $r_i < 0.5$, execute full dimensional CLS which is shown as Eq. (8):

$$y_{ij}^{(l+1)} = y_{ij}^{(l)} + \delta_t z_{ij}^{(l+1)} L_j, j \in [1, n] \tag{8}$$

else execute part dimensional CLS, $n/2$ index values are equiprobably chosed from $1 \sim n$ to constitute a set $\boldsymbol{P}_{ld}$, and new variable is generated according to Eq. (9).

$$y_{ij}^{(l+1)} = \begin{cases} y_{ij}^{(l)} + \delta_t z_{ij}^{(l+1)} L_j, j \in \boldsymbol{P}_{ld} \\ y_{ij}^{(l)}, \text{ others} \end{cases} \tag{9}$$

where $\boldsymbol{L}$ is the neighborhood ranges around $\boldsymbol{y}_i$ for PCLS that can be set from $0.01\boldsymbol{S}$ to $0.1\boldsymbol{S}$, and $\boldsymbol{S}$ is the scale of the variable, if $\boldsymbol{U}_b$ and $\boldsymbol{L}_b$ is the upper and lower boundary of the search space, $\boldsymbol{S} = \boldsymbol{U}_b\text{-}\boldsymbol{L}_b$; δ_t is the compressibility factor of chaotic local variable that can be evaluated:$\delta_t = e^{-C*t/T_{max}}$, where C is a predefined number usually set in the interval [1, 10]. Smaller value of C may result in a poor accuracy, but larger value of it may lead to a failure of search on later stage, so it should be determined according to the real problem. All of the new variables $\boldsymbol{y}_i^{(l+1)}$ generated by PCLS constitute a matrix $\boldsymbol{y}^{(l+1)}$ which is shown as Eq. (10).

$$\boldsymbol{y}^{(l+1)} = \begin{bmatrix} y_{11}^{(l+1)} & y_{12}^{(l+1)} & \cdots & y_{1n}^{(l+1)} \\ y_{21}^{(l+1)} & y_{22}^{(l+1)} & \cdots & y_{2n}^{(l+1)} \\ \cdots & \cdots & \cdots & \cdots \\ y_{N1}^{(l+1)} & y_{N2}^{(l+1)} & \cdots & y_{Nn}^{(l+1)} \end{bmatrix} \tag{10}$$

Step 3.3: Calculate objective function $f(\boldsymbol{y}^{*(l+1)})$ for all variables $\boldsymbol{y}_i^{(l+1)}(i = 1,\ldots,N)$, and find the optima $\boldsymbol{y}^{*(l+1)}$, if $f(\boldsymbol{y}^{*(l+1)}) < f(\boldsymbol{x}_{pg})$, then $f(\boldsymbol{x}_{pg}) = f(\boldsymbol{y}^{*(l+1)})$, and update the current optimal solution $\boldsymbol{x}_{pg} = \boldsymbol{y}^{*(l+1)}$, which replaces a random firefly $\boldsymbol{x}_i$ simultaneously.

Step 3.4: Update $l = l+1$, if $l < C_{max}$, go to Step3.1, else go to Step4.

Step 4: Update $t = t+1$, if $t < T_{max}$, go to Step2, else stop the algorithm and the final $\boldsymbol{x}_{pg}$ is global optima.

According to the above procedure, if the $\boldsymbol{x}_{pg}$ achieved in Step 2 is in the neighborhood of global optimal solution, it will converge to the global optima further by PCLS in Step 3. And if the $\boldsymbol{x}_{pg}$ achieved in Step 2 falls in local optima, there will be a certain probability to jump out of local optima and converge to the global optimal solution in Step 3. Therefore, an enhanced version of FA named as EFA can be obtained to search for better solution using PCLS.

4 The Proposed Hybrid Clustering Algorithm

In this paper, the objective function KHM(X,C) is used to denote the brightness I_i of firefly *i*, in order to combine KHM and EFA to build a hybrid clustering algorithm. A vector $\boldsymbol{X} = (x_{11},\ldots,x_{1n},\ldots,x_{k1},\ldots,x_{kn})$ represents a firefly individual of $k*n$ dimension, where k is the number of clusters and n is the number of one object's attribution, hence x_{iq} denotes the qth dimension of the ith clustering center. The hybrid clustering algorithm KHM-EFA can be summarized as follows.

1. Initialize the parameters of the algorithm such as γ, α, β, C_{max}, N, $\boldsymbol{L}$ and randomly generate the initial population of size P_{size}.
2. Set the initial iteration number $gen = 0$ and the max total iteration number of the algorithm is *IterCount*.
3. Execute EFA for *Gen*1 times, and find the current optimal individual *Gbest* with the fitness Fg.
4. Execute KHM for *Gen*2 times based on *Gbest* as the initial centers, and then the value of the objective function KHM(X,C) and the center $\boldsymbol{X}_{KHM}$ are obtained, if KHM(X,C) < Fg, Fg = KHM(X,C) and $Gbest = \boldsymbol{X}_{KHM}$, and also a firefly individual is chosen randomly to be replaced by $\boldsymbol{X}_{KHM}$.
5. Update $gen = gen + 1$, if $gen < IterCount$, go to Step 3, else stop.
6. Assign data point $\boldsymbol{x}_i$ to cluster j with the biggest $m_{KHM}(\boldsymbol{c}_j/\boldsymbol{x}_i)$.

5 Experiment and Analysis

To verify the effectiveness of KHM-EFA, three other algorithms KHM, KHM-FA and KHM-PSO [3] are adopted as the comparative methods, what needs to be pointed that almost the only difference between KHM-FA and KHM-EFA is step 3 of Sect. 4. Four real life data sets from UCI Repository are employed to evaluate the performance of these algorithms including Seeds, Wisconsin Breast Cancer (WBC), Wine and Image

Table 1. The feature of experimental data sets

Dataset	Class	Dimension	Size(size of each class)
Seeds	3	7	210(69,65,76)
WBC	2	9	683(444,239)
Wine	3	13	178(59,71,48)
Image Segmentation	7	19	210(30 × 7)

Segmentation. The details of these data sets are described in Table 1. The algorithms are simulated using MATLAB2010b and executed on a computer with Intel Core i5-4200 M CPU 2.5 GHz and 4 GB RAM.

KHM(X,C) is an internal cluster validity measure, and the smaller it is, the better the result is. F-measure is an external cluster validity measure, which employs ideas of precision and recall from retrieval, for each class i and cluster j, the definition of precision and recall is shown as: $p(i, j) = n_{ij}/n_j$, $r(i, j) = n_{ij}/n_i$, where n_i denotes the size of class i, n_j denotes the size of cluster j, n_{ij} denotes the number of objects of class i within cluster j. For class i and cluster j, F-measure can be defined as Eq. (11).

$$F(i,j) = \frac{2 \times p(i,j) \times r(i,j)}{p(i,j) + r(i,j)} \tag{11}$$

The global F-measure for dataset of size n is shown as: $F = \sum_i \frac{n_i}{n} \max_j\{F(i,j)\}$, and the bigger it is, the higher the quality of cluster is.

All the parameter settings are configured: the maximum iteration of KHM is $Maxgen = 100$; the parameters of KHM-PSO are $P_{size} = 18$, $w = 0.7298$, $c_1 = c_2 = 1.496$, $IterCount = 5$, $Gen1 = 8$, $Gen2 = 4$, what are the same as that in literature [6], and it should be noted that $Gen2$ is 10 for dataset Image Segmentation which has not yet appeared in literature [6]; the corresponding parameters of KHM-FA, KHM-EFA are set as the same as that in KHM-PSO, remainders such as $\gamma = 1$, $\alpha = 0.1$ and β is rewritten: $\beta = (\beta_{max} - \beta_{min})e^{-\gamma r_{ij}^2} + \beta_{min}$, where $\beta_{max} = 1$, $\beta_{min} = 0.2$, meanwhile, $C_{max} = 4$, $N = 5$ and $\boldsymbol{L} = 0.1\boldsymbol{S}$ only including in KHM-EFA. The experimental results are averages of 20 runs of simulation for p is 2.5, 3 and 3.5 respectively which is

Table 2. The result of four algorithms when p = 2.5

Dataset	Algorithm	KHM(X,C)	F-measure	Runtime(s)
Seeds	KHM	1611.21	0.891	0.272
	KHM-FA	1610.74	0.891	1.743
	KHM-PSO	1611.08	0.891	1.879
	KHM-EFA	1610.73	0.891	3.282
WBC	KHM	57168.99	0.962	0.614
	KHM-FA	56997.84	0.962	4.456
	KHM-PSO	57028.62	0.962	4.895
	KHM-EFA	56871.08	0.962	8.542
Wine	KHM	75338585.3	0.689	0.272
	KHM-FA	75335326.4	0.707	2.216
	KHM-PSO	75336097.5	0.703	2.232
	KHM-EFA	75333137.9	0.709	3.780
Image Segmentation	KHM	54412682.7	0.595	0.924
	KHM-FA	42302723.3	0.597	6.597
	KHM-PSO	39615172.4	0.598	6.869
	KHM-EFA	27756979.3	0.598	10.903

Table 3. The result of four algorithms when p = 3

Dataset	Algorithm	KHM(X,C)	F-measure	Runtime(s)
Seeds	KHM	2199.69	0.891	0.269
	KHM-FA	2196.35	0.891	1.747
	KHM-PSO	2197.14	0.891	1.881
	KHM-EFA	2195.85	0.891	3.268
WBC	KHM	113726.84	0.965	0.601
	KHM-FA	112383.21	0.967	4.453
	KHM-PSO	112568.20	0.965	4.831
	KHM-EFA	111498.07	0.972	8.495
Wine	KHM	1.0491e + 09	0.622	0.262
	KHM-FA	1.0491e + 09	0.656	2.193
	KHM-PSO	1.0490e + 09	0.632	2.136
	KHM-EFA	1.0490e + 09	0.659	3.776
Image Segmentation	KHM	1.7904e + 08	0.552	0.883
	KHM-FA	1.7883e + 08	0.558	6.563
	KHM-PSO	1.7876e + 08	0.560	6.710
	KHM-EFA	1.7710e + 08	0.559	10.814

Table 4. The result of four algorithms when p = 3.5

Dataset	Algorithm	KHM(X,C)	F-measure	Runtime(s)
Seeds	KHM	3097.43	0.882	0.273
	KHM-FA	3085.92	0.882	1.756
	KHM-PSO	3090.08	0.880	1.890
	KHM-EFA	3082.87	0.882	3.287
WBC	KHM	232191.25	0.966	0.615
	KHM-FA	226306.73	0.969	4.467
	KHM-PSO	226214.58	0.966	4.852
	KHM-EFA	222866.91	0.974	8.519
Wine	KHM	2.7175e + 10	0.630	0.273
	KHM-FA	1.4204e + 10	0.655	2.201
	KHM-PSO	1.4419e + 10	0.636	2.140
	KHM-EFA	1.4191e + 10	0.658	3.817
Image Segmentation	KHM	3.9979e + 09	0.537	0.937
	KHM-FA	1.9433e + 09	0.543	6.672
	KHM-PSO	1.5798e + 09	0.542	6.829
	KHM-EFA	1.4984e + 09	0.542	10.861

known to be a key parameter for the algorithms. All of the results including KHM(X, C), F-measure and runtime are shown in Tables (2, 3, 4).

As we can see from Tables 2, 3, 4, in terms of the KHM(X,C) and F-measure values, the result of three hybrid clustering algorithms KHM-FA, KHM-PSO and KHM-EFA are better than KHM for almost all the four data sets. The KHM-EFA algorithm proposed in this paper can achieve the best result in most cases. It should be noted for dataset Seeds, owing to its characteristic that it can be clustered easier relatively, all the three hybrid clustering algorithms achieve the F-measure values equal to that of KHM. Hence the superiority of the hybrid clustering algorithms can be merely indicated from KHM(X,C) values, and the proposed algorithm KHM-EFA obtains the lowest value that represents the best searching ability. To compare overall performance of the three hybrid clustering algorithms to that of KHM more obviously, the largest decreasing rate of KHM(X,C) and increasing rate of F-measure are computed and shown in Tables 5, 6 for all cases of p value.

Table 5. The largest decreasing rate of KHM(X,C)

	Seeds	WBC	Wine	Image segmentation
KHM-FA	0.37 %	2.53 %	47.73 %	51.39 %
KHM-PSO	0.24 %	2.57 %	46.94 %	60.48 %
KHM-EFA	0.47 %	4.02 %	47.78 %	62.52 %

Table 6. The largest increasing rate of F-measure

	Seeds	WBC	Wine	Image segmentation
KHM-FA	0 %	0.31 %	5.47 %	1.12 %
KHM-PSO	0 %	0 %	2.03 %	1.45 %
KHM-EFA	0 %	0.83 %	5.95 %	1.27 %

The results of Tables 5 demonstrate that KHM-EFA can get reduction of KHM(X, C) value to a largest degree for all data sets, which proves that the searching capability of the proposed algorithm is the best. For data sets Wine and Image Segmentation, with the result of being trapped in local optima when executing KHM in some runs of simulation which produce a large KHM(X,C) value and increase the mean value as well, a large decreasing rate can be obtained in some cases of p value. However, the hybrid clustering algorithms can effectively jump out of local optima and obtain a lower mean value than that of KHM(X,C). Meanwhile, KHM-EFA can get a better comprehensive increasing degree of F-measure which is shown in Table 6, even though in a few cases KHM-FA and KHM-PSO obtain a little higher increasing rate that may be achieved by chance. Therefore, it is concluded that the proposed algorithm can overcome the drawback of KHM effectively and improve the accuracy and stability of clustering.

6 Conclusion

In this paper, a novel hybrid k-harmonic means clustering based on the enhanced firefly algorithm is proposed. The clustering performance of our method is compared to conventional KHM and two other hybrid clustering algorithms KHM-FA and KHM-PSO for four data sets from UCI Repository, which shows that our method can escape from local optima effectively and has accurate and stable superiority to a certain degree. However, a main shortage of KHM-EFA is that it requires more runtime due to the process of the combination of swarm intelligence search with PCLS which needs more computational times of the objective function. Therefore, this method is not applicable when time is a vital factor in the system, and we will do some research to reduce the runtime in the future.

Acknowledgment. This research was supported by the Cooperative Industry-Academy-Research Innovation Foundation of Jiangsu Province (BY2013015-33).

References

1. Jain, A.K.: Data clustering: 50 years beyond K-means. PRL **31**(8), 651–666 (2010)
2. Zhang B, Meichun Hsu, Umeshwar Dayal.: K-Harmonic Means-a Data Clustering Algorithm. Technical report HPL-1999-124, Hewlett-Packard Labs (1999)
3. Hamerly G, Elkan C.: Alternatives to the k-means algorithm that find better clusterings. In: Proceedings of the Eleventh International Conference on Information and knowledge Management, ACM, 600–607(2002)
4. Güngör, Z., Ünler, A.: K-harmonic means data clustering with simulated annealing heuristic. Appl. Math. Comput. **184**(2), 199–209 (2007)
5. Güngör, Z., Ünler, A.: K-Harmonic means data clustering with tabu-search method. Appl. Math. Model. **32**(6), 1115–1125 (2008)
6. Yang, F.Q., Sun, T.E.L., Zhang, C.H.: An efficient hybrid data clustering method based on K-harmonic means and Particle Swarm Optimization. Expert Syst. Appl. **36**(6), 9847–9852 (2009)
7. Jiang, H., Yi, S., Li, J., et al.: Ant clustering algorithm with K-harmonic means clustering. Expert Syst. Appl. **37**(12), 8679–8684 (2010)
8. Alguwaizani, A., Hansen, P., Mladenović, N., et al.: Variable neighborhood search for harmonic means clustering. Appl. Math. Model. **35**(6), 2688–2694 (2011)
9. Yin, M., Hu, Y., Yang, F., et al.: A novel hybrid K-harmonic means and gravitational search algorithm approach for clustering. Expert Syst. Appl. **38**(8), 9319–9324 (2011)
10. Hung, C.-H., Chiou, H.-M., Yang, W.-N.: Candidate groups search for K-harmonic means data clustering. Appl. Math. Model. **37**(24), 10123–10128 (2013)
11. Abdeyazdan, M.: Data clustering based on hybrid K-harmonic means and modifier imperialist competitive algorithm. J. Supercomput. **68**(2), 574–598 (2014)
12. Yang, X.-S.: Nature-Inspired Metaheuristic Algorithms. Luniver press, Bristol (2010)
13. Senthilnath, J., Omkar, S., Mani, V.: Clustering using firefly algorithm: performance study. Swarm Evol. Comput. **1**(3), 164–171 (2011)
14. Adaniya, M.H.: Anomaly detection using met heuristic firefly harmonic clustering. J. Netw. **8**(1), 82–91 (2013)
15. Yuan, X., Zhao, J., Yang, Y., et al.: Hybrid parallel chaos optimization algorithm with harmony search algorithm. Appl. Soft Comput. **17**, 12–22 (2014)
16. He, Y., Yang, S., Xu, Q.: Short-term cascaded hydroelectric system scheduling based on chaotic particle swarm optimization using improved logistic map. Commun. Nonlinear Sci. Numer. Simul. **18**(7), 1746–1756 (2013)
17. He, Y., Xu, Q., Yang, S., et al.: A novel chaotic differential evolution algorithm for short-term cascaded hydroelectric system scheduling. Int. J. Electr. Power Energy Syst. **61**, 455–462 (2014)

A Feature Selection Method Based on Feature Grouping and Genetic Algorithm

Xiaohui Lin(✉), Xiaomei Wang, Niyi Xiao, Xin Huang, and Jue Wang

School of Computer Science & Technology, Dalian University of Technology, 116024 Dalian, China
datas@dlut.edu.cn

Abstract. Feature selection technique has shown its power in analyzing the high dimensional data and building the efficient learning models. This study proposes a feature selection method based on feature grouping and genetic algorithm (FS-FGGA) to get a discriminative feature subset and reduce the irrelevant and redundancy data. Firstly, it eliminates the irrelevant features using the symmetrical uncertainty between features and class labels. Then, it groups the features by Approximate Markov blanket. Finally, genetic algorithm is applied to search the optimal feature subset from the different groups. Experiments on the eight public datasets demonstrate the effectiveness and superiority of FS-FGGA in comparison with SVM-RFE and ECBGS in most cases.

Keywords: Feature selection · Symmetrical uncertainty · Feature grouping · Genetic algorithm

1 Introduction

As the quick development of genomic, proteomics and metabolomics techniques, they have been widely applied in the study of pathology, diagnostics and prognosis. Since the bioinformatic data are often high dimensional and contain noise and redundant variables, finding the interested features to get an efficient classification model is becoming very important. Many feature selection methods, such as Support Vector Machine-Recursive Feature Elimination (SVM-RFE) [1], Random Forests (RF) [2], Genetic Algorithm (GA) [3], Relief-F [4], and Mutual Information (MI) [5, 6], have been applied to select the meaningful feature subset from the high dimensional data to induce a classification model with a high performance [7, 8].

SVM [9] is a supervised machine learning technique. It is suitable to analyze the high dimensional data [10]. Originally, SVM was proposed for binary problems. And it could solve the multi-class problems by means of "one-versus-all" and "one-versus-one" methods[11], etc. SVM-RFE [12] is a popular feature selection approach based on SVM. It calculates the weights of the features according to the SVM learning model and removes the features with the smallest weights iteratively. GA is a stochastic global search technique [13] and has got a promising performance. Many feature selection techniques have been proposed based on GA [14, 15].

X. He et al. (Eds.): IScIDE 2015, Part II, LNCS 9243, pp. 150–158, 2015.
DOI: 10.1007/978-3-319-23862-3_15

To filter out noise and redundant data simultaneously, several techniques have been proposed, such as min-redundancy and max-relevance (mRMR) [16], a method combining SVM-RFE and correlation coefficient [17], a method where SVM-RFE and mRMR work together [18], and a dynamic weighting-based feature selection algorithm [19].

To select the meaningful feature subset from the high dimensional data, this paper proposes a new feature selection method based on feature grouping and GA (FS-FGGA). It removes the irrelevant data which has small relevance with the class label, groups the features, and applies GA to search the optimal combination feature subset from different feature groups. The applications on eight public data verify the effectiveness of FS-FGGA.

2 Methods

To improve the performance of the learning model, FS-FGGA selects the meaningful non-redundant features from the original data. It eliminates the irrelevant features by symmetrical uncertainty [20, 21] and groups the features according to the relevance among the features. The features lying in the same group have the similar information related to the class label. Hence each group contributes one feature to the final feature subset. But selecting different features from each group may induce different learning models which may have different classification performance. GA is adopted to search the optimal combination feature subset. Fig. 1 shows the main framework of FS-FGGA.

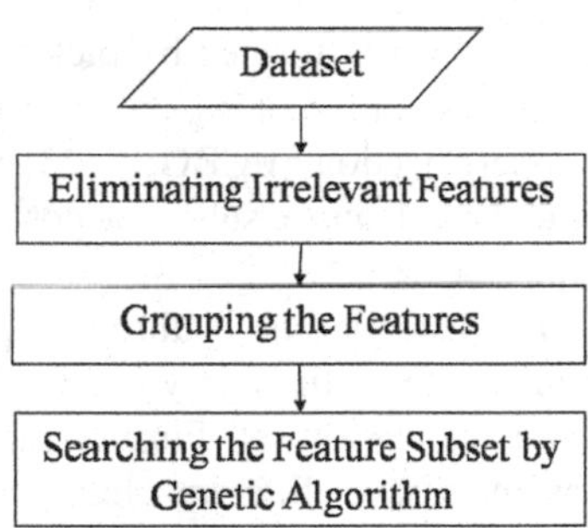

Fig. 1. Framework of FS-FGGA.

2.1 Symmetrical Uncertainty

Symmetrical Uncertainty (SU) [20, 21] is an effective technique to measure the correlation of two random variables. Let X and Y be two variables, their correlation $SU(X, Y)$ is defined as follows:

$$SU(X,Y) = 2 \cdot \frac{IG(X|Y)}{H(X) + H(Y)}. \quad (1)$$

H(*X*) is the entropy of *X*, *IG*(*X*|*Y*) is the information gain which reflects additional information about *X* provided by *Y*.

Let $F = \{f_1, f_2, \ldots, f_n\}$ denote the feature set, *C* denote the class label set. In order to filter out the irrelevant features, FS-FGGA adopts symmetrical uncertainty, $SU(f_i, C)$ ($1 \leq i \leq n$), to measure the relation between feature $f_i \in F$ and the class label *C*. If $SU(f_i, C)$ is low enough, i.e. it is lower than a threshold σ, feature f_i has little relevance with the class label, and is removed from the data [20, 21].

2.2 Grouping Features

Fast Correlation-Based Filter (FCBF) [21] is an efficient feature selection technique. It analyzes the relevance by symmetrical uncertainty, and removes the redundant data by means of Approximate Markov blanket (AMB). For two different features $f_i \in F$ and $f_j \in F$ ($1 \leq i \neq j \leq n$), f_i is an Approximate Markov blanket [21] of f_j, if and only if

$$SU(f_i, C) \geq SU(f_j, C) \ and \ SU(f_i, f_j) \geq SU(f_j, C). \quad (2)$$

FS-FGGA groups the features according to AMB. The features which are relevant to each other by FCBF [21] are put into the same group.

2.3 Searching the Optimal Feature Subset by GA

FCBF produces a feature subset which is formed by picking the center feature of the group [21]. But the center may be different as the training samples change [22]. Ensemble correlation-based gene selection (ECBGS) [23] method uses the different starting points and selects the best feature subset according to the corresponding classification performance.

Let $FG = \{FG_1, FG_2, \ldots, FG_k\}$ denote the feature group set. Since the features in the same group contain the similar information, only one feature is picked up from each group to constitute the selected feature subset. Further the combination of different features from different groups may have different classification performance. Hence FS-FGGA applies GA to search the optimal one. Initially, FS-FGGA randomly selects a feature from each group to form a feature subset as an individual and repeats this operation to get the initial population of GA. The flow chart of searching the optional feature subset is shown in Fig. 2.

The fitness of an individual in a population is assessed by the classification accuracy rate of SVM. Roulette wheel selection is adopted to select the parents from the population. A single-point crossover operation and a single-point mutation are also applied for the offspring individuals.

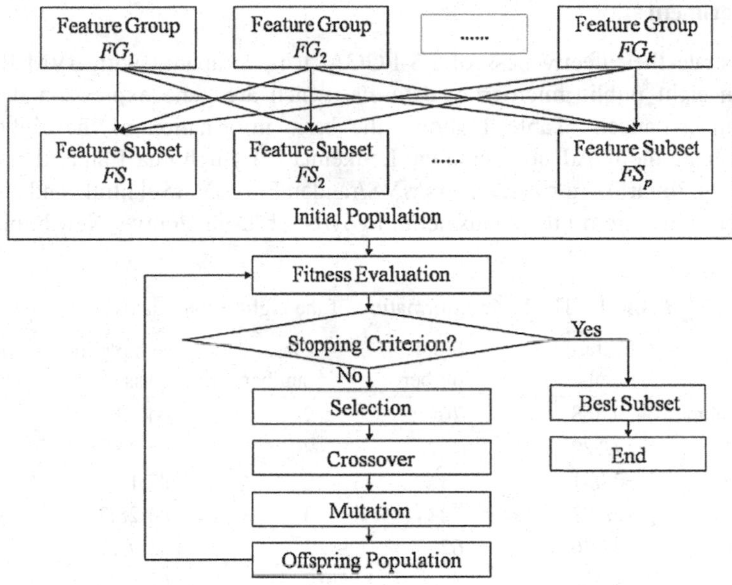

Fig. 2. Flow chart of searching the best feature subset.

3 Results and Discussion

3.1 Performance Metrics

Features selection technique aims at selecting a feature subset having the high classification ability. Meanwhile, the stability of the method is also very important. This study applied the classification accuracy and stability to evaluate the performance of the methods. The percentage of overlapping features related (POFR) [24] is used to measure the method stability. It is defined as follows [24]:

$$POFR_{F_1F_2} = \frac{|F_1 \cap F_2| + |R_{F_1F_2}|}{|F_1|}. \tag{3}$$

$$POFR_{F_2F_1} = \frac{|F_1 \cap F_2| + |R_{F_2F_1}|}{|F_2|}. \tag{4}$$

where F_1 and F_2 are two different feature subsets selected by the different running of a algorithm, $|F_1|$ (or $|F_2|$) is the number of the features in F_1 (or F_2), $R_{F_1F_2}$(or $R_{F_2F_1}$) is the set of the features in F_1 (or F_2) which are not in F_2 (or F_1) but have a strong correlation with at least one feature in F_2 (or F_1). The greater its value is, the more stable the feature selection algorithm is.

3.2 Experiment

To demonstrate the effectiveness of FS-FGGA, it is compared with SVM-RFE and ECBGS on eight public microarray datasets, which are gene expression data from various human cancers. Table 1 shows the basic information of the eight public datasets. Among them, Adenocarcinoma, Leukemia 2, Lymphoma 1 and Srbct datasets are from http://ligarto.org/rdiaz/Papers/rfVS/randomForestVarSel.html, and the other four datasets come from http://linus.nci.nih.gov/~brb/DataArchive_New.html.

Table 1. The basic information of the eight public datasets

Datasets	Feature number	Sample number	Class number	Sample number of every class
Adenocarcinoma	9,868	76	2	64:12
Leukemia 1	7,129	72	2	47:25
Leukemia 2	3,051	38	2	27:11
Leukemia MLL	12,582	72	3	24:28:20
Lymphoma 1	4,026	62	3	42:9:11
Lymphoma 2	4,026	96	2	62:34
Prostate	12,600	102	2	50:52
Srbct	2,308	63	4	23:20:12:8

Auto scaling is used to reduce the differences of the magnitude of different features. To calculate SU, equal width discretization (EWD) [25, 26] is adopted, where the real data is divided into h (h is set to 3 in the experiments) intervals with equal width between the minimum value and the maximum value.

Parameter σ for FS-FGGA and ECBGS is set as follows:

$$\sigma = 0.5 * (SU_{\max} - SU_{\min}). \tag{5}$$

where $SU_{\max}$ and $SU_{\min}$ are the maximal and the minimal relevant values of the features with the class label, respectively.

For FS-FGGA, the maximal number of iterations and the size of population are set to 50 and 100, respectively. The crossover probability and mutation rate are set to 0.8 and 0.01, respectively. When the generation is up to the maximal number of iterations or the best fitness comes to 0.95, the GA search procedure stops.

Ten-fold cross-validation was run ten times. SVM is adopted as the classification method, and the RBF kernel function and the LINEAR kernel function are used respectively. The source code of SVM is from http://www.csie.ntu.edu.Tw/~cjlin/libsvm and the other algorithms were written in C++.

3.3 Results and Discussion

Tables 2 and 3 show the comparison of the three methods on the average classification accuracy rates. The bold face means the largest accuracy rate among the three methods in a data set. The last row (W/T/L) of the two tables count the number of wins/ties/losses compared to the FS-FGGA over all data sets. It can be seen that FS-FGGA is superior to the other two feature selection methods in most cases.

In comparison with SVM-RFE, FS-FGGA ties with SVM-RFE on RBF kernel function (Table 2), but it shows a clear superiority over SVM-RFE on the LINEAR kernel function (Table 3), where FS-FGGA wins seven times to SVM-RFE. With LINEAR kernel function, the average classification accuracy rate of SVM-RFE is equal to that of FS-FGGA only on the Adenocarcinoma data, but the standard deviation of SVM-RFE is 1.55% higher than that of FS-FGGA.

In comparison with ECBGS, the average classification accuracy rates of FS-FGGA are higher than those of ECBGS on all the eight datasets with RBF kernel function (Table 2). While in Table 3, using the LINEAR kernel function, FS-FGGA wins ECBGS seven times. Only on the Leukemia 1 data, the average classification accuracy rate of ECBGS is higher than that of FS-FGGA a little.

Tables 4 and 5 show the average *POFR* of the three feature selection algorithms. From the two tables, FS-FGGA algorithm is more stable than the other two algorithms in the majority cases.

Table 2. The comparison on SVM with RBF kernel function

Datasets	SVM-RFE(%)	ECBGS(%)	FS-FGGA(%)
Adenocarcinoma	**82.37±3.11**	79.47±2.86	81.05±2.92
Leukemia 1	**94.44±2.45**	93.75±1.99	94.03±1.47
Leukemia 2	97.37±2.77	95.26±3.23	**98.16±2.50**
Leukemia MLL	93.61±2.55	92.50±2.19	**94.58±2.01**
Lymphoma 1	96.77±1.32	98.06±1.67	**99.84±0.51**
Lymphoma 2	**91.56±1.51**	90.52±1.04	91.35±1.56
Prostate	90.98±1.84	90.49±1.47	**91.18±1.46**
Srbct	**97.94±0.77**	94.60±3.76	95.40±1.58
W/T/L	4/0/4	0/0/8	-

Table 3. The comparison on SVM with LINEAR kernel function

Datasets	SVM-RFE(%)	ECBGS(%)	FS-FGGA(%)
Adenocarcinoma	**79.47±4.12**	76.84±2.99	**79.47±2.57**
Leukemia 1	91.94±2.91	**92.64±2.78**	92.36±1.76
Leukemia 2	95.79±3.96	96.58±2.50	**97.37±2.15**
Leukemia MLL	87.64±2.31	93.61±2.47	**95.28±2.09**
Lymphoma 1	93.87±2.82	97.90±1.87	**98.87±0.78**
Lymphoma 2	88.75±2.24	86.77±2.60	**89.38±2.90**
Prostate	88.73±2.70	89.61±1.74	**90.78±1.97**
Srbct	92.70±3.01	95.56±2.97	**96.83±1.06**
W/T/L	0/1/7	1/0/7	-

Table 4. The average *POFR* using RBF kernel function

Datasets	SVM-RFE	ECBGS	FS-FGGA
Adenocarcinoma	**0.6311**	0.3274	0.3590
Leukemia 1	**0.8320**	0.7409	0.8243
Leukemia 2	0.7316	0.6967	**0.9454**
Leukemia MLL	0.8419	0.8457	**0.8971**
Lymphoma 1	0.7360	0.7945	**0.9797**
Lymphoma 2	**0.7117**	0.5905	0.5653
Prostate	0.7792	0.7864	**0.8112**
Srbct	**0.9497**	0.6260	0.7466

Table 5. The average *POFR* using LINEAR kernel function

Datasets	SVM-RFE	ECBGS	FS-FGGA
Adenocarcinoma	**0.4904**	0.3131	0.3384
Leukemia 1	0.5501	0.7090	**0.8139**
Leukemia 2	0.6748	0.7101	**0.9457**
Leukemia MLL	0.6312	0.8296	**0.8931**
Lymphoma 1	0.5012	0.8274	**0.9797**
Lymphoma 2	0.4176	0.5667	**0.5757**
Prostate	0.4272	0.7514	**0.8051**
Srbct	**0.7769**	0.6034	0.7501

4 Conclusions

This paper proposes a new feature selection method based on feature group and genetic algorithm (FS-FGGA). The method can effectively eliminate the irrelevant features and reduce the redundant features. Applications on eight public microarray data show the effectiveness of FS-FGGA. It can select more discriminative feature subsets to build more efficient classification models than SVM-RFE and ECBGS in most cases.

Acknowledgments. The study has been supported by the State Key Science & Technology Project for Infectious Diseases (2012ZX10002011), the Sino-German Center for Research Promotion (GZ 753), National Natural Science Foundation of China (21375011).

References

1. Tang, Y.C., Zhang, Y.Q., Huang, Z.: Development of two-stage SVM-RFE gene selection strategy for microarray expression data analysis. IEEE/ACM Trans. Comput. Biol. Bioinform. **4**, 365–381 (2007)
2. Breiman, L.: Random forests. Mach. Learn. **45**, 5–32 (2001)
3. Holland, J.H.: Adaptation in Natural and Artificial Systems. MIT Press, Cambridge, MA (1992)

4. Kononenko, I.: Estimating attributes: analysis and extensions of RELIEF. Mach. Learn. **784**, 171–182 (1994)
5. Cover, T.M., Thomas, J.A.: Elements of Information Theory. Wiley, New York (1991)
6. Xu, J.C., Xu, T.H., Sun, L.: An efficient gene selection technique based on fuzzy C-means and neighborhood rough set. Appl. Math. Inf. Sci. **8**, 3101–3110 (2014)
7. Yassi, M., Moattar, M.H.: Robust and stable feature selection by integrating ranking methods and wrapper technique in genetic data classification. Biochem. Biophys. Res. Commun. **446**, 850–856 (2014)
8. Liu, X.M., Tang, J.S.: Mass classification in mammograms using selected geometry and texture features, and a new SVM-based feature selection method. IEEE Syst. J. **8**, 910–920 (2014)
9. Vapnik, V.N.: Statistical Learning Theory. Wiley, New York (1998)
10. Shen, L., Tan, E.C.: Dimension reduction based penalized logistic regression for cancer classification using micro-array data. IEEE/ACM Trans. Comput. Biol. Bioinform. **2**, 166–175 (2005)
11. Zhou, X., Tuck, D.P.: MSVM-RFE: extensions of SVM-RFE for multiclass gene selection on DNA microarray data. Bioinformatics **23**, 1106–1114 (2007)
12. Guyon, I., Weston, J., Barnhill, S., Vapnik, V.: Gene selection for cancer classification using support vector machines. Mach. Learn. **46**, 389–422 (2002)
13. Arunachalam, J., Kanagasabai, V., Gautham, N.: Protein structure prediction using mutually orthogonal Latin squares and a genetic algorithm. Biochem. Biophys. Res. Commun. **342**, 424–433 (2006)
14. Ram, R., Chetty, M.: A Markov-Blanked-Based model for gene regulatory network inference. IEEE-ACM Trans. Comput. Biol. Bioinform. **8**, 353–367 (2011)
15. Abbasnia, R., Shayanfar, M., Khodam, A.: Reliability-based design optimization of structural systems using a hybrid genetic algorithm. Struct. Eng. Mech. **52**, 1099–1120 (2014)
16. Maji, P., Garai, P.: On fuzzy-rough attribute selection: criteria of max-dependency, max-relevance, min-redundancy, and max-significance. Applied Soft Computing. **13**, 3968–3980 (2013)
17. Xie, Z.X., Hu, Q.H., Yu, D.R.: Improved feature selection algorithm based on SVM and correlation. Adv. Neyral Netw. **3971**, 1373–1380 (2006)
18. Mundra, P.A., Rajapakse, M.J.: SVM-RFE with mRMR filter for gene selection. IEEE transactions on nano bioscience. **9**(1), 31–37 (2010)
19. Sun, X., Liu, Y.H., Xu, M.T., Chen, H.L., Han, J.W., Wang, K.H.: Feature selection using dynamic weights for classification. Knowl.-Based Syst. **37**, 541–549 (2013)
20. Shen, L.L., Zhu, Z.X., Jia, S.: Discriminative Gabor feature selection for hyper spectral image classification. IEEE Geosci. Remote Sens. Lett. **10**, 29–33 (2013)
21. Yu, L., Liu, H.: Efficient feature selection via analysis of relevance and redundancy. J. Mach. Learn. Res. **5**, 1205–1224 (2004)
22. Liu, H.W., Liu, L., Zhang, H.J.: Ensemble gene selection by grouping for microarray data classification. J. Biomed. Inform. **43**, 81–87 (2010)
23. Piao, Y.J., Piao, M.H., Park, K.J., Ryu, K.H.: An ensemble correlation-based gene selection algorithm for cancer classification with gene expression data. Bioinformatics **28**, 3306–3315 (2012)
24. Zhang, M., Zhang, L., Zou, J.F., Yan, C., Xiao, H., Liu, Q.: Evaluating reproducibility of differential expression discoveries in microarray studies by considering correlated molecular changes. Bioinformatics **25**, 1662–1668 (2009)

25. Bennasar, M., Setchi, R., Hicks, Y.: Unsupervised discretization method based on adjustable intervals. In: 16th International Conference on Knowledge-Based and Intelligent Information and Engineering Systems, vol. 243, pp. 79–87, San Sebastian (2012)
26. Orhan, U., Hekim, M., Ozer, M.: Epileptic seizure detection using artificial neural network and a new feature extraction approach based on equal width discretization. J. Fac. Eng. Archit. Gazi Univ. **26**, 575–580 (2011)

Tuning GSP Parameters with GA

Wei Chi Cheng[1], Ping Yu Hsu[1], Ming Shien Cheng[1,2(✉)], and Shih Hsiang Huang[1]

[1] Department of Business Administration, National Central University, No. 300, Jhongda Road, Jhongli City 32001, Taoyuan County, Taiwan, ROC
984401019@cc.ncu.edu.tw,
mscheng@mail.mcut.edu.tw

[2] Department of Industrial Engineering and Management, Ming Chi University of Technology, No. 84, Gongzhuan Road, Taishan District, New Taipei City 24301, Taiwan, ROC

Abstract. In data mining, association rules can be shown when customers buy products, which products will be purchased at the same time. Scholars use this feature to develop market basket analysis to formulate marketing strategies for business. As we know, the data are changing all the time. When new data generate, the old data will be replaced. In the database, time become a very important attribute. And new data mining method have been proposed, called Generalized Sequential Patterns (GSP). GSP uses time stamp to find the product portfolio with sequential patterns. However, the GSP parameter is user-defined. The result of the operation may be unstable, because of the parameter setting incorrectly. Tuning the parameters used in this study combined GSP and Genetic Algorithm (GA) to improve the result continuously, to find the appropriate parameters. In the experiment, we use a medium-sized supermarket verify the results and found that after comparing with random input parameters, the parameters of the proposed method found significantly better than a random set of parameters.

Keywords: Sequential pattern mining · Generalized sequential patterns · Genetic algorithm

1 Introduction

Market basket analysis pertains to similar application of association rule, and frequently found in supermarket data analysis [1, 3]. However, for whole sale and retail business, back-buy behavior of customers related to the issue of whether the enterprise could be profitable continuously, and if the enterprise could correctly predict customers buying behavior, it will lead to enterprise continued sales increase and acquire growth [5]. However, market basket analysis can only be applied to products in the same basket, and unable to predict product relations among different basket. For further analyzing market basket data and finding potential customers, thereby sequence pattern mining developed by scholars.

However, data will change at all time, especially transaction data increasing speed is very fast, only after a while, new data set will be generated, and old data set will be

X. He et al. (Eds.): IScIDE 2015, Part II, LNCS 9243, pp. 159–170, 2015.
DOI: 10.1007/978-3-319-23862-3_16

updated, replaced. Obviously, time stamp plays a very important role in data set. Generalized sequential patterns (GSP) method is through two properties of time constraints and sliding windows, and through parameter settings (like maxGap, minGap, maxSpan, windowSize), to generate final results. After that, there have been many eastern and western scholars, specific to accuracy of its results and efficiency of calculation time, proposed some modifications [12, 13]. In addition, There have been related studies exploring GSP parameters specifically [4, 14]. Market basket analysis papers with GSP method on supermarket transaction data related issues, usually only work on one or two parameters among them, there has been no paper discussing multiple parameters simultaneously. However, among the four parameters, those are in fact inter-affected.

The study expects to use Genetic Algorithm (GA) properties to generate parameter combinations, through fitness function to determine good-bad of results, and continuously generate next generations until terminate condition reached, whereby feasible solution of a parameter configuration could be found. In addition, the study proposes a parameter base concept, where the parameter combinations solution generated through GA will be stored in the parameter base, so along with data changes, time and place changes, parameter combination solution can be updated, then use that parameter combinations to show final results.

This paper is organized as follow: Sect. 2 reviews of time series data issues, time series mining models and genetic algorithm. Section 3 describes the methodology of our study. Section 4 analyzed result and evaluated our research. Section 5 draws some conclusion and addresses the future work.

2 Related Work

Sequential patterns are item sets frequently happened and with specific rules, in sequential patterns, all items in itemsets are assumed as specific event of the same transaction time or within specific time interval. Usually, all transaction event of a customer will be viewed as a sequence, and known as customer-sequence, where each transaction will be attributed to the same itemsets, and transaction items will be sorted according to transaction time [7]. Sequential patterns mining is a process of mining for special sequential pattern under minimal support threshold satisfied situation [9]. Owing to data volume could be very huge, and users could have various interests and requirements, so threshold values are usually set by users on their own. Through self-defined minimal support threshold values, users can restrain uninterested situations by themselves, thus mining process becomes more efficient.

In early days, most sequential patterns mining are all based on association rule of Apriori property [2], wherein the property is like that as long as sub-sequences are all frequent, then the sequence is also frequent. For instance, three shopping combinations of beer and bread, bread and milk, beer and milk, if all appeared in market basket frequently, then the shopping combination of beer, bread and milk will be appeared in market basket frequently for sure. GSP [15] is a sequential pattern mining developed with Apriori as basis. GSP added time limits and broaden transaction definition, in addition, added taxonomy concept. In time limits, maxGap and minGap are defined as gap between any two adjacent records in designated sequence. If distance between two records were not within range of maxGap and minGap then those two cannot be treated as continuous records in the same sequence. GSP also defines a transaction through a sliding window, this also meaning that if maxGap and minGap were not greater than, it will be treated as in the same transaction event. The taxonomy also applied to generating multi-level sequential pattern, with these new factors, then sequential patterns mining can be defined as: Given a sequential data D, a classification T, user self-defined minGap and maxGap limits, user defined sliding window size, to inquire sequences where support greater than user defined minimum support.

However, GSP is still frequently be used in time sequence mining patterns [8, 11], because it is added with time constraint concept, so that data set spacing too long or too close situations can be avoided, leading to mining information equipped with more time effectiveness, such property let GSP show representativeness in cross market basket analysis research. However, among GSP parameter related literatures [4, 14], there is no discussions specific to GSP parameter configurations, only applied pre-process method to enhance calculation time, instead of process accuracy problem.

When using usual GSP, parameters settings must be input, thereby final results will be affected for the input parameters, leading to great variations in results of each time. Through the approach proposed by the study, user will be able to generate more stable parameter combination results through computer automated calculation, and product mix with sequential patterns can be found. And in the method proposed by the study, specific to three parameters in GSP of minGAP, maxGap and windowSize are all modified simultaneously, because the study believe that there will be interact among the three parameters, and through GA properties to help the study find out a set of better parameter combination feasible solution

3 Methodology

In the study, it is mainly through genetic algorithm properties, to solve problem in parameters setting of GSP method, and conduct comparison sort for the generated parameter combinations, select better solution to store into parameter base. When new data are added in days later, extract the stored parameter combinations from the parameter base, to enhance computational efficiency and maintain stability of calculation results. The study method is divided into 6 steps:

Step 1: Use GA to initialize settings and arbitrarily generate GSP parameters combination groups.

Step 2: Conduct GSP calculations with generated parameters combination groups according to user interested products and formulated threshold, and calculate fitness values of each group.

Step 3: Determine whether calculated results are better in fitness function value than predecessor.

Step 4: Compare and sort the generated fitness function values, and select better results to store into parameter base

Step 5: Based on high-low of fitness values, conduct selection from parameter base to evolve parent cluster of next generation, conduct GA chromosome crossover, mutation functions, to generate parameter combinations of new next generation.

Step 6: After user input settings, conduct step 5, 2, 3, 4, whereby show the present optimal parameter calculation results

3.1 Generate Parameter Combinations

When conducting GA, the most important is to determine individual coding approach, where gene is the most basic coding unit, in other words, that is to code the issue to be processed into gene mode. In the study, GSP patterns parameters are thus the issues to be processed. Generally speaking, gene coding can be divided into real number coding, binary coding, and symbolic coding. Binary coding was proposed by John Holland [6], It imitates nature creature evolution concept, and the study adopts binary coding as chromosome coding approach, binary coding is more matched with creature gene in gene crossover, mutation [10], as shown in Table 1.

Table 1. Binary coding of gene

Representing Value	7			5			1		
Location of Gene	1			2			3		
Gene Value	1	1	1	1	0	1	0	0	1

Table 2. Chromosome group

Gene Code	MinGap	MaxGap	WindowSize
1	35	70	10
2	12	38	5
......	......	......	......
P_{Size}	10	55	7

While chromosome is a set composed of by a group of gene [9], meanwhile each chromosome is a solution in the study, each gene position in chromosome represents a parameter in GSP. In addition, the set composed by chromosome is known as population [9]. In other words, population structure is as shown in Table 2. It is explained in Table 2 that chromosome population is generated from initial population size set by user, wherein, there are three genes in each chromosome. After each gene decoded, it will be represented in real number, and each chromosome is given a code.

3.2 GSP Method

GSP is a method proposed by scholar Agrawal and Srikant in 1996 [15], mainly to add time constraints so that user can find required results according to the settings. And the study is to through input user acceptable threshold value and sliding widow parameter and time constraint parameter in GSP, to conduct calculation.

Threshold value is user self-define acceptable minimum value, usually is support [9]. However, in the study, user will input product item to be inquired, and the standards to measure threshold, and set as confidence. Confidence is that under product A bought condition, the probability of buying product B. For instance, assume product A is bought, confidence of buying product B as 0.5, which shows that for every two people bought product A, one of them will be buying product B.

Sliding windows parameter, the sliding widows parameter represents all transactions generated within a certain time period, and all will be viewed as the same transaction event, and able to enhance flexibility of transaction event [15]. Normally, assume there is a data sequence d = $(d_1,\ldots,d_m)$, and included in s = $(s_1,\ldots,s_n)$, in addition, according to time stamp, window range is l_1 <= u_1 <= l_2 <= u_2 <= l_3 <= u_3 <= l_n <= u_n, showing every lower side must be less than upper side of sliding widow, in addition, window (n-1) upper side must be less than lower side of window (n). as shown below:

1. s_1 included $\mathrm{U}_{k=l_i}^{u_i} d_k, 1 \leq i \leq n$ in
2. transaction time (d_{ui}) - transaction time (d_{li}) <= windowSize, 1 <= i <= n

Time constraint parameter, time constraints is formed by constraining to a specific time interval of time stamp. User is able to set three time constraints of windowSize, maxGap, and minGap to find time interval of their own interest [15]. As shown below in followings.

1. s_i included $\mathrm{U}_{k=l_i}^{u_i} d_k, 1 \leq i \leq n$ in
2. transaction time (d_{ui}) - transaction time (d_{ll}) <= windowSize, 1 <= i <= n
3. transaction time (d_{li}) - transaction time (d_{ui-1}) > minGap, 2 <= i <= n
4. transaction time (d_{ui}) - transaction time (d_{li-1}) <= maxGap, 2 <= i <= n

3.3 The Process of Generating New Chromosome

The study adopts roulette wheel selection to select parent population, where the policy of roulette wheel selection is through fitness function value to judge good-bad of parent generation chromosome, whereby determines probability of selected for crossover execution. If fitness function value were higher, then selected probability will be higher; on the contrary, the selected possibility will be lower. The policy of roulette wheel selection can be divided into four steps:

Step 1: Sort chromosome of parent generation population according to fitness function values, and based on Eq. 1 to calculate weight of individual chromosome.

$$Total\_FitnessValue = \sum_{i=1}^{P_{size}} FV_i$$

$$FV\_Rate_j = \sum_{i=1}^{j} \frac{FV_i}{\text{Total\_FitnessValue}}$$

$$\text{Where j} = 1,2,\ldots,\text{P}_{size},\ FV = Fitness\ Value \tag{1}$$

Step 2: Randomly generate as set of value R between 0 and 1. *RandValue = R[0,1]*

Step 3: Based on the generated R values, determine chromosome of parent generation population for executing crossover. If $R < FV\_Rate_i$, then select chromosome i, for instance, if $FV\_Rate_9 < R < FV\_Rate_{10}$ then select chromosome 10 for executing crossover.

Step 4: Repeat Step 2, 3, until there are enough chromosome for generating next generation chromosome selected.

In GA, gene crossover is mainly to generate new next generation through crossover of gene values of two selected chromosome, so that the generated next generation can be provided with partial properties of good parents. About gene crossover policy, the prior study works proposed many crossover methods of single point, two points, uniform, and order etc.

In the study, it is configuration methods among three parameters. At gene crossover, it is not gene position to be considered, instead specific to parameter position to consider whether to conduct gene crossover, the steps are as following.

Step 1: Set probability of crossover (PC), and specific to chromosome randomly generate a set of R values between 0 and 1. *RandValue = R[0,1]*

Step 2: If $R < PC$, then prepare to conduct gene crossover; otherwise, don't conduct gene crossover.

Step 3: Generate $CrossoverSite = Rand[1,C_{size}],\ where\ size = 1,2,\ldots,C_{count}$

In mutation of GA, it will let gene values in chromosome generate changes arbitrarily, thereby chromosome jump out of the present searching space, in this way, generating range constrained better solution phenomenon can be avoided to find better solution, the implementation steps are as shown in the following.

Step 1: Set mutation rate (*MR*), and specific to each parameter position, randomly generate a set of *R* value between 0 to 1. *RandValue = R[0,1]*

Step 2: If $R < MR$ then choose to conduct mutation; otherwise, stop mutation.

Step 3: Generate MutationSite (*MS*) = $Rand[1,C_{size}],\ where\ size = 1,2,\ldots,C_{count}$

Step 4: Transfer gene at *MS* in step 3 from 0 to 1 or from 1 to 0.

4 Empirical Analysis

4.1 Data Description and Experiment Design

The data source adopted in the study is from M fresh supermarket chain, at present, the supermarket has 75 stores, mainly provides products of various kinds of fresh foods, beverages, daily supplies and groceries. The study uses daily transaction data of one year of one store of M supermarket chain as study object. The study employs Microsoft SQL server 2012 as main experiment tool. The master data adopted by the study is consumption details file, where data contents include not only company id, store no., cash register no, cashier no., but also transaction date, time, etc. However, the study aims at finding product mix with series relation, so must to identify whether consumption records at different time are by the same consumer. Hence data screening was conducted in SQL server, reserve these three fields and delete member card no. field as null data, the remaining data is totally 640,623 transaction records, and the final data form applied is as shown in Table 3.

Table 3. Data format

	transdat	memno	itemno	stamp
1	20011231	0200017903-00	4710088411785	4
2	20011231	0200017903-00	4710199035986	4
3	20011231	0200017903-00	4710063312168	4
4	20011231	0200017903-00	4710126170148	4
5	20011231	0200017903-00	4710126170148	4
6	20011231	0200017903-00	4710247006999	4
7	20011231	0200017903-00	9004080	4
8	20011231	0200017903-00	9004080	4
9	20011231	0200017903-00	9004224	4
10	20011231	0200017983-00	4711194164060	4
11	20011231	0200017983-00	4712934010074	4
12	20011231	0200017903-00	9005086	4

In the experiment, it is considered that chromosome must be stored into parameter base, for controlling chromosome volume in parameter base, so the experiment adopts the one with fewer population number which proposed by J.J. Grefenstette: crossover rate = 0.9, mutation rate as = 0.01, initial population number = 30. On comparison and analysis of experiment, the study conducts differentiation comparison between the method of the study and arbitrary generated parameter combinations, by comparing each time calculated average confidence and variance of results, to explore efficacy of the research method. Equation description is followed:

$$(\text{Avgerage Confidence}) = \sum_{i=1}^{N} C_i/_N \quad (\text{Variance}) = \sum_{i=1}^{N} (C_i - \bar{C})^2/_N$$

Where N:The Amount of Parameter C_i: Confidence Value of Each Chromosome, $i = 1,2,\ldots,N$

The study conducts experiment with random selected 10 products. Where parameter settings related information are synthesized as shown in Table 4.

Table 4. Parameter setting of experiment

Population	30
Probability of Crossover (PC)	0.9
Mutation Rate (MR)	0.01
Threshold- Confidence	0.25
Fitness Function	avg. *conf.*

4.2 Experiment Results

The study randomly selected 10 products and input the following products according to the process of methodology in Fig. 1 as followings:

1. Black bridge sausage (original flavor) 2. I-Mei traditional soy milk (no sugar) 1000 ml 3. Kuang-Chuan embryo rice milk - 1000 ml 4. PECOS meat ball noodle - 90 g 5. Chicken inner muscle 6. Organic Town fresh egg – 10 ea 7. Kyoho grapes (16 ea) 8. Water spinach 9. Cow-Head brand soup stock original flavor - 411 g 10. High class fine salt - 1 kg

Comparison conducted between experiment results and random input parameter results, and with average fitness function value and fitness function variance of each set of chromosome as comparison method, the results are as shown in Figs. 1 and 2.

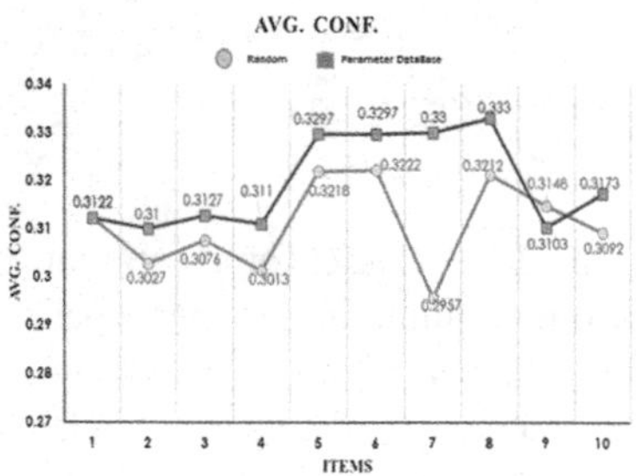

Fig. 1. Average fitness function values data source: synthesized by the study

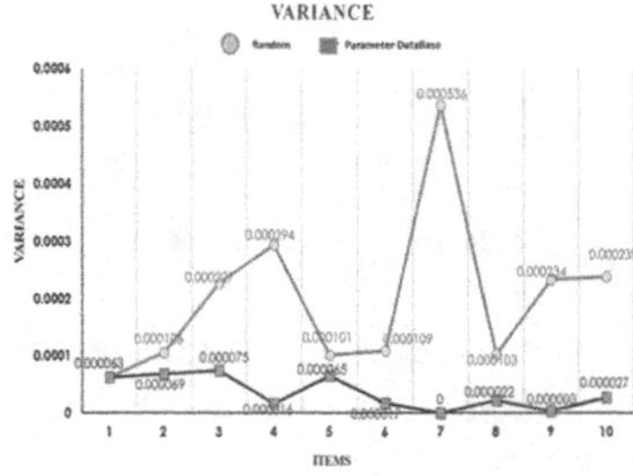

Fig. 2. Variance of fitness function value data source: synthesized by the study

In experiment results, the parameter combination not reaching threshold set by user will be shown as null value. And it is found in the results that in test with random generated parameters, about 6.5 null values generated from calculation results of each product; while with the experiment of the study, except 7 null values appeared from random generated parameters for the first product, there is no any null value generated with the other chromosome combinations generated from brought-in variables, and on fitness function value, performance also shown more superior.

Further, in observation of fitness function value combinations calculated from different product, it is found that the parameter combinations stored into parameter base after calculated for different products will alter continuously, so parameter base will not generate problem of over rigid, leading to over convergence, resulted in area best solution generated.

As shown in experiment results, comparing parameter establishment to arbitrary input, no matter on average value of calculation results, or on calculation results of input multiple parameter combinations, all provided with superior performance. In addition, by observations on parameter combinations of best performance in various products, it is found that there will be no dominated parameter combinations in parameter base, leading to area best solution problem. Therefore, in the experiment, parameter base establishment is able to let user more efficiently find proper product mix with sequence relation. The GSP conducted in the study is to explore configuration relations among three parameters of minGap, maxGap, and windowSize, the study uses upper-lower to change one of the parameters, specific to each product, to show best parameter configuration and conduct efficacy test, the experiment results are as shown in Figs. 3, 4.

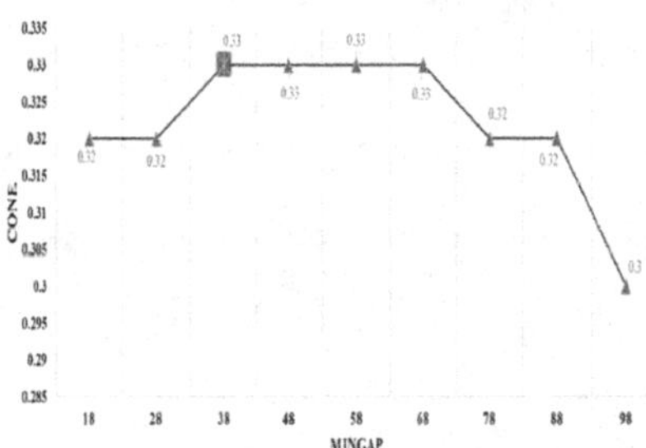

Fig. 3. Parameter configuration (maxGap)

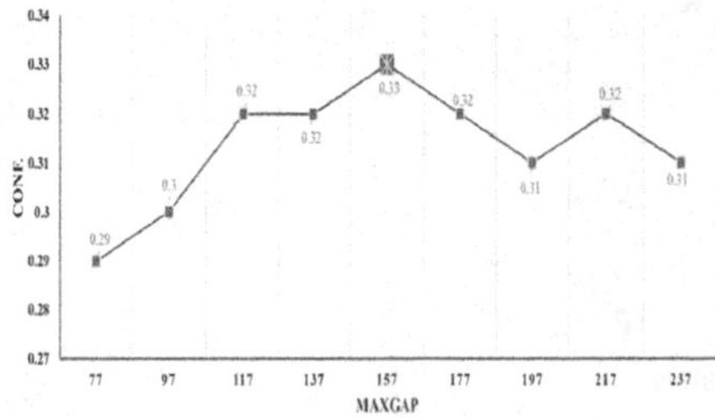

Fig. 4. Parameter configuration (minGap)

In addition, it is also proved that results listed in Table 5 have better performance in parameter configuration test. Finally, calculated results with GSP for different products will be explained, to judge whether there are time sequence related products. Table 5 shows results of the first product. The parameter combinations used for the first product are random input, and the results show that there are five products provided with sequence relation with it, it is about average every three people bought Black bridge sausage (original flavor) for a period of time, there will be one person back-buy the latter product.

Table 5. Consequence result of black bridge sausage

Antecedent	Consequent	Confidence
Black bridge sausage (original flavor)	Scallion	0.33
	Lin Fengmu Field 100% premium milk 1 / 2G	0.33
	High Agricultural egg	0.33
	Vacuum Packed mushroom	0.32
	Highland cabbage	0.32

Table 6. Consequence result of other products

Antecedent	Consequent Confidence	Parameter Combinations		
		maxGap	minGap	windowSize
Black bridge sausage (original flavor)	Vacuum Packed mushroom (0.49) Highland cabbage (0.39) Scallion (0.38)	157	45	2
I-Mei traditional soy milk (no sugar) 1000ml	Vacuum Packed mushroom (0.39) Highland cabbage (0.38) High Agricultural egg (0.36)	211	110	5
Kuang-Chuan embryo rice milk 1000ml	Vacuum Packed mushroom (0.39) Lin Fengmu Field 100% premium milk 1 / 2G(0.32) Scallion (0.32)	235	78	6
PECOS meat ball noodle - 90g	Highland cabbage (0.49) Vacuum Packed mushroom (0.43) Scallion (0.41)	196	31	6
Chicken inner muscle	Scallion(0.43) Vacuum Packed mushroom (0.41) Lin Fengmu Field 100% premium milk 1 / 2G(0.38)	157	47	5
Organic Town fresh egg – 10 ea	Highland cabbage (0.46) Scallion(0.44) Vacuum Packed mushroom (0.44)	235	17	6
Kyoho grapes (16 ea)	Scallion (0.54) Highland cabbage (0.51) Vacuum Packed mushroom (0.5)	196	16	6
Water spinach	Vacuum Packed mushroom (0.46) Scallion (0.41) Highland cabbage (0.38)	225	102	2
Cow-Head brand soup stock original flavor - 411g	Scallion (0.39) Lin Fengmu Field 100% premium milk 1 / 2G(0.39) High Agricultural egg (0.37)	157	38	6

After other product input, results are synthesized as shown in Table 6, and other detail results are as shown in attachment 1. Table 6 lists out top three products of fitness function values of calculation results, and marked out the input parameter combination values. As shown in the results, there are certain products with longer buying period interval, such as Kuang-Chuan embryo rice milk - 1000 ml, Cow-Head brand soup stock original flavor - 411 g, sequence product will be appeared after 100 days later. And there are other products take about 2–3 weeks buying interval, such as Kyoho grapes (16 ea), Water spinach.

5 Conclusion and Future Research

5.1 Conclusion

1. In average fitness function value, parameter calculation results with parameter base usage are higher than arbitrary parameter input. On calculating fitness function value of each product, the study method will input all calculated parameter combinations, the generated results through averaging to calculate superiority of performance of chromosome of the set.
2. On stability of calculated results after parameters input, parameter base is equipped with stable performance. GSP is through user self input parameters for settings to conduct calculation, so results will also be different owing to difference in input parameters, leading to different results shown each time, so solve the situation of unstable results owing to self input, the study proposed parameter base establishment, is able to go through continuous evolution, so that when parameters input, there could be a feasible solution of more stable results.
3. The best parameter combination of each kind of product is different, so there will be no such situation that one parameter combination dominated. GSP will have different best parameter configuration owing to different data, however, through calculation, GA will converge into a set of better feasible solution, while the set of feasible can only be used in one product.

5.2 Future Research

1. Explore multiple itemsets: In the study, sequence relation of a single product is explored. However, GSP is able to explore multiple products sequence relation, therefore, it should be capable for exploring problems in multiple itemsets.
2. Add maxSpan parameter: the GSP parameter discussions of the study, there are only three of maxGap, minGap and windowSize explored, without maxSpan mentioned. If this parameter could be added, the study could be more abundant.
3. Apply to other parameter issues: As shown in experiment results, parameter base is able to effectively reduce the situations of unstable results generated owing to improper parameter setting. Therefore, the method could be applied to other parameter issues, to explore feasibility of other different issues.

References

1. Agrawal, R., Srikant, R.: Fast algorithms for mining association rules. In: Proceedings of 20th International Conference very Large Data Bases VLDB, vol. 1215, pp. 487–499, September 1994
2. Brijs, T., Swinnen, G., Vanhoof, K., Wets, G.: Using association rules for product assortment decisions: a case study. In: Proceedings of the Fifth ACM SIGKDD International Conference on Knowledge Discovery and Data Mining, pp. 254–260. ACM, August 1999
3. Chiang, D.A., Wang, Y.H., Chen, S.P.: Analysis on repeat-buying patterns. Knowl.-Based Syst. **23**(8), 757–768 (2010)
4. De Jong, K.A., Spears, W.M.: An analysis of the interacting roles of population size and crossover in genetic algorithms. In: Schwefel, H.-P., Männer, R. (eds.) PPSN 1990. LNCS, vol. 496, pp. 38–47. Springer, Heidelberg (1991)
5. Grefenstette, J.J.: Optimization of control parameters for genetic algorithms. IEEE Trans. Syst. Man Cybern. **16**(1), 122–128 (1986)
6. Han, J., Pei, J., Mortazavi-Asl, B., Chen, Q., Dayal, U., Hsu, M.C.: FreeSpan: frequent pattern-projected sequential pattern mining. In: Proceedings of the Sixth ACM SIGKDD International Conference on Knowledge Discovery and Data Mining, pp. 355–359. ACM, August 2000
7. Wang, J.-T., Chern, M.-S., Yang, D.-L.: A two-machine multi-family flowshop scheduling problem with two batch processors. J. Chin. Inst. Ind. Eng. **18**(3), 77–85 (2001)
8. Kantardzic, M.: Data mining: concepts, models, methods, and algorithms. John Wiley & Sons, Hoboken (2011)
9. Lin, M.Y., Lee, S.Y.: Incremental update on sequential patterns in large databases. In: 1998 Proceedings of Tenth IEEE International Conference on Tools with Artificial Intelligence, pp. 24–31. IEEE, November 1998
10. Masseglia, F., Poncelet, P., Teisseire, M.: Pre-processing time constraints for efficiently mining generalized sequential patterns. In: 2004 Proceedings of 11th International Symposium on Temporal Representation and Reasoning, TIME 2004, pp. 87–95. IEEE, July 2004
11. Nakanishi, Y.: Application of homology theory to topology optimization of three-dimensional structures using genetic algorithm. Comput. Methods Appl. Mech. Eng. **190**(29), 3849–3863 (2001)
12. Oracle: Big Data for the Enterprise, An Oracle White Paper. (2013). http://www.oracle.com/us/products/database/big-data-for-enterprise-519135.pdf
13. Pei, J., Han, J., Mortazavi-Asl, B., Wang, J., Pinto, H., Chen, Q., Hsu, M.C.: Mining sequential patterns by pattern-growth: the prefixspan approach. IEEE Trans. Knowl. Data Eng. **16**(11), 1424–1440 (2004)
14. Qin, L.X., Shi, Z.Z.: Efficiently mining association rules from time series. Int. J. Inf. Technol. **12**(4), 30–38 (2006)
15. Zaki, M.J.: SPADE: an efficient algorithm for mining frequent sequences. Mach. Learn. **42** (1–2), 31–60 (2001)

Event Recovery by Faster Truncated Nuclear Norm Minimization

Debing Zhang, Long Wei[(✉)], Bin Hong, Yao Hu, Deng Cai, and Xiaofei He

Zhejiang University, Hangzhou, China
{debingzhangchina,weilongzju,
hongbinzju,yaoohu,dengcai,xiaofeihe}@gmail.com

Abstract. When we want to know an event we are concerned, it is likely that the collected information is incomplete which may severely affect the consequent analysis. In this paper, we focus on the event recovery problem that aims to discover missing historical information for a certain event based on the limited known information. We formulate an event as a two dimensional data matrix, which will be called the event matrix in this paper, and convert the original problem to matrix completion problem. We observe that the event matrix has low-rank structure due to the strong dependence between different event attributes. Then we adopt a recently proposed approach called Truncated Nuclear Norm Minimization (TNNM) to recover the event matrix. We also propose an early stopping strategy to further accelerate the optimization of TNNM. Experimental results on a collected event dataset demonstrate the effectiveness and the fast convergence rate of the proposed algorithm.

Keywords: Event evolution analysis · Matrix completion · Missing information · Optimization · Truncated nuclear norm

1 Introduction

Nowadays, it is convenient to collect information from the Internet for an event we are concerned. The information may come from different sources, describe different aspects about the event and record it at different time. However, in many real world applications, we are facing the problem of incomplete information, which makes it difficult to analyze historical evolution of the event. There are several possible reasons about it. First, different media may provide information from different viewpoints or at different time. Second, some indicators about an event may be calculated once a month while some may be calculated once every quarter year. Third, collecting all the information is very time consuming, and it is highly possible that one may miss some information. And fourth, some collected information may be false and should be discarded.

To address this problem, we propose event recovery to present a full historical picture of a certain event from all aspects at any time. It is an important topic in various real applications such as city expansion, disease spreading [8], cultural

X. He et al. (Eds.): IScIDE 2015, Part II, LNCS 9243, pp. 171–180, 2015.
DOI: 10.1007/978-3-319-23862-3_17

evolution, public security [9] and so on. It has been paid more and more attention with the rapid development of the Internet, which enables people to discover detailed information of an event more conveniently. Note that in this paper we only care about recovering missing historical information about an event.

It is natural to model the collected information about an event as a matrix $X \in \mathbb{R}^{n \times t}$ which will be referred to as the *event matrix*. Each row of X represents an attribute of the event at different time, and each column of X represents all the attributes of the event at a certain time. For example, suppose we are considering "economic crisis" as an event, then different rows of the event matrix may represent different economic indicators of different countries. And different columns represent all the indicators at different time, for example, from Jan. 2007 to Dec. 2012.

We observed that it is possible to recover the event matrix based on the available incomplete information about the event since all rows (attributes) of the event matrix belong to the same event and there should be strong connections and dependencies between them. So the rank of the event matrix should be low. As we test on the collected "economic crisis" dataset, the event matrix can be well approximated by a low rank matrix. The result is shown in Fig. 1. As can be seen, although the original event matrix's size is 88×161, it can be very well approximated by a low rank matrix whose rank is only 20. So the problem of event recovery can be viewed as a low rank matrix completion problem.

In this paper, we propose a novel method called Faster Truncated Nuclear Norm Minimization(FTNNM) to recover missing information of a certain event. We use truncated nuclear norm to capture the low-rank structure of the event matrix. And an early stopping strategy is proposed to accelerate the TNNM. The experimental results on a collected dataset show the effectiveness of the proposed algorithm.

The rest of the paper is organized as follows. In the next section, we first provide a brief description of matrix completion methods. In Sect. 3, we present our approaches to build the event matrix and recover missing information by FTNNM. We then conduct a series of experiments on a real-world dataset, and the results are shown in Sect. 4. Finally, we provide some concluding remarks in Sect. 5.

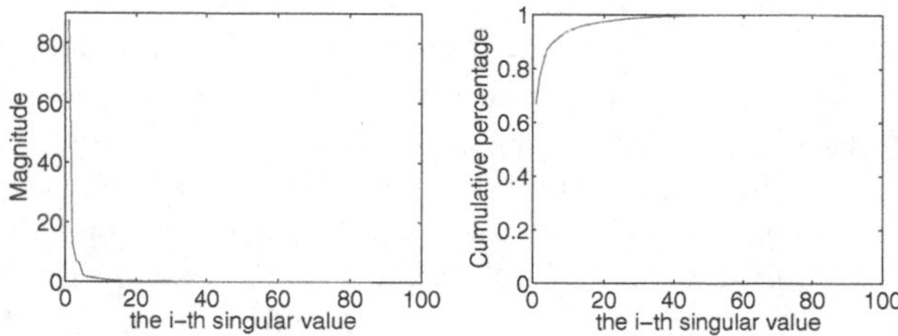

Fig. 1. The singular values' distribution of a normalized real-world event matrix. The size of the event matrix is 88×161, However, the top 20 singular values can keep most of the event information (more than 95 %).

Notations: Given a matrix $X \in \mathbb{R}^{n \times t}$, let σ_i denote X's i-th largest singular value. So $\sigma_1 \geq \sigma_2 \geq ... \geq \sigma_{\min(n,t)}$ are all the singular values of X. And the nuclear norm of X is defined as: $\|X\|_* = \Sigma_{i=1}^{\min(n,t)} \sigma_i$. let $\Omega \subset \{1, 2, ..., n\} \times \{1, 2, ..., t\}$ denote the set of indices of the observed information of X, and let Ω^c denote the indices of the missing entries. It is convenient to summarize the information available of X via symbol X_Ω, which is defined as

$$(X_\Omega)_{ij} = \begin{cases} X_{ij} & \text{if}(i, j) \in \Omega \\ 0 & \text{if}(i, j) \in \Omega^c. \end{cases} \tag{1}$$

2 Related Work

To recovery missing historical information about an event is a typical matrix completion problem. In this section, we briefly review three approaches of matrix completion, including rank minimization, nuclear norm minimization and truncated nuclear norm minimization.

2.1 Rank Minimization

Rank minimization problem is NP-hard due to the non-convex and non-smooth nature of the rank operator $rank(.)$. Meka et al. [1] proposed an iterated greedy algorithm (Singular Value Projection, SVP) to solve the rank minimization problem in Eq. (2) directly. Under the incoherence condition [1], SVP can converge to a low rank solution of problem. However, in real applications, the incoherence condition is usually not satisfied.

$$\begin{aligned} \min_X \quad & rank(X) \\ s.t. \quad & X_\Omega = M_\Omega \end{aligned} \tag{2}$$

2.2 Nuclear Norm Minimization

Inspired by compressed sensing which uses the L_1 norm to approximate the L_0 norm, researchers adopt the nuclear norm as a convex surrogate of the rank operator [2,3]. Nuclear norm which is defined by the summation of all the singular values is much easier to optimize than the original rank operator. Cai et al. [3] proposed a Singular Value Thresholding (SVT) algorithm to solve the nuclear norm based matrix completion problem like Eq. (3). It can be seen easily that when τ tends to ∞, the nuclear norm can be accurately approximated by the SVT's objective function.

$$\begin{aligned} \min_X \quad & \tau\|X\|_* + \frac{1}{2}\|X\|_F^2 \\ s.t. \quad & X_\Omega = M_\Omega \end{aligned} \tag{3}$$

2.3 Truncated Nuclear Norm Minimization

Compared with the rank operator in which all the non-zero singular values are evaluated equally as 1, the nuclear norm treats the singular values differently. So in general, the nuclear norm based approaches can not guarantee an accurate low rank solution. Zhang et al. [4] modified the traditional nuclear norm, and proposed a new objective function called truncated nuclear norm minimization as shown in Eq. (4). Suppose $X \in \mathbb{R}^{n \times t}$, the *truncated nuclear norm* $\|.\|_r$ is defined as the summation of the smallest $\min(n,t) - r$ singular values. By leaving the largest r singular values free, we can always achieve the low rank solution when $\|.\|_r$ tends to 0 during the optimization.

$$\begin{aligned} \min_X \quad & \|X\|_r \\ s.t. \quad & X_\Omega = M_\Omega \end{aligned} \tag{4}$$

3 Our Approach

3.1 Building Event Matrix From Collected Information

To recover a certain event, the first thing is to build the event matrix from information collected from different sources. Each piece of information describes an attribute about the event at some time. We gather all mentioned attributes about the event as an attribute set A, and gather all mentioned times as a time set T. Then each piece of information can be associated to a two dimensional tuple $I = (i, j)$, where $i \in A, j \in T$.

For those attributes describing numerical character about the event, we directly use the scalar to represent the corresponding piece of information. For example, if we regard the "economic crisis" as an event, then economic indicators like Gross Domestic Product(GDP), Consumer Price Index(CPI) can be used directly as attributes of this event. Unfortunately, we may be confronted with situations when no scalar is contained or explicitly given in the collected information. For these cases, we need to quantizate the attribute by calculating some important statistics such as variance and density or adopting some cross-modal processing techniques, such as [5,6,10]. Then the event can be represented by an *incomplete information matrix* $M \in \mathbb{R}^{n \times t}$ with missing values, where $n = |A|$ and $t = |T|$. We use Ω denote the indices of observed values in M, where each observed entry of M quantizes a piece of information $I = (i, j)$ for some $i \in A, j \in T$.

To recover the *complete event matrix* X, we need to consider the relationship between the complete event matrix and the incomplete information matrix. A natural requirement for the complete event matrix is to consistent with the incomplete information matrix on the observed values since the derived event matrix should respect the existed information of the event. In addition, as we mentioned in the Introduction section, each row of the event matrix describes some attribute or aspect of the event, thus there exists strong dependence among rows of the event matrix. And statistics test we conducted on real world event "economic crisis" also shows the low-rank structure of the event matrix. Therefore, our goal is to recover the complete event matrix X with the following two constraints: (1)$X_\Omega = M_\Omega$; (2)minimizing the rank of X.

3.2 Learning the Event Matrix Using Truncated Nuclear Norm Minimization

Inspired by matrix completion approaches proposed in [4], we will introduce how to recover the complete event matrix based on Truncated Nuclear Norm Minimization. And we will also show that Truncated Nuclear Norm Minimization can be further accelerated by adding an early stopping strategy to the original optimization.

The truncated nuclear norm of matrix X is the summation of the smallest $\min(n,t)-r$ singular values of X, denoted by $\|X\|_r$. Theoretical analysis in [4] shows that there exists close relationship between the truncated nuclear norm and the traditional nuclear norm:

$$\|X\|_r = \|X\|_* - \max_{AA^T=I, BB^T=I} Tr(AXB^T) \tag{5}$$

where $A \in \mathbb{R}^{r\times n}, B \in \mathbb{R}^{r\times t}$.

Considering the constraint that the complete event matrix should be consistent with the incomplete information matrix on observed values, we only need to optimize the following objective function:

$$\begin{aligned} \min_X \quad & \|X\|_* - \max_{AA^T=I, BB^T=I} Tr(AXB^T) \\ s.t. \quad & X_\Omega = M_\Omega. \end{aligned} \tag{6}$$

3.3 Optimization

Following [4], we also apply the Augmented Lagrangian Method(ALM) to optimize the Eq. (6) in an iterative scheme.

$$\begin{aligned} L(X,Y,W,\beta) = & \|X\|_* - Tr(AWB^T) + \frac{\beta}{2}\|X-W\|_F^2 \\ & + Tr(Y^T(X-W)), \end{aligned} \tag{7}$$

where $\beta > 0$ is the penalty parameter. Given the initial settings $X_1 = M_\Omega, W_1 = M_\Omega$ and $Y_1 = M_\Omega$,

we can recover the final event matrix by minimizing $L(X,Y,W,\beta)$ via iteratively performing the following four steps:

$$X_{k+1} = \underset{X}{\operatorname{argmin}}\, L(X, Y_k, W_k, \beta) \tag{8}$$

$$W_{k+1} = \underset{W}{\operatorname{argmin}}\, L(X_{k+1}, Y_k, W, \beta) \tag{9}$$

Algorithm 1. Faster Iterative Scheme for TNNM

Input: the incomplete information matrix M_Ω, the error ϵ, the early stopping parameters $\{\epsilon_i\}, i = 0, 1, 2, ...$
Initialization: $X_0 = M_\Omega, l = 0$
repeat
  Calculate the SVD of X_l as $X_l = U_l \Sigma_l V_l^T$
  Calculate A_l, B_l as [4]
  $\hat{X}_0 = M_\Omega, W_0 = M_\Omega, Y_0 = M_\Omega, k = 0$
  repeat
    $\hat{X}_{k+1} = \mathcal{D}_{\frac{1}{\beta}}(W_k - \frac{1}{\beta}Y_k)$
    $W_{k+1} = \hat{X}_{k+1} + \frac{1}{\beta}(A_l^T B_l + Y_k)$
    $Y_{k+1} = Y_k + \beta(\hat{X}_{k+1} - W_{k+1})$
    $W_{k+1} = (W_{k+1})_{\Omega^c} + M_\Omega$
    $k = k + 1$
  until $L(X_k, Y_k, W_k) - L(X_{k+1}, Y_{k+1}, W_{k+1}) < \epsilon_l$
  $X_{l+1} = \hat{X}_{k+1}$
  $l = l + 1$
until $\|X_{l+1} - X_l\|_F^2 \leq \epsilon$

$$Y_{k+1} = Y_k + \beta(X_{k+1} - W_{k+1}). \tag{10}$$

$$W_{k+1} = (W_{k+1})_{\Omega^c} + M_\Omega. \tag{11}$$

The iteration procedure stops when the decrease of the augmented lagrange function $L(X, Y, W, \beta)$ is small enough as follows

$$L(X_k, Y_k, W_k, \beta) - L(X_{k+1}, Y_{k+1}, W_{k+1}, \beta) < \epsilon. \tag{12}$$

3.4 Early Stopping Stategy

In [4], each iteration is required to be solved accurately. However, this requirement needs many iterations which will make the optimization inefficient. Inspired by the Contrastive Divergence strategy used in the training of Restricted Boltzmann Machines (RBMs) [7], we find that there is no need to accurately solve the optimization problem in each iteration to guarantee the convergence of the whole optimization. So we add a simple but effective early stopping strategy by adaptively changing the ϵ in the stop condition (12) as follows

$$L(X_k, Y_k, W_k, \beta) - L(X_{k+1}, Y_{k+1}, W_{k+1}, \beta) < \epsilon_l. \tag{13}$$

ϵ_l is empirically chosn to be $\epsilon_l = \min(10^{-(\alpha+l/\gamma)}, \epsilon), \alpha = 3, \gamma = 2$. And in this way, the optimization will stop earlier in the first several iterations which can greatly reduce the computational cost of TNNM.

We summarize the modified faster optimization of truncated nuclear norm minimization(TNNM) in Algorithm (1).

4 Experiments

In this section, we show the experimental results on a collected real-world event data matrix.

4.1 Data Set Description

The data set is collected from The World Bank, Global Economic Monitor (http://data.worldbank.org/data-catalog/global-economic-monitor). The information that how the economic indicators change from Jan. 2000 to May 2013 in many different countries has been collected. Specifically, the collected indicators contain "total reserves", "terms of trade", "stock market index", "official exchange rate", "real effective exchange rate", "industrial production", "Consumer Price Index(CPI)" and "Exports Merchandise". And the collected countries include USA, UK, Canada, France, Germany, Italy, Japan, Australia, Switzerland, Spain and Russia. The event matrix is built as described in Sect. 3.1. And the size of the event matrix is 88×161.

4.2 Compared Methods

Altogether, we compare the following five methods:

- Baseline. One simple intuitive idea is to use linear interpolation to estimate the missing values of each row seperately. And we consider it as the baseline method.
- SVP [1]. Singular Value Projection method estimates the missing values by solving the rank minimization directly by using an iterated greedy algorithm.
- SVT [3]. Singular Value Thresholding method estimates the missing values by solving the nuclear norm minimization problem.
- TNNM [4]. Truncated Nuclear Norm Minimization method estimates the missing values by solving a modified nuclear norm minimization.
- FTNNM. Faster Truncated Nuclear Norm Minimization is the proposed method.

4.3 Evaluation Criterion

We use the relative error on the missing values to evaluate the performance of different methods. Suppose the ground truth event matrix is X_{gt}, the observed indices is Ω, and the recovered matrix is X_{rec}. Then the relative error is defined as follows:

$$\frac{\|(X_{rec} - X_{gt})_{\Omega^c}\|_F^2}{\|(X_{gt})_{\Omega^c}\|_F^2}.$$

Clearly, the smaller the relative error is, the better the performance can be achieved.

4.4 Results

Firstly, we randomly cover some positions of the event matrix, and then compare the performance of the five algorithms. The observed ratio can vary from 60 % to 90 %. And the results are shown in Fig. 2. As can be seen, all the matrix completion based methods perform better than the simply linear interpolation. This is because linear interpolation can only use the local information of a single row (attribute) while matrix completion methods can take advantage of all the observed information. Also we can see that truncated nuclear norm based approach is better than nuclear norm based SVT due to the better approximation ability of truncated nuclear norm to the rank function. And FTNNM's performance is very similar to that of TNNM.

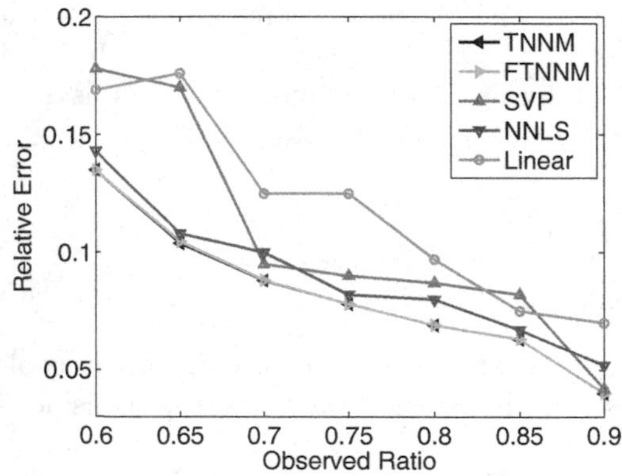

Fig. 2. The relative errors achieved by all the five compared methods. It can be seen that TNNM and FTNNM can recover the incomplete event matrix more accurately than other compared methods.

Secondly, for TNNM and FTNNM, the parameter r in truncated nuclear norm $\|.\|_r$ needs to be set first. We simply try all the small r and choose the one with best performance. How the performance changes with r is shown in Fig. 3 when the observed ratio is 90 %. We can see that, the performance will first decrease and then increase when the parameter r changes from small to large. And based on this property, we do not need to exhaustively search all the rs in order to find a good r in real applications.

Thirdly, we show the fast convergence property of the proposed FTNNM method. Suppose the observed ratio is 90 %. We can see that although we adopt an early stopping strategy, the convergence of problem (4) can still be guaranteed. And there is almost no performance drop. In this way, the TNNM based approach can be greatly accelerated. And computational cost comparison of TNNM and FTNNM is shown in Fig. 4. Overall, FTNNM only takes half time of TNNM to achieve similar performance.

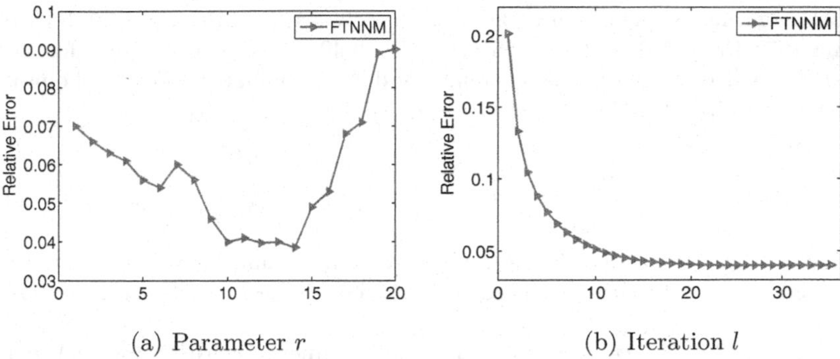

(a) Parameter r (b) Iteration l

Fig. 3. (a) shows the relative errors achieved by the proposed FTNNM with different choices of the parameter r. It can be seen that FTNNM can achieve very good results when the parameter r is between 10 and 15. And (b) shows how the relative error converges with respect to the iteration counter l. Although an early stop strategy is adopted to reduce the computational cost in each iteration, the convergence can still be guaranteed.

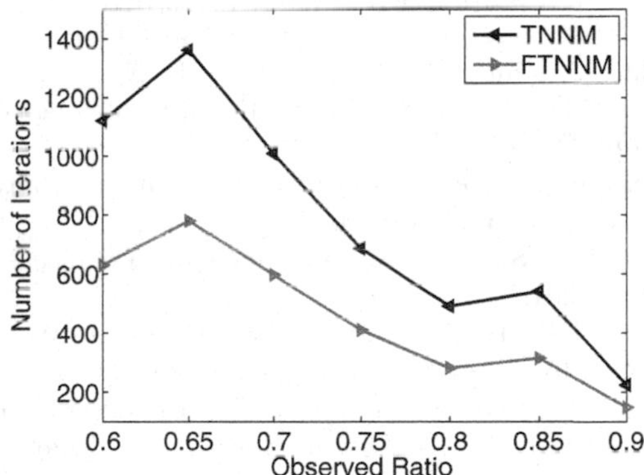

Fig. 4. The total number of iterations needed by TNNM and FTNNM. FTNNM only takes about half the iterations, and can achieve similar results (Fig. 2) compared with TNNM.

5 Conclusion

In this paper, we propose the idea of using matrix completion techniques to recover the event matrix. We adopt the recently proposed truncated nuclear norm minimization to achieve better recovery. What's more, an efficient early stopping strategy is proposed to greatly reduced the computational cost of truncated nuclear norm minimization. Experiments demonstrate the effectiveness and the efficiency of the proposed method. In the future, we will consider how to learn the event prediction based on the recovered event matrix.

Acknowledgements. This work was supported by National Basic Research Program of China (973 Program) under Grant 2012CB316404, National Program for Special Support of Top-Notch Young Professionals, and National Natural Science Foundation of China under Grant 61233011 and Grant 61125203.

References

1. Jain, P., Meka, R., Dhillon, I.S.: Guaranteed rank minimization via singular value projection. In: Advances in Neural Information Processing Systems, pp. 937–945 (2010)
2. Toh, K.C., Yun, S.: An accelerated proximal gradient algorithm for nuclear norm regularized linear least squares problems. Pac. J. Optim. **6**, 615–640 (2010)
3. Cai, J.F., Cands, E.J., Shen, Z.: A singular value thresholding algorithm for matrix completion. SIAM J. Optim. **20**(4), 1956–1982 (2010)
4. Zhang, D., Hu, Y., Ye, J., Li, X., He, X.: Matrix completion by truncated nuclear norm regularization. In: IEEE Conference on Computer Vision and Pattern Recognition (CVPR), pp. 2192–2199 (2012)
5. Rasiwasia, N., Costa Pereira, J., Coviello, E., Doyle, G., Lanckriet, G. R., Levy, R., Vasconcelos, N.: A new approach to cross-modal multimedia retrieval. In: Proceedings of the International Conference on Multimedia, pp. 251–260. ACM (2010)
6. Ngiam, J., Khosla, A., Kim, M., Nam, J., Lee, H., Ng, A. Y.: Multimodal deep learning. In: Proceedings of the 28th International Conference on Machine Learning, pp. 689–696 (2011)
7. Hinton, G.E.: Training products of experts by minimizing contrastive divergence. Neural Comput. **14**(8), 1771–1800 (2002)
8. Sadilek, A., Kautz, H.A., Silenzio, V.: Modeling spread of disease from social interactions. In: ICWSM (2012)
9. Jasso, H., Fountain, T., Baru, C., Hodgkiss, W., Reich, D., Warner, K.: Prediction of 9-1-1 call volumes for emergency event detection. In: Proceedings of the 8th Annual International Conference on Digital Government Research: Bridging Disciplines & Domains, pp. 148–154. Digital Government Society of North America (2007)
10. Mao, X., Lin, B., Cai, D., He, X., Pei, J.: Parallel field alignment for cross media retrieval. In: Proceedings of the 21st ACM International Conference on Multimedia, pp. 897–906. ACM (2013)

Detecting Fake Review with Rumor Model—Case Study in Hotel Review

Tien Chang[1], Ping Yu Hsu[1], Ming Shien Cheng[1,2(✉)], Chen Yao Chung[1], and Yi Liang Chung[1]

[1] Department of Business Administration, National Central University, No. 300, Jhongda Road, Jhongli City, Taoyuan County 32001, Taiwan, ROC
984401019@cc.ncu.edu.tw, mscheng@mail.mcut.edu.tw
[2] Department of Industrial Engineering and Management, Ming Chi University of Technology, No. 84, Gongzhuan Road, Taishan District, New Taipei City 24301, Taiwan, ROC

Abstract. With the development of the Internet economy, various websites accumulate numerous reviews about different products and services. Those reviews have become one major information source besides official product information, expert opinion, and automatically generated individualized advice. The survey shows that percentage of gathering buying information on Internet gradually increases by years, and the relevant researchers have also proven that consumers pay more attention to others' reviews, thus deeply affect consumers' shopping decision. Unfortunately, by taking advantage of such trend, some dealers manipulate reviews in order to exaggerate their own product or defame their rivals. Those behaviors have brought severe damage to consumers and commerce. This study takes Internet reviews as research object, using rumor model to detect the truth of these review. Our rumor model applied text mining technique and extract 3 major characteristic of review content: important attribute word, specific quantifier, and noun verb ratio to build the model. For testing our rumor model, we take hotel reviews on America website "TripAdvisor" and the comparison group "Fake reviews" as analysis objects. We try to automatically and easily classify true and fake reviews. The result, generated by developed model in this research, shows that the more unique vocabulary and specific quantifier and noun it contains, the less possibility it is fake.

Keywords: Text mining · Fake review · Rumor · Hotel review

1 Introduction

The development of the network and the emergence of Internet marketing have gradually become the backbone of product marketing. Various websites are filled with all types of reviews [11, 16, 17, 21], including different types of products such as 3C products, books, movies, as well as service providers such a restaurants, hotel, etc. Users will write their opinions about products or services based on their experience. The opinions are the important sources of information for sellers and producers. Dellarocas (2003) [6] and Chevalier et al. (2006) [4] mentioned in their research that reviews are the best tools for evaluating products, services, and companies. Opinions

X. He et al. (Eds.): IScIDE 2015, Part II, LNCS 9243, pp. 181–192, 2015.
DOI: 10.1007/978-3-319-23862-3_18

from users are more reliable than advertisement claims, thus ultimately affecting purchase decisions. Sellers and producers can use these reviews to analyze product performance and send feedbacks to market and R&D departments.

A survey from Cone Research shows that 80 % of consumers change their purchase decision after reading negative reviews, while 87 % of consumers make purchase decisions based on positive product reviews. [5] Certainly, many positive reviews can create miracles, and the power of Internet reviews is unprecedentedly enormous. The research of Michael Anderson and Jeremy Magruder [15], proposed to measure the relationship between customer purchase decisions and online star rating. In this research, the ratings of 328 restaurants in San Francisco (5-star rating system) and the amount of daily reservations were collected. It was found that if a restaurant whose rating increased from 3 stars to 3.5 stars. The occupancy rate of the restaurants increased from 13 % to 34 % during peak hours; the restaurants' rating increasing from 3.5 stars to 4 stars contributed to increase 19 % rate of occupancy during peak hours. The positive relationship between positive reviews and occupancy rates has become the motivation for online fraud.

Hu et al. (2008) [10] also found in their study that when consumers search product comments on the Internet, they trust reviews published and generated by other users. Producers have also noticed this phenomenon, so they deliberately create fake reviews, attempting to affect consumers' purchase decisions. Research findings of Hu et al. (2012) [9] further indicate that more and more consumers rely on Internet reviews to make decisions. The study covered tests on writing style, emotions, readability, and other attributes of reviews. It was found that about 10 % of reviews on the Internet are fake reviews. Thus, it is important for business review-based service websites to provide customers with real and valid information. The purpose of this study is to establish a review screening model through text mining technology.

This paper is organized as follow: Sect. 2 covers several scholars' researches on studies of rumor detection and detecting fake review. Section 3 describes the methodology and model of our research. Section 4 evaluated our research result and Sect. 5 draws some conclusion and addresses the future work.

2 Related Work

2.1 Rumor Detection Model

According to Oxford dictionary: the definition of rumor is "a piece of information, or a story, that people talk about, but that may not be true" [23] (http://oald8.oxfordlearnersdictionaries.com/dictionary/rumour). Some scholars tend to define rumors as unproven messages. Shibutani (1996) [22] mentioned that "rumors" start from important and unspecified events. Message receivers are likely to accept or believe a rumor due to the great importance of an event and their failure to obtain adequate knowledge or information to prove the authenticity of the rumor. Fisher (1998) [8] believes that rumors are the circulation of information for a certain purpose, a kind of collective behavior among the masses.

Allport and Postman (1947) [1, 2] as the pioneer of rumor research have developed a conceptual equation to calculate the degree of rumor for a given piece of message:

$$\text{Rumors} = \text{Importance of events} \times \text{Ambiguity of events} \tag{1}$$

According to the Eq. 1, if the importance of an event is zero, or if the event itself is not vague, rumors cannot be generated [20]. Schachter and Burdict [19] further approved Eq. 1 by experiment, in their study, intentionally make a rumor in a school, saying that some of the examinations would be canceled, this rumor quickly spread on the campus. Two days after the rumor spread, one female student suddenly was interviewed by the principal (experimentally arrangement), and transferred to another class. At the same time, various rumors about transferred girl were spread on campus. This study experiment confirmed the proposition of Allport and Postman research that two basic elements: "the importance of the events" and "the ambiguity of events" caused rumor be wide-spreading. We could reach the conclusion that when the message was relative important to the recipient and the message was still faint; various related messages were likely to be made and circulated.

The way people detecting rumors also depend on the content, senders and transmitting pipeline. But the Internet rumors are hard to detect sender and transmitting pipeline. Sender may create or spread rumors by changing their identity or through anonymity, so proofing would involves complexity and difficulty [6]. In recent years, blogs, web forums, bulletin board systems (BBS), email, or social media like Facebook and Twitter have gradually become pipelines for rumors. Pasted and shared information are forwarded through convenient function button before the information is proven true. Through the interpersonal networks, friendly reminders, and alarmist rumors, outdated and erroneous information continue to be passed to all corners of the world. Before a rumor can be refuted or confirmed, it may have been passed around for months or even years. According to the arguments we discussed, content analysis would be the direct and efficient way to detect rumors.

The research of Reyes and Rosso [3] pointed out that content analysis could classify into three approaches: 1. N-grams: the approach tries to find frequent sequence of words considering n-grams of different orders; 2. POS n-grams: this one endeavors to find morpho-syntactic templates given the Part Of Speech (POS) tags. 3. Profiling: this feature evaluates a selection of the best-performing rumor characteristics found in the articles. Profiling approach would be the most adopted model in recent text mining researches. The profiling of our research are based on the conceptual formula proposed by Allport and Postman [1] in Eq. (1).

2.2 Detecting Fake Review

The focus of this study lies in fake reviews published on websites. According to rumors classification [18], fake reviews fall under deliberate rumors, and the purpose is to deceive or mislead consumers to affect their purchase decisions. Positive reviews bring business entities good reputation and economic benefits. To a certain extent, it also leads to the production of reviews. Therefore, identification and filter of fake reviews have substantial significance and theoretical value. Based on the summary of previous scholars' research on fake review [12], fake reviews can be divided into two types, as follows:

1. Deceptive reviews: Deliberately provide positive reviews to facilitate the selection of a certain product or business entity; deliberately provide negative reviews to lower the reputation of a product or business entity in order to affect sales.
2. Destructive reviews: This type of reviews is generally non-review in nature, such as unrelated advertisements and viewpoints or messages not related to the review body.

The second type of destructive reviews causes relatively less harm because consumers can identify them with ease. On the other hand, deceptive reviews conform to Scholar Shibutani's definition of the characteristics of rumors: [22] Receivers lack sufficient knowledge or information to verify events. Thus, this type of fake reviews possesses the characteristics of seclusion and diversity, which are difficult for consumers to identify. This study falls under the second category–the identification of deceptive reviews.

The first to engage in research on related fields was Jindal and Liu (2007) [11] who discovered the widespread existence of fake reviews. With reviews on Amazon as the research participants, multiple product samples were obtained. The data set was sourced from the reviews' related eight attributes, namely, the original reviews on Amazon, extracting product ID, reviewer ID, rating, date, review title, review body, number of helpful feedbacks, and number of feedbacks to engage in model establishment and distinguish true reviews from fake reviews. This method is based on the meta-heuristic algorithm. Ott et al. (2011) [17] used AMT (Amazon Mechanical Turk) to develop the first data set for identifying fake reviews. Additionally, the 20 most popular hotels on TripAdvisor were collected for true review cross-comparison. The traditional text classification from the psychological perspective was adopted to identify fake reviews.

Li et al. [14] tagged the corpus using product review data obtained from the network. Co-training algorithm was used to identify fake reviews. The research of Ott et al. [17] showed that large a number of fake examples exist in the used manually tagged corpus. In the face of reviews for a huge amount of information, the use of manually tagged data sets to engage in the identification of fake reviews is not the most reasonable method. Feng et al. [7] used probability-based context-free grammar rules and SVM classifier to classify true and fake text. Certain validation pertaining to classification rates was also obtained from the standard data set.

In addition, some characteristics for identifying fake review selection come from reviewers' behaviors. For instance, Jindal and Liu (2008) [12] started from user rating behaviors to establish a model based on experience and targeting fake commentators. In addition, based on experience, the weights for various behavioral characteristics were set up. By finding fake review producers, the purpose of testing fake reviews was achieved. Lappas (2012) [13] believed writing fake review is a kind of deliberate attack. From the perspective of the attacker, the discussion focused on how to write fake reviews to increase reliability. Several contribution factors with high reliability were proposed, and different attack strategies determined the different impacts of the attacker. The characteristics of fake reviews and the weaknesses of online review corpus were put forward, although the effectiveness is yet to be verified.

3 System Design

The text mining technique was adopted in this study to engage in the preprocessing of original text and extract keywords of reviews. According to the concept formula proposed by Allport and Postman: rumor = importance of event × ambiguity of event as the basis, a model was developed to identify fake reviews. The importance of events in this study is defined as the ratio of number of words that represent the characteristics of reviews and the total number of words; event ambiguity was extended in this study into reverse meanings, which are deterministic events. Deterministic is defined as the expression of deterministic quantifiers, such as the time and date the customers in the data set stayed in hotels, the number of nights checked in, and the number of people. The more number of quantifiers that appeared in reviews, the more realistic the authors' hotel experience and the lower the likelihood of falsity will be. According to the research of the data set collectors Ott et al. [17], since the contents of fake reviews never took place, the authors could not produce accurate descriptions using words. Therefore, compared to real reviews, more verbs are used in fake contents. Hence, the ratio of nouns and total of verbs and nouns was adopted as another element of the model. The higher the ratio, the lower the likelihood of fake reviews will be.

The processes in this study involve the preprocessing of original data. Then, the relevant attributes were used for word frequency computation and term weight calculation, based on which a rumor prediction model was established. In the future, this model can be used as an initial filter of reviews to be determined and screen dubious reviews. The research processes are as shown in Fig. 1.

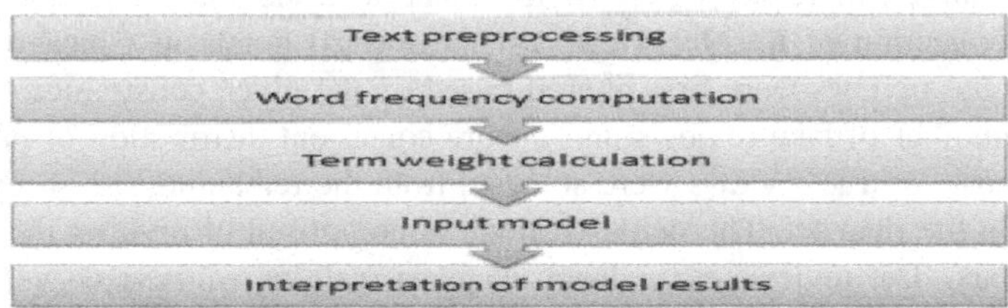

Fig. 1. System design process

3.1 Research Model

In this study, the number of terms served as an important indicator for evaluating reviews. Since the reviews that greatly varied in length resulted in greatly different terms, the logarithm was adjusted, leading to the development of a model. The smaller the evaluation value, the more likely it is for reviews to be fake reviews. The model formula is as follows. In this study, the precision, recall, and F-measure of fake review identification served as the bases for evaluating model effectiveness.

$$\begin{aligned} &\text{Evaluation Value of Fake Review} \\ &= \frac{\textit{Amount of Quantifier} + 1}{\textit{Amount of TerminReview}} \\ &\times \log(\textit{Amount of Characteristic Term} + 1) \\ &\times \frac{\textit{Amount of Noun}}{\textit{Amount of Noun and Verb}} \end{aligned} \tag{2}$$

4 Empirical Analysis

4.1 Data Description and Preprocessing

One major problem of fake review research is that it is difficult to obtain fake reviews and true reviews that can be used for comparative study. In this study, the review data set of Ott et al. [17] was adopted. It is the only publicly available tagged data set at present. The data set covers four parts: positive true reviews, positive fake reviews, negative true reviews, and negative fake reviews. The subjects were the 20 most popular hotels in Chicago listed on TripAdvisor website. With the fake reviews sourced from AMT (Amazon Mechanical Turk), reviews were written on the 20 above-mentioned hotels. Through the platform, the marketing department managers of the hotels requested users to write positive evaluations of their hotel to help in hotel development; the negative reviews on the other hand were intended to attack competitors. Each person was given 30 min to complete the mission for the price of $1. Through the platform, 400 positive reviews and 400 negative reviews were collected.

All the positive true reviews were reviews that rated the hotels five stars. Through the manual collection of 6,977 reviews on the top 20 hotels in Chicago listed on TripAdvisor, the reviews were first filtered based on the given constraints. In order to balance the number of fake reviews and ensure consistent distribution of fake review length, 400 positive true reviews were selected from the remaining 2,124 reviews and were stored in the data set. The negative reviews comprised of reviews that rated the hotels 1–2 stars. The reviews came from six most well-known review websites: Expedia, Hotels.com, Orbitz, Priceline, TripAdvisor, and Yelp. Through the aforementioned method, the same one used for positive true reviews, 300 negative true reviews were finally selected and stored in the data set.

4.2 Model Results

In this study, the true and fake reviews were randomly arrayed, 70 % of which were randomly extracted as testing data, and 30 % were used as training data. The study adopted Stanford Log-linear Part-Of-Speech Tagger to tag speech, count, and engage in TF-IDF calculation through StatSoft Statistical 10 software. Finally, after compilation, the data was outputted using the Excel format and imported using the Microsoft Excel software. There were six columns: total number of words in a review, number of quantifier, number of terms greater than the weight average value, number of verbs, number of nouns, and true or fake category. The corresponding values were used in the model formula in the study. The rumor evaluation values for individual articles were calculated using this formula. First, targeting the training data, the results were arranged from high to low to facilitate the calculation of precision, recall, F-measure, and threshold. The higher the evaluation value, the more likely it is for an article to be a rumor; on the contrary, the lower the evaluation value, the more likely it is for an article to be a normal article.

Findings show that the precision is a decline curve, while the recall is an incremental curve. Hence, the threshold for identifying fake review values took into account both the mean value of precision and that of recall, with F-measure as the evaluation indicator. Past studies show that positive reviews and negative reviews may differ [17]. Thus, assuming that positive and negative review data show differences in model results, the positive reviews and the negative reviews should first be discussed. Of the 800 negative review articles, 572 articles accounting for 70 % were selected as training data and were introduced into the model. The results are as shown in the Fig. 2: the maximum value of F-measure was 71.4 %, and the precision was 60.9 %. With the recall of 86.4 %, the fake evaluation value at this point was 0.016952338. The value as the threshold was used for 238 testing data entries, accounting for 30 %. The results are as shown in the Fig. 3: the threshold of 0.016952338 obtained from the training data was used as the classification standard. The maximum fake evaluation value obtained from the testing data lower than the threshold was 0.016734130277926. The precision was 59.4 %, the recall was 85.6 %, and the F-measure was 70.3 %.

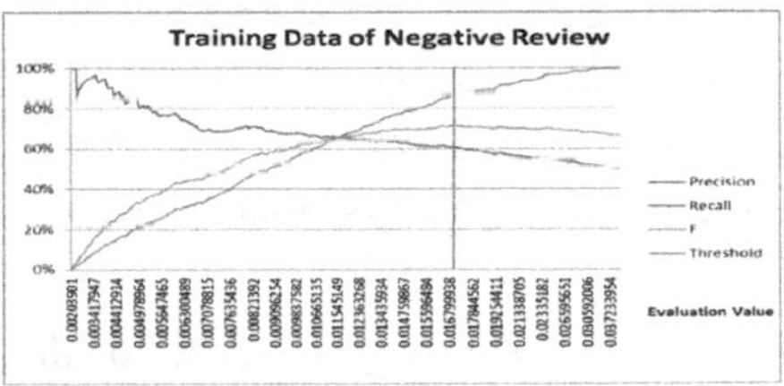

Fig. 2. Training data of negative review

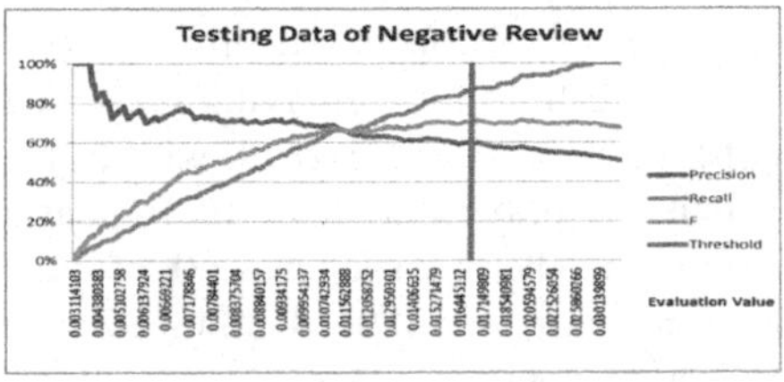

Fig. 3. Testing data of negative review

Of the 800 positive reviews, 562 articles accounting for 70 % were selected as training data and were introduced into the model. The results are as shown in the Fig. 4: the maximum value of F-measure was 69.8 %. The precision was 59.6 %; the recall was 84.2 %. The fake evaluation value at this point was 0.0169895244580622, which served as the threshold used for 238 testing data entries, accounting for 30 %. The results are as shown in Fig. 5: The threshold of 0.0169895244580622 obtained from the training data served as the classification standard. The maximum fake evaluation value obtained from the testing data lower than the threshold was 0.0168408389182647. The precision was 62.1 %, the recall was 78.7 %, and the F-measure was 69.4 %.

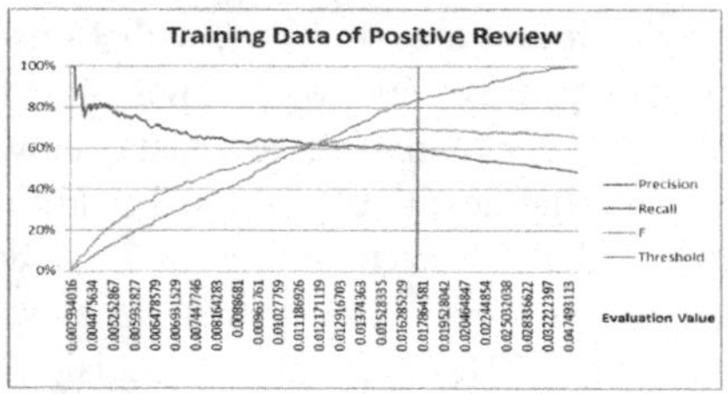

Fig. 4. Training data of positive review

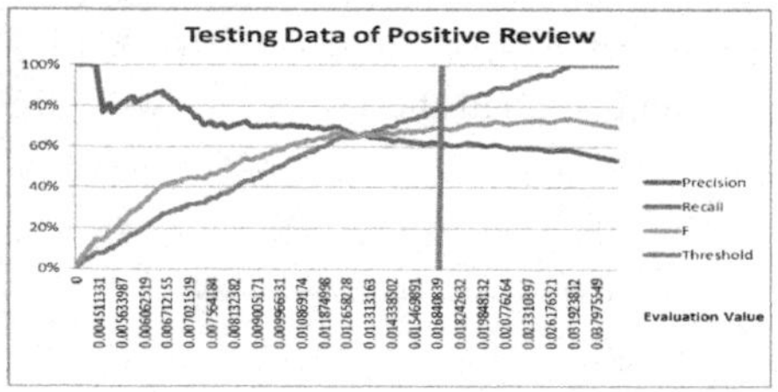

Fig. 5. Testing data of positive review

The result summary shows that the analysis results of the positive and negative reviews introduced into the model showed no significant differences. Hence, the model used for the research development was used to conduct fake value evaluation on 1,600 data entries. Among them, 1,158 articles were part of the training data. Through the model evaluation, the results are as shown in Fig. 6 follows.

As shown in the Fig. 6, the precision is a decline curve, while the recall is an incremental curve. Hence, the threshold for identifying fake evaluation values took into account the mean values of both precision and recall, with F-measure as the evaluation indicator. Findings show that the maximum value of F-measure in the data set was 70.6 %, which was the best result. The precision was 60.5 %; the recall was 84.8 %. At this point, the fake evaluation value was 0.0170733430376586, which served as the threshold for 30 % of the testing data.

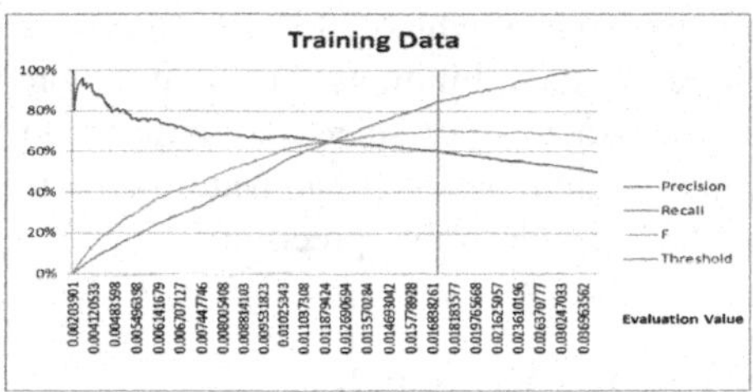

Fig. 6. Training data

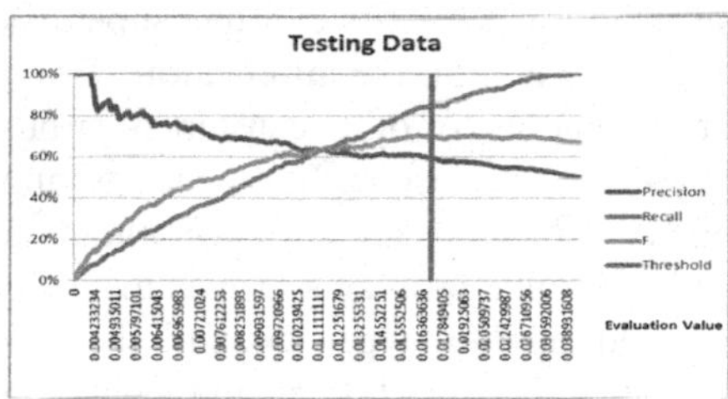

Fig. 7. Testing data

The results of 422 testing data entries are as shown in the Fig. 7. The threshold of 0.0170733430376586 obtained from the training data served as the classification standard. The maxim fake evaluation value lower than the threshold obtained from the testing data was 0.0169889746352642. The precision was 59.6 %, the recall was 84.6 %, and the F-measure was 69.9 %. The model prediction results put forth in this study are as following Table 1.

Table 1. Model result

Data Set	Precision	Recall	F--measure
All Reviews	59.6%	84.6%	69.9%
Negative Reviews	59.4%	85.6%	70.3%
Positive Reviews	62.1%	78.7%	69.4%

4.3 Model Comparison

The SVM algorithm was selected as the comparison model. This classification method is deemed one of the best classification models. The three elements for identifying fake reviews selected by the model were imported into the StatSoft Statistica 10 software. SVM method was adopted for classification and prediction. In particular, the SVM type selected Classification type 1 (capacity = 5.000); Kernel type selected RBF (Radial basis function), most commonly used in SVM. Classification and prediction were conducted on all the 1,600 reviews, of which 70 % were used as training data, while the other 30 % were used as testing data. Findings show that the precision and recall were both 50 %. With the same characteristics selected, compared to the cross-comparison model SVM algorithm, the prediction model had better predictive effectiveness, but it was also observed that it had less desirable precision'.

5 Conclusion and Future Research

5.1 Conclusion

The Internet has rich information, but it lacks effective supervision and filter, resulting in the failure to correctly convey and utilize information. True and reliable consumer experience reviews can bring benefits for other consumers or business operators, while fake evaluations will lead to negative results. In the face of huge review information, effective automatic identification methods are a worth-noting issue with practical value.

Former study attempted to resolve problems such as manual screening of terms and constraints of thesauruses and product types through the automatic screening and extraction of characteristic elements using text mining technology. Findings show that the elements selected in this study had certain effectiveness in determining fake reviews, but the identification effectiveness was undesirable.

According to relevant studies, it shows that the attributes of commentators in the identification model had a significant influence. The model that combined review contents and attributes of reviewer possessed better identification results. The study constraint pertaining to data sets is that only the review contents were analyzed. In the future, it is recommended that the data scope be expanded, and website operators may use this model along with attributes of reviewers to enhance the accuracy of fake review identification.

5.2 Future Research

1. In conjunction with other algorithms and research tools: A number of classification algorithms commonly used in text mining include CART classification trees, Bayesian classification, logistic regression, etc., based on different forms of data, selecting the appropriate algorithm.
2. Increase of different product type research: This study only focused hotel reviews, whether there are differences in the findings on other products? In other consumer electronic products, such as digital cameras, tablet PCs, or other products of e-WOM related such as cars, to be the research scope, will it affect the results? These objects can be explored in future researches.
3. Expand the breadth and depth of research data: In this study, the reviews setting ranges are Chicago hotel. Concerned broadly, the future research can extend to the reviews of other city's hotel or even include reviews of all hotels of American.

References

1. Allport, G.W., Postman, L.: An analysis of rumor. Publ. Opin. Q. **10**, 501–517 (1947)
2. Allport, G.W., Postman, L.: The Psychology of Rumor. Henry Holt, New York (1947)
3. Reyes, A., Rosso, P.: Making objective decisions from subjective data: detecting irony in customer reviews. Decis. Support Syst. **53**(4), 754–760 (2012)
4. Chevalier, J.A., Mayzlin, D.: The effect of word of mouth on sales: online book reviews. J. Mark. Res. **43**(3), 345–354 (2006)
5. Cone Research. Game changer: cone survey finds 4-out-of-5 consumers reverse purchase decisions based on negative online reviews. http://www.coneinc.com/negative-reviews-online-reverse-purchase-decisions (2011)
6. Dellarocas, C.: The digitization of word of mouth: promise and challenges of online feedback mechanisms. Manage. Sci. **49**(10), 1407–1424 (2003)
7. Feng, S., Banerjee, R., Choi, Y.: Syntactic stylometry for deception detection. In: Proceedings of the 50th Annual Meeting of the Association for Computational Linguistics (ACL 2012), pp. 171–175, Jeju Island, Korea. ACL, Stroudsburg, PA, USA, 8–14 July 2012
8. Fisher, D.R.: Rumoring theory and internet-a framework for analyzing the grass roots. Soc. Sci. Comput. Rev. **16**(2), 158–168 (1998)
9. Hu, N., Liu, L., Sambamurthy, V.: Fraud detection in online consumer reviews. Decis. Support Syst. **50**, 614–626 (2012)
10. Hu, N., Liu, L., Zhang, J.J.: Do online reviews affect product sales? The role of reviewer characteristics and temporal effects. Inf. Technol. Mgmt. **9**(3), 201–214 (2008)
11. Jindal, N., Liu, B.: Review spam detection. In: Proceedings of the 16th International Conference on World Wide Web, pp. 1189–1190. ACM, New York, NY, USA (2007)
12. Jindal, N., Liu, B.: Opinion spam and analysis. In: Proceedings of the 2008 International Conference on Web Search and Data Mining, pp. 219–230. ACM, New York, NY, USA (2008)
13. Lappas, T.: Fake reviews: the malicious perspective. In: Proceedings of the 17th International Conference on Applications of Natural Language Processing to Information Systems, pp. 23–34 (2012)
14. Li, F., Huang, M., Yang, Y., et al.: Learning to identify review spam. In: Proceedings of the 22nd International Joint Conference on Artificial Intelligence (IJCAI 2011), pp. 2488–2493, Bar celona, Spain. AAAI, Palo Alto, CA, USA, 16–22 July 2011
15. Anderson, M., Magruder, J.: learning from the crowd: regression discontinuity estimates of the effects of an online review database. Econ. J. **122**(563), 959–989 (2012)
16. Mukherjee, A., Venkataraman, V.: What yelp fake review filter might be doing?. In: Proceedings of the 7th International Conference on Weblogs and Social Media, pp. 409–418. AAAI Press, Palo Alto (2013)
17. Ott, M., Choi, Y.J., Cardie, C., et al.: Finding deceptive opinion spam by any stretch of the imagination. In: Proceedings of the 49th Annual Meeting of the Association for Computational Linguistics: Human Language Technologies, pp. 309–319. Association for Computational Linguistics, Stroudsburg, PA, USA (2011)
18. Rownow, R.L.: On rumor. J. Commun. **24**(3), 26–38 (1974)
19. Schacter, S., Burdick, H.: A field experiment on rumor transmission and distortion. J. Abnorm. Soc. Psychol. **50**, 363–371 (1955)
20. Schultz, C.K., Luhn, H.P.: Pioneer of Information Science - Selected Works. Macmillan, London (1968)
21. Sernovitz, A.: Word of Mouth Marketing. Kaplan Publishing, New York (2009)

22. Shibutani, T.: Improvised News: a Sociological Study of Rumor. Bobbs Merrill, Indianapolis (1966)
23. Oxford Dictionary. The definition of rumor. http://oald8.oxfordlearnersdictionaries.com/dictionary/rumor

Dynamic Multi-relational Networks Integration and Extended Link Prediction Method

Hong Wang[1,2(✉)] and Yanshen Sun[3]

[1] School of Information Science and Engineering, Shandong Normal University, Wenhua Road 88, Jinan 250014, China
1456029328@QQ.com, Wanghong106@163.com

[2] Shandong Provincial Key Laboratory for Distributed Computer Software Novel Technology, Wenhua Road 88, Jinan 250014, China

[3] Department of Earth Sciences, Zhejiang University, Xihu Road, Hangzhou 310000, China
969481221@QQ.com

Abstract. Link prediction is an effective method in complex networks analysis, not only in simple networks but also in dynamic multi-relational ones. How to integrate these multi-relational networks is of great importance to link prediction results. In this paper, we study the integration method of dynamic multi-relational networks and extend the link prediction method which estimates the likelihood of missing links and existent links disappearing in the future. Accordingly, we put forward an algorithm for building dynamic multi-relational weighted networks and an extended link prediction algorithm in integrated dynamic multi-relational networks, which is capable of predicting bi-directional links. We apply our method in a real multi-relational network. The experimental results show that our method can improve the link prediction performance in dynamic multi-relational networks.

Keywords: Dynamic multi-relational networks · Networks integration · Extended link prediction · Principle component analysis · Visibility graph

1 Introduction

With complex networks scales getting larger, network analysis becomes a crucial focus and attracts more and more researchers. One of the most important analyzing method is link prediction.

Examining these networks, we can observe that they have features of multi-dimension and dynamics. The multi-dimensional feature means a complex network always includes various kinds of nodes and multiple types of links between nodes. This kind of networks are referred as heterogeneous networks, a.k.a. multi-relational networks. If each type of interaction is described as a single dimensional network and then all of them compose multi-relational networks. Obviously, these sub-networks are not isolated but interrelated, we should integrate all of them and do comprehensive link prediction. The dynamic characteristics of complex networks

X. He et al. (Eds.): IScIDE 2015, Part II, LNCS 9243, pp. 193–203, 2015.
DOI: 10.1007/978-3-319-23862-3_19

indicates that networks usually change over time, enlarging or diminishing, which adds a new temporal dimension to the problem of link prediction.

2 Related Work

Main methods of link prediction are Markov chain, probabilistic method, maximum likelihood estimation, node-property-based method and network-structure-based method [1], in which the network-structure-based method is relatively ideal in complex networks. There are many research fruits in this area [2–4], which proves that the link prediction method has become a powerful means to analyze complex networks. In addition to theoretical research results, there are a lot of application findings [5–8]. In addition, the link prediction method is used in identifying not only missing links but also evanescent links [9–11].

Main contributions of this paper are as follows. Firstly, we present an integrated model of dynamic multi-relational networks based on visibility graph (VG) and further reduce the dimension of networks by using PCA to deal with increasing scale of networks. Secondly, we do link prediction in dynamic multi-relational networks. Thirdly, we show a kind of extended link prediction method, which predicts both missing links and error links. Most of the previous works focus on the former but neglect to the latter. In addition, some studies about spurious link prediction is different from us.

3 Multi-relational Networks Integration

3.1 Related Definition

As we know, a weighted network can be described as a graph, denoted as $H = (V, E, W)$, in which V is a node set, E is a link set and W is a set of weight. Similarly, we define a single-relational network (*SRN*) as a 5-tuple form, named $H = (ID, V, E, W)$, in which ID is a unique identification of the single-relational network, other elements are the same as in the weighted network. For instance, if $H = (ID, V, E, W)$ is used to present a single-relational network of kinship, then ID is the label of kinship relation, V is a set of persons, E is a set of kinship relation and W presents the closeness of a link. If person v_i and person v_j are relatives, then there is a link $e_{ij} \in E$.

Based on descriptions above, we define an integrated multi-relational network as a 6-tuple form $\hat{H} = (\hat{ID}, \hat{V}, \hat{E}, \hat{W})$, in which $\hat{H} = \{H_i | H_i = (ID_i, V_i, E_i, W_i) \in SRN, 1 \le i \le n\}$, $\hat{V} = \{\cup V_i | 1 \le i \le n\}$, $\hat{E} = \{\cup E_i | 1 \le i \le n\}$, $\hat{ID} = \{\cup ID_i | 1 \le i \le n\}$ and $\hat{W} = \{\cup W_i | 1 \le i \le n\}$. That is, $\hat{H}$ is composed by n types of relations. It is emphasized that $E_i \cap E_j = \phi (1 \le i, j \le n,\ i \ne j)$ as these n types of relations are different but $V_i \cap V_j$ may be null or not. When V_i and V_j present different sort of nodes, $V_i \cap V_j$ may be null. An integrated multi-relational network is abbreviated as *IMRN*.

Figure 1 shows how three *SRN*s forming an *IMRN*. Suppose three *SRN*s present movie recommendation network, book recommendation network and music recommendation network, respectively. Node 1 to node 7 present different persons. In (a), there is a link between two persons if they like the same kind of movies, similarly, in (b) and (c), there is a link between two persons if they like the same kind of books or music. According to the definitions above, we get the *IMRN* shown in (e). Noted that the width of links in (e) are different, which presents different weight value of links after integration.

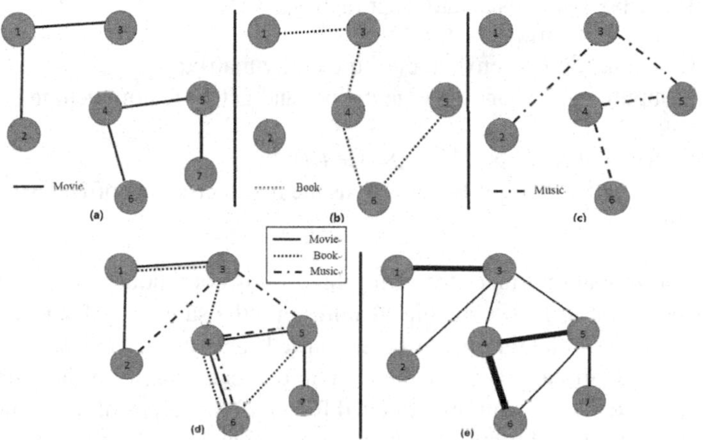

Fig. 1. Integrated multi-relational network. (a) Movie recommendation network. (b) Book recommendation network. (c) Music recommendation network. (d) Composed multi-relational network with single relational networks in (a), (b) and (c). (e) Integrated multi-relational network from (d) by accumulating weight values of the same links.

3.2 Integration Method

Heterogeneous multi-relational networks make the link prediction work complicated. It calls for new solutions to extract information in heterogeneous networks. It is noted that, one cannot reduce a heterogeneous network to several homogeneous ones, because the interaction information in one dimension might be too noisy to detect meaningful relations. It is generally helpful to utilize information from other modes or dimensions for more effective link prediction.

There are three types of integration strategies: network integration, utility integration and feature integration [12]. Given advantage in clustering performance, feature integration is preferred in this paper. The main idea of integrating multi-relational networks is as follows, shown in Algorithm 1.

Step 1: With each single-dimensional network, a utility matrix is constructed. Depending on the objective function, we can construct different utility matrices.

Step 2: After obtaining the utility matrix, we obtain the soft community indicator S that consists of the top eigenvectors with the largest (or smallest subject to formulation) eigenvalues. The selected eigenvectors capture the prominent relation patterns, so this step can also be considered as a de-noising process.

Step 3: Try to integrate soft community indicators extracted from each dimension of the network.

Algorithm 1. Integration algorithm

INPUT: $\hat{H} = (\hat{ID}, \hat{V}, \hat{E}, \hat{W})$
OUTPUT: IMRN
Integration-IMRN ( ) {
FOR (i =1 TO n DO) { // n is the number of *SRN*s
Describe a *SRN* as an adjacent matrix;
Compute its utility matrix;
Compute top l eigenvectors of the utility matrix;
Compute corresponding eigenvalues and get its feature denoted as $S^{(i)}$;
ENDFOR
Compute SVD of $X = [S^{(1)}, S^{(2)}, \ldots, S^{(p)}] = UDV^T$;
Obtain averaged structural features S^* as the first k columns of U;
RETURN S^*;
}

As the solution that optimizes the utility function is not unique, we have to apply a certain transformation to map structural features into the same coordinates. After that, the multi-dimensional integration can be conducted. In order to find out the transformation associated with each type of interaction, one can maximize the pairwise correlations between structural features extracted from different types of interactions. Once such a transformation is found, structural feature which is the averaged structural features after transformation can be shown.

3.3 Time Relation Networks

As mentioned in section one, multi-relational networks are dynamically changing, which obviously exert new impacts on the link prediction result. So we should consider the time effect during the construction of an integrated network.

Visibility Graph (VG) [13] is a kind of method which transform time series into a complex network whose topology keeps inherent characteristics of original time series. It has been proved that VG is an effective bridge between the time sequence and the complex networks [14, 15]. So we extend the VG and apply it to our multi-relational networks.

Suppose a dynamic multi-relational network is described as a sequence $S = \{(t_1, \tilde{H}_{t_1}), (t_2, \tilde{H}_{t_2}), \ldots, (t_c, \tilde{H}_{t_c})\}$, in which $\tilde{H}_{t_1}$ is a *IMRN* at time $t_i(i = 1, 2 \ldots c)$ (suppose the time series is divided into c parts). In order to define the concept of visibility in the extended VG, we should use a value to represent an *IMRN* $\tilde{H}_{t_1}(i = 1, 2, \ldots c)$. The value here is the average value of eigenvalues of the *IMRN* at time $t_i(i = 1, 2, \ldots c)$. Of cause, other values can be used. Then, we define the visibility between two nodes $(t_p, \tilde{H}_{t_p})$ and $(t_q, \tilde{H}_{t_q})$, if $\forall (t_x, \tilde{H}_{t_x}) \in S$, $(t_p < t_x < t_q)$, they fulfill the condition in formula 1.

$$\tilde{H}_{t_x} < \tilde{H}_{t_q} + (\tilde{H}_{t_p} - \tilde{H}_{t_q}) \times (t_x - t_q)/(t_p - t_q) \tag{1}$$

That means, if satisfying formula 1, there will be a link between two nodes $(t_p, \tilde{H}_{t_p})$ and $(t_q, \tilde{H}_{t_q})$. The algorithm of building a dynamic multi-relational network is shown in algorithm 2.

Algorithm 2: Construction algorithm of a Dynamic *IMRN*.
INPUT: *IMRN* series obtained from algorithm 1
OUTPUT: a Dynamic *IMRN*
Building-Dynamic-IMRN ( ) {
FOR (*i*=1 to *c* DO)
1. Compute the average value of eigenvalues in each *IMRN* $\tilde{H}_{t_i}$;
2. Compute the visibility between two nodes in $S=\{(t_1, \tilde{H}_{t_1}), (t_2, \tilde{H}_{t_2}), ..., (t_c, \tilde{H}_{t_c})\}$ according to formula 1 and finally get a complex network denoted as *Z*;
3. RETURN Z;
}

On line 1, we compute the average value of eigenvalues in each *IMRN*. And then we get a time relation network based on formula 1.

3.4 Reducing Integrated Network Dimension

After running algorithms above, we get the dynamic integrated network. Considering the time and space overhead, we need to find some more important relations to present the entire integrated multi-relational network. It is obvious that randomly selecting relations has no effect. We use the extended principal component analysis (PCA) [16] method to solve this problem. Then, the original data is approximated by data with much fewer dimensions, which summarizes well on the original data. The algorithm of reducing integrated network dimension is as follows.

Algorithm 3: Algorithm of reducing integrated network dimension
INPUT: an *IMRN*
OUTPUT: Low-dimensional integrated network
Reducing-Dimension-IMRN ( ) {
1. Describe IMRN as a standard adjacent matrix $M_{n\times l}$;
2. Find correlation coefficient matrix $A_{n\times n}$ based on $M_{n\times l}$;
3. Compute eigenvalues of $A_{n\times n}$, denoted as $\lambda_1 \geq \lambda_2 \geq \cdots \geq \lambda_n \geq 0$;
4. Compute the matrix $E_{n\times n}$ whose columns are the corresponding eigenvectors of $\lambda_i (i = 1, 2, 3...n)$;
5. Determining the value *k* based on the accumulate contribution rate of principal components;
6. Let $E_{n\times k}$ be the first *k* columns of $E_{n\times n}$;
7. Compute $B_{n\times k} = A_{n\times n} \times E_{n\times k}$;
8. RETURN $B_{n\times k}$;
}

Detailed explanations of algorithm 3 is as follows. On line 1, in order to compare link weight, we have to normalize the adjacent matrix to a standardized one named $M_{n\times l}$. On line 3, as the main idea of PCA is to find the eigenvectors for the matrix M^TM or MM^T, we get the correlation coefficient matrix $A_{n\times n} = MM^T/(p-1)$ based on $M_{n\times l}$, in which:

$$A[i,j] = \sum_{k=1}^{p} m_{ik} \cdot m_{jk}/(p-1),\ i,j = 1,2,\ldots n \tag{2}$$

On line 5, in order to reduce dimension, we should compute the value k based on the accumulate contribution rate of principal components. The accumulate contribution rate is expressed in formula 3. Generally speaking, the threshold value is set to 0.85, which can be set to other values.

$$\sum_{j=1}^{k} \lambda_j / \sum_{j=1}^{n} \lambda_j \geq 0.85 \tag{3}$$

On line 7, $B_{n\times k} = A_{n\times n} \times E_{n\times k}$ is the k-dimensional representation of $M_{n\times l}$. So the *IMRN* is expressed by k principal components, in which the ith component is presented as follows.

$$R_i = \sum_{j=1}^{k} w_j u_{ij}, \quad i = 1,2,\ldots n \tag{4}$$

In which:

$$w_j = \lambda_j / \sum_{i=1}^{k} \lambda_i,\ j = 1,2,\ldots,k \tag{5}$$

4 Extended Link Prediction

After getting a dynamic *IMRN* through previous algorithms, we will do extended link prediction process, which estimates not only the likelihood of the existence of missing links but also the possibility of links disappearing in the future.

4.1 Predicting Appearing of Missing Links

There are different kinds of similarity indexes in link prediction research area, such as Jaccard, Common Neighbor (CN), Resource allocation (RA), Katz, average commute time (ACT), random walk with restart (RWR), SimRank (SimR) and so on. The detailed explanations of these definitions are shown in reference [17]. We use path based similarity indexes Katz. Katz similarity is defined as follows.

$$Sim_{xy} = \sum_{l=1}^{\infty} a^l \cdot |paths_{x,y}^{<l>}| = a(A)_{xy} + a^2(A)_{xy}^2 + a^3(A)_{xy}^3 + \cdots \quad (6)$$

In which, a is an adjusted weight coefficient. $|paths_{x,y}^{<l>}|$ presents the number of links whose length is l between node x and node y, and A is the adjacent matrix of a complex network.

We can easily compute those similarity indexes with the adjacent matrix of the dynamic *IMRN* called $A_{N\times N}$. In fact, it has been proved that when $a < 1/\lambda_{\max}$, Sim_{xy} in formula 6 is convergent and can be described as formula 7, which is much simpler than formula 6.

$$Sim_{xy} = (I - a \cdot A)^{-1} - I \quad (7)$$

In which, X^{-1} presents the inverse matrix of X and I is a unit matrix. Now the link prediction algorithm for missing links is shown as follows.

Algorithm 4: Predicting appearing of missing links
INPUT: $A_{N\times N}$
OUTPUT: The most likely appearing links between nodes in the future
Forward-Link-Prediction ( ) {
1. Compute the eigenvalues of $A_{N\times N}$, denoted as $\lambda_1 \geq \lambda_2 \geq \lambda_3 \geq \ldots$;
2. Randomly select a value $a < 1/\lambda_1$;
3. Compute the Katz similarity values by formula 9;
4. List L={node pair (x,y)| Sort the Katz similarity value in descending order} ;
5. RETURN first k node pairs in L;

}

In algorithm 4, on line 4, we get node pairs' list in descending order according to the Katz similarity values computed on line 3. Then, the first k node pairs in L are returned, which means the larger the Katz similarity value, the more likely to exist a link between nodes. By the way, the Katz similarity is used in Algorithm 4, in fact, other similarity indexes can be used instead of it.

4.2 Predicting Disappearing of Spurious Links

The purpose of predicting spurious links is to estimate the possibility of some existed links that might disappear in the future. The traditional link prediction method regards that the more similar two nodes are, the more likely they are connected with a link. Based on its converse-negative proposition, we hold that less similar two nodes are, the more likely the link between them will disappear. So we change weight values in the dynamic *IMRN*, changing an original bigger value into a smaller one, similarly, an original smaller one to a bigger one. After this change, we do link prediction like algorithm 4. The output is just the links that is likely to disappear in the future.

5 Experimental Results and Analysis

5.1 Experimental Data Set

The experimental data used in this paper are from the logs of a video website of a university during last three months. There are many kinds of resources on the website, such as movies, books, TV show, music and so on. We carry out our link prediction methods by using three kinds of data, movie, book and music. First of all, we pre-process these data to build three single-relational networks, denoted as the movie network, the book network and the music network. The movie network is constructed as follows. Each student, denoted by a student No., is a node in the movie network. If two students see the same movie, there is a link between them. Then we get the movie network. The building process of two other networks is similar. After pre-processing, we obtain three single-relational networks whose features are shown in Table 1.

Table 1. Features of three networks: number of nodes (N) and edges (E), average degree (<k>), average shortest path length (<d>), radius (R), clustering coefficient (C), Modularity (Q), density (ρ)

Name	*N*	*E*	*<K>*	*<d>*	*R*	*C*	*Q*	*ρ*
Movie	3196	32446	20.30413016	3.005439937403983	4	0.086	0.17	0.007
Book	3328	24375	14.64843750	3.181040076740527	1	0.063	0.198	0.005
Music	2975	26233	17.63563025	3.0376148410096713	1	0.084	0.181	0.007

To evaluate the accuracy of our method, we use the standard metrics of the area under the receiver operating characteristic curve (AUC) and precision [12].

5.2 Experimental Results

In our experiments, we first build an *IMRN* with algorithms 1–3. Its features are shown in Table 2. Comparing with Table 1, we can see that the *IMRN* is much denser than any single-relational ones, but the radius is not.

Table 2. Features of multi-relational networks: number of nodes (N) and edges (E), average degree (<k>), average shortest path length (<d>), radius (R), clustering coefficient (C), Modularity (Q), density (ρ)

Name	*N*	*E*	*<K>*	*<d>*	*R*	*C*	*Q*	*ρ*
Multi-relation	3592	68733	34.459	2.609	1	0.135	0.123	0.014

Then we compare our link prediction method with other kinds of methods. Data sets are randomly divided into a training set and a test set, with a proportion of 80 and 20 % respectively. We compare our revised link prediction algorithms in an IMRN, abbreviated as DMRLP, with Common Neighbor based method, local naive Bayesian method, path-based method, local random walk (LRW) and Matrix forest Index method (MFI). In Common Neighbor based method, we choose CN, Jaccard, Adar-Adamic (AA) and Resourse Allocation (RA) indexes. In local naive Bayesian model, we select Local naïve bayes-Common Neighbor (LNBCN) and Local naive bayes-Resource Allocation Index (LNBRA) indexes. With the test data divided into ten groups, we carry out these link prediction methods independently in each group. The comparison results are shown in Fig. 2.

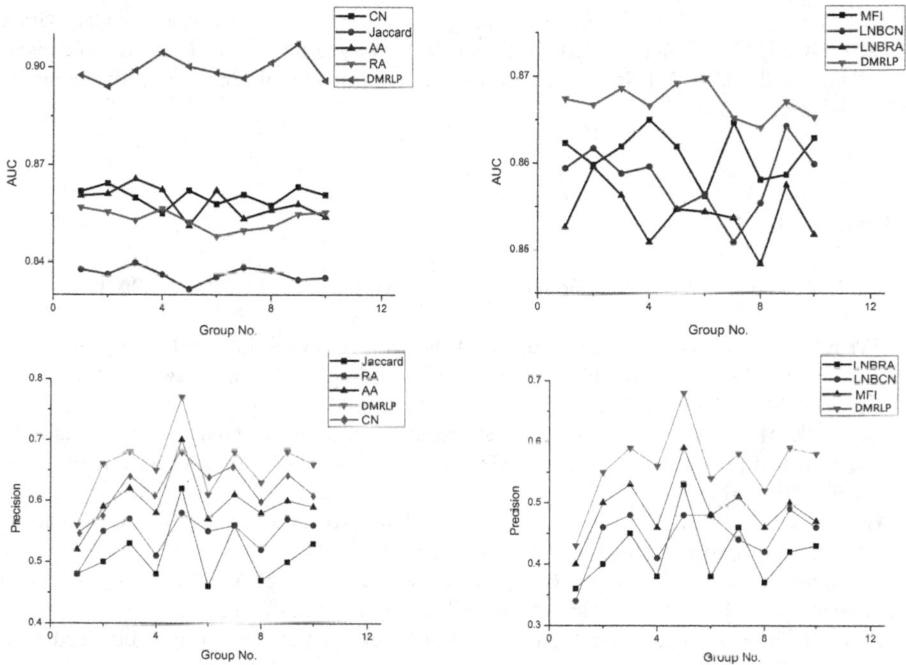

Fig. 2. Comparison of AUC and precision between our method and others

From the results, we can see that both AUC and precision value of DMRLP is higher than those of other methods. However the execution time of DMRLP is longer than that of LNBCN and LNBRA algorithm, as DMRLP has to spend time on constructing integrated multi-relational networks and computing its eigenvalues.

6 Conclusions

In this paper, we put forward algorithms of integrated dynamic multi-relational networks and present an extended link prediction method in integrated dynamic multi-relational networks, which not only predicts possibly existent links but also estimates links to disappear in the future. The experimental results show that our method can improve the link prediction performance. But our method needs to be improved. One problem is that we should further reduce the dimension of each single relational network. We think that its dimension will be reduced if these resources are clustered. Another problem is that the complexity of PCA is very high, which makes our algorithms spend more time. If we use SVD decomposition method to replace PCA, our algorithm will spend less time and cost. Hence, our further work mainly includes some meticulous work such as optimizing the algorithms and improving their efficiency.

Acknowledgments. Supported by National Natural Science Foundation under Grant (No.61373149, 61472233), Technology Program of Shandong Province under Grant (No. 2012GGX10118,2014GGX101026), Exquisite course project of Shandong Province under Grant (No. 2012BK294, 2013BK402), education research project of Shandong Province under Grant (No. ZK1437B010).

References

1. Lü, L.Y., Zhou, T.: Link prediction in complex networks: a survey. Phys. A **390**, 1150–1170 (2011)
2. Tylenda, T., Angelova, R., Bedathur, S.: Towards time-aware link prediction in evolving social networks. In: Proceedings of the 3rd Workshop on Social Network Mining and Analysis, SNAKDD ACM, Paris, France, pp. 1–10 (2009)
3. Allali, O., Magnien, C., Latapy, M.: Link prediction in bipartite graphs using internal links and weighted projection. In: IEEE Conference on Computer Communications Workshops, pp. 93–941 (2011)
4. Huang, L.W., Li, D.Y.: A meta path-based link prediction model for heterogeneous information network. Chin. J. Comput. 37(4), 8–857 (2014)
5. Gallagher, B., Tong, H., Eliassi-Rad, T., Faloutsos, C.: Using ghost edges for classification in sparsely labeled networks. In: ACM SIGKDD, pp. 25–264 (2008)
6. Zhou, T., Ren, J., Medo, M.: Bipartite network projection and personal recommendation. Phys. Rev. E 76–115 (2007)
7. Zhou, T., Jiang, L.L., Su, R.Q.: Effect of initial configuration on network-based recommendation. Europhys. Lett. 8–85 (2008)
8. Naji, G., Nagi, M., ElSheikh, A.M., Gao, S., Kianmehr, K., Özyer, T., Rokne, J., Demetrick, D., Ridley, M., Alhajj, R.: Effectiveness of social networks for studying biological agents and identifying cancer biomarkers. In: Wiil, U.K. (ed.) Counterterrorism and Open Source Intelligence. LNCS, pp. 285–313. Springer, Heidelberg (2011)
9. Guimera, R., Sales-Pardo, M.: Missing and spurious interactions and the reconstruction of complex networks. PNAS **106**(52), 22073–22078 (2009)
10. Kim, D.H., Noh, J.D., Jeong, H.: Scale-free trees: the skeletons of complex networks. Phys. Rev. E **70**, 046126 (2004)
11. An, Z., Giulio, C.: Removing spurious interactions in complex networks. Phys. Rev. E **85**, 036101 (2012)
12. Tang, L., Liu, H.: Community Detection and Mining in Social Media. Morgan & Claypool, San Rafael (2010)
13. Lacasa, L., Luque, B., Luque, J.: From time series to complex networks: the visibility graph. Proc. Nat. Acad. Sci. **105**, 4972–4975 (2008)

14. Luque, B., Lacasal, L.: Horizontal visibility graphs: exact result for random time series. Phys. Rev. E **80**, 046103 (2009)
15. Zhao, L.: Time series analysis based on complex network theory. J. Shanghai Univ. Sci. Technol. **33**(1), 47–51 (2011)
16. Jolliffe, I.T.: Principal Component Analysis, p. 487. Springer, New York (1986). doi:10.1007/b98835
17. Lü, L.Y., Zhou, T.: Link Prediction. Higher Education Press, Beijing (2013)

Matrix Factorization Approach Based on Temporal Hierarchical Dirichlet Process

Liang Chen(✉) and Peidong Zhu

College of Computer, National University of Defense Technology, Changsha, People's Republic of China
chenliang@nudt.edu.cn, Chl160@163.com, zpd136@gmail.com

Abstract. User-based collaborative filtering algorithms generally rely on the users' stationary preferences, yet user preferences in real world are seldom stationary. User preference patterns may have the time-evolving statistical properties in many social contexts. Motivated by this phenomenon, we propose a temporal collaborative filtering approach based on temporal Hierarchical Dirichlet Process (tHDP). This approach can capture the density changes on the time-evolving datasets. Experiments on large real world datasets demonstrate the superiority of our proposed approach.

Keywords: Matrix factorization · Hierarchical dirichlet process · Collaborative filtering · Temporal based approach

1 Introduction

Recommendation systems have served as an effective approach to recommend the items that users most preferred. With the increase of information on Internet and the growth of personalize needs of users, recommender systems have become the widespread used applications and one of the key technologies of e-commerce, social networking, video, music and other Web 2.0 services.

Collaborative filtering (CF) is one of the most important techniques that recommender system used [1]. The principle of CF is to recommend the items that users do not know but most likely be of interest. It based on the assumption that if two users have similar interests, it is likely that one user might prefer another user's favorite items. CF algorithms have the advantages of not being limited by the specific content of recommended items, closely integrated with social networks and the recommendation accuracy. But CF techniques face many challenges on how to deal with huge and growing amounts of data in the Internet to make accurate recommendations:

- Huge Amount of Data: the recommendation algorithm need to be able to respond in shortest possible time.
- Data Sparsity: which looks like contradictory with huge amount of data. But relative to the large number of users and items, the information that user rating on preferred items is actually sparse.

X. He et al. (Eds.): IScIDE 2015, Part II, LNCS 9243, pp. 204–212, 2015.
DOI: 10.1007/978-3-319-23862-3_20

- Dynamics of the Data: the users' preferences usually evolve over time and more and more data are continuous generate in the processes of users' using online social services.

For the first two challenges, many researchers have proposed a number of approaches based on data clustering, dimensionality reduction and etc. These include the probability-based approaches [2, 3], matrix factorization based approaches [4, 5], co-clustering approaches [6–9] and etc. These approaches can effectively reduce the dimension of the training data, and some can effectively reduce the sparsity of data. However, these approaches have some defects of the large offline computation overhead, difficulty on model update. So these approaches can not be effectively applied in some contexts. To address the third challenge, in this paper we propose a temporal collaborative filtering approach based on tHDP. This approach can capture the density changes on the time-evolving datasets. Thus the time-varying properties of the training data can be effectively integrated into the matrix decomposition. It is important for the CF systems to detect user preferences shifts in the context of real world datasets.

The rest of this paper is organized as follows. Related works are discussed in Sect. 2. Section 3 introduce some preliminaries about the Dirichlet process. Section 4 describe the details of our proposed approach. The experimental results are discussed in Sect. 5. Finally we conclude the paper in Sect. 6.

2 Related Work

Matrix factorization is a very effective method that collaborative filtering algorithms commonly used. It uses different mathematical or machine learning methods to decompose the latent feature from the user-item rating matrix. There is a lot of work on factorization methods for collaborative filtering, among which the most well-known one is Singular Value Decomposition (SVD), which is also called Latent Semantic Analysis (LSA) in the language and information retrieval communities. Based on the LSA, probabilistic LSA [10] was proposed to provide the probabilistic modeling. Along another direction, methods like [11–13] improve the SVD using more sophisticated factorization. Bayesian probabilistic matrix factorization [14] provides a Bayesian treatment for probabilistic matrix factorization to achieve automatic model complexity control. It demonstrates the effectiveness and efficiency of Bayesian methods in real-world large-scale data mining tasks. Temporal modeling has been largely neglected in the collaborative filtering community until Koren [15] proposed their award winning algorithm timeSVD++. The timeSVD++ method assumes that the latent features consist of some components that are evolving over time and some others that are dedicated bias for each user at each specific time point. This model can effectively capture local changes of user preference which the authors claim to be vital for improving the performance. Several algorithms aimed at learning the evolution of relational data were proposed. Yang and Lozano [16] proposed a nonparametric hierarchical Bayesian approach that improves the performance of the existing collaborative approach. Li et al. [17] considers the approximately orthogonal nonnegative matrix factorization problem and proposed a method based on the hierarchical alternating least squares and the accelerated proximate gradient. Yao et al. [18] synthesize the human mobility patterns with a tensor decomposition method. All these works reveal the dynamic nature of various problems.

The above mentioned approaches face the problem that how to adjust the recommendation when the user preferences changed. How to improve the practicality and flexibility of the recommender system is an important and challenging issue. To address this challenge, in this paper we propose a temporal collaborative filtering approach based on tHDP. This approach can capture the density changes on the time-evolving datasets. Thus the time-varying properties of the training data can be effectively integrated into the matrix decomposition. It is important for the CF systems to detect user preferences shifts in the context of real world datasets. We conduct several experiments on the Netflix [19] dataset to evaluate the accuracy and efficiency of the proposed approach.

3 Preliminaries

The Dirichlet Process (DP) is a family of stochastic processes in probability theory. Dirichlet process (DP) models were first introduced by Ferguson [20]. A DP is characterized by a base distribution G_0 and a positive scalar α, usually referred to as the innovation parameter, and is denoted as $\mathrm{DP}\left(\alpha, G_0\right)$. Let us suppose we randomly draw a sample distribution G from a DP, and we independently draw M random variables $\{\theta_m^*\}_{m=1}^M$ from G:

$$G|\alpha, G_0 \sim \mathrm{DP}\left(\alpha, G_0\right). \tag{1}$$

$$\theta_m^*|G \sim G, \quad m = 1, \ldots, M. \tag{2}$$

Integrating out G, the joint distribution of the variables $\{\theta_m^*\}_{m=1}^M$ can be shown to exhibit a clustering effect.

Let $\{\theta_c\}_{c=1}^C$ be the set of distinct values taken by the variables $\{\theta_m^*\}_{m=1}^{M-1}$. A characterization of the distribution of the random variable G drawn from a DP, $\mathrm{DP}\left(\alpha, G_0\right)$, is provided by the stick-breaking construction of Sethuraman [21]. Consider two infinite collections of independent random variables $v = [v_c]_{c=1}^{\infty}, \{\theta_c\}_{c=1}^{\infty}$, where the v_c are drawn from a Beta distribution, and the θ_c are independently drawn from the base distribution G_0. The stick-breaking representation of G is then given by:

$$G = \sum_{c=1}^{\infty} \varpi_c(v)\, \delta_{\theta_c}, \tag{3}$$

where δ_{θ_c} denotes the distribution concentrated at a single point θ_c,

$$p\left(v_c\right) = \beta\left(1, \alpha\right). \tag{4}$$

$$\varpi_c(v) = v_c \prod_{j=1}^{c-1} \left(1 - v_j\right). \tag{5}$$

Teh et al. [22] proposed a HDP model that allows for linking a set of group-specific Dirichlet processes, learning the model components jointly across multiple groups of data. Specifically, let us assume J groups of data; let the dataset from the jth group be denoted as $\{x_{ji}\}_{i=1}^{N_j}$. An HDP-based model considers that the data from each group are drawn from a distribution with different parameters $\{\theta_{ji}^{*}\}_{i=1}^{N_j}$, which are in turn drawn from group-specific DPs. In addition, HDP assumes that the base distribution of the group-specific DPs is a common underlying DP. Under this construction, the generative model for the HDP yields:

$$x_{ji} \sim F\left(\theta_{ji}^{*}\right), \tag{6}$$

$$\theta_{ji}^{*} \sim G_j, \tag{7}$$

$$G_j \sim \mathrm{DP}\left(\alpha_j, G_0\right), \tag{8}$$

$$G_0 \sim \mathrm{DP}\left(\gamma, H\right), \tag{9}$$

where $j = 1, \ldots, J$, and $i = 1, \ldots, N_j$.

In the context of the HDP, different observations that belong to the same group share the same parameters that comprise G_j. In addition, observations across different groups might also share parameters, probably with different mixing probabilities for each DP G_j; this is a consequence of the fact that the DPs G_j pertaining to all the modeled groups share a common base measure G_0, which is also a discrete distribution.

4 The tHDP-Based Approach

At the beginning of the approach, there are a set of users $U = \{u_1, u_2, \ldots, u_m\}$, a set of items $I = \{i_1, i_2, \ldots, i_n\}$ and a set of time slices $T = \{1, 2, \ldots, t\}$ Matrix $R = [R_{u,i}]_{m \times n}$ record the users' real ratings on items. The entry $r_{u,i}$ in matrix R denotes the rating expressed by user u on item i. $r_{u,i}$ can be any real number. In this paper, the ratings' in Netflix dataset range in [0, 5]. Let us also consider that the preference patterns of the users may change over time, due to different moods, contexts, or pop culture trends. In addition, we assume that user preferences actually exhibit strong temporal patterns, thus tending to evolve gradually over time. To obtain a model-based CF method capable of incorporating these assumptions into its inference and prediction mechanisms, we proceed as follows.

Let us denote $R^t = \{r_{ui}^t\}$ as the set of rankings obtained at time slice t. We consider

$$p\left(R^t | U^t, I; \sigma^2\right) = \prod_{i=1}^{N} \prod_{j=1}^{M} [p\left(r_{ui}^t | U^t, I; \sigma^2\right)]^{I_{ui}^t}, \tag{10}$$

where I_{ui}^t is an indicator variable equal to 1 if the user u rated the item i at time slice t. U^t is the set of user latent feature vectors at the time slice t. To introduce temporal dynamics into the model, and capture the evolution of the latent feature vectors of the users over time, we impose a HDP prior over the user latent feature vectors:

$$r_{ui}^{t}|z_{u}^{t} = k \sim \mathcal{N}\left(u^{k}i, \sigma^{2}\right), \tag{11}$$

where u^k is the latent feature vector of the user u, if we consider that user i belongs at time t to the kth group of users, then $z_u^t = k$. In other words, to model the transitive dynamics of user preference patterns, we consider that each user belongs at each time slice to some user group, users change groups over time, and the latent feature vectors of the users belonging to the same group are generated from the same underlying distribution.

Following the assumptions of the imposed tHDP, we subsequently have:

$$z_{u}^{t}|\phi_{u}^{t};\{\pi_{\tau}\}_{\tau=1}^{t} \sim \text{Mult}\left(\pi_{\phi_{u}^{t}}\right), \tag{12}$$

where $\pi_\tau = (\pi_{\tau l})_{l=1}^{\infty}$, we have:

$$\pi_{\tau l} = \tilde{\pi}_{\tau l}\prod_{h=1}^{l-1}\left(1 - \tilde{\pi}_{\tau h}\right), \tag{13}$$

$$\tilde{\pi}_{\tau l} \sim \beta\left(\alpha_{\tau}\beta_{l}, \alpha_{\tau}\left(1 - \sum_{m=1}^{l}\beta_{m}\right)\right), \tag{14}$$

$$\beta_{k} = \varpi_{k}\prod_{q=1}^{k-1}\left(1 - \varpi_{q}\right), \tag{15}$$

$$\varpi_{k} \sim \beta\left(1, \gamma\right), \tag{16}$$

and the latent variables ϕ_u^t of the tHDP yield:

$$\phi_{u}^{t}|\tilde{w} \sim \text{Mult}\left(w_{t}\right), \tag{17}$$

with $\boldsymbol{w}_t = (w_{tl})_{l=1}^{t}$,

$$w_{tl} = \tilde{w}_{l-1}\prod_{m=1}^{t-1}\left(1 - \tilde{w}_{m}\right), \ l = 1, \ldots, t, \tag{18}$$

$$\tilde{w}_{t}|a_{t}, b_{t} \sim \beta\left(\tilde{w}_{t}|a, b\right), \tag{19}$$

and $\tilde{w}_0 = 1$. Finally, we also consider:

$$p\left(\boldsymbol{u}^{k}|\boldsymbol{\mu}_{U}^{k}, \boldsymbol{\Lambda}_{U}^{k}\right) = \mathcal{N}\left(\boldsymbol{u}^{k}|\boldsymbol{\mu}_{U}^{k}, [\boldsymbol{\Lambda}_{U}^{k}]^{-1}\right), \tag{20}$$

$$p\left(\boldsymbol{i}^{k}|\boldsymbol{\mu}_{I}^{k}, \boldsymbol{\Lambda}_{I}^{k}\right) = \mathcal{N}\left(\boldsymbol{i}|\boldsymbol{\mu}_{I}^{k}, \boldsymbol{\Lambda}_{I}^{-1}\right), \tag{21}$$

with

$$p\left(\boldsymbol{\mu}_U^k, \boldsymbol{\Lambda}_U^k\right) = \mathcal{NW}\left(\boldsymbol{\mu}_U^k, \boldsymbol{\Lambda}_U^k | \lambda_U, \boldsymbol{m}_U, \eta_U, \boldsymbol{S}_U\right), \tag{22}$$

$$p\left(\boldsymbol{\mu}_I^k, \boldsymbol{\Lambda}_I^k\right) = \mathcal{NW}\left(\boldsymbol{\mu}_I^k, \boldsymbol{\Lambda}_I^k | \lambda_I, \boldsymbol{m}_I, \eta_I, \boldsymbol{S}_I\right), \tag{23}$$

Where $\mathcal{W}$ denotes the Wishart distribution.

Having derived the above model, we now proceed to derivation of its prediction algorithm. This consists in using the trained model to estimate the unknown rating a user with index $u \in \{1, \ldots, m\}$ would have rate to an item with index $i \in \{1, \ldots, n\}$ at some time slice $T \in \{1, \ldots, t\}$, if they had rated it at that time slice. For this purpose, we use block Gibbs sampling to generate multiple samples from the posterior distributions over the model parameters and hyper-parameters. Using these samples, the generated prediction is approximated as follows:

$$\hat{r}_{ui}^t \approx \frac{1}{\Psi} \sum_{\psi=1}^{\Psi} \boldsymbol{u}^{(\psi)} \boldsymbol{i}^{(\psi)}, \tag{24}$$

where Ψ is the number of drawn samples, and $\boldsymbol{u}^{(\psi)} \boldsymbol{i}^{(\psi)}$ is the ψth drawn sample.

5 Experiments and Analysis

In this section, we conducted experiments using the Netflix dataset. The Netflix dataset contains 100, 480, 507 ratings from N = 480, 189 users to M = 17, 770 movies between 1999 and 2005. Among these ratings, 1, 408, 395 are selected uniformly over the users as the probe set for validation. Time information is provided in days. This dataset comprises users ratings to various movies, given on a 5-star scale.

To obtain some comparative results, we also evaluate recently proposed dynamic relational data modeling approaches, namely the tensor factorization-based BPTF method [23], and the Bi-LDA-based RMGM-OT method [24]. We also compare to a related static modeling method, namely BPMF [25]. We run our experiments as a single-threaded MATLAB process on a 3.40 GHz Intel Core i7 CPU. In our experiments, we use source code provided by the authors of the BPTF and RMGM-OT papers. For all models, prediction results are clipped to fit in the interval (1, 5). Similar hyper-parameter values are selected for our model, to ensure fairness in our comparisons. For the RMGM-OT method, we heuristically determined that using 20 user and item groups, as also suggested in [23], and preset in the source code provided by the authors, yields optimal results. Finally, for the BPTF method, hyper-parameter selection is based on the suggestions of [23]. Convergence of the Gibbs sampler is diagnosed by monitoring the behavior of the Frobenius norms of the sampled model parameters and hyper-parameters. Performance is evaluated on the grounds of the root mean square error (RMSE) metric, a standard evaluation metric in CF literature.

In Fig. 1, we illustrate how performance changes for the BPMF, BPTF, and tHDP models by varying the number of latent features. These results are averages over the conducted 10 folds of our experiment. As we observe, in general, all models yield an RMSE decrease as the number of factors increases. We also note that increasing the

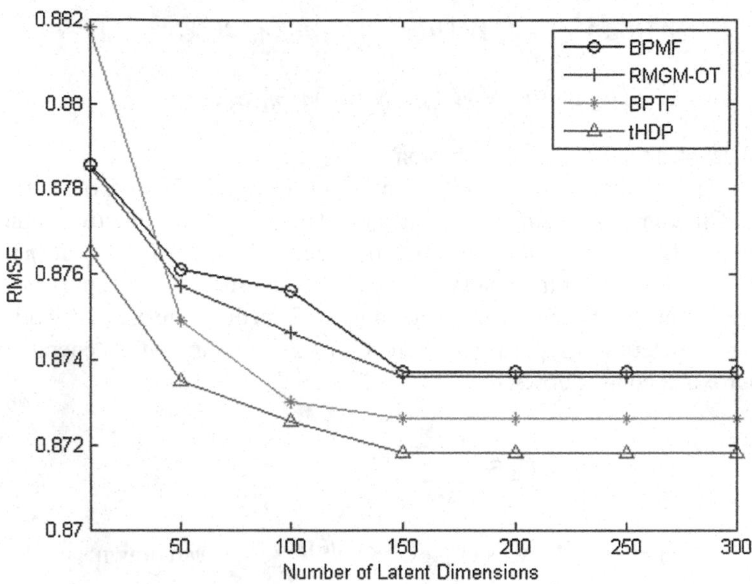

Fig. 1. Performance fluctuation as a function of latent dimensionality.

latent features to more than 100 does not result in substantial further performance improvements for the dynamic data modeling methods; in contrast, BPMF performance continues to yield some improvement. This fact indicates that the lack of temporal structure extraction mechanisms in the context of static models results in them relying on the derived latent subspace representations to make up for this inadequacy.

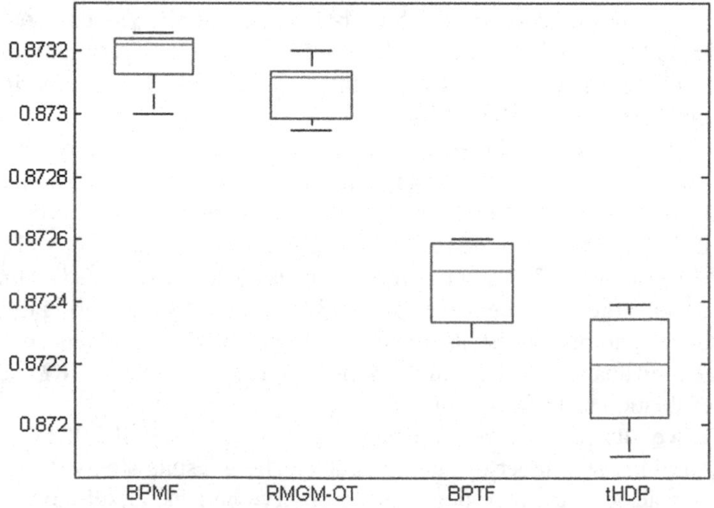

Fig. 2. Box plot of the obtained RMSEs for optimal number of latent features.

In Fig. 2, we also illustrate in detail the performances of all the evaluated algorithms for optimal latent space dimensionality, as determined from the previous illustration, in the form of box plots. As we observe, our algorithm works better than the competition, including the related BPTF method. Specifically, our approach output better result in terms of median performance, and the 25th and 75th percentiles of the set of the obtained performances. Further, we observe that RMGM-OT obtains a statistically significant improvement over the other methods, being the second best performing method.

6 Conclusions

In this paper, we presented a matrix factorization approach based on tHDP. Formulation of our method is based on an extension of the probabilistic matrix factorization framework under a Bayesian model treatment. To allow for effectively capturing preference pattern shifts in a dynamic fashion, we imposed a hierarchical Dirichlet process prior over the latent feature vectors of the users. We evaluated our method using the Netflix datasets. We observed that our approach obtain better performances in contrast with several related methods.

References

1. Chen, W.Y., Chu, J.C., Luan, J., Bai, H., Wang, Y., Chang, E.Y.: Collaborative filtering for Orkut communities: discovery of user latent behavior. In: Proceedings of WWW 2009, pp. 681–690 (2009)
2. Hofmann, T.: Latent semantic models for collaborative filtering. ACM Trans. Inf. Syst. **22**(1), 89–115 (2004)
3. Blei, D., Ng, A., Jordan, M.: Latent Dirichlet allocation. J. Mach. Learn. Res. **3**(3), 993–1022 (2003)
4. Netflix update: Try this at home (2006). http://sifter.org/~simon/journal/20061211.html
5. Zhang, S., Wang, W., Ford, J., Makedon, F.: Learning from incomplete ratings using non-negative matrix factorization. In: Ghosh J., (ed.) Proceedings of the 6th SIAM Conference on Data Mining, pp. 549–553. SIAM, Bethesda (2006)
6. Marlin, B.: Collaborative filtering: a machine learning perspective. MS, thesis, University of Toronto (2004)
7. Cheng, Y.Z., Church, G.M.: Biclustering of expression data. In: Bourne, P.E. (ed). Proceedings of the 8th International Conference on Intelligent Systems for Molecular Biology, pp. 93–103. AAAI Press (2000)
8. Cheng, G., Wang, F., Zhang, C.S.: Collaborative filtering using orthogonal nonnegative matrix tri-factorization. Inf. Process. Manage. **45**(3), 368–379 (2009)
9. Shan, H.H., Banerjee, A.: Bayesian co-clustering. In: Altman, R., (ed.) Proceedings of the ICDM 2008, pp. 530–539. IEEE Computer Society Press, Washington (2008)
10. Hofmann, T.: Probabilistic latent semantic analysis. In: Proceedings of Uncertainty in Artificial Intelligence, UAI 1999, pp. 289–296 (1999)
11. Salakhutdinov, R., Mnih, A.: Probabilistic matrix factorization. In: Advances in Neural Information Processing Systems (NIPS), vol. 20 (2007)

12. Rennie, J.D., Srebro, N.: Fast maximum margin matrix factorization for collaborative prediction. In: Proceedings of the 22nd International Conference on Machine Learning, pp. 713–719 (2005)
13. Chen, G., Wang, F., Zhang, C.: Collaborative filtering using orthogonal nonnegative matrix tri-factorization. Inf. Process. Manage. 2863–2875 (2009)
14. Salakhutdinov, R., Mnih, A.: Bayesian probabilistic matrix factorization using Markov chain Monte Carlo. In: Proceedings of the 25th International Conference on Machine Learning, pp. 880–887 (2008)
15. Koren, Y.: Collaborative filtering with temporal dynamics. Commun. ACM **53**(4), 89–97 (2010)
16. Yang, H., Lozano, A.: Multi-relational learning via hierarchical nonparametric Bayesian collective matrix factorization. J. Appl. Stat. **42**, 1113–1147 (2014)
17. Li, B., Zhou, G., Cichocki, A.: Two efficient algorithms for approximately orthogonal nonnegative matrix factorization. IEEE Sig. Process. Soc. **7**(22), 843–846 (2015)
18. Yao, D., Yu, C., Jin, H.: Human mobility synthesis using matrix and tensor factorizations. Inf. Fusion **23**, 25–32 (2015)
19. Netflix: http://archive.ics.uci.edu/ml/datasets/Netflix+Prize
20. Ferguson, T.: A Bayesian analysis of some nonparametric problems. Ann. Stat. **1**, 209–230 (1973)
21. Sethurasman, J.: A constructive definition of the Dirichlet prior. Stat. Sin. **2**, 639–650 (1994)
22. Teh, Y.W., Jordan, M.I., Beal, M.J., Blei, D.M.: Hierarchical Dirichlet processes. Technical report, Department of Computer Science, National University of Singapore (2005)
23. Xiong, L., Chen, X., Huang, T.K., Schneider, J., Carbonell, J.G.: Temporal collaborative filtering with Bayesian probabilistic tensor factorization. In Proceedings of SDM (2010)
24. Li, B., Zhu, X., Li, R., Zhang, C., Xue, X., Wu, X.: Cross-domain collaborative filtering over time. In: Proceedings of 22nd AAAI, pp. 2293–2298 (2011)
25. Salakhutdinov, R., Mnih, A.: Bayesian probabilistic matrix factorization using Markov chain monte carlo. In: Proceedings of ICML 2008 (2008)

Efficient Location-Based Event Detection in Social Text Streams

Xiao Feng[1(✉)], Shuwu Zhang[2], Wei Liang[2], and Jie Liu[2]

[1] Beijing Shu Shi Yu Tong Technologies, Beijing, China
soledad1030@gmail.com

[2] Institute of Automation, Chinese Academy of Sciences, Beijing, China
{shuwu.zhang,wei.liang,jie.liu}@ia.ac.cn

Abstract. Social networks provide a wealth of online sources about real-world events. Due to the large volume of data in social streams, the event detection suffers from high computational complexity. In this work, we present a location-based event detection approach using Locality-Sensitive Hashing to accelerate the similarity comparison. We use this approach to detect real-world events from Sina Weibo by clustering microblogs with high similarities. We propose a message-mentioned location extraction method based on the textual content based on Part-of-Speech tagging and a Support Vector Machine classifier and a novel similarity measurement considering content, location, and time between messages to improve the precision of event detection. We compare our approach with the state-of-the-art baselines on event detection, and demonstrate the effectiveness of our approach.

Keywords: Event detection · Location extraction · Social networks · Microblogs

1 Introduction

Currently, social networks, also known as the User Generated Content (UGC) platforms, provide a wealth of online sources about real-world events. Popular social media services, such as Twitter and Sina Weibo, allow people to report and share short messages (limited to 140 characters) about what is happening in their daily lives. Clearly, we can benefit from real-time event detection from social messages to support emergency managements and damage control.

The event, in this paper, is referring to an actual occurrence that happens at some specific time and place [1]. Under this definition, the social events range from widely known ones such as natural disasters or political affairs to local ones such as accidents or crimes. Our research interest is to detect real-world events including both widely-known ones and local ones via monitoring the social text stream.

The majority of works in event detection from social messages rely on clustering algorithms [2, 6, 7, 9, 11, 14], keyword co-occurrence graph [8, 13] and topic models [3, 15, 16]. Generally, clustering algorithms have relatively high computational complexity due to the large volume of data, and thus the process of event detection is inevitably time consuming. To address the problem of time delay, researchers applied

X. He et al. (Eds.): IScIDE 2015, Part II, LNCS 9243, pp. 213–222, 2015.
DOI: 10.1007/978-3-319-23862-3_21

hashing algorithm to accelerate the similarity comparison [4, 5]. The proposed algorithms based on Locality-Sensitive Hashing (LSH) only consider the cosine similarity of text contents between message pairs but fail to take spatial and temporal similarity into account.

In this paper, we present a location-based event detection approach using LSH to avoid pair wise similarity computation. We use this approach to detect real-world events from Sina Weibo by clustering microblogs with high content and location similarities under the time constraint. The research challenges of our work are: (1) the amount of social message is huge, and we need to process the data efficiently; (2) social messages are very noisy, and it is often difficult to identify whether they are truly describing a real event; (3) the event location extraction is a challengeable problem, because the GPS tags and the registered locations in user profiles can only indicate where the message was sent out and where the user often hung around, and meanwhile it is difficult to extract message-mentioned locations from the text.

Our main contributions include: (1) a real-time event detection approach using LSH to accelerate the similarity comparison, (2) a message-mentioned location extraction method based on Part-of-Speech (POS) tagging and a Support Vector Machine (SVM) classifier, (3) a novel similarity measurement considering content, location, and time to improve the precision of event detection. We compare our approach with two state-of-the-art baselines on event detection, and demonstrate the effectiveness of our approach.

The remainder of this paper is organized as follows. Section 2 introduces some related work. Section 3 introduces the scheme of event detection in social streams and the proposed methods. We evaluate the performance of proposed methods in Sect. 4 and we finally conclude our work in Sect. 5.

2 Related Work

2.1 Social Event Detection

Event detection in social networks has received considerable attention in the fields of data mining and knowledge discovery [3]. The most common approach for event detection is text-based clustering, and a variety of clustering algorithms are applied, such as hierarchical clustering [2], single-pass incremental clustering algorithm with threshold [6, 9, 11, 14], density-based clustering algorithm [7]. However, the pair-wise similarity comparison during the process of clustering is very expensive, and thus we need to limit the number of similarity comparison between messages by firstly finding candidate similar items.

Research [4] presented a first story detection method based on LSH to overcome the limitations of traditional method which relied on pair-wise similarity comparison. The proposed method used the hashing scheme [10] in which the probability of two massages colliding was proportional to the cosine of the angle between them. Besides, a variance reduction strategy was introduced to reduce the false alarm rate returned by LSH. Further, research [5] incorporated paraphrase information in the LSH scheme proposed in [4] to improve the performance of streaming first story detection. Both these two research only focused on the content text similarity between social messages without considering the location similarity and time similarity under the LSH scheme.

In this paper, we present our LSH scheme for streaming event detection based on the method proposed in [4], and furthermore extend the scheme to take both content and location similarities into consideration under the time constraint.

2.2 Event Location Extraction

An event location is a place where the event happened or is happening, and the message-mentioned location is the location mentioned in the text content of a message. According to research [2], compared to the GPS-tagged locations and the user profile locations, the message-mentioned locations are much more likely to be the actual event locations. Therefore, we use message-mentioned locations to identify event locations in this paper.

Researches on event location extraction are still relatively limited. [17] utilized a multinomial naive Bayes classifier to predict user-level location for each event-related tweet. Cheng et al. [18] proposed a method to automatically identify location keywords and further estimating a Twitter user's city-level location based purely on the textural contents. Li et al. [19] presented a method to predict the POI (Point Of Interest) tag of a tweet based on its textual content and time of posting by using ranking algorithm and web pages retrieved by search engines as an additional source of evidence. All approaches above focused on assigning one or more locations from the existing geo-name datasets to a social message, while our work attempts to extract the message-mentioned locations from the text content of a message. According to current knowledge, our work is the first try to address this problem.

3 Event Detection in Social Streams

3.1 Text Stream Clustering Based on LSH

In this paper, we detect real-world events from Sina Weibo by clustering similar microblogs. In order to accelerate the process of clustering and avoid pair-wise comparison between massages, for each newly arrived social message, we firstly use LSH to hash the message into the same bucket with its candidate similar messages, and then compute the similarity between the new message and each message in its candidate set. If the similarity between the new message and its nearest neighbor is higher than the pre-specified threshold, the new message is assigned to the same cluster to which its nearest neighbor belongs.

Basically, our method extends the LSH scheme proposed in [4] by using two kinds of hash functions instead of one. The first puts messages into the same bucket only if they have high cosine similarity in their textual content, which is the hash function utilized in [4]; the second put messages into the same bucket only if they have high Jaccard similarity [12] in their Message-Mentioned Location (MML) sets.

Under the LSH scheme for the cosine similarity in textual contents of messages ($LSH_{(content)}$, for short), the set of messages that collide with a newly arrived message $S_{content}(m)$ is defined as:

$$S_{content}(m) = \{m' : h_{ij}(m') = h_{ij}(m), \exists \in i[1...L], \forall j \in [1...k]\} \quad (1)$$

where L is the number of hash tables, k is the number of bits per key in the hashing scheme, and the hash functions h_{ij} are defined as:

$$h_{ij}(x) = \text{sgn}(u_{ij}^T x) \quad (2)$$

where x is the Vector Space Model (VSM) vector of a message with TFIDF term weights, and the random vectors u_{ij} are drawn independently for each i and j by sampling a Gaussian function with mean 0 and variance 1.

Under the LSH scheme for Jaccard similarity in MML sets ($LSH_{(MML)}$, for short), the set of messages that collide with a new arrived message $S_{MML}(m)$ is defined as:

$$S_{MML}(m) = \{m' : h_{ij}(m') = h_{ij}(m), \exists \in i[1...B], \forall j \in [1...r]\} \quad (3)$$

where B is the number of bands, r is the number of rows per band in the hashing scheme, and the hash function h is the Minhashing function which was introduced in [12].

Furthermore, for the candidate similar messages which are sent into the same bucket by the two LSH schemes ($LSH_{(content)}$, and $LSH_{(MML)}$), the pair-wise similarity between two messages is computed using the similarity measurement which will be described in more detail in Sect. 3.3.

The pseudo code shown in Algorithm 1 summarizes our text stream clustering approach based on LSH.

```
Algorithm 1: LSH-based text stream clustering approach
input: threshold t
foreach message m in social stream do
  add m to LSH_(content)
  S_content(m) ← set of messages that collide with m in LSH_(content)
  add m to LSH_(MML)
  S_MML(m) ← set of messages that collide with m in LSH_(MML)
  Sim_max(m) ← 0
  NearestNeighbor(m) ← Ø
  foreach message m' in S_content(m) ∪ S_MML(m)
    c=Sim(m,m')
    if c> Sim_max(m) then
      Sim_max(m) ← c
      NearestNeighbor(m) ← {m'}
    end
  end
  if Sim_max(m) <=t then
    m is the first message of a new event cluster
  end
  assign m to EventCluster(NearestNeighbor(m))
  add m to InvertedIndex_(content)
  add m to InvertedIndex_(MML)
end
```

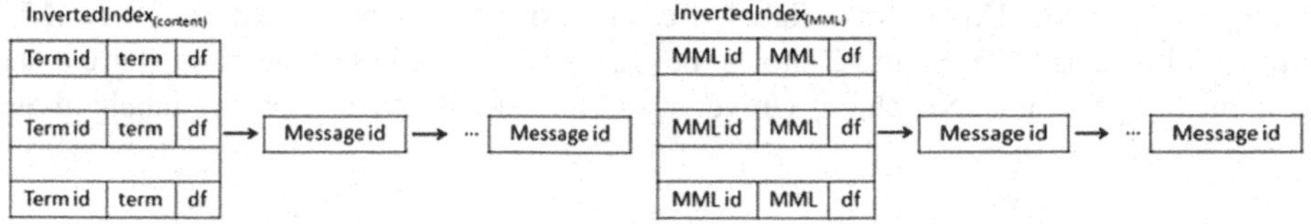

Fig. 1. The structures of $\text{InvertedIndex}_{(\text{content})}$ and $\text{InvertedIndex}_{(\text{MML})}$

Indices are built to avoid unnecessary computation of the messages that have been processed. As shown in Fig. 1, two indices are kept for all the saved messages: an inverted index of the textual content and an inverted index of the MML. The $\text{InvertedIndex}_{(\text{content})}$ has an entry for each term in the vocabulary, which is built though word segmentation and stop word elimination. The entry for term t is a linked list of the message ids of all the messages whose textual content contained t. The $\text{InvertedIndex}_{(\text{MML})}$ has an entry for each MML extracted from messages by using the method which will be described in more detail in Sect. 3.2. The entry for an MML is a linked list of the message ids of all the messages whose textual content mentioned it. The linked lists of message ids in both indices are sorted in descending order of messages' arrival time.

3.2 Message-Mentioned Location Extraction

The investigation in [2] shows that the message-mentioned locations are much more likely to be the actual event locations, compared to the GPS-tagged locations and the user profile locations. Therefore, we use message-mentioned locations to identify event locations in this paper. However, the location extraction from text is one challenging problem.

Generally, by using POS tagging and Named Entity Recognition (NER) tools, such as ICTCLAS[1] and FudanNLP[2], all country-level, province-level and city-level locations can be identified after POS-tagging. The locations identified by POS tagging and NER tools from textual contents are regarded as one part of the MMLs.

However, these tools often fail to identify street-level locations, which are the other important part of the MMLs. For example, here is a message, such as "I see a car accident happens near Nanchang University Commercial Street.". ICTCLAS can identify the city-level location "Nanchang", but it fails to identify the street-level location "Nanchang University Commercial Street". To address this problem, we present a novel method based on POS tagging and a SVM classifier to extract Street-Level Message-Mentioned Locations (SLMMLs) from textual contents.

The classifier is utilized to predict one label out of four {B,M,E,N} for each term in the text of a message to indicate that whether this term is the beginning of a SLMML phrase (labeled as 'B'), or in the middle of a SLMML phrase (labeled as 'M'),

[1] http://ictclas.org/.

[2] http://jkx.fudan.edu.cn/nlp/.

or the end of a SLMML phrase (labeled as 'E'), or nothing to do with any SLMML phrase (labeled as 'N'). Figure 2 shows the SLMML phrase in the message "I see a car accident happens near Nanchang University Commercial Street" and the labels given by the SVM classifier.

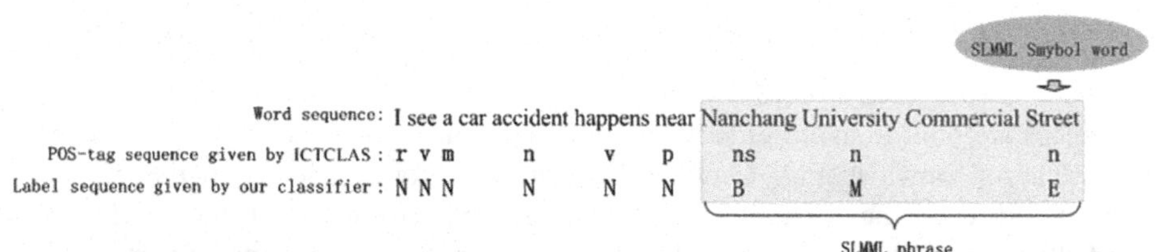

Fig. 2. The SLMML phrase and the labels given by our classifier

The classifier for labeling SLMML phrases is based on the POS tags given by POS tagging tools and a group of heuristic features. The features are described in Table 1.

Table 1. Heuristic features for labeling SLMML phrases

Feature	Definition
PosTag	The POS tag of term t
PosTagBefore	The POS tag of the term before t
PosTagAfter	The POS tag of the term after t
LabelAfter	The SLMML label of the term after t
IsSymWord	Whether t is a symbol word of SLMML phrases

We manually annotate 383 phrases of SLMML phrases from 370 event-related messages as the training set. Then, extract 65 words which are commonly used at the end of SLMML phrases in the training set, such as "Road", "Bridge", "Avenue", "Square", "Street", and so on. These words are designated as the symbol words which can indicate SLMMLs. We construct the feature vectors according to the definitions in Table 1, and train the SVM classifier using the training set. Finally, the extracted phrase are used as keywords to search the most similar geo-name in our Geo-Names dataset, and we take the most similar search result as the SLMML of the message. If there are not any similar geo-names in the dataset, the SLMML phrase are saved as a new entry after the manual review.

3.3 Similarity Measurement for Event Detection

In this section, we propose a novel similarity measurement considering content, location, and time similarities between social messages for pair-wise similarity comparison between candidate similar messages in the union of $S_{content}(m)$ and $S_{MML}(m)$ returned by $LSH_{(content)}$ and $LSH_{(MML)}$. The proposed formula combines content, spatial and temporal dimensions. Supposed that there is a newly arrived message m, and its candidate near neighbor m', the similarity between m and m' can be denoted as follows:

$$Sim(m, m') = \cos(m, m') \times sp(m, m') \times tp(m, m') \tag{4}$$

$$cos(m, m') = \sum_i \frac{m_i \times m'_i}{|m||m'|} \tag{5}$$

$$sp(m, m') = \frac{|MML(m) \cap MML(m')|}{|MML(m) \cup MML(m')|} \tag{6}$$

$$tp(m, m') = e^{\frac{\zeta|time(m) - time(m')|}{W}} \tag{7}$$

where the cosine similarity is used for content-based similarity measurement. To make the considerations of spatial and temporal similarities, the spatial penalty *sp(m,m)* is the Jaccard similarity between the MML sets of two messages, and the temporal penalty *tp(m,m')* is an exponential distribution which can reduce the similarity if the time distance between two massages is long. The parameter ζ can adjust the temporal decay rate, and *W* is the size of the sliding time window.

4 Experiments and Evaluation

4.1 Evaluation of Message-Mentioned Location Extraction

We manually annotate 383 SLMMLs from 370 microblogs from Sina Weibo as the training set, and another 272 SLMMLs from another 312 microblogs as the test set.

As described in this paper, the task of SLMML extraction is to extract out the phrases describing SLMMLs from the text content of a message. To evaluate the performance of proposed method, we compare the phrase extracted by our method and the phrase annotated manually. If the difference is no more than one word, then we consider the phrase extracted by our method is precise, because the difference of one word can be corrected by using geo-names dataset in practical application or applying fuzzy matching strategy for phrases. Moreover, we use the recall to measure whether our proposed method can extract each SLMML from the text content. Table 2 shows

Table 2. Evaluation of SLMML extraction

Feature	Precision	Recall	F-value
Pos	37.5 %	55.8 %	44.8 %
PosBefore	42.1 %	63.1 %	50.5 %
PosAfter	80.1 %	83.2 %	81.6 %
LabelAfter	25.7 %	33.5 %	29.0 %
IsSymWord	29.3 %	43.3 %	34.9 %
All proposed features	**90.3 %**	**87.6 %**	**88.9 %**

the Precision, Recall and F-value using different features for SLMML extraction described in Table 1.

The experiment results show that the highest F-value is achieved on the set test when all proposed features are used for SLMML label prediction, and the proposed method is effective in SLMML extraction.

4.2 Evaluation of Event Detection Based on LSH

In order to evaluate our approach, we collected microblogs between June 1, 2013 and June 3, 2013 from Sina Weibo to simulate a live social text stream. The search keywords used for data collection are “car accident”, “fire”, and “earthquake”, because messages containing these keywords may be related to actual important events. We collected 257872 messages containing the search keywords. Since it is impractical to manually label the overly large number of messages in the dataset, we labeled the messages as relevant and irrelevant in 5 event clusters, which are detected by our approach and both two baselines. The precision of one event cluster is defined as:

$$\text{precision(event cluster)} = \frac{|\text{relevant message set of an event}|}{|\text{detected message set of an event}|} \tag{8}$$

We also use the definition of the precision used in [2] to evaluate the ability of our approach to detect real-world events, which is defined as follows:

$$\text{precision(real-world event detection)} = \frac{\text{the number of real-world events}}{\text{the number of detected events}} \tag{9}$$

In the experiment, we compare our proposed approach with two state-of-the-art baselines. **Baseline 1** is the traditional 1-NN clustering which used cosine similarity and TFIDF weight document representation. **Baseline 2** is the LSH-based clustering approach proposed in [4], which also used cosine similarity between content text of messages and TFIDF weight scheme with $k = 13$ and $L = 100$ in Eq. (1). Both these two baselines and our approach utilized inverted indices to acculturate the similarity computation. The parameters B and r of our approach in Eq. (2) are set to $B = 5$, $r = 20$. The parameters ζ and W of our approach in Eq. (7) are set to $\zeta = -0.5$, $W = 24$ h. Table 3 shows the descriptions of the five event clusters detected by our approach and both two baselines and the frequent terms for each event. Table 4 shows the precision of the five event clusters detected by our approach and two baselines, and our approach achieved the highest precision for five event clusters. Table 5 shows the precision of real-world events detected by our approach and two baselines, and our approach achieves the highest precision for real-world event detection.

The experiment results demonstrate that the message-mentioned locations are likely to be the actual event locations and can improve the real-world event detection. Moreover, the similarity measure considering content, spatial and temporal dimensions is more effective than cosine similarity between textual contents.

Table 3. The five event clusters detected by our approach and two baselines

Event id	The frequent terms
1	Zhangzhou, traffic accident, family, child, truck, Taiwanese
2	Muxidi, car accident, AUDI, Beijing, driver, die,serious
3	Sinograin, fire, grain, barn, burn, Heilongjiang, last
4	Jilin, fire, explode, die, poutry, factory
5	Taiwan, earthquake, shake, feel, where

Table 4. The precision of five event clusters detected by our approach and two baselines

Event id	Number of messages in event cluster			Precision (event cluster)		
	Baseline 1	Baseline 2	Our approach	Baseline 1	Baseline 2	Our approach
1	526	483	567	78.3 %	79.5 %	**80.8 %**
2	247	196	223	75.4 %	76.3 %	**79.6 %**
3	1031	897	1192	72.6 %	76.8 %	**79.3 %**
4	723	573	649	80.2 %	81.3 %	**83.2 %**
5	1763	1488	2039	68.5 %	69.3 %	**73.8 %**

Table 5. The precision of real-world events detected by our approach and two baselines

Method	Num of detected events	Num of detected real-world events	Precision (real-world event detection)
Baseline 1	343	53	15.5 %
Baseline 2	421	52	12.4 %
Our Approach	287	49	**17.1 %**

5 Conclusions

In this paper, we presented a real-time event detection approach using LSH to accelerate the similarity comparison, a message-mentioned location extraction method, and a novel similarity measurement to improve the precision of event detection. We compare our approach with two state-of-the-art baselines on event detection, and the experiment results demonstrate the effectiveness of our approach.

Acknowledgements. The work is supported by the National Key Technology R&D Program of China under Grant No. 2012BAH75F03 and 2013BAH61F01.

References

1. TDT 2004: Annotation manual. http://www.ldc.upenn.edu/Projects/TDT2004
2. Unankard, S., Li, X., Sharaf, M.A.: Location-based emerging event detection in social networks. In: Ishikawa, Y., Li, J., Wang, W., Zhang, R., Zhang, W. (eds.) APWeb 2013. LNCS, vol. 7808, pp. 280–291. Springer, Heidelberg (2013)
3. Zhou, X., Chen, L.: Event detection over Twitter social media streams. VLDB J. **23**(3), 381–400 (2014)
4. Petrović, S., Osborne, M., Lavrenko, V.: Streaming first story detection with application to Twitter. In: NACL, pp. 181–189. ACL (2010)
5. Petrović, S., Osborne, M., Lavrenko, V.: Using paraphrases for improving first story detection in news and Twitter. In: NACL, pp. 338–346. ACL (2012)
6. Becker, H., Naaman, M., Gravano, L.: Learning similarity metrics for event identification in social media. In: WSDM, pp.291–300. ACM(2010)
7. Lee, C.: Mining spatio-temporal information on microblogging streams using a density-based online clustering method. Expert Syst. Appl. **39**, 9623–9641 (2012). Elseiver
8. Cataldi, M., Caro, L.D., Schifanella, C.: Emerging topic detection on twitter based on temporal and social terms evaluation. In: MDMKDD, no. 4. ACM(2010)
9. Ozdikis, O., Senkul, P., Oguztuzun, H.: Semantic expansion of hashtags for enhanced event detection in Twitter. In: VLDB (2012)
10. Charikar, M.S.: Similarity estimation techniques from rounding algorithms. In: STOC, pp. 380–388. ACM(2002)
11. Sankaranarayanan, J., Samet, H., Teitler, B.E., Lieberman, M.D., Sperling, J.: TwitterStand: news in Tweets. In: GIS, pp.42–51. ACM (2009)
12. Rajaraman, A., Ullman, J.D.: Mining of Massive Datasets. Cambridge University Press, Cambridge (2011)
13. Sayyadi, H., Hurst, M., Maykov, A.: Event detection and tracking in social streams. In: ICWSM, pp. 311–314 (2009)
14. Becker, H., Naaman, M.R, Gravano, L.: Beyond trending topics: real-world event identification on Twitter. In: AAAI, pp. 438–441 (2011)
15. Tan, Z., Zhang, P., Tan, J.: A multi-layer event detection algorithm for detecting global and local hot events in social networks. Procedia Comput. Sci. **29**, 2080–2089 (2014)
16. Wang, Y., Agichtein, E., Benzi, M.: TM-LDA: efficient online modeling of latent topic transitions in social media. In: KDD, pp. 123–131. ACM (2012)
17. Baldwin, T., Cook, P., Han, B., Harwood, A., Karunasekera, S., Moshtaghi, M.: A support platform for event detection using social intelligence. In: EACL, pp. 69–72. ACL (2012)
18. Cheng, Z., Caverlee, J., Lee, K.: You are where you tweet: a content-based approach to geo-locating Twitter users. In: CIKM, pp. 759–768. ACM (2010)
19. Li, W., Serdyukov, P., de Vries, A.P., Eickhoff, C., Larson, M.: the where in the Tweet. In: CIKM, pp. 2473–2476. ACM (2011)

Opposition-Based Backtracking Search Algorithm for Numerical Optimization Problems

Qingzheng Xu[1(✉)], Lemeng Guo[1], Na Wang[1], and Li Xu[2]

[1] Xi'an Communications Institute, Xi'an 710106, China
xuqingzheng@hotmail.com
[2] Unit 73689 of PLA, Nanjing 210042, China

Abstract. Backtracking Search Algorithm (BSA) is a novel global optimization algorithm for solving real-valued numerical optimization problems. In this paper, several opposition-based BSAs are proposed and compared comprehensively. Its key character is that a candidate solution and its corresponding opposite solution are considered simultaneously to achieve an optimal approximation. The simulation results on 58 widely used benchmark problems demonstrate that, the opposition-based learning method can significantly improve the performance of original BSA. In addition, the proposed algorithm performance has evident positive correlation with the utilization rate of opposite points.

Keywords: Backtracking Search Algorithm · Opposition-Based Learning · Number of function calls · Utilization rate of opposite points

1 Introduction

Real-world optimization problems are very important and often challenging to treat, especially when the objective function has a non-linear and non-differentiable form, and has an unknown and possibly large number of local minima within the feasible region [1]. To solve such NP-hard problems, it is difficult to find the global optimum with classical techniques and then optimization tools have to be used, though there is no guarantee that the optimal solution can be obtained within reasonable time.

As a relatively young nature-inspired algorithm, Backtracking Search Algorithm (BSA) was first introduced for the real-valued optimization problems in 2013 by Civicioglu [2]. The simulation and comparison results showed that, though it has a simple structure and less parameter to adjust, BSA has quite powerful local and global search abilities and it is easily adapt to different numerical optimization problems and real-world optimization problems. In just two years, it is widely used to solve parameter estimation [3], satellite formation flying control [4], optimal design of antenna arrays [5, 6] and induction magnetometer [7], optimal locations of distributed generations [8], and economic dispatch problems [9].

In general, choosing suitable control parameter values in evolutionary algorithm is frequently difficult. The robustness and reliability of BSA algorithm are affected mainly

X. He et al. (Eds.): IScIDE 2015, Part II, LNCS 9243, pp. 223–234, 2015.
DOI: 10.1007/978-3-319-23862-3_22

with different control parameters. In [3], in order to efficiently estimate the unknown parameters and obtain optimal solutions, a novel BSA algorithm was proposed and investigated. Due to its nonrepetitive nature, Burger's chaotic map is applied to adjust the scaling factor during mutation operation to balance the convergence rate and population diversity. An adaptive BSA was also proposed to reinforce the convergence performance of classical BSA in [7]. Its key idea is to adapt the amplitude control factor and mix rate based on the fitness statistics of population at each generation.

2 Opposition-Based Backtracking Search Algorithms

2.1 Backtracking Search Algorithm

Like many evolutionary algorithms, BSA suffers three basic genetic operators such as selection, mutation, and crossover to minimize the objective function. In general, BSA comprises five evolutionary mechanisms: initialization, selection-I, mutation, crossover, and selection-II. The general flowchart of BSA is illustrated in Fig. 1 and the detailed descriptions of BSA can be found in [2].

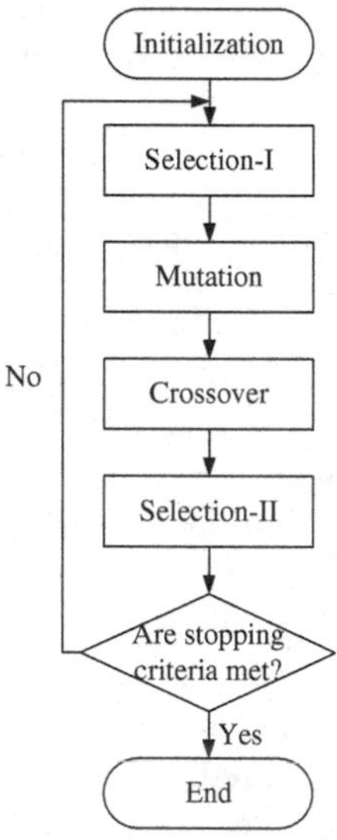

Fig. 1. General flowchart of Backtracking Search Algorithm (BSA).

2.2 Opposition-Based Learning

The basic concept of Opposition-Based Learning (OBL) was original introduced in 2005 by Tizhoosh [10]. The essential idea of this optimization strategy is the serious consideration of an estimate and its corresponding opposite estimate simultaneously to achieve an optimal approximation of the current candidate solutions. It is amazing that, it has been utilized in a vast majority of soft computing areas over a very short period of time. For a detailed overview of a range of opposition-based learning and its typical applications, we strongly recommend a recent review article by Xu etc., entitled "*A review of opposition-based learning from 2005 to 2012*" [11].

Until now, Opposition-Based Learning has been varied in different levels and successfully employed for solving continuous domain optimization problems. These variants include Quasi-Opposition-Based Learning (QOBL) [12], Quasi-Reflection Opposition-Based Learning (QROBL) [13], Generalized Opposition-Based Learning (GOBL) [14] and Opposition-Based Learning using the Current Optimum (COOBL) [15]. For further details please read the references listed above.

2.3 Opposition-Based BSAs

Similar to all population-based optimization algorithms, two main steps are distinguishable for BSA, namely, population initialization and producing new generations by basic evolutionary operations such as mutation, crossover, and selection. In this paper, both two steps are enhanced by using the OBL schemes of the Opposition-based Differential Evolution [16]: opposition-based population initialization and opposition-based generation jumping. As far as we know, this exact scheme seems to be a generally-accepted and widely used mode in many opposition-based soft computing algorithms. The basic procedure of OBSA algorithm is presented as follows in Fig. 2.

```
procedure OBSA (Problem, Opposition method)
    Randomly generate initial population, P
    Generate the opposite of initial population, OP
    Maintain the fittest amongst P and OP
    while Generation ≤gen limit do
        Perform BSA algorithm
        Remove duplicates from population
        Calculate the fitness of P
        if random ≤Opposition Jumping Rate then
            Create the opposite population, OP
            Calculate the fitness of OP
            Maintain the fittest amongst P and OP
        end if
        Restore Elite individuals
    end while
    return Best Individual
end procedure
```

Fig. 2. Opposition-based backtracking search algorithms.

It is clearly that several opposition-based backtracking search algorithms (OBSAs) can be established when using different opposition-based leaning schemes mentioned above. In order to distinguish them each other, we use different abbreviations in this paper, including OBSA which uses Opposition-Based Learning, QOBSA which uses Quasi-Opposition-Based Learning, QROBSA which uses Quasi-Reflection Opposition-Based Learning, GOBSA which uses Generalized Opposition-Based Learning, and COOBSA which uses Opposition-Based Learning using the Current Optimum.

3 Simulation Results and Discussion

3.1 Experimental Setup

A comprehensive set of numerical benchmark functions, including 58 different well-known global optimization problems [16], has been employed for performance verification of the proposed approach. The definition, the range of search space, and also the global minimum of each function are given in Appendix A in literature [16].

For all conducted experiments, parameter settings are kept the same as follows for all competing algorithms: population size, $N_p = 30$; mix rate in crossover process, DIM_RATE = 1; Jumping rate constant, $J_r = 0.3$; maximum NFCs, MAX-NFC = 2×10^6; value to reach, VTR = 10^{-8}. The termination criterion is to find a value smaller than the value-to-reach (VTR) before reaching the maximum number of function calls.

We compare the convergence speed of BSA and all OBSAs by measuring the number of function calls (NFC) which is the most commonly used metric in literature. A smaller NFC means higher convergence speed. In order to minimize the effect of the stochastic nature of the algorithms on the measured metric, the reported values for each function are the average over 100 independent trials. In order to maintain a reliable and fair comparison, extra fitness evaluation for the opposite points (both in population initialization and also generation jumping phases) are counted.

The number of times, for which the algorithm successfully reaches the VTR for each test function, is measured as the success rate (SR):

$$\mathrm{SR} = \frac{\text{number of times reached VTR}}{\text{total number of trials}} \tag{1}$$

Utilization rate of opposite points (UR) is another comparison criterion. A larger UR means that opposite points play more effects on algorithm performance during the evolution process.

$$\mathrm{UR} = \frac{\text{number of opposite points reserved as offspring in the next generation}}{\text{total number of points and opposite points}} \tag{2}$$

3.2 Simulation Results and Discussion

A random version of OBSA (called ROBSA in this paper) is introduced firstly. Instead of using opposite points for the population initialization and the generation jumping, the random points generated uniformly in entire search space, are employed in ROBSA algorithm. Results of applying BSA, OBSA, QOBSA, QROBSA, GOBSA, COOBSA and ROBSA to solve 58 benchmark functions are given in Table 1.

First of all, the performance of Backtracking Search Algorithm is tested on 58 benchmark problems. As seen from Table 1, the success rates for problems, F17 (20 %) and F50 (31 %), are both less than fifty percent. In addition, the original BSA is utterly fail to make optimize solution within MAXNFC when solve problems F22, F24, F33, F51, F55 and F56. For some problems (F17, F22, F24, F51, F55 and F56), based on our previous results [17], we can not also obtain the optimal solution using the Differential Evolution (DE) approach. According to frequently reported comprehensive studies, DE has been preferred in terms of accuracy, convergence speed, and robustness. From this point of view, BSA can provide good performance over well-known benchmark functions, when compared with many other heuristic algorithms.

From Table 1, we can also observe that the success rate of COOBSA is the lowest one among all tested algorithms. It can not solve other 10 problems successfully, including F3, F5, F10, F17, F18, F19, F31, F34, F50 and F58. Until now, we have no idea what causes the sharply decline in success rate. We suspects that the possible reasons are algorithm implementation, parameters selection, robustness of algorithm. It is our top issue faced and will be solved in the near future.

Furthermore, it is obvious from Table 1 that, NFC of each opposition-based BSA is less than that of BSA algorithm. For example, its mean NFC of 100 independent Monte Carlo simulations for solving F1 is 1.88 times (OBSA) to 4.45 times (GOBSA) higher than that of opposition-based BSAs. Now, we can conclude that opposition-based BSAs have already showed better convergence speed than classical BSA.

The next step is to find out which opposition-based method can demonstrate the best outstanding performance on NFC. As far as we know, pair comparison or group comparison of these algorithms is the simplest and most commonly used subjective rating approach, presented in Table 2, in soft computing research. Given number in each cell shows on how many functions the algorithm in the table's row outperforms the corresponding one in the table's column. The penultimate column of the table shows the total numbers, which the algorithm can outperform other competitors. In addition, the last column is the corresponding percentage ratio to all comparison objects. It is special note that COOBSA and other algorithms solved 41 problems and 51 problems, respectively. As a result, the all compared cases for COOBSA are 246, and the all compared cases for other algorithms are 296. By comparing these percentage ratios, the following ranking result is obtained: COOBSA (best), QROBSA, QOBSA, GOBSA, OBSA, ROBSA and BSA.

Based on the features of 58 benchmark functions, they can be categorized into two types [15]. Within this paper, Type I (including 25 functions) means that the global optimum is the midpoint of the range of function domain, while type II (including 33 functions) holds a contrary opinion.

The focus of this paragraph is to explore the reason behind the superior quality of COOBSA. QROBSA has a good lead with 72.0 % over other competitors, and can be considered as a suboptimal algorithm. So we will compare the performances of QROBSA and COOBSA in details. It is observed from Table 3 that, the contribution of the functions of type I to QROBSA dominant and type II to COOBSA goes beyond 50 % and 60 %, respectively. When solving the functions of type I by COOBSA, for

Table 1. Comparison of BSA and OBSAs.

Benchmark function	Type	Criteria	BSA	OBSA	QOBSA	QROBSA	GOBSA	COOBSA	ROBSA
F1	I	SR	1	1	1	1	1	0.92	1
		NFC	88269.6	46854.9	34596.3	42161.4	19823.4	29751.2	44163.9
		UR	-	11.95	24.80	24.82	19.03	17.31	10.48
F2	I	SR	1	1	1	1	1	0.93	1
		NFC	98155.2	52093.5	38484.9	47565.6	22141.2	33499.7	49063.5
		UR	-	12.00	24.84	24.72	18.98	17.27	10.46
F3	I	SR	1	1	1	1	1	0	1
		NFC	806910.6	468599.7	232203.6	298243.2	106762.5	0	548373.9
		UR	-	3.26	17.41	21.91	4.88	0	1.75
F4	II	SR	1	0	0	0	0	0	0
		NFC	1496516	0	0	0	0	0	0
		UR	-	0	0	0	0	0	0
F5	I	SR	1	0.98	1	0.98	1	0	1
		NFC	93587.4	81990.9	81144	58649.7	75554.1	0	103849.5
		UR	-	1.46	0.61	2.30	1.49	0	0.47
F6	I	SR	1	1	1	1	1	0.96	1
		NFC	93580.8	50054.7	36717	43785.3	20832	31273.4	46361.4
		UR	-	11.94	24.88	24.85	19.00	17.33	10.50
F7	I	SR	1	0.99	1	1	1	0.98	1
		NFC	19501.8	8204.9	2604	3030.6	5398.5	6590.2	6719.1
		UR	-	13.95	23.23	22.94	17.10	18.15	16.00
F8	I	SR	1	1	0.38	0.11	1	0.61	1
		NFC	174558.6	98577	69130.3	80091.8	37958.7	59813.6	90449.1
		UR	-	10.71	23.53	24.30	18.75	17.00	9.46
F9	II	SR	1	1	1	1	1	0.79	1
		NFC	14879.4	11222.4	6263.1	3888.9	8882.7	3505.1	7522.2
		UR	-	3.49	7.46	14.17	4.98	13.64	6.03
F10	II	SR	1	0.99	1	1	1	0	1
		NFC	80218.8	53039.7	108613.2	173920.8	70906.5	0	73368.9
		UR	-	2.46	10.34	17.64	0.59	0	0.51
F11	II	SR	1	1	1	1	1	0.87	1
		NFC	14542.8	7500.9	3777.6	3171.3	5945.1	4414.8	7615.5
		UR	-	7.38	13.69	15.82	9.86	12.33	7.26
F12	II	SR	1	1	1	1	1	0.94	1
		NFC	9771.6	5806.2	3383.7	3148.8	5489.4	3718.7	5556.6
		UR	-	10.71	16.52	19.58	12.10	16.02	12.12
F13	II	SR	1	1	1	1	1	0.61	1
		NFC	26184.6	15495.3	14671.8	10189.2	22364.7	7728.2	27750.6
		UR	-	8.45	8.75	15.71	3.31	15.64	1.32
F14	II	SR	1	1	1	1	1	0.95	1
		NFC	9903	4412.1	9270.3	7306.8	4024.8	2946.6	9080.4
		UR	-	13.53	2.37	3.50	14.33	15.09	2.61
F15	II	SR	1	0.96	0.7	0.5	1	0.57	1
		NFC	97160.4	55492.2	37882.3	45484.2	56245.2	32670.5	51960.3
		UR	-	10.35	22.27	23.76	8.41	16.82	9.16
F16	I	SR	1	1	1	1	1	0.8	1
		NFC	11713.2	3890.4	2249.1	2349.3	3891.6	3246	6102.6
		UR	-	13.88	17.17	18.25	14.59	14.48	7.47

(*Continued*)

Table 1. (*Continued*)

F17	II	SR	0.2	0.14	0.14	0.1	0.08	0	0.12
		NFC	1515066	1504667	1502919	1462620	1292198	0	1581355
		UR	-	0.009	0.003	0.004	0.012	0	0.012
F18	II	SR	0.96	0.99	0.99	0.97	0.98	0	0.99
		NFC	135848.8	151289.4	137224.5	126334	157790.8	0	157727.9
		UR	-	0.12	0.09	0.14	0.07	0	0.07
F19	II	SR	1	1	1	1	1	0	1
		NFC	1186103	888006.9	355393.8	359697.6	286618.5	0	750346.5
		UR	-	0.56	6.48	14.53	2.05	0	1.48
F20	II	SR	1	1	1	1	1	0.79	1
		NFC	14868.6	5886.6	15196.8	6992.7	13499.1	3214.6	14865.9
		UR	-	10.22	0.71	4.36	0.79	14.21	0.75
F21	I	SR	1	1	1	1	1	0.41	1
		NFC	166696.8	134848.5	62310.3	77529.9	40166.4	55502.2	135799.5
		UR	-	6.12	23.40	23.68	17.07	15.48	3.73
F22	I	SR	0	0	0	0	0	0	0
		NFC	0	0	0	0	0	0	0
		UR	-	0	0	0	0	0	0
F23	I	SR	1	1	0.29	0.24	1	0.52	1
		NFC	21352.2	16891.2	312663.1	178333.8	5521.8	75429.8	13812
		UR	-	6.05	8.08	9.35	16.59	10.64	6.91
F24	I	SR	0	0	0	0	0	0	0
		NFC	0	0	0	0	0	0	0
		UR	-	0	0	0	0	0	0
F25	II	SR	1	0.99	1	1	1	0.08	1
		NFC	81814.8	66029.7	39022.8	47414.4	65520.9	27322.5	57206.7
		UR	-	1.44	10.64	16.12	0.87	7.77	1.16
F26	II	SR	1	0.89	0.98	0.88	1	0.44	1
		NFC	33744.6	16239.1	19183.2	14964.2	28806	7018.0	35728.5
		UR	-	8.32	4.76	11.07	1.28	15.11	0.56
F27	II	SR	1	0.95	0.98	0.99	1	0.59	1
		NFC	36893.4	26530.4	14204.4	7734.8	23616.9	7243.2	27500.4
		UR	-	3.92	9.38	17.04	3.68	15.08	2.59
F28	II	SR	1	0.93	1	0.99	1	0.61	1
		NFC	35041.2	27086.8	11988	7107.9	22397.7	7347.0	26089.8
		UR	-	3.72	9.61	17.84	3.72	14.85	2.89
F29	II	SR	1	1	1	1	1	0.27	1
		NFC	56777.4	32250	24436.5	11758.5	29072.7	6572.2	47920.5
		UR	-	2.07	0.44	7.18	0.45	15.03	0.30
F30	I	SR	1	1	1	1	1	1	1
		NFC	1333.8	969.9	545.4	480	880.8	813	1006.8
		UR	-	14.49	20.35	21.05	15.21	15.83	13.93
F31	I	SR	1	1	1	1	1	0	1
		NFC	592862.4	661116.3	606311.4	305619	579159	0	663906.9
		UR	-	0.09	0.04	2.57	0.08	0	0.02
F32	I	SR	1	1	1	1	1	0.78	1
		NFC	13891.2	6860.7	3561.6	2421.3	4115.4	3303.5	8423.7
		UR	-	7.94	11.37	16.27	13.37	13.87	5.03

(*Continued*)

Table 1. (*Continued*)

F33	I	SR	0	0	0	0	0	0	0
		NFC	0	0	0	0	0	0	0
		UR	-	0	0	0	0	0	0
F34	I	SR	1	0.99	1	1	0.98	0	1
		NFC	60405.6	39255.2	23682.9	11032.2	20003.9	0	58773
		UR	-	2.91	2.74	10.61	8.18	0	0.51
F35	II	SR	1	1	1	1	1	0.97	1
		NFC	6540	3673.8	2872.2	2395.5	3453.9	2774.8	3919.5
		UR	-	12.74	14.13	17.82	13.53	15.75	12.62
F36	II	SR	1	1	1	1	1	0.92	1
		NFC	9540	4097.1	11289.6	10161.9	3941.1	2974.6	10702.5
		UR	-	13.98	0.77	0.82	14.25	14.17	1.15
F37	I	SR	1	1	1	1	1	0.95	1
		NFC	9286.8	4964.1	3352.8	2681.4	4265.1	3754.1	5766.6
		UR	-	12.85	16.38	19.67	14.81	15.53	11.13
F38	I	SR	1	1	1	1	1	0.93	1
		NFC	10762.2	4818	2838.6	2680.2	4341.3	3946.5	5766.6
		UR	-	13.37	18.93	19.88	14.99	15.35	11.23
F39	I	SR	1	1	1	1	1	0.96	1
		NFC	8652	3424.2	2171.4	1941.6	3247.5	2862.5	4311
		UR	-	13.63	17.78	20.02	14.73	15.37	10.78
F40	II	SR	1	1	1	1	1	0.86	1
		NFC	12870.6	5913	13743.3	11923.5	5689.2	4492.3	13603.8
		UR	-	14.10	1.65	2.12	14.86	15.18	1.80
F41	I	SR	1	1	1	1	1	0.96	1
		NFC	29616	16159.8	6372.6	7836.9	10051.8	10563.8	17084.4
		UR	-	12.10	24.08	24.41	17.09	16.23	11.32
F42	II	SR	1	1	1	1	1	0.91	1
		NFC	11309.4	6745.5	3270	2950.2	5085.3	3471.8	5220.9
		UR	-	7.61	15.80	18.08	11.36	15.11	11.14
F43	II	SR	1	1	1	1	1	0.34	1
		NFC	36774	22352.4	15296.7	19127.4	29311.2	10279.4	29064.6
		UR	-	4.84	8.86	11.95	1.41	9.33	1.41
F44	II	SR	1	1	1	1	1	0.91	1
		NFC	12911.4	6419.4	3415.2	3492.9	6139.5	5011.6	7296.3
		UR	-	13.77	20.44	21.11	14.40	15.93	12.74
F45	II	SR	1	1	1	1	1	0.98	1
		NFC	5272.8	3178.5	2202.3	2043.6	3284.7	2445.9	3179.1
		UR	-	12.40	17.71	19.25	12.67	16.05	12.88
F46	II	SR	1	1	1	1	1	0.96	1
		NFC	10749	5957.4	3612.3	3138.9	7171.2	4365	6990
		UR	-	11.81	16.82	19.81	9.40	15.36	9.77
F47	II	SR	1	1	1	1	1	0.97	1
		NFC	7329	3261.9	2158.5	2171.7	3759.3	2638.5	3765.3
		UR	-	13.22	18.46	19.60	11.49	15.59	11.76
F48	II	SR	1	1	1	1	1	0.83	1
		NFC	11041.8	3798.9	10319.7	13154.4	8369.4	4414.3	7981.8
		UR	-	10.76	22.71	21.59	5.77	11.43	6.22

(*Continued*)

Table 1. (*Continued*)

F49	II	SR	1	1	1	0.99	1	0.55	0.98
		NFC	16874.4	18116.4	18032.1	7322.7	15258	3432.5	9954.5
		UR	-	0.52	0.64	5.25	1.54	13.04	4.24
F50	II	SR	0.31	0.29	0.31	0.73	0.14	0	0.17
		NFC	1053030	1150814	922672.3	1175594	1050017	0	1221856
		UR	-	0.02	0.01	0.02	0.02	0	0.02
F51	II	SR	0	0	0	0	0	0	0
		NFC	0	0	0	0	0	0	0
		UR	-	0	0	0	0	0	0
F52	II	SR	1	1	1	1	1	0.91	1
		NFC	39798	28353.9	11579.7	14764.8	25092.3	13454.5	24694.5
		UR	-	7.47	23.04	23.57	9.42	15.95	9.73
F53	I	SR	1	0.98	1	1	1	0.56	1
		NFC	17169	6981.4	5865.6	3945	8335.5	3456.4	17514.6
		UR	-	9.23	4.82	8.99	7.84	13.14	0.62
F54	I	SR	1	1	1	1	1	0.23	1
		NFC	32391	10512.6	17193.9	31148.7	11542.5	11807.0	26561.1
		UR	-	11.83	11.74	12.11	10.12	10.41	2.22
F55	II	SR	0	0	0	0	0	0	0
		NFC	0	0	0	0	0	0	0
		UR	-	0	0	0	0	0	0
F56	I	SR	0	0	0	0	0	0	0
		NFC	0	0	0	0	0	0	0
		UR	-	0	0	0	0	0	0
F57	I	SR	1	1	1	1	1	0.45	1
		NFC	50320.8	17973.6	10679.1	9768.6	14795.4	13710	22070.4
		UR	-	11.25	16.36	17.79	13.99	13.97	9.34
F58	II	SR	1	1	1	1	1	0	1
		NFC	80167.2	54286.2	104824.8	191402.7	70470	0	73046.1
		UR	-	2.51	10.04	18.08	0.59	0	0.53

the long distance between the local optimum and the global optimum, the opposite points have little influence during the primary stage of evolution, and then the performance of COOBSA slightly worse than QROBSA. When solving the functions of type II by QROBSA, the opposite points may lapse from the global optimum. On the contrary, when COOBSA is utilized for the functions of type II, the opposite points encircle the global optimum, which is approximately equal to the local optimum, and keep high influence on the next generation during the later stage of evolution. As a result of replacing the quasi-reflection opposition-based points by opposition-based points using the current optimum, COOBSA successfully illustrate the better performance than QROBSA.

By computing and comparing the mean utilization rates of all opposition-based algorithms in Table 1, the following ranking result is obtained: QROBSA (highest, 15.06 %), COOBSA (14.68 %), QOBSA (12.30 %), GOBSA (8.12 %), OBSA (8.11 %) and ROBSA (5.85 %). It is clear and interesting that the ranking result is similar to that mentioned above, where the only difference is the sequence between QROBSA and COOBSA. Thus we have reason to hypothesize that, the higher the UR

Table 2. Pair comparison of all tested algorithms.

	BSA	OBSA	QOBSA	QROBSA	GOBSA	COOBSA	ROBSA	Total	Percentage ratio (%)
BSA	-	4	9	6	3	1	11	34	11.5
OBSA	47	-	10	10	14	3	26	110	37.5
QOBSA	42	41	-	21	30	16	42	192	64.9
QROBSA	45	41	30	-	34	18	45	213	72.0
GOBSA	48	37	21	17	-	12	41	176	59.5
COOBSA	40	38	25	23	29	-	40	195	79.3
ROBSA	40	25	9	6	10	1	-	91	30.7

value is, the better the opposition-based algorithm performance is. The relationship between UR and percentage ratio becomes very apparent, especially when they are recorded and illustrated in Fig. 3.

Table 3. Comparison of QROBSA and COOBSA.

Functions	QROBSA dominant	COOBSA dominant
Type I	9	8
Type II	9	15
Total	18	23

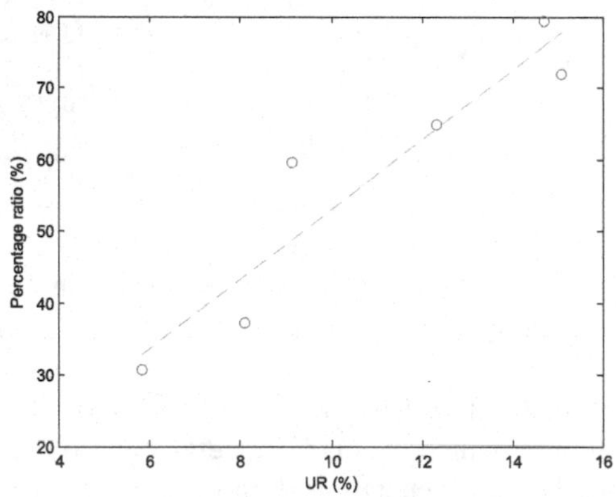

Fig. 3. The relationship between UR and percentage ratio.

In statistics, the Pearson's correlation coefficient between two variables X and Y was developed by Karl Pearson and widely used in the sciences as a measure of the degree of linear dependence between two variables [18]. With the help of software Statistical Product and Service Solutions (SPSS) 14.0 version, the Pearson's correlation coefficient between UR and percentage ratio can be calculated and obtained as equal to 0.943, and the significance with a two-sided is also obtained as equal to 0.005. These results of data statistic analysis indicate clearly that the percentage ratio has evident positive correlation with UR value. Typically, the least square fit is applied to their UR (x) and percentage ratio (y) of six opposition-based algorithms, and then the linear equation, $y = 4.8779x + 4.3248$, is obtained based on the experimental data in this paper. Therefore we can draw a conclusion intuitively that, the higher the utilization rate of opposite points is, the better the opposition-based algorithm performance is. However, it is should also be noted that this conclusion is valid only in statistical meaning. That is to say, for a given problem, an opposition-based algorithm, even with the greatest UR value, may show worse performance than other algorithms.

4 Conclusion

BSA is a newly-proposed global optimization algorithm and already proven competent over other soft computing algorithms. In this paper, some opposition-based learning methods used widely were applied into original BSA to improve its performance. We investigated comprehensively the effectiveness of opposition-based learning in solving some benchmark problems. The results of our simulation demonstrate that, the opposition-based learning can significantly improve the performance of BSA. Among all OBSA techniques, COOBSA is the best one, because the opposite points may always encircle the global optimum and keep high influence on the next generation during the later stage of evolution. Another useful and interesting conclusion is that, in statistical meaning, the algorithm performance has evident positive correlation with the utilization rate of opposite points. In the future, the proposed method should be compared with state-of-art BSA methods through various experiments on more benchmark data sets.

Acknowledgments. This work was supported in part by the National Natural Science Foundation of China (Nos. 61375089 and 61305083).

References

1. Jr, I.F., Yang, X.S., Fister, I., Brest, J., Fister, D.: A brief review of nature-inspired algorithms for optimization. Elektrotehniški Vestnik **80**, 116–122 (2013)
2. Civicioglu, P.: Backtracking search optimization algorithm for numerical optimization problems. Appl. Math. Comput. **219**, 8121–8144 (2013)
3. Askarzadeh, A., Coelho, L.S.: A backtracking search algorithm combined with Burger's chaotic map for parameter estimation of PEMFC electrochemical model. Int. J. Hydrogen Energy **39**, 11165–11174 (2014)
4. Kolawole, S.O., Duan, H.: Backtracking search algorithm for non-aligned thrust optimization for satellite formation. In: IEEE International Conference on Control and Automation, pp. 738–743 (2014)
5. Guney, K., Durmus, A., Basbug, S.: Backtracking search optimization algorithm for synthesis of concentric circular antenna arrays. Int. J. Antennas Propag. **2014**, 1–11 (2014)
6. Muralidharan, R., Athinarayanan, V., Mahanti, G.K., Mahanti, A.: QPSO versus BSA for failure correction of linear array of mutually coupled parallel dipole antennas with fixed side lobe level and VSWR. Adv. Electr. Eng. **2014**, 1–7 (2014)

7. Duan, H., Luo, Q.: Adaptive backtracking search algorithm for induction magnetometer optimization. IEEE Trans. Magn. **50**, 6001206 (2014)
8. El-Fergany, A.: Optimal allocation of multi-type distributed generators using backtracking search optimization algorithm. Electr. Power Ener. Syst. **64**, 1197–1205 (2015)
9. Modiri-Delshad, M., Rahim, N.A.: Solving non-convex economic dispatch problem via backtracking search algorithm. Energy **77**, 372–381 (2014)
10. Tizhoosh, H.R.: Opposition-based learning: a new scheme for machine intelligence. In: International Conference on Computational Intelligence for Modelling, Control and Automation, and International Conference on Intelligent Agents, Web Technologies and Internet Commerce, pp. 695–701 (2005)
11. Xu, Q.Z., Wang, L., Wang, N., Hei, X.H., Zhao, L.: A review of opposition-based learning from 2005 to 2012. Eng. Appl. Artif. Intell. **29**, 1–12 (2014)
12. Rahnamayan, S., Tizhoosh, H.R., Salama, M.M.A.: Quasi-oppositional differential evolution. In: IEEE Congress on Evolutionary Computation, pp. 2229–2236 (2007)
13. Ergezer, M., Simon, D., Du, D.W.: Oppositional biogeography-based optimization. In: IEEE International Conference on Systems, Man and Cybernetics, pp. 1009–1014 (2009)
14. Wang, H., Wu, Z.J., Liu, Y., Wang, J., Jiang, D.Z., Chen, L.L.: Space transformation search: a new evolutionary technique. In: ACM/SIGEVO Summit on Genetic and Evolutionary Computation, pp. 537–544 (2009)
15. Xu, Q.Z., Wang, L., He, B.M., Wang, N.: Opposition-based differential evolution using the current optimum for function optimization. J. Appl. Sci. **29**, 308–315 (2011). (in Chinese)
16. Rahnamayan, S., Tizhoosh, H.R., Salama, M.M.A.: Opposition-based differential evolution. IEEE Trans. Evol. Comput. **12**, 64–79 (2008)
17. Xu, Q.Z.: Research on the Artificial Co-computing Model and Its Applications. Xi'an University of Technology, Xi'an (2011)
18. Pearson, K.: Notes on regression and inheritance in the case of two parents. Proc. Roy. Soc. London **58**, 240–242 (1895)

Real Orthogonal STBC MC-CDMA Blind Recognition Based on DEM-Sparse Component Analysis

S.I.M.M. Raton Mondol, Bui Quang Chung(✉), and Zhang Tian Qi

Chongqing Key Laboratory of Signal and Information Processing,
School of Communication and Information Engineering,
Chongqing University of Posts and Telecommunications,
Nan'an District, Chongqing 400065, China
simmraton@gmail.com

Abstract. The method for blind recognition of real orthogonal STBC of underdetermined systems based on sparse component analysis is proposed. This algorithm first built the model of received signals and the virtual channel matrix, as the virtual channel matrix includes information of space time code and can be used for blind recognition, and then the virtual channel matrix is separated by using the DEM algorithm. Finally two characteristic parameters of correlation matrix of virtual channel matrix are extracted according to the characteristics of real orthogonal STBC for blind recognition such as sparsity and energy ratio of non-main and main diagonal elements energy. The simulation results and theoretical analysis indicates that the proposed algorithm can detect signals efficiently and also work well on the lower SNR input.

Keywords: Orthogonal space time block coding (OSTBC) · Space time block coding multicarrier code division multiple access (STBC MC-CDMA) · Virtual channel matrix · Sparse component analysis (SCA)

1 Introduction

The OFDM (orthogonal frequency division multiplexing) has received widespread interest for wireless broadband multimedia applications over the last decade and more recently in the railway context. The main advantages of this technique are its robustness in the case of frequency selective fading channels, its capability of portable and mobile reception and its flexibility. Spread spectrum has been successfully used by the military services for decades and nowadays takes a significant role in cellular and personal communications. Advantages of spread spectrum techniques are widely known: immunity against multi path distortion and jamming, low transmitted power, no need for complex frequency planning when using code division multiple access (CDMA) properties. The advantages and success of multicarrier modulation and spread spectrum technique motivated many researchers to investigate the suitable combination of both techniques, known as multicarrier spread spectrum (MC-SS) which benefits from the main advantages of both schemes. Space time coding (STC) is an effective coding technique that uses transmit diversity to combat the detrimental effects in

X. He et al. (Eds.): IScIDE 2015, Part II, LNCS 9243, pp. 235–246, 2015.
DOI: 10.1007/978-3-319-23862-3_23

wireless fading channels by combining signal processing at the receiver with coding techniques appropriate to multiple transmit antennas to achieve higher date rates. Thus, the combination of STC and MC-CDMA is one of the most promising schemes for next generation (next G) wireless communication [1–3]. Receive diversity receiver shortcoming is the computation load is high, may cause the power of mobile station in the downlink of the consumption of large. The transmitter uses space time coding can achieve diversity gain and decoding at the receiver only need simple linear processing. Space time codes to transmit diversity techniques channel coding and modulation techniques in combination can effectively improve the transmission performance in fading channels.

Recently, existing algorithms of multicarrier OFDM modulation signal recognition problems mostly embed inter class recognition of single carrier and OFDM signals. Young et al. [4] under the condition of only one receive antenna, using fourth-order cycle accumulation characteristics of Alamouti code to recognize Alamouti code and general spatial multiplexing scheme. Choqueuse et al. [5–8] proposed space time block codes (STBC) identification based on correlation matrix of methods. This method according to the different STBC correlation matrix at different time delays, Frobenius norm is zero or not, using a decision classifier tree to realize space-time block codes recognition. In [6] three kinds of STBC classifier is proposed based on maximum likelihood such as: the optimal classifier, the second-order statistics (SOS) classifier, the code parameter (CP) classifier, where code parameters classifier enables blind identification. The blind recognition of linear space time block codes is proposed in [8], this method first whitened the intercepted sample, then the characterization of the STBC is obtained from a time-lag correlation of the whitened process, where the correlation norm only depends on the construction matrices of the STBC, finally the automatic recognition of the STBC is realized by comparing the theoretical and experimental values of this correlation norm. However, these methods are not yet deeply researched on the type recognition of space-time codes and almost of them just working good in the determined systems. Furthermore, to decode the received signal, we need to know the used coding mode, so the type recognition of space-time codes methods need to deeply researched.

For the blind recognition orthogonal STBC of underdetermined systems problem, the method for blind recognition of real orthogonal STBC of underdetermined systems based on sparse component analysis is proposed. This algorithm first built the model of received signals and the virtual channel matrix, then using the DEM [9] algorithm to separate the virtual channel matrix, finally based on sparsity and energy of ratio of non-main and main diagonal elements energy to recognition orthogonal STBC signal.

The paper is organized as follows: Sect. 1 presents the introduction; Sect. 2 describes OSTBC baseband physical model MC-CDMA system, the signal structure and matrix; Sect. 3 describes characteristic parameters extraction; Sect. 4 describes the OSTBC code recognition algorithm; Sect. 5 is the computer simulation and analysis; Sect. 6 gives the conclusions and outlook of this article.

Notation $[\bullet]^*$, $[\bullet]^T$ and $[\bullet]^H$ denote the complex conjugate, the transpose and the conjugate transpose of matrix, respectively.

2 The Signal Model

The OSTBC- MIMO system is showed in the Fig. 1 with n_T transmits and n_R receives antennas. Where $\boldsymbol{S}(k) = [S_1(k), S_2(k), \ldots, S_N(k)]^{\mathrm{T}}$ is the $k-th$ block of N symbols to be transmitted by the system, where the symbols are belong to a finite alphabet set.

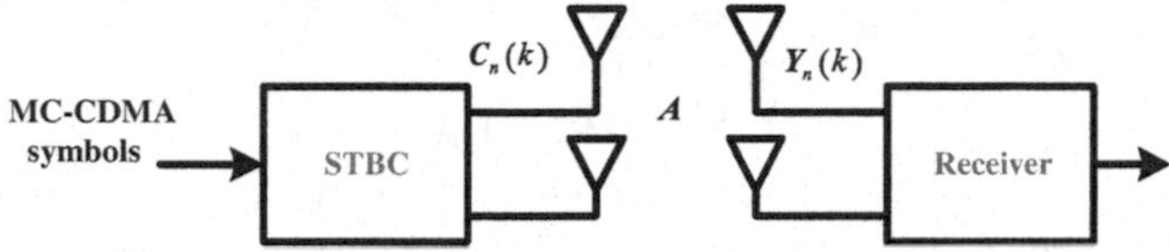

Fig. 1. MIMO system based on OSTBC.

For simplicity, we assume that the symbols belong to the real alphabet $\boldsymbol{\Omega} = \{\pm 1, \pm 3, \ldots\}$, although the results could be easily extended to the complex case. Let $\boldsymbol{c}_i(k)$ be the $n_T \times 1$ coded vector containing the signals transmitted by the antennas at the $i-th$ time instant of the $k-th$ block. As shown in [3], the N symbols represented in the vector $\boldsymbol{S}(k) = [S_1(k), S_2(k), \ldots, S_N(k)]^{\mathrm{T}}$ are mapped onto a $n_T \times N$ space time block coding matrix $\boldsymbol{C}(k) = [\boldsymbol{c}_1(k), \boldsymbol{c}_2(k), \ldots, \boldsymbol{c}_N(k)]$ by means of the following expression [10]:

$$\boldsymbol{C}(k) = \sum_{i-1}^{N} \boldsymbol{X}_i s_i(k) \tag{1}$$

where $\boldsymbol{X}_i$ is the $n_T \times L$ dimension of orthogonal coding matrix corresponding to the $i-th$ symbol $s_i(k)$, with the following properties [11]:

$$\begin{cases} \boldsymbol{X}_i \boldsymbol{X}_i^{\mathrm{T}} = \boldsymbol{I}_{n_T},\ i = 1, 2, \ldots, N \\ \boldsymbol{X}_j \boldsymbol{X}_i^{\mathrm{T}} + \boldsymbol{X}_i \boldsymbol{X}_j^{\mathrm{T}} = \boldsymbol{0},\ i \neq j \end{cases} \tag{2}$$

where $\boldsymbol{I}_{n_T}$ is $n_T \times n_T$ dimension of unit matrix.

The received signal can be expressed as:

$$\boldsymbol{Y}(k) = \boldsymbol{G}\boldsymbol{C}(k) + \boldsymbol{V}(k) \tag{3}$$

where $\boldsymbol{G} = \begin{bmatrix} h_{11} & h_{1n_T} \\ h_{21} & h_{2n_T} \\ \vdots & \vdots \\ h_{n_R 1} & h_{n_R n_T} \end{bmatrix} = \begin{bmatrix} \boldsymbol{b}_1 \\ \boldsymbol{b}_2 \\ \vdots \\ \boldsymbol{b}_{n_R} \end{bmatrix}$, $\boldsymbol{Y}(k) = \begin{bmatrix} y_1(k) \\ y_2(k) \\ \vdots \\ y_{n_R}(k) \end{bmatrix}$, $\boldsymbol{V}(k) = \begin{bmatrix} v_1(k) \\ v_2(k) \\ \vdots \\ v_{n_R}(k) \end{bmatrix}$

$\boldsymbol{G}$ is the $n_R \times n_T$ channel response matrix, $\boldsymbol{b}_m = [h_{m1} \ldots h_{mn_T}]$ $(m = 1, 2, \ldots, n_R)$ is the n_T dimension vector, $\boldsymbol{Y}(k)$ is $n_R \times L$ matrix, $\boldsymbol{y}_m(k)$ is L dimension at $m-th$ receiver antenna of vector, $\boldsymbol{V}(k)$ is a $n_R \times L$ matrix containing the Gaussian noise with zero mean, variance is σ_n^2.

From (1) and (3) we get:

$$\begin{aligned} \boldsymbol{y}_m(k) &= \boldsymbol{b}_m \boldsymbol{C}(k) + \boldsymbol{v}_m(k) \\ &= \boldsymbol{b}_m \sum_{i=1}^{N} \boldsymbol{X}_i s_i(k) + \boldsymbol{v}_m(k) \\ &= \boldsymbol{S}^{\mathrm{T}}(k) \boldsymbol{\Omega}_m + \boldsymbol{v}_m(k) \end{aligned} \tag{4}$$

where $\boldsymbol{\Omega}_m = \begin{bmatrix} \boldsymbol{b}_1 \boldsymbol{X}_1 \\ \boldsymbol{b}_2 \boldsymbol{X}_2 \\ \vdots \\ \boldsymbol{b}_m \boldsymbol{X}_N \end{bmatrix}$, $\boldsymbol{\Omega}_m$ is $N \times L$ matrix.

Transpose (4) we can get:

$$\boldsymbol{y}_m^{\mathrm{T}}(k) = \boldsymbol{\Omega}_m^{\mathrm{T}} \boldsymbol{S}(k) + \boldsymbol{v}_m^{\mathrm{T}}(k) \tag{5}$$

$$\begin{bmatrix} \boldsymbol{y}_1^{\mathrm{T}}(k) \\ \boldsymbol{y}_2^{\mathrm{T}}(k) \\ \vdots \\ \boldsymbol{y}_{n_R}^{\mathrm{T}}(k) \end{bmatrix} = \begin{bmatrix} \boldsymbol{\Omega}_1^{\mathrm{T}} \boldsymbol{S}(k) \\ \boldsymbol{\Omega}_2^{\mathrm{T}} \boldsymbol{S}(k) \\ \vdots \\ \boldsymbol{\Omega}_{n_R}^{\mathrm{T}} \boldsymbol{S}(k) \end{bmatrix} + \begin{bmatrix} \boldsymbol{v}_1^{\mathrm{T}}(k) \\ \boldsymbol{v}_2^{\mathrm{T}}(k) \\ \vdots \\ \boldsymbol{v}_{n_R}^{\mathrm{T}}(k) \end{bmatrix} = \begin{bmatrix} \boldsymbol{\Omega}_1^{\mathrm{T}} \\ \boldsymbol{\Omega}_2^{\mathrm{T}} \\ \vdots \\ \boldsymbol{\Omega}_{n_R}^{\mathrm{T}} \end{bmatrix} \boldsymbol{S}(k) + \begin{bmatrix} \boldsymbol{v}_1^{\mathrm{T}}(k) \\ \boldsymbol{v}_2^{\mathrm{T}}(k) \\ \vdots \\ \boldsymbol{v}_{n_R}^{\mathrm{T}}(k) \end{bmatrix} \tag{6}$$

Equation (6) can be expressed as:

$$\tilde{\boldsymbol{Y}}(k) = \boldsymbol{\Omega}^{\mathrm{T}} \boldsymbol{S}(k) + \tilde{\boldsymbol{V}}(k) = \boldsymbol{A}\boldsymbol{S}(k) + \tilde{\boldsymbol{V}}(k) \tag{7}$$

where $\boldsymbol{\Omega} = [\boldsymbol{\Omega}_1, \boldsymbol{\Omega}_2, \ldots, \boldsymbol{\Omega}_{n_R}]$, $\boldsymbol{A} = \boldsymbol{\Omega}^{\mathrm{T}}$ is a $n_R L \times N$ virtual channel matrix, it is a independent vectors composed by a statistically independent source.

3 Characteristic Parameters Extraction

3.1 The Characteristics of Virtual Channel Matrix

For analysis correlation matrix of virtual channel matrix, set $\boldsymbol{R} = \boldsymbol{A}^{\mathrm{T}}\boldsymbol{A} = \boldsymbol{\Omega}\boldsymbol{\Omega}^{\mathrm{T}}$ we get:

$$\boldsymbol{R} = \begin{bmatrix} \boldsymbol{b}_1\boldsymbol{X}_1 & \boldsymbol{b}_2\boldsymbol{X}_1 \ldots \boldsymbol{b}_{n_R}\boldsymbol{X}_1 \\ \boldsymbol{b}_1\boldsymbol{X}_2 & \boldsymbol{b}_2\boldsymbol{X}_2 \ldots \boldsymbol{b}_{n_R}\boldsymbol{X}_2 \\ \vdots & \vdots \\ \boldsymbol{b}_1\boldsymbol{X}_N & \boldsymbol{b}_2\boldsymbol{X}_N \ldots \boldsymbol{b}_{n_R}\boldsymbol{X}_N \end{bmatrix} \begin{bmatrix} \boldsymbol{b}_1\boldsymbol{X}_1 & \boldsymbol{b}_2\boldsymbol{X}_1 \ldots \boldsymbol{b}_{n_R}\boldsymbol{X}_1 \\ \boldsymbol{b}_1\boldsymbol{X}_2 & \boldsymbol{b}_2\boldsymbol{X}_2 \ldots \boldsymbol{b}_{n_R}\boldsymbol{X}_2 \\ \vdots & \vdots \\ \boldsymbol{b}_1\boldsymbol{X}_N & \boldsymbol{b}_2\boldsymbol{X}_N \ldots \boldsymbol{b}_{n_R}\boldsymbol{X}_N \end{bmatrix}^{\mathrm{T}} \tag{8}$$

For the orthogonal space-time block codes, the element (i, i) of $\boldsymbol{R}$ can be expressed as:

$$\boldsymbol{R}_{ii} = \boldsymbol{b}_1 \boldsymbol{X}_i \boldsymbol{X}_i^{\mathrm{T}} \boldsymbol{b}_1^{\mathrm{T}} + \boldsymbol{b}_2 \boldsymbol{X}_i \boldsymbol{X}_i^{\mathrm{T}} \boldsymbol{b}_2^{\mathrm{T}} + \ldots + \boldsymbol{b}_{n_R} \boldsymbol{X}_i \boldsymbol{X}_i^{\mathrm{T}} \boldsymbol{b}_{n_R}^{\mathrm{T}} \tag{9}$$

with Eq. (2) we can get:

$$\boldsymbol{R}_{ii} = \boldsymbol{b}_1 \boldsymbol{b}_1^{\mathrm{T}} + \boldsymbol{b}_2 \boldsymbol{b}_2^{\mathrm{T}} + \ldots + \boldsymbol{b}_{n_R} \boldsymbol{b}_{n_R}^{\mathrm{T}} \tag{10}$$

the element (i, i) of $\boldsymbol{R}$ can be expressed as:

$$\boldsymbol{R}_{ij} = \boldsymbol{b}_1 \boldsymbol{X}_i \boldsymbol{X}_j^{\mathrm{T}} \boldsymbol{b}_1^{\mathrm{T}} + \boldsymbol{b}_2 \boldsymbol{X}_i \boldsymbol{X}_j^{\mathrm{T}} \boldsymbol{b}_2^{\mathrm{T}} + \ldots + \boldsymbol{b}_{n_R} \boldsymbol{X}_i \boldsymbol{X}_j^{\mathrm{T}} \boldsymbol{b}_{n_R}^{\mathrm{T}} \tag{11}$$

Because $\boldsymbol{R}_{ij}$ is a scalar, so $\boldsymbol{R}_{ij} = (\boldsymbol{R}_{ij})^{\mathrm{T}} = \boldsymbol{b}_1 \boldsymbol{X}_j \boldsymbol{X}_i^{\mathrm{T}} \boldsymbol{b}_1^{\mathrm{T}} + \boldsymbol{b}_2 \boldsymbol{X}_j \boldsymbol{X}_i^{\mathrm{T}} \boldsymbol{b}_2^{\mathrm{T}} + \ldots + \boldsymbol{b}_{n_R} \boldsymbol{X}_j \boldsymbol{X}_i^{\mathrm{T}} \boldsymbol{b}_{n_R}^{\mathrm{T}}$, using (2) we get $\boldsymbol{R}_{ij} = -\boldsymbol{b}_1 \boldsymbol{X}_j \boldsymbol{X}_i^{\mathrm{T}} \boldsymbol{b}_1^{\mathrm{T}} - \boldsymbol{b}_2 \boldsymbol{X}_j \boldsymbol{X}_i^{\mathrm{T}} \boldsymbol{b}_2^{\mathrm{T}} - \ldots - \boldsymbol{b}_{n_R} \boldsymbol{X}_j \boldsymbol{X}_i^{\mathrm{T}} \boldsymbol{b}_{n_R}^{\mathrm{T}} = -\boldsymbol{R}_{ij}$ when $i \neq j, \boldsymbol{R}_{ij} = 0$. By using (2) and (10) we can get:

$$\boldsymbol{R} = \left(\boldsymbol{b}_1 \boldsymbol{b}_1^{\mathrm{T}} + \boldsymbol{b}_2 \boldsymbol{b}_2^{\mathrm{T}} + \ldots + \boldsymbol{b}_{n_R} \boldsymbol{b}_{n_R}^{\mathrm{T}}\right) \boldsymbol{I}_N = \left(\sum_{m=1}^{n_R} \|\boldsymbol{b}_m\|^2\right) \boldsymbol{I}_N \tag{12}$$

So, for the orthogonal space-time block codes, $\boldsymbol{R}$ is a $N \times N$ diagonal matrix. When $n_R = 1$, $\boldsymbol{R} = (\boldsymbol{b}_1 \boldsymbol{b}_1^{\mathrm{T}}) \boldsymbol{I}_N = \|\boldsymbol{b}_m\|^2 \boldsymbol{I}_N$. For the non-orthogonal space-time block codes, so (2) does not hold and $\boldsymbol{R}$ is not a diagonal matrix, so (11) does not hold.

3.2 Characteristic Parameters Extraction

3.2.1 The Sparsity θ Extraction

According to the characteristic of $\boldsymbol{R}$ matrix, the sparsity θ of $\boldsymbol{R}$ is given as:

$$\theta = \|\boldsymbol{R}\|^0 = \sum_{i,j} |\boldsymbol{R}_{ij}|^0 \tag{13}$$

θ is non zero number of $\boldsymbol{R}$.

For orthogonal space-time block codes, $\boldsymbol{R} = \left(\sum_{m=1}^{n_R} \boldsymbol{b}_m \boldsymbol{b}_m^{\mathrm{T}}\right) \boldsymbol{I}_N$ is a $N \times N$ diagonal matrix, so sparsity $\theta = N$; for non-orthogonal space-time block codes, sparsity of $\boldsymbol{R}$ $\theta > N$; so we can set sparsity $\theta = N$.

Due to the presence of noise and the estimation error, estimated $\hat{\boldsymbol{R}}$ can be not diagonal matrix, so first we can taking the absolute value of $\hat{\boldsymbol{R}}$, then just find every value of $\hat{\boldsymbol{R}}$ which less than γ and set them be equal to zero, where γ is the maximum value of the diagonal elements of $\hat{\boldsymbol{R}}$.

The sparsity θ value will be influence directly by the noise cancel parameter γ. For orthogonal space-time block codes, we set $\gamma = \gamma_1$ for make sure $\theta = N$ and we can get high recognition efficiency, for non-orthogonal space-time block codes, we set $\gamma = \gamma_2$ for make sure $\theta > N$, where $\gamma_1 \gg \gamma_2$.

3.2.2 The Energy Ratio of Non-main and Main Diagonal Elements

D **Extraction.** The elements of $\boldsymbol{R}$ matrix are divided into two parts: main diagonal elements and non-main diagonal elements. According to the characteristic of $\boldsymbol{R}$ matrix, we set the energy ratio of non-main and main diagonal elements energy D of $\boldsymbol{R}$ matrix:

$$D = \frac{\sum_{i \neq j} r_{ij}^2}{\frac{1}{F} \sum_k r_{kk}^2} \tag{14}$$

Where F is the number of main diagonal elements. $\boldsymbol{R}$ is a $N \times N$ diagonal matrix, so we can get $D = 0$, but due to the presence of noise and the estimation error, estimated can be not equal to zero. In order to solve the noise cancel parameter γ selection problem, we set the threshold D_{th} as the second characteristic parameter (see in the Figs. 6, 7).

Number of symbols N **estimation**

After get $\tilde{\boldsymbol{Y}}(k)$, we can get the correlation matrix $\boldsymbol{R}_y$, by using MDL criterion we can get the signal subspace dimension $\hat{M}$ of $\boldsymbol{R}_y$, because $\hat{M} = 2N$, so we can get number of symbols N.

MDL calculation [12] can be given as:

$$\begin{aligned} MDL(a) = -\log\left(\frac{\prod_{i=a+1}^{P} \lambda_i^{1/(P-a)}}{\sum_{i=a+1}^{P} \lambda_i/(P-a)}\right)^{K(P-a)} + \\ + \frac{a(2P-a)}{2} \log K \\ (a = 1, 2, \ldots, P) \end{aligned} \tag{15}$$

where K is observation time, thus transmitter signal block number; λ_i is element i th of autocorrelation matrix $\boldsymbol{R}_y$, which is decomposed by using SVD, $P = 2n_R L$. Signal subspace dimension is given as:

$$\hat{M} = \arg \min_{m=0,1,\ldots,-1} MDL(m) \tag{16}$$

Because restrictive conditions of MDL criterion was that, signal subspace dimension $\hat{M}$ will be less than the space dimension of received signal, thus $2N < P = 2n_R L$, so when bit rate $r = N/L < n_R$, we can use MDL to estimate N.

4 Recognition Algorithm

This algorithm first built the model of received signals and the virtual channel matrix, because the virtual channel matrix was including information of space time code, so we can using it for blind recognition space time code, then the virtual channel matrix is separated by using the DEM algorithm, after then two characteristic parameters of correlation matrix of virtual channel matrix are proposed by according the characteristics of real orthogonal STBC for recognition. The proposed algorithm is summarized as following steps (Fig. 2):

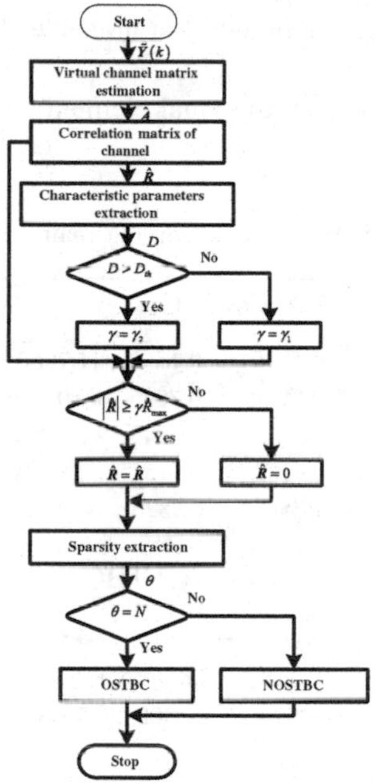

Fig. 2. Flowchart of the proposed algorithm

5 Simulation and Results

As a result the data symbol duration will be $3.2\mu s$. On the other side the number of cyclic prefix is 16 with a cyclic prefix duration equals to $0.8\mu s$. The used technique of modulation is QAM. Rayleigh fading is the used channel for broadcasting data on this system. WH codes are used. $L = N = 5$, $SNR = [-20:5]$ dB, for the performance simulations, 100 time of Monte Carlo simulations were run to highlight the behavior of the proposed algorithm in different environments. They were aimed at recognizing communications using OSTBC and NOSTBC [1–3] (Fig. 3).

Table 1 shown that, due to the presence of noise and the estimation error, estimated $\hat{\boldsymbol{R}}$ of OSTBC was not diagonal matrix. If compare with $\hat{\boldsymbol{R}}$ of NOSTBC, main diagonal

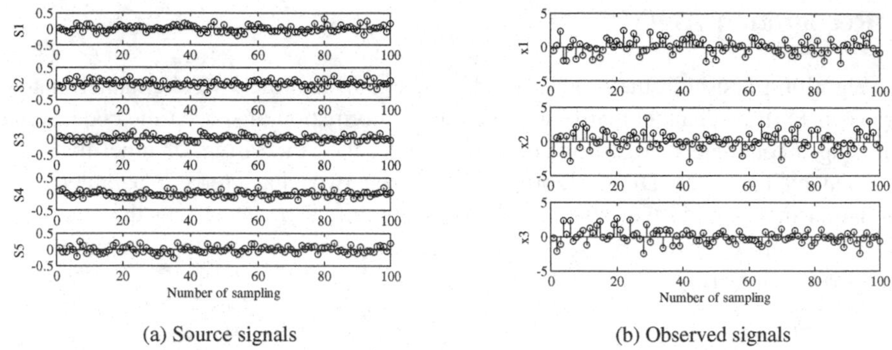

(a) Source signals (b) Observed signals

Fig. 3. The source and observed of insufficiently sparse signals

elements bigger more than non-main diagonal elements. So if set a suitably threshold, we can discern its.

Table 1. $\hat{R}$ of mixing matrix

$\hat{R}$ of OSTBC

$$\begin{bmatrix} 35.246 & -0.8810 & 0.1550 & -1.7560 & 0.2511 \\ 0.7300 & 34.850 & -0.7826 & 0.0829 & 1.6252 \\ 0.4030 & -0.2580 & 34.244 & -0.3284 & -0.7483 \\ -1.2850 & 0.6002 & 0.9152 & 33.698 & 0.0594 \\ -0.0180 & -1.7502 & -0.8820 & 0.8621 & 33.028 \end{bmatrix}$$

$\hat{R}$ of NOSTBC

$$\begin{bmatrix} 32.654 & 7.1065 & -3.6015 & 4.6750 & 2.3251 \\ -10.730 & 32.318 & 2.8782 & -9.2908 & 1.6252 \\ 3.0040 & -4.2580 & 32.008 & 7.0328 & -1.7483 \\ 1.2850 & 5.2608 & -6.9152 & 31.698 & 0.9598 \\ 2.8105 & -1.0777 & 3.0882 & -5.3186 & 31.082 \end{bmatrix}$$

5.1 *N* Estimation Analysis

If want to set a suitably sparsity θ threshold, first we should be know the number of symbol of each block N. We remember that, (n_T,n_R) is expressed different transmit antennas n_T and receive antennas n_R.

Figures 4 and 5 shown that, the error of N estimation of OSTBC and NOSTBC, respectively: when $SNR \geq -6\,\text{dB}$, the number of symbol N can be estimated quite exactly by using MDL algorithm; when the (n_T,n_R) is same, the effect of number of symbol of OSTBC is more better than NOSTBC's; when the transmit antenna have not changed, with extending of receive antenna, the effect of number of symbol is increased; when the received antenna have not changed, with extending of transmit antenna, the effect of number of symbol have also increased.

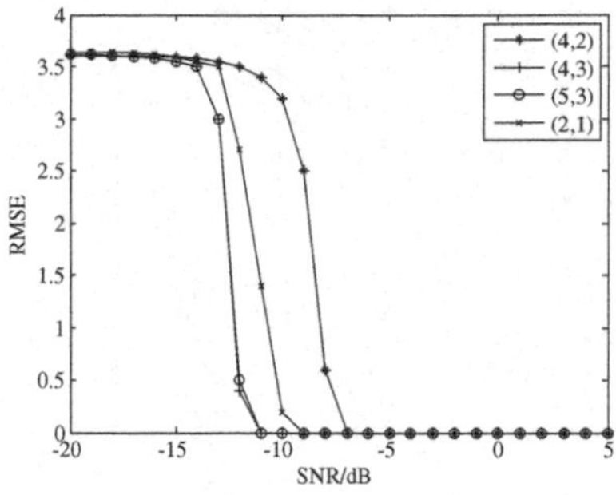

Fig. 4. Error of N estimation of OSTBC

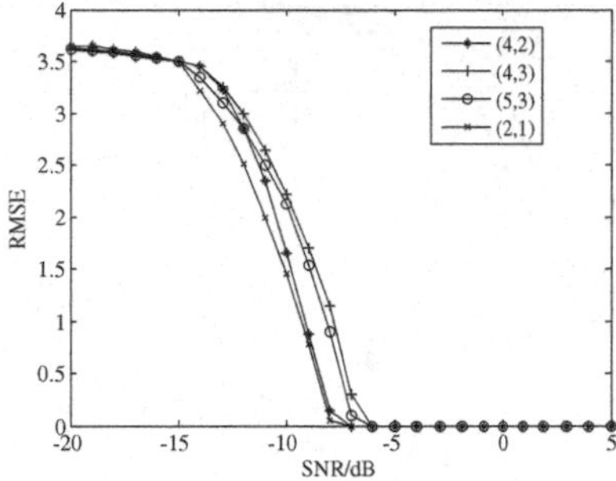

Fig. 5. Error of N estimation of NOSTBC

5.2 Influence of Noise Analysis

Figures 6 and 7 shown that, conversion of characteristic parameter D and θ, respectively: the values of D and θ of NOSTBC are more bigger than OSTBC's, they were not overlap; with the conversion of SNR, the values of D and θ of OSTBC have not changed but values of D and θ of NOSTBC have changed; when the received antenna have not changed, with extending of transmit antenna, values of D and θ of NOSTBC have increased; the Fig. 7 also shown that, when the symbol, (n_T, n_R) are same, with extending of SNR, values of D and θ of NOSTBC have also increased.

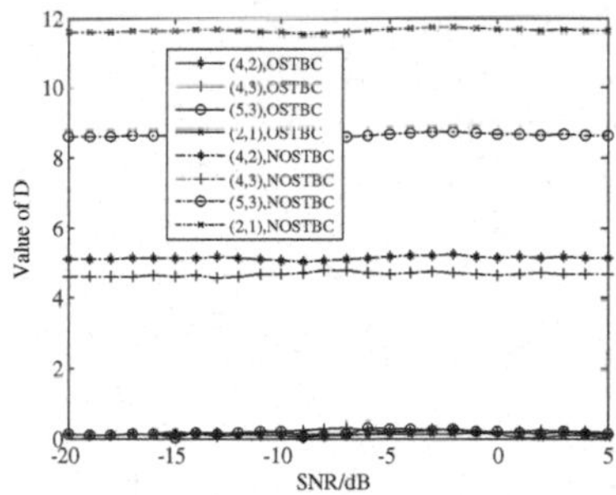

Fig. 6. Conversion of characteristic parameter Ds

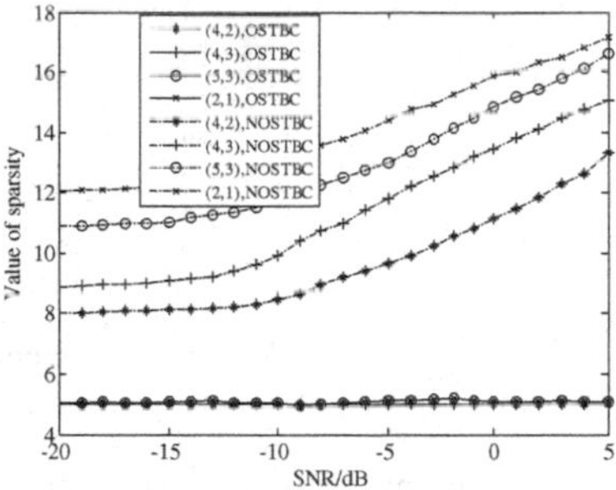

Fig. 7. Conversion of characteristic parameter θ

5.3 Recognition Efficiency of Algorithm Analysis

According to the results of last sections, we set $D_{th} = 1$, when $D > D_{th}$, set $\gamma = \gamma_2 = \frac{1}{40}$; when $D \leq D_{th}$, set $\gamma = \gamma_1 = \frac{1}{8}$. Set sparsity threshold $\theta_{th} = N = 5$.

Figures 8 and 9 shown the recognition efficiency of algorithm. In the same conditions, the recognition efficiency of OSTBC is better than NOSTBC's, especially in low SNR as $SNR < -10dB$; when $SNR \geq -5dB$, the recognition efficiency of OSTBC have reached about 99%, but for the NOSTBC, when only $SNR \geq -1dB$, the

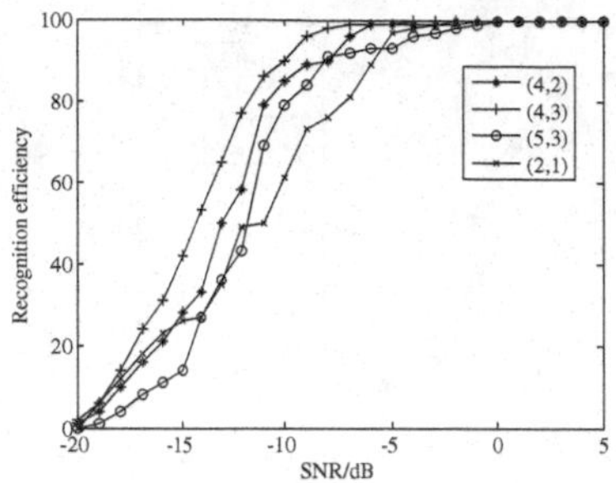

Fig. 8. Recognition efficiency of OSTBC

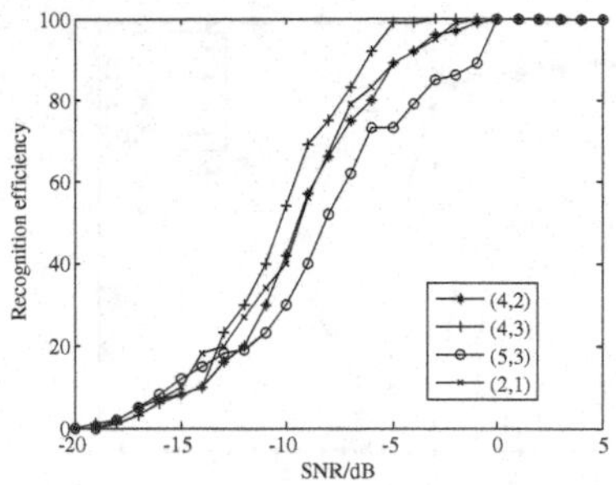

Fig. 9. Recognition efficiency of NOSTBC

recognition efficiency just have reached 100%; when the received antenna have not changed, with extending of transmit antenna, the recognition efficiency also have increased; when the transmit antenna have not changed, with extending of receive antenna, the recognition efficiency also have increased.

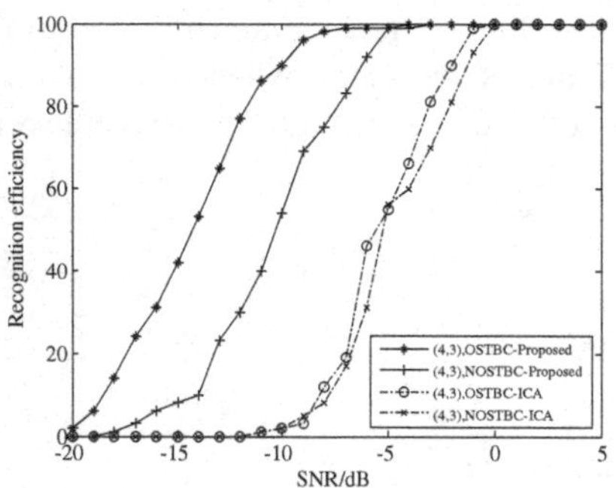

Fig. 10. Efficiency of algorithms compared

The efficiency of proposed algorithm and efficiency of DS-ICA algorithm [10] is compared in Fig. 10. For the DS-ICA algorithm, by using ICA algorithm to estimate the mixing matrix. The efficiency of DS-RCA algorithm is more better than DS-ICA's, especially in low SNR; when $SNR \geq -5dB$, the recognition efficiency of DS-ICA algorithm just have reached about 60%, when $SNR \leq -10dB$, the recognition efficiency of DS-ICA was very low, the main reason was that, for the underdetermined systems, efficiency of estimate the mixing matrix of this algorithm is not good; besides, the computational complexity of DS-ICA algorithm is quite high, including computational complexity of singular value decomposition computational complexity $\tilde{\boldsymbol{Y}}(k)$ is $O(K\hat{M}^2)$, pre whitening's is $O(KP\hat{M})$.

The computational complexity of proposed algorithm including: computational complexity of DEM algorithm [9]; computational complexity of $\hat{\boldsymbol{R}}$ estimation is

$O(P\hat{M}^2)$; computational complexity of θ estimation is $O(\hat{M}^2)$, computational complexity of D estimation is $O(\hat{M}^2)$.

6 Conclusion

To conclude we can say that, when $SNR \geq -5dB$, the recognition efficiency of OSTBC have reached about 99%, and for the NOSTBC signal when $SNR \geq -1dB$, the recognition efficiency just have reached to 100% for both same and different input and output antenna case.

Acknowledgment. This work is supported by the National Natural Science Foundation of China (No. 61371164, 61275099, 61102131), the Project of Key Laboratory of Signal and Information Processing of Chongqing (No.CSTC2009CA2003), the Chongqing Distinguished Youth Foundation (No. CSTC2011jjjq40002), the Natural Science Foundation of Chongqing (No. CSTC2012JJA40008), the Research Project of Chongqing Educational Commission (KJ120525, KJ130524) and Graduate Research and Innovation Projects of Chongqing (No. CYS14140).

The authors are grateful to the Chongqing University of Posts and Telecommunication, Chongqing Key Laboratory of Signal and Information Processing (CQKLS&IP) for providing the facility in carrying out this research.

References

1. Cho, Y.S., Kim, J., Yang, W.Y., et al.: MIMO-OFDM Wireless Communications with Matlab. Publishing House of Electronics Industry, Beijing (2013)
2. Alamouti, S.M.: A simple transmit diversity technique for wireless communications. IEEE J. **16**, 1451–1458 (1998)
3. Tarokh, V., Jafarkhani, H., Calderbank, A.R.: Space time block codes from orthogonal designs. IEEE Trans. Inf. Theor. **45**, 1456–1467 (1999)
4. Young, M.D., Health, R., Evans, B.L.: Using higher order cyclostationarity to indentify space-time block codes. In: 2008 IEEE Global Telecommunications Conference. New Orleans, Louisiana, USA, pp. 3370–3374 (2008)
5. Shi, M., Bar-ness, Y., Su, W.: STC and BLAST MIMO modulation recognition. In: 2007 IEEE Global Telecommunications Conference. Washington, DC, USA, pp. 3334–3339 (2007)
6. Choqueuse, V., Yao, K., Collin, L., et al.: Hierarchical space-time block code recognition using correlation matrices. IEEE Trans. Wireless Commun. **7**, 3526–3534 (2008)
7. Choqueuse, V., Yao, K., Collin, L., et al.: Blind recognition of linear space-time block codes. In: ICASSP 2008. Las Vegas, Nevada, USA, pp. 2833–2836 (2008)
8. Choqueuse, V., Yao, K., Collin, L., et al.: Blind recognition of linear space-time block codes: a likelihood-based approach. IEEE Trans. Signal Process. **58**, 1290–1299 (2010)
9. Lee, K.I., Woo, K.S., Kim, J.K., et al.: Channel estimation for OFDM based cellular systems using the DEM algorithm. Pimrc 2007, Athens, (2007)
10. Ju, L., Antonio, I.P., Miguel, A.L.: Blind separation of OSTBC signals using ICA neural networks. In: IEEE International Symposium on Signal Processing and Information Technology, Darmstadt, Germany, pp. 502–505 (2003)

11. Jafarkhani, H.: Space Time Coding: Theory and Practice. Cambridge University Press, Cambridge (2005)
12. Wax, M., Kailath, T.: Detection of signals by information theoretic criteria. IEEE Trans. Acoust. Speech Signal Process. **33**, 387–392 (1985)

Temporal Association Rule Mining

Ting-Feng Tan[1], Qing-Guo Wang[1(✉)], Tian-He Phang[1], Xian Li[1], Jiangshuai Huang[1], and Dan Zhang[2]

[1] Department of Electrical and Computer Engineering, National University of Singapore, Singapore 117576, Singapore
{a0123784, elewqg, lixian, jshuang}@nus.edu.sg
[2] Department of Automation, Zhejiang University of Technology, Hangzhou 310023, People's Republic of China
danzhang@zjut.edu.cn

Abstract. A modified framework, that applies temporal association rule mining to financial time series, is proposed in this paper. The top four components stocks of Dow Jones Industrial Average (DJIA) in terms of highest daily volume and DJIA (index time series, expressed in points) are used to form the time-series database (TSDB) from 1994 to 2007. The main goal is to generate profitable trades by uncovering hidden knowledge from the TSDB. This hidden knowledge refers to temporal association rules, which represent the repeated relationships between events of the financial time series with time-parameter constraints: sliding time windows. Following an approach similar to Knowledge Discovery in Databases (KDD), the basic idea is to use frequent events to discover significant rules. Then, we propose the Multi-level Intensive Subset Learning (MIST) algorithm and use it to unveil the finer rules within the subset of the corresponding significant rules. Hypothesis testing is later applied to remove rules that are deemed to occur by chance.

Keywords: Temporal data mining · Financial time series · Knowledge discovery · Events · DJIA · Hypothesis testing

1 Introduction

The state of art temporal association rule mining techniques are able to unveil the underlying association rules from the financial TSDBs. However, curse of dimensionality could arise when dealing with high-dimensional data. To overcome this, we propose the Multi-level Intensive Subset-based Training (MIST) algorithm and use it to extract the subset data of the significant rules in a recursive fashion. Hence, this breadth-first association rules searching mechanism enables efficient discovery of finer rules as the complexity and the processing time are greatly reduced by removing unrelated elements in the subset. Another shortcoming of the current rule mining techniques is that some rules might be useful but are ignored if they do not satisfy the support threshold as they occur rarely. To resolve this, we suggest using hypothesis testing to retain rules if the occurrences are sufficiently large compared to chance occurrences. These two major additions will be further explained in the modified framework after we have described the general framework.

X. He et al. (Eds.): IScIDE 2015, Part II, LNCS 9243, pp. 247–257, 2015.
DOI: 10.1007/978-3-319-23862-3_24

The rest of the paper is organized in the following manner. The general framework, based largely on the heuristic methodology for temporal association rule mining by Dante et al. [2], is introduced in Sect. 2. The limitations are discussed and will provide motivation for improvements. By making some modifications to the general framework, our proposed modified framework is elaborated in Sect. 3 and Sect. 4. Here, Multi-level Intensive Subset Learning (MIST) algorithm and hypothesis testing are introduced to overcome the limitations. In Sect. 5, the top four component stocks in terms of the highest daily volume from Dow Jones Industrial Average (DJIA) and DJIA itself are used as a case study. This section also analyses the rules and simulation results. Section 6 concludes the paper.

2 Temporal Association Rule Mining

Let p_i where $i = 1, 2, \ldots, n$ be the time-series data which is used to form a time-series database where sliding time windows are introduced as the timing constraints. The time window for antecedents will be referred to as Input Window and window width is given by IW, while the time window for consequents will be referred to as Output Window and window width is given by OW. The definitions for antecedents and consequents will be elaborated later in this section. Sliding time windows are shown in Fig. 1, where unit of t is in days and t_0 refers to a particular day.

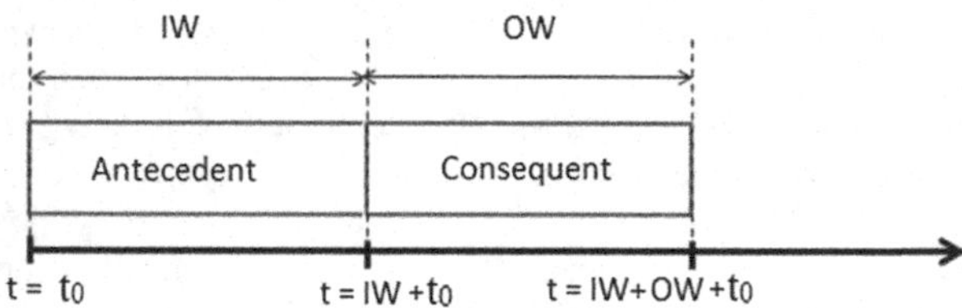

Fig. 1. Illustration of sliding time windows

p_i can be classified according to the respective range for each item. The Symbolic Aggregate Approximation (SAX) algorithm [3, 4] is chosen and used for this purpose because one of its key advantages is the ability to classify time series data based on its percentile position in the overall distribution.

To elaborate on SAX further, it is the first symbolic representation for time series that allows data points reduction and indexing with a lower-bounding distance measure. SAX is as well-known as representations such as Discrete Wavelet Transform (DWT) and Discrete Fourier Transform (DFT), except that it requires less storage space. One of the advantages of the SAX algorithm is its ability to convert time-series data of N size into time-series data of n size, where $n \leq N$ and $N/n \in Z^+$, by applying the Piecewise Aggregate Approximation (PAA) technique. Z^+ refers to the set of positive integers. The idea of PAA technique is to divide time-series data of N size into n segments with equal length and the average value of each segment is used. The equation of the PAA technique to convert a time-series data, C, of the length of N into length n, in which the *i*th element of C, is calculated by the following equation [5].

$$\tilde{p}_i = \frac{k}{N} \sum_{j=\frac{N}{k}(i-1)+1}^{\frac{N}{k}i} C_j \tag{1}$$

Figure 2 shows a time-series C that is represented by PAA (by the mean values of equal segments). In the example above, data points are reduced from N = 60 to k = 6.

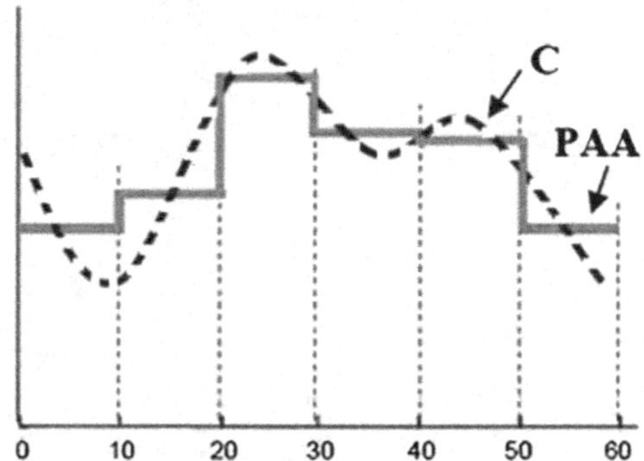

Fig. 2. Illustration of PAA

In order to obtain the string representation after a time-series data is transformed into the PAA representation, symbolization region should be determined. According to [4], by empirically testing more than 50 datasets, it is defined that normalized subsequences have distribution that highly resembles the Gaussian distribution. Therefore, by calculating the mean and the variance of the subsequences PAA data and symbolizing each of the PAA data based on its percentile value, the SAX algorithm is then able to convert the whole time series data into symbolized string representations. This quantization process transforms $\tilde{p}_i$ to $\bar{p}_i$. The implementation of the SAX algorithm is contributed by the researchers [4]. Their implementation is chosen because it is widely used in the related research field and is also recommended by the SAX main research website to interested researchers; hence, the correctness of the implementation is ensured. Figure 3 gives the illustration of SAX. The SAX program requires four parameters as shown in Table 1.

Table 1. Font sizes of headings. Table captions should always be positioned above the tables.

Parameter	Description
Data_Raw	The data to be processed by SAX algorithm
N	The number of entries in the provided Data_Raw
n	The number of output segments
alphabet_size	Number of discrete symbols used to represent the string

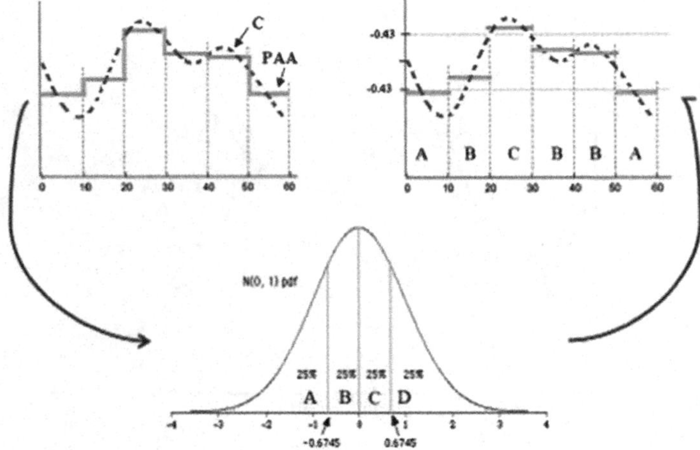

Fig. 3. Illustration of SAX: The N-size time series data was first converted into k-size data using PAA technique. The k-size data are then normalized and symbolized by mapping it in the Gaussian distribution.

Hence, the corresponding inputs and outputs can be given by $X_j = [\bar{p}_i, \bar{p}_{i+1}, \bar{p}_{i+2}, \ldots]$ and $Y_j = [\bar{p}_i]$ where i $= 1, 2, \ldots,$ n and j $= 1, 2, \ldots,$ m. The inputs are known as antecedents and the outputs are known as consequents. A typical association rule can be written as

$$X \Rightarrow Y \tag{2}$$

The "support" $S(X \Rightarrow Y)$ of the rule is defined as the fraction of observations where the rule holds and the formula is as follows

$$S(X \Rightarrow Y) = \frac{N(X, Y)}{N(T)} \tag{3}$$

where N(X,Y) is the number of times when both the antecedent and consequent are observed and N(T) is the number of transactions in the database. The "confidence" $C(X \Rightarrow Y)$ of the rule is its support divided by the support of the antecedent.

$$C(X \Rightarrow Y) = \frac{S(X \Rightarrow Y)}{S(X)} \tag{4}$$

This can also be viewed as the conditional probability that Y occurs, given that X occurs [8]. We can then generate all possible rules based on the below constraints

$$S(X \Rightarrow Y) \geq s \quad and \quad C(X \Rightarrow Y) \geq c \tag{5}$$

where "*s*" is the support threshold and "c" is the confidence threshold.

For the temporal association rules extraction algorithm, we have chosen to use MOWCALT algorithm by Harms et al. [6] as they provide a clear explanation with pseudo code for their algorithms and the implementation is relatively simple. The rules extracted in this manner are called significant rules. See Algorithm 1.

Algorithm 1

(1) Generate Antecedent Target Episodes of length 1 which we denote as ATE_1;
(2) Generate Consequent Target Episodes of length 1 which we denote as CTE_1;
(3) Record occurrences of ATE_1 and CTE_1 episodes;
(4) Prune unsupported episodes from ATE_1 and CTE_1 based on minimum support threshold;
(5) k − 1;
(6) while ($ATE_k \neq \emptyset$) do
(7) Generate Antecedent Target Episodes ATE_{k+1} from ATE_k;
(8) Record each occurrence of the episodes;
(9) Prune the unsupported episodes from ATE_{k+1};
(10) k++;
(11) Repeat Steps 5 – 11 for consequent episodes using CTE_{k+1};
(12) Generate combination episodes from antecedent episodes and consequent episodes;
(13) Record each occurrence of the combination episodes;
(14) Return the supported combination episodes that satisfy the minimum confidence threshold and these are the relevant rules;

3 MIST Algorithm

The number of possible temporal association rules, H, can be easily formulated as the following.

$$H = \sum_{j=2}^{Q} \left(M^j \times \binom{Q}{j} \right) \tag{6}$$

where M is the number of symbols, Q is the number of items, and j is the total number of both Antecedents' and Consequent's episodes in the corresponding rules. It can be observed that the number of possible temporal association rules is growing exponentially with the number of symbols and stocks. If there are only fifteen items to be symbolized with just five symbols, H will be 4.7018×10^{11}. Therefore, huge memory resources and processing time are required even for small values of M and Q. If items are required to be quantized with more symbols to achieve higher resolution, the hardware requirements might be a huge issue.

To tackle the problem of heavy computation costs, we propose the Multi-level Intensive Subset-Training (MIST) Algorithm as an efficient way to discover the finer temporal association rules with less symbols (breadth searching) first, and only perform a deeper searching with more symbols (depth searching) for the relevant rules recursively. This algorithm will continue to perform depth searching until there is insufficient data size or no more significant rules are needed to be diagnosed. The processing time and memory required will be greatly reduced because only the subset of the data is discretized with more symbols. The idea behind MIST algorithm is mainly based on the downward closure lemma, which states that the subsets of a frequent pattern are also frequent. See Algorithm 2 for the pseudo code.

Algorithm 2

(1) Check for rules where the numbers of occurrences satisfy the minimum data size;
(2) For each of the relevant rule, discretize with additional symbols;
(3) Create new time-transaction database using the occurrences of the rule
(4) Use Algorithm 1 to discover new finer rules;
(5) Repeat Steps 1–4 as long as there are rules to be analyzed;
(6) Return all new finer rules.

4 Hypothesis Testing

Some temporal association rules might have low occurrences and, hence are rejected by the support threshold. However, these could be useful rules if their occurrences are not by chance.

To overcome this limitation, we need to check if the occurrences are sufficiently large enough for us to believe that the rules do not take place by chance. To find this out, the idea is to perform hypothesis testing where null hypothesis is set up as H0: $O_{rule} = O_{random}$ and the alternative hypothesis is formulated as H1: $O_{rule} > O_{random}$ at the a% significance level. O_{random} is the number of times that a particular rule gets triggered successfully in the random walk case and O_{rule} is the recorded occurrence of the rule as observed during learning from the training data. We assume that distribution is normal since we will use a sample size that is large enough [7]. So, Z-test is used. If H0 is accepted, it means the rule occurs by chance. Alternatively, if H0 is rejected, it means the rule does not occur by chance. Table 2 shows the steps and Algorithm 3 shows the pseudo code.

Table 2. Explanation of the steps for hypothesis testing

#	Description
Step 1	H0: $O_{rule} = O_{random}$ H1: $O_{rule} > O_{random}$
Step 2	Select a significance level a%
Step 3	Use Z-test
Step 4	Compare the computed test statistic with critical value. If the computed value is within the rejection region, the null hypothesis will be rejected. Otherwise, the null hypothesis will not be rejected

Algorithm 3

(1) Generate normally-distributed pseudorandom numbers using randn in Matlab;
(2) Use results from Step 1 to create time-series database to mimic random walk;
(3) Follow the steps in Algorithm 1 to get the occurrences O_{random};
(4) Repeat, say k times for Steps 1–3 to get different samples of O_{random};
(5) Set the recorded occurrence of the rule we are interested in as O_{rule};
(6) Conduct Z-test where H0: $O_{rule} = O_{random}$ and H1: $O_{rule} > O_{random}$ at the a% significance level
(7) Repeat Steps 5 – 6 for the rest of the rules
(8) Return the rules that pass the Z-test;

By adding MIST algorithm and hypothesis testing, we get the modified framework in the below Fig. 4.

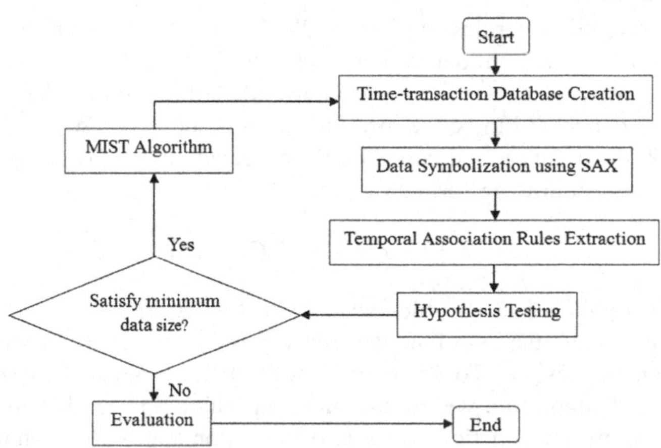

Fig. 4. Flowchart of modified framework

5 Case Study for US Stocks

We will use the top four component stocks in terms of the highest daily volume from Dow Jones Industrial Average (DJIA) and DJIA itself will be used as a case study. Financial data are configured with 5 financial time-series: CSCO, GE, INTC, MSFT, DJIA. Data is taken for January 1994 to December 2013 from Yahoo Finance [8]. Data for January 1994 to December 2007 is used as training data while data for January 2008 to December 2013 is used as test data. Note that the adjusted close prices take into considerations all corporate actions, such as stock splits, dividends/distributions and rights offerings.

Instead of using p_i as per Sect. 2, we use r_i which is rate of return (ROR) and formula is given below.

$$r_i = \frac{(P_t - P_{t+W})}{P_t} \times 100\% \tag{7}$$

where P_t represents the price at time t and P_{t+W} refers to the price after W number of days from time t. Here, we use W = 20 and IW = 3 and OW = 1.

After the stock prices database is converted into ROR database, the data can then be classified according to the range for ROR of each stock. One of the options is to classify ROR with a fixed range. For instance, a fixed range from 0 % to 1 % can be classified as weak positive trend with symbol "+", while 1 % to 3 % as semi-strong positive trend with symbol "++" for all the stocks. However, this method does not work well because every stock should have its own range due to its nature of business or certain unique characteristics. Hence, the SAX algorithm is used. The time series ROR data is first separated into three main categories, which are (negative gain) "–", (positive gain) "+", and (no gain) "0". Classification is then carried out by the SAX algorithm. Notice that the (no gain) category does not require to be symbolized again because its ROR is zero.

To discover significant rules, we use MOWCALT algorithm, and select support threshold to be 10 % and confidence threshold to be 50 %. Also, the consequent will have two sets of measurements: one for when the rule holds and the other for when the rule fails. Let us define TM to be the mean for ROR in the first case and FM to be the mean for the ROR in the second case. So, the expected gain (EG) of the rule can be calculated using the following formula.

$$EG = C \times TM + (1 - C) \times FM \tag{8}$$

where C is the confidence of the rule as per defined in Eq. (4).

Next, we implement the algorithm described in Sect. 4 with k = 10 and a = 1 % to carry out hypothesis testing. To get finer rules from the significant rules discovered using MOWCALT algorithm, we use the MIST algorithm and set data minimum size as 40. Lastly, we use multi-period portfolio optimization approach to simulate trading in the real world. We shall call this 'Rule Learning' model and will compare this against standard mean-variance model and equally-weighted model in terms of

Table 3. Simulation results

Rule	Antecedents	Consequent	Support (%)	Confidence (%)	Expected Gain (%)	Total Trades	Success Rate (%)	Average Return per Trade (%)
1	{CSCO2+, DJA3+, INTC1+}	{CSCO+}	0.73	96	3.02	21	62	**−0.13**
2	{DJA2+, INTC1+, MSFT3+}	{CSCO+}	0.61	95	3.57	32	**47**	**−0.16**
3	{CSCO2+, GE1+, MSFT3+}	{CSCO+}	0.58	95	3.30	16	69	1.27
4	{CSCO2+, MSFT1+, MSFT3+}	{CSCO+}	0.58	95	2.69	42	60	0.43
5	{CSCO1+++, CSCO3+, MSFT2+}	{GE+}	0.58	95	1.57	5	**40**	**−2.95**
6	{DJA3+, GE1 +, GE2+}	{GE+}	0.64	92	1.35	9	**44**	**−0.23**
7	{CSCO3+, GE2+, MSFT1+}	{INTC+}	0.67	96	2.60	10	70	1.89
8	{CSCO3+, DJA2+, MSFT1+}	{INTC+}	0.61	95	2.60	19	84	2.23
9	{CSCO3+, GE2+, MSFT1+}	{DJA+}	0.73	96	0.92	12	75	0.49
10	{GE2+, MSFT1+, MSFT3+}	{DJA+}	0.70	96	1.06	23	87	1.60
11	{DJA3+, GE1 +, GE2+}	{DJA+}	0.70	96	1.11	10	50	**-0.39**
12	{DJA3+, GE1 +, MSFT2 +}	{DJA+}	0.64	96	1.06	15	53	**-0.27**
13	{DJA2+, MSFT1+, MSFT3+}	{DJA+}	0.61	95	1.31	15	93	0.95
14	{CSCO2+, DJA3+, GE1+}	{DJA+}	0.61	95	1.07	12	83	1.08
15	{DJA3+, GE1 +, GE2+}	{DJA+}	0.61	95	1.17	11	55	**0.00**
16	{DJA3+, GE2 +, MSFT1 +}	{DJA+}	0.58	95	1.09	19	68	0.56

(Continued)

Table 3. (*Continued*)

Rule	Antecedents	Consequent	Support (%)	Confidence (%)	Expected Gain (%)	Total Trades	Success Rate (%)	Average Return per Trade (%)
17	{CSCO2+, DJA3+, MSFT1+}	{DJA+}	0.58	95	1.06	35	86	1.35
18	{DJA3+, GE1 +, MSFT2+ +}	{DJA+}	0.58	95	1.27	6	67	0.92

measurements such as terminal return and Sharpe ratio. This approach enables us to evaluate the usefulness of the rules.

Our trading strategy and some assumptions are as below.

- Trade according to the rule's consequent when all the antecedents are triggered
- Entry price is tomorrow's open price
- Close the position by the end of Output Window
- No limitation on short sell holding period
- No target profit (TP) and no stop loss (SL)
- Dividends, stock split and other corporate actions are ignored
- Trading costs (TC) are taken to be 0.18 % of the value of the position

In addition, we will only select rules, which have antecedents that contain episodes for each of the three time steps. This will allow the price trends to be sufficiently described before the possible occurrence of the consequent. From here on, we will focus only on the finer rules and will refer to them simply as rules. For ease of presentation, we take the top rules in terms of confidence levels from each of the assets and focus on those that satisfy a certain support threshold. This gives us 18 rules and the simulation results are presented in Table 3.

The rules in Table 3 can be interpreted in the following manner. For example, rule 18: {DJA3+, GE1+, MSFT2++}$\Longrightarrow${DJA+} means that when DJA3 in time step k + 2 and GE1 in time step k show "+" positive trends, and MSFT2 in time step k + 1 shows "++" positive trend, DJA in time step k + 3 will show "+" positive trend with support value of 0.58 % and confidence value of 95 %. The average "Success Rate" here is 66 % compared to the average "Confidence" value of 95 %. This is expected because the time period used for testing is different from the time period used to discover rules. Using entry price as tomorrow's open price has also contributed to this phenomenon. Also, it is observed that the number of times that each rule is triggered is small compared to the number of occurrences for a typical significant rule. This is reasonable as these rules in Table 2 were generated using MIST algorithm, which discovers finer rules with more symbols from the subset of the corresponding significant rules.

6 Conclusions

We have proposed a modified framework that applies temporal data mining technique to financial time series. After using the MOWCATL algorithm to discover significant rules, we found finer rules within the subset of the respective significant rules by applying our proposed MIST algorithm. Later, we use hypothesis testing to filter for rules that do not occur by chance. From the simulation results, the rules are found to be useful in trading individual positions and for portfolio management. The right mixture of rules would enable investors to enhance portfolio performance.

Acknowledgments. This research is funded by the Republic of Singapore's National Research Foundation through the grant "NRF2013EWT-EIRP001-025" under Energy Innovation Research Program and by the Republic of Singapore's National Research Foundation through the grant "NRF2013EWTEIRP001-025" under Energy Innovation Research Programme and a grant to the Berkeley Education Alliance for Research in Singapore (BEARS) for the Singapore-Berkeley Building Efficiency and Sustainability in the Tropics (SinBerBEST) Program. BEARS has been established by the University of California, Berkeley as a center for intellectual excellence in research and education in Singapore.

References

1. Malkiel, B.G.: The efficient market hypothesis and its critics. J. Econ. Perspect. **17**, 59–82 (2003)
2. Dante, C., Francisco, J.M.D.P., Alpha, P.: Finding temporal associative rules in financial time-series: a case of study in madrid stock exchange (IGBM). Adv. Comput. Intell. Man-Mach. Syst. Cybern **1**, 60–68 (2010)
3. Warasup, K., Nukoolkit, C.: Discovery association rules in time series data. **29**(4), 447–462 (2012)
4. Lin, J., et al.: A symbolic representation of time series, with implications for streaming algorithms. In: Proceedings of the 8th ACM SIGMOD Workshop on Research Issues in Data Mining and Knowledge Discovery. ACM (2003)
5. Lkhagva, B., Yu, S., Kyoji, K.: Extended sax: extension of symbolic aggregate approximation for financial time series data representation. DEWS2006 4A-i8 7 (2006)
6. Harms, S.K., Deogun, J.S., Tadesse, T.: Discovering sequential association rules with constraints and time lags in multiple sequences. In: Hacid, M.-S., Raś, Z.W., Zighed, D.A., Kodratoff, Y. (eds.) ISMIS 2002. LNCS (LNAI), vol. 2366, pp. 432–441. Springer, Heidelberg (2002)
7. http://www.investopedia.com/terms/c/central_limit_theorem.asp
8. https://sg.finance.yahoo.com/q?s=%5EDJI

Evaluating Diagnostic Performance of Machine Learning Algorithms on Breast Cancer

George Gatuha and Tao Jiang(✉)

College of Information and Communication Engineering,
Harbin Engineering University, Harbin China
ggrpb@yahoo.com, jiangtao@hrbeu.edu.cn

Abstract. This paper focuses on comparing performance of six data mining methods namely: Bagging, SVM (SMO), Decorate, C4.5 (J48), Naïve Bayes and IBK in analyzing Wisconsin Breast Cancer (WBC) datasets. The datasets were obtained from the UCI Machine Learning Repository and comprises of 699 instances and 11 attributes. A confusion matrix, based on a 10-fold cross validation technique was used in our experiment to provide the basis for measuring the accuracy of each algorithm. We introduce an idea of combining the algorithms at classification level to obtain the most ideal multi-classifier approach for the WBC data set. Waikato Environment Knowledge Explorer (WEKA), open source data mining software was used for the experimental analysis. The experimental results show that SMO offers the best accuracy (97 %) among the six algorithms, while merging SMO, Naïve Bayes, J48 and IBK offers the best accuracy (97.3 %) on the data set.

Keywords: Data mining · Classification · Open source · Confusion matrix · Breast cancer

1 Introduction

Breast cancer is one of the most common cancers affecting women both in the developed and the developing countries. It is estimated that over 508, 000 women died in 2011 due to breast cancer and over 50 % of breast cancer cases occur in the developing world [1]. Despite research conducted by leading oncological research institutes around the world, breast cancer etiology is still not known [2]. Several causative factors for breast cancer have been well documented, several theories have also been put forward to explain causative agents of cancer (carcinogens) such as gene mutation, age and heredity among others. However, for the majority of women presenting with breast cancer, it is difficult to identify specific risk factors [3]. Early detection not only leads to better management and treatment but also lowers the cost associated with treatment.

Data mining is the extraction of previously unknown, implicit and important information from databases [4]. It performs a set of processes automatically, whose aim is to discover and extract hidden patterns from large datasets. Manual analysis of large datasets would be an arduous and time consuming task and therefore computer software's have been developed to sift through the databases and obtain valuable information. Data mining

X. He et al. (Eds.): IScIDE 2015, Part II, LNCS 9243, pp. 258–266, 2015.
DOI: 10.1007/978-3-319-23862-3_25

combines statistical analysis, artificial intelligence, database management systems and machine learning techniques to extract the hidden patterns in databases. Several data mining functions exists, such as Association Rules, Concept descriptions, Classification, Clustering, Prediction and Sequence discovery to extract the useful patterns [5].

Classification is a supervised data mining technique that assigns labels or classes to different groups. Classification technique is a two-step process; the first step is model construction and is defined as the analysis of the training records of a database. Second step is model usage, the constructed model is used for classification. The classification accuracy is estimated by the percentage of test samples or records that are correctly classified [6].

The prediction of occurrence of breast cancer is possible by extracting the knowledge from the WBC datasets. The datasets source is the UCI machine learning repository, a unique, reliable and important resource for analysis of different aspects of breast cancer. The selected attributes for our experiment are: Normal Nuclei, Single Epithelial cell size, Clump thickness, Uniformity of Cell Size, Uniformity of Cell Shape, Mitoses, Bare Nuclei, Marginal adhesion, Bland Chromatin and Class [7]. Class attribute indicates benign or malignancy events.

In this paper, we analyze several data mining algorithms to predict occurrence of breast cancer in women. In our experiment, we have used the WBC datasets and have introduced a concept of combining the algorithms at classification level to get the most suitable multi-classifier approach for the analyzing WBC data set.

This paper is organized as follows. In Sect. 2 a review of literature is carried out. Section 3 outlines the materials and methods used. Section 4 outlines experimental results and discussion. Section 5 concludes the work and provides an idea about the future research direction.

2 Literature Review

Recently, medical data mining has gained prominence in the medical professional literature. Scientists have been looking for valuable patterns in the medical data to acquire valuable knowledge which may be useful in the diagnosis and treatment of diseases including breast cancer. Several data mining analytical methods for breast cancer have been put forward.

In [8], the performance comparison of five classification algorithms C4.5, SVM-RBF kernel, Naïve Bayes, CART and RBF neural networks was done on WBC breast cancer datasets. The experimental result showed that SVM-RBF kernel was more accurate than the other classifiers; it scored an accuracy of 96.84 % in WBC.

In [9], analysis of performance between four classification algorithms Naïve Bayes, C4.5, K- Nearest Neighbor (K-NN) and Support Vector Machine (SVM) was done to determine the best classifier using WBC datasets. SVM obtained an accuracy of 96.99 % and emerged the best classifier.

In [10], the decision tree classifier (CART) algorithm performance analysis was carried out on WBC datasets. The first analysis tested the classifier without feature selection and achieved an accuracy of 94.84 %. The second analysis was done using feature selection PrincipalComponentsAttributeEval method; it achieved an accuracy of 96.99 %. When the feature selection was changed to ChiSquaredAtrributeEval method, it scored an accuracy of 94.56 %.

All of the above data mining analyses are of no use to medical practitioners unless they are included in a robust medical decision support system. The Authors in [11] observe that breast cancer has suffered from specialists having inconsistencies in medical information and guidelines to strengthen their decisions; the author's research aimed at providing a means for easy acquisition of consistent medical data via a robust decision support system that incorporate data mining techniques.

3 Materials and Methods

The main purpose of our experiment was to compare the performance of various classification algorithms used in data mining on WBC breast cancer datasets obtained from UCI Machine Learning Portal. WEKA open source data mining software was used for the exercise. The data sets and the algorithm used are described below.

3.1 Wisconsin Breast Cancer Datasets (WBC) [12]

The description of the attributes is as follows: Clump thickness describes the grouping of cells in the breast; benign cells tend to be grouped in monolayers while malignant cells are often grouped in multiple layers. The uniformity of cell size/shape describes the variation in size and shape of cells, normal or benign cells are evenly distributed while malignant cells are not. Marginal adhesion describes the way cells appear together, benign cells tend to stick together while malignant cells lack this adhesion property. Single epithelial cell size is closely related to uniformity of cell shape/size; epithelial cells having significant enlargement maybe a sign of malignancy. Bare nuclei describes nuclei lacking cytoplasm, cells exhibiting this phenomena are highly likely to be malignant. Bland chromatin describes the texture of the nucleus; uniformity of the texture is an indication of benign and coarse texture is an indication of malignancy. The normal nucleoli describes small structures found in the nucleus, the nucleoli is usually very small if visible. In malignant cells the nucleoli become prominent and sometimes they are many in number. Mitosis is the cell division together with cytokines and produces two daughter cells at prophase stage. A cancer specialist can determine malignancy by counting the number of mitotic divisions. The independent variables values range from 1–10 and for class variable, benign has value of 2 and malignant class has value of 4. For consistency, the missing values were replaced with the mean value of the respective attribute.

3.2 Data Mining Algorithms

The classification algorithms used in our experiment for analysis of are outlined below.

3.2.1 Bagging Algorithm. It is an ensemble method, where a weak learning algorithm and a training set ((x1, y1), ..., (xm, ym)) are given, it generates a number of training sets randomly including some samples from the initial training set, that are almost identical in size. The classification algorithm is trained on each of the training sets and generates a predicted sequence g1, g2, ..., gn where the final function G can be obtained by voting [13].

3.2.2 DECORATE Algorithm. (Diverse Ensemble Creation by Oppositional Relabeling of Artificial Training Examples). It is a simple general meta-learner that directly makes diverse hypothesis by using artificially constructed training examples, it also uses diverse and strong learners as a base classifier to build diverse committees [14].

3.2.3 Naïve Bayes Algorithm. It denotes a Bayesian network implemented in WEKA as class NaïveBayes. The Bayesian network represents a joint probability distribution over a set of categorical attributes. Numerical attributes are discretized. It comprise of a directed acyclic graph, conditional probability tables, and allows the computation of the (joint) posterior probability distribution of any subset of unobserved assignments of values to attributes, which makes it ideal for classification [15].

3.2.4 Support Vector Machine Algorithm (SVM). It is a classification algorithm used in supervised learning models with associated learning algorithms that are used for data analysis and pattern recognition. It is a nonlinear classifier which is known to produce better classification result compared to other classification methods. SVM's have excellent theoretical basis and has been identified as one of the best in applications in bioinformatics [16]. It is represented by SMO in WEKA.

3.2.5 K-Nearest Neighbor Algorithm (KNN) [9]. It groups the instances based on their similarity, where an object is classified by a majority of its neighbors. It is represented by IBK in WEKA and is one of the most popular algorithms for data mining. It is a type of Lazy learning algorithm where the function is only approximated locally and all computation is deferred until classification. The neighbors are selected from a set of objects for which the correct classification is known.

3.2.6 C4.5 Algorithm. It is a decision tree classification algorithm represented as J48 in WEKA. It is an ID3 successor and it builds decision trees from training dataset, using the concept of information entropy [17]. The decision tree is constructed in a top-down recursive divide-and-rule manner. The decision tree has two types of nodes:

(a) The internal and root nodes that is associated to an attribute;
(b) The leaf nodes that is associated with a class.

Its creation is based on searching attributes for potential associations with nodes on the basis of information they bring; ideally, every non-leaf node has a branch for every possible categorical value/subset of numerical values of the attribute. A decision tree is used to determine the risk of a disease for a particular patient, starting with the root; all the successive internal nodes are visited until a leaf node is reached.

The supervised classification techniques described above are commonly referred to as prediction. They are also called learning algorithms. Learning means that the algorithm is given a set of instances, each provided with an outcome (class), the method operates as if it was supervised by this training set telling it what class should be assigned to a particular instance.

4 Results and Discussion

The basic phenomenon used to classify breast cancer using classification algorithms is its performance; this is predicted in terms of Sensitivity, Specificity and Accuracy. Classification confusion matrix is a visualization tool commonly used for presenting the performance of classifiers. It is used to show the relationships between outcomes and predicted classes. Effectiveness of a classification model is determined by the measure of correct and incorrect classification in each possible value of the variable being classified [18].
The confusion matrix (Table 1) values have the following meaning:

- *TN* represents the correct predictions that a given instance is negative,
- *FP* represents the incorrect predictions that a given instance is positive,
- *FN* represents the incorrect predictions that a given instance is negative,
- *TP* represents the correct predictions that a given instance is positive.

Table 1. Confusion matrix.

		Predicted	
		Negative	Positive
Actual	Negative	TN	FN
	Positive	FP	TP

The following methods below were used to measure performance of the classification algorithms:

Sensitivity is a measure of the proportion of true positives and it complements the false negative rates.

$$\text{Sensitivity (TPR)} = \frac{\text{TP}}{(\text{TP} + \text{FN})} \quad (1)$$

Specificity is a measure of the number of the true negatives that are correctly identified as such and is complimentary to the false positive rate.

$$\text{Specificity (TNR)} = \frac{\text{TN}}{(\text{TN} + \text{FP})} \tag{2}$$

Accuracy is the degree of correctness of a diagnostic classification and is determined from measures of specificity and sensitivity. The most efficient predictor would be described as 100 % sensitive and 100 % specific; in theory any predictor will possess a Bayes error rate. However for any test there is ideally a trade-off between the measures. Receiver operating characteristic curve can be used to represent this trade-off.

$$\text{Accuracy} = \frac{\text{TP} + \text{TN}}{(\text{TP} + \text{FP} + \text{FN} + \text{TN})} \tag{3}$$

The values FN, TP, FP, and TN are obtained during a 10-fold cross-validation feature in WEKA classification. In the 10-fold cross-validation, the data is divided into 10 patient folds. All the partitions are at random, and all folds contain almost similar number of non-recurrent and recurrent patients. To avoid over fitting, a single fold of the 10 folds is used as the testing data set for evaluation, and the remaining 9 folds are employed for learning. Analysis of the learning data is done by the algorithm for discovering the unknown patterns in data. Cross validation process is iterated 10 times, and each of the 10 folds is used once for testing.

The methods in Table 1 are evaluated so that models having high accuracy, sensitivity and specificity are preferred. Grievous situations occur when patient's tumor which is high risk is considered as low risk, ascertained by sensitivity. This risk should be reduced and therefore sensitivity should be maximized. Treatment costs are increased when low risk patients are classified as high risk, determined by specificity.

SMO algorithm gave the highest accuracy of 97.0 %. Table 2 illustrates the algorithms performance and Fig. 1 gives a graphical representation.

Table 2. Comparison of Algorithms according to performance.

Tally	Algorithm	TP	TN	FP	FN	SEN. (%)	SPE. (%)	ACC. (%)
1	SMO	232	446	12	9	96.2	97.3	97.0
2	DAGGING	228	447	11	13	94.6	97.5	96.6
3	DECORATE	231	443	15	10	95.9	96.7	96.4
4	Naïve Bayes	235	436	22	6	97.5	95.2	96.0
5	IBK	222	443	15	19	92.1	96.7	95.1
6	J48	223	438	20	18	92.5	95.6	94.6

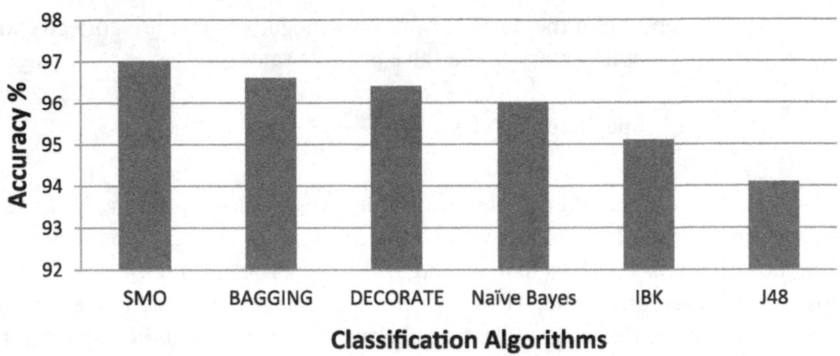

Fig. 1. Graphical representation of algorithms according to performance.

Fusion of classifiers is the combination of multiple classifiers to obtain the best accuracy. It is a set of classifiers whose individual predictions are combined in some way to classify new examples. In WEKA, the class for combining classifiers is called Vote. Different combinations of probability estimates for classification were made and the best combination was found to be SMO, J48, Naïve Bayes and IBK having an accuracy of 97.3 %. Results are presented in Table 3 and a graphical representation in Fig. 2.

Table 3. Comparison of merged algorithms according to performance.

No.	Merged algorithms	TP	TN	FP	FN	SPE. (%)	SEN.(%)	ACC. (%)
1	SMO + J48 + Naïve Bayes + IBK	235	445	13	6	97.2	97.5	97.3
2	SMO + BAGGING + Naïve Bayes + IBK	234	444	14	7	96.9	97.1	97.0
3	SMO + J48 + Naïve Bayes + DECO-RATE	232	446	12	9	97.4	96.3	97.0
4	SMO + J48 + BAGGING + Naï ve Bayes	234	443	15	7	96.7	97.1	96.9

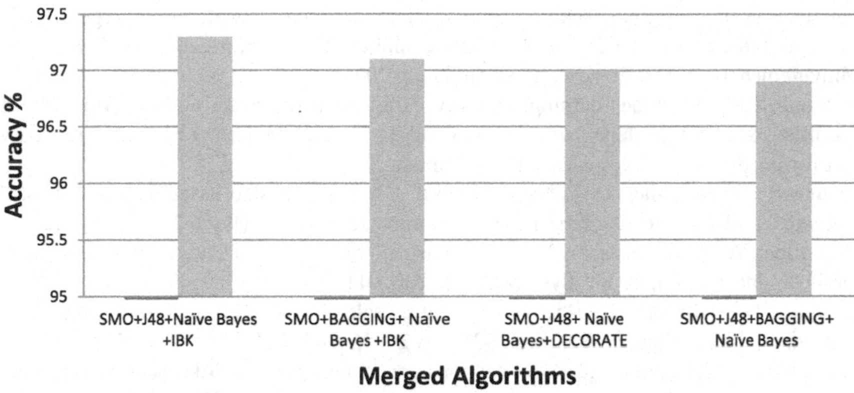

Fig. 2. Graphical representation of merged algorithms according to performance.

5 Conclusion

The aim of this project was to analyze performance of different classification data mining algorithms in terms of Sensitivity, Specificity and Accuracy in classifying WBC datasets. Classification algorithm SMO gave the highest accuracy of 97.0 %. We introduced the concept of merging the algorithms and found that the combination of SMO, J48, Naïve Bayes and IBK algorithms gave the best accuracy of 97.3 %.

In future we intend to design and implement a web based application, where cancer data can be collected via mobile communication gadgets and integrated in a server environment for quick analysis.

Acknowledgements. The authors would like to thank the Chinese Scholarship Council, Harbin Engineering University and the Kenyan Government for their support in these efforts.

We also acknowledge Dr. William H. Wolberg at the University of Wisconsin for availing the breast cancer dataset used in our analysis.

References

1. Ferlay, J., Soerjomataram, I., Ervik, M., Dikshit, R., Eser, S., Mathers, C., Rebelo, M., Parkin, D.M., Forman, D., Bray, F.: GLOBOCAN 2012 v1.0, cancer incidence and mortality worldwide: IARC cancer base no. 11 [Internet]. International Agency for Research on Cancer, Lyon, France (2013)
2. Danaei, G., et al.: Causes of cancer in the world: comparative risk assessment of nine behavioural and environmental risk factors. Lancet **366**, 1784–1793 (2005)
3. Lacey Jr., J.V., et al.: Breast cancer epidemiology according to recognized breast cancer risk factors in the prostate, lung, colorectal and ovarian (PLCO) cancer screening trial cohort. BMC Cancer **9**, 84 (2009)
4. Witten, H.I., Frank, E.: Data Mining: Practical Machine Learning Tools and Techniques, 2nd edn. Morgan Kaufmann Publishers, Burlington (2005)

5. Pei, J., Han, J., Wang, W.: Mining sequential patterns with constraints in large databases. In: Proceedings of 2002 International Conference on Information and Knowledge Management (CIKM 2002), Washington, D.C. (2001)
6. Mitchell, T.M.: Machine Learning. McGraw-Hill Science/Engineering/Math, Boston (1997)
7. Lichman, M.: UCI machine learning repository [http://archive.ics.uci.edu/ml]. University of California, School of Information and Computer Science, Irvine, CA
8. Aruna, S., Rajagopalan, D.S., Nandakishore, L.V.: Knowledge based analysis of various statistical tools in detecting breast cancer. Comput. Sci. Inf. Technol. **2**, 37–45 (2011)
9. Christobel, A., Sivaprakasam, Y.: An empirical comparison of data mining classification methods. Int. J. Comput. Inf. Syst. **3**(2), 24–28 (2011)
10. Lavanya, D., UshaRani, K.: Analysis of feature selection with classification: breast cancer datasets. Indian J. Comput. Sci. Eng. (IJCSE) **2**, 756–763 (2011)
11. Skevofilakas, M.T., Nikita, K.S.: A decision support system for breast cancer treatment based on data mining technologies and clinical practice guidelines. In: Proceedings of the 2005 IEEE Engineering in Medicine and Biology 27th Annual Conference. IEEE (2005)
12. Frank, A., Asuncion, A.: UCI machine learning repository. University of California, School of Information and Computer Science, Irvine, CA (2010)
13. Breiman, L.: Bagging predictors. Mach. Learn. **24**(2), 123–140 (1996). doi:10.1007/BF00058655. CiteSeerX: 10.1.1.121.7654
14. Melville, P., Money, R.: Constructing diverse classifier ensembles using artificial training examples. In: Proceedings of the Eighteenth International Joint Conference on Artificial Intelligence, pp. 505–510, Acapulco, Mexico (2003)
15. Han, J., Kamber, M.: Data Mining: Concepts and Techniques. Academic Press, San Francisco (2001). ISBN 1-55860-489-8
16. Vapnik, V.N.: The Nature of Statistical Learning Theory, 1st edn. Springer, New York (1995)
17. Wu, X., Kumar, V., Quinlan, J.R., et al.: Top 10 algorithms in data mining. Knowl. Inf. Syst. **14**, 1–37 (2008)
18. Matyja, D., Tuzinkiewicz, L.: Analysis of oncological data with use of MS BI SQL server. In: Proceedings of the Methods and Tools of Software Development Conference, pp. 293–306. Wroclaw University of Technology Publishing House (2007)

Application of TOPO to the Multistage Batch Process Optimization of Gardenia Extracts

Bing Xu, Fei Sun, Jianyu Li, Xianglong Cui, Xinyuan Shi(✉), and Yanjiang Qiao(✉)

Research Center of TCM Information Engineering, Beijing University of Chinese Medicine, No. 11, Northern Third Ring Road East, Chaoyang District, Beijing 100029, People's Republic of China
shixinyuan01@163.com, yjqiao@263.net

Abstract. In this paper, the target-oriented overall process optimization (TOPO) strategy is for the first time applied to a six-unit production process of gardenia extract, in order to improve the product quality during the multistage batch manufacturing process. The optimization action is performed actively and iteratively from the second stage as the process continued, giving each stage the maximum probability for the product quality meeting the specified requirements. Simulation results demonstrated that TOPO could lead the product quality towards the predefined target and mitigate the variations from the raw materials.

Keywords: Multistage batch process · Target-oriented overall process optimization (TOPO) · Quality control · Probability trajectory · Herbal medicine

1 Introduction

As an important role in the modern industry, batch process is preferred in a plenty of manufacturing systems. In practice, most batch processes are carried out in the sequential form, and can therefore be called multistage batch processes [1]. Generally, a multistage batch process refers to a production system that consists of multiple processing units [2]. Through the chained processing stages, the desired quality is transformed from the starting raw material into the final product step by step. The main objective of the multistage batch process is to ensure the consistency, reproducibility and high level of product quality that is vital to the business success [3].

Nevertheless, there are great challenges to the quality oriented investigations for multistage batch processes, due to the complex nature of the multistage manufacturing systems. Firstly, the interrelationships among different stages are complex. The output of one stage is the input of its following stage, and the product quality of the downstream stages is influenced by the upstream stages [4]. Therefore, variations from the raw material or one stage could be propagated along with the proceeding of the process, causing run-to-run variations of the final product [5]. Secondly, the role and the importance of process variables and stages in relation to the product quality may change as the process progressed. There exist temporal correlations between different variables and stages.

X. He et al. (Eds.): IScIDE 2015, Part II, LNCS 9243, pp. 267–276, 2015.
DOI: 10.1007/978-3-319-23862-3_26

In order to tackle the forementioned difficulties, systematic methodologies including multistage process modeling, monitoring, control and optimization, are necessary to effectively control and improve the product quality during the multistage manufacturing process. Some recent articles provide comprehensive surveys of the existing technologies on this research area [6–9]. Among these technologies, there are two active research directions. One is the analysis of the stage wise casual relationships, which belongs to the multistage modeling aspects. PLS path modeling [10] and Bayesian network [11] are two recently reported techniques accounting for this purpose, and they are very useful to identify the underlying interactions of different stages. The other research direction refers to active control or predictive control [12–14], which means the product quality is maintained or improved by process adjustment in the feed-forward mode. By active control, manipulated variables are adjusted to compensate for the quality error. As a result, the quality of the product is maintained at an acceptable level, while the variations can be mitigated to the minimum.

Lately, a new approach named target-oriented overall process optimization (TOPO) that's linked to the above two research directions has been proposed, which is suitable for a batch production system with several processing units [15]. The assumption of TOPO is that the number of process variables in every stage does not change over time. After several batches of the production, each stage can form a data block. Based on the historical production data, the expanding PLS process modeling method inspired by the path modeling techniques is employed in TOPO to build the mathematical models of the multistage manufacturing system. With the help of the established series of PLS models, the quality oriented process control and optimization are designed to start from the second stage to the last one. The Bayesian approach is integrated into optimization procedures. For a certain stage to be optimized, all the historical data, as well as measurements from its previous stages are utilized. The optimization goal is to provide the continuously optimal assurance of product qualities meeting the defined specifications.

In this paper, the TOPO strategy is applied to the six-unit natural product manufacturing process which suffers from the raw material's variability and unclear mechanisms of quality transferring, to enhance the process understanding as well as improve the final product quality. The remainder of this paper is organized as follows. First, a brief illustration is made about mathematical fundamentals of TOPO. Then, a real industrial case study applying the TOPO strategy is described. It aims at mitigating the impact of crude drugs' variation on the final product. Finally, a conclusion of this paper is provided.

2 TOPO Fundamentals

In general, TOPO is composed of the optimization target definition, data pretreatment, process modeling, and overall process optimization. Details of the TOPO methodology are illustrated in the following sections.

2.1 The Optimization Target Setup

TOPO begins with the explanation of the optimization problem, as well as the target that must be met with special purposes. Generally speaking, the possible goals of optimization could be minimum, maximum or target values. TOPO mainly focuses on the optimization towards the target response region. Such target region should contain an upper and a lower limit, or at least one of them. The value of the target limit should be within the acceptable range of the response and be characterized by the process or product specifications. In real applications, the optimization problem often involves multiple responses, each of which has a predefined target to be optimized.

2.2 Expanding Process Modeling

An expanding PLS process modeling technique is employed in TOPO modeling step. The word "expanding" means a series of PLS models are built at the end of any stages, where the quality variables can also be predicted. Considering that an overall production process is composed by k sequential unit operations, the meaning of unit here involves the summing up of several similar processes that can cause a physical or chemical change and are functionally the same. After a collection of the available historical batch production data, each unit can form a data block Xi. The rows of Xi correspond to the number of batches collected, while the columns correspond to the process parameters in the i-th unit. Thus the whole production process can be represented by a series of data blocks:

$$X = (X_1, X_2, \ldots, X_i, \ldots, X_k). \tag{1}$$

where $1 < i \leq k$.

The vector of quality value y forms the last $k + 1$ data block $\mathbf{Y}$.

At the i-th stage of the production process, all the previous data blocks can make a joint block $\mathbf{X}_{(i-1)}$ which could be seen as the input data:

$$X_{(\mathrm{i}-1)} = (X_1, X_2, \ldots, X_{\mathrm{i}-1}). \tag{2}$$

And $\mathbf{X}_\mathrm{i}$ stands for the controllable data that need to be optimized.

Before the optimization task is performed, the PLS regression model is used to grasp the underlying relationship between the process parameters and the product quality. Because the X-block is expanded along with the proceeding of the production process, a series of PLS models can be established at different stages of the overall production process:

$$\mathrm{XY}_1{:}X_{(1)} \rightarrow y, \mathrm{XY}_2{:}X_{(2)} \rightarrow y, \ldots, \mathrm{XY}_\mathrm{i}{:}X_{(\mathrm{i})} \rightarrow y, \ldots, \mathrm{XY}_\mathrm{K}{:}X_{(\mathrm{k})} \rightarrow y. \tag{3}$$

XY_i denotes that the data block $\mathbf{X}_{(\mathrm{i})}$ is mapped to $\mathbf{y}$. $\mathbf{X}_{(\mathrm{k})}$ equals the whole data block $\mathbf{X}$. These PLS models can then be applied in the prediction of quality values at each stage of the overall process. It is particular helpful in making decisions about process parameters of the incoming i-th stage, which will be illustrated in the next section.

2.3 Overall Process Optimization

Let's begin the overall process optimization problem with a unit operation optimization. Assuming that the i-th stage of a new batch process is going to be optimized, the optimization objective is to find the best process parameters of the i-th stage (i.e. $\mathbf{x}_\mathrm{i}$) in order to meet the target with specified boundaries.

Using the historical data, the PLS regression XY_i relating the dependent variable $\mathbf{y}$ and independent variables $\mathbf{X}_{(i)}$ is considered as:

$$y = X_{(i)}b_{(i)} + \varepsilon. \quad (4)$$

where $\mathbf{b}_{(i)}$ is the regression coefficient and $\boldsymbol{\varepsilon}$ the prediction error. $\mathbf{b}_{(i)}$ is actually a joint vector of coefficients and can be divided into two parts, $\mathbf{b}_{(i-1)}$ and $\mathbf{b}_i$, which are associated with the data block $\mathbf{X}_{(i-1)}$ and $\mathbf{X}_i$, respectively. Therefore, the prediction of the quality y for a new batch process at the i-th stage can be written as:

$$\hat{y} = x_{(i)}b_{(i)}^T = x_{(i-1)}b_{(i-1)}^T + x_i b_i^T. \quad (5)$$

The right part of above linear prediction model consists of two terms: a fixed term determined by the previous $i-1$ stages and a free term that need to be optimized. To find the optimal solution of $\mathbf{x}_i$, the optimization problem is transformed as:

$$\max P(\hat{y} \in T|data, x_i) \quad \mathbf{x_i} \in \mathbf{L_i}. \quad (6)$$

where $P(\cdot)$ refers to the probability of the prediction y meeting the target of interest; data represents all the available information, including the historical data and the accomplished $i-1$ stages of the ongoing process; $\mathbf{L}_i$ is all the acceptable combinations of process parameters in the i-th stage, namely the parameter space; max $P(\cdot)$ is solved via Bayesian predictive approach. Given certain process variables, the Bayesian posterior probability $P(\cdot)$ reflects the relationships between the current state of the process and the optimization target. Large value of probability provides high level of assurance for the quality variables meeting the target. The maximization of $P(\cdot)$ gives the highest guarantee that satisfies the process objective.

Furthermore, the unit optimization scheme can be extended to the overall process in a sequential way from the second stage to the last k stage. The first stage is treated as the initial input and cannot be optimized. Along with the proceeding of a production process, the probabilities meeting the target at different stages could be connected to form a line, called the probability trajectory, which is very useful to visualize and evaluate the performance of the multistage production process. In this way, TOPO is to provide the consistent optimal assurance for the product quality meeting the target during the course of multi-stage optimization.

3 Case Study

3.1 Process Description

The object studied in this paper refers to the production system of gardenia extract involving six processing units $(\mathbf{X}_1, \mathbf{X}_2, \ldots, \mathbf{X}_6)$, which represents a typical multistage production process of the natural product. Each unit is treated as one stage, the function of which is briefly described in Table 1. For example, Stage 3 represents the ethanol precipitation process, during which a majority of unwanted substances like starch and proteins are eliminated. Totally, 22 critical process variables $(x_1, x_2, \ldots, x_{22})$ are included in the research and are distributed in the six stages, as shown in Table 1. Each process variable has its own meaning. For example, x_1 represents the quality of the input material, which is the content of Geniposide (Ge) in gardenia fruit.

Table 1. Description of stages and corresponding parameters

Stage	Name of unit operation	Process variables	
		Observable	Controllable
1	Water extraction	x_1–x_5	None
2	Concentration	None	x_6
3	60 % alcohol precipitation	x_7–x_9	x_{10}–x_{13}
4	Alcohol Recovery	x_{14}	x_{15}–x_{16}
5	Water precipitation	x_{17}	x_{18}–x_{21}
6	Purification	None	x_{22}

Ge is considered as one of the most effective components in gardenia extract [16]. However, until now, this compound cannot be industrially synthesizedand has been mainly obtained from the fruit of gardenia (Gardenia Jasminoides Ellis) [17]. During the production process researched in this study, Ge is separated from the starting material to the gardenia liquid extract stage by stage. The quality of gardenia extract is evaluated by the content of Ge quantified by HPLC.

In practice, it is expected that the maximum yield of Ge could be obtained in the final extract. After a collection of 162 historical batch production records from 2008 to 2011, it was found that Ge concentrations spread in a wide range from about 10.0 mg·mL^{-1} to 22.0 mg·mL^{-1}. Low concentrations indicated loss of Ge in the production process. Therefore, there is an urgent need to optimize the whole production process to improve and stabilize the quality of gardenia extract. In this paper, the optimization target in line with the TOPO requirements is defined as an interval ranging between 19.0 mg·mL^{-1} and 22.0 mg·mL^{-1} to test the proposed approach of optimization.

3.2 Preprocessing of the Data

The collected historical production data were organized into the data blocks **X** and **Y**, respectively. All the process variables and quality variables were scaled between −1 and 1. The optimization target for the concentration of Ge was denoted as [0.5, 1]. The whole dataset was partitioned into the calibration set with 114 batches and the validation set with 48 batches, using the Kennard and Stone algorithm. The validation set was used to validate the PLS models in the phase of process modeling, and was treated as the control set in the phase of process optimization. Then, the process variables of the calibration set were arranged into six joint data blocks from $\mathbf{X}_{(1)}$ to $\mathbf{X}_{(6)}$.

3.3 Development of Process Models

By relating the process variables block $\mathbf{X}_{(i)}$ and the quality response block $\mathbf{Y}$, six process PLS models were developed with the expanding of the $\mathbf{X}$ block, and the results were shown in Table 2. The latent variables (LVs) were investigated via the leave-one-out cross validation method. The chemometric indicators, such as RMSEC, RMSEP, which explained the percent of X-variance and Y-variance, were applied to select the best number of LVs. Taking the model XY_6 for example, the RMSEC and RMSEP values stabilized around 7 LVs (see Fig. 1), by which about 85 % of X-variance and 74 % of Y-variance were explained. Since more LVs involved may lead to over-fitting problems, seven LVs were used to establish the PLS model during the sixth stage.

Moreover, it could be seen that the latent variables changed as the multistage process progressed. The RMSE and Bias values for both the calibration set and the validation set showed a decreasing trend, while the correlation coefficient r and the RPD values showed an increasing trend. This indicated that the precision and the predictive ability of the models were improved with the increasing number of the process variables. In other words, the final concentrations of Ge in gardenia extract were influenced by these process variables and stages. The established PLS models could be deemed as the summary of the production knowledge and would be applied to guide the process optimization phase.

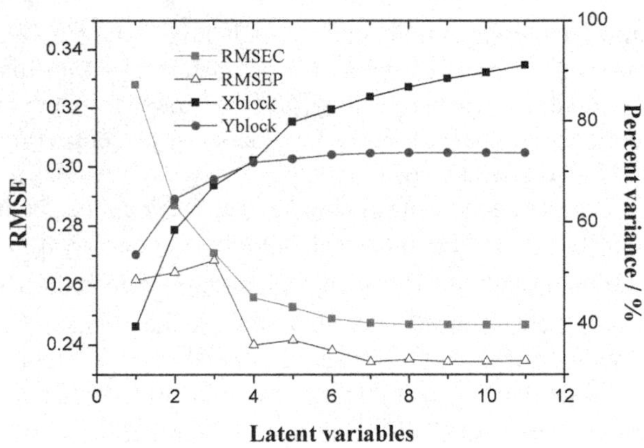

Fig. 1. Selection of the optimal latent variables for the PLS model established at the last stage

3.4 Simulated Optimization for the Control Set

Every batch in the control set was investigated and optimized according to TOPO. For a single batch, the optimization was simulated to adjust the values of the controllable variables from the second stage to the last stage, and the values of the observable variables were kept unchanged. All the possible combinations of variables during a certain stage were employed to form $\mathbf{L}_j$ in Eq. 6. For example, $\mathbf{L}_i$ was generated from a grid of $11 \times 10 \times 11 \times 37$ data points according to the control precision of the four manipulated variables at the third stage.

The effects of optimization were assessed from two aspects, i.e. the course of one batch optimization and the final target achievement of all batches in the control set. Taking Batch 25 for example, the original and optimized process variables were depicted in Fig. 2. The concentrations of Ge predicted by the original process variables gradually approached the reference value determined by HPLC. While, the concentrations of Ge predicted by the optimized process variables showed an increasing trend and finally stopped at 0.7240 which was within the predefined target interval. Under either the optimized or original process variables, the Bayesian posterior predictive distributions of Ge concentrations after each stage optimization were visualized by the box plot shown in Fig. 3. The probabilities meeting the target differed from the second stage and reached at 18.30 % and 56.71 % at the last stage for the original process and optimized process respectively.

Table 2. Performance of the PLS models established at different stages

Stage	LVs	Calibration				Validation			
		RMSEC	RMSECV	$BIAS_{cal}$	r_{cal}	RMSEP	RPD	$BIAS_{val}$	r_{val}
1	4	0.3493	0.3650	0.2615	0.5650	0.2720	1.33	0.2044	0.7110
2	4	0.3370	0.3558	0.2547	0.5847	0.2800	1.29	0.2209	0.6674
3	6	0.2729	0.3059	0.2073	0.7434	0.2563	1.41	0.1841	0.7270
4	6	0.2705	0.3102	0.2064	0.7482	0.2589	1.40	0.1929	0.7216
5	7	0.2598	0.3182	0.2071	0.7709	0.2389	1.51	0.1788	0.7661
6	7	0.2473	0.3060	0.1939	0.7948	0.2342	1.54	0.1844	0.7673

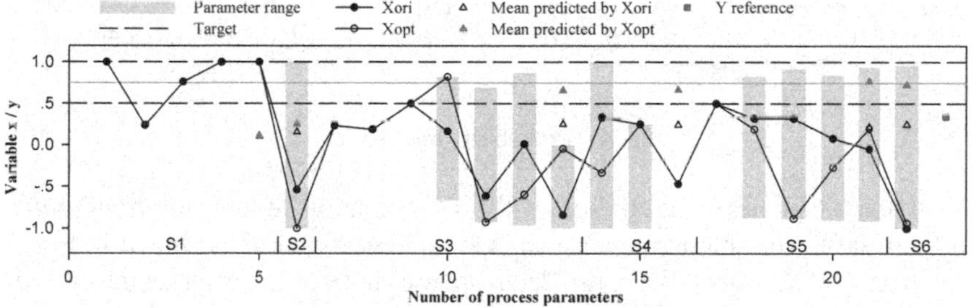

Fig. 2. Optimization results for batch 25

To make a clear illustration, four batches (i.e. Batches 1, 15, 30 and 44) with different starting materials were selected from the control set as examples, and the probability trajectories of the four batches were shown in Fig. 4. It was obvious that the probability trajectories for processes without optimization evolved irregularly (see Fig. 4A). By contrast, the probability trajectories for optimized processes gradually rose up and reached the highest at the final stage (see Fig. 4B). These results demonstrated that the proposed approach of optimization could navigate the quality response moving towards the target stage by stage.

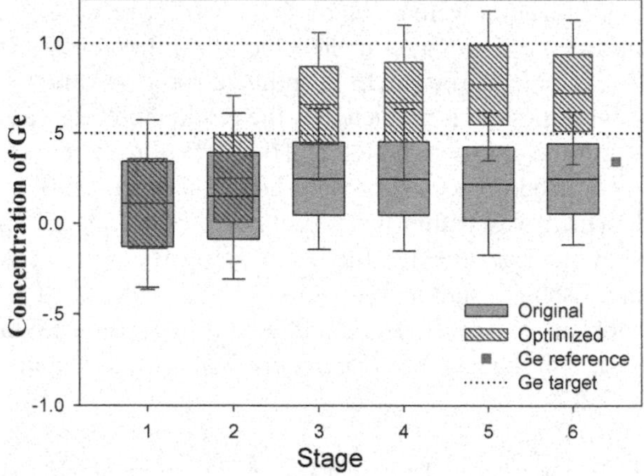

Fig. 3. The Bayesian posterior predictive distributions of Ge concentrations calculated under the original and optimized process variables of batch 25

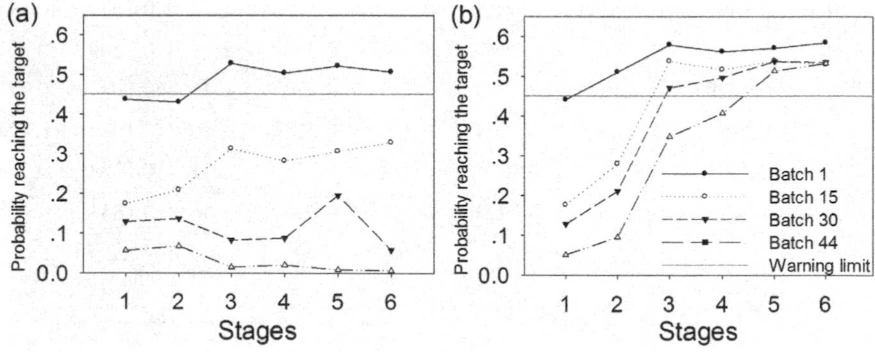

Fig. 4. Probability trajectories

For all the 48 batches in the control set, the histograms of Ge concentrations predicted at the final stage were plotted against the Ge reference values, as shown in Fig. 5. It could be seen that almost all the optimized responses lay within the target interval. After optimization, the mean value of Ge concentration was shifted from 0.2527 to 0.6811, and the standard deviation was reduced from 0.3612 to 0.0356. These results were satisfactory and revealed that the proposed method could lead the concentrations of Ge toward the predefined target and reduce the quality variability of gardenia extracts. Moreover, it was also noticed that the optimized responses were distributed close to the lower limit side of the target interval. The reason may be attributed to that the predictive ability of the PLS model at the upper limit side of the response range was weak, and the Bayesian approach of optimization tended to choose the results with relatively high reliability.

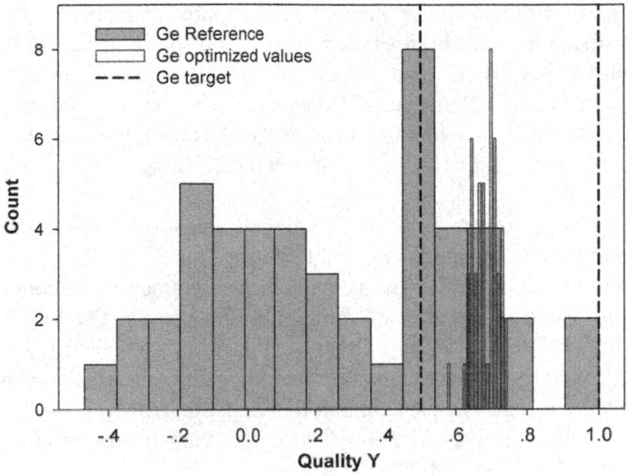

Fig. 5. Target achievement

4 Conclusions

In this paper, applicability of TOPO was demonstrated by a six-unit production process of gardenia extract. The probability trajectory was drawn to help the process engineer understand and optimize the multistage manufacturing processes. The final target achievement of product quality was satisfactory. However, there is still room for further research. If more controlled variables are involved in a process unit, the optimization by brute force search can be computationally challenging. This challenge may be tackled by combining the optimization solvers (e.g. genetic algorithm, GA) with the Bayesian predictive density.

Acknowledgement. This work is supported by the National Natural Science Foundation of China (No. 81403112) and the Joint Development Program Supported by Beijing Municipal Education Commission – Key Laboratory Construction Project (Study on the Integrated Modeling and Optimization Technology of the Chained Pharmaceutical Process of Chinese Medicine Products. And the historical production data used in this study were kindly provided by the assistant manager Haiyan Zhou of Yabao Beizhongda (Beijing) Pharmaceutical Co., Ltd.

References

1. Yao, Y., Gao, F.: Phase and transition based batch process modeling and online monitoring. J. Process Control **19**, 816–826 (2009)
2. Zhao, C., Fu, J., Xu, Q.: Real-time dynamic hoist scheduling for multistage material handling process under uncertainties. AIChE J. **59**, 465–482 (2013)
3. Vaghefi, A., Sarhangian, V.: Contribution of simulation to the optimization of inspection plans for multi-stage manufacturing systems. Comput. Ind. Eng. **57**, 1226–1234 (2009)

4. Li, H., Huang, J.: Process goose queue (PGQ) approaches toward plantwide process optimization with applications in supervision-driven real-time optimization. Ind. Eng. Chem. Res. **51**, 10848–10859 (2012)
5. Palumbo, B., De Chiara, G., Sansone, F., Marrone, R.: Technological scenarios of variation transmission in multistage machining processes. Qual. Reliab. Eng. Int. **27**, 651–658 (2011)
6. Shi, J., Zhou, S.: Quality control and improvement for multistage systems: a survey. IIE Trans. **41**, 744–753 (2009)
7. Yao, Y., Gao, F.: A survey on multistage/multiphase statistical modeling methods for batch processes. Annu. Rev. Control **33**, 172–183 (2009)
8. Liu, J.: Variation reduction for multistage manufacturing processes: a comparison survey of statistical-process-control vs stream-of-variation methodologies. Qual. Reliab. Eng. Int. **26**, 645–661 (2010)
9. Shetwan, A.G., Vitanov, V.I., Tjahjono, B.: Allocation of quality control stations in multistage manufacturing systems. Comput. Ind. Eng. **60**, 473–484 (2011)
10. Höskuldsson, A., Rodionova, O., Pomerantsev, A.: Path modeling and process control. Chemometr. Intell. Lab. Syst. **88**, 84–99 (2007)
11. Li, J., Shi, J.: Knowledge discovery from observational data for process control through causal Bayesian networks. IIE Trans. **39**, 681–690 (2007)
12. Jin, R., Shi, J.J.: Reconfigured piecewise linear regression tree for multistage manufacturing process control. IIE Trans. **44**, 249–261 (2012)
13. Jiao, Y., Djurdjanovic, D.: Joint allocation of measurement points and controllable tooling machines in multistage manufacturing processes. IIE Trans. **42**, 703–720 (2010)
14. Jiao, Y., Djurdjanovic, D.: Compensability of errors in product quality in multistage manufacturing processes. J. Manuf. Syst. **30**, 204–213 (2011)
15. Xu, B., Lin, Z., Wu, Z., Shi, X., Qiao, Y., Luo, G.: Target-oriented overall process optimization (TOPO) for reducing variability in the quality of herbal medicine products. Chemometr. Intell. Lab. Syst. **128**, 144–152 (2013)
16. Ma, T., Huang, C., Zong, G., Zha, D., Meng, X., Li, J., Tang, W.: Hepatoprotective effects of geniposide in a rat model of nonalcoholic steatohepatitis. J. Pharm. Pharmacol. **63**, 587–593 (2011)
17. Zhang, M., Ignatova, S., Hu, P., Liang, Q., Wang, Y., Sutherland, I., Jun, F., Luo, G.: Cost-efficient and process-efficient separation of geniposide from Gardenia jasminoides Ellis by high-performance counter-current chromatography. Sep. Purif. Technol. **89**, 193–198 (2012)

An Efficient String Searching Algorithm Based on Occurrence Frequency and Pattern of Vowels and Consonants in a Pattern

Kwang Sik Chung[1(✉)], Soo Young Kim[2], and Heon Chang Yu[3]

[1] Department of Computer Science, Korea National Open University, Seoul, Republic of Korea
kchung0825@knou.ac.kr

[2] Department of Computer Science, Korea Advanced Institute of Science and Technology, Daejeon, Republic of Korea
sooyoungkim@kaist.ac.kr

[3] Department of Computer Science, Korea University, Seoul, Republic of Korea
yuhc@korea.ac.kr

Abstract. Information and communication technologies enable people to access to various documentations and information. Huge documents and information in the Internet or storage disks have made search time more important. Especially as the volume size and the number of documents on the Internet increase, string search times and costs increase have become big burden to search service. But, most of string searching algorithms have not consider lexical structures nor vowels' occurrence frequency. Formal documents (articles, news, novels, etc.) have important characteristic that is 'well-formed written' English. And words of formal documents have 'limit number of words and alphabets' that are listed in a dictionary. The 'limit number of words and alphabets' has predictable occurrence probability in real world's documentations.

We try to use the alphabet occurrence probability as first search condition. We analyze all the words in the dictionaries (dictionary of free dictionary project, scrabblehelper – Revision 20, Winedit dictionary) and calculate each alphabet occurrence probability of repeated vowels, repeated consonants, not-repeated vowels and not-repeated consonants. In this paper, we define and propose the search rules and string searching algorithm, based on occurrence frequency and patterns of vowels and consonants. We use only the occurrence patterns and repeated positions of vowel and consonant in a text. Therefore, in the real world, proposed string searching algorithm (OFRP algorithm) is based on occurrence frequency and repetition pattern of vowels and consonants and is usefully and effectively applied to string search service and web search engine.

Keywords: String search · Vowel and consonant-based string search · Occurrence frequency of vowels · Occurrence frequency of consonants · Repetition pattern of vowels · Repetition pattern of consonants

X. He et al. (Eds.): IScIDE 2015, Part II, LNCS 9243, pp. 277–286, 2015.
DOI: 10.1007/978-3-319-23862-3_27

1 Introduction

As demands for document keyword search and document search increase, search time also increases and most portion of total documents working time and a spelling checking time is string search time. Traditionally string searching algorithm is used for word processor software, editors and web search engines. Nowadays string search or information search based text in the web documentations and text files have become the most important goal.

String searching algorithm finds a string in given text file or web documentation, and returns search results (the string position, and search success or fail). Brute-Force algorithm [4, 5] is naïve. KMP (Knuth-Morris-Pratt) algorithm [2–6] is based on finite automata. Boyer-Moore algorithm [1, 4, 5] and Rabin-Karp algorithm [4, 5] have the complex preprocess. Lastly Vowel based algorithm [6] focused at vowels in a pattern.

Since previous methods did not consider characteristics of English word composition, previous algorithms made unnecessary overload and non-effective performance if they were applied to real text files. An English word is made of alphabets which is combination of vowels and consonants. We consider frequency of vowels and consonants, and same vowel's occurrence repetition at two or more position in a pattern in order to search a pattern in text file. And in order to use these characteristics of a pattern, we analyze a pattern structure of vowels and consonants, extract vowels and repetition position of vowels and after that, search and compare the pattern in text file. In other words, we decide a pivot vowel that occurs most frequently in a given pattern, and the pivot vowel is firstly compared with an alphabet of text after mismatch happens. We compare the pivot vowel firstly and then compare the other vowels.

In this paper, we propose string searching algorithm based on repeated vowels and consonants, and their repetition positions, and evaluate the proposed OFRP algorithm. The first search step of the propose string searching algorithm is 'repeated vowels and consonants' match. The second search step of the propose string searching algorithm is 'repetition position' match. The proposed OFRP algorithm is compared with other string searching algorithm in the view of time complexity. Finally we show the other study plan.

2 Related Works

Brute-Force algorithm is the simplest string searching algorithm. Brute-Force algorithm compares first one alphabet of the string with one alphabet of text file as the bubble sort algorithm. If one alphabet of the string matches one alphabet of text file, then the next alphabet of the string is compared with the next alphabet of text file. And when last alphabet of the string matches one alphabet of text file, string searching algorithm successes to find out the string and comparison stops. Comparisons stop when all alphabets of text file are consumed or all alphabet of the string are found in text file. At the worst case, number of comparison between the string and a text file is $(m\text{-}n)\bullet n$ (m is pattern and n is alphabet numbers in text file) and the worst time complexity of Brute-Force algorithm is $O(n^2)$. Time complexity of Brute-Force algorithm is determined by length of text file.

Knuth-Morris-Pratt (KNP) string searching algorithm [2, 6] can find out 'incorrect start position' in a text file when a mismatch happens on the comparison of string and text file. Therefore, after mismatching, Knuth-Morris-Pratt string searching algorithm can decide more advanced comparison restart position in a pattern. And it is very effective to search a binary pattern in a binary string. But, in the real world, Boyer-Moore string searching algorithm [4] basically has efficiency derived from the fact that it avoids each unsuccessful comparison to find a match between a pattern and a text. Boyer-Moore string searching algorithm uses the information gained from that comparison in order to rule out as many positions of a text as possible where a pattern cannot match.

To make it more efficient, Boyer-Moore string searching algorithm makes two kinds of tables. One table calculates 'how many positions' ahead to start the next search based on the identity of the character that caused the match attempt to fail. The other table makes a similar calculation based on how many characters were matched successfully before the match attempt failed. But Boyer-Moore string searching algorithm needs consecutive repetition of a alphabet unit with vowels and consonants. But those kinds of repetition of a alphabet unit are few cases in a real text.

Lastly vowel based algorithm [6] dealt with only vowels. Since consonants in a pattern were not dealt with, it could be efficient if the consonants and vowels would be dealt with together.

3 Description of OFRP Algorithm

The basic idea behind the OFRP algorithm is that alphabets in a text have different frequencies and occurrence probabilities. The time complexity of string searching can be reduced if the searching processes are based on alphabet occurrence frequencies.

For fast pattern matching, we propose three stages of string matching: pattern preprocessing, text preprocessing, and matching. Based on the frequency, less frequent collections of alphabets in a pattern, a pivot, is decided by the pattern preprocessing. Text is also preprocessed in order to erase words whose length are different from those of the pattern. Then, the pattern are compared with only the words of the same length in the text.

3.1 Detailed Description of OFRP Algorithm

To find out the frequencies of each alphabet in a pattern, we examine the frequencies of repeated consonants, repeated vowels, consonant tuples, and vowel tuples in a pattern. The repeated vowels or the repeated consonants are the one that occurs most frequently. For an example, 'b' for the pattern 'baby' is the most frequently repeated consonant. Consonant tuple has the information about position of all consonant. For example the consonant tuple of 'realist' is 'r__l__st'. Similarily, vowel tuple includes the information of vowels.

By counting the number of words that have the same repetition or same tuple, we found out that the frequency of repeated vowel and consonant is less than the vowel or consonant tuples. Since most English words have at least one vowel, we believed that the tuples of consonants have more number of comparisons than the tuples of vowels.

p_p__... Develo**p** **p**rocess...jum**p** **p**robability...

Based on the analysis, the following pattern preprocessing rules are constructed: (1) repeated vowels or consonants, (2) a tuple of vowels, or (3) no pivot. In addition to the alphabet frequencies in a pattern, the position data of each alphabet in a pattern is also included in the pivot of a pattern.

The pattern preprocessing rule (1) is applied to patterns with an alphabet repetition. The pattern 'consonant' has a certain vowel or consonant that occurs more than once. In this case, 'n' is the most frequent alphabet, so that '__n__n_n_' is the pivot of 'consonant'. For patterns with no repetition of consonants or vowels, rule (2) is applied. The pattern 'realist' has no alphabet repetition, and has the pivot '_ea_i__'. Although pattern preprocessing rule (1) and (2) can cover most of English words, there are words that do not have a vowel. For those words, no pivot will be selected. For example, 'crwth', the longest word that have neither repeated alphabets nor vowels, is compared without pivot selection.

The text preprocessing stage checks the length of each word. For efficient matching, words in a text, which have the same length as the pattern, are marked. If the length of a certain word, which is the same as pattern's length, is unique in the whole text, the OFRP algorithm becomes more efficient. Also, without text preprocessing, a blank would be considered as an alphabet, which makes unnecessary comparison cost. Suppose 'paper' is searched in text. Figure 1 shows the unnecessary result of matching the pivot 'p_p__'. Since the pivot can be detected over several words, pivot matching can produce unintended overhead when blanks are not handled.

Row Text	The Nymphs, unfortunately, reflects her conviction that 1) the world and love are no longer young and 2) truthfulness, especially when trying to convince his lover to be with him, is not the shepherd's greatest attribute. The stanza also employs an important poetic technique--alliteration, the repetition usually of initial consonants --to move a line smoothly along: "pretty pleasures might me move."
Preprocessed Text	... especially ... repetition ... consonants...

Fig. 1. Preprocessing based on pivot consonants

When the pattern and text preprocessing are finished, the matching stage begins. The first step of matching stage is to compare the preprocessed pattern and the text. Before comparing alphabets, the length of each word, marked in the text preprocessing stage, is checked. When the word in the text has the same length as the pattern, the pivot of the pattern are compared.

The pattern preprocessing decided the pivots of 'consonants', 'realist', and 'crwth' in the previous example. For the pattern 'consonants', the pivot '__n__n_n__' is compared with the preprocessed text. In Fig. 2, only one word in the preprocessed text matches with this pivot. Then, after a pivot of the pattern equals to a word of the text, remaining alphabets in the pattern is compared with the word. When the remaining alphabets, c, o, s, o, a, t, s, and their relative position is matched, the matching stage ends.

Row Text	Someone who work in leasing office often dislike a realist.
Preprocessed Text	Someone … leasing … dislike … realist

Fig. 2. Preprocessing based on pivot realist

The pivot of pattern 'realist' is '_ea_i__', a tuple of vowels. This pivot is compared with the preprocessed text in Fig. 3. 'Someone', 'leasing', 'dislike', and 'realist' are the words to be compared with the pivot. The word 'Someone' and 'dislike' is not equal to the pivot, but 'realist' and 'leasing' matches with the pivot. Then, the remaining alphabets of the pattern ('r', 'l', 's', and 't') are compared with the pivot-matched words, 'realist' and 'leasing'. In this way, the word 'realist' is found and the matching stage succeeds.

The patterns without pivot uses brute-force algorithm. Suppose the pattern 'crwth' is compared with the preprocessed text 'clear … crwth … class …'. Firstly, the pattern is compared with 'clear'. By comparing from 'c' to 'h', the word 'clear' is excluded after comparing 'r' in the pattern 'clear', and 'l' in the word 'clear'. Then, the pattern 'crwth' and the word 'crwth' in the preprocessed text are compared. After comparison of the fifth alphabet 'h' of the pattern and the text, the matching stage succeeds.

3.2 OFRP Algorithm

We now specify the OFRP algorithm. Figure 3 shows data structure for OFRP algorithm. *temp_vowels* stores vowels and their positions in the pattern. *temp_consonants* stores consonants and their positions in the pattern lastly, pivot_position is a node for consonants and vowels position (Fig. 4).

The first stage of the OFRP algorithm is pattern preprocessing. Pattern preprocessing is composed of two steps, pivot finding and pivot producing. In the pivot finding step, the most repeated alphabet is determined. pivot_finding function returns *pivot_array*. And *pivot_array* is constructed from analysis of *temp_vowels* and *temp_consonants* (Figs. 5 and 6).

Depending on the existence of repeated alphabets, a pivot array is constructed by using the pattern preprocessing rules.

Rule 1: If there is repetition of vowels or consonants, the pivot array for the information of the most repeated alphabet is constructed.

```
struct {
char temp_pivot;
int Num_vowel;
  pivot_position * next_pivot_position;
} temp_vowels;

struct {
char temp_pivot;
int Num_consonants;
struct * pivot_position;
} temp_consonants;

struct {
int pivot_position;
pivot_position * next_pivot_position;
} pivot_position;
```

Fig. 3. Data structure for OFRP algorithm

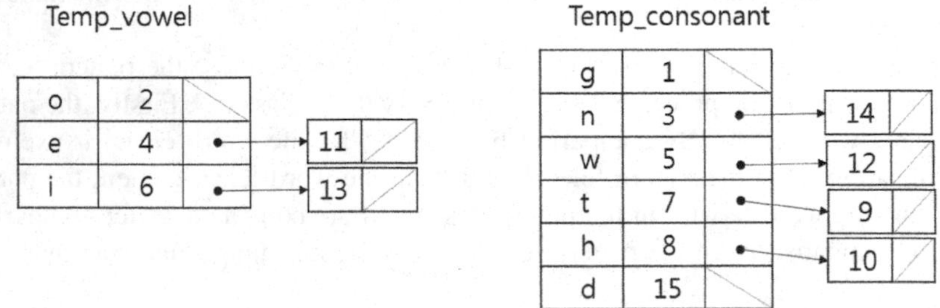

Fig. 4. Temp_vowels and Temp_consonants

pivot_array

e	e	i	i	o
4	11	6	13	2

Fig. 5. Pivot_array for pivot of pattern

Rule 2: If there is not repetition of vowels or consonants, a vowel tuple is formulated in an array.

In both cases, the position of pivot alphabets is represented as index in the array. The worst-case time complexity of pattern preprocessing is *O(n)*, where *n* is the length of the pattern (Fig. 7).

```
Pivot_find() {
  for(i=0; i<= pattern_length; i++) {
    if(pattern[i] in set_vowels) {
      temp_vowels[pattern[i].Num_vowel++;
      temp_next_pivot_position =
                          (struct pivot_position*)maloc(sizeof(pivot_position));
      temp_next_pivot_position.pivot_position = i;
      temp_next_pivot_position.next_pivot_position =
                          temp_vowels[pattern[i].next_pivot_position;
      temp_vowels[pattern[i].next_pivot_position = temp_next_pivot_position
     }
   if(pattern[i] in set_consonants) {
     temp_consonants[pattern[i]]. Num_consonants++;
     temp_next_pivot_position =
                          (struct pivot_position*)maloc(sizeof(pivot_position));
     temp_next_pivot_position .pivot_position = i;
     temp_next_pivot_position.next_pivot_position =
                             temp_consonants[pattern[i].next_pivot_position;
     temp_consonants[pattern[i].next_pivot_position =
                                           temp_next_pivot_position;
    }
  }

  temp_most_frequent_pivot_1 = 0;
  temp_most_frequent_pivot_2 = 0;

  for(i=0; i < num_vowels; i++) {
    if(temp_vowels[pattern[i]] > temp_most_frequent_pivot_1)
       temp_most_frequent_pivot_1 = temp_vowels[pattern[i]];
 }

  for(i=0; i < num_consonants; i++) {
    if(temp_consonants[pattern[i]] > temp_most_frequent_pivot_2)
       temp_most_frequent_pivot_2 = temp_ consonants [pattern[i]];
  }
}
```

Fig. 6. Finding pivot

For the fast comparison between a pattern and a text, OFRP algorithm also preprocesses the text in Fig. 8. Text preprocessing simply checks the length of words in text. Alphabets of each word in the text are counted and the length of each word is notated in the blank space in front of UTF-8 encoded words.

The worst-case time complexity for text preprocessing in Fig. 8 is $O(n)$, where n is the length of text.

```
pivot_position * next_pivot;
  if(pattern_pivot in set_vowels && temp_most_frequent_pivot_1 >= 2) {
    next_pivot = temp_vowels[pattern_pivot].pivot_position;
    for(i=0; next_pivot != NULL, next_pivot = next_pivot.pivot_position) {
          pattern_arrau[i][0] = pattern_pivot;
          pattern_array[i][1] = next_pivot.pivot_position;
    } }
  else if(pattern_pivot in set_vowels && temp_most_frequent_pivot_1 = 1) {
    pivot_array <= temp_vowel;
    }
    else if(pattern_pivot in set_consonants && temp_most_frequent_pivot_2 >=
2)
        {
          next_pivot = temp_vowels[pattern_pivot].pivot_position;
          for(i=0; next_pivot != NULL, next_pivot = next_pivot.pivot_position) {
             pattern_arrau[i][0] = pattern_pivot;
          pattern_array[i][1] = next_pivot.pivot_position;
   } }
        else if((pattern_pivot in set_consonants &&
                                             temp_most_frequent_pivot_2 = 1) {
             pivot_array <= temp_consonants;
       }
  }
```

Fig. 7. Making pivot array

```
word_length = 0;
while(text[i] != EOF) {
word_length = word_length + 1;
if(text[i] == BLANK) {
   text[i-word_length] = word_length - 1;
  word_length = 0;
  }
}
```

Fig. 8. Finding blanks in text

The matching stage starts by checking the length of the first word in the text. If the length of each word in the text is compared with the length of the pattern, after comparison success, the pattern is compared with the word of the text by the pivot array. After comparison success of pivot array, remaining alphabets in the pattern are compared with the word of the text. The matching stage searches fixed number of same length words as the pattern so that finding same length words in the text does not affect the whole algorithm's time complexity in the real life and only comparison between the pattern and the word affects the whole algorithm's time complexity. The OFRP algorithm has worst-case time complexity *O(n)*, where m is the length of the pattern. It means that the OFRP algorithm will not be affected by the pattern (Fig. 9).

```
i = 0;
int num_pivot;

while(text[i] != EOF) {
if(pattern_length < text[i]) {
 pattern_pivot_IN_Text_word = 0;
   for(j=0; j < num_pivot; j++) {
     if(text[i+pivot_array[j][1]] == pivot_array[j][0])
       pattern_pivot_IN_Text_word++ ;
   }
   if(pattern_pivot_IN_Text_word == num_pivot)
     CALL matching_whole_word();
       }
       i = i+text[i];
     }

         position_IN_Word = i+j;
         }
       }

       if(pattern_pivot_IN_Text_word == TRUE) {
        for(num_pivot_temp = 0; num_pivot_temp <
                                        num_pivot; num_pivot_temp ++)
       {
         pivot_move = pivot_array[num_pivot];
           for( ; pattern_pivot == text[pivot_move.position+i+j] &&
                                                pivot_move != NULL;
)

             pivot_move = pivot_move.link;

           if(pivot_move != NULL)
          BREAK;
     }
 }
 i=i+text[i];
```

Fig. 9. Finding a pattern in a text

4 Conclusion

In this paper, we consider the characteristics of English alphabets and text structures, and propose efficient new string searching algorithm based on vowel structure of a pattern and a text. We compare the time complexity of proposed vowel-based string searching algorithm with others. Proposed vowel-based string searching algorithm considers the frequency of vowels in a pattern and compares a pattern with a text mainly by the most frequent vowel in the pattern. For frequency of vowels and selection of a pivot vowel or pivot consonant (the most frequent vowel or consonant in the pattern), we define pivot_array as data structure that stores the most frequent vowel or consonant and their positions in the pattern. Proposed vowel and consonant-based string searching can avoid un-matching vowels in a text and unnecessary comparison between a pattern and a word in a text. Therefore in reality, proposed vowel and consonant-based string searching algorithm is usefully and effectively applicable to string searching function of a pattern and a text. Proposed vowel and consonant-based string searching algorithm has $O(n)$ time complexity and reasonable process time in the real world.

In the future, we will apply automata theory and tree structure to vowel and consonant-based string searching algorithm. It will reduce the formalization overhead of pattern preprocessing step. And we consider text preprocessing step that will extract vowels of text and formalize vowels structure of a text.

Acknowledgements. This work was supported by 2014 Korea National Open University Research Fund.

References

1. Boyer, R.S., Moore, J.S.: A fast string searching algorithm. Commun. ACM **20**, 762–772 (1977)
2. Knuth, D., Moris, J.H., Pratt, V.: Fast string searching in strings. SIAM J. Comput. **6**(2), 323–350 (1977)
3. Baeza-Yates, R., Gonnet, G.H.: A new approach to text searching. Commun. ACM **35**, 74–82 (1992)
4. Baase, S., Van Gelder, A.: Computer Algorithms: Introduction to Design and Analysis. Addis-Edisson-Wesley Pub., Reading (1999)
5. Sedgewick, R.: Algorithms in C++. Addison-Wesley Pub., Reading (1998)
6. Chung, K.S., Yu, H.-C., Jin, S.H.: An efficient string searching algorithm based on vowel occurrence pattern. In: Park, J.J., Yang, L.T., Lee, C. (eds.) FutureTech 2011, Part II. CCIS, vol. 185, pp. 379–386. Springer, Heidelberg (2011)

Neural Network Based PID Control for Quadrotor Aircraft

Dandan Zhao[1,2], Changyin Sun[1,2(✉)], Qingling Wang[1,2], and Wankou Yang[1,2,3]

[1] School of Automation, Southeast University, Nanjing 210096, China
cysun@seu.edu.cn
[2] Key Lab of Measurement and Control of Complex Systems of Engineering, Ministry of Education, Southeast University, Nanjing 210096, China
[3] Key Laboratory of Child Development and Learning Science of Ministry of Education, Southeast University, Nanjing 210096, China

Abstract. A back propagation neural network (BPNN) based PID control strategy for the attitude of quadrotor is proposed in this paper. Firstly, the architecture and dynamic model of quadrotor are analyzed according to the Newton-Euler Equation. Secondly, a nonlinear attitude model is established on the basis of the mathematical analysis. Thirdly, by eliminating the inverse error adaptively, a BPNN based PID controller is introduced to improve the robustness. Furthermore, PID parameters are adaptively adjusted through the training of neural network weighted coefficients. Finally, numerical examples demonstrate the performance of the designed BPNN based PID controller in terms of precision, adaptability and robustness.

Keywords: Neural network · PID controller · Attitude control · Quadrotor

1 Introduction

Quadrotor has great practical value and wide development prospects as a kind of vertical take-off and landing (VTOL) aircraft with excellent performance. Rotorcraft has interested the robotics and control community in recent years. Quadrotor, as the representative of the rotorcraft, has always been a hotspot in research of rotorcraft. Featured with multi-variable, nonlinearity, strong coupling and sensitivity of interference, quadrotor achieves six degrees of freedom (position and attitude) of aircraft movement by adjusting the four motor speeds [1].

The autonomous flight control circuit can generally be divided into three parts, namely position control, speed control and attitude control (Fig. 1). Attitude stability control (internal loop control) is the basis of micro flying platform for autonomous flight, the control result of which is of great importance to the aircraft flight characteristics. Thus the performance of attitude control is the key to the overall flight control. In these days, many researchers have being conducted on this field. The typical methods include PID control [1], adaptive control [2, 3], the sliding mode variable structure control [4, 5], back-stepping control [6, 7], robust control [8] and neural network control [9].

X. He et al. (Eds.): IScIDE 2015, Part II, LNCS 9243, pp. 287–297, 2015.
DOI: 10.1007/978-3-319-23862-3_28

The BPNN based PID control proposed in this article combines the advantages of PID control and neural network control. With the identification results of the BPNN, the selection and setting of proportion, integral and differential parameters are conducted.

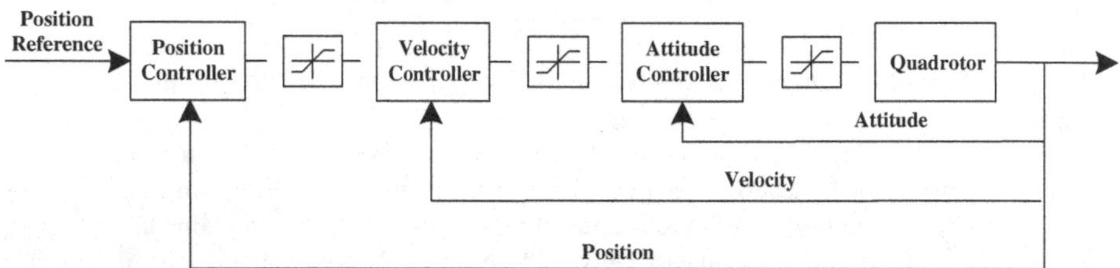

Fig. 1. The control structure of quadrotor

2 Dynamic Modeling for a Quadrotor

A quadrotor conventional structure is composed of four perpendicular arms which have motor and rotor, respectively (Fig. 2). The four rotors provide upwards propulsion as well as the direction control. In order to balance the effect of torque, the micro quadrotor system is divided into two opposite rotor pairs (rotor 1, 3 and rotor 2, 4) where one pair rotates clockwise while the other pair counters clockwise (Fig. 3) [1]. Vertical ascending (descending) flight is obtained by increasing (decreasing) thrust forces. By applying speed difference between front and rear rotors, we will have pitching motion which is defined as a rotation motion around Y axis. Using the same analogy in rolling motion, we will have a rotation motion around X axis. Yaw rotation, which is defined as a rotation motion around Z axis, is obtained through increasing (decreasing) speed of rotors (1.3) while decreasing (increasing) speed of rotors (2.4). Unlike pitch and roll motions, yaw rotation is the result of reactive torques produced by rotors rotation.

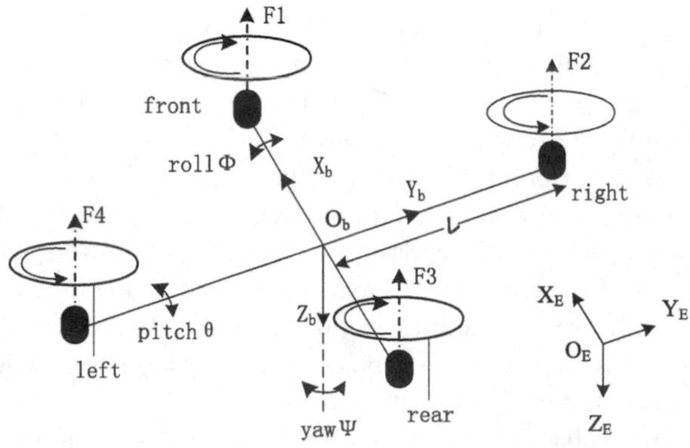

Fig. 2. Inertia and body coordinate system of the quadrotor

2.1 System Dynamics

The derivation of the dynamics is based on inertial and body frames. The quadrotor has six degrees of freedom, Cartesian (x, y, z) and angular (ϕ, θ, ψ). Using the formalism of Newton-Euler and assuming that the quadrotor has a symmetrical structure, the thrust and the drag forces are proportional to the square of the rotors speed, the dynamic model of quadrotor can be expressed as [10]:

$$\begin{cases} I_{xx}\dot{p} = qr(I_{yy} - I_{zz}) + J_r q\Omega_r + lU_1 \\ I_{yy}\dot{q} = pr(I_{zz} - I_{xx}) - J_r p\Omega_r + lU_2 \\ I_{zz}\dot{r} = pq(I_{xx} - I_{yy}) + J_r\dot{\Omega}_r + U_3 \\ m\ddot{x} = (\sin\psi\sin\phi + \cos\psi\sin\theta\cos\phi)U_4 - K_1\dot{x} \\ m\ddot{y} = (-\cos\psi\sin\phi + \sin\psi\sin\theta\cos\phi)U_4 - K_1\dot{y} \\ m\ddot{z} = \cos\psi\cos\phi U_4 + K_2\dot{z} + mg \end{cases} \tag{1}$$

Where:

- p, q, r is the angular velocity around the x, y, z axis in body coordinate respectively
- $I_{xx,yy,zz}$ is the inertia of the x, y, z axis respectively
- J_r is the inertia of the quadrotor
- l is the lever length
- $\Omega_r = \Omega_2 + \Omega_4 - \Omega_1 - \Omega_3$, $\Omega_{1,2,3,4}$ is angular velocity of rotor 1, 2, 3, 4, respectively
- U_1, U_2, U_3, U_4 are the defined control signals instead of use rotor speeds as control inputs, b is the thrust coefficient, d is the drag coefficient and $F_{1,2,3,4}$ is the rotor lift, $F_i = b\Omega_i^2$

$$\begin{cases} U_1 = b(\Omega_4^2 - \Omega_2^2) \\ U_2 = b(\Omega_1^2 - \Omega_3^2) \\ U_3 = d(-\Omega_1^2 + \Omega_2^2 - \Omega_3^2 + \Omega_4^2) \\ U_4 = \sum_{i=1}^{4} F_i = -b(\Omega_1^2 + \Omega_2^2 + \Omega_3^2 + \Omega_4^2) \end{cases} \tag{2}$$

From the above analysis, we can safely draw the conclusion that the attitude control of quadrotor is independent and the displacement control is affected by the attitude control.

2.2 Attitude Modeling

Due to the symmetrical structure of quadrotor, the design of the roll controller and the pitch controller can be exactly the same, so this paper just considers the specific design method of roll control.

If quadrotor is close to hover state, then the dynamic can be linearized as the gyroscopic effect and air resistance can be ignored [11]. Meanwhile, the attitude angular is very small in which case $(p, q, r) \approx (\dot{\phi}, \dot{\theta}, \dot{\psi})$. The linearized attitude model is:

$$\begin{cases} I_{xx}\ddot{\phi} = lU_1 \\ I_{yy}\ddot{\theta} = lU_2 \\ I_{zz}\ddot{\psi} = U_3 \end{cases} \tag{3}$$

It can be seen from the linear attitude model that there is no coupling between the attitude channels, the controllers can be designed separately.

Extract the attitude equation from Eq. (3):

$$I\ddot{\varphi} = \tau \tag{4}$$

τ is proportional to the difference of the square of the rotor angular velocity, I is the inertia of the quadrotor and φ is the attitude angular. Assuming the transform function of command signal to the motor and the actual speed difference output is first-order inertia link [12], the transfer function of motor instruction signal and the attitude angular can be obtained as

$$G_{\varphi u} = \frac{\varphi(s)}{u(s)} = \frac{k_{\varphi}}{s^2(\tau s + 1)} \tag{5}$$

The unknown parameters can be identified through manual flight inputs and outputs data fitting. The roll channel transfer function is identified as:

$$G_{\phi u} = \frac{53.46}{s^3 + 12.05s^2 + 101.8s} \tag{6}$$

And the yaw channel transfer function is identified as:

$$G_{\psi u} = \frac{1.6762}{0.1256s^3 + 0.7087s^2 + s} \tag{7}$$

3 BPNN Based PID Control Strategy

Attitude control is the core of the quadrotor control and its control performance will greatly affect the flight stability. BPNN based PID is introduced as attitude control method due to its effective transient and steady performance. The BPNN based PID controller can reduce the computation load and improve the system response speed [13]. According to the decoupled linear model, the roll channel, pitch channel and yaw channel BPNN based PID controllers can be designed independently.

The BPNN based PID controller (Fig. 3) is introduced to eliminate the inverse error adaptively and improve the robustness of the controller.

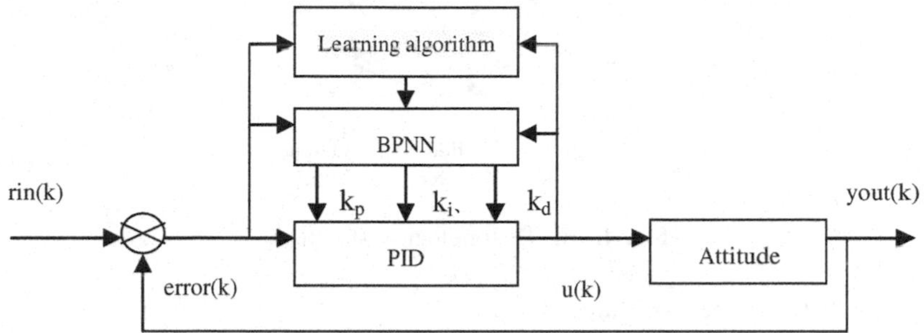

Fig. 3. BPNN based PID control structure

The typical PID control algorithm is as follow:

$$u(t) = k_p\left[e(t) + \frac{1}{T_I}\int_0^t e(t)dt + T_D\frac{de(t)}{dt}\right] \tag{8}$$

For the convenience of simulation, using the typical incremental digital PID control equation:

$$u(k) = u(k-1) + kp(e(k) - e(k-1)) + kie(k) + kd(e(k) - 2e(k-1) + e(k-2)) \tag{9}$$

kp, *ki*, *kd* stands for proportional, integral, differential coefficient respectively. Regard *kp*, *ki*, *kd* as dependent coefficients based on the adjustments of system operating status. Therefore Eq. (9) can be described as:

$$u(k) = f[u(k-1), kp, ki, kd, e(k), e(k-1), e(k-2)] \tag{10}$$

$f(.)$ is a nonlinear function of *kp*, *ki*, *kd*, u(k-1), e(k). Such a control law $u(k)$ can be obtained by BPNN training and learning as it has the ability to approximate any nonlinear function.

BPNN is one of the most widely used neural network model currently. BPNN is the multilayer feedforward network based on the training of error back propagation algorithm. BPNN can learn and store a lot of input - output model mapping without the knowledge of mathematical equations of the mapping relationship. The learning rule is that using the steepest descent method to minimize the error sum of squares of the network by back propagation, which can constantly adjust the network weights and threshold [14]. BPNN topological structures include input layer, hidden layer and output layer (Fig. 4).

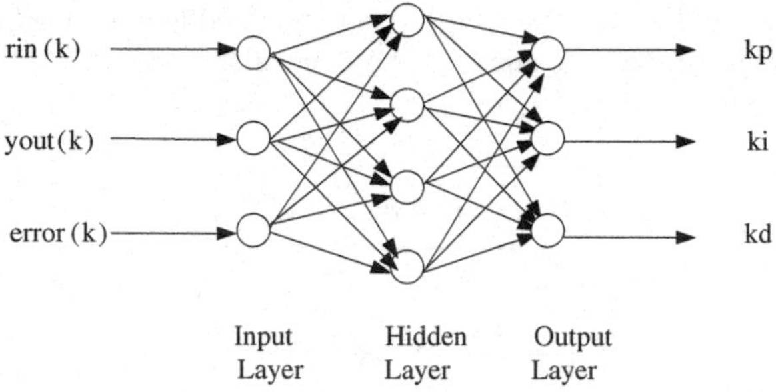

Fig. 4. BPNN topological structure

The inputs of the network input layer:

$$o_j^{(1)} = x(j), j = 1, 2 \ldots m \tag{11}$$

The inputs and outputs of the network hidden layer:

$$net_i^{(2)}(k) = \sum_{j=1}^{m} w_{ij}^{(2)} o_j^{(1)}, j = 1, 2 \ldots m, i = 1, 2 \ldots q \tag{12}$$

$$o_i^{(2)}(k) = f(net_i^{(2)}(k)), i = 1, 2 \ldots q \tag{13}$$

$\left\{w_{ij}^{(2)}\right\}$ stand for the hidden layer weighted coefficient, superscript (1), (2), (3) represent the input layer, hidden layer and output layer, $f(x) = (e^x - e^{-x})/((e^x + e^{-x})$.

The inputs and outputs of the network output layer:

$$net_l^{(3)}(k) = \sum_{i=1}^{q} w_{li}^{(3)} o_i^{(2)}(k) \tag{14}$$

$$o_l^{(3)}(k) = g(net_l^{(3)}(k)), l = 1, 2, 3 \tag{15}$$

$$o_1^{(3)}(k) = kp, o_2^{(3)}(k) = ki, o_3^{(3)}(k) = kd \tag{16}$$

$\{w_{li}^{(3)}\}$ stand for the output layer weighted coefficient, the output layer neuron activation function is the nonnegative function Sigmoid $g(x) = e^x/(e^x + e^{-x})$.

Take a performance index function:

$$E(k) = \frac{1}{2}(rin(k) - yout(k))^2 \tag{17}$$

Modify the weights of the network according to the gradient descent method with an inertia item that speeds up the convergence search:

$$\Delta w_{li}^{(3)}(k) = -\eta \frac{\partial E(k)}{\partial w_{li}^{(3)}} + \gamma \Delta w_{li}^{(3)}(k-1) \tag{18}$$

Where η is the learning coefficient and γ is the Inertia coefficient.

$$\frac{\partial E(k)}{\partial w_{li}^{(3)}} = \frac{\partial E(k)}{\partial y(k)} \cdot \frac{\partial y(k)}{\partial u(k)} \cdot \frac{\partial u(k)}{\partial o_l^{(3)}(k)} \cdot \frac{\partial o_l^{(3)}(k)}{\partial net_l^{(3)}(k)} \cdot \frac{\partial net_l^{(3)}(k)}{\partial w_{li}^{(3)}} \tag{19}$$

Using $\operatorname{sgn}\left(\frac{y(k)-y(k-1)}{u(k)-u(k-1)}\right)$ to approximately representative $\frac{\partial y(k)}{\partial u(k)}$.
According to Eq. (9):

$$\begin{aligned} \frac{\partial u(k)}{\partial o_1^{(3)}(k)} &= e(k) - e(k-1) \\ \frac{\partial u(k)}{\partial o_2^{(3)}(k)} &= e(k) \\ \frac{\partial u(k)}{\partial o_3^{(3)}(k)} &= e(k) - 2e(k-1) + e(k-2) \end{aligned} \tag{20}$$

The output layer weight can be obtained:

$$\Delta w_{li}^{(3)}(k) = \eta \delta_l^{(3)} o_i^{(2)}(k) + \gamma \Delta w_{li}^{(3)}(k-1) \tag{21}$$

$$\delta_l^{(3)} = e(k) \frac{\partial \hat{y}(k)}{\partial u(k)} \frac{\partial u(k)}{\partial o_l^{(3)}(k)} g'(net_l^{(3)}(k)), l = 1, 2, 3 \tag{22}$$

Similarly, the hidden layer weight can be obtained:

$$\Delta w_{ij}^{(2)}(k) = \eta \delta_i^{(2)} o_j^{(1)}(k) + \gamma \Delta w_{li}^{(2)}(k-1) \tag{23}$$

$$\delta_i^{(2)} = f'(net_i^{(2)}(k)) \sum_{l=1}^{3} \delta_l^{(3)} w_{li}^{(3)}(k), i = 1, 2, \ldots q \tag{24}$$

The steps of BPNN based PID algorithm:
Step 1: Determine the structure of BP network, the input layer node number M, the number of hidden layer nodes Q; the weighted coefficient of each layer and the selected learning rate and inertia coefficient are initialized; k = 1 for this step
Step 2: Sampling to get $rin(k)$, $yout(k)$; Calculate $error(k) = rin(k) - yout(k)$
Step 3: Calculate the inputs and outputs of each layer, the outputs of output layer are the parameters *kp*, *ki*, *kd*

Step 4: Calculate u(k) of the PID controller
Step 5: Do neural network learning to adjust the weighted coefficient to realize the adaptive adjustment of PID control parameters; Back to Step 2.

4 Simulations

4.1 Attitude Model Validations

The compared results of the model outputs and the real flight outputs are presented in Fig. 5. The figures demonstrate the validation of the attitude model.

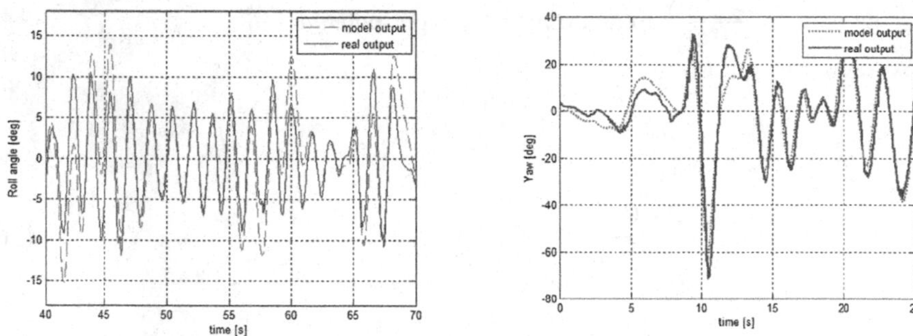

Fig. 5. Attitude model validations

4.2 Attitude Control Simulations

The neural network structure of the roll channel simulations consists of 3 input layer nodes, 4 hidden layer nodes and 3 output layer nodes. The initial values of the rate and inertia coefficient are 0.3 and 0.05 respectively.

The roll channel simulation results are in the following Figs. 6 and 7. Figure 6 shows that roll channel can follow the step signal with quite low overshot of 5 % and shorter rise time of 0.4 s in comparison of typical PID control of 1 s.

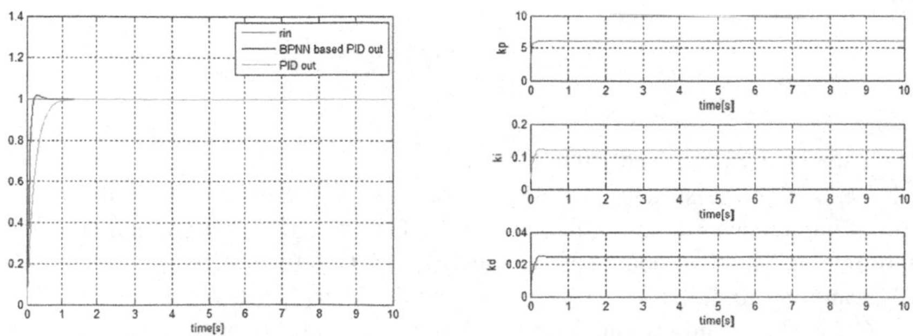

Fig. 6. The step response of roll channel

Furthermore, in order to test the capacity of resisting model uncertainty, the transfer function is slightly changed in the simulation of Fig. 7. The typical PID control can not follow the step signal with expected performance. Nevertheless, the BPNN based PID method can still follow the step signal with desired overshot and rising time due to the adaptability of the PID coefficient values. The theory presented in the paper is tested as having good robustness.

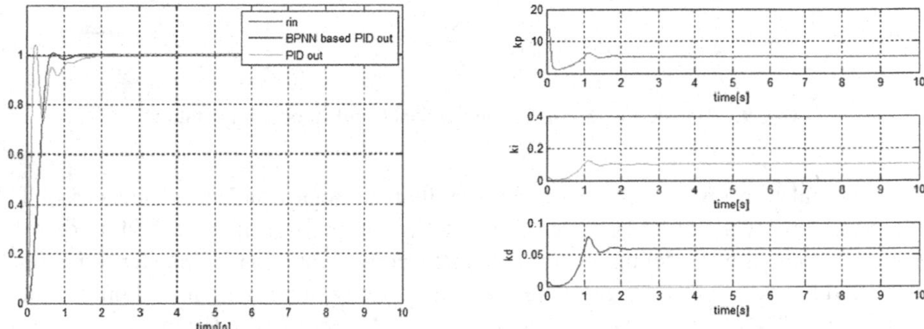

Fig. 7. The step response of roll channel with model uncertainty

The neural network structure of the yaw channel simulations consists of 3 input layer nodes, 4 hidden layer nodes and 3 output layer nodes. The initial values of the rate and inertia coefficient are 0.25 and 0.05 respectively.

Similarly, the yaw channel simulation results are in the following Figs. 8 and 9. Figure 8 shows that roll channel can follow the step signal with quite low overshot of 5 % and shorter rise time of 0.3 s in comparison of typical PID control of 0.9 s.

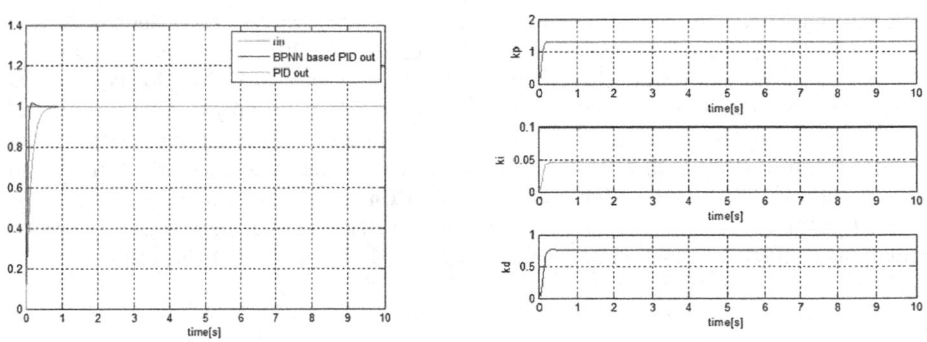

Fig. 8. The step response of yaw channel

The BPNN based PID method can still follow the step signal with desired overshot and rising time due to the adaptability of the PID coefficient values while the typical PID control performs not so well (Fig. 9).

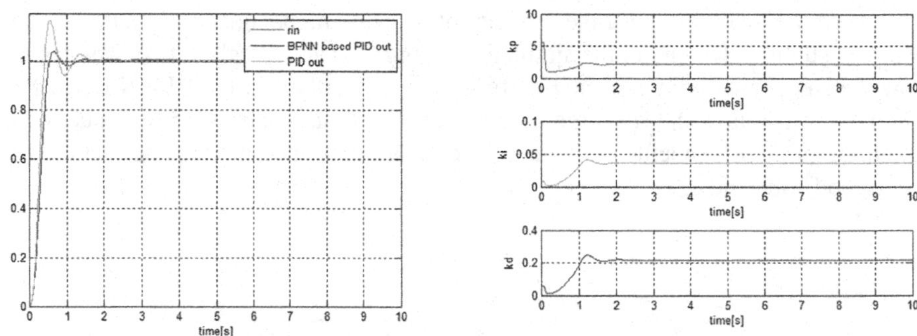

Fig. 9. The step response of yaw channel with model uncertainty

The simulation results demonstrate that attitude angular can follow the step signal with quite low overshot and shorter rise time in comparison of typical PID control. Besides, taking the modeling uncertainty into consideration, the transfer functions indentified before may have a certain error. The capacity of resisting model uncertainty through the BPNN based PID method is tested in the above simulations.

5 Conclusions

This paper proposed a BPNN based PID control of quadrotor attitude. Quadrotor dynamic model and attitude model were established. Furthermore, BPNN based PID controller was designed toachieve the adaptive adjustment of PID parameters by neural network back propagation. Considering the precision, adaptability and robustness of the control law, Matlab/Simulink based simulations were conducted. The performances of the control law were validated by simulation results.

Acknowledgments. The work was supported in part by Natural Science Foundation of Jiangsu Province of China under Grant No. BK20130471 and No. BK20140638, China Postdoctoral Science Foundation under grant No. 2013M540404, Jiangsu Planned Projects for Postdoctoral Research Funds under grant No. 1401037B, open fund of Key Laboratory of Measurement and Control of Complex Systems of Engineering, Ministry of Education under Grant No. MCCSE2013B01, the Open Project Program of Key Laboratory of Child Development and Learning Science of Ministry of Education, Southeast University (No. CDLS-2014-04), and A Project Funded by the Priority Academic Program Development of Jiangsu Higher Education Institutions (PAPD), and the Fundamental Research Funds for the Central Universities.

References

1. Cavalcante Sa, R., De Araujo, A.L.C, Varela, A.T, et al.: Construction and PID control for stability of an unmanned aerial vehicle of the type quadrotor. In: 2013 Latin American, Robotics Symposium and Competition (LARS/LARC), pp. 95–99. IEEE (2013)

2. Lee, T.: Robust adaptive attitude tracking on with an application to a quadrotor UAV. IEEE Trans. Control Syst. Technol. **21**(5), 1924–1930 (2013)
3. Mohammadi, M., Shahri, A.M.: Adaptive nonlinear stabilization control for a quadrotor UAV: theory, simulation and experimentation. J. Intell. Robot. Syst. **72**, 105–122 (2013). doi:10.1007/s10846-013-9813-y
4. Besnard, L., Shtessel, Y.B., Landrum, B.: Quadrotor vehicle control via sliding mode controller driven by sliding mode disturbance observer. J. Franklin Inst. **349**, 658–684 (2012)
5. Efe, M.: Battery power loss compensated fractional order sliding mode control of a quadrotor UAV. Asian J. Control **14**(2), 413–425 (2012)
6. Huang, B.B., Chun-Juan, L.I., Yong, H.E., et al. Back-stepping control for an under-actuated micro-quadrotor. J. Luoyang Inst. Sci. Technol. **23**(2), 56–61 (2013)
7. Honglei, A., Jie, L., Jian, W., et al.: Backstepping-based inverse optimal attitude control of quadrotor. Int. J. Adv. Robot. Syst. **10** (2013)
8. Raffo, G.V., Ortega, M.G., Rubio, F.R.: An integral predictive/nonlinear H ∞ control structure for a quadrotor helicopter. Automatica **46**, 29–39 (2010)
9. Wang, D.L, Qiang, L., Liu, F.: Quadrotor attitude control based on L_1 neural network adaptive control method. Comput. Eng. Des. **33**(12), 4758–4761 (2012)
10. Li, J., Li, Y.: Dynamic analysis and PID control for a quadrotor. In: 2011 International Conference on Mechatronics and Automation (ICMA), pp. 573–578. IEEE (2011)
11. Kendoul, F., Lara, D., Fantoni-Coichot, I., et al.: Real-time nonlinear embedded control for an autonomous quadrotorn helicopter. J. Guid. Control Dyn. **4**, 1049–1061 (2007)
12. Wang, W., Ma, H., Xia, M., et al.: Attitude and altitude controller design for quad-rotor type MAVs. Math. Prob. Eng. **58**(7), 98–107 (2013)
13. Dierks, T., Jagannathan, S.: Neural network output feedback control of a quadrotor UAV[C]. In: 47th IEEE Conference on Decision and Control, pp. 3633–3639. CDC 2008, IEEE (2008). doi:10.1109/CDC.2008.4738814
14. Chang, X.U., Jian-Hong, L., Cheng, M., et al. Neural network PID adaptive control and its application. Control Eng. China **14**(3), 284–286 (2007)

GPU-Based Parameter Estimation Method for Photovoltaic Electrical Models

Jieming Ma[1(✉)], T.O. Ting[2], Huiqing Wen[2], Baochuan Fu[1], and Jianmin Ban[1]

[1] School of Electronics and Information Engineering, Suzhou University of Science and Technology, 1 Ke Rui Road, Suzhou 215009, People's Republic of China
jieming84@gmail.com

[2] Xi'an Jiaotong-Liverpool University, 111 Ren'ai Road, Suzhou 215123, People's Republic of China

Abstract. Parameter estimation (PE) is one of the most challenging problems in photovoltaic (PV) system modeling. Owing to the ability to handle nonlinear functions regardless of the derivatives information, meta-heuristics have attracted many researchers. Recently, many implementations of particle swarm optimization (PSO) based PE method have been proposed in the literature. However, these algorithms utilize multiple agents or particles in the search process, and are normally compute intensive. In this paper, we describe our implementation of PSO on graphic processing units (GPUs) using open computing language (OpenCL). The proposed method has been specifically designed and entirely executed on the GPUs to provide a reduction of computational costs. Results show that the GPU-based PE is faster in comparison with its sequential implementation of PSO, and this proves the efficacy of the GPU framework.

Keywords: Photovolatic · Modeling · Parameter estimation · Particle swarm optimization

1 Introduction

Photovoltaic (PV) cell prices fell by 74 % from 1995 to 2011 [1], yet the initial cost of a PV system is relatively high. An accurate assessment of the electrical characteristics is therefore indispensable in the system design [2]. Significant research efforts have been made to develop electrical models of PV generation systems [3]. These models include analytical models based on PV cell physics, empirical models, and a combination of these two approaches [4]. Their mathematical expressions formulate the terminal current I with the most crucial technical characteristics and environment variables, such as terminal voltage V, the ambient temperature T, and the irradiance G.

However, these models cannot be directly used because of the lack of proper model parameters characterizing the PV cells. Parameter estimation (PE) is a discipline that provides tools for estimating constants appearing in the model [5]. With the parameters obtained in such a way, the difference between the simulated and experimental data can be minimized.

X. He et al. (Eds.): IScIDE 2015, Part II, LNCS 9243, pp. 298–307, 2015.
DOI: 10.1007/978-3-319-23862-3_29

In the literature [6,7], conventional PE methods are classified into two categories: analytical techniques and numerical extraction techniques. The former utilizes mathematical equations to describe the parameters of PV electrical models. There is much research on addressing the PE problem by analytical expressions in terms of the physical parameters, such as the coefficient of diffusion of electrons in the semiconductor, lifetime of minority carriers, the intrinsic carrier density, etc. [8]. However, the values of these physical parameters are normally not provided by manufacturers. This scenario impels researchers to explore an alternative way of formulating the parameters by using the information available in datasheet (e.g. short circuit current coefficient K_i, open circuit voltage coefficient K_v, short circuit current I_{sc}, open circuit voltage V_{oc} etc.).

Albeit simple, they are generally dependent on the key points on the *I-V* curve. The errors can be significant and cannot be further improved if these key points are incorrectly specified. Assisted by a statistical method, numerical extraction techniques fit a great many operating points on the *I-V* curves to obtain a more accurate solution [9,10]. The numerical extraction techniques are normally considered as accurate approaches in parameter estimation since all the measured data can be used in calculation. However, the non-linear curve-fitting procedures are quite complicated both mathematically and in terms of computer code [11]. Moreover, the algorithms can be computationally expensive as the size of required data is considerably large.

Evolutionary Algorithm (EA) techniques are very efficient in optimizing real-valued multi-modal objective functions [12,22]. To date, genetic algorithm (GA) [13], particle swarm optimization (PSO) [14], bacterial foraging algorithm (BFA) [16], simulated annealing (SA) [17], pattern search [18], differential evolution [7,19] have been employed for estimating parameters of various PV electrical models due to their ability to handle non-linear functions without requiring derivatives information. Although the EA techniques may obtain accurate solution if their initial points and algorithm parameters are set properly, most of these methods apply multiple agents or particles in random search and do not provide a significant improvement in computational efficiency.

With respect to the PSO-based PE approach, this work considers a means of computation, in which the PE process is carried out simultaneously to increase the computational efficiency. The proposed method is implemented in OpenCL, which is a heterogeneous programming framework that supports a wide range of levels of parallelism and efficiently maps to general graphic processing units (GPUs) [20]. Computational performance is evaluated by identifying the parameters of a most widely applicable PV electrical model.

2 Parameter Estimation of a Photovoltaic (PV) Electrical Model

In this section, a solution to the problem of PE for a single-diode model (SDM) is proposed. We provide a formalization of the modeling framework and a proper fitness function for this problem. Then the proposed GPU-based PE method will be exploited.

2.1 Problem Formulation

The SDM is usually considered to offer a good compromise between simplicity and accuracy, and thus it is widely used to predict the I-V electrical characteristics of PV generation systems. The SDM is described by a modified Shockley diode equation incorporating a diode quality factor to account for the effect of recombination in the space-charge region [15]. For a PV cell, its I-V relations can be described via the expression of SDM:

$$I = I_{ph} - I_o(e^{\frac{q(V+IR_s)}{AkT}} - 1) - \frac{V + IR_s}{R_p} \tag{1}$$

In the above equation, the temperature T, terminal voltage V and current I are the quantities that can be measured. The k and q represent the Boltzmann constant ($1.3806488 \times 10^{-23}\ J/K$) and the electron charge ($1.602176 \times 10^{-19}\ C$), respectively. The five unknown parameters are series resistance R_s, shunt resistance R_p, photocurrent I_{ph}, saturation current I_o, diode ideality constant A, series resistance R_s, and shunt resistance R_p.

Based on a typical optimization algorithm, the PE method minimizes the differences between calculated values and measured data by generating a population of trial solutions, and moving these particles around in the search-space in accordance with simple mathematical formulas over the particles' positions and velocities. Normally, the fitness value of a trial solution is evaluated by the root mean square (RMS) error f_{rms} which serves to aggregate absolute differences into a single measure of predictive power. If the size of experimental data is denoted by N, the RMS error can be mathematically described by the following equation [21]:

$$f_{rms} = \sqrt{\frac{1}{N}\sum_{d=1}^{N}\left(f_d(\widehat{V}, \widehat{I}, \mathbf{X})\right)^2}, \tag{2}$$

where the $\widehat{V}$ and $\widehat{I}$ denote the measured voltage and current, respectively. The vector $\mathbf{X}$ involves the unknown parameters $\mathbf{X} = (I_{ph}, I_o, A, R_s, R_p)$. The function $f_d(\widehat{V}, \widehat{I}, \mathbf{X})$, representing the differences between directly measured and calculated values for terminal current in d^{th} sample, and is basically a homogeneous form of (1):

$$f_d(\widehat{V}, \widehat{I}, \mathbf{X}) = I_{ph} - I_o(e^{\frac{\widehat{V}+\widehat{I}R_s}{AV_t}} - 1) - \frac{\widehat{V} + \widehat{I}R_s}{R_p} - \widehat{I}, \tag{3}$$

2.2 Parameter Estimation Using Particle Swarm Optimization (PSO)

Particle swarm optimization (PSO) [22] is one of the prominent algorithms in the category of nature-inspired algorithms and has been one of the most successful metaheuristic optimization methods applied in many fields. Compared with many evolutionary algorithms such as GA and DE, the PSO usually obtains faster convergence speed.

The basic idea behind the PSO is that a potential solution to the optimization problem is treated as a particle flying like a bird in multi dimensional space, adjusting its position in search space according to its own previous experience and that of its neighbors. Its movement is guided by the essentially important ingredient formulas:

$$x_{i,j}^{t+1} = x_{i,j}^{t} + v_{i,j}^{t+1}, \tag{4}$$

where $v_{i,j}^{t+1}$ is the velocity, expressed as:

$$v_{i,j}^{t+1} = wv_{i,j}^{t} + \alpha\epsilon_1(x_{i,j}^{t} - gbest^t) + \beta\epsilon_2(x_{i,j}^{t} - pbest_i^t). \tag{5}$$

where by:

$v_{i,j}^t$	Velocity for i^{th} particle in j^{th} dimension at time t.
w	Inertia weight, usually set to 0.5.
$X_{i,j}^t$	Current position of ith dimension at time t.
$gbest^t$	The best solution among all participating particles for i^{th} dimension at time t, also known as global best.
$pbest_i^t$	The best position for i^{th} dimension at time t of a particle, also known as personal best.
$\epsilon_1,\ \epsilon_2$	Independent uniform random numbers within [0, 1].
$\alpha,\ \beta$	Acceleration coefficients towards $pbest_i^t$ and $gbest^t$ respectively.

By careful inspection of Eqs. (4) and (5), the following interpretations can be concluded with regard to the PSO [23]:

i. the velocity somehow acts as short-term memory retention and plays a crucial role in the update process;
ii. the update of a dimensional value is guided by the $pbest_i^t$ and $gbest^t$. Simply, this means that a particle explores positions in between the $pbest_i^t$ and $gbest^t$;
iii. the independent random numbers ϵ_1 and ϵ_2 control the ratio of movement towards the $pbest_i^t$ and $gbest^t$.

In this work, the five parameters characterizing SDM will be predicted by PE method. To illustrate the application of the PSO in PE, these parameters are described by the set $\mathbf{X} = [I_{ph}, I_o, A, R_s, R_p]$. The $\mathbf{X}$, describing the Nth particle at t^{th} generation, is determined as $\mathbf{X_i^t} = [\mathbf{X_1^t},\ \mathbf{X_2^t},\ ..., \mathbf{X_N^t}]$, whose velocity is $\mathbf{V_i^t} = [\mathbf{V_1^t},\ \mathbf{V_2^t},\ ..., \mathbf{V_N^t}]$. Algorithm 1 shows the estimation process. First, a swarm of particles $\mathbf{X}$, as well as their velocities $\mathbf{V}$ are seeded onto the search space in a random manner. These particles then move through the problem space. At generation t, the historical best solution of the ith particle is recorded as local best $pbest^i$. Particles then communicate with each other to find out the global best $gbest^t$ among the personal bests and direct themselves towards the global best position. The estimation process will not end until it satisfies the stopping criterion. In the next section we show how to exploit a PSO to estimate the vector of parameters $\mathbf{X}$ in a parallel manner.

Initialize PSO parameters;
Initialize particles $\mathbf{X}_\mathrm{i}^\mathrm{t}$ and velocities $\mathbf{V}_\mathrm{i}^\mathrm{t}$;
while *Stopping criterion is not satisfied* **do**
  for $i = 1$ *to* N *(particle)* **do**
    Evaluate the f_{rms_i} for each trial via (2);
    Update the $pbest_i^t$ for each particle;
  end
  Update the $gbest^t$;
  Update the $\mathbf{V}_\mathrm{i}^\mathrm{t}$ via (5);
  Update the $\mathbf{X}_\mathrm{i}^\mathrm{t}$ via (4);
end

Algorithm 1: Pseudocode for the PSO.

2.3 Implementation and Discussion of the GPU-Based PSO

The PSO described so far is computationally expensive due to the huge number of fitness evaluations performed by the algorithm. In order to reduce the computation cost, the performances of PSO is improved by using the GPU architecture, which exploits the great computational power of modern video cards. The proposed method has been implemented with OpenCL, a GPU computing framework supporting a wide range of levels in parallelism and efficiently maps to a variety of computing devices [20]. In this implementation, particles/velocities update is followed by two kernels: i. fitness evaluation; ii. global best compu-

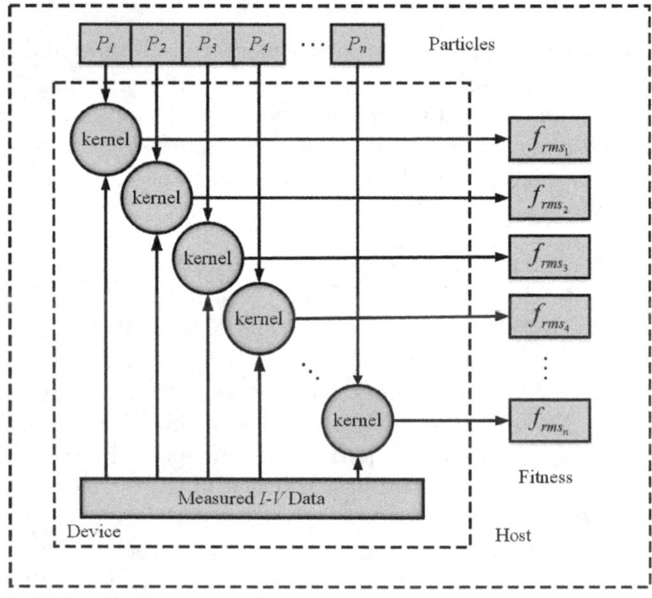

Fig. 1. Parallel implementation of fitness evaluation functions.

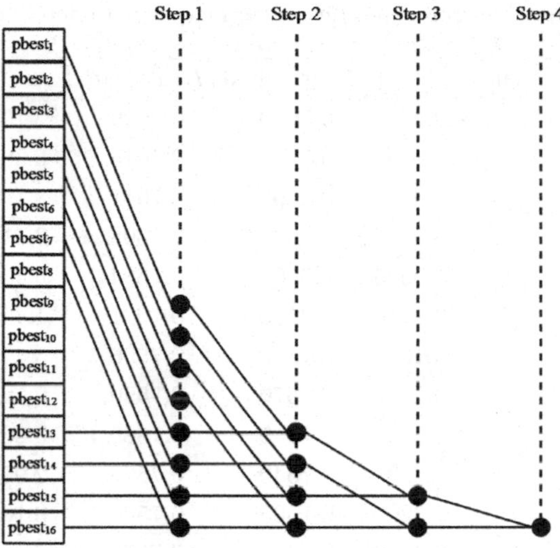

Fig. 2. A reduction tree implemented in OpenCL.

Table 1. Best solutions obtained by the GPU-based PSO.

I_{ph}	I_o	A	R_s	R_p
0.7609	4.82E-06	1.8161	0.0204	180.0000

tation. The former generates a separate execution instance to perform fitness evaluation for each particle, since fitness evaluations are independent. Figure 1 depicts the concurrent process. With the measured *I*-*V* data, the RMS errors f_{rms} are calculated concurrently in the kernel. By comparing the new fitness values, the particles' personal bests can be updated then. In the second kernel, the $gbest^t$ is obtained by parallel reduction as illustrated in Fig. 2. The input local bests are divided into $N/2$ different work groups, where each group is responsible for computing a single element. Within a work group, the reduction is performed over multiple stages. At each stage, work-items compare the values between $pbest_i^t$ and $pbest_{i+N/2}^t$. The smaller one will be placed in $pbest_i^t$. The number of work groups then is reduced by half. In this manner, the smallest f_{rms} can be stored in $pbest_1^t$ after log_2N stages.

3 Experimental Results and Analysis

The experimental platform for this paper is based on Intel Core i7-4770 k CPU, 8.0 G RAM, NVIDIA GTX 760, and Windows 7 operating system.

To study the accuracy and efficiency of the proposed GPU-based PE method, a 57 mm diameter commercial (R.T.C. France) silicon solar cell is modeled by the SDM. During the PE process, the parameters are evaluated by using the

Table 2. Relative error based on the estimated parameters in Table 1.

Measurement	$\widehat{V}$	$\widehat{I}$ measured	I calculated	e
1	−0.2057	0.7640	0.7619	2.70E-03
2	−0.1291	0.7620	0.7615	6.36E-04
3	−0.0588	0.7605	0.7611	8.19E-04
4	0.0057	0.7605	0.7608	3.41E-04
5	0.0646	0.7600	0.7604	5.45E-04
6	0.1185	0.7590	0.7601	1.40E-03
7	0.1678	0.7570	0.7596	3.49E-03
8	0.2132	0.7570	0.7590	2.70E-03
9	0.2545	0.7555	0.7580	3.36E-03
10	0.2924	0.7540	0.7562	2.92E-03
11	0.3269	0.7505	0.7529	3.18E-03
12	0.3585	0.7465	0.7470	7.12E-04
13	0.3873	0.7385	0.7372	1.71E-03
14	0.4137	0.7280	0.7215	8.90E-03
15	0.4373	0.7065	0.6984	1.15E-02
16	0.4590	0.6755	0.6652	1.53E-02
17	0.4784	0.6320	0.6212	1.71E-02
18	0.4960	0.5730	0.5652	1.36E-02
19	0.5119	0.4990	0.4975	2.91E-03
20	0.5265	0.4130	0.4174	1.07E-02
21	0.5398	0.3165	0.3265	3.16E-02
22	0.5521	0.2120	0.2245	5.89E-02
23	0.5633	0.1035	0.1144	1.06E-01
24	0.5736	−0.0100	−0.0022	7.76E-01
25	0.5833	−0.1230	−0.1291	4.93E-02
26	0.5900	−0.2100	−0.2251	7.18E-02

I-V experimental data in [24]. Table 1 lists the best solutions obtained by the GPU-based PSO by using 19200 particles after 500 generations. The relative error, $e = |\widehat{I} - I|/\widehat{I}$, for each measurement is shown in Table 2. The error is within the domain [3.41E-04, 7.76E-01] and the mean absolute error is 4.61E-02. The calculated value of the MAE denotes the high accuracy of the identification process.

In Fig. 3, it is possible to see how execution time depends on the number of swarms: plotted data refer to the mean time, and corresponding standard deviations, over 100 consecutive runs. The PSO-based PE is performed both on a GPU (AMD R9 200) and a CPU(Intel i7 4770 k). The execution time and

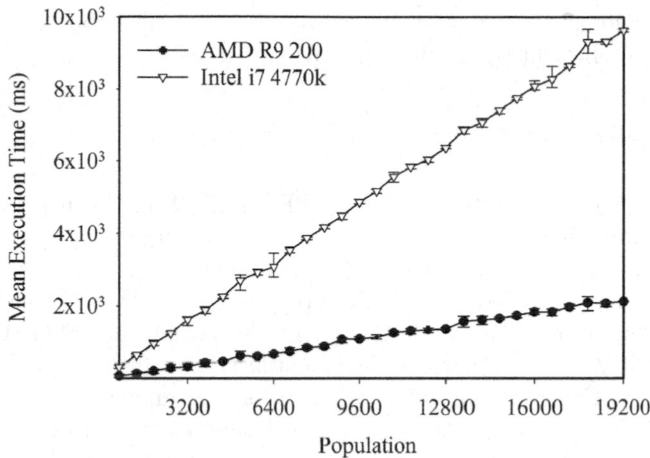

Fig. 3. Comparison between the mean execution time identified by PSO on CPU (Intel i7 4770 k) and GPU (AMD R9 200).

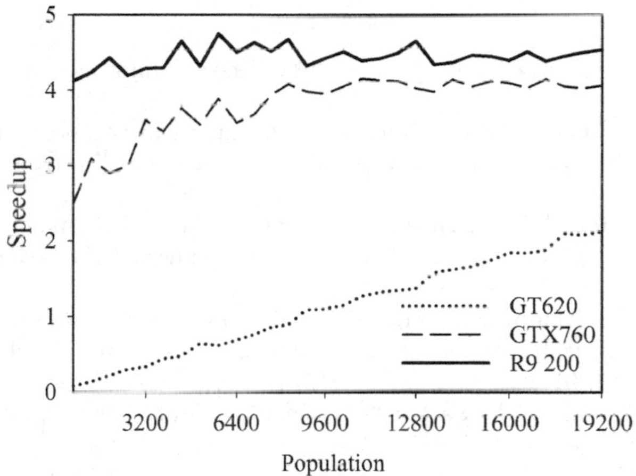

Fig. 4. Comparison between the speed up obtained by PSO on various GPUs.

swarm population size take a linear relationship on GPU. It takes less than 2 ms for GPU to optimize PE process.

Another experiment is created to compare the speedup on various GPUs:

1. NVIDIA OpenCL on GT 620 in Window 7;
2. NVIDIA OpenCL on GTX 760 in Window 7;
3. AMD OpenCL on Radeon HD5850 in Windows 7.

The results are as shown in Fig. 4. The population size varied from 640 to 19200 with constant iterations count of 640, and the number of generations is 500. The speedup on R9 200 fluctuates around 4.3 in our experiments. As for

NVIDIA GT 620 and GTX 760, it can be observed that the GPU-based PE method become efficient when the population increases.

4 Conclusion

In this paper, a novel way to implement PE on GPU has been presented. The proposed method distributes the workload of a PSO algorithm appropriately to computing devices in parallel mode. The results show that the PSO, as well as the objective functions used in the PE, is probably one of the algorithms that is most suitable for parallelization on GPUs. The execution time of GPU-based PE method is greatly shortened over its sequential counterpart, while maintaining similar accuracy. Our planned future work includes comparing the OpenCL performance with CUDA, studying more parallel intelligent optimization algorithms and their applications in PE, as well as exploring more aggressive CPU/GPU sharing on more recent hardware that has improved memory bandwidth.

References

1. DOE, U.S.: Annual Energy Review 2011. Energy Information Administration (EIA) (2012)
2. Villalva, M.G., Gazoli, J.R., Filho, E.R.: Comprehensive approach to modeling and simulation of photovoltaic arrays. IEEE Trans. Power Electron. **24**, 1198–1208 (2009)
3. Jung, J.-H., Ahmed, S.: Real-time simulation model development of single crystalline photovoltaic panels using fast computation methods. Sol. Energy **86**, 1826–1837 (2012)
4. Siddiqui, M.U., Abido, M.: Parameter estimation for five- and seven-parameter photovoltaic electrical models using evolutionary algorithms. Appl. Soft Comput. **13**, 4608–4621 (2013)
5. Beck, J.V., Arnold, K.J.: Parameter estimation in engineering and science parameter. Wiley Series in Probability and Mathematical Statistics, New York (1977)
6. Ishaque, K., Salam, Z., Taheri, H., Shamsudin, A.: A critical evaluation of EA computational methods for photovoltaic cell parameter extraction based on two diode model. Sol. Energy **85**, 1768–1779 (2011)
7. Ishaque, K., Salam, Z., Mekhilef, S., Shamsudin, A.: Parameter extraction of solar photovoltaic modules using penalty-based differential evolution. Appl. Energy **99**, 297–308 (2012)
8. Nishioka, K., Sakitani, N., Uraoka, Y., Fuyuki, T.: Analysis of multicrystalline silicon solar cells by modified 3-diode equivalent circuit model taking leakage current through periphery into consideration. Solar Energy Mater. Solar Cells **91**, 1222–1227 (2007)
9. Gottschalg, R., Rommel, M., Infield, D.G., Kearney, M.J.: The influence of the measurement environment on the accuracy of the extraction of the physical parameters of solar cells. Meas. Sci. Technol. **10**, 796 (1999)
10. Mullejans, H., Hyvarinen, J., Karila, J., Dunlop, E.D.: Reliability of the routine 2-diode model fitting of PV modules. In: 19th European Photovoltaic Solar Energy Conference, pp. 2459 (2004)

11. Yordanov, G., Midtgrd, O., Saetre, T.: Two-diode model revisited: parameters extraction from semi-log plots of IV data. In: 25th European Photovoltaic Solar Energy Conference (2010)
12. Yang, X.-S.: Nature-Inspired Metaheuristic Algorithm. Luniver Press, Bristol (2010)
13. Joseph, A.J., Hadj, B., Ali, A.-L.: Solar cell parameter extraction using genetic algorithms. Meas. Sci. Technol. **12**, 1922 (2001)
14. Huang, W., Jiang, C., Xue, L., Song, D.: Extracting solar cell model parameters based on chaos particle swarm algorithm. In: 2011 International Conference on Electric Information and Control Engineering, pp. 398–402 (2011)
15. Ye, M., Wang, X., Xu, Y.: Parameter extraction of solar cells using particle swarm optimization. J. Appl. Phys. **105**, 094502–094508 (2009)
16. Rajasekar, N., Krishna Kumar, N., Venugopalan, R.: Bacterial foraging algorithm based solar PV parameter estimation. Sol. Energy **97**, 255–265 (2013)
17. El-Naggar, K.M., AlRashidi, M.R., AlHajri, M.F., Al-Othman, A.K.: Simulated annealing algorithm for photovoltaic parameters identification. Sol. Energy **86**, 266–274 (2012)
18. AlHajri, M.F., El-Naggar, K.M., AlRashidi, M.R., Al-Othman, A.K.: Optimal extraction of solar cell parameters using pattern search. Renew. Energy **44**, 238–245 (2012)
19. da Costa, W.T., Fardin, J.F., Simonetti, D.S.L., Neto, L.d.V.B.M.: Identification of photovoltaic model parameters by differential evolution. In: 2010 IEEE International Conference on Industrial Technology (ICIT), pp. 931–936 (2010)
20. Gaster, B., Howes, L., Kaeli, D.R., Mistry, P., Schaa, D.: Heterogeneous Computing with OpenCL. Elsevier, Waltham (2012)
21. Ma, J., Ting, T.O., Man, K.L., Zhang, N., Guan, S.-U., Wong, P.W.H.: Parameter estimation of photovoltaic models via cuckoo search. J. Appl. Math. **2013**, 8 (2013)
22. Eberhart R.C., Yuhui S.: Particle swarm optimization: developments, applications and resources. In: Proceedings of the 2001 Congress on Evolutionary Computation, vol. 81, pp. 81–86. IEEE Press (2001)
23. Ting, T.O., Man, K.L., Guan, S.-U., Nayel, M., Wan, K.: Weightless swarm algorithm (WSA) for dynamic optimization problems. In: Park, J.J., Zomaya, A., Yeo, S.-S., Sahni, S. (eds.) NPC 2012. LNCS, vol. 7513, pp. 508–515. Springer, Heidelberg (2012)
24. Easwarakhanthan, T., Bottin, J., Bouhouch, I., Boutrit, C.: Nonlinear minimization algorithm for determining the solar cell parameters with microcomputers. Int. J. Solar Energy **4**, 1–12 (1986)

MAD: A Monitor System for Big Data Applications

Mingruo Shi(✉) and Ruiping Yuan

Beijing Wuzi University, Beijing 100123, China
shimingruo@163.com, yuanruiping@bwz.edu.cn

Abstract. A big data application usually needs to build a pipeline on the top of workflow engine which connects relevant periodic workflow jobs. It's crucial to timely alert pipeline issues, provide an issue diagnosis subsystem to find out root cause from a variety of sources, and measure pipeline/service by predefined metrics. In this paper, we identify three indispensable qualities monitor systems must fulfill namely timeliness, accuracy and flexibility. We find that the conventional monitoring tools lack at least one of three qualities, and introduce a general purpose MAD (Monitoring, Alerting and Diagnosis) system for big data applications to keep data freshness, collect measurement metrics to meet SLA.

Keywords: MAD (Monitoring Alerting and Diagnosis) · Hadoop · Oozie · SLA

1 Introduction

A MAD system becomes increasingly important in big data applications falling in a wide family of application scenarios: from online advertising to financial securities exchange, from social networks to medical information systems. MAD contains three subsystems i.e. measurement, alerting and diagnosis. Measurement measures if Service Level of Agreement (SLA) is achieved; if Key Performance Indicator (KPI) is met; if system resource is within budget; and other internal measurement indicators such as usage/adoption/coverage/precision/recall of individual components and prediction models. Alerting is targeted to alert about system abnormal situations such as pipeline/service error/over SLA. The purpose of alerting is to shorten MTTD (mean time to detection). Diagnosis provides tools to better understand the whole system and find the root cause of a system fault more quickly. The target of diagnosis is to shorten the MMTR (mean time to repair). Without MAD system faults are difficult to detect and hard to track the KPI, hence more efforts and time are required to fix a business system issue. In 2006, Khanna et al. [1] developed an external monitor by analyzing external message exchanges. In 2007, Khanna et al. [2] proposed a rule based diagnosis for distributed IT infrastructures. In 2010, Haifeng et al. [3] proposed an invariants

This work is specially supported by the Science and Technology Plan General Program of Beijing Municipal Education Commission (KM201510037001), Chinese Mountaineering Association (CMA2014-B-A04) and Intelligence Logistics System Beijing Key Laboratory (NO:BZ0211)

X. He et al. (Eds.): IScIDE 2015, Part II, LNCS 9243, pp. 308–315, 2015.
DOI: 10.1007/978-3-319-23862-3_30

based failure diagnosis method for distributed computing systems. In 2010, Joshi et al. [4] proposed a probabilistic model-driven recovery for distributed systems. Some individual software packages e.g. Ganglia, Nagios and Splunk [5–7] are provided some functionalities for monitor, alerting and diagnosis for distributed systems such as Hadoop [8]. None of these systems is general purpose for big data applications.

MapReduce is the hot distributed and parallel programing paradigm for processing over big data in IT industry. In 2004, Jeffery and Sanjay [9] proposed MapReduce to simplify data processing on large clusters and widely used in Google. In 2006, Hadoop [10] is an open source implementation of MapReduce which is a subproject of Apach Lucene. Hadoop jobs could be aggregated by Pig and Hive. SCOPE [11] like Hive is a SQL like language to script Map/Reduce job under COSMOS, Map/Reduce and storage system of Microsoft. They are widely used in Search engines, data analytics and so forth. It's vital important to provide a general purpose MAD for jobs running on both these MapReduce platforms and corresponding online systems.

Workflow engine orchestrates the running of pipeline jobs. These jobs consist of MapReduce job, DB load job, timer job, stream monitoring job or other jobs. A job could be run in hourly, daily, weekly or monthly. Workflow engine is the key components to make job done in the expected way. Oozie [12] is a scalable open source workflow engine on top of Hadoop.

In this paper, we proposed a general purpose MAD system for big data applications in order to keep data freshness, to shorten SLA and to provide corresponding metrics to corresponding product system. Current implementation is based on Hadoop and Oozie. It's easy to extend to support COSMOS or google MapReduce by implementing the predefined interfaces.

The paper is organized as follows: in Sect. 2, we describe in detail the architectural attributes of MAD that enable timeliness, accuracy and flexibility. Section 3 describes its components in detail. In Sect. 4 discusses the experimental setup to measure the performance. We conclude and summarize our ongoing work in Sect. 5.

2 Architecture

MAD is defined on top of big data platform and workflow engine. The current implementation is based on Hadoop and Oozie respectively. MAD also get signal from log files, SQL DBs and performance counters defined in production environment. MAD aims at measuring, alerting and diagnosing big data applications. There concerns problems in the architecture collaborated by monitoring service, measurement subsystem, alerting subsystem and diagnosis subsystem. Figure 1 shows the architecture of MAD system.

Monitor service is responsible for monitoring production pipeline and measurement pipeline. Measurement subsystem is responsive for measurement metrics calculation. Alerting subsystem deals with alert collection, maintenance and alert email sending. Diagnosis subsystem provides the insights of target application details, dependency of production workflow/measurement workflow pipeline. Diagnosis subsystem also provides enough information for issue diagnosis.

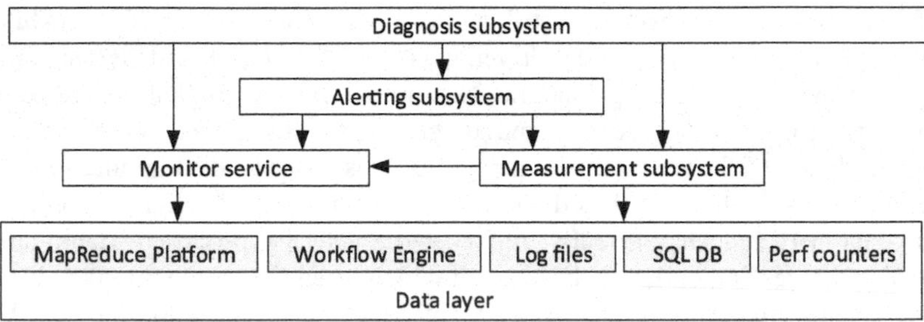

Fig. 1. MAD architecture.

3 MAD Components

3.1 Monitor Service

Monitor service collects and caches pipeline latest status with which pipeline owners could get enough information for diagnosis subsystem and alerting subsystem about production and measurement pipelines. Also collect performance information of product services by leveraging service's performance counters. It provides Web Service interface from which alerting subsystem and diagnosis subsystem could get required information. Its major components are shown in Fig. 2.

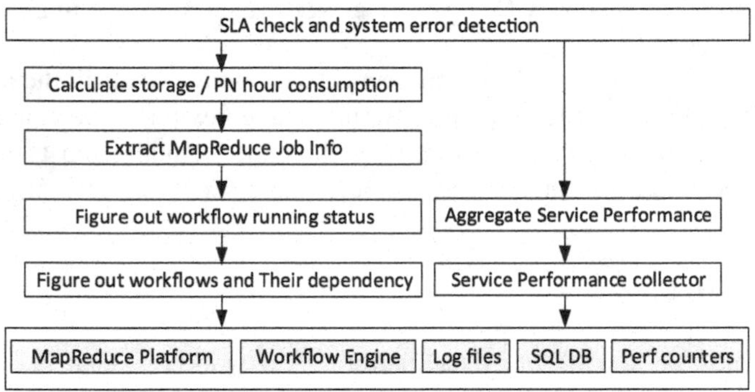

Fig. 2. Monitor service.

For product services, monitor service collect the availability and percentile metrics by service log, service performance collector and service performance aggregator. For backend workflow job, monitor service figures out workflows and their dependency by workflow group information defined in workflow engine. Then figure out workflow running status. For scheduled jobs, monitor service can figure out map-reduce job information including input/output streams, storage consumption and PN hours. Both product services and map-reduce jobs, monitor service will check if they meet SLA or workflow/service errors occur. Monitor service caches workflows/service latest information for diagnosis, alerting and metrics. Cache contains last 15 day workflow/job information. Monitor service supports Hadoop and workflow engine Oozie.

3.2 Measurement

Measurement pipeline is a key component to make the business system measurable to meet business goal. The Fig. 3 shows the general workflows and dependency for a typical big data application. Generally, raw log monitor, common log cooking, core measurement, and DbLoad workflows are required.

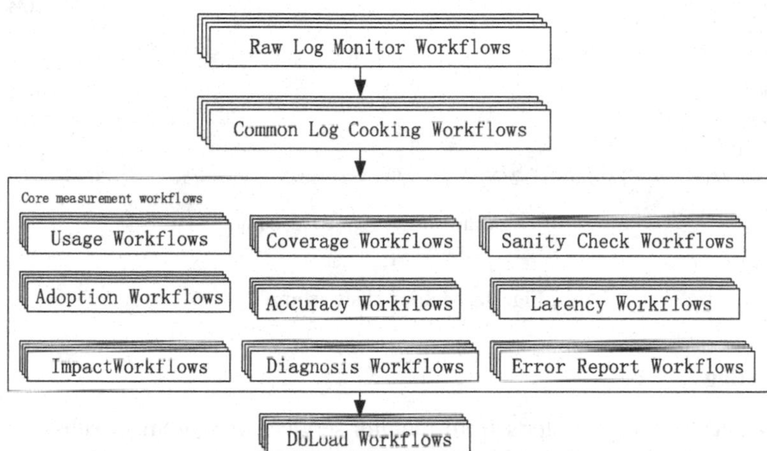

Fig. 3. Measurement pipeline for a typical big data application

Raw log monitor workflow checks if raw log upload is complete. If done, MAD will generate a signal file or post an event to indicate its completion. Common log cooking workflow cooks different raw logs into single one to be consumed by downside workflows. Take service quality metrics as example. Some service stores usage log, performance log and error log separately, they could be cooked into single log i.e. each row for a usage with error information and performance information. Service performance metrics and other metrics could be based on the cooked log instead of several separate logs with better performance and simpler process logic. The performance for a XXX service is shown in Fig. 4. Core measurement workflow calculates a variety of measurement metrics which includes and not limited to usage, coverage, sanity check, adoption, accuracy, latency, impact, diagnosis, and error report workflows. Some of these metrics are defined as KPI. DbLoad workflow loads metrics from HDFS to Database.

Measurement pipeline workflows with predefined SLA are also monitored by Monitor service. In order to keep KPI fresh, the MapReduce jobs related to KPI metrics should take higher priority. Figure 4 shows the availability and call count for some product service. It's hard to ensure the correctness of metrics in big data analytics since data is from heterogeneous data sources. Sanity check introduced in Sect. 3.3 is a good way to reduce metrics calculation mistakes.

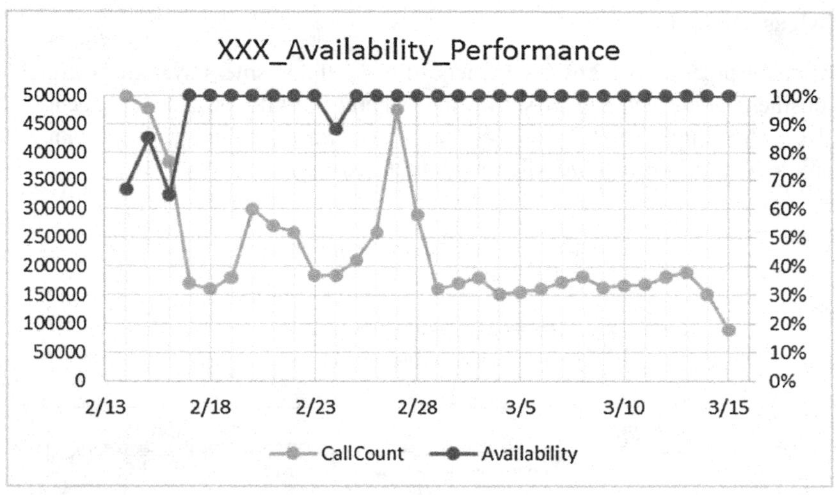

Fig. 4. Metrics chart example

3.3 Alerting

Alert subsystem aggregate alerts from monitor service, measurement subsystem. The major types of alerts include E2E check alerts, sanity check alerts, over SLA of backend workflow job or product service and workflow job errors shown in Fig. 5.

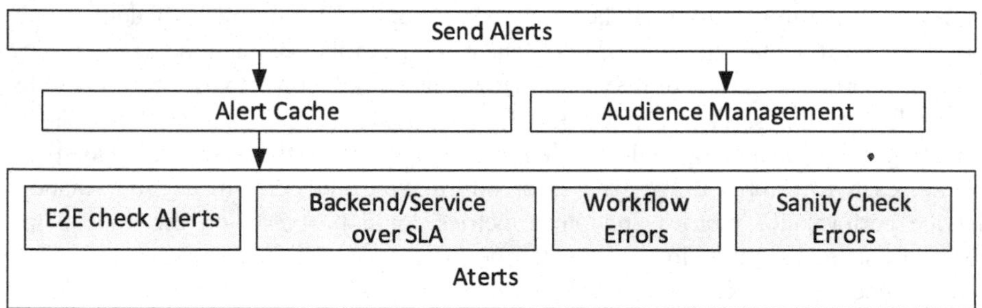

Fig. 5. Alerting subsystem

Measurement subsystem provides flexibility for data quality check i.e. sanity check. Both measurement result and the streams generated by product pipelines are required to satisfy some predefined rules. We call such rule check process sanity check. We provide a general sanity check framework accordingly. Feature owner just need write sanity check workflow. Empty result stream means sanity check is OK. Otherwise framework will send alert email for detailed check results to feature owners.

E2E check is done by operators in daily basis by which MAD know if the product system does work as expected. Alerting subsystem leverages monitor service's results for backend/service over SLA and workflow errors. Alert subsystem create an alert cache to void duplicated alerts. Alerting subsystem also provide a user interface to set alert audience.

3.4 Diagnosis

Diagnosis subsystem provides detailed info about target system. Take backend pipeline as an example in Fig. 6, it shows the dependency about feature area *XXX*.

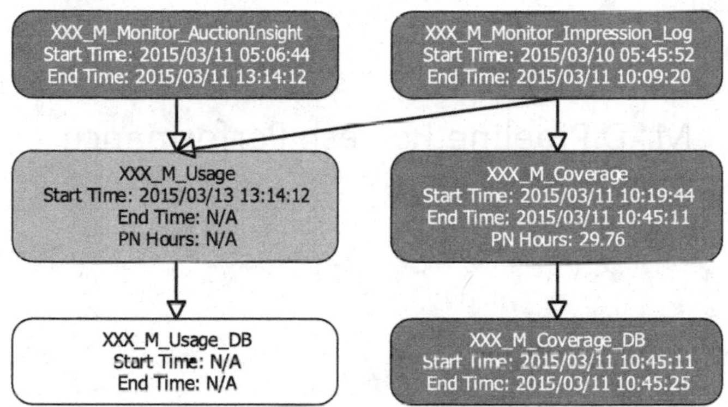

Fig. 6. Workflow pipeline dependency graph by feature area XXX

The feature area includes six workflows. Workflow color in workflow means different running status. XXX_M_Monitor_AuctionInsight, XXX_M_Monitor_Inpression_Log, XXX_M_Coverage and XXX_M_Coverag_DB are colored blue means workflow are finished. The green colored i.e. XXX_M_Usage is in running state. The white colored i.e. XXX_M_Usage_DB is not started to run. The start time is shown for scheduled jobs. The end time is shown for the finished jobs. The PN hours are only available for the finished map-reduce jobs. The more detailed information including error message will be shown to users when user double click one workflow.

4 Performance

Accuracy, timeliness and flexibility are the most important identified qualities of MAD. For evaluation purpose, we apply our MAD to our online Ads test platform. We setup 1000 product workflow jobs per day which is divided into 4 feature areas. Twenty workflows are hourly scheduled. Five hundred and twenty workflows are daily scheduled. These workflows has some predefined dependency. So the start time of these workflows are potentially different. Ten hourly workflow jobs has issue. Ninety daily workflow jobs has issues. The root causes of these issues are: (1) map or reduce function bug; (2) lack dependent jar package; (3) lack required input HDFS file; and (4) workflow configuration error. We running these workflows 2 days.

We found that 101 hundred alerts received of which 1 alerts is false-alarms i.e. job resubmitting message instead of workflow error. So recall achieves 100 % and precision is 98 %. For timeliness, we refresh the pipeline status every 15 min. The actual refresh time per round is shown in Fig. 7. We found the refresh time per round is increased in initial stage then keep stable. This is due to the number of running workflows is increasing in initial rounds due to workflow dependency. Then the number of running workflows keep stable then. Note that we will analyze log files of failed jobs as well as log files of running jobs. The average refresh time is less than 5 min per round.

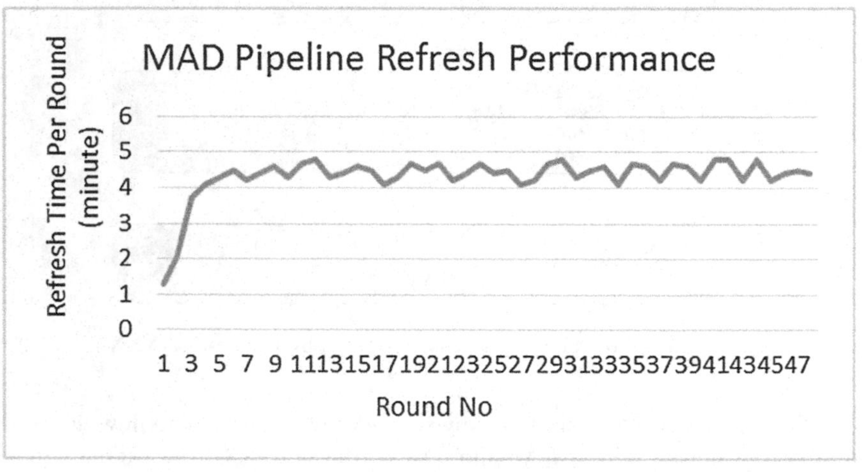

Fig. 7. MAD pipeline experiment refresh performance

As discussed in previous section, we could verify sanity check result with a simple and common rule. So alerting for sanity check failures is general by leveraging our sanity check framework. This means data correctness verification is flexible.

5 Conclusion

A general-purpose MAD system is proposed to maintain an online/offline big data application in efficient and easy way. The MAD system can alert timely with 100 % recall rate and very low false alarm (less than 2 %) and provide fine-grained diagnosis capacity such that the data freshness is kept and maintenance cost is greatly reduced. A general metrics framework is provided for KPI slice-dice by measurement pipelines and UI tools, which provides the indicator for continuous improvement. However, our MAD system doesn't resist on single point of failure. ZooKeeper [13] is a good candidate to conquer this issue. Another future work is to support other MapReduce implementations such as SCOPE and other workflow engine though the design itself provide corresponding interfaces.

References

1. Khanna, G., Varadharajan, P., et al.: Automated online monitoring of distributed applications through external monitors. IEEE Trans. Dependable Secure Comput. **3**(2), 115–129 (2006)
2. Khanna, G., Cheng, M.Y., et al.: Automated rule-based diagnosis through a distributed monitor system. IEEE Trans. Dependable Secure Comput. **4**(4), 266–279 (2007)
3. Chen, H., Jiang, G., et al.: Invariants based failure diagnosis in distributed computing systems. In: IEEE Symposium on Reliable Distributed Systems, pp: 160–166 (2010)
4. Joshi, K.R., Hiltunen, M.A., et al.: Probabilistic model-driven recovery in distributed systems. IEEE Trans. Dependable Secure Comput. **8**(6), 913–928 (2011)
5. Ganglia - a scalable distributed monitoring system for high-performance computing systems. http://ganglia.sourceforge.net/
6. Nagios - the industry standard in IT infrastructure monitoring. http://www.nagios.org/
7. Splunk - the leading platform for Operational Intelligence. http://www.splunk.com/
8. Apache Hadoop. http://wiki.apache.org/hadoop
9. Jeffery, D., Sanjay, G.: MapReduce: simplified data processing on large clusters (2004). http://labs.google.com/papers/mapreduce.html
10. Hadoop - Yahoo! Lauches world's largest hadoop production applications. http://developer.yahoo.com/blogs/hadoop/posts/2008/02/yahoo-worlds-largest-product-hadoop/
11. Ronnie, C., Bob, J., et al.: SCOPE: easy and efficient parallel processing of massive data sets. In: VLDB 2008, pp. 24–30 (2008)
12. Mohammad, I., Angelo, K.H. Oozie: torwards a scalable workflow management system for hadoop. In: SWEET 2012, 20 May 2012
13. Patrick, H., Mahadev, K., et al.: ZooKeeper: wait-free coordination for internet-scale. In: Usenix (2010)

Research on SQLite Database Encryption Technology in Instant Messaging Based on Android Platform

Aite Zhao, Zhiqiang Wei, and Yongquan Yang(✉)

Ocean University of China, Qingdao 266100, Shandong, China
tiddyzhao@hotmail.com, weizhiqiang@ouc.edu.cn,
i@yangyongquan.com

Abstract. Due to the characteristics of the current open-source Android system, and looser user rights management mechanism, information security issues become a problem of common concern to the user. At the present stage, there are many measures encrypting data for Android database, which are divided into several aspects of hardware layer, core layer, virtual machine layer and application framework layer. This paper presents a method which achieves the database encryption by modifying SQLite source code in Libraries and Android Runtime Layer. The encryption function can not only encrypt the data itself, but also be invoked by the upper application though JNI interface, so that the application can access more secure database, and improve the security of the instant messaging application. Experimental results show feasibility and effectiveness of the SQLite database encryption which is used in the instant messaging application.

Keywords: Android · Encryption · SQLite · Security · Instant messaging application

1 Introduction

Since the emergence of Android operating system, it has become one of the most popular mobile platforms in the world [1]. It is used in single binary phones, tablet PCs, and other devices. Moreover, the number of users who use smart phones and tablet PCs that have Android operating system has more than one billion [2]. This situation indicates that the Android mobile phones have been favored by a large number of users. Instant messaging application described here can make the majority of users exchange ideas easily. The growing number of users is bound to get a lot of data, how to protect these data, establish and improve the security mechanism have become a major issue [3].

SQLite [4] is an embedded lightweight database in Android operating system, SQLite supports SQL syntax of standard relational database, and can do many complex queries, you only need to define the SQL statements for creating and updating databases, and SQLite will manage your Android platform automatically. The instant messaging application which is described will put the personal chat log and the group chat log into local SQLite database. This article will use SQLite data encryption technology to encrypt the chat log and modify the SQLite source code, in order to avoid the risk of storing in plain text, so that the local database information is protected.

X. He et al. (Eds.): IScIDE 2015, Part II, LNCS 9243, pp. 316–325, 2015.
DOI: 10.1007/978-3-319-23862-3_31

Experiments will use the Android 4.4 version of the operating system for testing. We adopt the Android platform to develop the whole instant messaging system, and use Openfire as the server, use SQLServer 2008 as the database.

2 Related Work

Market share of the Android system is very high, but the security issue is serious. This makes mobile security market become an important field among competing security software vendors. After the offensive and defensive combat between attackers and security vendors, and fierce competition among security vendors, the work of Android security enhancement has made a lot of progress and got a lot of important research results. The following paragraphs describes the current situation of the Android system security research in the internal hardware layer of Android device, the core layer, the virtual machine layer and application framework layer.

2.1 Hardware Layer

In hardware layer, methods for realizing security protection are the transformation of mobile phone hardware and the expansion of the security mechanism,In the field of security studies of hardware layer, David Kleidermiacher [5] proposed to enhance the security of Android devices by using ARM TrustZone. The Android system implemented a set of Java Level DRM API, and also implements a C/C++ Level DRM Manager. The implementation of each specific DRM strategy is connected to the DRM Manager by plugging in. The specific DRM needs to be implemented by the system vendors according to actual needs, and the application developer can develop relevant applications according to the standard of DRM API.

2.2 Core Layer

Android operating system is an embedded operating system which is based on Linux kernel development for the mobile device. Linux security block is the main module to protect the system security, which is a lightweight universal access control framework of the Linux kernel. It makes a variety of security access control model can be achieved in the form of Linux loadable kernel module. Users can select the appropriate security module loaded into the kernel according to their needs, thus greatly improving the flexibility and availability of the security access control mechanism. At present, there are many well-known enhanced access control system achieved in the LSM, including SELinux [6], Domain and Type Enforcement (DTE) and the Linux Intrusion Detection System (LIDS) etc.

2.3 Virtual Machine Layer

Dalvik virtual machine is an important part of the Android platform. According to Android Dalvik, researchers proposed a security enhancement scheme based on virtualization technology, which allowed Dalvik virtual machine to have a higher authority to manage and monitor the behavior of the top programs. Currently, in the field of mobile security, virtualization products mainly include L4Android [7], vmware [8], Xen [9] etc.

2.4 Application Framework Layer

Application framework layer is the foundation of the developer in Android application development. Security enhancements of this layer focused on improvement of Activity Manager to form dynamic monitoring method to observe the runtime permissions, currently it also have mature commercial products. For example, the 360 mobile guard etc.

3 Overall Security Mechanism of the Instant Messaging

The instant messaging application is designed for the social users. To protect the security of chat data is the most important work. Figure 1 is the overall security architecture diagram of the instant messaging.

- The system is based on XMPP protocol. XMPP uses Transport Layer Security (TLS) to prevent the sending of data between servers from being tampered or overheard. This mechanism ensures the data security during transmission.
- The server uses the DES encryption algorithm to encrypt the user's login password database query. It will use the DES algorithm to decrypt again to protect the security of user data.
- The chat data stored in local uses SQLite database encryption technology to protect the data security.

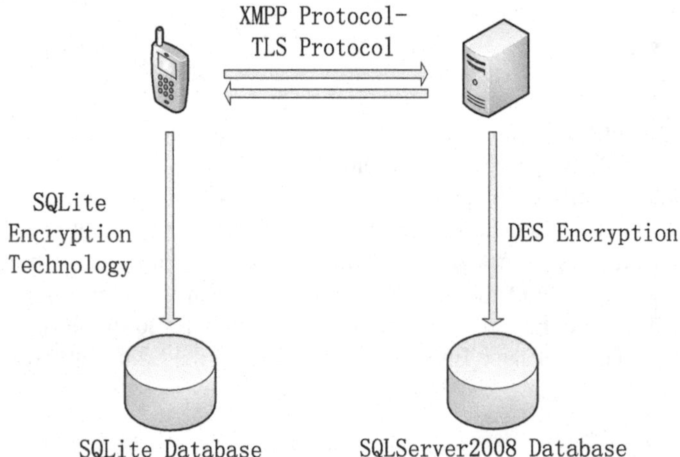

Fig. 1. Security mechanisms of instant messaging

4 SQLite Encryption Design

The design idea of SQLite encryption is to find the encryption interface in SQLite source code, and then implement the key setting function in the interface to complete the encryption of the data itself, and finally, provide the application with encryption function through JNI (Java Native Interface) interface to make the application more secure access to Android SQLite database.

4.1 The Choice of Encryption Method

When the server is running, SQLite data and virtual database engine are both running in the same process. In a stored procedure, they are also stored in the same physical file. So SQLite database is a file type embedded database. For this type of database, there are generally three kinds of encryption methods: application layer encryption, operating system and DBMS encryption layer encryption.

Encryption in Application Layer. If you encrypt some data in the application, the general approach is encrypted storage and decrypted read, and the database will receive and store the cipher text of the application. When reading data, the application can read the cipher text and decrypt it. This method is inflexible and inefficient that can increase the developer's difficulty, and increase the application load.

Encryption in Operating System Layer. Operating system layer encryption is to regard the SQLite database as an ordinary file and encrypt it, and regardless of the relationship between the data SQLite database. Inserting data during operation, the operating system firstly encrypt the data in RAM, and then the file system writes the encrypted data to the SQLite database file. The system will do inverse decryption when reading data. Database administrators need to properly manage the key.

Encryption in DBMS Layer. Encryption in DBMS layer needs to operate the SQLite database itself. This method has completed the encryption and decryption in the process of writing and reading data of database, and has done the transparent processing for the application layer, users and the storage layer. Despite the encryption method increases the DBMS load; its encryption function also has strong and high efficiency. Therefore, in order to ensure the operational efficiency of the customized Android system, so we use high efficient DBMS layer encryption to encrypt the SQLite database.

4.2 Encryption Interface Options

Encryption interface method in SQLite reserved mainly include SQLite3_key(), SQLite3_rekey(), SQLite3CodecGetKey() and SQLite3CodecAttach(), SQLite3_key() is mainly used for the database to set the key, SQLite3_rekey() is used to reset the key, SQLite3CodecGetKey() is used to return the current key, SQLite3CodecAttach() is used to relate the page encode function and keys to the database.

4.3 Encryption Algorithm Selection

Encryption algorithm is divided into symmetric encryption algorithm and asymmetric encryption algorithm according to whether encryption and decryption key is consistent. When the symmetric encryption algorithm encrypts data and files, it has fast speed, high key strength, the short key and is suitable for mobile devices. Usually, symmetric encryption algorithm is AES [10], DES, IDEA, etc. In comparison, non-symmetric encryption algorithm is more complex, the performance of encrypting data has low efficiency, long strength etc. Therefore, the instant messaging application uses the AES encryption algorithm to encrypt data. The key length is 256 bits.

AES encryption algorithm flow chart is shown in Fig. 2.

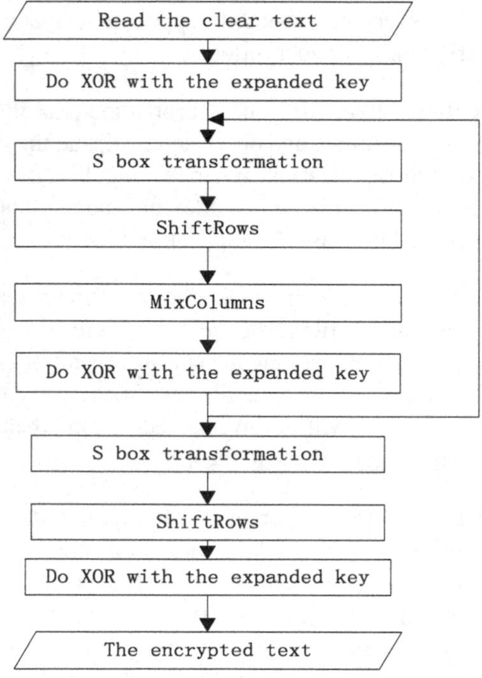

Fig. 2. AES encryption algorithm flow chart

5 SQLite Encryption Process

Because the amount of Android source code is large and the structure is complex, in the entire source code to modify and debug database module SQLite source code is very complex. The specific flow chart of encryption is shown in Fig. 3.

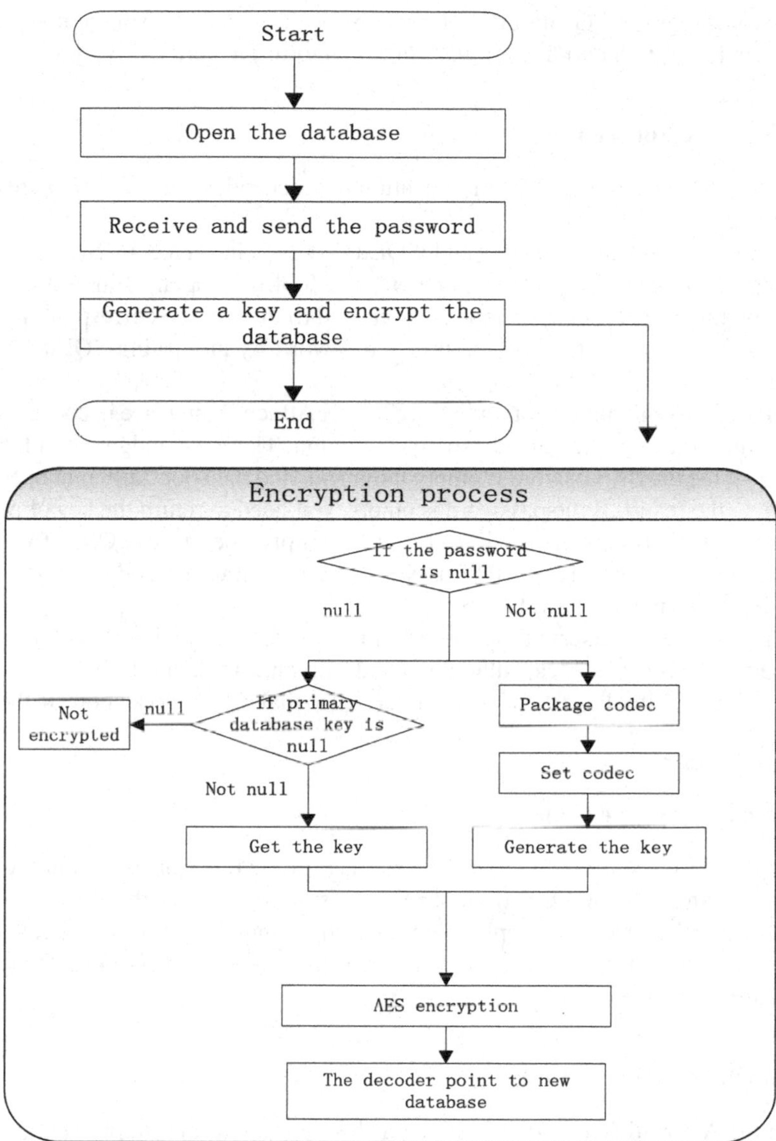

Fig. 3. Implementation of SQLite encryption flow chart

5.1 Separation of The Source Code

Firstly, I separate all the SQLite source code from Android source code. Source code located in/external/SQLite/dist, which has some more Android profiles and "make" files than open source SQLite official code. So we take out the source code of this directory and establish another software program to implement encryption.

5.2 Modify the Source Code

Modify the source code that is the modification of independent source code in reserved JNI interface.

First, modify SQLite3_open() and SQLite3_key() interface in SQLite3.c file. According to the encoding and parameters, the SQLite3_open () interface can be divided into three different open modes. You need to modify the corresponding code, and then to receive and transmit the user's password by modifying SQLite3_key () interface.

This part focus on implementing SQLite3CodecAttach () interface in codecext.c file to achieve generation of key and database encryption. There is a judgment in Interface that judges if the user's password is empty judgment, and then does different operations according to the result. If the password is empty, you need to return the key of primary database and don't do the encryption, if it is not empty, then to use CodecCopy () to return the key and encrypt the MD5 password by rewriting the codec.c and codec.h header files, and finally return the key.

Using the AES256 encryption algorithm to encrypt returned key. Rijndael algorithm used in AES256 is encapsulated in rijndael.h and rijndael.c file.

Modifying SQLite3PagerSetCodec () interface to make the decoder point to the latest database.

5.3 Merge the Source Code

To put the SQLite source code which has encryption function back into Android source code, and then modify the entire open source to open the path of SQLite encryption interface from the application layer to the inner layer to make developers call database encryption API directly in application layer, just like call the official Android API.

5.4 Compile the Source Code

Compile the Android source code which has been modified and improved, and form the Android system and SDK that has underlying database encryption feature.

5.5 Transplant the Source Code

You can also put the SQLite source code with encryption into applications. You can use the full encryption after adding the required libSQLite3_jni.so package.

6 Analysis of Experimental Results

Create a database called native.db in instant messaging application to save chat record, and then create a data table called History including numbers, user name, age and chat record three fields, number is primary key. Now insert six records and encrypt data by the achieved method, database. key (String key) is the method for data encryption.

The result displays in UI is shown in Fig. 4:

- The key in Fig. 4 comes from the project. It is implemented by internal code, and encrypted key can be displayed on the front screen. It's easy to observe the result.
- If the database which stores local chat records has been stolen maliciously, he could not open the database because he didn't have user' password.
- The encryption method encrypted the entire database file. Users need to know encrypted key that can open the database. Otherwise, they cannot see any available data through the editor.
- According to the project set, every time a chat record is inserted, encrypted key will change.

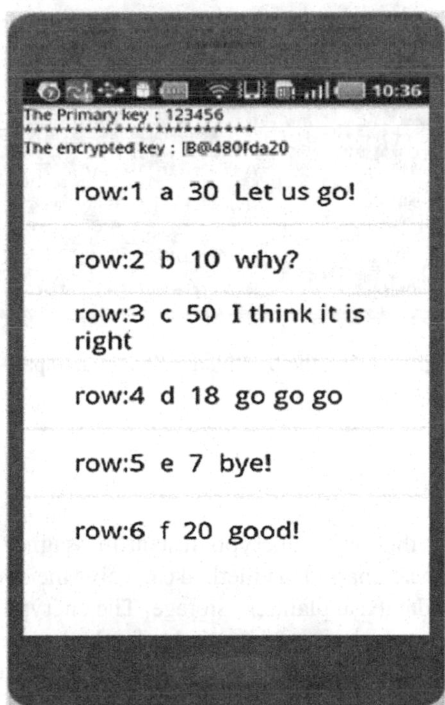

Fig. 4. The experimental result

Figures 5 and 6 display the result that the encrypted native.db is opened by SQLite Developer and Notepad.

Fig. 5. Open the SQLite database by SQLite Developer

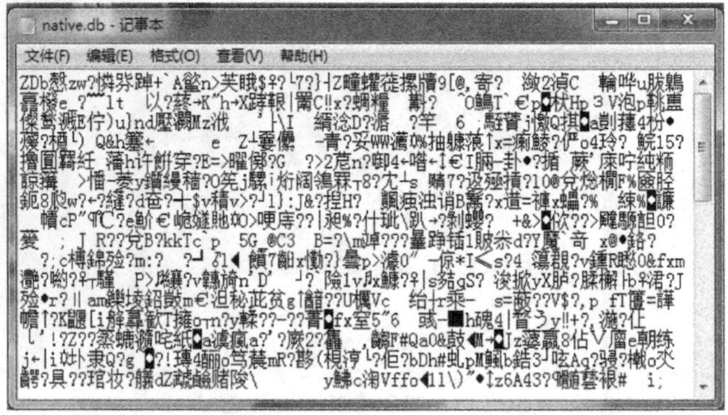

Fig. 6. Open the SQLite database by notepad

7 Conclusion

The method presented in this paper encrypts instant messaging chats stored in local, ensuring the security of local chats. This method can solve the Android system security issues caused by SQLite database plaintext storage. The encryption method belongs to a database management system layer and is suitable for Android mobile phone, also can be ported to other applications that need to be encrypted. Experimental results show that this method is feasible and effective, and the encrypted database is more secure.

Acknowledgment. This work is supported by the Fundamental Research Funds for the Central Universities (Grant No. 201413065); National Key Technology R&D Program (Grant No. 2012BAH17F03).

References

1. Android, the world's most popular mobile platform (2014). http://developer.android.com/about/index.html
2. The Android Story (2014). http://www.android.com/history
3. Enck, W., Octeau, D., McDaniel, P., et al.: A study of android application security. In: USENIX Security Symposium, vol. 2, p. 2 (2011)
4. Owens, M., Allen, G.: The Definitive Guide to SQLite. Apress, Berkeley (2006)
5. Kleidermacher, D.: Bringing security to Android-based devices. http://www.igmagazine-online.coin/current/pdf/Pg56-58_IQ_32-Bringing—Security—to—Android-based—Devices.pdf. Accessed 22 October 2010
6. Shabtai, A., Fledel, Y., Elovici, Y.: Securing Android-powered mobile devices using SELinux. IEEE Secur. Priv. **8**(3), 36–44 (2010)
7. Lange, M., Liebergeld, S., Lackorzynski, A., et al.: L4Android: a generic operating system framework for secure smartphones. In: Proceedings of the 1st ACM Workshop on Security and Privacy in Smartphones and Mobile Devices, pp. 39–50. ACM (2011)
8. Barr, K., Bungale, P., Deasy, S., et al.: The VMware mobile virtualization platform: is that a hypervisor in your pocket? ACM SIGOPS Oper. Syst. Rev. **44**(4), 124–135 (2010)
9. Barham, P., Dragovic, B., Fraser, K., et al.: Xen and the art of virtualization. ACM SIGOPS Oper. Syst. Rev. **37**(5), 164–177 (2003)
10. Daemen, J., Rijmen, V.: The Design of Rijndael: AES-the Advanced Encryption Standard. Springer, Berlin (2002)

Predicting Protein-Protein Interactions with Weighted PSSM Histogram and Random Forests

Zhi-Sen Wei[1], Jing-Yu Yang[1], and Dong-Jun Yu[1,2(✉)]

[1] School of Computer Science and Engineering, Nanjing University of Science and Technology, Xiaolingwei 200, Nanjing 210094, China

[2] Changshu Institute, Nanjing University of Science and Technology, Changshu 215513, People's Republic of China

njyudj@njust.edu.cn

Abstract. The prediction of protein-protein interactions is one of the most important and challenging problems in computational biology. Because position specific scoring matrix (PSSM) encodes the evolutionary conservation information of a protein, the PSSM-derived features have been widely used to predict protein-protein interaction residues in previous studies. In this paper, we developed a novel method to extract feature, called weighted PSSM histogram, from the PSSM of a protein by introducing the concept of histogram in digital image processing field. Based on the extracted weighted PSSM histogram and several traditional features, we trained a random forests prediction model. Experiment results on benchmark datasets demonstrated the efficacy of the proposed method.

Keywords: Protein-protein interaction · Weighted PSSM histogram · Random forests

1 Introduction

Proteins play important roles in all biological systems. They interact with each other to perform various biochemical activities that are essential to life. To understand the mechanisms of various biochemical activities, it is necessary to study the process of protein-protein interactions.

For its potential value in biological research and medical treatment, prediction of protein-protein interaction sites has been an object of intense research in computational biology. A number of papers proposing various methods have been published. Among them, predicting protein-protein interaction residues based on sequence information is studied more and more. Porollo and Meller [1] applied SVM and neural network (NN) to identify protein-protein interaction sites with relative solvent accessibility (RSA). Their results revealed that RSA is more discriminating than other considered features. Murakami and Mizuguchi [2] applied kernel density estimation method to feature vectors combining position specific scoring matrix (PSSM) and Predicted accessibility (PA) calculated from protein sequences, followed by training a naïve Bayesian classifier. Dhole et al. combined position specific scoring matrix (PSSM), predicted relative solvent accessibility (PRSA) and averaged cumulative hydropathy (ACH) to form

X. He et al. (Eds.): IScIDE 2015, part II, LNCS 9243, pp. 326–335, 2015.
DOI: 10.1007/978-3-319-23862-3_32

a 186-D feature vector, based on which a l1-regularized logistic regression predictor [3] and a artificial neural networks predictor [4] were developed to perform protein-protein interaction residues prediction.

Sequence-based prediction of protein-protein interaction residues is a challenge problem. Despite that much progress in this field has been achieved, this problem is yet to be solved. In this paper, a novel PSSM-derived feature is proposed and a random forests predictor is trained for protein-protein interaction residues prediction.

2 Materials and Methods

Our method is to train a random forests (RF) predictor based on a new designed feature combined with two previous used features. The study is composed of the training phase and the testing phase. In the training phase, a random forests predictor was trained using the features calculated from the training dataset. We searched the optimal parameters for random forests through the leaving-one-out crossing-validation. In the testing phase, the trained random forests predictor was used to predict samples with the features calculated from two independent testing datasets. The details of benchmark datasets, feature representation and random forests are described in the following sections.

2.1 Benchmark Datasets

In this study, three datasets developed to evaluate the performance of protein-protein interaction residues predictors were utilized to demonstrate the efficacy of the proposed method, and to compare it with recently released state of art protein-protein interaction predictors. These datasets has been utilized by several recently published literatures [2–4]. One of them is used as training dataset, and the other two are the independent testing datasets for evaluation.

The training dataset Dset186 was developed by Murakami and Mizuguchi [2]. Dset186 selected 186 heterodimeric, non-transmembrane, transient protein sequences from Protein Data Bank (PDB) [5]. The 186 sequences is non-redundant with sequence identity <25 %. These chains were found by filtering with structures solved by X-ray crystallography with a resolution of ≤3.0 Å. A residue was defined as an interacting one that lost absolute solvent accessibility of <1.0 $Å^2$ on complex formation.

The independent testing datasets are expected to evaluate the generalization of the proposed method. In this study, two datasets were utilized. The first one, denoted as Dtestset72 [2], was part of the protein–protein docking benchmark set version 3.0 [6], in which any sequences showing ≥25 % sequence identity over a 90 % overlap with any of the sequences in Dset186 were removed by using BLASTClust [7]. After the removing procedure, 72 protein sequences from 36 protein complexes, which are non-overlapping with sequences in Dset186, were obtained as Dtestset72.

The second testing dataset, denoted as PDBtestset164, was recently released by Singh et al. [4]. Because the number of proteins deposited in PDB [5] increases over time, there has been a large number of proteins newly added to PDB since the construction of Dset186 and Dtestset72. Singh et al. [4] extracted non-redundant 164 protein sequences from these newly annotated proteins (from June 2010 to November 2013) in the PDB with the same filters that have been used to create Dset186 and Dtestset72. Then, the interaction residues was identified by using Software PSAIA [8].

2.2 Feature Representation

An effective feature representation of protein sequence properties is crucial for the prediction of the protein-protein interaction residues using machine learning. Three types of features that consider evolutionary conservation, hydropathy and predicted structural information of proteins have been demonstrated to be useful for predicting the protein-protein interaction sites [3]. They are position specific scoring matrix (PSSM), averaged cumulative hydropathy (ACH) [9], and predicted relative solvent accessibility (PRSA). Here, we developed a novel PSSM-derived feature, called weighted PSSM histogram and denoted as wHPSSM. The wHPSSM, ACH and PRSA are combined as the feature representations of samples for our method.

2.2.1 Weighted PSSM Histogram (WHPSSM)

Position specific scoring matrix (PSSM) represents the evolutionary conservation information of a protein. PSSM-derived features have been found to be useful in different protein related prediction problems [10, 11] including the protein-protein interactions prediction [2–4].

In this paper, we generate the PSSM of a protein sequence using the tool BLAST+ [12] to search the NCBI non-redundant protein sequence database through three interactions with 0.001 as E-value cutoff for multiple sequence alignment against the query sequence. Furthermore, we normalize each element in the obtained PSSM to range [0, 1] with the logistic function $f(x) = 1/(1 + e^{-x})$, where x is the original value in PSSM.

According to previous methods [2–4], for a query residue in a given protein sequence, the normalized PSSM elements of the sequence corresponding to a sequence segment centered on the residue were concatenated. Instead of that, we propose a novel feature represent for utilizing PSSM. We calculated a weighted PSSM histogram (wHPSSM) corresponding to a sequence segment centered on the query residue. Specifically, given the normalized PSSM of a query protein sequence, the wHPSSM of the residue in position i is a 20D histogram defined as

$$H_{ij} = \sum_{t=i-(N-1)/2}^{i+(N-1)/2} W_t PSSM_{tj} \quad j = 1 \cdots 20 \tag{1}$$

Where W is a weighting vector and W_t is the weight for the residue in position t, $PSSM_{tj}$ is the j-th elements of the normalized PSSM corresponding to the residue in position t, and N is the segment size. In this study, N is set to 9 since previous studies [2] have found that a nine-residue segment size would be optimal for protein-protein interaction prediction. The calculating process of wHPSSM is shown in Fig. 1.

2.2.2 Averaged Cumulative Hydropathy (ACH)

The hydropathy index, proposed by Kyte and Doolittle [13], represents the hydrophobic or hydrophilic properties of the side chain of a residue. It is a scalar quantity, for which,

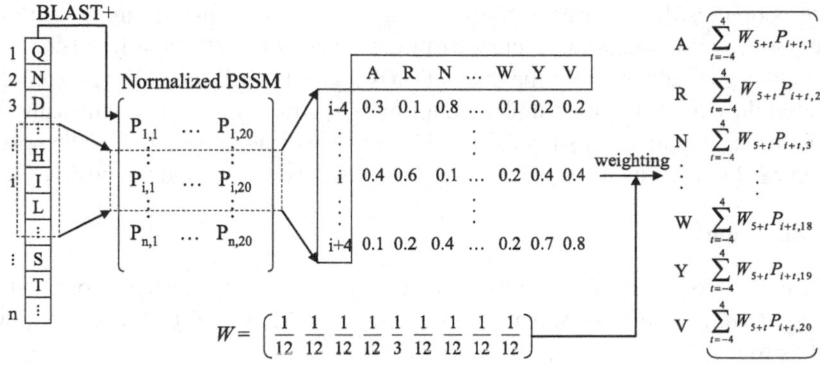

Fig. 1. The calculation of wHPSSM for the residue in position *i*.

larger value stands for higher hydrophobic property, while smaller value means higher hydrophilic property.

In this study, averaged cumulative hydropathy (ACH) feature is used, which have been used to predict catalytic residues [9]. We calculated hydropathy indices of a residue and its neighborhood to extract its averaged cumulative hydropathy (ACH) feature. For a query residue in a given protein sequence, its ACH feature is the concatenation of the averaged cumulative hydropathy indices over a window size varying between 1, 3, 5, 7 and 9 centered on the residue. In this study, the ACH feature of a residue, which is a 5-D vector, was calculated by using the Python codes provided by Dhole et al. [3].

2.2.3 Predicted Relative Solvent Accessibility (PRSA)

The solvent accessibility, introduced by Lee and Richards [14], is found to be closely related to the spatial structure of a protein, i.e. the spatial arrangement, the packing of residues on protein folding, which influence the characteristics of the protein–protein interactions [14]. Considering this, we took the predicted relative solvent accessibility (PRSA) of residues as one of our features. The PRSA features of residues in a protein sequence were calculated by feeding the sequence to the online server, i.e., SANN [15], which is freely available at http://lee.kias.re.kr/~newton/sann/. In this study, we just utilized the continuous value of the solvent accessibility predicted by SANN.

In all, a residue is represented as a 26-D vector by serially combining its PSSM feature (20-D), ACH feature (5-D), and PRSA feature (1-D).

2.3 Random Forests Classifier

Random forests [16] are ensembles of decision trees. Each tree is independently grown based on bootstrap samples from the training data. When growing a tree, each split in the tree is determined based on a small random selected subset of the feature space. Moreover, different from the training of general decision trees that needs post-pruning, the trees do not need post-pruning in random forests since random sample guarantees no over-fitting. Random forests aggregate over the predictions of all trees as the final prediction.

In this study, because the training data is unbalanced in which the number of negative samples (the non-interaction residues) is more than 5 times that of positive samples (the interaction residues), directly training a random forest using all samples will tend to focus more on the prediction accuracy of the majority class (the non-interaction one) thus results in poor accuracy for the interaction residues [17]. To deal with this problem, we utilized the commonly used under-sampling [18] strategy. We randomly sampled a subset from the negative set such that the size of the subset equals that of the positive set. Then random forests were trained using the negative subset and the positive set.

2.4 Evaluation

Six routinely used measures were explored to assess the performance of our method, including *Recall*, *Precision*, *Specificity*, *Accuracy*, *MCC*, and *F-measure*. They were defined as follows,

$$Recall = TP/(TP + FN) \tag{2}$$

$$Precision = TP/(TP + FP) \tag{3}$$

$$Specificity = TN/(TN + FP) \tag{4}$$

$$Accuracy = (TP + TN)/(TP + FN + TN + FP) \tag{5}$$

$$MCC = ((TP \times TN) - (FP \times FN))/\sqrt{(TP + FP) \times (TP + FN) \times (TN + FP) \times (TN + FN)} \tag{6}$$

$$F - measure = 2 \times (Recall \times Precision)/(Recall + Precision) \tag{7}$$

Where, *TP*: Residues correctly predicted as interacting, *FP*: Residues incorrectly predicted as interacting, *TN*: Residues correctly predicted as non-interacting, *FN*: Residues incorrectly predicted as non-interacting.

Among the six measures, *MCC* and *F-measure* reflect the overall performance. Matthews correlation coefficient (*MCC*) [19] is essentially a correlation coefficient between the observed and predicted classes of the samples. *F-measure* [20] is the weighted harmonic mean of *Recall* and *Precision*.

In this paper, leave one out crossing validation was adopted to assess the performance of our method on the training Dset186 and explore impact of different parameter values. One protein sequence was used as test data. The remaining 185 sequences were used to train a model. Then, of this model on the test data, the six measures mentioned above were calculated. This process was repeated 186 times with distinct sequences as test data. Averaging measures over 186 iterations gave an assessment to the model with specific parameters. Different parameter configurations were explored and the optimal setting was achieved to maximize *MCC*. In addition, the optimal threshold of the classifier was obtained with leave one out validation.

After optimal parameter configuration and threshold were found, a model was trained on all examples in Dset186. Dtestset72 and PDBtestset164 were used to independently test the performance of the model. The performance measures were computed for each protein sequence using the optimal threshold, and the average measures were exhibited over all sequences in Dtestset72 and PDBtestset164 respectively. The averages within each subcategory of Dtestset72 were also calculated.

Table 1. Performance comparisons between the proposed RF^{hist} and $RF^{non\text{-}hist}$ on dataset Dset186 over leave-one-out cross-validation *.

Method	MCC	Precision %	Recall %	Specificity %	Accuracy %	F-measure %
RF^{hist}	0.224(0.005)	29.8(0.2)	60.8(0.6)	65.4 (0.3)	64.6 (0.3)	37.7 (0.3)
$RF^{non\text{-}hist}$	0.219(0.007)	29.1(1.4)	60.7(6.1)	64.5 (6.2)	63.7 (3.9)	37.0 (0.4)

*For each method, experiments were performed five times and the averaged results followed by standard deviations were reported. RF^{hist} is composed of 250 trees and $RF^{non\text{-}hist}$ consists of 400 trees

3 Result and Discussion

3.1 The WHPSSM Feature Is Effective

In this section, we will experimentally show the efficacy of the proposed wHPSSM feature. The performance on the training dataset Dset186 over leave-one-out cross-validation will be explored.

The input of a random forest is a 26-D feature vector obtained by serially concatenating a residue's wHPSSM feature (20-D), ACH feature (5-D), and PRSA feature (1-D). Considering the prediction for a residue, intuitionally the PSSM elements related to the residue itself should make more important influence than the ones related to its neighbors. So we explore several weighting schemes for wHPSSM, finding that setting the weighting vector W to $\left[\frac{1}{12}\ \frac{1}{12}\ \frac{1}{12}\ \frac{1}{12}\ \frac{1}{3}\ \frac{1}{12}\ \frac{1}{12}\ \frac{1}{12}\ \frac{1}{12}\right]^T$ is feasible for this study. The parameters of random forest are optimized through leave one out crossing validation (refer to Sect. 2.4).

Since previous studies [2–4] have constructed a non histogram type of PSSM-derived feature (denoted as non-hist-PSSM, refer to Sect. 2.2.1) and have demonstrated its efficacy, we compare the performance between our method and a random forest trained with a 186-D feature vector (combining non-hist-PSSM, ACH and PRSA as [3]). The difference between this method and our method is only in the different PSSM-derived features. The parameters of this method are also optimized through leave one out crossing validation (refer to Sect. 2.4).

Table 1 presents the performance comparison between the proposed method (RF^{hist}) and random forest with non-hist-PSSM ($RF^{non\text{-}hist}$) on dataset Dset186 over leave-one-out cross-validation. As shown in the table, RF^{hist} performs slightly better but not significantly than $RF^{non\text{-}hist}$. It proofs that the proposed wHPSSM is as effective as the existing non-histogram type of PSSM-derived feature. Furthermore, wHPSSM is only of 20-D length, while the non-histogram type of PSSM-derived feature is a 180-D vector. As shown in Fig. 2, RF^{hist} only needs 250 trees to achieve best performance, while $RF^{non\text{-}hist}$ needs 400 trees. These make our method more efficient.

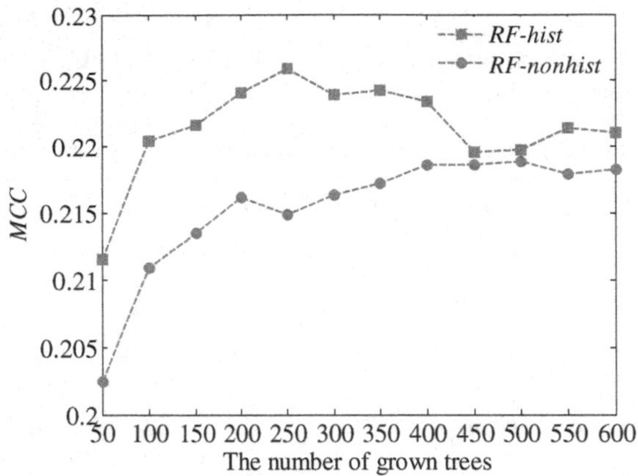

Fig. 2. The curves of *MCC* along with the number of trees to be grown for RF^{hist} and $RF^{non\text{-}hist}$. For each setting of the number of trees, the leave one out crossing validation was performed five times and the averaged *MCC* was calculated.

3.2 Comparisons with Existing Methods

To demonstrate the efficacy of the proposed method (RF^{hist}), we compare RF^{hist} with three recent reported state of the art methods [2–4] on the training dataset and two independent testing dataset.

3.2.1 Crossing Validation Comparisons on Dset186

In this section, we explored the performance on Dset186 over leave-one-out cross-validation. The performances of LORIS [3] and PSIVER [2] are gained from their published papers. SPRINGS [4] is not presented because its author did not report the performance in detail. The comparisons are listed in Table 2. From the table, we can find that RF^{hist} achieves an overall performance close to LORIS [3] and outperforms PSIVER [2]. The standard deviations of all measures of RF^{hist} are observably lower than that of LORIS, demonstrating that RF^{hist} performs more robustly than LORIS.

Table 2. Performance comparisons between RF^{hist} and existing methods on Dset186 over leave-one-out cross-validation*

Method	MCC	Precision %	Recall %	Specificity %	Accuracy %	F-measure %
RF^{hist}	0.224 (0.005)	29.8 (0.2)	60.8 (0.6)	65.4 (0.3)	64.6 (0.3)	37.7 (0.3)
LORIS [3]	0.221 (0.025)	28.7 (2.7)	69.8 (6.5)	58.6 (9.4)	60.4 (6.3)	38.4 (1.4)
PSIVER[2]	0.151	30.6	41.6	74.3	67.3	35.3

*For RF^{hist}, experiments were performed five times and the averaged results followed by standard deviations were reported. For LORIS, experiments were performed three times and the averaged results followed by standard deviations were reported. The performances of LORIS and PSIVER are based on previous studies [2, 3]

3.2.2 Comparisons on Two Independent Test Datasets

To compare the generalization ability of the proposed method with that of other methods, we made an evaluation on two independent test datasets, i.e. Dtestset72 and PDBtestset164. The results are listed in Tables 3 and 4, respectively.

To explore the relation between performance and conformation changes, Dtestset72 was categorized to three subcategories [2], i.e. rigid body (27 complexes), medium cases (6 complexes) and difficult cases (3 complexes). The performances on these subcategories and the whole dataset were respectively explored. The comparisons between RF^{hist} and other methods are listed in Table 3.

Table 3. Performance comparisons on Dtestset72*

Method	MCC	Precision %	Recall %	Specificity %	Accuracy %	F-measure %
Rigid body (27)						
RF^{hist}	0.191	24.3	62.9	62.9	63.0	32.5
LORIS [3]	0.175	23.2	63.8	60.3	60.9	32.0
SPRINGS [4]	0.167	23.5	59.2	62.5	62.1	31.3
PSIVER [2]	0.127	23.9	46.5	68.8	65.5	27.3
Medium cases (6)						
RF^{hist}	0.219	26.6	64.5	65.1	64.8	34.2
LORIS	0.187	25.0	60.9	63.4	63.3	32.9
SPRINGS	0.197	26.2	59.1	65.6	64.9	33.7
PSIVER	0.171	28.9	43.5	75.3	70.2	27.1
Difficult cases (3)						
RF^{hist}	0.226	31.2	56.9	71.9	68.5	38.4
LORIS	0.174	26.5	61.1	62.7	61.8	35.5
SPRINGS	0.143	24.9	57.7	62.3	60.3	32.8
PSIVER	0.139	26.9	53.2	61.9	62.8	33.2
Overall average performance (72)						
RF^{hist}	0.198	25.3	62.7	64.1	63.8	33.3
LORIS	0.177	23.8	63.1	61.0	61.4	32.4
SPRINGS	0.170	24.1	59.0	63.0	62.4	31.8
PSIVER	0.135	25.0	46.5	69.3	66.1	27.8

*The performances of methods (other than RF^{hist}) are sourced from previous studies [2–4]

Table 4. Performance comparisons on PDBtestset164*

Method	MCC	Precision %	Recall %	Specificity %	Accuracy %	F-measure %
RF^{hist}	0.134	31.1	52.7	63.6	60.4	35.8
LORIS [3]	0.111	26.3	53.8	60.9	58.8	32.3
SPRINGS [4]	0.108	26.8	40.7	64.8	60.6	31.1
PSIVER [2]	0.078	25.3	46.4	63.4	59.6	29.5

*The performances of methods (other than RF^{hist}) are sourced from previous studies [3, 4]

As shown in Table 3, the proposed method outperformed all other three methods on the whole dataset and each of three subcategories. Specially, overall averaged improvements of 2.1 % and 0.9 % on *MCC* and *F-measure*, respectively were obtained compared with the second-best predictor (LORIS), which is the most recently released sequence-based PPI predictor.

From Table 4, we can find that the proposed method obtained more significant improvements on PDBtestset164. The proposed method also performed best among all mentioned methods. Compared with LORIS, the proposed method also made improvements of 2.3 % on *MCC* and 3.5 % on *F-measure*.

The results listed in Tables 3 and 4 demonstrate that the generalization capability of the proposed predictor is better than that of existing sequence-based PPI predictors.

4 Conclusion

In this study, we have proposed a PSSM-derived feature and have trained a random forests predictor for protein-protein interaction residues prediction. Experiment results demonstrate the efficacy of the proposed method. Though improvement is achieved, predicting protein-protein interaction residues is still a challenging problem and need more efforts, among which looking for a more effective feature representation will be one of the most important work.

Acknowledgements. The authors would like to thank the anonymous reviewers for suggestions and comments which helped improve the quality of this paper. This work was supported by the National Natural Science Foundation of China (No. 61373062 and 61233011), the Natural Science Foundation of Jiangsu (No. BK20141403), the Jiangsu Postdoctoral Science Foundation (No. 1201027C), and the China Postdoctoral Science Foundation (No. 2013M530260, 2014T70526).

References

1. Porollo, A., Meller, J.: Prediction-based fingerprints of protein–protein interactions. Proteins Struct. Funct. Bioinf. **66**, 630–645 (2007)
2. Murakami, Y., Mizuguchi, K.: Applying the Naïve Bayes classifier with kernel density estimation to the prediction of protein–protein interaction sites. Bioinformatics **26**, 1841–1848 (2010)
3. Dhole, K., Singh, G., Pai, P.P., Mondal, S.: Sequence-based prediction of protein–protein interaction sites with L1-logreg classifier. J. Theor. Biol. **348**, 47–54 (2014)
4. Singh, G., Dhole, K., Pai, P.P., Mondal, S.: SPRINGS: prediction of protein-protein interaction sites using artificial neural networks. PeerJ PrePrints **1**, 7 (2014)
5. Berman, H.M., Westbrook, J., Feng, Z., Gilliland, G., Bhat, T., Weissig, H., Shindyalov, I.N., Bourne, P.E.: The protein data bank. Nucleic Acids Res. **28**, 235–242 (2000)
6. Hwang, H., Pierce, B., Mintseris, J., Janin, J., Weng, Z.: Protein–protein docking benchmark version 3.0. Proteins Struct. Funct. Bioinf. **73**, 705–709 (2008)
7. Altschul, S.F., Madden, T.L., Schäffer, A.A., Zhang, J., Zhang, Z., Miller, W., Lipman, D.J.: Gapped BLAST and PSI-BLAST: a new generation of protein database search programs. Nucleic Acids Res. **25**, 3389–3402 (1997)
8. Mihel, J., Šikić, M., Tomić, S., Jeren, B., Vlahoviček, K.: PSAIA–protein structure and interaction analyzer. BMC Struct. Biol. **8**, 21 (2008)
9. Li, B.-Q., Feng, K.-Y., Chen, L., Huang, T., Cai, Y.-D.: Prediction of protein-protein interaction sites by random forest algorithm with mRMR and IFS. PLoS ONE **7**, e43927 (2012)
10. Yu, D., Hu, J., Yang, J., Shen, H., Tang, J.: Designing template-free predictor for targeting protein-ligand binding sites with classifier ensemble and spatial clustering. IEEE/ACM Trans. Comput. Biol. Bioinf. **10**, 15 (2013)
11. Yu, D.J., Hu, J., Huang, Y., Shen, H.B., Qi, Y., Tang, Z.M., Yang, J.Y.: TargetATPsite: a template-free method for ATP-binding sites prediction with residue evolution image sparse representation and classifier ensemble. J. Comput. Chem. **34**, 974–985 (2013)
12. Camacho, C., Coulouris, G., Avagyan, V., Ma, N., Papadopoulos, J., Bealer, K., Madden, T.L.: BLAST+: architecture and applications. BMC Bioinf. **10**, 421 (2009)
13. Kyte, J., Doolittle, R.F.: A simple method for displaying the hydropathic character of a protein. J. Mol. Biol. **157**, 105–132 (1982)
14. Lee, B., Richards, F.M.: The interpretation of protein structures: estimation of static accessibility. J. Mol. Biol. **55**, 379–IN4 (1971)
15. Joo, K., Lee, S.J., Lee, J.: Sann: solvent accessibility prediction of proteins by nearest neighbor method. Proteins Struct. Funct. Bioinf. **80**, 1791–1797 (2012)
16. Breiman, L.: Random forests. Mach. Learn. **45**, 5–32 (2001)
17. Gallet, X., Charloteaux, B., Thomas, A., Brasseur, R.: A fast method to predict protein interaction sites from sequences. J. Mol. Biol. **302**, 917–926 (2000)
18. He, H., Garcia, E.A.: Learning from imbalanced data. IEEE Trans. Knowl. Data Eng. **21**, 1263–1284 (2009)
19. Matthews, B.W.: Comparison of the predicted and observed secondary structure of T4 phage lysozyme. Biochim. Biophys. Acta (BBA)-Protein Struct. **405**, 442–451 (1975)
20. Hripcsak, G., Rothschild, A.S.: Agreement, the f-measure, and reliability in information retrieval. J. Am. Med. Inf. Assoc. **12**, 296–298 (2005)

Research on Multiple Files Input Programming Method Based on MapReduce

Jing Zhang[1(✉)], Xiaoyuan Li[2], Yanmei Huo[1], and Siqi Li[3]

[1] College of Computer Science and Technology, Jilin University, Changchun, China
583505399@qq.com, huoym@jlu.edu.cn

[2] 65 Southampton Avenue, Berkeley, CA 94707, USA
Xiaoyuanli66@gmail.com

[3] College of Economics and Management, Yanbian University, Yanbian, China
741316485@qq.com

Abstract. Hadoop is a software framework that allows for the distributed processing of large data sets across clusters of computers using simple programming models. Hadoop is widely used due to its scalability, high reliability, low-cost, high efficiency and so on. Hadoop Distributed File System (HDFS) and MapReduce programming model are respectively used for storing and processing of the data. This paper firstly carries depth research and detailed introduction on HDFS and MapReduce, then proposing a programming method that can sort output according to the order input. Experimental results demonstrate that the proposed method is feasible and effective. Through such output, it will bring great convenience in data processing on the later work.

Keywords: Hadoop · Mapreduce · Multiple files · Programming method

1 Introduction

Hadoop is an open-source implementation of Google's Map/Reduce framework to support distributed applications. It has been widely used not only by the Internet but also received widespread attention on the other industries such as electric power industry, intelligent transportation, judicial system etc. Hadoop mainly consists of two core components: MapReduce framework and the Hadoop Distributed File System (HDFS, renamed by NDFS) [1]. HDFS is a distributed file system with high reliability and high scalability, it can provide storage capacity of mass files across multiple machines [2]. MapReduce is a programming model of parallel processing for massive data sets [3]. MapReduce has become the most popular framework for processing and analyzing of large-scale data, mainly because of its simplicity, highly effective and fault tolerant for large-scale data analysis, parallel programming and so on [4–6]. MapReduce often fails to exhibit acceptable performance for various processing tasks. Quite often this is a result of weaknesses related to the nature of MapReduce or the applications and use-cases it was originally designed for [7]. There are many limitations of the processing model adopted in MapReduce. Hadoop is a software framework that developed for distributed processing of large data sets, and it is works well with large

X. He et al. (Eds.): IScIDE 2015, Part II, LNCS 9243, pp. 336–342, 2015.
DOI: 10.1007/978-3-319-23862-3_33

files. There's a doubt whether it also works well with small files [8]. However many articles have reported the study on the performance of handling small files [9–11]. They respectively prove the improvement of performance by merging small files, indexing, compressing the block and using the different input format. Seen, there is no problem in dealing with a large number of small files using Hadoop. In this paper, we do not discuss the performance but to solve the problem when handing multiple files. This paper mainly proposes a programming method based on MapReduce programming model that can allows the order output in accordance with the order input when processing large numbers of small files. It has been proved in the experiment that the method is feasible and effective. And currently, there is no article that having been referred to such issues.

The rest of this article is organized as follows: first, we provide an overview of Hadoop mainly focusing on HDFS and MapReduce. Then, an outline of our method to handle multiple files input. Finally, we show the process and results of experiments and conclude our main works.

2 Hadoop Core Components

2.1 Distributed File System – HDFS

HDFS is highly scalable and distributed and is designed to run on commodity hardware. Rather than other existing distributed file systems, HDFS is designed for storing large files and has a high degree of fault-tolerance and is suitable for applications of large datasets, such as machine learning, data mining, etc. HDFS firstly cut the file into slices, and stored on three nodes in the form of blocks (a block is 64 M). And then it adds a node to record the mapping relationship between block and each node, this process is transparent for users. It is an easily scalable distributed file system, by adding nodes, it can apply on a large number of ordinary low-cost machines, providing fault tolerance mechanism and providing file access services of well performance to large numbers of users.

HDFS has the master/slave architecture with a single NameNode and numbers of DataNodes. NameNode manages the file system namespace and the metadata of the file system. DataNodes store the actual data and provide block storage and serve I/O requests from clients. Client accesses the file system through the interaction between DataNodes and NameNode. Client Contacts NameNode to obtain metadata of file, but the real file I/O operations directly interact with DataNodes.

2.2 Distributed Computing Framework - Mapreduce

MapReduce framework architecture provides a parallel processing to the massive data. In the case of unfamiliar with distributed parallel programming, it is convenient for programmers to run their own process on distributed systems. MapReduce programs are easy to parallelize, thus putting very large-scale data analysis into the hands of anyone with enough machines at their disposal. Its implementation is to split parallel execution into two phases: Map assemble one set of a set of key/value pairs, and returns a new set of key/value pairs. Reduce ensures that all key/value pairs in the mapping for each of the group to share the same key. MapReduce is useful in applications like log analysis, machine learning, document clusterings and so on.

It is still a Master/slave structure, there are two main services: JobTracker and TaskTracker. The JobTracker responses to client's commands, and monitors the running state of TaskTracker, MapTask and ReduceTask. TaskTracker is installed on each node, receiving the JobTracker's command, and then executes MapTask or ReduceTask. After startup, TaskTrackers report the status of these tasks to JobTracker through the heartbeat mechanism. JobTracker initializes job, manage all job and decomposes the job into a set of tasks, communicate with TaskTrackers, assigns tasks to TaskTracker, coordinates the entire job, monitor job/task, handles errors etc. However, JobTracker exists is single point of failure, once occurs, the whole cluster is unavailable. Fortunately, this problem has been resolved in the second generation of Hadoop [14].

3 Programming Model

Dean and Ghemawat [3] described the MapReduce programming model as follows:

The computation takes a set of input key/value pairs, and produces a set of output key/value pairs. The user of the MapReduce library expresses the computation as two functions: Map and Reduce. The Map turns the input key/value pairs into a set of intermediate key/value pairs. The Reduce produces a set of intermediate values which share a key to a smaller set of values. The MapReduce programming model is shown in Fig. 1.

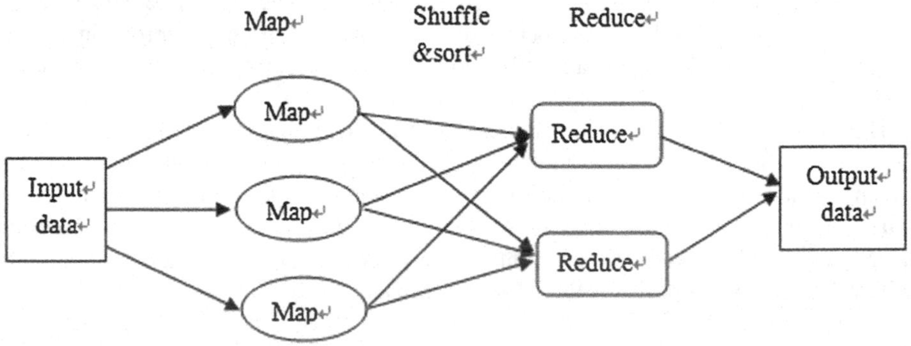

Fig. 1. MapReduce programming model

The input data is firstly splitted into a number of small slices, and each slice is handled by a Map. After the Map's processing, then divide it into partitions by the same key. Each reduce will receive a partition to process. Finally merge each reduce output into one output.

However, the MapReduce output is not in accordance with the specified order input when handling of multiple files. Thus, we design a programming method that can make the output according to the order input.

First, acquire the content of each file we want to handle:

```
String content = value.toString();
```

Then, the following code is how to achieve the file name:

```
FileSplit fileSplit = (FileSplit)context.getInputSplit();
int filename = Integer.parseInt(fileSplit.getPath().getName
());
```

Using the properties of MapReduce sort by the file name, the result of MapReduce output actually is sorted according to the key [12]. Convert the filename from "String" to "int", the output can be sorted by the order of numbers rather than dictionary. So output the content with the file name as the key:

```
Context.write(new IntWritable (filename), new Text
(content))
```

Finally, delete the file name and output the file contents.
In this program, we use Reduce(): reduce (IntWritable key, Iterable <Text> values, Context context) to iteratively output thc values.

```
String valueString = new String();
For(Text value : values)
{
  valueString += value.toString() + "\r\n";
}
```

4 Experiments

We set up Hadoop with stand-alone pseudo-distributed mode on VMware, and its experimental environment is shown in Table 1. The Hadoop version is 1.0.3. We use the data is the scores of Electronic and Information Engineering major of College of Electronic Science and Engineering of Jilin University from 2002–2010 (except 2005).

Table 1. Experimental environment

Hardware environment	Hadoop experimental environment
CPU: Intel Pentium CPU	RAM: 1 GB
RAM: 4G	System: Ubuntu12.04
System: windows 7 32 bit	Disk: 20 GB

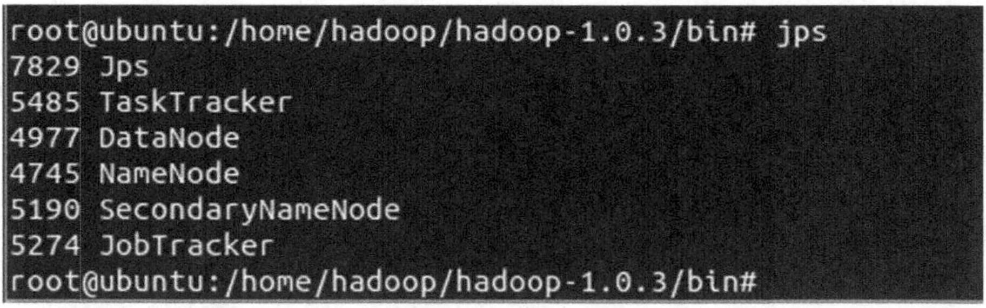

```
root@ubuntu:/home/hadoop/hadoop-1.0.3/bin# jps
7829 Jps
5485 TaskTracker
4977 DataNode
4745 NameNode
5190 SecondaryNameNode
5274 JobTracker
root@ubuntu:/home/hadoop/hadoop-1.0.3/bin#
```

Fig. 2. Hadoop daemons

Experimental Setup [13]:

(1) First to install Java 1.6 or higher versions, we install Java 1.7.0.
(2) ssh must be installed and check that you can ssh to the localhost without a passphrase. "sshd" must be running to use the Hadoop scripts that manage remote Hadoop daemons.

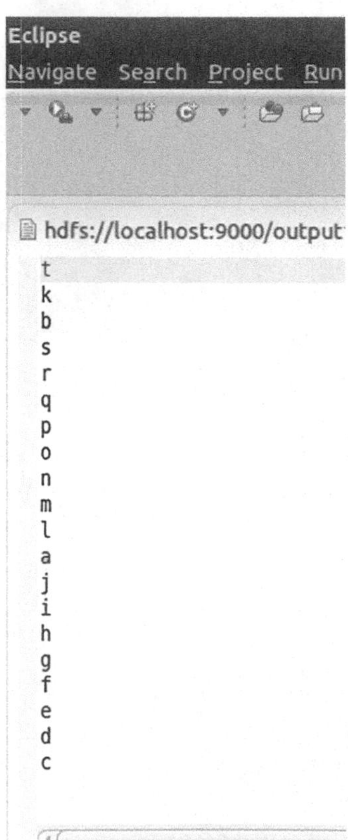

Fig. 3. The result of common program that outputting the contents of the 20 files.

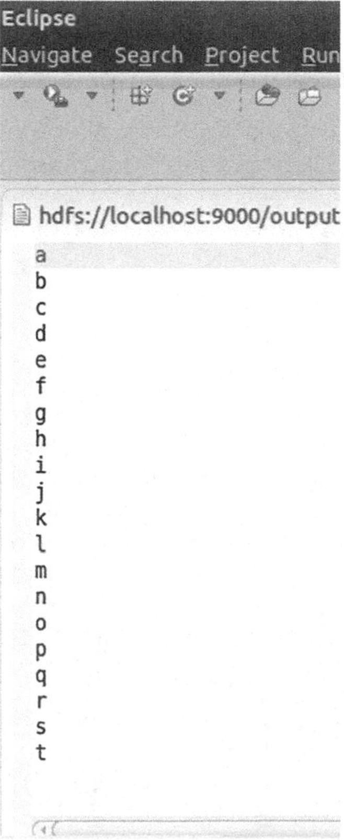

Fig. 4. The result of the program using our method

(3) Unpack the downloaded Hadoop distribution. In the distribution, edit the file conf/hadoop-env.sh to define at least JAVA_HOME to be the root of your Java installation and core-site.xml, hdfs-site.xml, mapred-site.xml.
(4) Initialization: after start the pseudo distributed servers, format namenode(./hadoop fs namenode -format) and start Hadoop(./start-all.sh), and then close security mode of the distributed file system(./hadoop dfsadmin –safemode leave), allowing other servers to access. Then we can see the daemons in Fig. 2.

Since we want to demonstrate the feasibility of this approach and not to study the performance issue, so we randomly create 20 files named from 1 to 20, and the content of the file named by "1" is a, the content of the file named by "2" is b, and so on, for the remaining files . By"./hadoop fs -copyFromLocal/home/q/*/input", copy data files to the HDFS. We use Eclipse to run mapreduce program.

Experimental Results:
Figure 3 is the result of common program that outputting the contents of the 20 files. Figure 4 is the result of the program using our method, we can see that the content output according to the order input.

Since our input file name is in order, then put the file name in the output and finally delete the file name, so the output is in order according to the input. Currently, we are studding on the influence on students' performance by student' scores, and we want to filter out the scores of freshman in each year sequentially using the method. The result is shown in Fig. 5, start from the second column, every column represents the scores of the top 20 of freshman from 2002 to 2010 (except 2005) year. Once again, it proves that this method is effective in such applications.

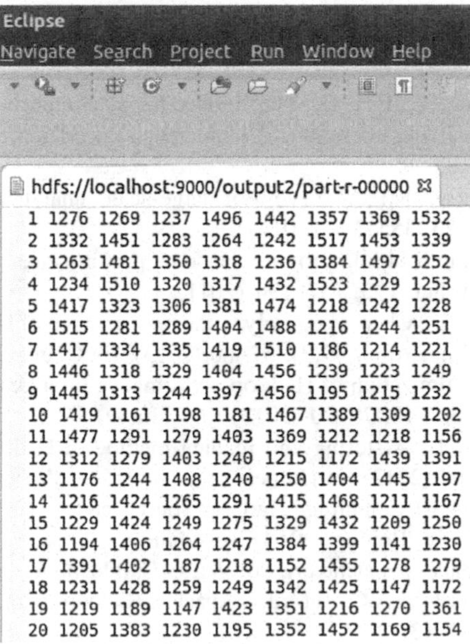


```
1 1276 1269 1237 1496 1442 1357 1369 1532
2 1332 1451 1283 1264 1242 1517 1453 1339
3 1263 1481 1350 1318 1236 1384 1497 1252
4 1234 1510 1320 1317 1432 1523 1229 1253
5 1417 1323 1306 1381 1474 1218 1242 1228
6 1515 1281 1289 1404 1488 1216 1244 1251
7 1417 1334 1335 1419 1510 1186 1214 1221
8 1446 1318 1329 1404 1456 1239 1223 1249
9 1445 1313 1244 1397 1456 1195 1215 1232
10 1419 1161 1198 1181 1467 1389 1309 1202
11 1477 1291 1279 1403 1369 1212 1218 1156
12 1312 1279 1403 1240 1215 1172 1439 1391
13 1176 1244 1408 1240 1250 1404 1445 1197
14 1216 1424 1265 1291 1415 1468 1211 1167
15 1229 1424 1249 1275 1329 1432 1209 1250
16 1194 1406 1264 1247 1384 1399 1141 1230
17 1391 1402 1187 1218 1152 1455 1278 1279
18 1231 1428 1259 1249 1342 1425 1147 1172
19 1219 1189 1147 1423 1351 1216 1270 1361
20 1205 1383 1230 1195 1352 1452 1169 1154
```

Fig. 5. Students' scores output according to the order input.

5 Conclusion

Hadoop has become a popular data processing platform with the data growing at an unprecedented rate. And MapReduce has been adopted by more and more programmers because of its easy programming model and fast processing speed. In this paper, we firstly introduce the frameworks of MapReduce and HDFS in detail. Then we carry depth research on MapReduce programming model and propose a programming method that can allows the order of output contents in accordance with the order of input files when processing large numbers of multiple files. Finally, it has been proved in the experiment that the method is feasible and effective, meanwhile it brings great convenience for the follow-up work of our application. However, we take two programs to complete the function. The future work is supposed to be more focused on optimizing the program, and to attempt to change the framework of MapReduce to implement the function.

References

1. White, T.: Hadoop: The Definitive Guide. Tsinghua university Publications, Beijing (2012). (in Chinese)
2. Borthakur, D.: The hadoop distributed file system: architecture and design. Hadoop Proj. Website **11**, 21 (2007)
3. Dean, J., Ghemawat, S.: MapReduce: simplified data processing on large clusters. Commun. ACM **51**(1), 107–113 (2008)
4. Dean, J., Ghemawat, S.: MapReduce: a flexible data processing tool. Commun. ACM **53**(1), 72–77 (2010)
5. Lammel, R.: Google's MapReduce programming model — revisited. Sci. Comput. Program. **70**(1), 1–30 (2008)
6. Ekanayake, J., Pallickara, S., Fox, G.: MapReduce for data intensive scientific analyses. In: Proceedings of Fourth IEEE International Conference on eScience, Indianapolis, Indiana, USA, pp. 277–284 (2008)
7. Doulkeridis, C., Nørvåg, K.: A survey of large-scale analytical query processing in MapReduce. VLDB J. **23**, 355–380 (2014)
8. Mohandas, N., Thampi, S.M.: Improving Hadoop performance in handling small files. Commun. Comput. Inf. Sci. **193**, 187–194 (2011)
9. Yuan, Y., Cui, C., Wu, Y., Chen, Z.: Analyze the performance of Hadoop small files under stand-alone. Comput. Eng. Appl. **49**(3), 57–60 (2013). (in Chinese)
10. Liu, X.: Research on optimization of Hadoop performance for a large number of small files. Comput. CD Softw. Appl. **18**, 78–80 (2013) (in Chinese)
11. Zhang, C., Rui, J., He, T.: A storing and reading method of Hadoop small files. Comput. Appl. Softw. **11**, 95–100 (2012) (in Chinese)
12. Zhang, Xin: Deep Cloud Computing. Hadoop Source code analysis, revised edn. China Railway Publishing House, Beijing (2014). (in Chinese)
13. Information on http://hadoop.apache.org/docs/r1.0.4/single_node_setup.html
14. Vavilapalli, V.K., Murthy, A.C., et al.: Apache Hadoop yarn: yet another resource negotiator. In: SOCC (2013)

Fuzzy C-means Based on Cooperative QPSO with Learning Behavior

Ping Lu[1(✉)], Husheng Dong[1,2], Huanhuan Zhai[2], and Shengrong Gong[2]

[1] Department of Information, Suzhou Institute of Trade and Commerce, Suzhou, China
{2218688723,85592637}@QQ.com

[2] School of Computer Science and Technology, Soochow University, Suzhou, China
{85592637,457607111}@QQ.com, shrgong@suda.edu.cn

Abstract. In this paper, we propose an improved fuzzy C-means clustering algorithm based on cooperative quantum-behaved particle swarm optimization with learning behavior. Though FCM is a widely used clustering method, it has the inherent limitation of being sensitive to initial value and prone to fall in local optimum. To address this problem, we utilize the widely used global searching algorithm—QPSO, and employ new strategies to enhance its performance. First, we use the cooperative evolution strategy to improve the global searching capacity. Second, for each particle, the behavior of learning from others is granted, which effectively boosts the local searching capability. Furthermore, a gene pool is constructed to share information among all subgroups periodically. Since the iteration process is replaced by the improved version of QPSO, FCM no longer depends on the initialization values. Our experiments show that the proposed algorithm outperforms FCM and its improved versions significantly. The convergence and clustering accuracy are both improved effectively.

Keywords: Fuzzy C-means · Clustering · Quantum-behaved particle swarm optimization · Cooperative evolution · Learning behavior

1 Introduction

Clustering analysis is an important technology used in many areas, such as data mining, machine learning and pattern recognition. Clustering is the process of grouping similar objects together, so that the similarities of the objects' properties in the same category are as large as possible, and vice versa. Fuzzy C-means (FCM) clustering algorithm employs the fuzzy set theory based on Hard C-Means and uses membership degree to determine the classification of each sample. It is the one of the most widely used fuzzy clustering algorithms. However, it has one similar shortcoming as K-means clustering [11] algorithm because they both use iterative gradient descent method to obtain the optimal solution. As a result, it is easy to fall in local optimum and sensitive to the clustering centers initially assigned. In order to get stable clustering results, some intelligent evolutionary algorithms are employed, such as genetic algorithm [1],

X. He et al. (Eds.): IScIDE 2015, Part II, LNCS 9243, pp. 343–351, 2015.
DOI: 10.1007/978-3-319-23862-3_34

artificial ant colony algorithm [2], artificial fish algorithm [3] and particle swarm optimization. As the Particle Swarm Optimization (PSO) algorithm has relatively fewer parameters and faster computation, it has been received more attentions [4–7]. Wang et al. [4] adopted the standard PSO algorithm to optimize the searching for the initial clustering centers and fuzzy weighted index of FCM. This could get an optimum fitness and a relatively stable value at the same time. Chen and Zhang [5] used the K-Means clustering algorithm with the population's center of gravity. It can effectively avoid the local optimum; however, the algorithm is still sensitive to the initial value selection. Izakian et al. [6] introduced a hybrid particle swarm algorithm to improve FCM, trying to get rid of the dependence on the initial value. Li et al. [8] and Long et al. [9] both used a Quantum-behaved particle swarm optimization (QPSO) to improve K-Means and FCM respectively. However, all these efforts do not consider the limitation of the PSO (or QPSO) itself, i.e., it has the problem of premature convergence.

In this paper, we leverage some new strategies of multi-subgroup co-evolution and particle's learning behavior to improve QPSO, and then apply the improved QPSO which we named LCQPSO to substitute the iteration of FCM clustering. Experiments showed that the proposed algorithm can reduce the sensitivity to initial centers effectively and achieve a much higher clustering accuracy.

The remaining of this paper is organized as follows: Sect. 2 introduces the fuzzy C-means algorithm. Section 3 reviews the QPSO and presents our improved version—LCQPSO. Section 4 describes the proposed algorithm of LCQPSO—FCM. Experimental results are given in Sects. 5 and 6 concludes the paper.

2 Fuzzy C-means Clustering Algorithm

Fuzzy C-means clustering (FCM) algorithm was proposed by Bezdek in 1973. It uses the membership of (0, 1) to determine the degree of each sample belongs to each cluster. Suppose FCM needs to categorize n samples $x_i (i = 1, 2, \ldots, n)$ to c clusters, u_{ij} is the degree of membership of sample j belongs to the ith cluster in membership matrix U, u_{ij} should satisfy $\sum_{i=1}^{c} u_{ij} = 1, \left(\forall j = 1, \ldots, n,\ u_{ij} \in [0, 1]\right)$. The target is to minimize the sum of intra-class weighted square distances:

$$J(U, c_1, \ldots, c_c) = \sum_{i=1}^{c} J_i = \sum_{i=1}^{c} \sum_{j=1}^{n} u_{ij}^m d_{ij}^2 \tag{1}$$

Where c_i is the center of cluster i, $d_{ij} = \|c_i - x_j\|$ is the Euclidian distance between sample j and clustering center i, $m \in [1, +\infty)$ is the weighted exponential, usually set to $m = 2$. According to the Lagrange multiplier method, we can obtain the following result:

$$c_i = \sum_{j=1}^{n} u_{ij}^m x_j \Big/ \sum_{j=1}^{n} u_{ij}^m, \quad u_{ij} = 1 \Big/ \sum_{k=1}^{c} \left(d_{ij}/d_{kj}\right)^{\frac{2}{m-1}} \tag{2}$$

By iterating (2), we will get the final clustering result when the termination criterion is met. FCM clustering is essentially a gradient descent optimization progress. Due to the inherent limitation, it is sensitive to the initial value and prone to local optimal. Therefore, a global optimization searching algorithm is supposed to be employed to address this problem.

3 Cooperative Quantum-Behaved Particle Swarm Optimization with Learning Behavior

3.1 Quantum-Behaved Particle Swarm Optimization

In QPSO, let the solution space dimension is D, the population size is N, the current position of the particles is $X_i = (x_{i1}, x_{i2}, \ldots, x_{iD})\; i = 1, 2, \ldots, N$, their best historical position is $p_i = (p_{i1}, p_{i2}, \ldots, p_{iD})$, and the best position of the population is noted as $p_g = (p_{g1}, p_{g2}, \ldots, p_{gD})$. Then, each particle updates its position as follows:

$$q_i = \phi P_i(t) + (1 - \phi) P_g(t) \tag{3}$$

$$mbest(t) = \sum_{i=1}^{N} \frac{p_{i\cdot}^{(t)}}{N} = \left(\sum_{i=1}^{N} \frac{p_{i1}^{(t)}}{N}, \ldots, \sum_{i=1}^{N} \frac{p_{iD}^{(t)}}{N} \right) \tag{4}$$

$$X_i(t+1) = q_i(t) \pm \beta |mbest(t) - X_i(t)| \ln(1/u) \tag{5}$$

Where q_i is a local attractor of each particle, which is decided by particle's p_i and population's p_g; mbest is the mean value of the current p_i; φ and u are random numbers in (0, 1). β is the contraction coefficient, usually linearly descends from 1 to 0.5 to get better convergence.

As particles have no speed parameters in QPSO, and they are granted quantum behavior instead, the convergence and accuracy are improved significantly.

3.2 "Survival of the Fittest" Model and Learning Behavior

However, QPSO still has the deficiency of premature convergence. In order to further enhance the searching capacity, inspired by [10], we employ "Survival of the fittest" strategy to reduce the probability of local optimal. The multi-subgroup co-evolution strategy simulates the social organizations' behavior, it divides the population into subgroups, and each subgroup searches in the solution space following the rule of QPSO independently. In order to share information among subgroups, we create a "gene pool" to exchange information among subgroups periodically. We use the strongest particles in "gene pool" to substitute the inferior particles in each subgroup periodically, and select the new optimal particles from subgroups to update the "gene pool". This mechanism not only ensures the robustness of the population, but also promotes the searching power. The model is illustrated in Fig. 1.

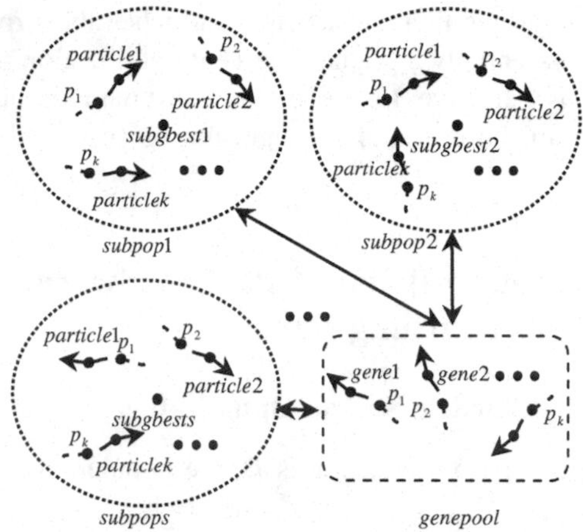

Fig. 1. "Survival of the fittest" model, see text for details

Individuals in society groups not only learn abilities from their own experience, but also from others actively. But in PSO and QPSO, particles are always following the individual optimal and population's optimal to update their state passively, and their ability to learn from other particles is inhibited. In order to enhance the individual's learning ability, we use the following method to grant particles the behavior of learning from the others:

$$q_{ij}(t) = \begin{cases} \phi_j \cdot p_{s,kj}(t) + (1 - \phi_j) \cdot p_{gj}(t) & l_{rand} < l_c \\ \phi_j \cdot p_{ij}(t) + (1 - \phi_j) \cdot p_{gj}(t) & l_{rand} \geq l_c \end{cases} \tag{6}$$

Where l_{rand} is a random value of (0, 1), l_c is a leaning probability parameter, s and k are the other subgroup index and the particle index respectively. The local attractor will partly learn the current dimension of the optimal particle's position of subgroup s based on the parameter l_c. Therefore, the learning probability parameter l_c will determine particles' ability of learning from others. In order to achieve a tradeoff between particles' individual development and the population's searching capability, particle i's value of l_c within a subgroup is determined as follows:

$$l_{ci} = l_{cMin} + (l_{cMax} - l_{cMin}) \cdot (i/s)^{\alpha} \tag{7}$$

Where α is a constant greater than 0, l_{cMax} and l_{cMin} are the maximum and minimum respectively, all assigned manually.

We name our improved version of QPSO by the strategies of multi-subgroup evolution and particles' learning behavior "LCQPSO".

4 FCM Based on LCQPSO

4.1 Particle Encoding Method and Fitness Function

Suppose the sample's dimension is D, the clusters number is c, the current clustering centers will be noted as $\{(C_{i1}C_{i2}\ldots C_{iD})\}_{i=1}^{c}$. We encode the configuration of all clustering centers as one particle, just as the illustration of Q in Fig. 2. Our target is to find the best setting of all centers, i.e. the best particle in the population.

$$Q\left\{\begin{array}{cccc} C_{11} & C_{12} & \cdots & C_{1D} \\ C_{21} & C_{22} & \cdots & C_{2D} \\ & & \vdots & \\ C_{c1} & C_{c2} & \cdots & C_{cD} \end{array}\right.$$

Fig. 2. Particle encoding method

According to the particle encoding method, we choose the following fitness function:

$$fit(Q) = \sum_{i=1}^{c}\sum_{j=1}^{n} u_{ij}^{m} \left\| Q_i - x_j \right\|^2 \tag{8}$$

Where Q is one particle, x_j is the jth sample.

4.2 Detailed Procedure

Given all samples, clusters number, fuzzy index, population size, we initialize all the following parameters: the membership degree matrix U, the whole population with each dimension, the max iteration number, the iterative termination threshold, etc. The detailed procedure is as follows:

Step 1: The population is divided into s subgroups, each with the scale of N/s. For each subgroup, we calculate the fitness value of each particle and obtain the optimal solution noted as $p_g^i(i = 1, 2, \ldots s)$. The optimal solution of the entire population is $p_{g_{pop}} = p_g^k$, where $k = \arg\min_{1 \le i \le s} fit\left(p_g^i\right)$. We also select the optimal particles from each subgroup by gene rate P_{gene} to create one gene pool.

Step 2: According to the current iteration index t, we calculate contraction coefficient β_t and learning probability parameter l_c for each particle. l_c and l_{rand} will codetermine q_i, then all particles update according to (5). When all particles' state updated, those cross-border dimensions are constrained to the boundary values.

Step 3: ReCompute fitness value for each particle and update the following parameters: each particle's individual optimal, subgroups' optimal $p_g^i(i = 1, 2, \ldots s)$, the entire population's optimal $p_{g_{pop}}$ and membership matrix U by the fitness function.

Step 4: Check whether the evolution period is arrived, if true, we replace each subgroup's inferior particles with particles of "gene pool", and then update "gene pool" with better particles from subgroups.

Step 5: Check whether the error termination criterion δ is satisfied, if true, we output the clustering result. Otherwise, we repeat Step 2 ~ Step 4.

5 Experiments

This section presents the experimental results of the proposed LCQPSO-FCM. The experiment is conducted on 4 UCI benchmark datasets: Iris, Wine, Breast Cancer, and Vowel. Table 1 shows the relevant information for each dataset.

Table 1. Dataset information

Dataset	Samples	Attributes	Classes
Iris	150	4	3
Wine	178	13	3
Breast cancer	683	9	2
Vowel	990	10	11

5.1 Performance Test

In order to evaluate the clustering effect of the algorithm, we choose improved FCM's variants as baselines. They are improved by gene algorithm (GA) [1], PSO, QPSO respectively and noted as GA-FCM, PSO-FCM and QPSO-FCM. All of them use the same fitness function as (8). The Measurements of performance used are as follows: objective function value, accuracy, mean intra-cluster distances and mean inter-cluster distances. We repeat each algorithm 10 times on each dataset, and take the average as the last result.

The experimental results are shown in Table 2. Compared to the standard FCM, all improved versions of FCM obtain higher clustering accuracy. Nevertheless, from the mean inter-class distance and intra-class distance, we can find that GA-FCM and PSO-FCM have similar results that only a little better than FCM. We believe they may be sucked in some local minimums. On the contrary, QPSO-FCM and LCQPSO both have larger inter-class distance, and smaller intra-class distance, this reveal that they have significant improvements on searching for global optimal solution. But our LCQPSO-FCM uses multi-subgroup co-evolution strategy and the particles can learn from others, the searching ability is the strongest: it has the smallest objective function value and the highest accuracy on all the 4 dataset.

Table 2. Comparison of clustering results

Dataset	Clustering method	Objective function value	Mean inter-cluster distances	Mean intra-cluster distances	Accuracy (%)
Iris	FCM	5.4892	0.7827	9.8535	90.00
	GA	5.2336	0.7827	9.7584	91.65
	PSO	5.2336	0.7827	9.7584	91.33
	QPSO	5.2335	0.7827	9.7584	91.67
	LCQPSO	5.2335	0.7853	9.7529	92.72
Wine	FCM	1.7961e+006	0.5077e+003	4.5707e+003	61.80
	GA	1.7922e+006	0.5082e+003	4.3228e+003	62.92
	PSO	1.7922e+006	0.5082e+003	4.3228e+003	62.92
	QPSO	1.7887e+006	0.5096e+003	4.2941e+003	65.83
	LCQPSO	1.7830e+006	0.5110e+003	4.2764e + 003	69.13
Breast cancer	FCM	1.5323e+004	14.0712	1.5407e+003	76.82
	GA	1.5322e+004	14.0721	1.5399e+003	80.65
	PSO	1.5322e+004	14.0726	1.5399e+003	80.70
	QPSO	1.5317e+004	14.0734	1.5381e+003	85.72
	LCQPSO	1.5310e+004	14.0790	1.5333e+003	88.59
Vowel	FCM	475.2976	0.5770	177.7012	77.93
	GA	473.1706	0.5783	175.7222	81.20
	PSO	472.9823	0.5788	175.7370	82.01
	QPSO	469.4541	0.5791	175.5287	83.52
	LCQPSO	469.4239	0.6159	173.2352	87.76

The fitness evolution curves in log space on each dataset are shown in Figs. 3. We can see that due to Iris's lower dimensions, all algorithms can effectively converge to the optimal solution, but LCQPSO-FCM has better performance than others obviously. On the Breast Cancer and Vowel datasets, we find that when the dataset is larger, the dimension is also higher, GA-FCM and PSO-FCM perform poorly, especially on Vowel, the data dimension is as high as 10, classes number is 11, they are both trapped into local optimal, and convergence are slow too. On the contrary, LCQPSO-FCM has the fastest convergence, and the fitness value still drops even in the later stage of evolution, reflecting the LCQPSO's powerful searching ability.

5.2 Complexity Test

We show the times each algorithm consumed in Table 3. We can find that all FCM's variants consume more time than the original version, for all of them substitute FCM's iteration with more complex searching steps which will incur more computation time inevitably. The time efficiency of all algorithms listed are basically consistent on all datasets. Among all of the algorithms, GA-FCM is the slowest for it has to process complex operations of selection, cross and mutation. Though PSO-FCM is fastest, it

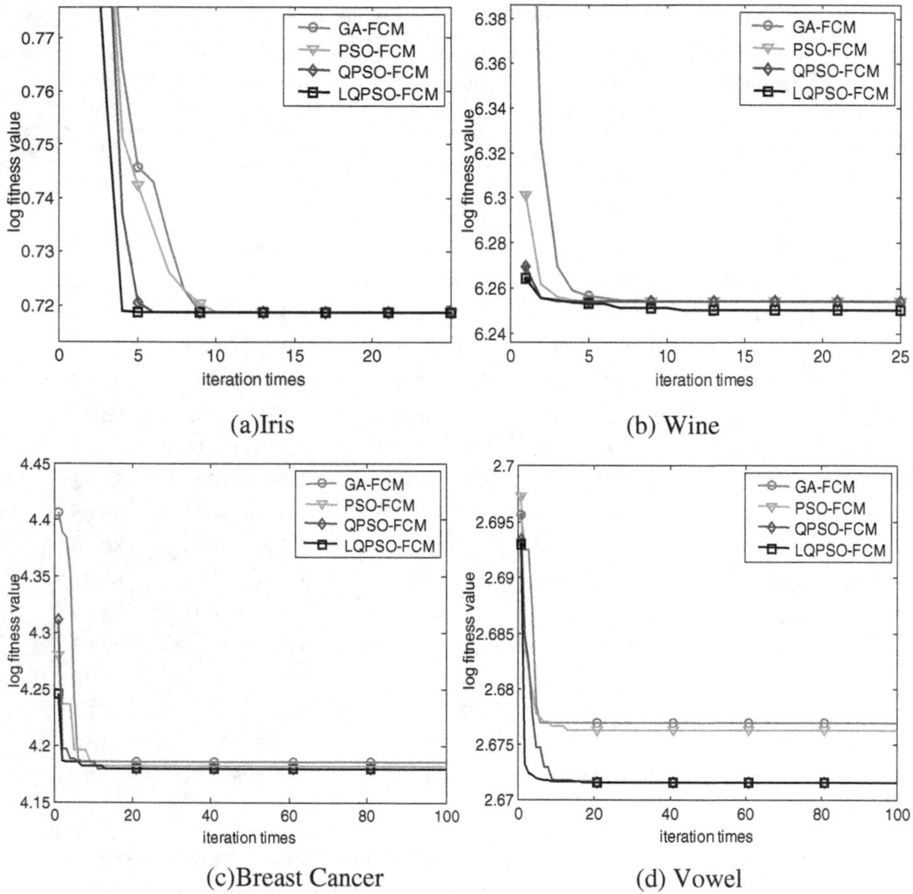

Fig. 3. Fitness curves on each dataset

has the poorest searching ability. LCQPSO-FCM has little larger time consumption than PSO/QPSO-FCM, but it is worthy, for it has the strongest searching ability.

Table 3. Run time comparison (seconds)

	FCM	GA-FCM	PSO-FCM	QPSO-FCM	LCQPSO-FCM
Iris	1.7474e-002	4.9284e+000	2.3624e-001	3.6912e-001	3.9445e-001
Wine	4.4583e-002	8.9796e+000	3.9191e-001	8.2144e-001	8.8227e-001
Breast cancer	4.2735e-002	1.0823e+001	5.1983e-001	6.7213e-001	6.4136e-001
Vowel	4.3441e-001	4.7272e+001	2.1623e+000	2.3278e+000	3.1450e+000

6 Conclusion

In this paper, we propose an improved FCM clustering algorithm based on the enhanced version of QPSO, which employs multi-subgroup co-operation evolution strategy and learning behavior to raise the searching capacity. As a result, the algorithm has faster convergence speed and better convergence accuracy. The evaluations on 4 UCI datasets show that compared to the original FCM and its improved versions by GA, PSO, QPSO, the proposed algorithm can overcome the initial value sensitivity problem, and obtain much better clustering effect.

Acknowledgments. This work is supported by National Natural Science Foundation of China (NSFC Grant No. 61272258, 61170124, 61301299, 61272005), and a prospective joint re-search projects from joint innovation and research foundation of Jiangsu Province (BY2014059-14).

References

1. Ye, A., Deng, D.: Clustering algorithm based on improved quantum genetic algorithm. J. Comput. Simul. 30(4), 275–278, 307 (2013)
2. Lv, Y.: Research of text clustering based on improved ant colony algorithm. J. Microelectron. Comput. **29**(3), 31–34 (2012)
3. Yu, H., Jia, M., Wang, H., Shao, G.: K-means clustering algorithm based on artificial fish swarm. J. Comput. Sci. **39**(12), 60–64 (2012)
4. Wang, Z., Liu, Z., Chen, D.: Research of PSO-based fuzzy C-means clustering algorithm. J. Comput. Sci. **39**(9), 165–169 (2012)
5. Chen, X., Zhang, J.: Clustering algorithm based on improved particle swarm optimization. J. Comput. Res. Dev. **49**(Suppl.), 287–291 (2012)
6. Izakian, H., Abraham, A., Snasel, V.: Fuzzy clustering using hybrid fuzzy c-means and fuzzy particle swarm optimization. In: 2009 IEEE World Congress on Nature and Biologically Inspired Computing, pp. 1690–1694 (2009)
7. Li, C., Zhou, J., Kou, P., et al.: A novel chaotic particle swarm optimization based fuzzy clustering algorithm. J. Neurocomputing **83**, 98–109 (2012)
8. Li, Y., Mao, L., Xu, W.: Research of improved fuzzy C-means algorithm based on quantum-behavior particle swarm optimization. J. Comput. Eng. Appl. 48(35), 151–155, 173 (2012)
9. Long, H., Xu, W., Sun, J.: Data clustering based on quantum-behaved particle swarm optimization. J. Appl. Res. Comput. 23(12), 40–42, 45 (2006)
10. Zhou, D., Sun, J., Xu, W.: An advanced quantum-behaved particle swarm optimization algorithm utilizing cooperative strategy. In: 2010 Third International Workshop on Advanced Computational Intelligence (IWACI), pp. 344–349 (2010)
11. Jain, A.K.: Data clustering: 50 years beyond K-means. Pattern Recogn. Lett. **31**(8), 651–666 (2010)

Design for an Interference-Suppressing DMX512 Protocol Expansion and Repeater

Xueli Zhu(✉), Shuxian Zhu, Yongjun Zhu, Shenghui Guo, and Hanwen Gao

Department of Mechanical and Electrical Engineering, Suzhou University of Science and Technology, Suzhou 215009, Jiangsu, China
zhuxueli77@163.com

Abstract. In view of lack of checking devices for the frame structure of DMX512 protocol, this paper presents a design scheme of pulse interference suppressor. The interference suppression repeater is composed of PIC micro-controller, level conversion and isolating circuit, and through the repeater, local network data transmission rate can be doubled and the check of the data transmission is realized. The design scheme can effectively reduce more than 95 % of the given communication interference environment operation error, and improve the reliability of the DMX512 application system.

Keywords: Protocol · Repeater · Pulse interference · Frame structure

1 Introduction

DMX512 protocol is a kind of general control protocol of digital entertainment lighting equipment, which is widely used in entertainment lighting industry. Because the communication protocol has simple, practical and efficient characteristics, so it has been widely used in of various stage effect lightings equipment such as computer lights, controller, console, color-changing device, electric hanger etc. [1–3]. However, because DMX512 protocol frame structure has no checking device, so under some complex electromagnetic environment situation, accidental flashing or loss of control information will occur, seriously affecting the overall visual effect.

Aimed at the shortcomings of the DMX512 protocol, this paper designs a kind of relay equipment which can suppress the interference pulse to improve the application effect of DMX512 protocol.

2 DMX512 Protocol

DMX512 serial communication protocol includes the electrical characteristics, data protocol, data format etc.. The protocol was proposed by United State Institute for Theatre Technology in August 1986. The protocol provides a data transmission standard that in the EIA–485 differential channels on a single frame transmission of up to 512 bytes control information, and then in 1990 and 2004 the protocol was further modified. The latest version is DMX512–A Digital Data Transmission Standard for

X. He et al. (Eds.): IScIDE 2015, Part II, LNCS 9243, pp. 352–360, 2015.
DOI: 10.1007/978-3-319-23862-3_35

Lighting Control (ANSI E1.11–2004). The domestic version of the protocol is the DMX512–A light control data transmission protocol, which is drafted by Performing Venue and Equipment Committee of CETA, and register information management.

2.1 Working Principle of DMX512 Protocol

DMX is the abbreviation of Digital Multiplex, which means the digital multiplexing protocols in the name of 512 representing a single frame of the longest data bytes, and in theory, 512 minimum control units can be mounted on a single DMX512 bus (such as a single light source brightness, motor rotation angle, etc.). Protocol should comply with the EIA–485 serial way to send and receive data; standard baud rate is 250 kbps; a single bit transmission time is 4 s; The single frame completing the longest transmission time is about 22 ms; data refresh rate can reach more than 40 Hz [4].

DMX512 uses asynchronous communication format, and the initial part of the protocol is made up of reset segments and data segments. The reset segment consists of BREAK (reset signal), MAB (After reset mark) and START code. Including the start code, each control information byte consists of 11 bits, including 1 start bit, 8 bits of data and 2 stop bits. Each byte of control information can be expressed in different values between 0 and 255. The frame structure of DMX512–A is shown in Fig. 1. The time requirements of DMX512 protocol refer to Table 1.

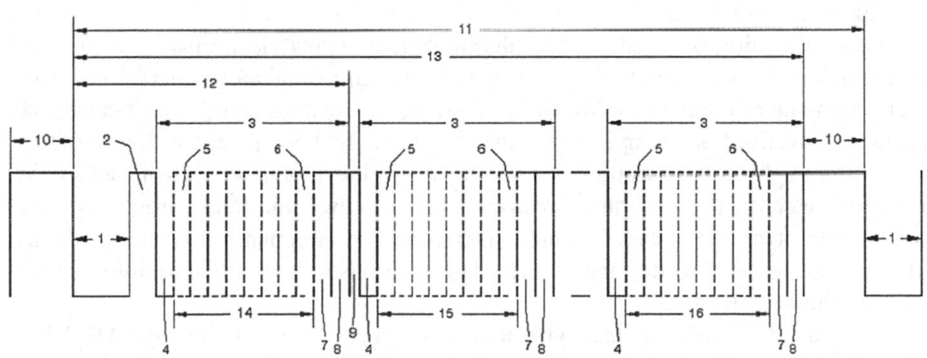

Fig. 1. The frame structure of DMX512–A 1–"SPACE" for BREAK 2–"MARK" after BREAK (MAB) 3–slot time 4–START time 5–LEAST SIGNIFICANT data bit 6–MOST SIGNIFICANT data bit 7–STOP bit 8–STOP bit 9–"MARK" time between slots 10–"MARK" before BREAK (MBB) 11–BREAK to BREAK time 12–RESET sequence (BREAK, MAB, START Code) 13–DMX512 packet 14–START CCODE (slot 0 data) 15–SLOT 1 DATA 16–SLOT nnn DATA (maximum 512)

Table 1. The time requirements of DMX512 protocol

Designation	Description	Min	Typical	Max	Unit
–	Bit rate	245	250	255	Kbp/s
–	Bit time	3.92	4	4.08	μs
–	Minimum update time for 513 slots	–	22.7	–	ms
–	Maximum refresh rate for 513 slots	–	44	–	updats/s
1	"SPACE" for BREAK	92	176	–	μs
2	"MARK" after BREAK	12	–	<1.00	μs s
9	"MARK" time between slots	0	–	<1.00	s
10	"MARK" before BREAK	0	–	<1.00	s
11	BREAK to BREAK time	1204	–	— 1.00	μs s
13	DMX512 packet	1204	–	— 1.00	μs s

2.2 Shortcomings of DMX512 Protocol

One of the shortcomings of DMX512 protocol is lacking of the checking devices for frame structure and the data can be sent only one-way from host to all slave machines. Slave machines cannot identify errors in the process of communication, so feedback on error status can not be done.

The shortcoming of the design results in some complex electromagnetic environment situation, the data sent by host is vulnerable to differential mode interference from other environment coupled to the data bus, so the instructions sent by host cannot be executed correctly. For example, when the adjustable light equipment such as building facade/stage lighting collocation is used to adjust the motor which can adjust the irradiation angle of the lamp and because of public power lines and wiring ways, the start or stop of motor is easy to cause interference to communication lines, and the pulse interference will lead to unexpected flashing lights or loss of control information, affecting the overall visual effect.

There are two ways to deal with the interference. One is to use the DMX512 protocol reserved EF extension, the standby communication line can be used as a redundant data lines, or use the backup line to send control information with a higher rate to realize some functions such as the data redundancy sending and verifying/feedback, to achieve the purpose of communication interference suppression. The advantages of this method is to provide a flexible way of expansion and good comprehensive effect for designer, the disadvantage is also obvious: high complexity of systems and communication protocols, long development cycle, and poor general degree, low reusability of equipment, difficult to popularize. In addition, the cost for line materials and installation costs will raise sharply. Therefore, purchasing the complementary equipment is needed.

Another solution is to use the repeater for custom protocol extension in the local or branch end of the DMX512 network. This scheme also has the problem such as complex, poor popularity, but relative to the first solution, the complexity of communication protocols can be reduced greatly, low cost repeater with simple structure can be used for protocol transfer, and line material and installation costs have no obvious increase.

This paper uses the latter solution, and has carried on the simple extension to DMX512 protocol. The relay equipment chosen can work in the general DMX512 mode or enhancement mode, and through lower cost and by using the relatively simple way, the ability to suppress line interference is improved.

3 Expansion of Protocol and Relay Equipment Design

Interference suppression repeater is a kind of communication system based on MCU, its effect is to extend DMX512 specification, and the working principle is to use the MCU with level conversion and isolation circuit to promote local DMX512 network transmission rate, so as to realize the transmission of data redundancy check and improve the reliability of data transmission.

3.1 Expansion of DMX512 Protocol

The expansion of the protocol described in this paper, relative to DMX512 specification, is mainly to improve the communication rate of subnet and redundancy to send control information.

Telecommunication lines interference leads to the abnormal operation of the equipment in the communication line, even check device is added and under the situation the receiver can determine the error data, what can only be guaranteed is that the error control information is not implemented. In certain circumstances, when lights flashing, beating direction etc. occur, control information frame with lost and wrong information will cause the short-term action not to be performed. The way of error correction coding is limited by MCU operation ability, and interfered with change control bit can not be more than 1–2, otherwise control information can not be restored correctly. This paper uses the following methods to make the expansion of the information frame structure, namely, join the frame number CRC-16 check-sum and repeat forwarding the frame number at the end of the original MDX512 frame.

3.2 Design of Interference Suppression Repeater

By adopting a simple DMX512 protocol extensions, the interference suppression repeater (hereinafter referred to as the repeater) can improve subnet or local communication rate to send control information redundancy, which minimizes the influence of the short pulse interference randomly generated by electrical devices.

The basic working principle of repeater is: in subnet or the local communication rate to 500 kbps, all subnets devices should support the doubled communication rate and can adjust and set the subnet equipment by ASC (Alternate START Code) retained by the DMX512 specifications. The equipment can work in ordinary DMX512 protocol mode and also switch to the anti-interference enhanced mode under complex interference circumstances.

In order to reduce the delay effect, the repeater will start forwarding by extending the protocol way with double rate at the signal output end when the repeater receives half single frame information from the console. When the repeater receives last letters of the console data frames, according to the definition of extended protocol repeater will place check section and the forwarding number in the end, at same time to start sending the received data frame. When the data frame is sent second time, half of the new console data frame has arrived. Specific signal timing of pulse interference suppressor is as shown in Fig. 2.

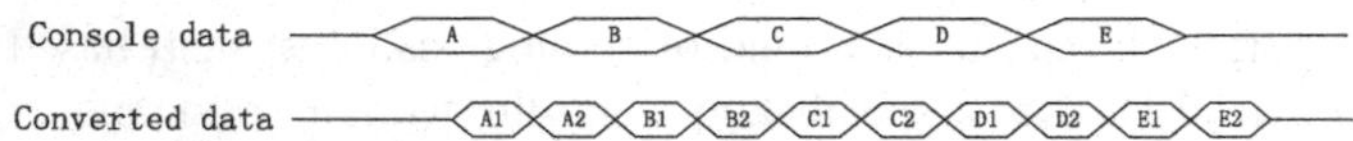

Fig. 2. Signal process of the pulse interference repeater

The receiver works in the following modes. After it receives in turn the two frame forwarding numbers A1 and A2, according to the following logic, the receiving end choose whether to refresh data or instructions:

(1) When the check of two frames is correct, use the data or instructions.
(2) When the check of two frames is wrong, continue to use the previous data or instructions.
(3) When only one frame of the two frames is correct, use the right frame data and discard the calibration error of the data.

The signal input of the repeater is RS485 data cables of the console, and signal output terminal is RS485 data line of subnet. In a variety of applications, interference is usually through the power line coupled to the communication line, so the repeater should be placed as close as to the control equipment of signal output port in order to reduce the possibility of the original control signal to be interfered.

3.3 Hardware Implementation of Repeaters

The short pulse interference suppressor developed in this study is a kind of communication system based on MCU. By using PIC24FV32KA304 as main control MCU, according to the basic requirement of DMX512 protocol, the power supply and communication lines will be isolated by 6N137 and DCV010505, the principle diagram of system is shown in Fig. 3.

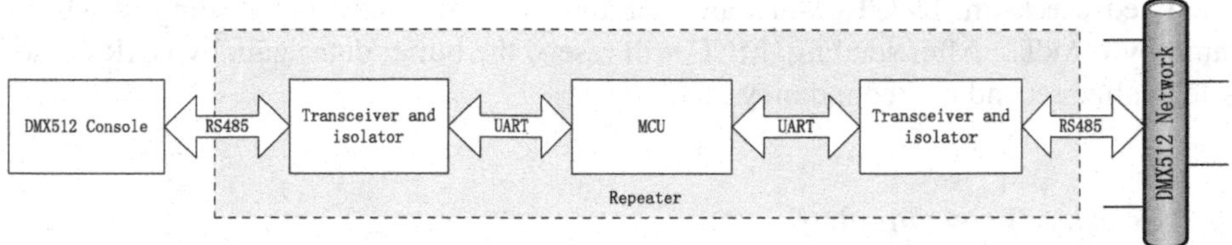

Fig. 3. Principle block diagram of suppressor

The master PIC24FV32KA304 is a low cost 16-bit RISC microcontroller designed by MicroChip company. The MCU uses modified Harvard architecture, with 32 KB Flash and 2 KB RAM memory, and it can achieve the highest processing capacity of 16 MIPS when the system clock is 32 MHz. The MCU has abundant peripheral resources, especially suitable for the application of CRC cyclic redundancy check generator and with 4 deep FIFO UART module of receiving/sending buffer. Part schematic diagram of UART1 and isolation of power are shown in Fig. 4.

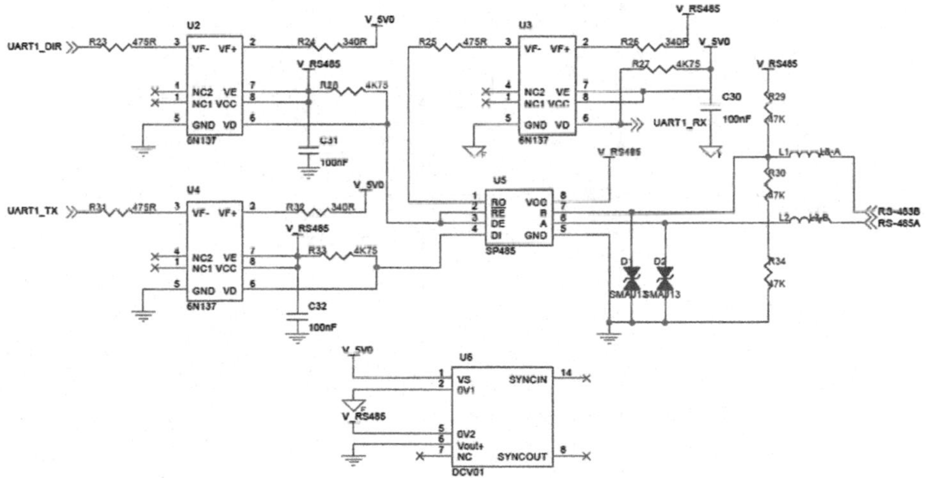

Fig. 4. Part schematic diagram of UART1 and isolation of power

UART1 is responsible for receiving control information of the console, communication rate of 250 kbps. UART2 is responsible for forwarding the data in the form of doubled original data rate (500 kbps) to the subnet. After detected frame starts, MCU will store the data one by one in a 516 byte buffer, and update and calculate the CRC check sum–16 data. When 258 bytes information is sent from the console, MCU will start to send the data received before 258 bytes in double rate through the UART2 and UART1 continues to receive the console data at the same time. By the time when the console data frames is sending the data to lower 512 bytes, MCU will place the checksum data and the forwarding number at the end of the buffer according to the calculated checksum by CRC - 16 and continue to complete of the sending of whole frame by UART2. After sending, MCU will resend the buffer data again by UART2, so as to realize secondary redundancy.

3.4 Software Programming

Software contains two parts, the repeater and the receiving device. The program flow of repeater section is shown in Fig. 5. After having detected the right start signal (BREAK) through the external interrupt and timer, MCU will enter the UART1 to receive UART1 interrupt. The delay part, sending length and the buffer length in the flow chart can be retained by PC interface using ASC Settings and stored in a block of repeater EEPROM in order to adapt to different application environment.

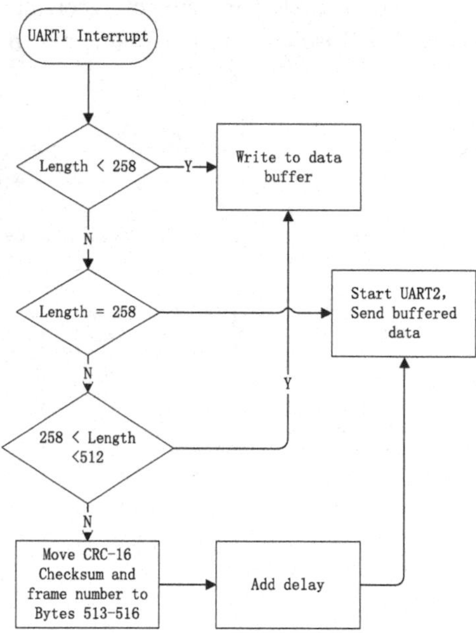

Fig. 5. The forwarding function of the repeater software flow chart

The program flow of receiving unit is shown in Fig. 6. The receiving terminal gives an auxiliary judgment according to the frame forwarding number at the end of the control information. Receiving length and buffer length can also be set and stored in EEPROM using the above method.

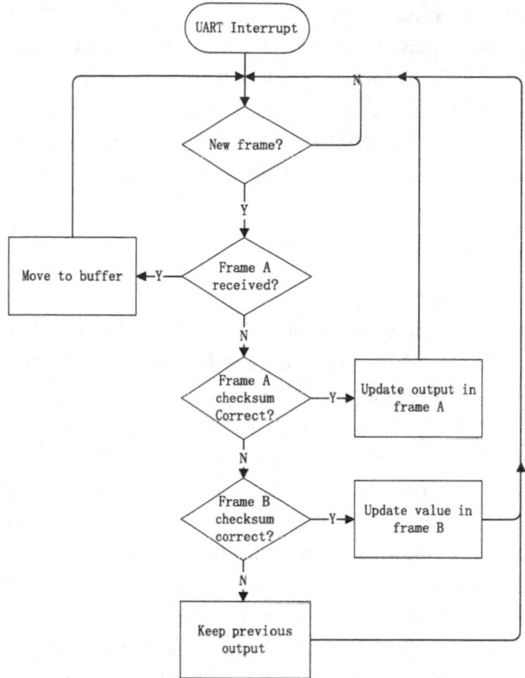

Fig. 6. The software flow chart of slave machine

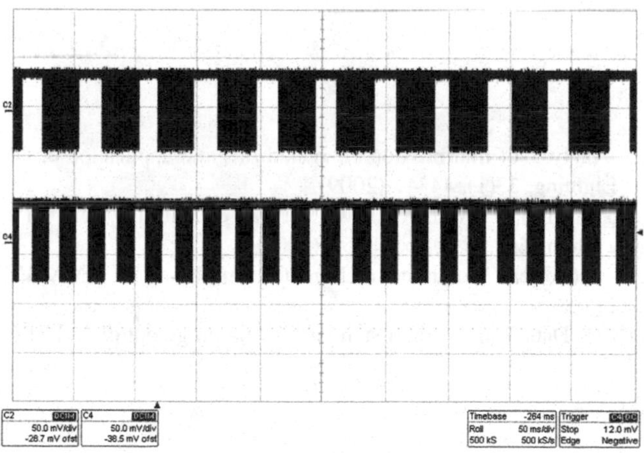

Fig. 7. Bus actual data diagram

3.5 Comprehensive Evaluation of Design Scheme

Figure 7 shows the bus actual data diagram when the repeater is working through differential probes. Above the figure it can be seen that the above channel is the original DMX512 control information frame sent by the controller, and information is 512 full bytes sent. Below is the channel of the control information frame forwarded by suppressor. So we can see the repeater can achieve the design requirement very well. In the case of different length of data, the repeater can forward data in accordance with the requirement. In practical anti-interference test, we can use a DC motor and lamp to test equipment. With the statistic, the scheme can effectively reduce more than 95 % of the operation error in the specific communication interference environment.

4 The Conclusion

The paper presents a design scheme of short pulse interference suppressor, by using this device, the short-time pulse interference can be greatly reduced, and the reliability of DMX512 application system can be improved.

The improved scheme of the protocol has the following advantages: (1) Small changes are needed to the original DMX512 protocol which is simple and easy to be implemented. (2) Realizing redundant data receiving which can greatly reduce the frame loss phenomenon when pulse interference occurs. (3) With relatively strong commonality, the manufacturer can provide the repeater with its complementary equipment, and can configure the repeater and lamps working in ordinary DMX512 mode or protocol extension mode through the PC interface. (4) The scheme is low cost, simple structure and easy for popularization, and has engineering application value.

Acknowledgements. This project is funded by Construction System Program of Jiangsu Provincial Department of Housing and Urban-Rural Development (Project Number: 2013ZD47).

References

1. Han, Z., Qi, L.: Analysis of transmission characteristics and application of DMX512 control protocol. Light Lighting. **33**(1), 44–47 (2009)
2. Huang, Y., Liao, S., Liu, Y., Cai, R.: DMX512 based control system for full color LED luminaire. China Illum. Eng. J. **20**(4), 48–53 (2009)
3. Yu, M.: Design and research of laser intelligent beams based on DMX512 protocol. Jilin University (2012)
4. DMX512-A Digital Data Transmission standard for lighting control (WH/T 32-2008)

A Convolutional Deep Neural Network for Coreference Resolution via Modeling Hierarchical Features

Xue-Feng Xi[1,3], Guodong Zhou[1(✉)], Fuyuan Hu[2], and Bao-chuan Fu[2]

[1] School of Computer Science & Technology, Soochow University, Suzhou, China
{20114027008,gdzhou}@suda.edu.cn
[2] School of Electronic & Information Engineering, Suzhou University of Science & Technology, Suzhou, China
[3] Suzhou Key Laboratory of Mobile Networking and Applied Technologies, Suzhou, China

Abstract. Coreference resolution is a major task of natural language processing (NLP) identifying which noun phrases (or mentions) refer to the same real-world entity or concept. The state-of-the-art methods applied to coreference resolution are mainly based on statistical machine learning, and their performance strongly depends on the quality of the extracted features. The extracted features are usually shallow features by artificial selection, which leads to the loss of unknown useful deep semantic information and becomes an obstacle for improving system performance. We explored a convolutional deep neural network (CDNN) to extract discourse level features automatically. Our method utilized all of the word tokens as input without complicated pre-processing. To begin with, the word tokens were transformed to vectors by looking up word embeddings. Secondly, mention-pair level features were extracted according to the given mentions. In the meanwhile, distance features were computed easily. Moreover, discourse level features were learned using a convolutional approach. Finally, these features were fed into a softmax classifier to predict the equivalence between two marked mentions. The experimental results demonstrate that our approach obtains a competitive score of average F1 over MUC, B3, and CEAF, which places it above the mean score of other systems on the dataset of CoNLL-2012 Shared Task.

Keywords: Deep learning · Convolutional deep neural network · Coreference resolution · Mention-pairs

1 Introduction

Coreference resolution is a major task of natural language processing identifying which Noun Phrases (or mentions) refer to the same real-world entity or concept. Traditional one of the most influential learning-based coreference resolver is mention-pair mode, which was operated by training a model for classifying whether two mentions were co-referring or not (Aone and Bennett [1], 1995).

X. He et al. (Eds.): IScIDE 2015, Part II, LNCS 9243, pp. 361–372, 2015.
DOI: 10.1007/978-3-319-23862-3_36

In this kind of architecture, the performance of the entire coreference system strongly depends on the quality of the extracted features. Numerous studies are concerned with feature extraction, typically trying to enrich the classifier with more linguistic knowledge and/or more world knowledge (Ng and Cardie [2], 2002; Kehler et al. [3], 2004; Ponzetto and Strube [4], 2006; Bengtson and Roth [5], 2008).

Despite their initial successes, these mention-pair models have two commonly-cited problems to hinder the performance of these systems. First, since mention-pairs independency, these models only determine how well a candidate antecedent of a pair is relative to the active mention, but not how well a candidate antecedent is relative to other candidates. Second, the information extracted from the two mentions alone is not be sufficient for making an informed coreference decision. In other words, the extracted features are usually some shallow features by artificial selection, that leads to the loss of unknown useful deep semantic information and that becomes an obstacle for improving system performance.

To address these weaknesses, we claim that mention-pairs should not only be processed at the local clause or sentence level, but also be treated in the global discourse. Our argument is therefore based on global discourse level considerations, rather than on purely single mention-pair at sentence level. Furthermore, we are interested in learning how to construct feature representations of input.

Motivated in part by these recently developed models (Collobert et al. [6], 2011; Yang et al. [7], 2013; Zeng et al. [8], 2014), we propose in this paper a deep learning approach to noun phrase coreference resolution based on mention-pair model that combines the strengths of hierarchical feature space. To the best of our knowledge, this work is the first example of using a convolutional DNN for coreference resolution.

In sum, we believe our work makes two main contributions to coreference resolution: (1) at the discourse level, we consider dependencies between pairs to construct a global processing mechanism; (2) based on mention-pairs, constructing a deep learning method with word tokens as raw source input is a novel approach for coreference resolution.

The rest of this paper is organized as follows. In Sect. 2, we describe this deep learning approach based on mention-pair model and explain how to model features hierarchically with it. Next, Sect. 3 provides the experiment description and evaluates the various models on CoNLL-2012 English datasets. Finally, Sect. 4 concludes our paper.

2 A Convolutional Deep Neural Network for Coreference Resolution

Motivated in part by some neural network architectures (Collobert et al. [6], 2011, Zeng et al. [8], 2014), we advocate a convolutional deep neural network (CDNN) for NP coreference resolution. Our architecture is summarized in Fig. 1. This system takes an input discourse and discovers multiple levels of feature

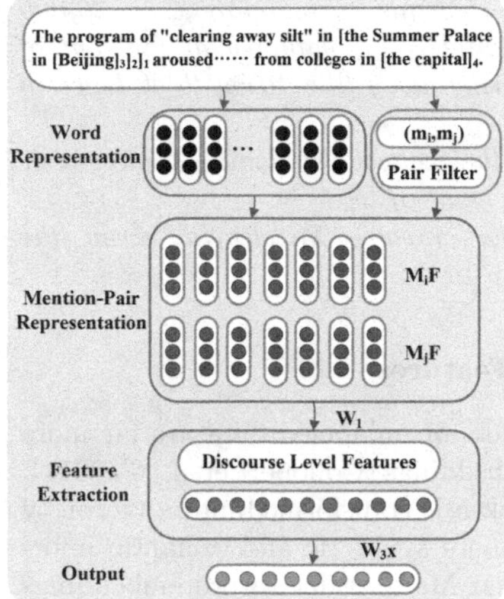

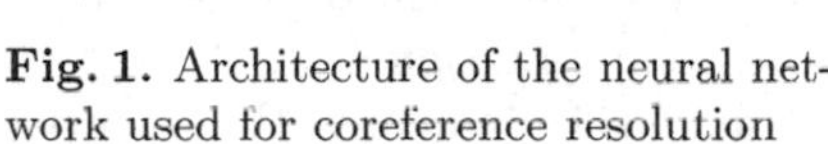
Fig. 1. Architecture of the neural network used for coreference resolution

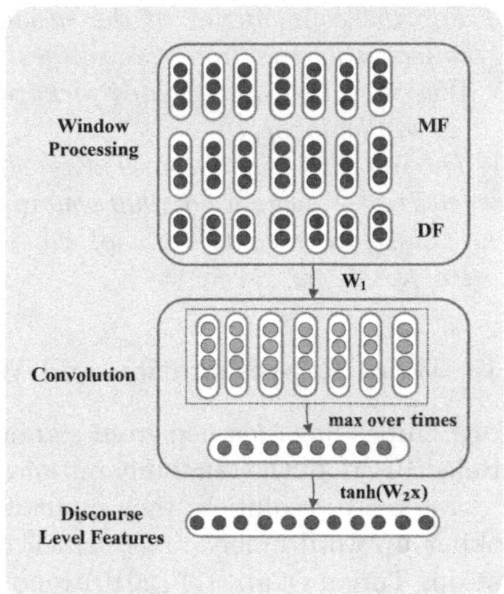

Fig. 2. The framework used for extracting discourse level features (Motivated by Zeng et al. [8], 2014)

extraction, where higher levels represent more abstract sides of the inputs. It primarily consists of the following five components: Word Representation, Mention-pair Representation, Pair Filter, Feature Extraction and Output. The system does not need any complicated syntactic or semantic preprocessing, and the input of the system is a discourse with marked mentions. Then, the word tokens were transformed to vectors by looking up word embeddings. Meanwhile, we get the mention-pairs selected by pair filter. Secondly, mention-pair level features are extracted according to the given mentions. Moreover, distance features are computed easily. In succession, discourse level features are learned using a convolutional approach. Finally, these features are fed into a softmax classifier to predict the equivalence between two marked mentions. The value of output from the classifier is the confidence score of the mention-pair. We will describe each layer we use in our networks as below. Assume that we have the following sequence of words with marked mention in a discourse.

(1) [The program of "clearing away silt" in [the Summer Palace in [Beijing]$_3$]$_2$]$_1$ aroused an enthusiastic response from colleges in [the capital]$_4$.

(2) College [students]$_{19}$ coming back from the work, at yesterday's forum, spoke openly of the experience, concluding that [volunteer work]$_5$ enables patriotic hearts to become a force for national service and that [it]$_6$ is very meaningful.

(3) [A large scale clearing away of silt from [the Kunming Lake in [the Summer Palace]$_7$]$_8$, which was the first time in the 240 years since[the lake]$_9$ came into being]$_{10}$, began late last year.

(4) *For the convenience of the mechanical clearing work, [more than 20,000 college students from 22 colleges in [capital]$_{11}$]$_{12}$, gathered at the foot of Wanshou Hill and wielded pickaxes clearing away [the 10 cm thick layer of ice on[the lake]$_{13}$]$_{14}$.*
(5) *During the eight consecutive days, [the college students]$_{15}$ altogether [cleared]$_{16}$ away a 660,000 square meter layer of ice.*
(6) *[This]$_{17}$ represents* 70 % *of the total work involved in clearing away [the ice]$_{18}$.*

2.1 Word Representation and Word Features

Word embeddings learned from significant lots of unlabeled data are far more satisfactory than the randomly initialized embeddings (Collobert et al. [6], 2011). In coreference resolution, each input word token is transformed into a vector by looking up word embeddings, which carry more syntactic and semantic information. Turian et al. [13] (2010) reported that Many trained word embeddings are freely available. So Our experiments directly use the trained embeddings provided by Turian et al. [13] (2010) as the word representation.

Distributional hypothesis theory (Harris [14], 1954) indicates that words occuring in the same context tend to have similar meanings. To capture this characteristic, the WF (word feature) combines a words vector representation and the vector representations of the words in its context. An example is shown below.

(6) This$_0$ *represents*$_1$ 70 %$_2$ *of*$_3$ *the*$_4$ *total*$_5$ *work*$_6$ *involved*$_7$ *in*$_8$ *clearing*$_9$ *away*$_{10}$ *the*$_{11}$ *ice*$_{12}$.

Each word is also associated with an index into the word embeddings. All of the word tokens are represented as a list of vectors$(x_0, x_1, \cdots, x_{12})$, where x_i corresponds to the embedding of the i-th word in this word sequence. To use a context size of w, we combine the size w windows of vectors into a richer feature. For example, when we take $w = 3$, the WF of the first word "this" and the last word "ice" in this sequence are expressed as $[x_s, x_0, x_1]$ and $[x_{11}, x_{12}, x_e]$, respectively. (x_s and x_e are special word embeddings that correspond to the beginning and end of the sequence respectively.)

2.2 Mention Representation and Mention Features

In Coreference Resolution based mention-pair model, we considers all pronouns (PRP, PRP$), noun phrases (NP) and heads of verb phrases (VP) as mentions. In other words, a mention is composed of one or more words. We convert a mention into a vector motivated partly by Socher's method [11] (2011), which one learns vector space representations for multi-word phrases.

For instance, the mention [*the ice*] in sequence (6) is composed of the word "the" and the word "ice". One word feature is expressed as $[x_{10}, x_{11}, x_{12}]$, the other one is expressed as $[x_{11}, x_{12}, x_e]$ (see Sect. 2.1). Based on the recursive auto-encoder method (Socher et al. [11], 2011), the words are first mapped into a vector space. Secondly they are recursively merged by the auto-encoder network into a fixed dimension vector, x_k. In the end, we get the MF (mention feature)of the mention [*the ice*] with representation similar to WF below.

$$mention[the\ ice] = m_j = [x_9, x_k, x_e]$$

2.3 Mention-Pair Representation

Every pair of mentions m_i and m_j is modeled by a random variable (more formally referring to Lassalle and Denis [15], 2013):

$$P_{ij} : \Omega \to X \times Y$$

$$\omega \mapsto (x_{ij}(\omega), y_{ij}(w))$$

where Ω classically represents randomness, X is the space of objects (mention-pairs) that is not directly observable and $y_{ij}(\omega) \in Y = \{1, 0\}$ are the labels indicating whether m_i and m_j are co-referring or not. The antecedent is m_i, and the anaphor is m_j.

One mention-pair is composed by two mentions, m_i and m_j. Since the mention is expressed by a vector space, similarly, considering the whole pair, the mention-pair can be represented as follows:

$$[m_i,\ m_j]^T$$

2.4 Pair Filter

The classical mention-pair level characteristics that can be found in details in previous work (Soon et al. [9], 2001; Bengtson and Roth [5], 2008; Rahman and Ng [10], 2011): grammatical type and subtype of mentions, string match and substring, apposition and copula, distance (number of separating mentions/sentences/words), gender/number match, synonymy/hypernym and animacy, family name (based on lists), named entity types, syntactic features (gold parse) and anaphoricity detection and so on. From these characteristics, we can find some very clear constrains that can help us eliminate negative cases.

Considering Number match's effect, in sequence (2), the mention $[students]_{19}$ and the mention $[it]_6$ are not co-referring absolutely because their Number is not same. The former is plural, but the later is singular.

There are some characteristics primarily include the mentions themselves, the types of the pairs of nominals and word sequences between the entities, the quality of which strongly depends on the results of existing NLP tools. Alternatively, we uses those that are not too difficult to compute as the source of constrain. Table 1 presents the constrains with selected pair characteristics that are related to the marked mentions. All of these constrains are used to construct the filtering rules for our pair filter.

Table 1. Constrains of the markable pair (m_i,m_j)

Constrains	Comments
C1	Number: if ((m_i is singular) and (m_j is plural)) then Y$(m_i,m_j)=0$
C2	Number: if ((m_i is plural) and (m_j is singular)) then Y$(m_i,m_j)=0$
C3	Gender: if ((m_i is male) and (m_j is female)) then Y$(m_i,m_j)=0$
C4	Gender: if ((m_i is female) and (m_j is male)) then Y$(m_i,m_j)=0$

2.5 Discourse Level Features

As mentioned in Sect. 2.1, all of the tokens are represented as word vectors, which have been demonstrated to correlate well with human judgments of word similarity. Despite their success, single word vector models are severely limited because they do not capture long distance features and semantic compositionality, the important quality of natural language that allows humans to understand the meanings of a longer expression. In this section, we propose a max-pooled convolutional neural network to offer discourse level representation and automatically extract discourse level features. Figure 2 shows the framework for discourse level feature extraction. In the Window Processing component, each mention-pair is further represented as two Mention Features (MF) (m_i,m_j), and Distance Features (DF) (see sects. 2.5.1 and 2.5.2). Then, the vector goes through a convolutional component. Finally, we obtain the discourse level features through a non-linear transformation.

2.5.1 Mention-Pair Features As an example, the mention-pair(m_{17}, m_{18}) in sequence (6), the antecedent m_{17}, "This", the anaphor m_{18}, "the ice". The mention features of "This" and "the ice" in this sequence are expressed as $[x_s, x_0, x_1]$ and $[x_9, x_k, x_e]$, respectively (see Sect. 2.2). According to mention-pair representation (see Sect. 2.3), the mention-pair features of (m_{17}, m_{18}) can be represented below:

$$[x_s, x_0, x_1, x_9, x_k, x_e]^T$$

2.5.2 Distance Features There are many important knowledge sources useful for coreference. One important factor is the distance between the two markables. McEnery (et al. [16], 1997) have done a study on how distance affects coreference, particularly for pronouns. One of their conclusions is that the antecedents of pronouns do exhibit clear quantitative patterns of distribution. The distance feature has different effects on different noun phrases. It is necessary to specify impact of distance between the m_i and the m_j.

For this purpose, DF (Distance Feature) are proposed for coreference resolution.In this paper, the DF is the combination of two distance features, D_1 and D_2. On the one hand, D_1 is the relative distances to m_1 and m_2. It's possible values are 0,1,2,3 ..... If m_i and m_j are in the same sentence,the value is 0; if they are one sentence apart, the value is 1; and so on. On the other hand, D_2 is the word distance. It's value represents the number of words between two mentions. In our method, the two distances also are mapped to a vector of dimension d_e (a hyperparameter); this vector is randomly initialized. Then, we obtain the distance vectors d_1 and d_2 with respect to the distances of the current mention in the discourse, and $DF = [d_1, d_2]$. Combining the MF and DF, the mention-pair is represented as $[MF_i, MF_j, DF]^T$, which is subsequently fed into the convolution component of the algorithm.

2.5.3 Convolution As we previously mentioned, word representation approach can capture contextual information through combinations of vectors in a window. But it only produces local features around each word of the sentence. Thus, using all of the local features and predict a coreference resolution relation globally might be necessary. When utilizing neural network, the convolution approach is a natural method to merge all of the features. Motivated by Collobert et al. [6] (2011), we first process the output of *Window Processing* via a linear transformation.

$$\mathbf{Z} = \mathbf{W_1 X} \tag{1}$$

$\mathbf{X} \in \mathbb{R}^{\mathbf{n_o \times t}}$ is the output of the *Window Processing* task, where $n_o = w \times n$,n (a hyperparameter) is the dimension of feature vector, and t is the mention-pair number of the input discourse. $W_1 \in \mathbb{R}^{n_1 \times n_0}$,where n_1 (a hyperparameter) is the size of hidden layer 1, is the linear transformation matrix. We can see that the features share the same weights across all times, which chiefly reduces the number of free parameters to learn. After the linear transformation is applied, the output $\mathbf{Z} \in \mathbb{R}^{\mathbf{n_1 \times t}}$ is dependent on t. To determine the most useful feature in the each dimension of the feature vectors, we perform a max operation over time on $\mathbf{Z}$.

$$m_i = maxZ(i, \cdot) \ \ 0 \leqslant i \leqslant n_1 \tag{2}$$

where $Z(i, \cdot)$ denote the i-th row of matrix $\mathbf{Z}$. Finally, we obtain the feature vector $m = m_1, m_2, \cdots, m_{n_1}$, the dimension of which is no longer related to t.

2.5.4 Discourse Level Feature Vector To learn more complex features, we designed a non-linear layer and selected hyperbolic tanh as the activation function. One useful property of tanh is that its derivative can be expressed in terms of the function value itself:

$$\frac{d}{dx} \tanh x = 1 - \tanh^2 x \tag{3}$$

It is easy to compute the gradient in the backpropagation training procedure via this approach. In due form, the non-linear transformation can be written as

$$g = \tanh(W_2 m) \tag{4}$$

$W_2 \in \mathbb{R}^{n_2 \times n_1}$ is the linear transformation matrix, where n_2 (a hyperparameter) is the size of hidden layer 2. Compared with $m \in \mathbb{R}^{n_1 \times 1}$, $g \in \mathbb{R}^{n_2 \times 1}$ can be considered higher level features (discourse level features).

2.6 Output

In this component, a single feature vector $f = [g]$ is generated by means of concatenating the automatically learned lexical to discourse level features mentioned above. To compute the confidence of each mention-pair, the feature vector $f \in \mathbb{R}^{n_2 \times 1}$ is fed into a softmax classifier.

$$o = W_3 f \tag{5}$$

Here, the transformation matrix is $W_3 \in \mathbb{R}^{n_3 \times n_2}$ and the ultima output of the network is $o \in \mathbb{R}^{n_3 \times 1}$, where n_3 is equal to the number of mention-pairs of the input discourse. Each output can be interpreted as the confidence score of the mention-pair. This score can be used as a conditional probability by applying a softmax operation (see Sect. 2.7)

2.7 Backpropagation Training

For the CDNN based Coreference resolution method proposed here, it could be stated as a quads $\theta = (X, W_1, W_2, W_3)$. We consider that each input discourse is independently. Given an input example E, the network with parameter θ outputs the vector o, where the i-th component o_i contains the score for mention-pair i. In order to obtain the conditional probability $p(i|x, \theta)$, we apply a softmax operation over all mention-pairs in a discourse:

$$p(i|x, \theta) = \frac{e^{o_i}}{\sum\limits_{k=1}^{n^3} e^{o_k}} \tag{6}$$

Given all T training examples$(x^{(i)}; y^{(i)})$, we can then infer the log likelihood of the parameters as follows:

$$J(\theta) = \sum_{i=1}^{T} \log p(y^{(i)}|x^{(i)}, \theta) \tag{7}$$

To get the value of the network parameter θ, we maximize the log likelihood $J(\theta)$ via a simple optimization technique called stochastic gradient descent (SGD). W_1,W_2 and W_3 are initialized randomly and X is initialized via the word embeddings. Because of the parameters located in different layers of the neural network, we carry out the backpropagation algorithm: the differentiation chain rule is applied through the network until the word embedding layer is reached by iteratively selecting an example (x,y) and applying the following update rule.

$$\theta \leftarrow \theta + \lambda \frac{\partial \log p(y|x, \theta)}{\partial \theta} \tag{8}$$

3 Experiments

3.1 Dataset and Evaluation Metric

We evaluated the system on the English part of the corpus provided in the CoNLL-2012 Shared Task. The corpus considers all pronouns (PRP, PRP$), noun phrases (NP) and heads of verb phrases (VP) as potential mentions. It contains 7 categories of documents (over 2 K documents, 1.3 M words). We used the official train/dev/test data sets. Evaluating our system in this closed mode requires the consideration of using that only provided data.

In our experiments, the gold mentions are only considered by us. This is a rather idealized setting but our focus is on comparing various pairwise models rather than on building a full coreference resolution system. We show the three metrics as below that are most commonly used in the CoNLL-2011 and CoNLL-2012 Shared Tasks. In addition, these campaigns use an un-weighted average over the F1 scores given by the three metrics, **MUC** (Vilain et al. [17], 1995), **B-Cubed** (Bagga and Baldwin [18], 1998), **CEAF** (Luo [19], 2005).

3.2 Results of Comparison Experiments

In this section,we experimentally study the effects of the three parameters in our proposed method: the window size in the convolutional component w, the layer size s, the number of hidden layer 1, and the number of hidden layer 2. We respectively vary the number of hyper parameters w,s,n_1 and n_2 and compute the $F1$. Table 2 reports all the hyperparameters used in the following experiments.

Table 2. Hyperparameters used in our experiments

Parameter	Win. size	Word dim	Distance dim	Layer size	Hidden layer1	Hidden Layer2	Learning rate
Value	$w = 3$	n=50	$d_e = 8$	s=2	$n_1 = 300$	$n_2 = 200$	$\lambda = 0.01$

To obtain the final performance of our automatically learned features, we select three approaches as competitors to be compared with our method in Table 3. These systems are Soons Closest-First algorithm (Soon et al. [9], 2001), NGCARDIEs Best-First algorithm (Ng and Cardie [2], 2002) and Lassalle-Deniss Closest-First algorithm (Lassalle and Denis [15], 2013). The first competitor uses 12 traditional features and employs HMM as the classifier. It learns from a small, annotated corpus (namely, the MUC-6 and MUC-7 coreference corpora) and obtains encouraging results, indicating that on the general noun phrase coreference task, the learning approach holds promise and achieves accuracy comparable to that of non-learning approaches. The second competitor extends the work of Soon et al., which had achieved good improvement by means of additional 43 traditional features,in addition to adopting the 12 Soon's features. The similar results are achieved when we construct the baseline system on new data set using these two approaches.

These systems design a series of features and take advantage of a variety of resources. The third competitor (Lassalle and Denis [15], 2013) learns how to best separate types of mention pairs into equivalence classes for which it construct distinct classification models. Their experiments on the CoNLL-2012 Shared Task English datasets (gold mentions) indicate that this method is robust relative to different clustering strategies and evaluation metrics, showing large and consistent improvements over a single pairwise model using the same base features. However,this method had only considered standard,heuristic linking strategies like Closest-First,without considering more sophisticated decoding strategies.

The CDNN model builds a single compositional semantics for the minimal constituent.It is almost certainly too much to expect a single fixed transformation to be able to capture the meaning combination effects of all natural language operators.

Table 3. CoNLL-2012 test (gold mentions))

	MUC			B^3			CEAF			
Models	P	R	F1	P	R	F1	P	R	F1	Mean
Soon	79.49	93.72	86.02	26.23	89.43	40.56	49.74	19.92	28.45	51.68
NGCARDIE	81.02	93.82	86.95	23.33	93.92	37.38	40.31	18.97	25.80	50.04
LassalleDenis	83.23	73.72	78.19	73.50	67.09	70.15	47.30	60.89	53.24	67.19
Our Model	83.33	91.67	**87.30**	64.39	78.29	**70.66**	48.26	58.73	**52.98**	**70.32**

Table 3 illustrates the macro-averaged F1 measure results for these competing methods along with the resources, features and classifier used by each method. Based on these results, we make the following observations:

(1) Richer feature sets lead to better performance when using traditional features. This improvement can be explained by the need for semantic generalization from training to test data. The quality of traditional features relies on human ingenuity and prior NLP knowledge. It is almost impossible to manually choose the best feature sets.

(2) Our method using CDNN start with a word embedding phase, which maps words into a fixed length, real valued vectors. CDNN can learn suitable features automatically with raw input data, given a training objective. Multilayer neural networks are trained with the standard back propagation algorithm. As the networks are non-linear and the task specific objectives usually contain many local maximums, special techniques such as layerwise pre-training (Bengio et al. [12], 2007) and many tricks (LeCun et al. [20], 1998)have been taken in the optimization process to train better neural networks. Compared with some shallow learning methods, the CDNN model can capture the meaning combination effectively and achieve a higher performance.

(3) Our method reaches a score of 70.32, which would place it above the mean score (66.41) of the systems that took part in the CoNLL-2012 Shared

Task[1] (gold mentions track). Except for the first at 77.22, the best performing systems have a score around 68-70. Considering the automatical feature extracting strategy we employed, our current system sets up a strong baseline.

4 Conclusion

In this paper, we described a method for modeling feature hierarchically in pairwise coreference resolution by using a convolutional deep neural network (DNN) architecture to extract mention and discourse level features. In this architecture, the first layer extracts features for each mention. The second layer extracts features from the whole discourse of containing mention, treating it as a sequence with local and global structure. The following layers are standard neural network layers. We applied this method to optimize the pairwise model of a coreference resolution system. Instead of exploiting hand-craft input features carefully optimized for coreference resolution, our system learns internal representations on the basis of raw features of mention-pairs from the whole discourse. Our experiments on the CoNLL-2012 Shared Task English datasets (gold mentions) indicate that our method has a competitive result.

Acknowledgments. This work was supported partly by the National Natural Science Foundation of China (No.61472267) and Suzhou key laboratory of mobile networking and applied technologies (Open Project).

References

1. Aone, C., Bennett, S.W.: Evaluating automated and manual acquisition of anaphora resolution strategies. In: 33rd Annual Meeting of the Association for Computational Linguistics, pp. 122–129. ACL Press, Massachusetts (1995)
2. Ng, V., Cardie, C.: Improving machine learning approaches to coreference resolution. In: 40th Annual Meeting of the Association for Computational Linguistics, pp. 104–111 (2002)
3. Kehler, A., Appelt, D.E., Taylor, L., Simma, A.: The (non) utility of predicate-argument frequencies for pronoun interpretation. In: HLT-NAACL, vol. 4, pp. 289–296 (2004)
4. Ponzetto, S.P., Strube, M.: Exploiting semantic role labeling, WordNet and wikipedia for coreference resolution. In: Proceedings of the Main Conference on Human Language Technology Conference of the North American Chapter of the Association of Computational Linguistics. Association for Computational Linguistics, pp. 192–199 (2006)
5. Bengtson, E., Roth, D.: Understanding the value of features for coreference resolution. In: Proceedings of the Conference on Empirical Methods in Natural Language Processing, pp. 294–303. Association for Computational Linguistics (2008)
6. Collobert, R., Weston, J., Bottou, L., Karlen, M., Kavukcuoglu, K., Kuksa, P.: Natural language processing (almost) from scratch. J. Mach. Learn. Res. **12**, 2493–2537 (2011)

[1] http://conll.cemantix.org/2012/.

7. Yang, N., Liu, S., Li, M., Zhou, M., Yu, N.: Word alignment modeling with context dependent deep neural network. In: 51st Annual Meeting of the Association for Computational Linguistics, pp. 166–175. ACL Press, Sofia (2013)
8. Zeng, D., Liu, K., Lai, S. et al.: Relation classification via convolutional deep neural network. In: Proceedings of COLING, pp. 2335–2344 (2014)
9. Soon, W.M., Ng, H.T., Lim, C.Y.: A machine learning approach to coreference resolution of noun phrases. Comput. Linguist. **27**(4), 521–544 (2001)
10. Rahman, A., Ng, V.: Narrowing the modeling gap: a cluster-ranking approach to coreference resolution. JAIR **40**(1), 469–521 (2011)
11. Socher, R., Lin, C.C., Ng, A.Y., Manning, C.D.: Parsing natural scenes and natural language with recursive neural networks. In: Proceedings of the 26th International Conference on Machine Learning (ICML), vol. 2, p. 7 (2011)
12. Bengio, Y., Lamblin, P., Popovici, D., Larochelle, H.: Greedy layer-wise training of deep networks. Adv. Neural Inf. Process. Syst. **19**, 153 (2007)
13. Turian, J., Ratinov, L., Bengio, Y.: Word representations: a simple and general method for semi-supervised learning. In: Proceedings of the 48th Annual Meeting of the Association for Computational Linguistics, pp. 384–394 (2010)
14. Harris, Z.: Distributional structure. Word **10**(23), 146–162 (1954)
15. Lassalle, E., Denis, P.: Improving pairwise coreference models through feature space hierarchy learning. In: Proceedings of the 51st Annual Meeting of the Association for Computational Linguistics, pp. 497–506, Sofia (2013)
16. McEnery, A., Tanaka, I., Botley, S.: Corpus annotation and reference resolution. In: Proceedings of a Workshop on Operational Factors in Practical, Robust Anaphora Resolution for Unrestricted Texts, pp. 67–74. Association for Computational Linguistics (1997)
17. Vilain, M., Burger, J., Aberdeen, J., Connolly, D., Hirschman, L.: A model-theoretic coreference scoring scheme. In: Proceedings of the 6th conference on Message understanding, pp. 45–52. Association for Compu-tational Linguistics (1995)
18. Bagga, A., Baldwin, B.: Algorithms for scoring coreference chains. In: The First International Conference on Language Resources and Evaluation Workshop on Linguistics Conference, vol. 1, pp. 563–566 (1998)
19. Luo, X.: On coreference resolution performance metrics. In: Proceedings of the Conference on Human Language Technology and Empirical Methods in Natural Language Processing, pp. 25–32. Association for Com-putational Linguistics (2005)
20. LeCun, Y., Bottou, L., Bengio, Y., Haffner, P.: Gradient-based learning applied to document recognition. In: Proceedings of the IEEE, vol. 86(11), pp. 2278–2324 (1998)

Linear Feature Sensibility for Output Partitioning in Ordered Neural Incremental Attribute Learning

Ting Wang[1,2(✉)], Sheng-Uei Guan[3], Jieming Ma[4], and Fangzhou Liu[5]

[1] State Key Laboratory of Intelligent Technology and Systems, Department of Computer Science and Technology, Tsinghua University, Beijing 100084, China
tingwang@tsinghua.edu.cn

[2] Research Center of Web Information and Social Management, Wuxi Research Institute of Applied Technologies, Tsinghua University, Wuxi 214072, China

[3] Department of Computer Science and Software Engineering, Xi'an Jiaotong-Liverpool University, Suzhou 215123, China
steven.guan@xjtlu.edu.cn

[4] School of Electronics and Information Engineering, Suzhou University of Science and Technology, Suzhou 215011, China
jieming84@gmail.com

[5] Department of Computer Science, University of Liverpool, Liverpool L69 3BX, UK
fangzhou.liu@liverpool.ac.uk

Abstract. Feature Ordering is a special training preprocessing for Incremental Attribute Learning (IAL), where features are trained one after another. Since most feature ordering calculation methods, compute feature ordering in one batch, no matter, this study presents a novel approach combining input feature ordered training and output partitioning for IAL to compute feature ordering with considering whether the output of the classification problem is univariate or multivariate. New metric called feature's Single Sensibility (SS) is proposed to individually calculate features' discrimination ability for each output. Finally, experimental benchmark results based on neural networks in IAL show that SS is applicable to calculates feature's discrimination ability. Furthermore, combined output partitioning can also improve further the final classification performance effectively.

Keywords: Fisher linear discriminant · Incremental attribute learning · Pattern classification · Machine learning · Neural networks

1 Introduction

Incremental Attribute Learning (IAL) is a "Divide-and-Conquer" machine learning strategy, which does not train features, also called attributes, in one batch, but incrementally trains them in one or more size. It has been shown as an applicable approach

X. He et al. (Eds.): IScIDE 2015, Part II, LNCS 9243, pp. 373–383, 2015.
DOI: 10.1007/978-3-319-23862-3_37

for solving machine learning problems in regression and classification using Genetic Algorithm (GA) [1, 2], Neural Network (NN) [3, 4], Support Vector Machine (SVM) [5], Particle Swarm Optimization (PSO) [6], Decision Tree (DT) [7], and so on. These previous studies also showed that IAL can outperform conventional batch-training methods, because IAL can reduce the interference brought by different input features [8]. If features are trained together by conventional methods in one batch, the interference between each other cannot be erased.

Previous studies investigated two different ways to reduce interference in IAL. One is Feature Ordering in terms of inputs [9–16], and the other is Output Partitioning [17]. Previous studies have successfully and independently verified that these two methods are applicable for final result improvement. Thus in this study, these two interference reduction approaches will be used simultaneously to investigate whether the integrated method is feasible to improve the final classification performance.

In this paper, feature's Single Sensibility (SS), a novel IAL Feature Ordering metric, is proposed for Output Partitioning based on feature's Single Discriminability (SD) [11] with considering whether the output of the classification problem is univariate or multivariate. New approach which combines feature ordering and output partitioning is also studied in this paper. As a neural network algorithm of IAL, incremental neural network training with an increasing input dimension (ITID) [9] is employed to test the feasibility and accuracy of this new integrated approach. This paper is divided into 5 sections. The Background of IAL and its preprocessing is introduced in next Section. Section 3 introduces the approach of Output Partitioning with feature ordered training based on feature's sensibility. Benchmarks are tested in Sect. 4 with some experimental result analysis, and the conclusions are drawn in the last section.

2 Preprocessing in Incremental Attribute Learning

2.1 Neural Incremental Attribute Learning

IAL gradually imports features one by one. At present, based on some intelligent predictive methods like NN, new approaches and algorithms have been presented for IAL. For example, ITID was shown applicable for classification. It divides the whole input space into several sub spaces, each of which corresponds to an input feature. Instead of learning input features altogether as an input vector in a training instance, ITID learns input features one after another through their corresponding sub-networks while the structure of NN gradually grows with an increasing input dimension based on Incremental Learning in terms of Input Attributes (ILIA) [4]. During training, information obtained by a new sub-network is merged together with the information obtained by the old network. Such architecture is based on ILIA1. After training, if the outputs of NN are collapsed with an additional network sitting on the top where links to the collapsed output units and all the input units are built to collect more information from the inputs, this results in ILIA2 as shown in Fig. 1. Finally, a pruning technique is adopted to find out the appropriate network architecture. With less internal interference among input features, ITID achieves higher generalization accuracy than conventional methods [9].

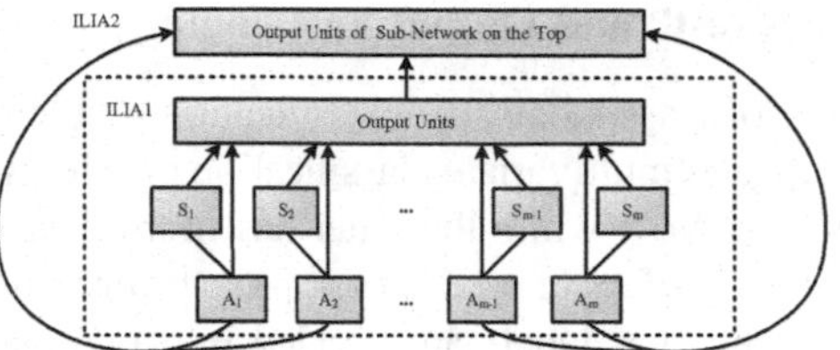

Fig. 1. The network structure of ITID

2.2 Fisher Linear Discriminant for Feature Ordered Training

Feature Ordered Training is unique in IAL. Because IAL imports and trains features one after another in a given sequence, Feature Ordering becomes a necessary pre-processing step in IAL. Previous studies have applied Fisher's Linear Discriminant (FLD) [10, 15], Entropy [12], Correlation Coefficient [13], mRMR [16, 18] and wrappers [3, 9] to IAL Feature Ordering computation, and the approach based on FLD shows higher speed and lower error rates in final results.

FLD provides simple ways to estimate the accuracy of classification. It firstly assumes that the datasets used in FLD are Gaussian conditional density models, where data have normal distributed classes or equal class covariance. The Fisher criterion aims to search a direction where the distance between different classes is the farthest and the distance of each pattern within every class is the closest. Thus, in this direction, the ratio of distance between-classes and within-classes is the largest compared with other directions. Such a direction often leads to the simplest classification. FLD in two-category classification is

$$J(w) = \frac{(\tilde{\mu}_2 - \tilde{\mu}_1)^2}{s_1^2 + s_2^2} \tag{1}$$

where $\tilde{\mu}_1$ and $\tilde{\mu}_2$ are two means of projected classes, and s_1 and s_2 are within-class variances. FLD aims to search the matrix w for maximum $J(w)$. The larger the $J(w)$, the easier the classification. FLD with such an objective can be treated as traditional linear discriminant, which creates new features to classify original datasets.

In previous research, Fisher Score (FS) [10] was improved based on FLD, which was successfully applied in IAL feature ordering. However, different from FLD, FS keeps the directions stable. It makes the current existing feature's direction as $J(w)$'s direction. Thus feature's discrimination ability can be calculated, and feature ordering can be obtained with the descending order of FS by $J(w)$. Moreover, FS is able to compute feature ordering before training based on features' discrimination ability for multivariate classification problems in one batch, while FLD is unable to do.

2.3 "One-Against-the-Rest" and Output Partitioning

According to the number of class labels, the classification problems can be divided into two categories: univariate and multivariate classification problems. The former has only two class labels: "Yes" or "No"; while the latter has more than two labels.

"One-Against-The-Rest" refers to a kind of approach for multivariate classification problems where all classification labels are divided into two categories: "Belonging-To" or "Un-Belonging-To". Thus a multivariate classification problem can be transformed to an univariate problem, when outputs are divided into two partitions. Previous research has show that Output Partitioning is feasible to reduce the classification error rates for each individual category in classification [17, 19, 20]. Therefore, in this study, Output Partitioning will be firstly employed, and Feature Ordering will be calculated by a linear discriminant based on FLD for IAL.

3 Feature Sensibility for Output Partitioning

3.1 Feature Discriminability for Multivariate Classification Problems

Discriminability is a metric for feature's discrimination ability based on FLD [11], which is often used to rank feature's contribution to outputs. In IAL, each feature's discriminability can be estimated in this feature's one-dimensional space. Features can be ordered by the ranking value of feature discriminability. For univariate classification problems, based on Eq. (1), discriminability of feature f_i can be given by

$$D(f_i) = \frac{(\mu_2 - \mu_1)^2}{s_1^2 + s_2^2} \tag{2}$$

where μ_1 and μ_2 are two means of classes, and s_1 and s_2 are within-class variances.

However, Eq. (2) is too simple-minded to cope with multi-category classifications, because the between-class scatter is difficult to describe merely by distance between patterns. Here, the difference between the centers of these multiple classes can be replaced by standard deviations of centers and standard deviations of patterns, so that the influence brought by classes whose mean is not the smallest or the largest among all means of classes can be measured.

Single Discriminability (SD) is a ratio of a feature by the standard deviation of all class centers and the sum of standard deviations of all patterns in each class. SD for both two-category and n-category classification problems can be unified as

$$SD(f_i) = \frac{std\left[\left(\mu_{f_i,C_j}\right)_{j=1}^{j=n}\right]}{\sum_{j=1}^{j=n} std(f_i)_{C_j}} \tag{3}$$

where n is the total number of classes, and two stds are standard deviations, one for all patterns belonging to class C_j in feature i, and the other for the vector consisting of the means of all classes.

3.2 Feature Sensibility

For multivariate classification problems, although SD is able to compute feature orderings for all outputs simultaneously, and acceptable classification results also can be obtained based on these feature orderings, it is still not very clear whether the final result performance still can be improved or not. Previous studies have not employed Feature Ordering and Output Partitioning together. However, these two approaches have been confirmed as effective methods to reduce error rates. Therefore, this study attempts to employ Feature Ordering and Output Partitioning simultaneously to improve the final classification results.

Generally, univariate classification is easier than multivariate classification. Based on the thinking of "Divide-and-Conquer", a multivariate classification can be easily transformed to a univariate classification according to a so-called "One-Against-The-Rest" strategy. In this strategy, data belonging to the target class are put into one group, and everything else in the other. Then an n-dimension ($n > 2$) output classification problem will be converted to n univariate classification problems. Thus if feature discriminability is used in these two-output classification problems and the number of input features is m, then the results of feature discriminability will be a $n \times m$ matrix. In comparison, feature discriminability in univariate classification problems with "One-Against-The-Rest" strategy will be replaced as Sensibility. Sensibility for each single feature is called feature's Single Sensibility (SS). Feature Single Discriminability (SD) only refers to the discrimination ability in the whole multiple output classification problems.

Definition 1. Single Sensibility (SS) refers to the discriminating capacity of one input feature f_i for distinguishing a given output C_j from all the other features $C_1, \ldots, C_{j-1}, C_{j+1}, \ldots, C_n$ where f_i is the ith feature in the set of inputs, n is the number of output features.

Similar to SD, two kinds of metrics can be employed in the computation of SS. One is the standard deviation of sample means from selected output feature and unselected output features in one input dimension, and the other is the sum of standard deviation of selected output feature and unselected output features in one input dimension. If C_j ($1 \leq j \leq n$) is the jth output feature, SS can be calculated by

$$SS(f_i, c_j) = \frac{std\left[\mu_{f_i,C_j}, \mu_{f_i,\bar{C}_j}\right]}{std(f_i)_{C_j} + std(f_i)_{\bar{C}_j}} \tag{4}$$

where $\bar{C}_j$ is the group of all the classes except C_j. It is manifest that if the number of classes is two, SD equals to SS.

4 Benchmarks

The proposed IAL with Output Partitioning based on Feature Ordering derived by feature's SS were tested on four classification benchmarks from University of California, Irvine (UCI) Machine Learning Repository. The datasets are Diabetes, Cancer, Glass and Thyroid, where Diabetes and Cancer are univariate classification problems, while Glass and Thyroid are multivariate problems. This experiment aims to confirm that the results derived by SS and SD in univariate classification problems are the same, while in multivariate problems, they are different. Moreover, results based on orderings derived by SS and Output Partitioning method usually show better performance than those derived by other feature orderings and classified together.

All patterns of these benchmarks were randomly divided into three sets: training (50 %), validation (25 %) and testing (25 %), where training data were firstly used to rank feature ordering based on SS or SD. The performance of the utilization of SS and Output Partitioning was evaluated based on the comparison of error rates with other three approaches: SD feature ordering, original feature ordering and the conventional batch training method. ITID was employed in all of these experiments. The input feature rank results with the values of SS and SD are showed from Tables 1, 2, 3 and 4, respectively, corresponding to these four relative datasets.

Table 1. Feature sensibility and discriminability with ranks of diabetes

Feature	SS of class 1 (rank)	SS of class 2 (rank)	SD (rank)
1	0.1437 (5)	0.1437 (5)	0.1437 (5)
2	0.3694 (1)	0.3694 (1)	0.3694 (1)
3	0.0509 (8)	0.0509 (8)	0.0509 (8)
4	0.1047 (6)	0.1047 (6)	0.1047 (6)
5	0.0960 (7)	0.0960 (7)	0.0960 (7)
6	0.2630 (2)	0.2630 (2)	0.2630 (2)
7	0.1567 (4)	0.1567 (4)	0.1567 (4)
8	0.1844 (3)	0.1844 (3)	0.1844 (3)

Table 2. Feature sensibility and discriminability with ranks of cancer

Feature	SS of class 1 (rank)	SS of class 2 (rank)	SD (rank)
1	0.7039 (5)	0.7039 (5)	0.7039 (5)
2	0.9566 (2)	0.9566 (2)	0.9566 (2)
3	0.9888 (1)	0.9888 (1)	0.9888 (1)
4	0.6604 (7)	0.6604 (7)	0.6604 (7)
5	0.6260 (8)	0.6260 (8)	0.6260 (8)
6	0.8725 (3)	0.8725 (3)	0.8725 (3)
7	0.7793 (4)	0.7793 (4)	0.7793 (4)
8	0.6895 (6)	0.6895 (6)	0.6895 (6)
9	0.3534 (9)	0.3534 (9)	0.3534 (9)

Table 3. Feature sensibility and discriminability with ranks of glass

Feature	SS of class 1 (rank)	SS of class 2 (rank)	SS of class 3 (rank)	
1	0.4691 (5)	0.1367 (4)	0.0622 (7)	
2	0.4738 (4)	0.0563 (9)	0.2038 (2)	
3	1.0344 (1)	0.5228 (1)	0.1355 (4)	
4	0.6229 (3)	0.4145 (2)	0.0829 (5)	
5	0.3797 (6)	0.0724 (7)	0.0245 (9)	
6	0.1318 (9)	0.0997 (5)	0.0284 (8)	
7	0.1356 (8)	0.0811 (6)	0.0405 (6)	
8	0.7330 (2)	0.2477 (1)	0.2312 (1)	
9	0.2186 (7)	0.0420 (3)	0.1561 (3)	
Feature	SS of class 4 (rank)	SS of class 5 (rank)	SS of class 6 (rank)	SD (rank)
1	0.0796 (7)	0.0141 (9)	0.1912 (7)	0.3226 (8)
2	0.0572 (8)	0.2381 (4)	0.4818 (1)	0.2605 (4)
3	0.4153 (1)	0.5930 (1)	0.3164 (4)	0.1716 (1)
4	0.1133 (5)	0.4418 (2)	0.0451 (8)	0.1566 (3)
5	0.2119 (4)	0.1488 (6)	0.2457 (5)	0.1514 (6)
6	0.0423 (9)	0.3321 (3)	0.4608 (2)	0.0976 (5)
7	0.0952 (6)	0.1587 (5)	0.0285 (9)	0.0802 (9)
8	0.2258 (2)	0.0874 (7)	0.2258 (6)	0.0764 (2)
9	0.2211 (3)	0.0527 (8)	0.4186 (3)	0.0542 (7)

Table 4. Feature sensibility and discriminability with ranks of thyroid

Feature	SS of class 1 (rank)	SS of class 2 (rank)	SS of class 3 (rank)	SD (rank)
1	0.0070 (19)	0.0335 (16)	0.0276 (13)	0.5890 (16)
2	0.0685 (8)	0.0593 (13)	0.0696 (9)	0.5272 (13)
3	0.1330 (5)	0.0275 (17)	0.2724 (1)	0.3816 (5)
4	0.0198 (14)	0.0806 (9)	0.0050 (17)	0.2883 (14)
5	0.0328 (12)	0.0750 (11)	0.0221 (14)	0.1067 (15)
6	0.0042 (21)	0.1441 (6)	0.0290 (12)	0.0727 (7)
7	0.0891 (7)	0.0867 (8)	0.0879 (7)	0.0672 (6)
8	0.0235 (13)	0.0231 (18)	0.0853 (8)	0.0648 (12)
9	0.0101 (17)	0.0189 (19)	0.0047 (18)	0.0551 (21)
10	0.0945 (6)	0.0755 (10)	0.0979 (6)	0.0487 (11)
11	0.0155 (15)	0.0481 (15)	0.0001 (21)	0.0478 (17)
12	0.0062 (20)	0.0493 (14)	0.0042 (19)	0.0420 (18)
13	0.0674 (9)	0.0657 (12)	0.0665 (10)	0.0373 (9)
14	0.0097 (18)	0.0079 (21)	0.0175 (15)	0.0321 (19)
15	0.0123 (16)	0.0119 (20)	0.0121 (16)	0.0316 (20)
16	0.0559 (10)	0.1625 (5)	0.0295 (11)	0.0278 (8)
17	0.3429 (3)	0.5429 (4)	0.1843 (4)	0.0247 (3)
18	0.2830 (4)	0.5569 (3)	0.1513 (5)	0.0209 (4)
19	0.4143 (2)	0.9917 (2)	0.2302 (3)	0.0120 (2)
20	0.0340 (11)	0.0899 (7)	0.0028 (20)	0.0100 (10)
21	0.4434 (1)	1.0780 (1)	0.2591 (2)	0.0095 (1)

The compared results of SS and SD with conventional classification method are presented from Tables 5, 6, 7 and 8 relative to the datasets of Diabetes, Cancer, Glass and Thyroid respectively. In this study, the experimental results were compare with other two approaches, which are Original Ordering method and conventional methods. The Original Ordering method is based on the feature's original order shown in datasets while the conventional methods is a traditional method which has no feature ordering and trains all features in one batch and is not an IAL approach [13].

Table 5. Result comparison of diabetes

Approach	Feature ordering	Error rate (ILIA1)	Error rate (ILIA2)
ITID-SS	2-6-8-7-1-4-5-3	21.84896 %	22.39583 %
ITID-SD	2-6-8-7-1-4-5-3	21.84896 %	22.39583 %
Original ordering	1-2-3-4-5-6-7-8	22.86458 %	23.80209 %
Conventional method		By batch: 23.93229 %	

Table 6. Result comparison of cancer

Approach	Feature ordering	Error rate (ILIA1)	Error rate (ILIA2)
ITID-SS	3-2-6-7-1-8-4-5-9	1.69541 %	1.72414 %
ITID-SD	3-2-6-7-1-8-4-5-9	1.69541 %	1.72414 %
Original ordering	1-2-3-4-5-6-7-8-9	2.90230 %	2.18391 %
Conventional method		By batch: 1.86782 %	

Table 7. Result comparison of glass

Approach	Feature ordering	Error rate (ILIA1)	Error rate (ILIA2)
ITID-SS (class 1)	3-8-4-2-1-5-9-7-6	18.67922 %	19.71696 %
ITID-SS (class 2)	3-4-8-1-6-7-5-2-9	29.15096 %	28.49057 %
ITID-SS (class 3)	8-2-9-3-4-1-7-6-5	7.54717 %	7.54717 %
ITID-SS (class 4)	3-8-9-5-4-7-1-2-6	0.00000 %	0.84906 %
ITID-SS (class 5)	3-4-6-2-7-5-8-9-1	0.75472 %	0.75472 %
ITID-SS (class 6)	2-6-9-3-5-8-1-4-7	8.96228 %	7.54717 %
ITID-SD	3-8-4-2-6-5-9-1-7	34.81133 %	28.96228 %
Original ordering	1-2-3-4-5-6-7-8-9	45.18870 %	36.03775 %
Conventional method		By batch: 41.22641 %	

For the datasets of Diabetes and Cancer, the number of outputs is 2, which present a univariate classification problem. According to Tables 1 and 2, it is obvious that Feature Ordering derived by SS and SD in univariate classification problems are the same. As the compared results showed in Tables 5 and 6, the final classification results of these two methods are naturally the same, as ITID-SS and ITID-SD shows exactly the same feature ordering and the same error rate in both ILIA1 and ILIA2 cases, which is smaller than the conventional method error rate and Original Ordering method error rate. Therefore, SS is SD in univariate classification problems, which has been verified by the experimental results of Diabetes and Cancer.

While for the datasets of Glass and Thyroid, the number of output is 3 and 6 separately, which indicated the number of class is 3 and 6 respectively, showing a multivariate classification problem. In multivariate classification problems, the results for SS and SD are different for Glass and Thyroid datasets, according to the input feature rank results with the values of SS and SD in Tables 3 and 4, as the SS for each class is considered separately. Based on SS and "One-Against-The-Rest" strategy, SS is employed to rank feature orderings for univariate classification problems, where some of them were transformed from multivariate classification problems. SS is no longer equal to SD in multivariate classification. SS values of different outputs are also different. As the output number is reduced, less targets brought about less interference. According to the compared experimental results from Tables 7 and 8, most of the SS results in Glass and Thyroid in both ILIA1 and ILIA2 cases are better than those derived by the other approaches except the second class of Thyroid. The reason why some class has larger error rate, for example Class 2 of Thyroid is an exception, it would be an important issue in future research. However, because the number of patterns belonging to Thyroid Class 2 is only 368 (about 10 % in the total), thus the overall result is still very good. Thus Feature Ordering based on SS and "One-Against-The-Rest" method are feasible to obtain better classification results based on IAL.

Table 8. Result comparison of thyroid

Approach	Feature ordering	Error rate (ILIA1)	Error rate (ILIA2)
ITID-SS (class 1)	21-19-17-18-3-10-7-2-13-16-20-5-8-4-11-15-9-14-1-12-6	0.50556 %	0.44444 %
ITID-SS (class 2)	21-19-18-17-16-6-20-7-4-10-5-13-2-12-11-1-3-8-9-15-14	2.62501 %	1.96944 %
ITID-SS (class 3)	3-21-19-17-18-10-7-8-2-13-16-6-1-5-14-15-4-9-12-20-11	1.61667 %	1.45278 %
ITID-SD	21-19-17-18-3-7-6-16-13-20-10-8-2-4-5-1-11-12-14-15-9	1.92778 %	1.52500 %
Original ordering	1-2-3-4-5-6-7-8-9-10-11-12-13-14-15-16-17-18-19-20-21	2.05000 %	1.59167 %
Conventional method		By batch: 1.86389 %	

5 Conclusion

This study presented a novel integrated IAL data preprocessing method which is combined with SS, a new Feature Ordering metric for multivariate classification problems, and "One-Against-The-Rest" Output Partitioning strategy. Experimental results using UCI datasets confirmed that such a method is able to cope with both univariate and multivariate classification problems. Moreover, SS is always equal to SD in univariate classification. In addition, experimental results also illustrated this novel integrated approach can outperform those conventional batch-training methods.

Acknowledgments. This research is supported by National Nature Science Foundation of China under Grant No. 61332007, China Jiangsu Provincial Science and Technology Foundation under Grant No. BK20131182, and China Postdoctoral Science Foundation under Grant No. 2015M571042.

References

1. Guan, S.U., Zhu, F.M.: An incremental approach to genetic-algorithms-based classification. IEEE Trans. Syst. Man Cybern. Part B Cybern. **35**, 227–239 (2005)
2. Zhu, F., Guan, S.U.: Ordered incremental training with genetic algorithms. Int. J. Intell. Syst. **19**, 1239–1256 (2004)
3. Guan, S.U., Liu, J.: Incremental ordered neural network training. J. Intell. Syst. **12**, 137–172 (2002)
4. Guan, S.U., Li, S.: Incremental learning with respect to new incoming input attributes. Neural Process. Lett. **14**, 241–260 (2001)
5. Liu, X., Zhang, G., Zhan, Y., Zhu, E.: An incremental feature learning algorithm based on least square support vector machine. In: Preparata, F.P., Wu, X., Yin, J. (eds.) FAW 2008. LNCS, vol. 5059, pp. 330–338. Springer, Heidelberg (2008)
6. Bai, W., Cheng, S., Tadjouddine, E.M., Guan, S.-U.: Incremental attribute based particle swarm optimization. In: 2012 8th International Conference on Natural Computation, ICNC 2012, pp. 669–674. IEEE Computer Society, 29 May 2012–31 May 2012
7. Chao, S., Wong, F.: An incremental decision tree learning methodology regarding attributes in medical data mining. In: 2009 International Conference on Machine Learning and Cybernetics, pp. 1694–1699. IEEE Computer Society, 12 July 2009–15 July 2009
8. Ang, J.H., Guan, S.U., Tan, K.C., Al Mamun, A.: Interference-less neural network training. Neurocomputing **71**, 3509–3524 (2008)
9. Guan, S.U., Liu, J.: Incremental neural network training with an increasing input dimension. J. Intell. Syst. **13**, 45–69 (2004)
10. Wang, T., Guan, S.U.: Feature ordering for neural incremental attribute learning based on fisher's linear discriminant. In: 5th International Conference on Intelligent Human-Machine Systems and Cybernetics, IHMSC 2013, vol. 2, pp. 507–510. IEEE Computer Society, Hangzhou, China (2013)
11. Wang, T., Guan, S.-U., Liu, F.: Feature discriminability for pattern classification based on neural incremental attribute learning. In: Wang, Y., Li, T. (eds.) ISKE2011. AISC, vol. 122, pp. 275–280. Springer, Heidelberg (2011)
12. Wang, T., Guan, S.-U., Liu, F.: Entropic feature discrimination ability for pattern classification based on neural IAL. In: Wang, J., Yen, G.G., Polycarpou, M.M. (eds.) ISNN 2012, Part II. LNCS, vol. 7368, pp. 30–37. Springer, Heidelberg (2012)
13. Wang, T., Guan, S.U., Liu, F.: Correlation-based feature ordering for classification based on neural incremental attribute learning. Int. J. Mach. Learn. Comput. **2**, 807–811 (2012)
14. Wang, T., Guan, S.U., Man, K.L., Ting, T.O., Lisitsa, A.: Optimized neural incremental attribute learning for pattern classification based on statistical discriminability. Int. J. Comput. Intell. Appl. **13**, 1450019 (2014)
15. Wang, T., Guan, S.-U., Ting, T., Man, K.L., Liu, F.: Evolving linear discriminant in a continuously growing dimensional space for incremental attribute learning. In: Park, J.J., Zomaya, A., Yeo, S.-S., Sahni, S. (eds.) NPC 2012. LNCS, vol. 7513, pp. 482–491. Springer, Heidelberg (2012)

16. Wang, T., Wang, Y.: Pattern classification with ordered features using mRMR and neural networks. In: 2010 International Conference on Information, Networking and Automation, ICINA 2010, pp. V2128–V2131. IEEE Computer Society, 17 October 2010–19 October 2010
17. Guan, S.U., Li, P.: Incremental learning in terms of output attributes. J. Intell. Syst. **13**, 95–122 (2004)
18. Peng, H., Long, F., Ding, C.: Feature selection based on mutual information: criteria of max-dependency, max-relevance, and min-redundancy. IEEE Trans. Pattern Anal. Mach. Intell. **27**, 1226–1238 (2005)
19. Guan, S.U., Li, S.C.: Parallel growing and training of neural networks using output parallelism. IEEE Trans. Neural Netw. **13**, 542–550 (2002)
20. Guan, S.U., Wang, K.: Hierarchical incremental class learning with output parallelism. J. Intell. Syst. **16**, 167–193 (2007)

Survey on Visualization Layout for Big Data

Pengju Teng[1], Hongjun Li[1(✉)], and Xiaopeng Zhang[2]

[1] College of Science, Beijing Forestry University, Beijing 100083, China
lihongjun69@bjfu.edu.cn
[2] NLPR-LIAMA, Institute of Automation, CAS, Beijing 100190, China

Abstract. A reasonable visualization is helpful for the representation and analysis of big data, and an optimal layout improves the effect of visualization and provides a nice and direct platform for visualization analysis. In this paper surveys visualization literatures focusing on graph topology structure analysis or optimal layout, especially on those for big data visualization. Typical algorithms are classified into four categories, namely parallel coordinates, scatter diagram, tree map and other layout. We also propose the feasible directions for future research.

Keywords: Information visualization · Graph layout · Large data sets

1 Introduction

A large data set called "big data", not only means that the data set is huge, but also means that the data format is complex. Modern data are generated quickly, and the processing speed of data is fast [1] also. One of the greatest challenge of processing big data is how to help users better understand and analyze the big data, and to show the result in a nice pattern [2]. Data visualization focuses on visual rendering methods and techniques for the data, and it is an interdisciplinary research field closely related to graph theory, statistics, data mining, machine learning, psychology and other disciplines [3].

Two basic problems of data visualization are graph structure analysis and optimal layout. Graph structure analysis contributes to optimizing the layout and improving the efficiency of algorithm. Because graphic layout optimization problem not only has great research significance in academic field, but also has extensive and eventual applications in industrial productions, for example electronics circuit. We briefly analyze some important results in this paper focusing on the graph structure analysis and the optimal layout algorithms.

2 Graph Structure Analysis

2.1 The Aesthetic Standard

Some visual systems give users a platform to play their visual apperception and to analyze the complex network structure. Although aesthetic standards are subjective, it is an important principle to design a visual system.

X. He et al. (Eds.): IScIDE 2015, Part II, LNCS 9243, pp. 384–394, 2015.
DOI: 10.1007/978-3-319-23862-3_38

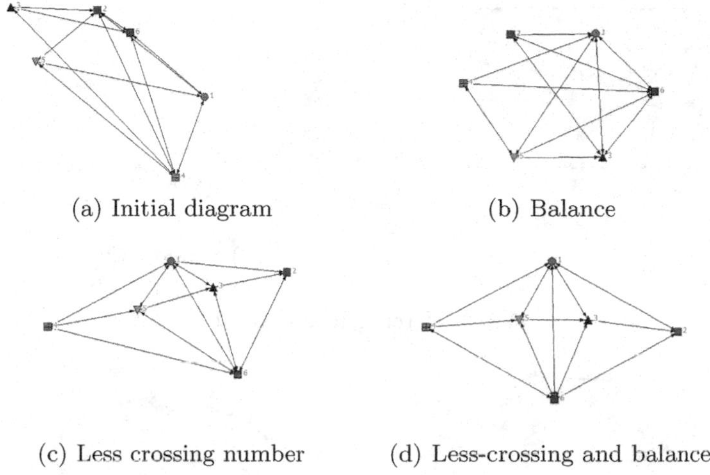

(a) Initial diagram (b) Balance

(c) Less crossing number (d) Less-crossing and balance

Fig. 1. Graph structure designing with different aesthetic standards.

Aesthetic standards include balance, less crossing number et al. Fig. 1(a) shows an initial graph drawn by FDA. The effects can be improved by employing the balance techniquc (Fig. 1(b)), or by decreasing crossing number (Fig. 1(c)). Figure 1(d) illustrates a comprehensive effect by combining the less crossing number and balance standard.

In 1981, Sugiyama et al. [4] summarized the earliest classical aesthetic standards, as "Close" layout of vertices connected to each other, "Balanced" layout of lines coming into or going from a vertex, "Straightness" of lines, and "Less crossing number" between lines (edges). Among those, "Less crossing number" gets the most widely popular. Then Sindren et al. [5] and Purchase et al. [6] put forward the new layout aesthetic standards based on graph readability. Among those standards, the thought that minimum the regional and maximum angle of connection is adopted. There are argues [7] that the connectivity principle more effective than the edges cross, which of course needs more experiment to validate its feasibility. The connectivity principle means that the network path should be a straight line as far as possible.

In recent years, a very hot research topic is heterogeneous layout which is also based on the aesthetics perspective making the graph more beauty and improving the readability. Some researchers now employ the rivers [8] (Fig. 2(a)), sunflower [9] (Fig. 2(b)) to complete the visualization of complex data and streaming data. But these layout aesthetics methods still need more practical and effective quantitative evaluation.

2.2 The Calculation of Edge Crossing Number

Minimum the number of edge crossing is one of the most important standards in graph aesthetics layout. The crossing number is an important topological

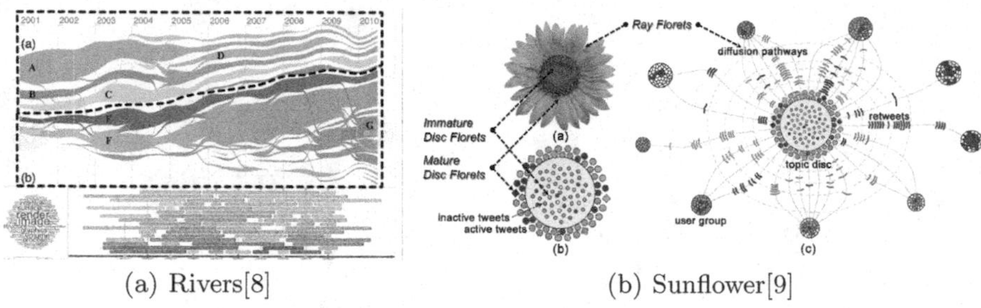

(a) Rivers[8] (b) Sunflower[9]

Fig. 2. Heterogeneous layout.

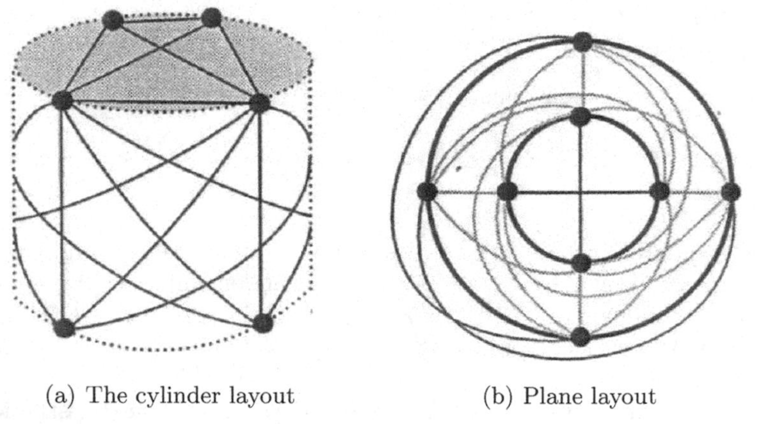

(a) The cylinder layout (b) Plane layout

Fig. 3. The plane layout and the cylinder layout for K_8.

parameter and is often taken into account in practice as an index, such as the circuit wiring problems of electronic circuit boards in industry. Although the calculation of edge crossing number is a NP complete [10], the boundary of the crossing number can be estimated for some kinds of figure. For example, M. Klesc identified the crossing number of all 4-order and some 5-order connected graph and the cartesian product [11,12] graphs. Concerning the complete graph K_p, D.R.Woodall [13] also gave a formula (1).

$$cr(K_p) = \left(\frac{1}{4}\right)\left(\frac{p}{2}\right)\left(\frac{p-1}{2}\right)\left(\frac{p-2}{2}\right)\left(\frac{p-3}{2}\right) 1 \leq \text{p} \leq 10 \qquad (1)$$

where p means the number of vertex, $cr(K_p)$ the minimum crossing number. The layout for graph of cylinder (Fig. 3(a)) and plane (Fig. 3(b)) can get the least number of crossing edges by solving the equation (1).

2.3 Graph Clustering Method

If vertices have high correlation, their distances will be small in the layout space by employing the graph clustering algorithm. At the same time, the number of edges between different clusters is decreased. In the past decades, a lot of graph clustering algorithms are put forward. According to the different main ideas, the clustering method can be divided into four categories: partitioning method, hierarchical method, density-based method, grid-based method.

Partitioning method uses similarity to cluster data. Given data set D with n objects, all individuals split into $k(k \leq n)$ [14]. Both K-means and k-mediods [15] methods work well. They show good performance even for dealing with big data. Figure 4 illustrates a cluster result by K-means method with a data generated from a normal distribution.

Hierarchical method uses the tree map to do stratification analysis for data [16]. The data is represented by a tree structure and a node graph which are sometimes combined with other cluster method such as BIRCH. This is a bottom-up algorithm. An advantage of the algorithm is that the algorithm can be terminated at any a point or meet a specific criterion what is given in advance, such as specifying clustering coefficient or optimizing the function. Now the mainly improvement for hierarchical method is the calculation process of nodes similarity. Figure 5(a) shows the tree structure of the tree map. Figure 5(b) displays the result of the hierarchical method with scatter diagram.

Density-based method is to filter some low density areas and to find the dense degree of sample points. A famous density-based method is DBSCAN (Density-Based Spatial Clustering of Applications with Noise)[17]. An advantage of the method is to find the cluster shown as any shape (Fig. 6). Figure 6(a) shows a result of K-mean for the data which have two cluster, and each cluster meets a condition: $y = sin(x) + z$ and $z \in (-1, 1)$. Figure 6(b) displays a result of

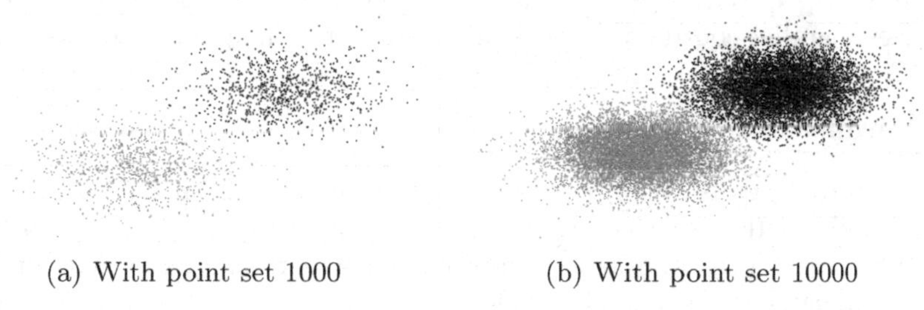

(a) With point set 1000 (b) With point set 10000

Fig. 4. The k-mean cluster.

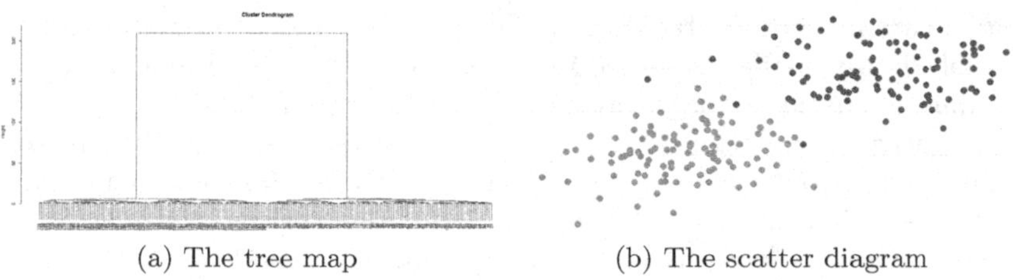

(a) The tree map (b) The scatter diagram

Fig. 5. The hierarchical method with 100 points

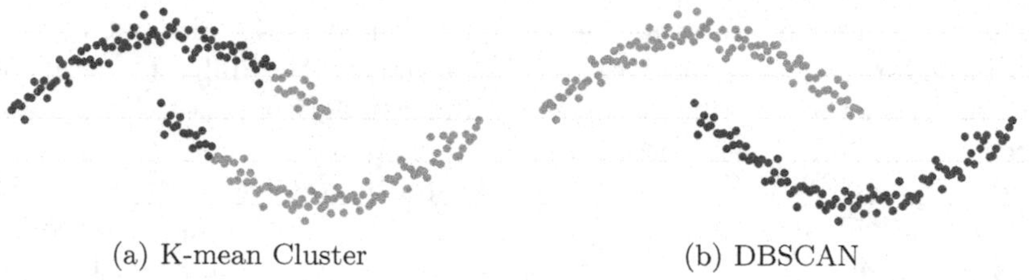

(a) K-mean Cluster (b) DBSCAN

Fig. 6. The data with 200 points.

DBSCAN for the data that have two cluster, and each cluster meets the condition also.

Grid-based method adopts multi-resolution grid data structure. It separates the screen into several rectangular elements. Different rectangular element size is corresponding to the different resolution. Advantages of the method are good locality, small computation, simplicity and fast. One of the most classical algorithms is STING that stores information in the grid cell [18]. The grid-based method is very effective to big data, and it is currently a hot research direction. Another new development is CLIQUE(clustering In QUEst) which combines the density-based method and grid-based method.

3 Graph Layout

In this section, we focus on the layout algorithms of static graphs which is vary popular [19]. The static graphs layout algorithm can be classified to four categories: parallel coordinates, scatter diagram, tree map and other layout.

3.1 Parallel Coordinates

Parallel coordinates is an important visualization expression method for high-dimension data. It uses the two-dimension graph to display multi-dimension data. The main principle is to represent each variable dimension of high-dimensional data by a series of parallel axes, and each variable is corresponding to a axis, see Fig. 7. The variable's value is corresponding to the position on the axis. Each line of high-dimensional data entry is corresponding to the coordinate at a line of parallel coordinates [20,21]. Therefore, parallel coordinates can make the variable data visually presented in one picture at the same time, and it does well for quickly checking and analyzing the variable data [22].

For Negative Correlation (NC) data visualization, a straight line in plane-coordinate system on the map in the parallel coordinates system is multiple line segments which intersect at a point; As for the Positive Correlation (PC) data visualization, a straight line in plane-coordinate system on the map in the parallel coordinates is radial parallel (Fig. 8). In addition, the hierarchical parallel coordinates [21] displays the set with many layers of massive module.

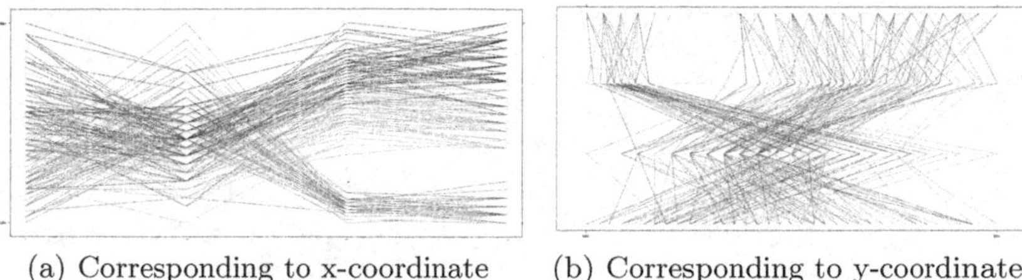

(a) Corresponding to x-coordinate (b) Corresponding to y-coordinate

Fig. 7. Parallel coordinates.

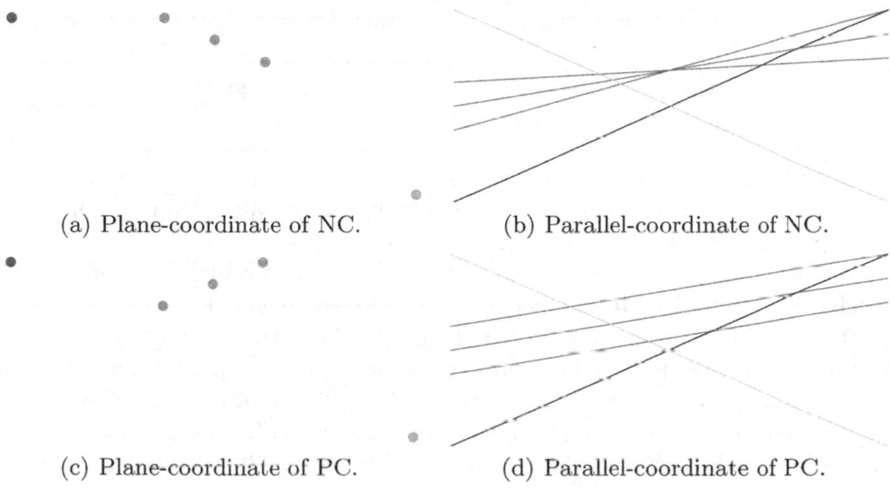

(a) Plane-coordinate of NC. (b) Parallel-coordinate of NC.

(c) Plane-coordinate of PC. (d) Parallel-coordinate of PC.

Fig. 8. The map of plane-coordinate to parallel-coordinate.

3.2 Scatter Diagram

Scatter diagram uses a node to represent an entry, using the connection between nodes to represent the relationship between two entries. Through the layout algorithm, a lot of points and lines will be systematically placed. The method rendering big data with relation information is fast. Scatter diagram method plays an important role in social network, sequence analysis etc.

Force directed layout algorithm called spring layout comes from the spring principle, a physics concept. By calculating the repulsive force and the gravitational force of each spring, the energy of the whole system can be estimated. In theory, the ideal state is the system in the equilibrium state (minimum total energy) [23]. Kamada et al. [24] improves the energy model and the ideal distance of the adjacent nodes. The ideal distance between two adjacent nodes is positive correlation to the shortest path between them.

Davidson and Harel based on the FDA, put forward the DH algorithm [25], and presented a new energy function which is widely used in the algorithm now (Fig. 9). In the algorithm, by adjusting the weight of energy function, the

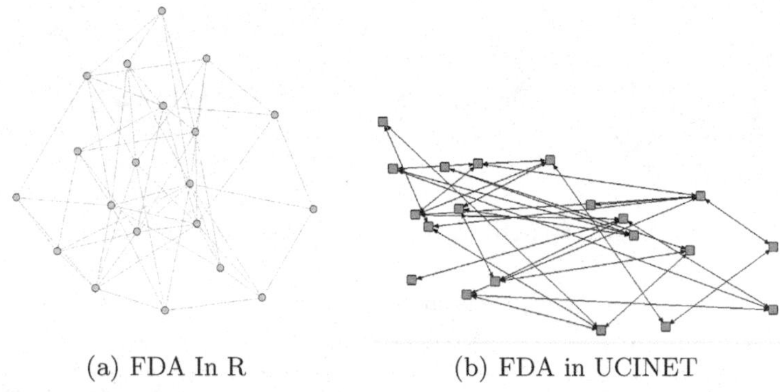

(a) FDA In R (b) FDA in UCINET

Fig. 9. The graph of force directed layout algorithm.

different layout which meets different aesthetic standard can be obtained. But in the balance, due to the inherent complexity of the algorithm, it will spend more time than FDA, especially for big data.

Another progress is to separate the drawing space into many grid layouts. In the FR algorithm (Fig. 10) named after Fruchterman and Reingold, annealing algorithm is simulated with an optimal solution [26]. But the gravity setting for nodes is not always reasonable and shows the overlap of points, and the time complexity of an iteration reaches $O(k * (n^2 + e))$, where n means the number of nodes, and e the number of edges. There are two obvious improvements in this direction. The one is as the Social Action System [27] and Matrix Explorers [28]. The other is based on the characteristics of the data itself and combines the theory of visual analysis to improve the algorithm.

The traditional FDA algorithm is hard to meet the demands of researchers for further mining and analyzing. It reduces the readability of a graph [29]. By improving FDA algorithm, SAL based on cluster is proposed in [30]. It is helpful to find the cluster on graph and decreases the time complexity.

As for data amount, about 1000 nodes can be managed [31] by employing the multi-scale technology in 2001. Using the maximum independent set as the standard of the multi-scale can get a great layout for big data [32]. Another improvement of the algorithm is from the drawing strategy. But it did not change the time complexity of the algorithm until Hu proposed the concept of super node [33].

Circular layout is easy to understand, as its name talks about. At initial of the algorithm, the shape of the layout is limited to a circular. This is also the most different between the circular with others. For example, Breitkreutz uses the circular layout to build a visual system called Osprey [34]. With the development, the layout not only limits on single circular but also expands to concentric circles. We compare the FDA with the circular layout in Fig. 11.

When the size of data set is too big, the layout shows a poor display. The solution for this problem is to use the visual analysis theory [35]. Zhu introduced an improve algorithm based on SOM [29] which used cluster algorithm

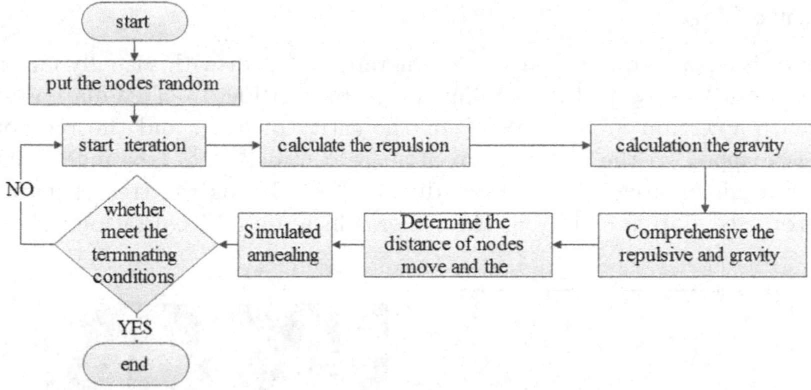

Fig. 10. The flow chart of FR.

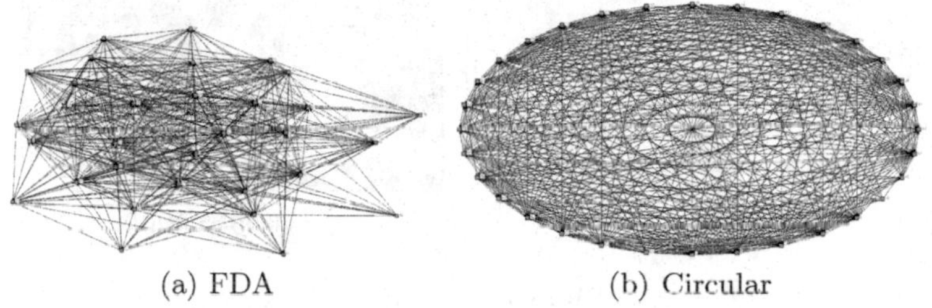

(a) FDA (b) Circular

Fig. 11. The layout of 45 nodes.

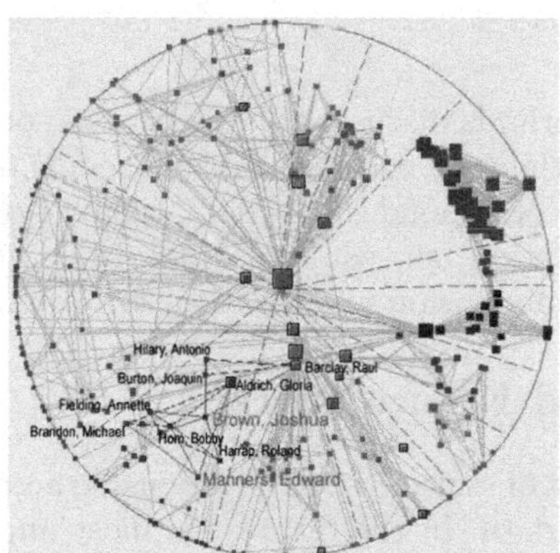

Fig. 12. The interface of the NetVizer system [29].

in advance, then put nodes on circular (Fig. 12). It takes full advantage of the dimensions of circular and has a nice visual effect.

3.3 Tree Map

Tree map is a space-filling method for hierarchical data with visually encoding the hierarchical property by containment, as seen in Fig. 13. Tree map uses the block with a certain area to represent the entity of data, and the relation of entity is mapped to the relationship of space contains [36]. Tree map is useful even for a million items on a single display [37]. But as the tree structure in computer, tree map is mainly used for hierarchical data visualization.

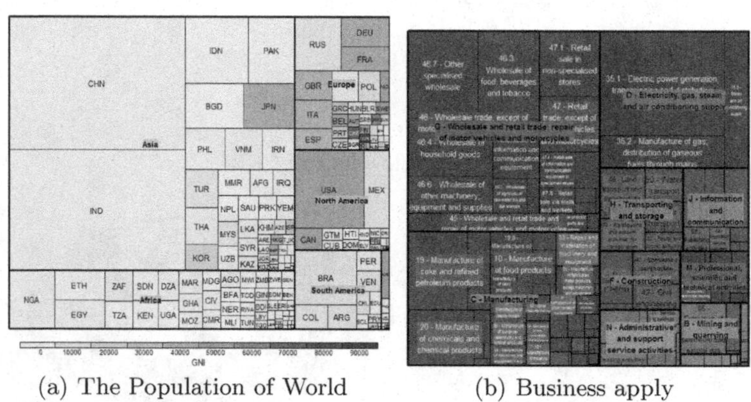

(a) The Population of World (b) Business apply

Fig. 13. Tree map.

3.4 Other Layouts

Besides visualization methods discussed above, various special layout algorithms have been presented in the past decades. One of them is map layout which puts the data on a real map correspond to the data, then we can draw some lines on the map to show the relation between nodes [38]. In addition, some researchers place emphasis on high-dimension visualization, for example, 3D layout [39].

4 Conclusion and Future Work

The visualization layouts of static graphs are discussed according to layout styles, especially those for big data. In the future, the most important problem could be usability, visual scalability, integrated analysis of heterogeneous data, in-situ visualization, errors and uncertainty [40]. In addition, we should consider visual systems combined with man-machine interaction techniques, and the optimizing layout should be connected to the clustering method and graph structure analysis.

Acknowledgment. This research work is supported by National Natural Science Foundation of China with projects Nos. 61372190, 61370193, and 61331018.

References

1. Gorodov, E.Y., Gubarev, V.V.: Analytical review of data visualization methods in application to big data. J. Elec. Comp. Eng. **22**(2–22), 2 (2013)
2. Sun, G.-D., Ying-Cai, W., Liang, R.H., Liu, S.X.: A survey of visual analytics techniques and applications: state-of-the-art research and future challenges. J. Comput. Sci. Technol. **28**(5), 852–867 (2013)
3. Debortoli, S., Mller, O., vom Brocke, J.: Vergleich von kompetenzanforderungen an business-intelligence- und big-data-spezialisten. WIRTSCHAFTSINFORMATIK **56**(5), 315–328 (2014)
4. Sugiyama, K., Shojiro, T.: Methods for visual understanding of hierarchical system structures. IEEE Trans. Syst. Man Cyber. **11**(2), 109–125 (1981)
5. Sindre, G., Gulla, B., Jokstad, H.: Onion graphs: aesthetics and layout. In: Proceedings on Visual Languages, pp. 287–291 (1993)
6. Purchase, H., Cohen, R., James, M.: Validating graph drawing aesthetics. In: Brandenburg, F.J. (ed.) GD 1995. LNCS, vol. 1027, pp. 435–446. Springer, Heidelberg (1996)
7. Ware, C., Purchase, H., Colpoys, L., Mcgill, M.: Cognitive measurements of graph aesthetics. Inf. Vis. **1**, 103–110 (2002)
8. Cui, W., Liu, S., Tan, L., Shi, C.: Textflow: Towards better understanding of evolving topics in text. TVCG **17**(12), 2412–2421 (2011)
9. Cao, N., Lin, Y., Sun, X., Lazer, D., Liu, S., Qu, H.: Whisper: tracing the spatiotemporal process of information diffusion in real time. TVCG **18**(12), 2649–2658 (2012)
10. Garey, M.R., Johnson, D.: Crossing number is np complete. SIAM J. Algebraic Discrete Meth. **4**(3), 312–316 (1983)
11. Kle, M.: The crossing numbers of products of paths and stars with 4-vertex graphs. J. Graph Theo. **6**, 605–614 (1994)
12. Kle, M.: The crossing numbers of cartesian products of paths with 5-vertex graphs. Discrete Math. **233**, 353–359 (2001)
13. Woodall, D.R.: Cyclic-order graphs and zarankiewicz's crossing-number conjecture. J. Graph Theo. **6**, 657–671 (1993)
14. Pothen, A., Simon, H.D., Liou, K.P.: Partitioning sparse matrices with eigenvectors of graphs. SIAM J. Matrix Anal. Appl. **11**(3), 430–452 (1990)
15. Kaufman, L., Rousseeuw, P.J.: Agglomerative Nesting (Program AGNES), pp. 199–252. Wiley, New York (2008)
16. de Abreu, N.M.M.: Old and new results on algebraic connectivity of graphs. Linear Algebra Appl. **1**, 53–73 (2007)
17. Ester, M., Kriegel, H-P., Sander, J., Xu, X.: A density-based algorithm for discovering clusters in large spatial databases with noise. AAAI Press, pp. 226–231 (1996)
18. Wang, W., Yang, J., Muntz, R.: Sting : a statistical information grid approach to spatial data mining. In: VLDB, pp. 186–195 (1997)
19. Ma, K.L., Muelder, C.W.: Large-scale graph visualization and analytics. Computer **46**(7), 39–46 (2013)
20. Inselberg, A., Dimsdale, B.: Parallel coordinates: a tool for visualizing multidimensional geometry. In: Visualization, pp. 361–378 (1990)
21. Fua, Y.H., Ward, M.O., Rundensteiner, E.A.: Hierarchical parallel coordinates for exploration of large datasets. In: Visualization 1999, pp. 43–50 (1999)

22. Dasgupta, A., Chen, M., Kosara, R.: Conceptualizing visual uncertainty in parallel coordinates. Comput. Graph. Forum **31**(3pt2), 1015–1024 (2012)
23. Battista, G.D., Eades, P., Tamassia, R., Tollis, I.G.: Graph Drawing: Algorithms for the Visualization of Graphs. Prentice Hall, Upper Saddle River (1998)
24. Kamada, T., Kawai, S.: An algorithm for drawing general undirected graphs. Inf. Process. Lett. **31**(1), 7–15 (1989)
25. Davidson, R., Harel, D.: Drawing graphs nicely using simulated annealing. ACM Trans. Graph. **15**(4), 301–331 (1996)
26. Fruchterman, T.M.J., Reingold, E.M.: Graph drawing by force-directed placement. Softw. Pract. Exper. **21**(11), 1129–1164 (1991)
27. Perer, A., Shneiderman, B.: Balancing systematic and flexible exploration of social networks. TVCG **12**(5), 693–700 (2006)
28. Henry, N., Fekete, J.-D.: Matrixexplorer: a dual-representation system to explore social networks. TVCG **12**(5), 677–684 (2006)
29. Zhu, B., Watts, S., Chen, H.: Visualizing social network concepts. Decis. Support Syst. **49**(2), 151–161 (2010)
30. Wu, P., Li, S.K.: Layout algorithm suitable for structural analysis and visualization of social network. J. Softw. **22**(10), 2467–2475 (2011)
31. Hadany, R., Harel, D.: A multi-scale algorithm for drawing graphs nicely. Discrete Appl. Math. **113**, 3–21 (2001)
32. Walshaw, C.: A multilevel algorithm for force-directed graph drawing. In: Marks, J. (ed.) GD 2000. LNCS, vol. 1984, pp. 171–182. Springer, Heidelberg (2001)
33. Hu, Y.: Efficient, high-quality force-directed graph drawing. Mathematica J. **10**, 37–71 (2005)
34. Ho, Y., Gruhler, A., Heilbut, A., Bader, G., Moore, L., Adams, S., Millar, A., Taylor, P., Bennett, K., Boutilier, K.: Systematic identification of protein complexes in saccharomyces cerevisiae by mass spectrometry. Nature **415**, 180–183 (2002)
35. Kohonen, T.: Self-Organizing Maps, 3rd edn. Springer, New York (2001)
36. Zhao, S., McGuffin, M., Chignell, M.H.: Elastic hierarchies: combining treemaps and node-link diagrams. In: INFOVIS 2005, pp. 57–64 (2005)
37. Bederson, B.B., Shneiderman, B., Wattenberg, M.: Ordered and quantum treemaps: making effective use of 2d space to display hierarchies. ACM Trans. Graph. **21**(4), 833–854 (2002)
38. Becker, R., Eick, S., Wilks, A.: Visualizing network data. TVCG **1**, 16–28 (1995)
39. Rekimoto, J., Green, M.: The information cube: Using transparency in 3d information visualization. In: WITS 1993 (1993)
40. Liu, S., Cui, W., Wu, Y., Liu, M.: A survey on information visualization: recent advances and challenges. Vis. Comput. **30**(12), 1373–1393 (2014)

Common Latent Space Identification for Heterogeneous Co-transfer Clustering

Liu Yang[1,2], Liping Jing[1(✉)], and Jian Yu[1]

[1] Beijing Key Lab of Traffic Data Analysis and Mining, Beijing Jiaotong University, Beijing, China
{yangliubjtu,lpjing,jianyu}@bjtu.edu.cn

[2] College of Mathematics and Information Science, Hebei University, Baoding, Hebei, China

Abstract. With the rapid development of collection techniques, it is easy to gather various data which come from different domains, such as images, videos, documents, and etc., how to group these heterogeneous data becomes a research issue. Traditional techniques handle these clustering tasks separately, that is one task for one domain, so that they ignore the interactions among domains. In this paper, we present a co-transfer clustering method to deal with these separate tasks together with the aid of co-occurrence data which contain some instances represented in different domains. The proposed method consists of two steps, one is to learn the subspace of different domains which uncovers the latent common topics and respects the intrinsic geometric structure, the next is to simultaneously cluster the instances in all domains via the symmetric nonnegative matrix factorization method. A series of experiments on real-world data sets have shown the performance of the proposed method is better than the state-of-the-art methods.

Keywords: Heterogeneous feature spaces · Co-transfer clustering · Collective matrix factorization · Symmetric nonnegative matrix factorization

1 Introduction

Clustering algorithm is a fundamental technique in unsupervised learning, it partitions a set of instances into clusters according to some similarity or distance strategies. Traditional clustering methods, such as k-means [16] and non-negative matrix factorization [13], are based on the original feature space. However, their performances are not satisfied when the feature space is nonlinear. Thus, some researchers study to map the original feature space into kernel space, such as spectral clustering [22], symmetric non-negative matrix factorization [12], etc. These linear or nonlinear clustering methods are focused on the single domain. They may work well in some domains, such as text domain, because text words contain more semantic information which is beneficial to the text understanding. While in some applications, such as image clustering, the performance may

X. He et al. (Eds.): IScIDE 2015, Part II, LNCS 9243, pp. 395–406, 2015.
DOI: 10.1007/978-3-319-23862-3_39

be poor due to the gap of visual words and the semantic information. Under such situations, some researches proposed transfer learning [15,20,24] to help improve the performance of target domain (such as image) by making use of the information from source domain (such as text).

Based on the feature spaces of source and target domains, transfer learning methods can be divided into two categories. One is homogeneous transfer learning [3,19], that is source and target data come from the same feature space. The other is heterogeneous transfer learning [25] which can handle heterogeneous data, and it has been applied to such as text-image [25,26] and cross-language problems [17]. The difficulty of heterogeneous transfer learning is how to connect different domains. Researchers have proposed various strategies to handel this problem, such as canonical correlation analysis (CCA) [5,11], collective matrix factorization (CMF) method [23], latent semantic analysis (aPLSA) [25] and correlational spectral clustering (CSC) [2] etc. They can learn the projections to simultaneously map the data from each feature space to one common space, they aim to improve the performance of the learning task in one target domain with the aid of source domain. Thus, they focus on one task in target domain and cannot simultaneously obtain the clustering results of multiple heterogeneous tasks.

In fact, some independent tasks may be related and they can be connected by learning the shared information for improving respective performance. Multi-task learning [4] is an approach to learn multiple tasks together at the same time by using a shared representation. But it is often under the supervised setting and the tasks are based on the same feature space. With the tasks in heterogeneous domains, there also may have some connections. For example, although the features are different, they may describe the same object, i.e. they belong to the same category. In this case, two tasks based on the heterogeneous feature spaces can share the same category space. Based on this observation, some researchers tried to combine these two tasks together to enhance performance of the singe task, such as co-transfer learning [17,18]. It can complete all tasks by using the knowledge cross different domains. But it focuses on the supervised problem by combining labeled data in multiple domains.

In this paper, we propose a Heterogeneous Co-Transfer Clustering framework (HCTC) to simultaneously handle multiple clustering tasks in different domains. Like heterogeneous transfer learning [25,26] and co-transfer learning [17,18], we build a bridge to transfer knowledge across domains based on the co-occurrence data which are represented in different domains. We assume that it is possible to obtain co-occurrence information between different domains. For instance, in the text-image mining problem, the images co-occur with texts on the same web page [7], and the linkage between texts and images in social networks [21,26], can be used for co-occurrence information. HCTC makes use of the co-occurrence data to form a bridge to connect different domains and learn the common latent subspace of different domains to keep the intrinsic geometric structure. In the new subspace, all separate clustering tasks are combined together to improve the performance of each task.

HCTC has the following merits. (i) HCTC has ability to co-transfer knowledge across different heterogeneous domains and respect the intrinsic geometric structure. (ii) It can simultaneously perform more than one unsupervised learning (clustering) tasks. (iii) It outperforms the state-of-the-art algorithms in the real-world benchmark data sets. The rest of this paper is organized as follows. In Sect. 2, the common latent space identification and heterogeneous co-transfer clustering framework are presented. The optimization analysis and convergence analysis are shown in Sect. 3. In Sect. 4, a series of experimental results are listed to illustrate the effectiveness of HCTC. Finally, some concluding remarks are given in Sect. 5.

2 Common Latent Space Identification for Co-transfer Clustering

In this section, we will show how to learn the common latent space of heterogeneous domains, and complete all clustering tasks in the new space.

2.1 Problem Statement

Given K heterogeneous domains, the k-th domain contains n_k instances $\{\widetilde{\mathbf{x}}_i^k\}_{i=1}^{n_k}$ (where $\widetilde{\mathbf{x}}_i^k \in \mathbb{R}^{m_k}$), and the data matrix is denoted as $\widetilde{X}^k = [\widetilde{\mathbf{x}}_i^k, \cdots, \widetilde{\mathbf{x}}_{n_k}^k] \in \mathbb{R}^{m_k \times n_k}$. The feature spaces of heterogeneous domains are different, such as image and text. The situation we focused on is that $\widetilde{X}^k$ and $\widetilde{X}^j$ share the same category space. The goal is to complete the clustering tasks in K domains, where each task is to group $\widetilde{X}^k$ into c clusters. Traditional clustering methods, such as k-means and spectral clustering, handle these tasks separately, then the interactions among domains are ignored.

The co-occurrence data, which can be easily obtained in real application, can be used to be a bridge to link separated tasks [7,25,26]. The co-occurrence data set is $O = \{X^k\}_{k=1}^K$ which contains n_o instances appear in K feature spaces ($X^k = [\mathbf{x}_1^k, \cdots, \mathbf{x}_{n_o}^k] \in \mathbb{R}^{m_k \times n_o}$), where X^k and $\widetilde{X}^k$ are in the same domain. $\mathbf{x}_i^1, \cdots, \mathbf{x}_i^K$ correspond to the representations of the i-th instance in K domains. We expect to simultaneously accomplish K clustering tasks together by using the co-occurrence data to improve the single clustering performance.

2.2 Common Latent Space Identification

Our motivation is to study the instance interactions from heterogeneous domains. However, $\widetilde{X}^k$ and $\widetilde{X}^j$ are in different feature spaces with different dimensional feature vectors, the relationship cannot be computed directly by using traditional similarity or distance methods. $\widetilde{X}^k$ and $\widetilde{X}^j$ should be firstly mapped to the common space, and then the similarity matrix can be computed in the new space. Based on the co-occurrence data, we propose a unified framework which is able to find the projections W^k from K feature spaces to one common space, and learn the new representation H^k. It should be guaranteed the data similarity

$\mathcal{K}_{H^k}$ in the common space is close to the similarity $\mathcal{K}_{X^k}$ in the original space. As a result, our model can be formulated as an optimization problem as follows,

$$\min_{W^k,H^k} \sum_{k=1}^{K} \alpha^k \ell(X^k, W^k, H^k) + \beta^k \text{Dist}(\mathcal{K}_{X^k}, \mathcal{K}_{H^k}) \tag{1}$$

where $W^k \in \mathbb{R}^{m_k \times r}$ represents the mapping matrix from the k-th feature space to the new r-dimensional common space $(1 \leq r \leq \min\{m_1, m_2, \cdots, m_K\})$, and $H^k = [\mathbf{h}_1^k, \cdots, \mathbf{h}_{n_o}^k] \in \mathbb{R}^{r \times n_o}$ is the new representation of n_o instances in the co-occurrence data set. The first term in (1) is the empirical risk functional of the decision function on the original data X^k, and the new representation H^k with the mapping matrix W^k. $\ell(\cdot)$ is the loss function which can be selected according to the application. The second term is the regularizer to describe the difference of representations X^k and H^k. Parameters $\alpha^k, \beta^k \geq 0$ are introduced to control the weights of different domains and balance the regularizer respectively. Hence, the K mapping functions $\{W^k\}_{k=1}^K$ can be learned by solving Problem (1).

Design of Loss Function: The empirical loss function on the data in the k-th domain represents the cost on replacing the original data X^k with the new representation H^k by the mapping matrix W^k. We consider the following widely used square loss functions for the k-th domain [13]:

$$\begin{aligned} \ell(X^k, W^k, H^k) &= \|X^k - W^k H^k\|_F^2 \\ \text{s.t. } \; W^k, H^k &\geq 0. \end{aligned} \tag{2}$$

where $\|\cdot\|_F$ denotes the Frobenius norm. Here the non-negative constraints on W^k and H^k are introduced due to its simple and interpretable part-based representation. Although the representations of the i-th instance ($\mathbf{x}_i^k \in \mathbb{R}^{m_k}$ and $\mathbf{x}_i^j \in \mathbb{R}^{m_j}$) correspond to different domains, they describe the same instance, that is $\mathbf{h}_i^k$ should be consistent with $\mathbf{h}_i^j$. Then, all representations $(\mathbf{h}_i^1, \cdots, \mathbf{h}_i^K)$ of the i-th instance should be unified as $\mathbf{h}_i$. Thus, we can constrain the new representations $H^1 = H^2 = \cdots = H^K = H$. Hence, (2) becomes

$$\begin{aligned} \ell(X^k, W^k, H) &= \|X^k - W^k H\|_F^2 \\ \text{s.t. } \; W^k, H &\geq 0. \end{aligned} \tag{3}$$

The loss function of all K domains is combined by $\sum_{k=1}^K \alpha^k \ell(X^k, W^k, H)$. The parameter α^k is the weight of each domain, we can set $\sum_{k=1}^K \alpha^k = 1$.

Difference of the Original and the New Representations: Give the co-occurrence data set O, we can construct a nearest neighbor graph G^k for each domain to model the local manifold structure with n_o vertices in the k-the domain, where each vertex represents an instance. Let $S^k \in \mathbb{R}^{n_o \times n_o}$ be the weight matrix of G^k. If $\mathbf{x}_i^k$ is among the e-nearest neighbors of $\mathbf{x}_j^k$ or $\mathbf{x}_j^k$ is among the e-nearest neighbors of $\mathbf{x}_i^k$, $S_{i,j}^k = \mathcal{K}(\mathbf{x}_i^k, \mathbf{x}_j^k)$, otherwise, $S_{i,j}^k = 0$. $\mathcal{K}(\mathbf{x}_i^k, \mathbf{x}_j^k)$ is kernel function which is selected by the application.

We define $d_i^k = \sum_{j=1}^{n_o} S_{i,j}$, and $D^k = \text{diag}(d_1^k, \cdots, d_{n_o}^k) \in \mathbb{R}^{n_o \times n_o}$. If two points are sufficiently close in the original feature space, then we expect that they have similar representations in the latent subspace. Consider the consensus of the new representation H with the weighted graph G^k, a reasonable criterion [1] for choosing a map to minimize the following objective function,

$$\frac{1}{2}\sum_{i=1}^{n_o}\sum_{j=1}^{n_o}(\mathbf{h}_i - \mathbf{h}_j)^2 S_{i,j}^k = \text{Trace}(HL^kH^T) \tag{4}$$

where $L^k = D^k - S^k$ is the Laplacian matrix.

Final Formulation: Combining (3) and (4) in all K domains, we arrive at the following minimization problem:

$$\min_{\{W^k\}_{k=1}^K, H} \sum_{k-1}^{K} (\alpha^k \|X^k - W^k H\|_F^2 + \beta^k \text{Trace}(HL^kH^T))$$
$$\text{s.t. } \{W^k\}_{k=1}^K, H \geq 0 \tag{5}$$

It can learn the mapping matrix W^k from the k-th domain to the common subspace and new representation H.

2.3 Heterogeneous Co-transfer Clustering

Given K data sets $\{\widetilde{X}^k\}_{k=1}^K$ from K different feature spaces, we aim to divide each data set $\widetilde{X}^k$ into c clusters. Because K data sets share the same category space, we expect to put them together for better clustering performance. We first use the matrices $\{W^k\}_{k=1}^K$ to map the data to the common subspace. Then we can obtain the new representation via $\widetilde{Q}^k = (W^k)^T\widetilde{X}^k \in \mathbb{R}^{r \times n_k}$ of the data in the k-th domain. By combining the instances in all domains, we can get the data matrix by $\widetilde{U} = [\widetilde{Q}^1\widetilde{Q}^2 \cdots \widetilde{Q}^K] \in \mathbb{R}^{r \times n_a}$ (where $n_a = \sum_{k=1}^K n_k$).

After getting the representation matrix $\widetilde{U}$, the goal becomes to group $\widetilde{U}$ into different clusters. We can use the clustering methods for single domain to get the results. Traditional methods such as k-means, non-negative matrix factorization, spectral clustering, symmetric non-negative matrix factorization (SNMF) have been widely used in clustering application. Among them, symmetric non-negative matrix factorization have been reported with better performance due to its ability of capturing the cluster structure [12]. Thus, we adopt SNMF to get the clustering results.

3 Optimization and Convergence Analysis

In this section, we will show the optimization of Problem (5) and the corresponding convergence analysis.

3.1 Optimization

We present the detailed optimization strategy for solving Problem (5). Following the iteratively optimization method, solving (5) can be separated into two steps: learning mapping matrix W^k by fixing the new representation H, and learning the representation H by fixing W^k. The multiplicative update approach [13] based on the coordinate descent is adopted due to its simple and easy to implement.

Computing W^k: Given the fixed H, the step of computing the optimal W^k becomes the form of non-negative quadratic programming, which can be solved by multiplicative updates. When fixing $H = H^{(t)}$ (where t is the current iteration number), the optimization problem that involves W^k is given by

$$\min_{W^k} ||X^k - W^k H||_F^2 \ \text{ s.t. } W^k \geq 0. \tag{6}$$

The optimization problem (6) is in form of non-negative quadratic programming which is same with the traditional NMF model. Thus, W^k can be updated via the strategy in [13], the updated rule is as follows,

$$(W^k)_{i,j}^{(t+1)} \leftarrow \left[\frac{\left(X^k H^T\right)_{i,j}}{\left((W^k)^{(t)} H H^T\right)_{i,j}} \right] (W^k)_{i,j}^{(t)}. \tag{7}$$

Computing H: For the fixed $W^k = (W^k)^{(t+1)}$, the objective function that involves H is given by

$$\min_{H} \sum_{k=1}^{K} \alpha^k ||X^k - W^k H||_F^2 + \beta^k \text{Trace}(H L^k H^T)$$
$$\text{s.t. } \ H \geq 0. \tag{8}$$

We can derive the updated rule for H as follows:

$$H_{i,j}^{(t+1)} \leftarrow \left[\frac{\sum_{k=1}^{K} \alpha^k \left((W^k)^T X^k\right)_{i,j}}{\sum_{k=1}^{K} \left(\alpha^k \left((W^k)^T W^k H^{(t)}\right)_{i,j} + \beta^k (H^{(t)} L^k)_{i,j}\right)} \right] H_{i,j}^{(t)}. \tag{9}$$

Obviously, if the initial $(W^k)^{(0)}$ and $H^{(0)}$ are non-negative, the update rules for W^k and H can guarantee the non-negativeness of the corresponding variables. Note that the objective value in (5) decreases in each step as proven in the next subsection.

3.2 Convergence Analysis

When H is fixed and W^k is updated, the optimization of (5) becomes the traditional non-negative matrix factorization. Then the proof of the convergence is similar with the traditional method [14]. When update H with fixed W^k, the decrease of the objective function value can be proved by the following theorem.

Theorem 1. *Suppose $\{W^k\}_{k=1}^K$ is fixed. When we update $H^{(t)}$ to $H^{(t+1)}$ by using (9), the objective function value of (5) monotonically decreases, i.e.,*

$$\sum_{k=1}^{K}\Big(\alpha^k(||X^k - W^kH^{(t+1)}||_F^2 - ||X^k - W^kH^{(t)}||_F^2) \\ +\beta^k \mathit{Trace}\left(H^{(t+1)}L^k(H^{(t+1)})^T - H^{(t)}L^k(H^{(t)})^T\right)\Big) \leq 0. \tag{10}$$

Proof. Let $P(H)$ be

$$P(H) = \sum_{k=1}^{K}(\alpha^k||X^k - W^kH||_F^2 + \beta^k\mathrm{Trace}(HL^kH^T)) \\ = \sum_{k=1}^{K}\mathrm{Trace}\,(\alpha^k(X^k)^TX^k - 2\alpha^kH^T(W^k)^TX^k + \alpha^kH^T(W^k)^TW^kH \\ +\beta^kH^THL^k)\,.$$

Then, we can reformulate (10) to show the following inequality:

$$P(H^{(t+1)}) - P(H^{(t)}) \leq 0, \tag{11}$$

In order to prove (11), we introduce an auxiliary function about H from [14] as

$$F(H,\hat{H}) = \sum_{k=1}^{K}\mathrm{Trace}\,(\alpha^k(X^k)^TX^k - 2\alpha^kH^T(W^k)^TX^k) \\ +\sum_{k=1}^{K}\left(\sum_{i=1}^{r}\sum_{j=1}^{n_o}\left(\frac{\alpha^k((W^k)^TW^k\hat{H})_{i,j}}{\hat{H}_{i,j}}H_{i,j}^2 + \frac{\beta^k(\hat{H}L^k)_{i,j}}{\hat{H}_{i,j}}H_{i,j}^2\right)\right).$$

According to [8], we know that the following matrix inequality holds.

$$\mathrm{Trace}(H^TAHB) \leq \sum_{i,j}\frac{(A\hat{H}B)_{i,j}}{\hat{H}_{i,j}}H_{i,j}^2, \tag{12}$$

where A, B and H are non-negative matrices, $A = A^T$ and $B = B^T$. Let $A = (W^k)^TW^k$, $B = I_{n_o}$, where I_{n_o} is an $n_o \times n_o$ identity matrix. It can be seen that

$$\mathrm{Trace}(H^T(W^k)^TW^kH) \leq \sum_{i=1}^{r}\sum_{j=1}^{n_o}\frac{((W^k)^TW^k\hat{H})_{i,j}}{\hat{H}_{i,j}}H_{i,j}^2. \tag{13}$$

If $A = I_r$ is an $r \times r$ identity matrix, and $B = L^k$, then it can be obtained

$$\mathrm{Trace}(H^T H L^k) \leq \sum_{i=1}^{r} \sum_{j=1}^{n_o} \frac{(\hat{H} L^k)_{i,j}}{\hat{H}_{i,j}} H_{i,j}^2. \tag{14}$$

Thus, we can get $P(H) \leq F(H, \hat{H})$, where $P(H) = F(H, \hat{H})$ if and only if $H = \hat{H}$. The optimal H can be obtained by minimizing $F(H, \hat{H})$. Let $f(H) = F(H, \hat{H})$, it can be seen that $f(H)$ is a convex function and there is a unique global minima. By setting $\frac{\partial f(H)}{\partial (H)_{i,j}} = 0$, we can get

$$H_{i,j} = \left[\frac{\sum_{k=1}^{K} \alpha^k ((W^k)^T X^k)_{i,j}}{\sum_{k=1}^{K} (\alpha^k ((W^k)^T W^k \hat{H})_{i,j} + \beta^k (\hat{H} L^k)_{i,j})} \right] \hat{H}_{i,j}, \tag{15}$$

When setting $H^{(t+1)} = H$ and $H^{(t)} = \hat{H}$, (15) becomes the updating rule (9). Also, $f(H^{(t+1)}) \leq f(H^{(t)})$, i.e., $F(H^{(t+1)}, H^{(t)}) \leq F(H^{(t)}, H^{(t)})$, then we can have

$$\begin{aligned} P(H^{(t+1)}) &= F(H^{(t+1)}, H^{(t+1)}) \leq F(H^{(t+1)}, H^{(t)}) \\ &\leq F(H^{(t)}, H^{(t)}) = P(H^{(t)}), \end{aligned} \tag{16}$$

which means that $P(H)$ monotonically decreases, i.e., the inequality (11) is hold. Hence Theorem 1 can be established.

4 Experimental Results

In this section, a series of experiments are conducted to demonstrate the effectiveness of the proposed algorithm.

4.1 Methodology

The proposed algorithm is tested on two data sets. The first is NUS-WIDE [6] which comes from Flickr and contains 81 categories. We randomly pick out 6 binary classification tasks from [17] by selecting 4 categories (flower, rock, sun, and tree). For each task, 600 images, 1,200 texts and 1,600 co-occurred image-text pairs are sampled. 500 visual words and 1,000 text terms are extracted to build the image and document features. The second data set is Cross-language Data which is crawled from Google and Wikipedia [18] and contains two categories of text documents about birds and animals. Among them, there are 3,415 English documents in the first domain, 2,511 French documents in the second domain, 3,113 Spanish documents in the third domain, and 2,000 co-occurred English-French-Spanish documents. 5,000 important terms are extracted from each domain are used to form a feature vector.

We compare the proposed HCTC with seven methods. Among them, k-means [10], spectral clustering (SC) and symmetric non-negative matrix factorization (SNMF) only use the instances in single domain, the other four methods

(CCA [11], CMF [23], aPLSA [25] and CSC [2]) focus on one task in target domain. The clustering result is evaluated by comparing the obtained label of each instance with the label provided by the data set. The clustering performance is measured via accuracy which is defined by $\sum_{i=1}^{n_k} \delta(y_i, \widehat{y_i})/n_k$, where y_i denotes the predicted cluster of instance $\widetilde{\mathbf{x}}_i^k$, $\widehat{y_i}$ denotes its true label, n_k is the number of data in the k-th task. $\delta(y_i, \widehat{y_i}) = 1$ if $y_i = \widehat{y_i}$, otherwise $\delta(y_i, \widehat{y_i}) = 0$. The larger accuracy is, the better clustering performance is.

4.2 Results and Discussion

In this subsection, we discuss the clustering performance of the proposed model. Before experimental comparison, it would be important to note the kernel function used in the following results. For image data, we use Gaussian kernel function [9] to yield the non-linear version of similarity in the intrinsic manifold structure of data. For text data, we adopt the cosine similarity.

NUS-WIDE data set contains image-text data, then the number of domains K is 2. Since CCA, CMF, aPLSA and CSC only consider to migrate the knowledge from a source domain to a target domain, experiment records the results of all methods on one clustering task (image clustering). Then the target domain is image, and the source domain is text.

The parameters of all methods will affect their performances, we randomly select one binary image clustering task (with 600 images, 300 for flower and 300 for tree, and 1,600 image-text pairs) as an example to see the effects of parameters. The parameter r is the number of bases or canonical variables in HCTC, CCA, CMF, aPLSA and CSC. Figure 1(a) gives the image clustering performance in terms of varying the number of r. We can see that HCTC can

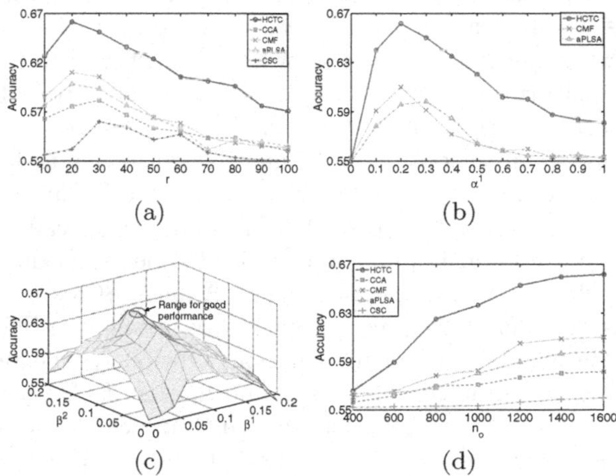

Fig. 1. Accuracy effect of (a) number of canonical variables or bases r, (b) transferred weight α^1, (c) regular parameters β^1 and β^2, and (d) co-occurred data size n_o on clustering accuracy.

get the best performance with 20 bases which is much smaller than the number of original image and text features. Then, the computational complexity of clustering can be efficiently reduced with the new image representation.

Since $K = 2$ in this data set, the transferred weights in HCTC, CMF and aPLSA are α^1 and α^2. For $\alpha^1 + \alpha^2 = 1$, we can only tune α^1 and set $\alpha^2 = 1 - \alpha^1$. Here α^1 represents the weight of image domain, it is tuned from 0 to 1 with step 0.1, and the results are shown in Fig. 1(b). The good prediction performance is obtained when α^1 is in the range of [0.1, 0.2], which confirms that the related text information is helpful to image clustering and it is reasonable to build a bridge between text domain and image domain for migrating the useful knowledge.

For HCTC, we test the effects of two regular parameters (β^1 and β^2) on the final clustering performance. Figure 1(c) gives the detailed performance for different values of β^1 and β^2. The performance becomes better and better with the increasing of β^1 and β^2 values, while decreases when they are larger than 0.1. The good performance can be achieved around the values close to $\beta^1 = 0.06$ and $\beta^2 = 0.1$. Since β^2 is larger than β^1, it is coincident with α^1 and α^2, which indicates that the text information is more important with the image clustering.

Usually, the performance is affected by the co-occurrence data size (i.e., n_o) on transfer learning. We run the image clustering task with different sizes, and the accuracies are recorded in Fig. 1(d). When the number of co-occurrence instances increases, the accuracies of all methods increase as well. It indicates that more co-occurrence instances make the representation more precise and helpful for the clustering data in transfer learning.

Moreover, we show the accuracy results of 6 binary image clustering tasks in the upper part of Table 1. For each task, the related 1,600 image-text pairs are used as the co-occurrence data. As expected, HCTC is better than other methods. On the average, HCTC achieves accuracy improvement of 0.05, 0.10, 0.13, 0.12, 0.12, 0.13 and 0.15 against CCA, CMF, aPLSA and CSC, SNMF, SC and k-means on all image clustering tasks respectively. This indicates that the image clustering benefits from the knowledge transferred from text domain. Because HCTC can utilize the text information more effective and reasonable, it gains better performance than other methods.

HCTC also can simultaneously handel multiple clustering tasks (i.e. $K > 2$), we conduct the experiments on three-cross-language data and the clustering results are shown in lower part of Table 1. CMF and aPLSA can be extended to transfer knowledge from multiple source domains to one target domain, but they handle the three clustering tasks separately. CCA and CSC cannot be extended to deal with this three-cross-language data. SC, k-means and SNMF are three baselines using the single domain information. Obviously, the performance is better when taking advantage of all domain information. HCTC can combine all tasks together and obtain the best accuracy which indicates that documents in different domains can help each other for single language clustering task.

Table 1. Clustering results of 6 image clustering tasks and 3 cross-language-clustering tasks.

	HCTC	CCA	CMF	aPLSA	CSC	SNMF	SC	k-means
Flower-tree	**0.66**±0.02	0.58±0.02	0.61±0.02	0.60±0.02	0.56±0.02	0.56±0.01	0.55±0.00	0.54±0.00
Flower-rock	**0.69**±0.02	0.61±0.02	0.54±0.02	0.52±0.02	0.60±0.02	0.59±0.01	0.59±0.00	0.53±0.00
Flower-sun	**0.74**±0.01	0.71±0.01	0.67±0.02	0.64±0.02	0.62±0.01	0.62±0.02	0.61±0.00	0.60±0.00
Tree-rock	**0.63**±0.01	0.60±0.02	0.51±0.01	0.51±0.01	0.52±0.02	0.52±0.01	0.51±0.00	0.50±0.00
Tree-sun	**0.72**±0.02	0.68±0.01	0.58±0.02	0.56±0.02	0.57±0.01	0.57±0.01	0.56±0.00	0.56±0.00
Rock-sun	**0.75**±0.01	0.72±0.02	0.68±0.02	0.59±0.01	0.60±0.02	0.60±0.02	0.59±0.00	0.58±0.00
English	**0.65**±0.01	-	0.62±0.01	0.61±0.01	-	0.59±0.01	0.58±0.00	0.57±0.01
French	**0.67**±0.02	-	0.64±0.02	0.64±0.02	-	0.63±0.01	0.59±0.00	0.58±0.00
Spanish	**0.68**±0.01	-	0.66±0.02	0.65±0.01	-	0.64±0.02	0.58±0.00	0.57±0.01

5 Conclusion

In this paper, a heterogeneous co-transfer clustering framework has been proposed to simultaneously complete more than one clustering task in different domains with the aid of co-occurrence data. The common subspace of all heterogeneous domains is learned by collective nonnegative matrix factorization via the co-occurrence data. After mapping all data into the subspace, the symmetric nonnegative matrix factorization is used to simultaneously obtain all clustering results. The proposed algorithm can be used to effectively perform heterogeneous co-transfer clustering tasks like image-text clustering problems and cross-language clustering problems. As for the future work, how to automatically identify the transferred weights from different domains will be considered.

Acknowledgement. This work was supported in part by the National Natural Science Foundation of China under Grant 61375062, Grant 61370129, the Ph.D Programs Foundation of Ministry of Education of China under Grant 20120009110006, the Fundamental Research Funds for the Central Universities under Grant 2014JBM029 and Grant 2014JBZ005, the Program for Changjiang Scholars and Innovative Research Team (IRT 201206), the Planning Project of Science and Technology Department of Hebei Province under Grant 13210347, and the Project of Education Department of Hebei Province under Grant QN20131006.

References

1. Belkin, M., Niyogi, P.: Laplacian eigenmaps and spectral techniques for embedding and clustering. In: NIPS, pp. 585–592 (2002)
2. Blaschko, M., Lampert, C.: Correlational spectral clustering. In: CVPR (2008)
3. Blitzer, J., Crammer, K., Kulesza, A., Pereira, F., Wortman, J.: Learning bounds for domain adaptation. In: NIPS (2008)
4. Caruana, R.: Multitask learning: a knowledge-based source of inductive bias. Mach. Learn. **28**, 41–75 (1997)
5. Chaudhuri, K., Kakade, S., Livescu, K., Sridharan, K.: Multi-view clustering via canonical correlation analysis. In: ICML, pp. 129–136 (2009)

6. Chua, T., Tang, J., Hong, R., Li, H., Luo, Z., Zheng, Y.: NUS-WIDE: a real-world web image database from National University of Singapore. In: ICICR (2009)
7. Dai, W., Chen, Y., Xue, G., Yang, Q., Yu, Y.: Translated learning: transfer learning across different feature spaces. In: NIPS, pp. 299–306 (2008)
8. Ding, C., Li, T., Jordan, M.: Convex and semi-nonnegative matrix factorizations. TPAMI **32**(1), 45–55 (2010)
9. Gartner, T.: A survey of kernels for structured data. KDD **5**(1), 49–58 (2003)
10. Hartigan, J., Wong, M.: A k-means clustering algorithm. Appl. Stat. **28**(1), 100–108 (1979)
11. Hotelling, H.: Relations between two sets of variates. Biometrika **28**, 321–377 (1936)
12. Kuang, D., Ding, C., Park, H.: Symmetric nonnegative matrix factorization for graph clustering. In: ICDM, pp. 106–117 (2012)
13. Lee, D., Seung, H.: Learning the parts of objects by non-negative matrix factorization. Nature **401**, 788–791 (1999)
14. Lee, D., Seung, H.: Algorithms for non-negative matrix factorization. In: NIPS, pp. 556–562 (2001)
15. Lu, Z., Zhu, Y., Pan, S., Xiang, E., Wang, Y., Yang, Q.: Source free transfer learning for text classification. In: AAAI (2014)
16. MacQueen, J.: Some methods for classification and analysis of multivariate observations. In: Proceedings of the Fifth Berkeley Symposium on Mathematical Statistics and Probability, vol. 1, pp. 281–297 (1967)
17. Ng, M., Wu, Q., Ye, Y.: Co-transfer learning via joint transition probability graph based method. In: KDD Workshop on CDKD, pp. 1–9 (2012)
18. Ng, M., Wu, Q., Ye, Y.: Co-transfer learning using coupled markov chains with restart. IEEE Intelligent Systems (2013)
19. Pan, S., Tsang, I., Kwok, J., Yang, Q.: Domain adaptation via transfer component analysis. TNN **22**(2), 199–210 (2011)
20. Pan, S., Yang, Q.: A survey on transfer learning. TKDE **22**(10), 1345–1359 (2010)
21. Qi, G., Aggarwal, C., Huang, T.: Towards semantic knowledge propagation from text corpus to web images. In: WWW, pp. 297–306 (2011)
22. Shi, J., Malik, J.: Normalized cuts and image segmentation. TPAMI **22**(8), 888–905 (2000)
23. Singh, A., Gordon, G.: Relational learning via collective matrix factorization. In: KDD, pp. 650–658 (2008)
24. Tan, B., Zhong, E., Ng, M., Yang, Q.: Mixed-transfer: transfer learning over mixed graphs. In: ICDM (2014)
25. Yang, Q., Chen, Y., Xue, G., Dai, W., Yu, Y.: Heterogeneous transfer learning for image clustering via the social Web. In: ACL/AFNLP, pp. 1–9 (2009)
26. Zhu, Y., Chen, Y., Lu, Z., Pan, S., Xue, G., Yu, Y., Yang, Q.: Heterogeneous transfer learning for image classification. In: AAAI (2011)

A Trusted Third Party-Based Key Agreement Scheme in Cloud Computing

Ying Li(✉), Liping Du, Guifen Zhao, and Fuwei Feng

Beijing Key Laboratory of Network Cryptography Authentication, Beijing Municipal Institute of Science and Technology Information, No. 138 Xizhimenwai StreetXicheng District, Beijing, China
happyliying2000@163.com

Abstract. In order to protect authentication information of cloud users transmitted between RPs and OPs, a trusted third party-based key agreement scheme is proposed. In the scheme, encryption card and advanced cipher graph are used to generate one-time encryption key to encrypt one-time process key. The process key is used to encrypt interactive authentication information to prevent it from being leaked out. Analyses show that the scheme is efficient and safe.

Keywords: Cloud computing · Key agreement · Trusted third party

1 Introduction

With the development of cloud computing technology, various popular network applications are moving into cloud. Cloud users can enjoy great profit from cloud services. For example, once a cloud users registers an account in one cloud service, he can login in multiple cloud services by the account whether these cloud services are provided by one cloud provider or not [1]. By this means, it's not necessary to remember multiple account and password for users to access different applications. But with the reduction of operation complexity of user', computing complexity of applications' increases. In the process of user' access to different cloud services, a large amount of authentication information are exchanged between different service providers [2], which risk leakage of user information because of frequent cooperation between different cloud services in complex cloud environment.

Single sign on (SSO) technology is popular used to realize cloud service authentication [3], e.g. a SSO based on SAML used in salesforce [4], a SSO based on federated identity active directory in Microsoft cloud services [5], and a SSO based on OpenID in Google cloud [6]. In Google cloud, when a user submit a login request, the request will be transmitted by RP (relying party) to Google OP (OpenID Provider). Google OP verified identities of users and return verification results to RP. According to verification results, RP can decide to allow or deny user's access to resources. The similar method is utilized in Microsoft cloud. In order to protect the security of interactive authentication information between RP and OP, in Microsoft cloud it is encrypted by a fix process key while in Google cloud it is encrypted by a varied process key, but the process of key agreement

X. He et al. (Eds.): IScIDE 2015, Part II, LNCS 9243, pp. 407–412, 2015.
DOI: 10.1007/978-3-319-23862-3_40

is encrypted by a fixed pair of private key and public key [7]. That is, when a cloud user gets access to a cloud resource or switches between several cloud services, an interactive process will be run between OP and RP which is encrypted by fixed key. Therefore, due to millions of login affairs carried out in cloud services, it's easy to get large number of cipher text encrypted by a fixed key which made it possible to break out both of the fixed key and authentication information.

In this case, a trusted third party key agreement scheme is proposed. In the scheme, special hardware equipment is deployed to solve the problem of symmetric key distribution, and a advanced CSK (combined symmetric key) technology is utilized to generate one-time encryption key in place of a fixed key to encrypt process key which guarantees secure transmission of interactive information between RP and OP.

2 System Architecture

In recent years, with the emergence of cloud computing, users can get access to resource of many cloud services by a client (browser, the user agent). In Google cloud, when a user try to get the resource, firstly a OID is submitted to RP by the user agent. Secondly, Diffie-Hellman key exchange protocol is used to negotiate a shared key between RP and OP. Lastly, the authentication information is protected in methods of encryption or signature in RP and OP. In Microsoft cloud, the second step have even been omitted [8]. According to the analyses of Sect. 1, the both method of key usage is not safe. In addition, in Google cloud, due to asymmetric encryption algorithm applied to exchange key, the key agreement process is relatively slow. Therefore, the cloud entrance will be blocked when a large number of users login the cloud at the same time (Fig. 1).

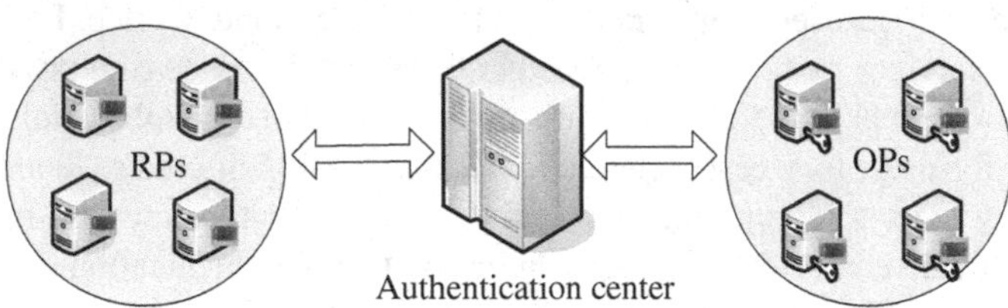

Fig. 1. In this paper, a trusted third party key agreement scheme is proposed to solve above problem. In the scheme a trusted third party, authentication center, is set up.

The authentication center, all RPs and OPs of cloud services are connected by internet. The function of authentication center is to distribute key pool and encryption device and to implement key agreement.

In authentication center, RPs and OPs, PCI encryption cards are deployed in, which is a encryption hardware with the features of high speed and safety. In encryption cards of RPs and OPs, identity information, key agreement protocol and key pool is filled. The key pool filled in devices is different from each other. In encryption card of authentication center, key agreement protocol and storage key are filled, while in databases of authentication center, cipher text of key pools of all OPs and RPs encrypted by storage key and encryption card are stored.

3 Key Agreement Scheme

3.1 Register

When an new application joins a existing cloud services group, it should register in authentication center of the cloud services group firstly. Authentication center receives register information including OPs' and RPs' of the new application and reviews the information, then delivers the new application encryption cards with identities and key pools. Then, authentication center adds the new application into its cloud services list and broadcasts this message to members of cloud services group.

3.2 Key Agreement Process

Key Agreement Process of RP. Process key is generated by hardware random number generator in encryption card of RP end. It is composed of 16 elements of one byte length and is utilized to encrypt authentication information transmitted between RP and OP.

RP gets timestamp and random number of 16 bytes generated by hardware random number generator in encryption card. Timestamp and random number are input into CSK key generation protocol to select and combine 16 bytes from key pool as an encryption key to encrypt process key.

Encrypt process key by encryption key. The process key cipher text, RP id, OP id, timestamp and random number are combined into one string and are send to authentication center (Fig. 2).

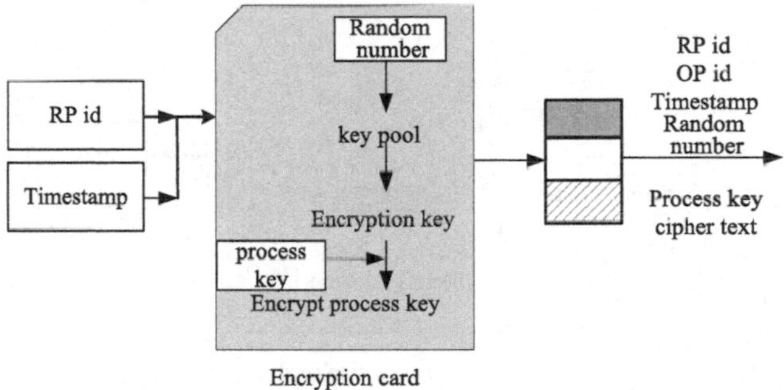

Fig. 2. Key agreement process of RP.

Key Agreement Process of Authentication Center. After Authentication center receives RP identity, OP identity, timestamp, random number and process key cipher text, the corresponding key pool cipher text of RP and OP stored in database are took out and decrypted to clear text by storage key in encryption card. Timestamp and random number are input into CSK key generation protocol to generate an decryption key of 16 bytes from key pool of RP to decrypt process key cipher text to process key, then timestamp and random number are input into CSK key generation protocol to generate an encryption key of 16 bytes from key pool of OP to encrypt process key to process key cipher text. RP id, timestamp and random number and the new process key cipher text are combined into one string and send to OP (Fig. 3).

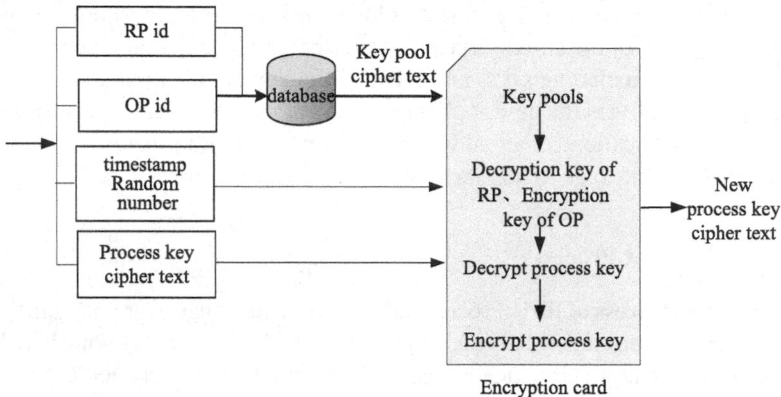

Fig. 3. Key agreement process of authentication center.

Key Agreement Process of OP. Timestamp and random number received from authentication center are input into CSK key generation protocol to select and combine 16 bytes from key pool in encryption card as an decryption key to decrypt process key.

Decrypt process key cipher text by decryption key, then key agreement process finishes (Fig. 4).

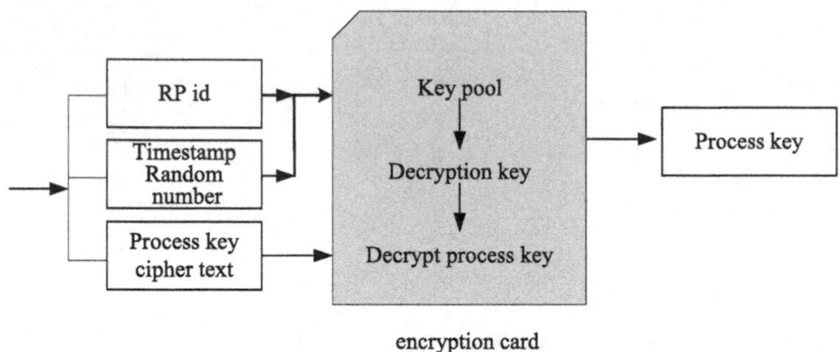

Fig. 4. Key agreement process of OP.

3.3 Unregister

When a cloud service is deemed unsafe, in order to ensure the security of other cloud services, this cloud service should be removed from the cloud service group, which is called as unregister. It's easy to unregister a cloud service. All need to do are deleting the key pools of RPs and OPs of the cloud service from authentication center database, excluding the cloud service from cloud service list and broadcasting message to other cloud services. Thus, the cloud service can't achieve key agreement and interact data correctly with other cloud services, this is, the unregister is successful.

4 Security Analysis

To protect the security of authentication information of cloud users, some important steps are taken as below:

- Encryption device of high security is used to ensure the secure storage of identity, key pools and key agreement protocol. All of the above can't be read out of encryption device. In this way, symmetric key need no longer updated at regular time.
- In the process of key agreement, decryption and encryption are operating in encryption device, using symmetric cryptography fixed in chips of encrypt device. Hence, symmetric key appears only in encryption device, and is impossible for hackers to trace and analyses.
- In authentication center, time stamp test and authentication parameters test are executed, which can prevent the server from replay attack.
- Encryption key is dynamic generated by key pool co-operating with parameters. By this means, it possesses a property of one-time and some attack like interception and fabrication is avoided.

5 Conclusion

In popular cloud services, fixed keys are utilized to encrypted transmitted authentication information or key agreement information between RPs and OPs, which risk the leakage of cloud user information. In the article, a trusted third party key agreement scheme is presented. In the scheme, by using encryption hardware and advanced cipher graph, the one-time encryption key is generated to encrypt one-time process key, which ensure that the interaction data is safe.

Acknowledgments. The authors wish to thank the helpful comments and suggestions from my director and colleagues in Beijing Key Laboratory of Network Cryptography Authentication. This work is supported by Innovation Project II2: Research and Development of Cryptographic Authentication System in Cloud Computing Security (No. PXM2014_178214_000011).

References

1. OWASP: Category:OWASP Top Ten Project [EB/OL]. https://www.owasp.org/index.php/Category:OWASP_Top_Ten_Project
2. Celesti, A., Tusa, F., Villari, M.: Three-phase cross-cloud federation model: the cloud SSO authentication. In: 2nd International Conference on Advances in Future Internet (AFIN 2010), pp. 94–101. IEEE CPS, Venice (2010)
3. Yu, N.-H., Hao, Z., Xu, J., Zhang, W.-M., Zhang, C.: Review of cloud computing security. Acta Electronica Sin. **2**, 371–381 (2013)

4. Lewis, K.D., Lewis, J.E.: Web single sign-on authentication using SAML. IJCSI Int. J. Comput. Sci. Issues **2**, 41–48 (2009)
5. Chalandar, M.E., Darvish, P., Rahmani, A.M.: A centralized cookie-based single sign—on in distributed systems. In: Information and Communications Technology, ICICT 2007, pp. 163–165. IBA, Karachi (2007)
6. Mather, T., Kumaraswamy, S., Latif, S.: Cloud Security and Privacy. O'Reilly Media Inc, Sebastopol (2009)
7. Zhu, R-H., Gao, N., Xiang, J.: Research on authentication mechanisms in cloud computing. In: 28th National Symposium on Computer Security, pp. 54–56. University of Science and Technology of China Press, Anhui (2013)
8. Jiang, W-Y., Gao, N, Liu, Z-Y., Lin, X-Y.: A multi-identities authentication and authorization schema in cloud computing. In: 27th National Symposium on Computer Security, pp. 7–10. University of Science and Technology of China Press, Anhui (2012)

Semi-supervised Learning Based on Improved Co-training by Committee

Kun Liu, Yuwei Guo, Shuang Wang(✉), Linsheng Wu, Bo Yue, and Biao Hou

Key Laboratory of Intelligent Perception and Image Understanding of Ministry of Education of China, Xidian University, Xi'an 710071, China
shwang@mail.xidian.edu.cn

Abstract. As a popular machine learning technique, semi-supervised learning can make full use of a large pool of unlabeled samples in addition to a small number of labeled ones to improve the performance of supervised learning. In co-training by committee, a semi-supervised learning algorithm, the class probability values predicted by committee may repeat, which brings a negative influence on the improvement of the classification performance. We propose a method to deal with this problem, which assign different class probability estimations for different unlabeled samples. Naïve Bayes is employed to help estimate the class probabilities of unlabeled samples. To prove that our method can reduce the introduction of noise, a data editing technique is employed to make a comparison with our method. Experimental results verify the effectiveness of our method and the data editing technique, and also indicate that our method is generally better than the data editing technique.

Keywords: Semi-supervised learning · Ensemble learning · Ensemble co-training

1 Introduction

Unlabeled instances have become abundant, but to obtain their labels is expensive and time consuming. Thus, semi-supervised learning is developed to deal with this problem [1, 2].

Co-training [3] is a multi-view and iterative semi-supervised learning algorithm, which has been widely applied to practical problems [4–7]. And a lot of works have been proposed for co-training [8–13]. In [8], a co-training approach is presented to deal with the case which only few labeled data available. Authors in [9] proposed another co-training algorithm, which combines with active learning. LapCo algorithm [10] added Laplacian regularization into the classifiers of co-training for increasing the algorithm performance. In [11], the co-training algorithm is improved in two aspects: the segmentation of the feature set and the selection of examples. Co-training by committee is a committee-based single-view method introduced by Hady and Schwenker [12]. In this algorithm, the class probability values predicted by committee may repeat, which results in the problem that some unlabeled samples share the same probability and will be selected randomly. This brings a negative effect on the

X. He et al. (Eds.): IScIDE 2015, Part II, LNCS 9243, pp. 413–421, 2015.
DOI: 10.1007/978-3-319-23862-3_41

improvement of the classification performance. Therefore, a more accurate ranking of class probability values is acquired. In [13], two distance metrics are used to put different probability value on different unlabeled example.

In this paper, we propose a method to adjust the class probability values predicted by committee. Different example with different class probability values is good at the selection of unlabeled example. Naïve Bayes is used to help estimate the class probabilities of unlabeled examples. Once two examples own the same class probability value predicted by committee, the one that owns larger class probability value predicted by Naïve Bayes should have the larger chance to be selected. To prove that our method can reduce the introduction of noise, a data editing technique is employed to make a comparison with our method. Experimental results on UCI data sets [20] verify the effectiveness of our method and the data editing technique, and also indicate that our method is generally better than the data editing technique.

The rest of this paper is organized as follows. Section 2 introduces the proposed method and the data editing technique. Experimental Results are shown in Sect. 3. Finally, conclusions are drawn in Sect. 4.

2 Proposed Method

2.1 The Method Combining with Naïve Bayes

The main goal of our work is to improve the classification performance of co-training by committee [13]. A committee can be constructed by training multiple classifiers and combining their predictions. At each iteration of co-training by committee, we select a fixed number of most confident samples according to the ranking of class probability values predicted by committee. Therefore, we could find that an accurate ranking of class probability values is very important. For an ideal case, each unlabeled sample is assigned a unique class probability value. However, the committee may not work as the ideal case, namely the ability of differentiating different samples is limited. Thus, the class probability values may repeat, which leads to the problem that some unlabeled samples own the same probability and will be selected randomly. Thus, a more accurate ranking of class probability values is acquired. The proposed method will be described in the following.

Naïve Bayes is an algorithm, which can generate different class probability estimations for different unlabeled examples directly, it is chosen to combine with ensemble learner to make predictions in this paper.

In co-training by committee, the confidence, namely the class probability, is computed as follows:

$$\text{Confidence}(x_u, w_c) = H(x_u, w_c) \tag{1}$$

where $H(x_u, w_c)$ is the probability predicted by committee that unlabeled example x_u belongs to class w_c.

However, the class probabilities computed by (1) may repeat. Now we assume that two different unlabeled samples own the same class probability predicted by committee, in order to distinguish them, Naïve Bayes is employed to predict their posterior probabilities, the one that owns larger value should have the larger chance to be picked out. Consequently, we get the following equation:

$$\text{Confidence}(x_u, w_c) = H(x_u, w_c) + P_{NB}(w_c|x_u)/c \tag{2}$$

where $P_{NB}(w_c|x_u)$ is the posterior probability predicted by Naïve Bayes that unlabeled example x_u belongs to class w_c and c is a constant.

Each unlabeled sample can be assigned a unique confidence by (2). The pseudo-code of how to select the most confident unlabeled samples at each round is shown in Table 1. Note that, Naïve Bayes is trained only on the original labeled data set, as the newly labeled samples may contain noise.

Table 1. The pseudo-code of selecting the most confident unlabeled examples

Input: U' unlabeled examples pool
H ensemble learner trained last time
$\{n_c\}_{c=1}^{C}$ number of unlabeled examples to be selected per class
C number of classes
NB Naïve Bayes trained on the original labeled data set
Output: L' the most confident unlabeled examples selected

$L' \leftarrow \varnothing$
For each $x_u \in U'$ do
Calculate the confidence that x_u belongs to class w_c as the following equation:
$$\text{Confidence}(x_u, w_c) = H(x_u, w_c) + P_{NB}(w_c \mid x_u)/c$$
End for
Rank the samples in U' according to the confidence
For $c = 1$ to C do
Select n_c examples with the maximum confidence for class w_c
Assign class label w_c to the n_c examples
Put the n_c examples into L'
End for
Return L'

2.2 The Data Editing Technique

Data editing is just the technique that identifying and dealing with the noise. To prove that our method can reduce the introduction of noise, we exploit a data editing technique to make a comparison with our method. A KNN classifier is employed to deal with the noise considering the time complexity. Firstly, a KNN classifier is trained on the original labeled data set. Then the classifier predicts the labels of the newly labeled samples at each round. The samples whose labels are not equal to their current labels are eliminated. The committee is retrained on the updated labeled data set.

3 Experiments

3.1 Experimental Setting

Twelve UCI data sets [18] are employed in the experiments. The detailed information of these data sets is given in Table 2.

Table 2. Description of data sets

Data set	Attributes	Size	Classes
Australian	14	690	2
Colic	22	368	2
Diabetes	8	768	2
German	24	1000	2
Glass	9	214	6
Heart-c	13	303	2
Hepatitis	19	155	2
Ionosphere	34	351	2
Iris	4	150	3
Sick	29	3772	2
Splice	60	3190	3
Wdbc	30	569	2

Bagging [15], Random Forest [14], AdaBoost [16], Random Subspace [17] and some other ensemble learners can all be exploited to evaluate our proposed method. However, considering the limited space, only AdaBoost [16] is employed as the ensemble learner in this paper.

C4.5 pruned decision tree [19] is used as the base learning algorithm of the ensemble learner. A KNN classifier is employed for the data editing technique. All of them are implemented by the WEKA library [20]. For each data set, 5 runs of 4-fold cross-validation are performed and the final result is obtained by averaging their results. Concretely, 25 % of the total examples are randomly chosen as the test data set, the rest examples are employed as the training data set. The training examples are further randomly divided into the labeled examples and the unlabeled examples under labeling rates 10 %. For the 10 % labeling rate, the algorithm stops when the number of labeled examples reaches 60 % of the training examples. The number of unlabeled examples selected at each round is decided to ensure that the number of rounds is not more than 50.

The constant c in the method proposed has an impact on the classification performance. The value of constant c is considered as $\{10, 20, 30, 40, 50, 60, 70\}$. In the data editing technique, the parameter k of the KNN classifier is suggested as $\{1, 3, 5\}$. Their optimal parameters are acquired based on 5-fold cross-validation.

We compare the three methods from different aspects in the following. The classification performance under different ensemble sizes and labeling rates is given in Tables 3 and 4. The effect of constant c on the classification performance is shown in Fig. 1.

In the experimental results, CoAdaBoost [13] indicates the original method, our method is referred to as CoAdaBoost-NB, and the data editing technique is denoted by CoAdaBoost-DE.

3.2 Comparison Under Different Ensemble Sizes and Labeling Rates

Tables 3 and 4 tabulate the test error rates of the three methods under different ensemble sizes. The classification performance of the hypotheses trained only on the original labeled data set is represented as *initial*. The final test error rates after using the unlabeled samples are denoted by *final*. The relative improvement is also given (*improve* = (*initial* − *final*) / *initial* × 100). The highest improvement for each data set has been boldfaced. The row *ave* indicates the average test error rates over all the experimental data sets.

Table 3. Test error rates for different methods under 10 % labeling rate and ensemble size 3

Data set		CoAdaBoost		CoAdaBoost-DE		CoAdaBoost-NB	
	Initial	Final	Improve	Final	Improve	Final	Improve
Australian	19.54	18.23	6.68	15.80	**19.14**	16.26	16.76
Colic	23.53	20.82	11.55	19.35	**17.78**	20.76	11.78
Diabetes	30.05	29.97	0.26	28.75	4.33	27.66	**7.97**
German	33.60	34.22	−1.85	31.74	**5.54**	31.78	5.42
Glass	57.01	52.24	8.37	52.81	7.37	50.91	**10.70**
Heart-c	26.54	28.39	−6.96	24.75	6.77	24.35	**8.25**
Hepatitis	27.59	27.21	1.37	27.86	−0.98	25.80	**6.48**
Ionosphere	18.79	18.00	4.20	17.26	**8.16**	17.26	**8.16**
Iris	19.80	18.05	8.83	13.92	29.74	12.31	**37.83**
Sick	3.26	3.07	5.85	3.17	2.76	2.77	**14.96**
Splice	14.04	14.31	−1.92	11.38	18.94	9.45	**32.69**
Wdbc	9.07	8.79	3.09	8.26	**8.93**	8.33	8.17
ave	23.57	22.78	3.29	21.25	10.71	20.64	**14.10**

From Table 3, under 10 % labeling rate and ensemble size 3, CoAdaBoost benefits from the unlabeled samples for 9 out of the 12 data sets. CoAdaBoost-DE performs better than CoAdaBoost on 9 out of the 12 data sets. CoAdaBoost-NB outperforms CoAdaBoost on all the 12 data sets. Also compared with CoAdaBoost-DE, CoAdaBoost-NB performs better on 7 out of the 12 data sets. The average test error rates over all the data sets demonstrate that CoAdaBoost-NB is the best.

From Table 4, under 10 % labeling rate and ensemble size 6, the performance of CoAdaBoost could be improved with unlabeled samples for only 6 out of the 12 data sets. One possible explanation is that overfitting may happen as the number of the original labeled samples is small and the ensemble size is large. Compared with CoAdaBoost, CoAdaBoost-DE performs better on 10 out of the 12 data sets.

CoAdaBoost-NB outperforms CoAdaBoost on all the 12 data sets. Also compared with CoAdaBoost-DE, CoAdaBoost-NB performs better on 9 out of the 12 data sets. The average test error rates over all the data sets demonstrate that CoAdaBoost-NB obtains the best performance.

Table 4. Test error rates for different methods under 10 % labeling rate and ensemble size 6

Data set		CoAdaBoost		CoAdaBoost-DE		CoAdaBoost-NB	
	Initial	Final	Improve	Final	Improve	Final	Improve
Australian	19.19	17.22	10.28	15.31	**20.23**	16.49	14.04
Colic	25.00	19.95	20.22	19.24	**23.04**	19.73	21.09
Diabetes	30.42	29.61	2.65	28.57	6.08	27.45	**9.76**
German	33.76	34.22	−1.36	32.02	5.15	30.84	**8.65**
Glass	55.03	52.52	4.57	52.43	4.73	51.39	**6.63**
Heart-c	25.74	25.87	−0.47	26.20	−1.76	24.62	**4.37**
Hepatitis	28.39	28.64	−0.87	29.15	−2.67	25.81	**9.09**
Ionosphere	18.45	19.81	−7.36	18.67	−1.21	17.66	**4.28**
Iris	20.59	18.05	12.33	13.92	32.43	12.04	**41.51**
Sick	3.05	3.12	−2.26	3.04	0.35	2.95	**3.30**
Splice	11.37	10.40	8.54	9.23	18.85	7.91	**30.43**
Wdbc	8.51	9.56	−12.42	8.61	**−1.24**	8.65	−1.66
ave	23.29	22.41	2.82	21.37	8.67	20.46	**12.62**

The results verify the effectiveness of our method, and also indicate that our method is generally better than the data editing technique.

3.3 Comparison with Different Values of Constant C

The value of constant c in the proposed method has an effect on the performance, thus an investigation on six data sets is given in Fig. 1. The results are obtained with 20 % labeling rate and ensemble size 6. Whatever the value of constant c, our method CoAdaBoost-NB outperforms CoAdaBoost for *Australian*, *German*, and *heart-c*. For *Colic*, the influence of the value of constant c on the performance is very large. Even so, our method obtains better performance with any value of c except the value 70. The similar case with *Hepatitis* and *glass*, concretely, CoAdaBoost-NB performs better than CoAdaBoost with 4 out of the 7 values for *Hepatitis*, also it outperforms CoAdaBoost with any value of c except the value 30 for *glass*. It can be concluded that our method could enhance the performance of CoAdaBoost with almost any value of constant c. However, the relative improvement is sometimes sensitive to the concrete value of constant c. If the value is chosen properly, the relative improvement can be very high.

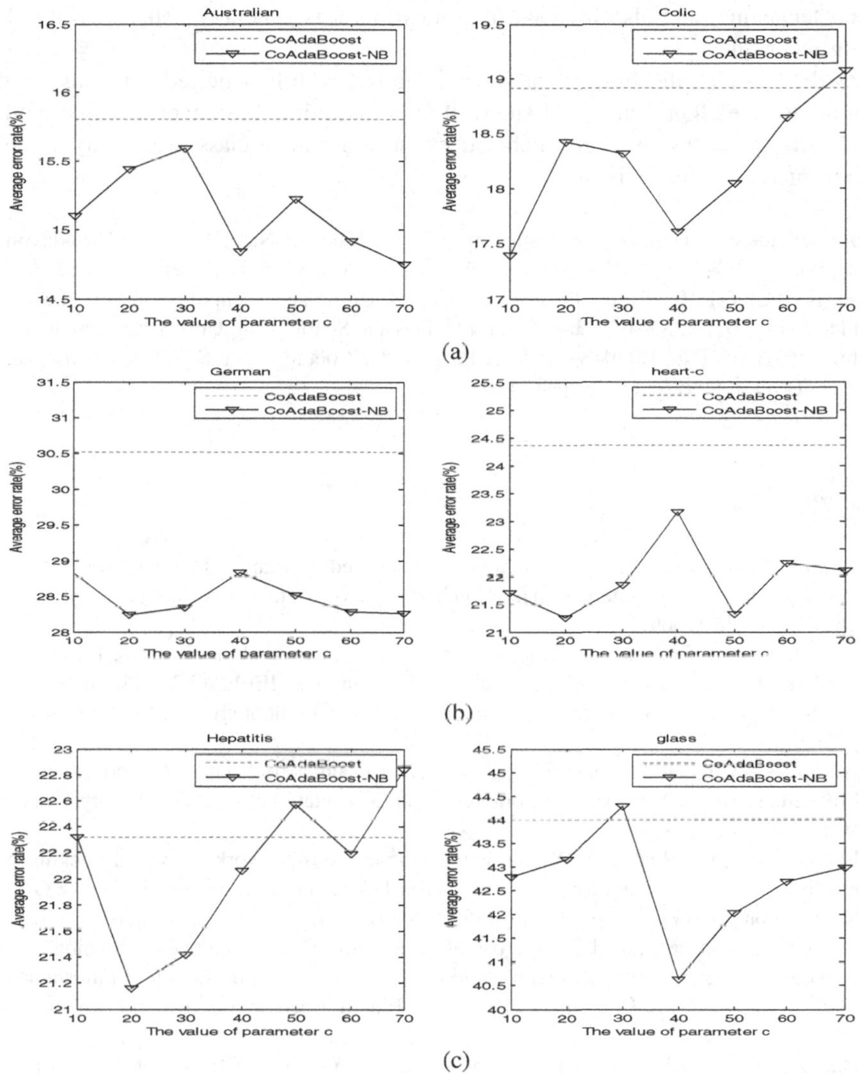

Fig. 1. The effect of constant c on the classification performance

4 Conclusion

In this paper, a method is proposed to deal with the problem that the class probability values predicted by committee may repeat. Naïve Bayes, which can generate different class probability estimations for different unlabeled examples, is used to help estimate the class probabilities of unlabeled examples. To prove that our method can reduce the introduction of noise, a data editing technique is used to make a comparison with our method. Experimental results verify the effectiveness of our method and the data editing technique, and also indicate that our method is generally better than the data editing technique.

As the best classification performance is not necessarily acquired at the last round, when to stop the algorithm is an issue well-worth studying. Moreover, to explore other better methods to assign each unlabeled example a unique class probability value is also an interesting future issue.

Acknowledgments. This work was supported by the National Natural Science Foundation of China (No. 61173092, No. 61271302), the Program for New Century Excellent Talents in University (No.NCET-11-0692), the Program for New Scientific and Technological Star of Shaanxi Province (No. 2013KJXX-64), the Fund for Foreign Scholars in University Research and Teaching Programs (No. B07048), and the Program for Cheung Kong Scholars and Innovative Research Team in University(No. IRT1170).

References

1. He, Z., Li, X., Hu, W.: A boosted semi-supervised learning framework for web page filtering. In: IEEE International Conference on Systems, Man and Cybernetics, pp. 2133–2136 (2009)
2. Sun, Z., Ye, Y., Zhang, X., Huang, Z., Chen, S., Liu, Z.: Batch-mode active learning with semi-supervised cluster tree for text classification. In: IEEE/WIC/ACM International Conferences on Web Intelligence and Intelligent Agent Technology (WI-IAT), pp. 388–395 (2012)
3. Blum, A., Mitchell, T.: Combining labeled and unlabeled data with co-training. In: Proceedings of the 11th Annual Conference on Computational Learning Theory, Madison, WI, pp. 92–100 (1998)
4. Lu, H., Zhou, Q., Wang, D., Xiang, R.: A co-training framework for visual tracking with multiple instance learning. In: IEEE International Conference on Automatic Face & Gesture Recognition and Workshops (FG), pp. 539–544 (2011)
5. Carneiro, G., Nascimento, J.C.: The use of on-line co-training to reduce the training set size in pattern recognition methods: application to left ventricle segmentation in ultrasound. In: IEEE Conference on Computer Vision and Pattern Recognition (CVPR), pp. 948–955 (2012)
6. Dai, P., Liu, K., Xie, Y., Li, C.: Online co-training ranking SVM for visual tracking. In: IEEE International Conference on Acoustics, Speech and Signal Processing (ICASSP), pp. 6568–6572 (2014)
7. Liu, B., Feng, J., Liu, M., et al.: Predicting the quality of user-generated answers using co-training in community-based question answering portals. Pattern Recogn. Lett. **58**, 29–34 (2015)
8. Fan, M., Qian, T., Chen, L., Liu, B., Zhong, M., He, G.: Authorship attribution with very few labeled data: a co-training approach. In: Li, F., Li, G., Hwang, S.-W., Yao, B., Zhang, Z. (eds.) WAIM 2014. LNCS, vol. 8485, pp. 657–668. Springer, Heidelberg (2014)
9. Zhang, Y., Wen, J., Wang, X., et al.: Semi-supervised learning combining co-training with active learning. Expert Syst. Appl. **41**(5), 2372–2378 (2014)
10. Li, Y., Liu, W., Wang, Y.: Laplacian regularized co-training signal processing (ICSP). In: 12th International Conference on IEEE, pp. 1408–1412 (2014)

11. Katz, G., Shabtai, A., Rokach, L.: Adapted Features and Instance Selection for Improving Co-training. In: Holzinger, Andreas, Jurisica, Igor (eds.) Interactive Knowledge Discovery and Data Mining in Biomedical Informatics. LNCS, vol. 8401, pp. 81–100. Springer, Heidelberg (2014)
12. Hady, M., Schwenker, F.: Co-training by committee: a new semi- supervised learning framework. In: Proceedings of the IEEE International Conference on Data Mining Workshops, pp. 563–572 (2008)
13. Wang, S., Wu, L., Jiao, L., et al.: Improve the performance of co-training by committee with refinement of class probability estimations. Neurocomputing **136**, 30–40 (2014)
14. Breiman, L.: Random forests. Mach. Learn. **45**(1), 5–32 (2001)
15. Breiman, L.: Bagging predictors. Mach. Learn. **24**(2), 123–140 (1996)
16. Freund, Y., Schapire, R.: A decision-theoretic generalization of online learning and an application to boosting. In: Proceedings of the 2nd European Conference on Computational Learning Theory, Barcelona, Spain, pp. 23–37 (1995)
17. Ho, T.: The random subspace method for constructing decision forests. IEEE Trans. Pattern Anal. Mach. Intell. **20**(8), 832–844 (1998)
18. Blake, C., Keogh, E., Merz, C.: UCI repository of machine learning databases. Department of Information and Computer Science, University of California, Irvine, http://www.ics.uci.edu/~mlearn/MLRepository.html (1998)
19. Quinlan, R.: C4.5: Programs for Machine Learning. Morgan Kaufmann Publishers, San Mateo (1993)
20. Witten, I.H., Frank, E.: Data Mining: Practical Machine Learning Tools and Techniques with Java Implementations. Morgan Kaufmann, San Francisco (1999)

A New Algorithm for Discriminative Clustering and Its Maximum Entropy Extension

Xiao-bin Zhi[1(✉)] and Jiu-lun Fan[2]

[1] School of Science, Xi'an University of Posts and Telecommunications, Xi'an 710121, China
Xbzhi@163.com

[2] School of Communication and Information Engineering, Xi'an University of Posts and Telecommunications, Xi'an 710121, China
jiulunf@xupt.edu.cn

Abstract. Discriminative clustering (DC) can effectively integrates subspace selection and clustering into a coherent framework. It performs in the iterative classical Linear Discriminant Analysis (LDA) dimensionality reduction and clustering processing. DC can effectively cluster the data with high dimension. However, it has complex form and high computational complexity. Recent work shows DC is equivalent to kernel k-means (KM) with a specific kernel matrix. This new insights provides a chance of simplifying the optimization problem in the original DC algorithm. Based on this equivalence relationship, Discriminative K-means (DKM) algorithm is proposed. When the number of data points (denoted as n) is small, DKM is feasible and efficient. However, the construction of kernel matrix needs to compute the inverse of a matrix in DKM, when n is large, which is time consuming. In this paper, we concentrate on the efficiency of DC. We present a new framework for DC, namely, Efficient DC (EDC), which consists of DKM and the whitening transformation of the regularized total scatter matrix (WRTS) plus KM clustering (WRTS+KM). When m (dimensions) is small and n far outweighs m, namely, $n \gg m$, EDC can carry out WRTS+KM on data, which is more efficient than DKM. When n is small and m far outweighs n, namely, $m \gg n$, EDC can carry out DKM on data, which is more efficient. We also extend EDC to soft case, and propose Efficient Discriminative Maximum Entropy Clustering (EDMEC), which is an efficient version of maximum entropy based DC. Extensive experiments on a collection of benchmark data sets are presented to show the effectiveness of the proposed algorithms.

Keywords: Dimension reduction · Discriminative clustering · Kernel clustering · Whitening transformation · Maximum entropy clustering

1 Introduction

Clustering is a fundamental data analysis step that has been widely used in pattern recognition [1], data mining [2], bioinformatics [3] and other fields. Recently, with the increasing of the data dimension man are dealing with in image processing, information

X. He et al. (Eds.): IScIDE 2015, Part II, LNCS 9243, pp. 422–432, 2015.
DOI: 10.1007/978-3-319-23862-3_42

retrieval and bioinformatics and many other application domains, the clustering algorithms that can adaptively reduce the data dimension have become the hot spot of current research on clustering [4–11]. LDA is one of the most classical statistical methods for feature extraction and dimensionality reduction, and has received wide attention [12–16]. LDA aims to compute the optimal linear transformation (projection), which minimizes the within-class distance (of the data set) and maximizes the between-class distance simultaneously, thus achieving maximum discrimination. Recently, combining LDA and the clustering process into a joint framework to improve the performance of the clustering is becoming one of hotspot in clustering research [6–11]. The algorithm, called Discriminative Clustering (DC) in the following discussion, works in an iterative fashion, alternating between LDA subspace selection and clustering, and can perform clustering and LDA dimensionality reduction simultaneously. However, the integration between subspace selection and clustering in DC is not well understood, due to the intertwined and iterative nature of the algorithm [8]. Ye et al. interpreted LDA-KM from the "kernel" trick point of view and proposed discriminative KM (DKM) algorithm [8]. This new insights provides a chance of simplifying the optimization problem in the original DC algorithm. However, DKM has the following shortcoming: The construction of kernel gram matrix needs to compute the inverse of some matrix, when the size of data is large, this is time consuming. In addition, the soft extended version of DC, namely, fuzzy DC has becomes the hot topic of DC research [9–11]. In [9], a new LDA-based fuzzy clustering (LDA-FC) algorithm is proposed based on a extension of classical scatter matrices. The interior point method is used to solve the corresponding optimization problem. However, no detailed algorithm procedure has been presented in the paper. And from the results of experiment presented in the paper, we can know the LDA-FC proposed is inefficient and time consuming. In [10, 11], fuzzy extended vision of LDA, viz. fuzzy LDA (FLDA), and maximum entropy clustering (MEC) algorithm [17–20] are combined into a coherent framework for adaptive dimension reduction and clustering.

Since DC and its fuzzy version have complex form and high computational complexity, in this paper, we focus on the efficiency of the DC and its fuzzy extended version. We present efficient algorithms for DC and its fuzzy (soft) extended version respectively. The new DC algorithms, called Efficient Discriminative Clustering (EDC) and Efficient Discriminative Maximum Entropy Clustering (EDMEC) respectively. EDC consists of DKM [8] and the whitening transformation of regularized total scatter matrix (WRTS) plus KM clustering (WRTS+KM). When the number of data far outweigh the dimensionality of data, namely, $n \gg m$, EDC can carry out WRTS+KM on data, which is more efficient. When the dimensionality of data far outweigh the number of data, namely, $m \gg n$, EDC can carry out DKM on data, which is more efficient. At the same time, EDMEC consists of Discriminative Maximum Entropy Clustering (DMEC) and WRTS+MEC, where DMEC can be obtain by applying the kernel used in DKM [8] to kernel MEC. Experimental results on seven real-world data sets show the effectiveness of EDC and EDMEC.

The remainder of the paper is organized as follows. We review LDA, DC and DKM algorithms in Sect. 2. EDC and EDMEC are presented in Sect. 3 and Sect. 4 respectively, Sect. 4 also explores the relations of EDMEC to other earlier approaches. Section 5 includes the experimental results and the conclusions are given in Sect. 6.

For convenience, Table 1 lists the important notations used in the rest of this paper. For simplicity, the data is centered in the preprocessing step, so that the total mean of data set $\bar{x} = \boldsymbol{X}\boldsymbol{e}^{\mathrm{T}}/n = \boldsymbol{0}$, where $\boldsymbol{e} = [1, 1, \ldots, 1]^{\mathrm{T}} \in \mathbb{R}^n$.

Table 1. The important notations used in the paper

Notation	Description	Notation	Description
n	Number of data points	$\boldsymbol{S}_{\mathrm{w}}$	Within-class scatter matrix
m	Dimensions	$\boldsymbol{S}_{\mathrm{b}}$	Between-class scatter matrix
c	Number of clusters	$\boldsymbol{S}_{\mathrm{t}}$	Total scatter matrix
l	Reduced dimensionality	$\boldsymbol{S}_{\mathrm{fw}}$	Fuzzy within-class scatter matrix
$\boldsymbol{X}$	Data matrix of size m by n	$\boldsymbol{S}_{\mathrm{fb}}$	Fuzzy between-class scatter matrix
C_j	The j-th cluster in X	$\boldsymbol{S}_{\mathrm{ft}}$	Fuzzy total scatter matrix
n_j	Size of the j-th cluster X_j	$\boldsymbol{H} = \{h_{ij}\}_{n \times c}$	Cluster indictor matrix
$\boldsymbol{P}$	Transformation matrix	$\boldsymbol{U} = \{u_{ij}\}_{n \times c}$	Membership matrix

2 LDA, DC and DKM

In this section, we first review LDA and DC [6, 7], and then present DKM proposed by Ye et al. [8].

Let $\boldsymbol{H} \in \mathbb{R}^{n \times c}$ be the cluster indicator matrix, the weighted cluster indicator matrix can be defined as follows [8]:

$$\boldsymbol{L} = [\boldsymbol{L}_1, \boldsymbol{L}_2, \ldots, \boldsymbol{L}_c].. \tag{1}$$

Where $\boldsymbol{L}_j = (0, \ldots, \overbrace{1, \ldots, 1}^{n_j}, 0, \ldots, 0)^{\mathrm{T}}/n_j^{\frac{1}{2}}$. Denote $\boldsymbol{v}_j = \sum_{\boldsymbol{x} \in C_j} \boldsymbol{x}/n_j$ as the mean of the j-th cluster C_j. In linear discriminant analysis (LDA), the within-cluster, between-cluster, and total scatter matrices are defined as $\boldsymbol{S}_{\mathrm{w}} = \sum_{j=1}^{c} \sum_{\boldsymbol{x}_i \in C_j}^{n} (\boldsymbol{x}_i - \boldsymbol{v}_j)(\boldsymbol{x}_i - \boldsymbol{v}_j)^{\mathrm{T}}$, $\boldsymbol{S}_{\mathrm{b}} = \boldsymbol{X}\boldsymbol{L}\boldsymbol{L}^{\mathrm{T}}\boldsymbol{X}^{\mathrm{T}}$ and $\boldsymbol{S}_{\mathrm{t}} = \boldsymbol{X}\boldsymbol{X}^{\mathrm{T}}$, respectively. Given class labels as specified by the indicator matrix $\boldsymbol{F}$ (or $\boldsymbol{L}$), the goal of LDA is to find an optimal linear transformation $\boldsymbol{P}^{\mathrm{T}} : \boldsymbol{x}_j \in \mathbb{R}^m \to \boldsymbol{y}_j \in \mathbb{R}^l$, such that the following objective function is maximized [12].

$$\operatorname{trace}\left(\left(\boldsymbol{P}^{\mathrm{T}}\boldsymbol{S}_{\mathrm{t}}\boldsymbol{P}\right)^{-1}\left(\boldsymbol{P}^{\mathrm{T}}\boldsymbol{S}_{\mathrm{b}}\boldsymbol{P}\right)\right) \tag{2}$$

Commonly, the optimization problem (2) can be optimally solved by applying an eigen decomposition on the matrix $\boldsymbol{S}_t^{-1}\boldsymbol{S}_{\mathrm{b}}$, which requires the total scatter matrix $\boldsymbol{S}_{\mathrm{t}}$ is non-singular. However, in many applications, the dimension of data set is often high, and $\boldsymbol{S}_{\mathrm{t}}$ may be singular, thus classical LDA cannot be used. The regularization technique [14] is commonly applied to prevent singularity as follows:

$$\tilde{S}_t = S_t + \lambda I_m = XX^T + \lambda I_m. \quad (3)$$

In (DC) [6, 7], the transformation matrix P and the cluster indicator matrix H are computed by maximizing the following objective function:

$$J(H, P) = \text{trace}\left(\left(P^T (XX^T + \lambda I_m)P\right)^{-1} (P^T XLL^T X^T P) \right). \quad (4)$$

The algorithm alternatively optimizes H and P. For a given H, P is given by the standard LDA procedure. For a given P, the optimal H can be computed by applying the gradient descent strategy [6] or by solving K-means problem in the lower-dimensional space resulting from the transformation P [7].

Experiments in [6, 7] have shown the effectiveness of DC in comparison with several other popular clustering algorithms. However, repeated calling to the LDA is very time-consuming in DC. By using kernel trick, Ye et al. showed DC is equivalent to kernel KM with a specific kernel Gram matrix and proposed Discriminative K-means (DKM) algorithm [8]. DKM can be formulated as the following optimization problem:

$$F^* = \arg\max_F \ \text{trace}\left(L^T \left(I_n - (I_n + G/\lambda)^{-1} \right) L \right), \quad (5)$$

which can be computed by solving a kernel KM problem with the kernel matrix given by $\tilde{G} = I_n - (I_n + G/\lambda)^{-1}$ [8].

Computing $G = X^T X$ takes $O\ (n^2 m)$ time, computing the inverse of $(I_n + G/\lambda)$ takes $O\ (n^3)$ time, and implement kernel KM takes $O\ (n^2 ct)$. So, the total time complexity of DKM is $O\ (n^2 m) + O\ (n^3) + O\ (n^2 ct)$. When n is small, DKM can handle high-dimension data sets efficiently. However, for applications with the large data sets, DKM begins to become infeasible.

3 An Efficient Framework for DC

In this section, we first present WRTS+KM, then we propose a flexible framework for DC: when m is small and $n \gg m$, we can implement DC by using WRTS+KM; however, when n is small and $m \gg n$, WRTS+KM is time-consuming, and then we can implement DC by using DKM [8]. This framework is more efficient.

First, based on the fact for LDA:

$$\max_P J(P) = \text{trace}\left(\left(P^T \tilde{S}_t P\right)^{-1} (P^T S_b P) \right) = \text{trace}\left(\tilde{S}_t^{-1} S_b \right), \quad (6)$$

we can know the objective function of DC (4) is equivalent to

$$J(H) = \text{trace}\left(\tilde{S}_t^{-1} S_b \right) = \text{trace}\left((XX^T + \lambda I_m)^{-1} (XLL^T X^T) \right). \quad (7)$$

Let $\tilde{S}_t = \boldsymbol{W\Sigma V}^T$ be the Singular Value Decomposition (SVD), let $\boldsymbol{M} = \boldsymbol{W\Sigma}^{-1/2}$.

Proposition 1. DC is equivalent to WRTS+KM.

Based on the Proposition 1 above, we can present the following WRTS+KM clustering algorithm.

Algorithm1 WRTS+HCM

Input: $\boldsymbol{X}$, c, ε, λ

Output: $\boldsymbol{H}$

Step1 Whitening the regularized total scatter matrix (WRTS)

a. Compute the regularized total scatter matrix as $\tilde{\boldsymbol{S}}_t = \boldsymbol{XX}^T + \lambda \boldsymbol{I}_m$. Computing the SVD of $\tilde{\boldsymbol{S}}_t$ as $\tilde{\boldsymbol{S}}_t = \boldsymbol{W\Sigma V}^T$, and let $\boldsymbol{M} = \boldsymbol{W\Sigma}^{-1/2}$.

b. Transform data set by using $\boldsymbol{Y} = \boldsymbol{M}^T\boldsymbol{X}$.

Step2 Implement KM in the transformation space resulting from the transformation $\boldsymbol{M}$.

Line 2 takes $O(m^2n)$ time for the matrix multiplication $\boldsymbol{XX}^T$, $O(m^3)$ time for the SVD computation, $O(m^3)$ time for the matrix multiplication $\boldsymbol{W\Sigma}^{-1/2}$. Line 3 takes $O(m^2n)$ time for the matrix multiplication. Line 4 takes $O(nrct)$ time for KM clustering in the transformation space resulting from the transformation $\boldsymbol{M}$, where r is the rank of data matrix $\boldsymbol{X}$, t is the iteration number of KM. Thus, the total time complexity is $O(m^2n) + O(m^3) + O(nrct)$. So, when m is small and n far outweighs m, namely, $n \gg m$, WRTS+KM is more efficient than DKM. However, when n is small and m far outweighs n, namely, $m \gg n$, WRTS+KM is more time-consuming than DKM. Then we can implement DC by using DKM [8]. In conclusion, we can implement DC by running WRTS+KM or DKM flexibly according to different scenarios of data for efficiency. For convenience, the proposed clustering framework is called Efficient Discriminative Clustering (EDC) in the following discussion.

4 Efficient Discriminative Maximum Entropy Clustering

Since EDC is still a hard clustering and soft (fuzzy) versions of DC have been proposed [9–11], which have been proved more effective than hard versions [10, 11]. In this section, in order to improve the performance of EDC further, based on Maximum Entropy Clustering (MEC) [17–20], we extend EDC to soft clustering and propose Efficient Discriminative Maximum Entropy Clustering (EDMEC).

4.1 WRTS + MEC and Discriminative Maximum Entropy Clustering

Replacing KM by MEC [17–20] in WRTS+KM, we can obtain WRTS+MEC algorithm. The total time complexity of WRTS+MEC is $O(m^3) + O(m^2n) + O(mnct)$. So when m is small, WRTS+MEC can efficiently perform clustering on data.

Kernel maximum entropy clustering (KMEC) can be formulated as the following optimization problem.

$$\begin{aligned} &\max J_{\text{KMEC}}(\boldsymbol{U}) = \sum\nolimits_{j=1}^{c}\sum\nolimits_{i=1}^{n} u_{ij}\|\Phi(\boldsymbol{x}_i) - \Phi(\boldsymbol{v}_j)\|^2 - \beta^{-1}\sum\nolimits_{j=1}^{c}\sum\nolimits_{i=1}^{n} u_{ij}\log u_{ij} \\ &\text{s.t. } u_{ij} \in [0,1], \sum\nolimits_{j=1}^{c} u_{ij} = 1, \ \sum\nolimits_{i=1}^{n} u_{ij} > 0, \quad 1 \le i \le n, 1 \le j \le c. \end{aligned} \tag{8}$$

This can be solved by using the following algorithm 2.

Applying the kernel matrix $\boldsymbol{I}_n - (\boldsymbol{I}_n + \boldsymbol{G}/\lambda)^{-1}$ [8] to KMEC above, We can extend DKM to Discriminative Maximum Entropy Clustering (DMEC). The total time complexity of DMEC is $O(m^2n) + O(n^3) + O(n^2ct)$. When n is small and $m \gg n$, DMEC is efficient. After we obtain WRTS+MEC and DMEC, we can present EDMEC: When m is small and $n \gg m$, we can use WRTS+MEC to perform clustering on data, while n is small and $m \gg n$, we can use DMEC to perform clustering on data.

Algorithm 2 Kernel MEC

Input: $\boldsymbol{X}, \boldsymbol{K}, c, \varepsilon, \lambda, \beta$

Output: $\boldsymbol{U}$

Step1 Initialization: initialize the matrix fuzzy membership matrix $\boldsymbol{U} = \{u_{ij}\}_{n\times c}$;

Step2 Optimization: Optimize the objective in Equation (22) by iterating between step a and step b as follows until convergence;

a. Compute the distance matrix $\boldsymbol{D}_{XV}$, where $\boldsymbol{D}_{XV}(i,j) = \boldsymbol{K}_{XX}(i,i) - 2\boldsymbol{K}_{XV}(i,j) + \boldsymbol{K}_{VV}(j,j)$, $\boldsymbol{K}_{XX}(i,j) = \boldsymbol{K}(\boldsymbol{x}_i,\boldsymbol{x}_j), i,j = 1,2,\cdots,n$,

$\boldsymbol{K}_{VV}(j,j) = \sum\nolimits_{k=1}^{n}\sum\nolimits_{l=1}^{n} u_{kj}u_{lj}\boldsymbol{K}(\boldsymbol{x}_k,\boldsymbol{x}_l) / (\sum\nolimits_{k=1}^{n} u_{kj})^2$,

$\boldsymbol{K}_{XV}(\boldsymbol{x}_i,\boldsymbol{v}_j) = \sum\nolimits_{k=1}^{n} u_{kj}\boldsymbol{K}(\boldsymbol{x}_i,\boldsymbol{x}_k) / \sum\nolimits_{k=1}^{n} u_{kj}$;

b. Update fuzzy membership matrix $\boldsymbol{U}' = \{u'_{ij}\}_{n\times c}$ by computing

$u'_{ij} = \exp(-\boldsymbol{D}_{XV}(i,j)/\beta) / \sum\nolimits_{k=1}^{c} \exp(-\boldsymbol{D}_{XV}(i,k)/\beta)$

c. If $\|\boldsymbol{U}' - \boldsymbol{U}\| < \varepsilon$, break, otherwise let $\boldsymbol{U} = \boldsymbol{U}'$, go to back to step a.

4.2 Relationships to Earlier Approaches

In this subsection, we discuss the relation of EDMEC to LDA-based Fuzzy Clustering (LDA-FC) Algorithm [9] and FLDA-MEC [10, 11].

In [9–11], the fuzzy within-cluster, between-cluster and total scatter matrices are defined as $\boldsymbol{S}_{\mathrm{fw}} = \sum_{j=1}^{c}\sum_{i=1}^{n} u_{ij}(\boldsymbol{x}_i - \boldsymbol{v}_j)(\boldsymbol{x}_i - \boldsymbol{v}_j)^{\mathrm{T}}$, $\boldsymbol{S}_{\mathrm{fb}} = \sum_{j=1}^{c}\sum_{i=1}^{n} u_{ij}\boldsymbol{v}_j\boldsymbol{v}_j^{\mathrm{T}}$ and $\boldsymbol{S}_{\mathrm{ft}} = \sum_{j=1}^{c}\sum_{i=1}^{n} u_{ij}\boldsymbol{x}_i\boldsymbol{x}_i^{\mathrm{T}}$, respectively. In fact, based on the fuzzy membership matrix above, we can define the objective function of WRTS + MEC as follows:

$$\max J(\boldsymbol{U}) = \mathrm{trace}\left((\boldsymbol{S}_{\mathrm{ft}} + \lambda \boldsymbol{I}_m)^{-1}\boldsymbol{S}_{\mathrm{fb}}\right) - \beta^{-1}\sum_{j=1}^{c}\sum_{i=1}^{n} u_{ij}\log u_{ij}. \tag{9}$$

At the same time, the objective function of LDA-FC [9] is formulated as

$$\max J(\boldsymbol{U}) = \mathrm{trace}\left(\tilde{S}_{\mathrm{fw}}^{-1}\boldsymbol{S}_{\mathrm{fb}}\right). \tag{10}$$

Where $\tilde{S}_{\mathrm{fw}} = \lambda \boldsymbol{S}_{\mathrm{fw}} + (1-\lambda)diag(\boldsymbol{S}_{\mathrm{fw}})$ is the regularized within-cluster scatter matrix. If within-cluster scatter matrix $\boldsymbol{S}_{\mathrm{fw}}$ is regularized as $\tilde{S}_{\mathrm{fw}} = \boldsymbol{S}_{\mathrm{fw}} + \lambda \boldsymbol{I}_m$, the objective function of LDA-FC turns out to be max $\mathrm{trace}\left((\boldsymbol{S}_{\mathrm{fw}} + \lambda \boldsymbol{I}_m)^{-1}\boldsymbol{S}_{\mathrm{fb}}\right)$, it is essentially equivalent to max $\mathrm{trace}\left((\boldsymbol{S}_{\mathrm{ft}} + \lambda \boldsymbol{I}_m)^{-1}\boldsymbol{S}_{\mathrm{fb}}\right)$, which is exactly objective function of WRTS + MEC.

The objective function of FLDA-MEC [10, 11] is

$$\max J_{\mathrm{FLDA-MEC}}(\boldsymbol{P}, \boldsymbol{U}) = \mathrm{trace}\left(\left(\boldsymbol{P}^{\mathrm{T}}(\boldsymbol{S}_{\mathrm{ft}} + \lambda \boldsymbol{I}_m)\boldsymbol{P}\right)^{-1}\boldsymbol{P}^{\mathrm{T}}\boldsymbol{S}_{\mathrm{fb}}\boldsymbol{P}\right) - \beta^{-1}\sum_{j=1}^{c}\sum_{i=1}^{n} u_{ij}\log u_{ij}. \tag{11}$$

Since $\max\limits_{\boldsymbol{P}} \mathrm{trace}\left(\left(\boldsymbol{P}^{\mathrm{T}}(\boldsymbol{S}_{\mathrm{ft}} + \lambda \boldsymbol{I}_m)\boldsymbol{P}\right)^{-1}\boldsymbol{P}^{\mathrm{T}}\boldsymbol{S}_{\mathrm{fb}}\boldsymbol{P}\right) = \mathrm{trace}\left((\boldsymbol{S}_{\mathrm{ft}} + \lambda \boldsymbol{I}_m)^{-1}\boldsymbol{S}_{\mathrm{fb}}\right)$, then maximizing $J_{\mathrm{FLDA-MEC}}(\boldsymbol{P}, \boldsymbol{U})$ is equivalent to maximizing $J(\boldsymbol{U}) = \mathrm{trace}\left((\boldsymbol{S}_{\mathrm{ft}} + \lambda \boldsymbol{I}_m)^{-1}\boldsymbol{S}_{\mathrm{fb}}\right)$ $-\beta^{-1}\sum_{j=1}^{c}\sum_{i=1}^{n} u_{ij}\log u_{ij}$. This is exactly the objective function of WRTS+DMEC. So WRTS+MEC is equivalent to FLDA-MEC. In fact, the proposed EDMEC can be seen as an efficient version of FLDA-MEC.

5 Experimental Results and Analysis

To verify the effectiveness of EDC and EDMEC, we present an empirical study to evaluate the EDC and EDMEC in comparison with classical KM, LDA-KM [7], DKM [8], MEC [20] and FLDA-MEC [10, 11].

Table 2. Summary of the test data sets used in our experiment

Data sets	n	m	c
Wine	178	13	3
Zoo	101	18	7
Letter	2291	16	3
USPS	800	256	4
Satimage	6435	36	6
Leukemia	72	7129	2
GCM	43	16063	3

5.1 Experiment Setup

We compare EDC and DA-EDC with KM, LDA-KM [7], DKM [8], MEC [20] and FLDA-MEC [10, 11] on seven real world datasets. They are four UCI data sets [21]: Wine, Zoo, Letter (a-c) and Satimage; one handwritten image data set: USPS [22] and two gene expression data sets: Leukemia and Global Cancer Map (GCM) (http://www.upo.es/eps/aguilar/data-sets.html, the three classes: Lymphoma, Bladder and Melanoma are selected). The information on the seven test data sets is summarized in Table 2. We use the accuracy (Acc) as the clustering performance measure. The Acc is defined as $\left(\sum_{j=1}^{c} c_j\right)/n$, where c_j is the number of the data points clustered correctly in j-th class. We randomly select the clustering center from the original data sets for 20 times and generate 20 groups clustering centers. Each algorithm is run for 20 times respectively using the same 20 groups clustering centers, and the mean of Acc (still denoted as Acc) and the runtimes of each algorithm(denoted as T) of twenty runs are taken as the final experimental results. In the experiments, the maximum number of iterations and the threshold value of iteration stop are set to 100 and 10^{-5} respectively. Our hardware configuration is 2.20 GHz CPU and 2G RAM.

5.2 Clustering Evaluation

Table 3 presents the clustering accuracies and the execution time of KM, LDA-KM, DKM, EDC, MEC, FLDA–MEC and EDMC on all seven test data sets, where regularization parameter λ of LDA-KM, DKM, EDMEC, FLDA-MEC and EDMC is chosen from the range $\lambda \in \{0, 0.1, 1, 10, 10^2, 10^3, 10^4, 10^5, 10^6, 10^7, 10^8\}$, parameter β involved in MEC, FLDA-MEC and EDMEC is chosen from the range $\beta \in \{1, 10^{-1}, 10^{-2}, 10^{-3}, 10^{-4}\}$. The best results of clustering algorithms with respect to different parameters on each dataset are reported in Table 3. We can observe from the Table 3 that, for hard clustering, the clustering accuracies of LDA-KM, DKM and EDC are almost equal to each other, and they are superior to the ones of KM on almost all seven data sets. The execution time of EDC is less than the ones of LDA-KM and DKM. This is more obvious on data sets Letter, USPS and Satimage. Especially for Satimage, The execution time of LDA-KM is up to 90.6765 s, for DKM, the phenomenon of "out of memory" occurs. However, EDC has less execution time. We can observe from the

Table 3. The clustering results of seven algorithms on seven real world data sets

	KM	LDA-KM	DHCM	EDC	MEC	FLDA-MEC	EDMEC
Wine							
Acc	0.6702	0.8638	0.8567	0.8553	0.7191	**0.9719**	**0.9719**
T(s)	0.0047	0.0508	0.5008	0.0219	0.0375	1.1227	0.2188
Zoo							
Acc	0.6955	0.7079	0.7272	0.7010	0.7099	**0.8025**	0.7762
T(s)	0.0094	0.8634	0.4422	0.0305	0.0164	0.5359	0.2859
Letter							
Acc	0.7177	0.9092	0.9092	0.9092	0.7172	**0.9110**	**0.9110**
T(s)	0.0476	1.9148	42.1812	0.6093	0.4023	7.0538	0.9657
USPS							
Acc	0.9195	0.9290	0.9291	0.9289	0.9275	0.9333	**0.9345**
T(s)	0.0555	2.9242	5.9368	0.7593	0.8040	10.2360	0.7672
Satimage							
Acc	0.6099	0.6868	×	0.6675	0.6769	0.7034	**0.7044**
T(s)	0.3088	90.6765	×	11.6977	9.9648	10.7327	6.5405
Leukemia							
Acc	0.6347	0.6354	0.6354	0.6354	0.6396	**0.6528**	**0.6528**
T(s)	0.0633	0.6742	0.2062	0.2062	2.5539	0.8437	0.1429
GCM							
Acc	0.6570	0.6539	0.6663	0.6663	0.7221	**0.7314**	0.7244
T(s)	0.2813	0.7399	0.3976	0.3976	9.8625	6.6102	0.3523

Table 3 that, for soft (or fuzzy) clustering, the clustering accuracies of FLDA-MEC and EDMEC are almost equal to each other, and they are superior to the ones of MEC on all nine data sets. The execution time of EDMEC is less than the ones of FLDA-MEFCA, which is more obvious on data sets Letter, USPS. In addition, the soft clustering algorithms are competitive with their hard versions in terns of clustering accuracy.

6 Conclusions

In this paper, we propose new algorithms for Discriminative Clustering (DC) and its soft version, namely, EDC and EDMEC respectively. Because EDC and EDMEC can flexibly adopt different execution strategy for m ≪ n or n ≪ m, they can obtain dramatic computational advantage while still achieving the comparable accuracy. These are not only shown by our theoretical analysis, but also supported by our empirical results. Relatively speaking, EDMEC has advantage over EDC in terms of clustering accuracy. With efficiency, EDC and EDMEC are promising in application involving large, high-dimensional data.

Acknowledgements. This work is supported by the National Science Foundation of China (No. 61102095), the Science Plan Foundation of the Education Bureau of Shaanxi Province (No. 2010JK835, No. 14JK1661), the Natural Science Basic Research Plan in Shaanxi Province of China (No. 2014JM8307) and The Science and Technology Plan in Shaanxi Province of China (No. 2014KJXX-72).

References

1. Webb, A.: Statistical Pattern Recognition. Wiley, New Jersey (2002)
2. Tan, P.N., Steinbach, M., Kumar, V.: Introduction to Data Mining. Addison-Wesley, Boston (2005)
3. Yeung, K.Y., Medvedovic, M., Bumgarner, R.E.: Clustering gene-expression data with repeated measurements. Genome Biol. **4**(5), R34 (2003)
4. Mitra, P., Murthy, C., Pal, S.: Unsupervised feature selection using feature similarity. IEEE Trans. Pattern Anal. Mach. Intell. **24**(3), 301–312 (2002)
5. Law, M.H.C., Figueiredo, M.A.T., Jain, A.K.: Simultaneous feature selection and clustering using mixture models. IEEE Trans. Pattern Anal. Mach. Intell. **26**(9), 1154–1166 (2004)
6. De la Torre, F., Kanade, T.: Discriminative cluster analysis. In: Proceedings of the 23th International Conference on Machine Learning, pp. 241–248. ACM Press, New York (2006)
7. Ding, C., Tao, L.: Adaptive dimension reduction using discriminant analysis and K-means clustering. In: Proceedings of the 24th International Conference on Machine Learning, pp. 521–528. ACM Press, New York (2007)
8. Ye, J.P., Zhao, Z., Wu, M.R.: Discriminative K-means for clustering. In: 21th Advances in Neural Information Processing Systems, pp. 1649–1656. MIT Press, USA (2007)
9. Li, C.H., Kuo, B.C., Lin, C.T.: LDA-based clustering algorithm and its application to an unsupervised feature extraction. IEEE Trans. Fuzzy Syst. **19**(1), 152–163 (2011)
10. Yin, X.S., Chen, S.C., Hu, E.L.: Regularized soft K-means for discriminant analysis. Neurocomputing **103**, 29–42 (2013)
11. Zhi, X.B., Fan, J.L., Zhao, F.: Fuzzy linear discriminant analysis-guided maximum entropy fuzzy clustering algorithm. Pattern Recogn. **46**(6), 1604–1615 (2013)
12. Fukunaga, K.: Statistical Pattern Recognition, 2nd edn. Academic Press, San Diego (1990)
13. Belhumeur, P.N., Hespanha, J.P., Kriegman, D.J.: Eigenfaces vs. Fisherfaces: recognition using class specific linear projection. IEEE Trans. Pattern Anal. Mach. Intell. **19**(7), 711–720 (1997)
14. Friedman, J.H.: Regularized discriminant analysis. J. Am. Stat. Assoc. **84**(405), 165–175 (1989)
15. Raudys, S., Duin, R.P.W.: On expected classification error of the fisher linear classifier with pseudo-inverse covariance matrix. Pattern Recogn. Lett. **19**(5–6), 385–392 (1998)
16. Ye, J., Xiong, T.: Computational and theoretical analysis of null space and orthogonal linear discriminant analysis. J. Mach. Learn. Res. **7**, 1183–1204 (2006)
17. Li, R.P., Mukaidono, M.: A maximum entropy approach to fuzzy clustering. In: Proceedings of the Fourth International Conference on Fuzzy Systems, pp. 2227–2232 (1995)
18. Karayiannis, N.B.: MECA: maximum entropy clustering algorithm. In: Proceedings of the Third IEEE International Conference on Fuzzy Systems, pp. 630–635 (1994)
19. Zhang, Z.H., Zheng, N.N., Shi, G.: Maximum-entropy clustering algorithm and its global convergence analysis. Sci. China Ser. E: Technol. Sci. **44**(1), 89–101 (2001)

20. Ren, S.J., Wang, Y.D.: A proof of the convergence theorem of maximum-entropy clustering algorithm. Sci. China Ser. F Inf. Sci. **53**(6), 1151–1158 (2010)
21. Newman, D.J., Hettich, S., Blake, C.L., Merz, C.J.: UCI repository of machine learning databases. http://www.ics.uci.edu/mlearn/MLRepository.html (2015)
22. Breitenbach, M., Grudic, G.: Clustering through ranking on manifolds. In: Proceedings of the 22nd International Conference on Machine Learning, pp. 73–80. ACM Press, New York (2005)

Multi-attribute Decision Making Under Risk Based on Third-Generation Prospect Theory

Yu Xiang[1(✉)] and Li Ma[2]

[1] School of Information Science and Technology, Yunnan Normal University, Kunming 650091, China
iamlionx@126.com
[2] Library of Kunming University of Science and Technology, Kunming 650091, China

Abstract. A method of multi-attribute decision making under risk (MADMUR) based on third-generation prospect theory (PT3) is presented to solve the decision making problems by allowing reference points of each attribute to be uncertain. First, some important properties of PT3 are defined. Second, the adaptation of uncertain reference point is represented specifically. Then, the MADMUR method and its relevant computing steps are illustrated including all required definitions of MADMUR, an psychological inference procedure with relative value function and relative cumulative decision weighting function, and ranking and ordering decision alternatives etc. Finally, an illustrative example is given to show this decision making method in detail.

Keywords: Third-generation prospect theory · Multi attribute decision making under risk · Uncertain reference point

1 Introduction

For decades, decision-making-theoretic methods have been considered inapplicable for general problem solving because they require agents to process a utility function which provides preference ordering over outcomes of actions, and to access a probability distribution over outcomes associated with each decision alternative [1]. Most previous studies of decision making have already investigated the methods to maximize the utility in reasoning expert system and given limitations in computational abilities and information processing technologies. Particularly, in this paper, we have explored the problem of computing probability distributions under uncertainty and risk on the basis of uncertain reference point, and to a lesser extent, we have studied the assessment and custom tailoring of decision weights for third-generation prospect theory utility model.

Practically performing inference to an acceptable probability distribution under uncertainty and risk can greatly delay an agent's action. Sequentially, inference-related delays can lead to losses stemming from competition based on limited information resources, contradictory reasoning conclusions and problems with coordination among independent decision makers and their confused minds [1, 2]. Endowing an agent with the ability to trade off the accuracy or precision of an analysis for more psychologically responses can increase the expected value of that agent's behavior. Recent work by

X. He et al. (Eds.): IScIDE 2015, Part II, LNCS 9243, pp. 433–442, 2015.
DOI: 10.1007/978-3-319-23862-3_43

several investigators has addressed such tradeoffs in reasoning and ranking systems, and presented two types of applicable decision making mechanisms which are compensatory integrative models and noncompensatory heuristic models [2–4].

In this paper, we introduced a multi-attribute decision making support method to experiment with the use of metareasoning procedure on the basis of third-generation prospect theory to dominate inference approximation methods [5]. The main idea of this method is to allow reference point to be uncertain, and reference point determines the critical point of gains and losses from the point of view of the decision maker on an actual inference problem in which he or she will face uncertain information about the world. That is to say, reference point psychologically reflects a risk sensitive estimate of the expected value of computation by balancing the cost of taking action immediately with the benefits expected from additional refinement of the probability used in the same decision problem in the future. The method makes use of information about the psychological expected results to extract decision propensity which controls the decision weighting function, and about the rank-dependent change of the utility of outcomes.

In the following parts of this paper we review background on third-generation prospect theory, describe the semantics and assessment procedures for rank-dependent utility, discuss the custom-tailoring of default rank-dependent utility model on the basis of uncertain reference point, and discuss some aspects of our work on the consideration of this model. Finally, we describe the practical operation and behavior of this decision making support method by presenting an illustrative example.

2 Prospect Theory and Rank-Dependent Utility

2.1 Prospect Theory

In 1944, for the first time, Von Neumann and Morgenstern introduced their Expected Utility Theory (EUT) which describes Expected Utility (EU) by probabilities of possible outcomes when certain decision alternative has been chosen and their Utility Values calculated by an Objective Utility Function [6]. Suppose that when decision maker chooses alternative X_k as the final answer to the inference problem, there will be a probability p_i that possible outcome x_i, where $i = 1, 2, \ldots, n$, may occur as the action consequence. Then Utility Value of x_i can be denoted as $U(x_i)$ and Expected Utility of alternative X_k can be computed by following formula:

$$EU(X_k) = \sum_{i=1}^{n} p_i U(x_i) \tag{1}$$

The key idea of EUT is to rank all alternatives by their EUs so that it becomes easier to decide which one is the best bet.

Not long after EUT had been widely used, a series of theories derived from EUT had been published. For example, Savage introduced subjective probability into EUT to make it more practical for problem solving, and his achievement was called Subjective Expected Utility Theory (SEUT) [7]. Instead of using objective or subjective probability, Edwards proposed to acquire EU by using a subjective weight function, and named it Subjectively Weighted Utility Theory (SWUT) [8]. Machina presented Generalized Expected Utility Theory (GEUT) [9], Bell, Loomes and Sugden gave their Regret Theory (RT) [10, 11]. And the most significant among them may be the Prospect Theory (PT) which was originally established on the basis of psychology theory of decision making behavior [12].

2.2 Rank-Dependent Utility and Cumulative Prospect Theory

Like all other decision making theories at that time, investigators found that even PT could not explain a psychology anomaly known as Stochastic Dominance on the inference procedure. To solve this problem, Quiggin utilized the Rank-Dependent method in the first place to represent utility function of EUT, and published his well-known Rank Dependent Utility Theory (RDUT) [13]. Not too soon after that, Tversky and Kahneman remodeled PT with a Rank-Dependent and Sign-Dependent method, and proposed Third-Generation Prospect Theory known as the Cumulative Prospect Theory (CPT) [14].

In CPT, decision value function was reconstructed by cumulative probability and a more complex decision weight function. For $X_k = \{(x_1, p_1), (x_2, p_2), \ldots, (x_n, p_n)\}$ which is a n-outcome alternative, where p_i denotes the subjective probability of outcome x_i and $x_1 \leq x_2 \leq \cdots \leq x_k \leq 0 \leq x_{k+1} \leq x_{k+2} \leq \cdots \leq x_n$, according to formula (1) the decision value of X_k, expressed as $V(X_k)$, can be acquired by following equations:

$$V(X_k) = V(X_k^+) + V(X_k^-) = \sum_{i=k+1}^{n} \pi_1^+ v(x_i) + \sum_{i=1}^{k} \pi_i^- v(x_i)$$

$$V(X_k) = \sum_{i=k+1}^{n} \left(w^+ \left(\sum_{j=i}^{n} p_j \right) - w^+ \left(\sum_{j=i+1}^{n} p_j \right) \right) v(x_i) + \sum_{i=1}^{k} \left(w^- \left(\sum_{j=1}^{i} p_j \right) - w^- \left(\sum_{j=1}^{i-1} p_j \right) \right) v(x_i) \tag{2}$$

Where w^+ and w^- indicate decision weight functions of gains and loses respectively, and $v(x_i)$ is a value function.

2.3 Third-Generation Prospect Theory

2.3.1 Theoretical Background and Axiom System

In 2005 Schmidt, Starmer and Sugden met great obstacle while using CPT to explain Preference Reversal phenomenon and willing-to-accept/willing-to-pay disparity in decision making practical activities. They found that PT and CPT have a common limitation: the reference points from which prospects are evaluated are assumed to be certainties. Then they generalized CPT so that it can encompass uncertain reference points and named the new model Third-Generation Prospect Theory (PT3) [5].

In PT3, preferences of decision makers are defined over Savage acts. An *Act* is defined as an assignment of consequences to state S_k of decision alternative X_s, and a reference act, denoted by Act_{rp}, is the original point between gain and loss. Whether a particular *Act* will produce positive or negative consequences is relative to such reference points. PT3 uses a rank-dependent framework to assign decision weights to any act, viewed relative to any reference act. For any state S_k, PT3 uses $Act(S_k)$ and $Act_{rp}(S_k)$ to denote the outcomes of *Act* and Act_{rp} in state S_k. The ex-post net gain from choosing *Act* rather than Act_{rp} is the value of $NG(Act, Act_{rp}) = Act(S_k) - Act_{rp}(S_k)$. If $NG(Act, Act_{rp}) \geq 0$ then construct gain states rankings for alternative X_s on state S_k. If $NG(Act, Act_{rp}) < 0$ then construct loss states ranking for alternative X_s on state S_k. These rankings are used to determine decision weights.

For a decision making under risk, $S = \{S_1, S_2, \ldots, S_n\}$ is a finite state space, the occurrence probability of state S_k in S is π_k, then we know that $\pi_k > 0$ and $\sum_{k=1}^{n} \pi_k = 1$. Let $X_s^S = \{x_1, x_2, \ldots, x_n\}$ be a set of all possible outcomes of alternative X_s on all possible states S, and $AS = \{Act_1, Act_2, \ldots Act_p\}$ is the set of all acts, then Act_f in AS can be modeled as a function $S \to X_s^S$. Namely, for every state S_i we know that $Act_f(S_i) \in X_s^S$. As in other versions of PT, preferences over acts are Reference-Dependent. PT3 formalizes this following the approach of Sugden's Reference-Dependent Subjective Expected Utility Theory (RDSEUT) [15]. For any three acts Act_f, Act_g and Act_{rp} in AS, $Act_f \succsim_{Act_{rp}} Act_g$ denotes that Act_f is weakly preferred to Act_g viewed from Act_{rp}, the reference act which can be interpreted as the status quo position. $Act_f \sim_{Act_{rp}} Act_g$ denotes that Act_f is as preferred as Act_g viewed from Act_{rp}, the corresponding relations of indifference. $Act_f \succ_{Act_{rp}} Act_g$ denotes that Act_f is strictly preferred to Act_g viewed from Act_{rp}. PT3 doesn't require Act_{rp} to be a constant act, i.e. an act which gives the same consequence in every state.

2.3.2 Relative Value Function, Relative Weight Function and Relative Decision Function

According to PT3, we use a relative value function $v(Act_f(S_i), Act_{rp}(S_i))$ to represent the magnitude deviation of decision expectation assigned to Act_f viewed from Act_{rp} on the same state S_i. Thus we draw some important conclusions as follow:

$$\begin{cases} Act_f(S_i) \sim_{Act_{rp}(S_i)} Act_{rp}(S_i) \Leftrightarrow v(Act_f(S_i), Act_{rp}(S_i)) = 0 \\ Act_f(S_i) \succsim_{Act_{rp}(S_i)} Act_g(S_i) \Leftrightarrow v(Act_f(S_i), Act_{rp}(S_i)) \geq v(Act_g(S_i), Act_{rp}(S_i)) \\ Act_f(S_i) \succ_{Act_{rp}(S_i)} Act_g(S_i) \Leftrightarrow v(Act_f(S_i), Act_{rp}(S_i)) > v(Act_g(S_i), Act_{rp}(S_i)) \end{cases}$$

We also use $w(S_i; Act_f, Act_{rp})$ as the relative weight function to represent the decision weight assigned to Act_f viewed from Act_{rp} on the same state S_i, i.e. consider any act pair, denoted as Act_f and Act_{rp}, on the same state S_i meets the criteria that $Act_f(S_i) \geq Act_{rp}(S_i)$, then there is a weak gain in the state S_i, and a strict loss if $Act_f(S_i) < Act_{rp}(S_i)$. Since the number of states in set S is n, for alternative X_s, let k^+ be the number of states in which there are weak gains, known that $0 \leq k^+ \leq n$, and let $k^- = n - k^+$ be the number of states in which there are strict losses. For all subscripts i, j, we have $i \geq j$ if and only if $v(Act_f(S_i), Act_{rp}(S_i)) \geq v(Act_f(S_j), Act_{rp}(S_j))$, and so that the states with weak gains are indexed $i \in \{1, 2, \ldots, k^+\}$ and the states with strict losses are indexed $i \in \{-k^-, -k^- + 1, \ldots, -1\}$. The relative weight function $w(S_i; Act_f, Act_{rp})$ is then defined as follow:

$$w(S_i; Act_f, Act_{rp}) = \begin{cases} w^+(\pi_i) & i = k^+ \\ w^+\left(\sum_{j=i}^{k^+} \pi_j\right) - w^+\left(\sum_{j=i+1}^{k^+} \pi_j\right) & 1 \leq i \leq k^+ - 1 \\ w^-\left(\sum_{j=-k^-}^{i} \pi_j\right) - w^-\left(\sum_{j=-k^-}^{i-1} \pi_j\right) & -k^- + 1 \leq i \leq -1 \\ w^-(\pi_i) & i = k^- \end{cases} \quad (3)$$

For Act_f and uncertain reference point Act_{rp}, PT3 implies maximization of the relative decision function:

$$V(Act_f(S), Act_{rp}(S)) = \sum_{i=-k^-}^{k^+} w(S_i; Act_f, Act_{rp}) v(Act_f(S_i) Act_{rp}(S_i)) \quad (4)$$

3 Multi-attribute Decision Making Under Risk Based on PT3

A multi-attribute decision making under risk problem is to find a best compromise solution from all feasible alternatives assessed on multiple attributes. Suppose the decision maker has to choose one of or rank n efficient non-inferior alternatives which are represented by a set $X = \{X_1, X_2, \ldots X_p\}$ and based on m attributes. Let $A = \{A_1, A_2, \ldots A_m\}$ be the attribute set and $\Omega = \{\omega_1, \omega_2, \ldots, \omega_m\}$ be the relative weight set over A, where $0 < \omega_k < 1$. Let $S = \{S_1, S_2, \ldots S_n\}$ be a finite set of possible states with occurrence probability distribution indicated by $\Pi = \{\pi_1, \pi_2, \ldots, \pi_n\}$, where $\pi_k > 0$ and $\sum_{k=1}^{n} \pi_k = 1$. Then the multi-attribute decision making under risk problem can be concisely expressed as a group of decision matrix $D_k = (d_{ijk})_{p \times n}$ on each state, where d_{ijk} is the decision value, which takes the form of positive real number, of alternative X_i with respect to attribute A_k on state S_j. Specifically, the decision making procedure of our method consists of following steps:

Step 1. Calculate the normalized decision matrix

In general, attributes can be classified into two types: benefit attributes and cost attributes. Moreover, attributes are incommensurable and contradictory, and can't be measured in unified dimension unit. In other words, to measure all attributes in dimensionless unit and to facilitate inter-attribute comparisons, we must normalize

decision matrix $D_k = \left(d_{ijk}\right)_{p\times n}$ into a corresponding decision matrix $\tilde{D}_k = \left(\tilde{d}_{ijk}\right)_{p\times n}$ by following formulas:
For benefit attributes

$$\tilde{d}_{ijk} = \frac{d_{ijk}}{\max_{i=1}^{p}\left(\max_{j=1}^{n}(d_{ijk})\right)} \tag{5}$$

For cost attributes

$$\tilde{d}_{ijk} = \frac{\min_{i=1}^{p}\left(\min_{j=1}^{n}(d_{ijk})\right)}{d_{ijk}} \tag{6}$$

Step 2. Choose reference points
In 2000, Ordonez, Conolly and Coughlan had studied multiple-reference-point technique and how it effects decision maker's satisfaction and fairness assessment. Soon after that, they developed a comprehensive satisfaction evaluation model based on two reference points as follow [16]:

$$u(x, y, z) = f(x) + g(x - y) + h(x - z) \tag{7}$$

In formula (7), $u(x, y, z)$ is the magnitude of comprehensive satisfaction evaluation that how decision maker feels about his or her income. y and z are two referential incomes, and $f(x)$ is the decision maker's satisfaction assessment about his or her income without knowing someone else's. Furthermore, $g(x - y)$ and $h(x - z)$ are decision maker's two satisfaction assessments after he or she compares his or her own income with referential incomes. In the same way, we can take $f(x)$ as the satisfaction assessments after decision maker compares with his or her own. Taken all together, whether it is a gain or loss largely depends on the comparison between what we have got and all other feasible outcomes.
We generalize formula (7) so that it can measure the comprehensive satisfaction evaluation on the basis of p reference points, where $p \geq 2$. Namely, we use comprehensive satisfaction evaluation technique to compare $\tilde{d}_{ijk}$ with all other $\tilde{d}_{ljk}$, where $l = 1, 2, \ldots, p$, which with respect to the same attribute A_k and in the same decision matrix $\tilde{D}_k$ and indicates the same alternative X_i on state S_j. Consequently, we develop a method to calculate reference point $RP_j^k(X_i)$ based on decision matrix $\tilde{D}_k$ of alternative X_i on state S_j as follow:

$$\tilde{d}_{ijk} - RP_j^k(X_i) = \left(\tilde{d}_{ijk} - \tilde{d}_{1jk}\right) + \left(\tilde{d}_{ijk} - \tilde{d}_{2jk}\right) + \cdots + \left(\tilde{d}_{ijk} - \tilde{d}_{ijk}\right) + \cdots + \left(\tilde{d}_{ijk} - \tilde{d}_{pjk}\right)$$

$$\tilde{d}_{ijk} - RP_j^k(X_i) = p\tilde{d}_{ijk} - \sum_{l=1}^{p} \tilde{d}_{ljk} \tag{8}$$

$$RP_j^k(X_i) = \sum_{l=1}^{p} \tilde{d}_{ljk} - (p-1)\tilde{d}_{ijk} \tag{9}$$

By using formula (9), we can calculate a corresponding reference point matrix $\tilde{D}_k^{rp}$ of decision matrix $\tilde{D}_k$.

Step 3. Calculate decision value

In this step, we can calculate decision value for each alternative using relative value function and relative weight function defined by PT3. For decision matrix $\tilde{D}_k$, we define its relative value function $v\left(\tilde{d}_{ijk}, RP_j^k(X_i)\right)$ of alternative X_i on state S_j as follow:

$$v\left(\tilde{d}_{ijk}, RP_j^k(X_i)\right) = \begin{cases} \left(\tilde{d}_{ijk} - RP_j^k(X_i)\right)^{\alpha} & \text{if } \left(\tilde{d}_{ijk} - RP_j^k(X_i)\right) \geq 0 \\ -\lambda\left|\left(\tilde{d}_{ijk} - RP_j^k(X_i)\right)\right|^{\beta} & \text{if } \left(\tilde{d}_{ijk} - RP_j^k(X_i)\right) < 0 \end{cases} \tag{10}$$

Utilize formula (10) to obtain relative value $v\left(\tilde{d}_{ijk}, RP_j^k(X_i)\right)$ of each alternative X_i in decision matrix $\tilde{D}_k$ on all feasible state S_j, where $j = 1, 2, \ldots, n$. Then rank there relative values, if $v\left(\tilde{d}_{ijk}, RP_j^k(X_i)\right) < 0$ then reassign subscript of S_j to $j \in \{-k^-, -k^- + 1, \ldots, -1\}$, if $v\left(\tilde{d}_{ijk}, RP_j^k(X_i)\right) \geq 0$ then reassign the subscript of S_j to $j \in \{1, 2, \ldots, k^+\}$. Accordingly, calculate relative weight value $w\left(S_j; \tilde{d}_{ijk}, RP_j^k(X_i)\right)$ of alternative X_i corresponding to decision matrix $\tilde{D}_k$ and state S_j by using formula (3), where:

$$w^+(\pi_j) = \frac{\pi_j^{\gamma}}{\left(\pi_j^{\gamma} + (1-\pi_j)^{\gamma}\right)^{\frac{1}{\gamma}}} \tag{11}$$

$$w^-(\pi_j) = \frac{\pi_j^{\delta}}{\left(\pi_j^{\delta} + (1-\pi_j)^{\delta}\right)^{\frac{1}{\delta}}} \tag{12}$$

More importantly, In formulas (10), (11) and (12), parameters α, β, γ and δ can be assigned by the values stemming from the psychology experiment researches, designed by Tversky and Kahneman, on American subjects, which are 0.88, 0.88, 0.61 and 0.69 [14, 17], and also can be assigned by the similar values from researches on Chinese subjects done by Zeng, which are 1.18, 1.01, 0.53 and 0.50 [18]. Moreover, in both cases, parameter λ can be assigned as $1 \leq \lambda \leq 2.5$.

Calculate the decision value $V(X_i)$ of each alternative X_i on the basis of corresponding decision matrix $\tilde{D}_k$ by using formula (4). After that, then convert $\tilde{D}_k$ into relative decision vector denoted as $\tilde{\tilde{D}}_k = \left(\tilde{\tilde{d}}_{ik}\right)_{p\times 1} = (V(X_i))_{p\times 1}$.

Step 4. Aggregate relative decision vectors and make decision
To rank the preference order of all decision alternatives, we aggregate elements in each relative decision vector corresponding to alternative X_i by following approach:

$$VA(X_i) = \sum_{k=1}^{m} \omega_k \tilde{\tilde{d}}_{ik} = \sum_{k=1}^{m} \omega_k V(X_i) \tag{13}$$

Here ω_k is the relevant attribute weight. Thereupon, we select the most desirable alternative(s) in accordance with the maximum decision value $V(X_i)$ as the ideal solution to the multi-attribute decision making under risk problem.

4 Numerical Example

In this section, we utilize a practical multi-attribute decision making under risk problem to illustrate the application of the developed method.

An electronics technology company wants to purchase a shipment of electronic components from the best option. There is a panel with four possible providers denoted as $X = \{X_1, X_2, X_3, X_4\}$, and three attributes denoted as $A = \{A_1, A_2, A_3\}$ which are considered here by this company in selection of the four possible providers are: ① circulating fund; ② purchase price; ③ quality of component, and $\Omega = \{\omega_1, \omega_2, \omega_3\} = \{0.7, 0.9, 0.8\}$ are decision weights of these three attributes. Based on the current market quotation, there are four feasible states denoted as $S = \{S_1, S_2, S_3, S_4\}$ of the purchase, which are: ① delivery in advance; ② delivery in time; ③ delivery partially delayed; ④ totally backordered, and the probability distribution of these four states is $\Pi = \{\pi_1, \pi_2, \pi_3, \pi_4\} = \{0.1, 0.3, 0.4, 0.2\}$. Then experts in the company establish three decision matrixes according to three attributes as follow:

$$D_1 = \begin{bmatrix} 7.4 & 7.0 & 6.8 & 6.5 \\ 6.0 & 5.9 & 5.7 & 5.3 \\ 6.5 & 6.2 & 5.5 & 5.4 \\ 6.9 & 6.6 & 6.4 & 6.0 \end{bmatrix}, \quad D_2 = \begin{bmatrix} 0.42 & 0.41 & 0.38 & 0.35 \\ 0.40 & 0.39 & 0.32 & 0.30 \\ 0.42 & 0.40 & 0.35 & 0.33 \\ 0.41 & 0.40 & 0.37 & 0.34 \end{bmatrix},$$

$$D_3 = \begin{bmatrix} 0.95 & 0.95 & 0.90 & 0.80 \\ 0.95 & 0.90 & 0.85 & 0.85 \\ 0.95 & 0.95 & 0.90 & 0.80 \\ 0.90 & 0.90 & 0.90 & 0.85 \end{bmatrix}$$

The procedure of the proposed decision making model can be shown as the following steps:

Step 1. Normalize decision matrixes by using formulas (5) and (6):

$$\widetilde{D}_1 = \begin{bmatrix} 1.00 & 0.95 & 0.92 & 0.88 \\ 0.81 & 0.80 & 0.77 & 0.72 \\ 0.88 & 0.84 & 0.74 & 0.73 \\ 0.93 & 0.89 & 0.86 & 0.81 \end{bmatrix}, \quad \widetilde{D}_2 = \begin{bmatrix} 0.71 & 0.73 & 0.79 & 0.86 \\ 0.75 & 0.77 & 0.94 & 0.00 \\ 0.71 & 0.75 & 0.86 & 0.91 \\ 0.73 & 0.75 & 0.81 & 0.88 \end{bmatrix},$$

$$\widetilde{D}_3 = \begin{bmatrix} 1.00 & 1.00 & 0.95 & 0.84 \\ 1.00 & 0.95 & 0.89 & 0.89 \\ 1.00 & 1.00 & 0.95 & 0.84 \\ 0.95 & 0.95 & 0.95 & 0.89 \end{bmatrix}$$

Step 2. Calculate reference points in each decision matrixes by using formula (9):

$$\widetilde{D}_1^{rp} = \begin{bmatrix} 0.62 & 0.63 & 0.53 & 0.50 \\ 1.19 & 1.08 & 0.98 & 0.98 \\ 0.98 & 0.96 & 0.07 & 0.95 \\ 0.83 & 0.81 & 0.71 & 0.71 \end{bmatrix}, \quad \widetilde{D}_2^{rp} = \begin{bmatrix} 0.77 & 0.81 & 1.03 & 1.07 \\ 0.65 & 0.69 & 0.58 & 0.65 \\ 0.77 & 0.75 & 0.82 & 0.92 \\ 0.71 & 0.75 & 0.97 & 1.01 \end{bmatrix},$$

$$\widetilde{D}_3^{rp} = \begin{bmatrix} 0.95 & 0.90 & 0.89 & 0.94 \\ 0.95 & 1.05 & 1.07 & 0.79 \\ 0.95 & 0.90 & 0.89 & 0.94 \\ 1.10 & 1.05 & 0.89 & 0.79 \end{bmatrix}$$

Step 3. Calculate both relative value and relative decision weight of each matrixes on the basis of formulas (10), (11) and (12). In this purchase problem we assign values 0.88, 0.88, 0.61, 0.69 and 1.8 to parameters $\alpha, \beta, \gamma, \delta$ and λ in relevant formulas. Subsequently, calculate relative decision vectors by using formula (4) as follow:

$$\widetilde{D}_1 = \begin{bmatrix} 0.41 \\ -0.55 \\ -0.45 \\ 0.14 \end{bmatrix}, \quad \widetilde{D}_2 = \begin{bmatrix} -0.34 \\ 0.25 \\ -0.01 \\ -0.17 \end{bmatrix}, \quad \widetilde{D}_3 = \begin{bmatrix} 0.00 \\ -0.17 \\ 0.00 \\ 0.06 \end{bmatrix}.$$

Step 4. According to formula (13), aggregate all corresponding vectors to determine decision value of each alternative and acquire results as: $VA(X_1) = -0.02, VA(X_2) = -0.3, VA(X_3) = -0.32, VA(X_4) = 0$. So, the company should select provider X_4 which has the maximum decision value $VA(X_4) = 0$.

5 Conclusions

In this paper, we have introduced a method of multi-attribute decision making under risk (MADMUR) based on third-generation prospect theory (PT3) to solve the decision making problems by allowing reference points of each attribute to be uncertain. We also have represented an illustrative numerical example of electronics technology company purchasing components from four providers to demonstrate the feasibility and practicability of the proposed model. Results show that the proposed multi-attribute decision making model can effectively deal with the uncertainty and risk of the world and does not cause more computational burden than other decision making method. The proposed group decision making model is not only efficient and robust, but also more realistic and reasonable for real-world applications.

The methodology of the proposed decision making model can be easily extended to various areas with similar decision problem, e.g., the decision making under risk problems contain multiple attributes. So the further research may focus on the application of the developed method to the fields of medical diagnosis, personnel dynamic examination and military system efficiency dynamic evaluation, etc.

References

1. Simon, H., Dantzig, G., Hogarth, R., Plott, C., Raiffa, H., Shelling, T., Shepsle, K., Thaler, R., Tversky, A., Winter, S.: Decision making and problem solving. Interfaces **17**, 11–31 (1987)
2. Wang, Z.J., Ou, C.W., Li, S.: Integrative model or the priority heuristic? A test from the point of view of the equate-to-differentiate model. Acta Psychol. Sin. **42**(08), 821–833 (2010)
3. Birnbaum, M.H.: New paradoxes of risky decision making. Psychol. Rev. **115**, 463–501 (2008)
4. Johnson, E.J., Schulte, M., Willemsen, M.C.: Process models deserve process data: comment on Brandstatter, Gigerenzer and Hertwig. Psychol. Rev. **115**, 263–273 (2008)
5. Schmidt, U., Starmer, C., Sugden, R.: Third-generation prospect theory. J. Risk Uncertainty **36**, 203–223 (2008)
6. Von Neumann, J., Morgensterm, O.: The Theory of Games and Economic Behavior. Princeton University Press, Princeton (1944)
7. Savage, J.: The Foundations of Statistics. Wiley, New York (1954)
8. Edwards, W.: The theory of decision making. Psychol. Bull. **51**, 380–417 (1954)
9. Machina, M.J.: Expected utility analysis without the independence axiom. Econometrica **50** (2), 277–323 (1982)
10. Bell, D.E.: Regret in decision making under uncertainty. Oper. Res. **30**, 961–981 (1982)
11. Loomes, G., Sugden, R.: Regret theory: an alternative theory of rational choice under uncertainty. Econ. J. **92**, 805–824 (1982)
12. Kahneman, D., Tversky, A.: Prospect theory: an analysis of decision under risk. Econometrica **47**(2), 263–291 (1979)
13. Quiggin, J.A.: Theory of anticipated utility. J. Econ. Behav. Organ. **3**(4), 323–343 (1982)
14. Tversky, A., Kahneman, D.: Advances in prospect theory: cumulative representation of uncertainty. J. Risk Uncertain. **5**(4), 297–323 (1992)
15. Sugden, R.: Reference-dependent subjective expected utility. J. Econ. Theor. **111**, 172–191 (2003)
16. Ordonez, L.D., Connolly, T., Coughlian, R.: Multiple reference points in satisfaction and fairness assessment. J. Behav. Decis. Mak. **13**, 329–344 (2000)
17. Tversky, A., Kahneman, D.: Advances in Prospect Theory: Cumulative Representation of Uncertainty, pp. 44–65. Cambridge University Press, Cambridge (2000)
18. Zeng, J.M.: An experimental test of cumulative prospect theory. J. Jinan Univ. (Nat. Sci.) **28** (1), 44–47 (2007)

Research of Massive Data Caching Strategy Based on Key-Value Storage Model

Lei Wang[1(✉)], Gongxin Chen[2], and Kun Wang[1]

[1] Faculty of Information Engineering, East China Institute of Technology, Nanchang 330013, China
Wlei598@163.com

[2] Faculty Water Resource and Environmental Engineering, East China Institute of Technology, Nanchang 330013, China

Abstract. The development trend of Internet application and software is needed to read-write and access the massive data efficiently and quickly. In order to improve the performance which Web access huge amounts of data and analyze the SQL of data caching strategy, this paper proposes a strategy of the massive data cache based on Key-Value storage model according to the characteristics of massive data access. This strategy can optimize the semantic analysis of SQL for the user's query, then it extracts data objects which are involved in the query, at last it calculates the cost of cache by Key characteristics. These data will be stored in the cache server in the form of object. Thereby, it can reduce the access to the main database and improve the performance of data access. The experiments show that the caching scheme can effectively reduce the average response time and increase the throughput capacity of system.

Keywords: Analysis of SQL · Data cache · Key-value storage model · Data access

1 Introduction

With the rapid development of Internet technology, and network accessed data volume growing exponentially. And the high concurrent access to produce the enormous pressure on the Web site. At present, the development of network applications falls short of highly concurrent read and write in the network database, mass data storage and access and the database of high scalability and high availability. On the other hand, user's demand for real-time interactive network are enhanced, and the response time of system requirements more stringent. Thus, In order to reduce user waiting time, improve the management capability for massive data access and increase the concurrent processing capability of system are getting more and more urgent.

In this paper, according to the demand for high efficiency of mass data storage and access, we plan to analysis of SQL semantic optimization for data caching scheme to achieve a method that can reduce the access to the database but obtain data quickly. In order to acquire this demand, we propose a massive data caching strategy based on Key-Value storage model. According to the difference of the user requests to SQL for databases, analysis on the SQL semantic optimization and combines with the

X. He et al. (Eds.): IScIDE 2015, Part II, LNCS 9243, pp. 443–453, 2015.
DOI: 10.1007/978-3-319-23862-3_44

characteristics of the stored data objects. We can use different Key-Value cache handling mode to make full use of the memory resources of Web server contains application server and speed up the processing the system. Experimental results show that the caching scheme can effectively improve the speed of the website or software data access, reduce the system response time, and significantly improve system performance.

2 The System Architecture Analysis and Improvement of Massive Data Access Caching Strategy

The traditional Service process of WEB server is that client sends a HTTP request to the WEB server and the WEB server needs to respond to HTTP requests. Before the respond WEB server needs to do a series of business logic processing and the data need to be checked out from the DB database server directly. Every request of service requires to establish a separate connection of database. It also need to operate of SELECT to database repeatedly. Therefore, in the whole process of service, It cost a lot of time and reduce the performance inevitably because of frequent connection operation. The traditional architecture shown in Fig. 1.

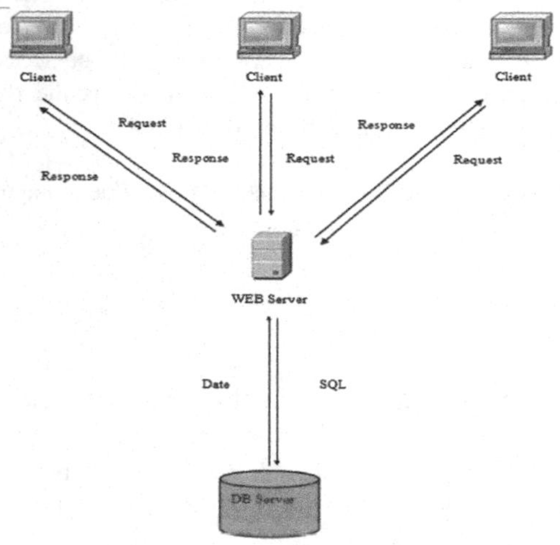

Fig. 1. The architecture of Web server program

For the above, we analyze that the architecture of this cache program will improve the traditional architecture. There are two modes in cache Scheme. One is the local cache of WEB server, the other is that set Cache server individually. The processes of specific Scheme s are as follows:

We still take the HTTP request for instance. When it was in the process of service, the first data were checked out from the DB database server directly. In this way caching server should be processed. In other words, it need to build a caching scheme based on Key-Value storage model on both Web server and Cache server. When there are access to those date secondly, there is no need to check out and gain data from the DB database directly. It should check the cache server of WEB server or Cache server. If there are in the cache, then it were returned to the service process directly. However, if there are not here, the service process will check out and gain data from the DB database server again. So there have Cache server between WEB server and DB database server, it can improve the performance of access for massive data effectively. Figure of the cache scheme and Figure of data cache Flow data are respectively showed in Figs. 2 and 3.

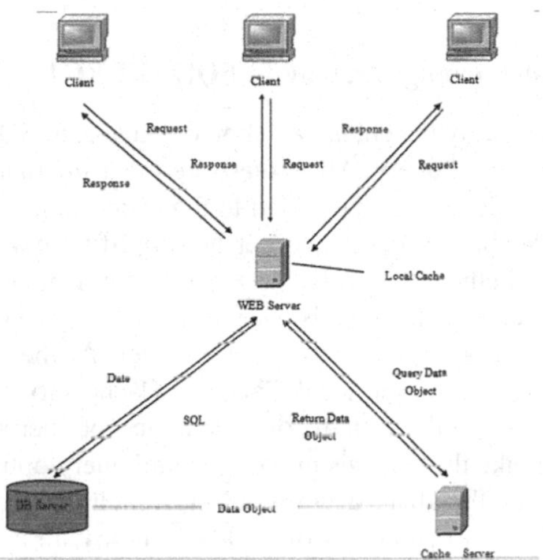

Fig. 2. WEB service architecture diagram added caching scheme

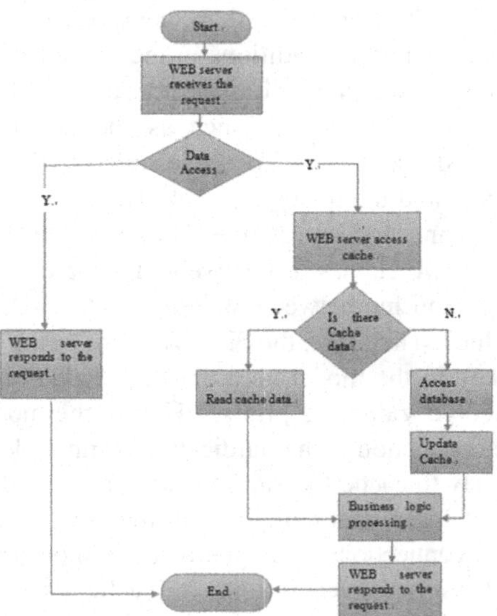

Fig. 3. The flowchart added caching scheme

3 The Analysis and Design of Massive Data Caching Strategy Based on Key-Value Storage Model

The Key-Value storage model is a data model which store the data based on the key value. It can store large amounts of semi-structured and unstructured data, and it takes the key value as identification to storage and access the data. The advantage of this model is that can cope with the amount of data and the scale of user continues to expand.

3.1 The Analysis of Caching Strategy in SQL-SELECT

For the queries which sent by the client, we should analysis to SQL semantic and judge and sort for the data object of query. All of these are to ensure that data of most recently and largest frequently accessed can be added to the cache space. When we need query of Select data from database server. In order to simplify the whole process of technological, no matter whether it was the first query, both directly to determine web server or local Cache server if the data is cached. If there are in cached and it returned directly. Otherwise it will access to the database server. At the same time the data by query saved to a Web server in the local Cache or Cache server as the way stored of Key-Value. Control nodes in the system continue to accept queries which submitted by network clients, they make the analysis processing and query optimization according to the cache strategy. When the query data by Select, it include some types of data query and the specific optimization strategies of SQL semantics are as follows:

- Form of Single - table access: the primary key way as objects to data cache, Key-Value storage model of Key as the primary key, Value for the object table.
- Form of Single table access with conditions (primary key): A primary key way to cache, as the same time there are more than one query conditions in cached, so it only returns to comply with the conditions in the object of Value.
- Form of Single table access with conditional (non-primary key): The data object as the query cache by query condition, such as the key of non-primary key in Key-Value storage model and value for table objects. We should consider combining the primary key and the query as whole Key.
- Form of Multi-table connection access: the primary key way to cache data objects. The cache object have two classes respectively. They are both Value1 and Value2. There is a kind of relationship between Value1 and Value2, and each Value1 may contain multiple Value2. Therefore, the Key-Value of cache object 1 stored Value1 as key of primary key in the model, and the Key-Value of cache object 2 stored Value2 that may involve Value1 as primary Key in the model.
- Form of Multi-table connection with qualification (primary key) access: on the basis of the primary key way to cache for more than a query conditions in cache. It only return to the object of Value that comply with the conditions from the cache.
- Form of Multi-table connection with qualification (non-primary Key) access: If statement of Select as query conditions in cache data, the query conditions

(non-primary Key) is not the only. So we should consider it join with the primary Key as the Key in Key-Value model at the same time.

- Form of Multiple tables join with fuzzy conditional access: we can join fuzzy query conditions with the primary Key as the key in Key-Value storage model.

3.2 SQL-UPDATE and SQL-DELETE Analysis of Caching Strategy

If the data in database server has been save to the Web local Cache or Cache server, and the user update operation on the data in the DB, we consider the problem of data synchronization and data consistency that the data should consistent in the database and web local cache servers or data cache. If the data object from the primary database changes in the current database, we should solve the problem of data consistency between the primary database and cache based on consistency maintenance strategy. The statements of update and delete are usually on one or more records in line with the conditions of the update or delete operation. It is not any means if we update and delete on all of the data. Analysis related to update and delete statements can be concluded that in the cache scheme using the key update and delete statements that are objects from update and delete statements. They obtain the value object that need to update the field and the new values from the update statement.

When data are be updated or deleted in the database server, we consider the corresponding data update or delete from the database directly. At the same time the data in local WEB server cache or Cache server should updated or deleted. There is an example by using to delete the data:

(1) Cascade delete: delete records of the specified primary key Table 1 in a database. Because of Table 1 connect with Table 2, we deleted Table 1 records and Table 2 records that corresponding foreign key. At the same time the data cache also need to remove the corresponding Value 1 and Value 2 object data.

(2) With delete condition: delete all tables that match conditions of a record, they also need to remove objects that meet specified criteria Value 1 data from data cache.

3.3 The Design of Data Caching Strategy Based on Query Condition

Most of the current data caching mechanism is based on the primary key caching. This caching program is designed for caching data based on common query conditions. The main idea is join the primary key with the query conditions as the key in Key-Value storage model and the data object as the Value in the Key-Value data storage model.

Query Q can be represented as triples Q (FS, TS, RP (fn, rr, rv)), in this triples, FS is the collection of query field, TS is a collection of relational tables, RP is the scope of the query, fn is the query field name, rr is a relationship between the query fields and values, the value of rr is {<, <=, >, >=, ==}, rv is the query field values. Let's analysis to the SQL statement as follows: "Select * from user_info where user_name = xiaohua", This query semantics Q ('*','user_info', RP('use_name','xiaohua')). Based on this related queries conditions, because of user_name is not the primary key in user_info, and the cache scheme table-object-mapping.xml profiles configured user_-info table the primary key is user_no, so caching strategy can be combined according to the user_no user_name together as key to store. Key is stored as: 'user_no', 'user_-name'. User_info cases recorded in Table 1:

Table 1. Records of user_info table

ID	user_no	user_name	user_pwd
1	1001	Xiaoming	123456
2	1002	xiaohua	123456

When we used the SQL statement: "select * from user_info where user_name = xiaohua" when key to a string: "user_info_1001, xiaohua". The problem need to consider is how "user_name = xiaohua". This query condition to obtain data from the cache.

When it used the query as a key, caching scheme requires traversal the key in CacheMap. It used a method of custom search Key (K key) and determined whether it include the Key "xiaoming" in this string. If a Key found in the string, it indicated that the data objects corresponding to the Value of Key had be found. So we can call method of user Info Cache Map. Set ("user_info_1001, xiaoming", userInfo) to cache the data objects by SQL statement to query. The userInfo is example for UserInfo class, which is declared as follows:

```
UserInfo1 =new UserInfo("user_info_1001", "xiaoming",
"123456")
```

3.4 The Problem of Hit Rate in Cache Strategy

Hit rate is an important factor to judge the merits of the cache, and the cache hit rate depends on many factors:

(1) Application of scene

Any application of scene, we must care for the frequency of data access. It is not any effect for a rarely access to the cache server. In general, the Internet web site is very suitable for cache applications of scene.

(2) The size of cache

If cache size is smaller, the hit rate is higher. Now the object cache is the smallest cache size and a higher chance of being hit. For example: If we access to this page in current, there need send to N SQL statements for ORM and take for their each user object. If the user object had in cache server, we can take it from the cache object directly at the time.

(3) The design of Update mode for caching scheme

The design of architecture for data caching model have a crucial impact on the cache hit rate. How to avoid the problem of cache invalidation, it provide caching accessed data frequently and need to design for cache. For a forum site, if we need to record the number of visits for each topic. When someone visit these topics every time, they must have on the topic data table is updated. That's means it is invalidated for update the cache when access the topic object of cache every time. We can use lazying update method for caching scheme. We can add an intermediate variable that save the number of visits. When it accumulated a certain number of visits, the database is updated. Thereby it reduced the frequency of cache invalidation.

4 Test and Evaluation on Massive Data Caching Scheme Based on Key-Value Storage Model

4.1 Design of Test Method for Caching Scheme

In order to verify the feasibility and the validity of cache scheme, we must carried out relevant tests on the caching scheme. The caching scheme's test results for the time value and milliseconds (ms) as a unit. It is a benchmark for testing. We get the same data and compared the time both access to database directly and access to cache of Key-Value storage model. If we get time to use the data from the cache is less than access to database directly. It prove the caching scheme has certain advantages.

Test caching scheme will use jUnit4.0 and a small test class to complete. It can get execution time on each method of performing by using JUnit4.0 and and a small test class respectively. jUnit4.0 can use the new features of Java annotations. The simple example of using jUnit4.0 as follows:

```
import org.junit.Test
public class TestClass {@Test
public void testAddUserInfo() {//Here call the need to
method of test}}
```

In this paper, design of small test class is using the callback and test method, it implements interface of TestService test and overwrite its test () method. The class of TimeTester calls TestService interface that give related methods of testing time directly. It is accomplished as follows:

```
public interface TestService
        {public void test();}
public class TimeTester {
public static long testTime(TestService testService)
  {long start = System.currentTimeMillis();
   testService.test();
   long end = System.currentTimeMillis();
   return end - start; }     }
```

4.2 Test Performance Evaluation of Caching Scheme

Test of caching scheme is based on data table user info and topic. Both of their fields and sample data are shown in Tables 2 and 3 as follows.

Table 2. Structure of user_info table

Number	Field name	Field description	Data type	Data length	Remark
1	user_no	User ID	varchar	20	Not null, PK
2	user_name	User name	varchar	50	Not null, unique
3	user_pwd	User password	varchar	20	Not null

Table 3. Structure of topic table

Number	Field name	Field description	Data type	Data length	Remark
1	topic_no	Topic number	varchar	20	Not null, PK
2	topic_name	Topic name	varchar	50	Not null, unique
3	topic_content	Thematic content	text	/	Not null
4	user_no	User ID	varchar	20	Not null, FK

Seven groups of test are done according to these two data tables, each of them is contrast-based test as well as carried out 10 tests. At last we calculate the result of an the average. In order to appear no conflicting and no reluctant, the design integrated directly both the Key and the primary key as the query conditions in the cache scheme based on the primary key caching scheme in this paper. While the value is a data object. The forms of key stored in User Info objects are: "user_no, user_name", Such as "user_no_5000, user_name_5000". While the forms of key stored in Topic object are: "topic_no, topic_name, user_no", such as "topic_no_4800, topic_name_4800".

We inserted 50000 records into table of user_info, and the time of inserted 50000 records in user_info is 79350 ms. Data table topic inserted 50000 records in total, and the time is 81651 ms.

(1) The test results of first group: Compare with the cost time that obtain 50000 records in user-info (table number 1) directly from the database and obtain 50,000 records from the data cache. The test results are shown in Table 4.

Table 4. The test results of first group

Number	Test 1	Test 2	Test 3	Test 4	Test 5	Test 6
1	748	746	748	736	744	748
2	356	376	392	355	368	366
Number	Test 7	Test 8	Test 9	Test 10	Mean value	
1	765	730	727	715	740.7	
2	390	352	355	339	364.9	

As we can show from the test results: The time that get 50,000 records of user_info from the cache have best performance advantages than from the database, the improvement of performance nearly 1 times.

(2) The test results of second group: Compare with the cost time that query the designated user_no record from database (table number 1)and the primary key as query condition designated user_no record from data cache (table number 2).The test results are shown in Table 5.

Table 5. The test results of second group

Number	Test 1	Test 2	Test 3	Test 4	Test 5	Test 6
1	275	274	279	281	280	279
2	45	46	44	44	44	44
Number	Test 7	Test 8	Test 9	Test 10	Mean value	
1	275	280	275	279	252.5	
2	46	44	44	44	44.5	

As it can be seen from the test results: get the specified user_no records from the cache have best performance advantage than query specifies user_no recorded from the database. The improvement of performance nearly 5 times.

(3) The test results of third group: There are compare with cost time on the records of the specified user_name use non-primary key as a query condition from the database directly (table number 1) and from the cache (table number 2). The test results shown in Table 6.

Table 6. The test results of third group

Number	Test 1	Test 2	Test 3	Test 4	Test 5	Test 6
1	270	273	270	282	274	273
2	41	42	42	41	41	41
Number	Test 7	Test 8	Test 9	Test 10	Mean value	
1	256	271	276	273	271.8	
2	41	41	41	41	41.2	

As it can be seen from the test results, the records of the specified user_name from cache have best performance advantages than from database. The improvement of performance nearly 6 times.

(4) The test results of fourth group: To Compare with cost time on query records in user_no from the database use foreign key as a condition(table number 1) and query records in user_no from the cache (table number 2). The test results shown in Table 7.

Table 7. The test results of fourth group

Number	Test 1	Test 2	Test 3	Test 4	Test 5	Test 6
1	259	231	234	233	236	234
2	7	8	7	7	7	8
Number	Test 7	Test 8	Test 9	Test 10	Mean value	
1	231	234	237	239	236.8	
2	10	7	8	7	7.6	

As it can be seen from the test results, the records of the specified user_no from cache have best performance advantages than from database. The improvement of performance nearly 31 times.

From test data of four groups above, it can be concluded that the design of strategy on data cache can improve massive data access performance effectively. It is a feasible and effective caching strategy.

5 Conclusion

With the development of Internet applications and software, how to improve the performance of accessing to massive data on web and how to store and access the massive data efficiently. That is become a hot research in the industry. In this paper, we analysis traditional architecture of network services based on web. There are using cache technology connect with key-Value storage model. So we propose a design of cache strategy on massive data based on Key-Value storage model. According to different requests of SQL to database from users, we should analysis for optimization of SQL semantic to data caching scheme in further. Consider the characteristics of data storage and improve the speed of process in system, we propose a data caching strategy using different form in Key-Value storage models. Experimental results show that the cache strategy can not only enhance the data access throughput of web site or software effectively but also reduce response time of the system.

Acknowledgment. This work is supported by project of teaching reform in Jiangxi Province (JXJG-12-8-15) and project of the Education Department of Jiangxi province science and technology projects (No. GJJ12752).

References

1. Yue, L.: Research on Key Technologies for Virtual Geographic Environment Based on Distributed Storage. PLA Information Engineering University, Zhengzhou (2011)
2. Wei, Li: Research and Implementation of Cache Technology in Grid Database [M]. Nanjing University of Aeronautics and Astronautics, Nanjing (2011)
3. Lu, C.-J.: On cache mechanism and its application model in data access layer. Appl. Res. Comput. **12**, 172–174 (2008)

4. Shen, X.-P.: Research on P2P data caching policy. Comput. Eng. Des. **8**, 2636–2638 (2011)
5. Liu, X.: The research and design of distributed data cache mechanism. Hunan University, Hunan (2013)
6. Ren, G.-Q., Yang, J.-M.: Content-based dynamic load-balancing algorithm of web server. Comput. Eng. **7**, 82–86 (2010)
7. Cao, W., Ying, J.: The research of hibernate cache mechanism and application. J. Hangzhou Dianzi Univ. **10**, 158–161 (2013)
8. Liu, W.-X., Yu, S.-Z.: Selective caching in content-centric networking. Chin. J. Comput. **2**, 275–287 (2014)

Trajectory Optimization for Cooperative Air Combat Engagement Based on Receding Horizon Control

Chengwei Ruan(✉), Lei Yu, Zhongliang Zhou, and An Xu

Aeronautics and Astronautics Engineering College,
Air Force Engineering University, Xi'an 710038, China
rcwrff@126.com

Abstract. Trajectory optimization for cooperative air combat engagement is studied. The optimization problem of cooperative air combat is established based on the analysis of vertical tactical engagement, target functions and terminal constraints through three different tactical processions are proposed. The receding horizon control model and the numerical solution based on Simpson-direct-collocation are put forward. A BP neural network based approximation of the performance measures is proposed In order to improve the online performance. Finally, a simulation shows that this method is feasible in cooperative air combat engagement.

Keywords: Cooperative attack · Air combat decision · Trajectory optimization · Artificial neural network (ANN) · Nonlinear programming (NLP)

1 Introduction

Cooperative air combat for airplane formation has become the main form of modern air combat at the background of NCW (network centered warfare) and cooperative attack plays a more and more important role in systemic confrontation. In cooperative combat dual flight formation can be described as a very important unit which is the foundation of systemic confrontation [1]. Tactical maneuver is an important process of multi-target-attack air combat, and the combat superiority can be approached with appropriate tactical maneuvers.

Tactical maneuver is a traditional trajectory optimization problem with given expected tactical requirements and the pursuit and evasion model is still the focus in current times [2]. The common methods we used to solve the problem are differential game [3], diagram game model [4], dynamic programming method [5], neural network [6], fuzzy reasoning [7] and optimum control [8], etc. Among all these methods, the differential game is too conservative and complex to apply to the problem, and the curse of dimensionality cannot be avoided if the diagram game model and dynamic programming method are used, a large amount of different academic knowledge is needed when neural network and fuzzy reasoning method are realized, and various difficulties will appear when using optimum control method to solve the problem.

X. He et al. (Eds.): IScIDE 2015, Part II, LNCS 9243, pp. 454–466, 2015.
DOI: 10.1007/978-3-319-23862-3_45

Speed and stealth are emphasized in tactical maneuver of cooperative air combat, concentrated control method is adopted in this study to pay much attention to the effectiveness of cooperative air combat as well as the characters mentioned above. The multi-stage optimum control model of cooperative air combat for dual fighters is put forward according to different mission process. Then the optimum control problem is transformed into parameter optimization problem by using RHC (receding horizon control) method in order to simplify the real-time calculation. A new neural network method is put forward to approximate index function in order to solve the calculation problem in nonlinear programming environment. And in the end, all the methods are put into practice and used in simulation.

2 Concept of Cooperative Tactical Maneuver for Dual Fighters

Cooperative tactical maneuver for dual fighters [9] can be described as the expansion of traditional one to one tactical maneuver style, both of the two styles are aiming at getting access to fulfill the attack condition. A series of constraint conditions and performance parameters have to be taken into consideration in the former combat style, so it has different decision method compared to the latter combat style. The main function of cooperation in air combat is to control the situation of combat as soon as possible.

The definition of cooperative tactical maneuver for dual fighters can be presented as below.

A tactical maneuver decision method and tactical execution process which is aiming at improving attack efficiency with integration of information, firepower, flight track and interference and the choice optimization from time domain and space field. Tactical maneuver is the most important stage in medium or long distance air combat, which may affect the attack efficiency and the final result directly. In this study, tactical maneuver is divided into three processes, namely tactical spread process (I), tactical transition process (II) and tactical synthesis process (III) (Fig. 1).

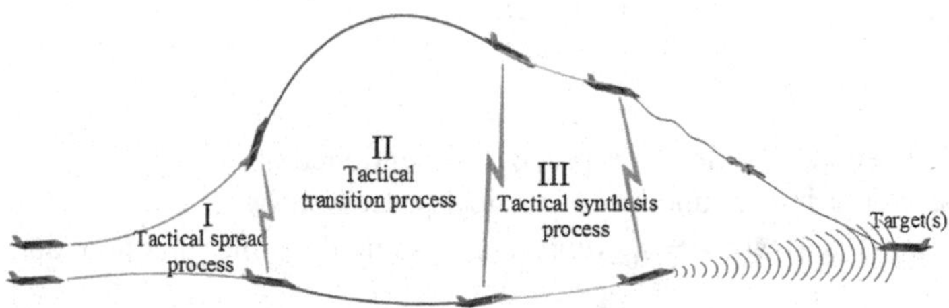

Fig. 1. Sketch map of vertical spread tactic for dual fighters

The vertical spread tactic [10] is mainly used in small formation air combat, which requires the formation to spread quickly in vertical plane. The leader fighter should climb to the expected height in order to occupy the first shot advantage and increase the missile attack distance. The wing fighter with stealth should descend to get close to the enemy, supervise the enemy and support the leader fighter to launch an attack. The tactic can be executed as follow steps:

(1) Decide the enemy's basic expectation according to current combat situation and determine the tactical organization as well as the general time of formation conversion.
(2) Form the optimum flight track and flight guidance command according to the tactical intent in order to guide the dual fighters with high security level to get close to the enemy in an expected flight track as soon as possible.
(3) Calculate real-time attack parameters of dual fighters and once the cooperative attack conditions are satisfied, the attack stage will be activated.

The main purpose of the three steps above is to make sure that the dual fighters can get into the attack area at the same time and they can also compose an efficient attack position from the high and relative low height in order to improve the formation attack probability. During the process of tactical maneuver, coordinated cooperative behaviors are needed to complete the whole mission, and a real-time tactical programming is also essential to help to execute the corresponding tactical maneuvers according to the policies.

3 Optimum Control Model of Cooperative Maneuver for Dual Fighters

According to the three processes, we can conclude that cooperative tactical maneuver problem can be solved by optimum control model, the functions are listed below:

$$\begin{aligned} J^k(\boldsymbol{u}) &= \int_{t_0^k}^{t_f^k} L^k(\boldsymbol{s},\boldsymbol{u},t)\mathrm{d}t + \varphi(\boldsymbol{s}_f^k, t_f^k) \\ \text{s.t.}\dot{\boldsymbol{s}} &= f(\boldsymbol{s},\boldsymbol{u},t), \boldsymbol{s}^k(t_0) = \boldsymbol{s}_0^k \\ & g(\boldsymbol{s},\boldsymbol{u},\dot{\boldsymbol{u}},\ddot{\boldsymbol{u}}) \le 0 \\ & h^k(\boldsymbol{s}_f^k) = 0 \end{aligned} \tag{1}$$

k = 1, 2, 3 represents the three processes mentioned above.
$J^k(\boldsymbol{u})$ means the index function of tactical process k.
$\dot{s} = f(\boldsymbol{s}, \boldsymbol{u}, t)$ and $\boldsymbol{s}(t_0^k) = \boldsymbol{s}_0^k$ represent system dynamic function and its initial condition.
$g(\boldsymbol{s},\boldsymbol{u},\dot{\boldsymbol{u}},\ddot{\boldsymbol{u}}) \le 0$ shows the control constraints.
$h^k(\boldsymbol{s}_f^k) = 0$ represents the time boundary condition of process k.
t_f^k and $\boldsymbol{s}_f^k$ are described as the initial time and status of process k.
t_f^k and $\boldsymbol{s}_f^k$ are described as the terminal time and status of process k.
$\boldsymbol{u}$ represents the control parameter of system input.

3.1 System Dynamic Equations

Here we definite $\boldsymbol{s} = [\boldsymbol{s}_l, \boldsymbol{s}_w]^T$ as state vector which $\boldsymbol{s}_l$ and $\boldsymbol{s}_w$ represent the state vectors of the leader fighter and the wing fighter relative to the target. We define $\boldsymbol{s}_v = [\varphi_v, \theta_v, d_v, \Delta h_v, \beta_v]^T$ as state variable of single fighter, while φ_v represents the lead angle of fighter's velocity, θ_v represents the aspect angle, d_v shows the relative distance, Δh_v represents the height difference, β_v represents the angular separation between the direction of the velocity of fighter and target. Finally, we can easily find that. $v = l$ *and* w are symbols of leader fighter and wing fighter.

The concrete computing method can be described as below:

$$\begin{aligned}
\varphi_v &= \arccos\{[(x_t - x_v)\cos\psi_v \cos\phi_v + \\
&\quad (y_t - y_v)\sin\psi_v \cos\phi_v + (h_t - h_v)\sin\phi_v]/d_t\} \\
\theta_v &= \arccos\{[(x_v - x_t)\cos\psi_t \cos\phi_t + \\
&\quad (y_v - y_t)\sin\psi_t \cos\phi_t + (h_v - h_t)\sin\phi_t]/d_t\} \\
d_v &= \sqrt{(x_t - x_v)^2 + (y_t - y_v)^2 + (h_t - h_v)^2} \\
\Delta h_v &= h_t - h_v \\
\beta_v &= \arccos(\cos\psi_v \cos\phi_v \cos\psi_t \cos\phi_t + \\
&\quad \cos\psi_v \sin\psi_t \cos\phi_v \sin\phi_t + \sin\psi_v \sin\psi_t)
\end{aligned} \tag{2}$$

x_v, y_v compose the location of fighter, while h_v represents the height and ψ_v, ϕ_v represent the track yaw angle and the track pitch angle.

3.2 Control Vector and Constraint

We define $\boldsymbol{u} = [\boldsymbol{u}_l, \boldsymbol{u}_w]^T$ and $\boldsymbol{u}_v = [\alpha_v, \mu_v, \eta_v]^T$ as system control vectors, we also define α_v, μ_v, η_v as attack angle, bank angle and throttle position. $g(\boldsymbol{x}, \boldsymbol{u}, \dot{\boldsymbol{u}}, \ddot{\boldsymbol{u}}) \leq 0$ represents the control constraint [11].

3.3 Index Function

When getting close to the enemy, our fighters should control the maneuver directions precisely to arrive in their own targets' area as soon as possible in order to reduce the probability of exposure and compose the launch condition. During this process, stealth and speed should be put into the first place, and the control loss should be taken into consideration as well. Under the condition of cooperation, the requirements of situation between dual fighters and situation between our fighters and targets at different stage are of a great difference. All the differences should be reflected in the index function.

We define two kinds of indexes during the process of tactical maneuver, they are general performance index and specific performance index. The former one is a common one which will be used through the whole combat, while the latter one can only be applied to the relative process.

(1) General Performance Index

According to the stealth requirement, index function is given as below:

$$J_s(\boldsymbol{u}) = \int_{t_0}^{t_f} \lambda(t)\mathrm{d}t \tag{3}$$

where $\lambda(t) = \begin{cases} 1, & \text{if } p_d(t) > p_T \\ 0, & \text{else} \end{cases}$, and p_T is regarded as the threshold probability of exposure, p_d can be calculated by current situation of both sides and RCS of the fighters.

According to the speed requirements, index function is given as below:

$$J_t(\boldsymbol{u}) = \int_{t_0}^{t_f} \mathrm{d}t = t_f - t_0 \tag{4}$$

Where t_f denotes the terminal time, which represents the time when the fighter arrives the targets' area.

According to the requirements of control loss, the index function can be composed by the angular velocity of the fighter.

$$J_e(\boldsymbol{u}) = \int_{t_0}^{t_f} (\dot{a}(t) + k\dot{\mu}(t))\mathrm{d}t \tag{5}$$

$\dot{a}(t)$ and $\dot{\mu}(t)$ are lateral acceleration and angular velocity in roll, k is weighting factor.

(2) Specific Performance Index

During the tactical spread process, the situation between dual fighters should be focused on, so we define the index below to describe the situation.

$$J_{ts}(\boldsymbol{u}) = \int_{t_0^1}^{t_f^1} (|\Delta H_l| - H_l^*)(|\Delta H_w| - H_w^*)|\Delta\beta|\mathrm{d}t \tag{6}$$

where ΔH_l is regarded as the attitude difference between leader fighter and the target while ΔH_w is regarded as the attitude difference between wing fighter. H_l^* and H_w^* are regarded as expected attitude difference, $\Delta\beta$ is the angle between the velocity direction of leader fighter and that of the wing fighter.

During the process of the tactical transition process, approach velocity and angle of the single fighter compared to the target can be described as below:

$$J_{td}(\boldsymbol{u}) = \int_{t_0^2}^{t_f^2} (\Delta H - \Delta H^*)(|\Delta H_l| - H_l^*)(|\Delta H_w| - H_w^*)|\Delta\beta|\mathrm{d}t \tag{7}$$

ΔH is the height deviation between both fighters, ΔH^* is the expected height deviation. The other variables are defined the same meaning as that of formula 7.

During the tactical synthesis process, relative situation of dual fighters is mainly put forward in order to make sure that the cooperative formation is able to compose an advanced position to attack the target (one of the dual fighters can track the target steadily and the other one can launch the missile or both of them can be able to launch missiles at the end).

$$J_{te}(\boldsymbol{u}) = \int_{t_0^3}^{t_f^3} \left\| \boldsymbol{s}_l - \boldsymbol{s}_l^* \right\| \left\| \boldsymbol{s}_w - \boldsymbol{s}_w^* \right\| \mathrm{d}t \tag{8}$$

$\boldsymbol{s}_l^*$ and $\boldsymbol{s}_w^*$ are expected situation of leader fighter and wing fighter, and the expected situation can be calculated by the tactical performance of both fighters. $\|\cdot\|$ represents F-2 norm. Figure 2 is the sketch map of the expected situation of leader fighter, the mathematical description of expected situation is given in the next paragraph.

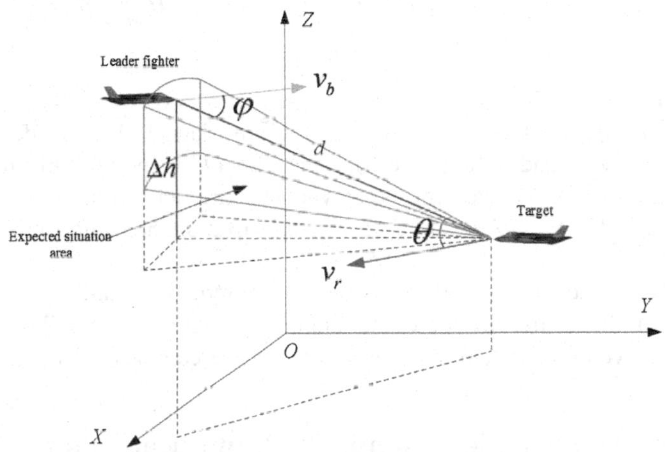

Fig. 2. Sketch map of expected situation of leader fighter

The index function of cooperative tactical maneuver for dual fighters can be defined as the sum of all the indexes in the whole cooperative system, the index functions in different process are given as below:

$$J^k(\boldsymbol{u}) = \begin{cases} \sum_v (J_{ts}^v + w_1^1 J_s^v(\boldsymbol{u}) + w_2^1 J_t^v(\boldsymbol{u}) + w_3^1 J_e^v(\boldsymbol{u})), \text{ if } k = 1 \\ \sum_v (J_{td}^v + w_1^2 J_s^v(\boldsymbol{u}) + w_2^2 J_t^v(\boldsymbol{u}) + w_3^2 J_e^v(\boldsymbol{u})), \text{ if } k = 2 \\ \sum_v (J_{te}^v + w_1^3 J_s^v(\boldsymbol{u}) + w_2^3 J_t^v(\boldsymbol{u}) + w_3^3 J_e^v(\boldsymbol{u})), \text{ if } k = 3 \end{cases} \tag{9}$$

Where $k = 1, 2, 3$ is regarded as different tactical processes, w_i^k is the weight factor of performance and $\sum_{i=1}^{3} w_i^k = 1$.

3.4 Terminal Condition

An expected cooperative attack situation is needed in the combat according to the definition of the goal set and different processes of the combat may have different requirements of terminal condition, so we define the value of time boundary to describe the terminal situation.

$$h^1(\boldsymbol{x}_f) = \{\boldsymbol{x} | \Delta H_i \geq H_i^*, \Delta\theta_i \leq \theta_i^*, i = l, w\} \tag{10}$$

$$h^2(\boldsymbol{x}_f) = \{\boldsymbol{x} | d_i \leq d_i^*, \theta_i \leq \theta_i^*, i = l, w\} \tag{11}$$

$$h^3(\boldsymbol{x}_f) = \left\{ \boldsymbol{x} | d_i \in [r_i, R_i], \varphi_i \in [o_i, O_i], \theta_i \in [l_i, L_i], |\Delta h_i| < H, \sum_i \beta_i \leq \pi/2, i = l, w \right\} \tag{12}$$

Where $[r_i, R_i]$ denote the minimum and maximum launch distance decided by the situation requirements and fighter's performance, $[o_i, O_i]$ is the limit of off-axis angle, $[l_i, L_i]$ is the limit of lead angle, The other variables are defined the same meaning as that of formula 9. The variables can be calculated based on the tactical project or given by the expert.

The terminal time of this optimum control problem is so hard to confirm that it brings a lot of difficulties to solve the engineering problem. So RHC method is introduced to solve this problem in order to avoid some complex calculating problems.

4 Receding Horizon Control and Its Numerical Solution

4.1 The Basic Idea of RHC

RHC [12] is an online calculating method which can solve the optimum control problem continuously based on the current changing system situation. In RHC method, the status at time t_d can be regarded as $\boldsymbol{x}(t_d)$, we can get the open-loop optimal control function $\boldsymbol{u}^*(\boldsymbol{x}(t_d))$ by calculating the following model during the period $t \in [t_d, t_d + T]$:

$$\begin{aligned} \tilde{J}_d(\boldsymbol{u}) &= \int_{t_d}^{t_d+T} L(\boldsymbol{x}, \boldsymbol{u}, t)\mathrm{d}t + V(\boldsymbol{x}(t_d + T), t_d + T) \\ \text{s.t. } & \dot{\boldsymbol{x}} = f(\boldsymbol{x}, \boldsymbol{u}, t), \boldsymbol{x}(t_d) = \boldsymbol{x}_d \\ & g(\boldsymbol{x}, \boldsymbol{u}, \dot{\boldsymbol{u}}, \ddot{\boldsymbol{u}}) \leq 0 \\ & h(\boldsymbol{x}(t_d + T)) = 0 \end{aligned} \tag{13}$$

In formula 13, the integral part equals to the accumulated value of index functions in time-domain T, $V(\cdot)$ is used as a cost function in order to describe the optimal cost from time $t_d + T$ to the final state $\boldsymbol{x}_f$. The current measured value is set to be the initial value, and the problem is regarded as the trajectory optimization problem with solution of Euler Lagrange method to carry on online computation and get the optimal control value $\boldsymbol{u}^*$, then $\boldsymbol{u}^*$ is applied to current system in time-domain T until the next measured value of a new state is acquired. This is a cyclic process which can be at work all the time until the requirements are satisfied and a group of state feedback control rate is acquired. The direct collocation method is introduced to solve this problem.

4.2 Third Order Simpson Direct Collocation Method

The classic optimal method is the indirect method [13], but in recent years, the direct collocation method has been used broadly to solve the trajectory optimization problem, and the core idea of this method is to transform the optimum control problem to NLP problem, then NLP algorithm can be used to calculate the control variables. Compared to the indirect optimum algorithm, the direct collocation method is insensitive to initial variable estimation and the result is sable, the process of calculation needs a real-time environment.

All the process can be divided into N parts from time-domain with direct collocation method, two extreme points of every part are regarded as nodes. The polynomial used between the two nodes represent the changing relationship between time and state. We suppose that the changing curve of control variables are linear, and all of this polynomials belong to Guass-Lobatto polynomial family. We adopt third order Simpson method [13] in this study.

The continuous time can be divided into N parts in expected time area $[t_d, t_d + T]$, and during every subinterval $[t_i, t_{i+1}]$, $i = 0, 1, \ldots, N - 1$. We define $h_i = t_i - t_{i-1}$, $r = (t - t_i)/h_i$, $r \in [0, 1]$, and every state variable can be described by cubic Hermite polynomial as below:

$$x = c_0 + c_1 r + c_2 r^2 + c_3 r^3 \tag{14}$$

The boundary condition is given as following:

$$x_1 = x(0),\, x_2 = x(1),\, \dot{x}_1 = \left.\frac{dx}{dr}\right|_{r=0},\, \dot{x}_2 = \left.\frac{dx}{dr}\right|_{r=1} \tag{15}$$

Then we can get:

$$\begin{bmatrix} x_1 \\ \dot{x}_1 \\ x_2 \\ \dot{x}_2 \end{bmatrix} = \begin{bmatrix} 1 & 0 & 0 & 0 \\ 0 & 1 & 0 & 0 \\ 1 & 1 & 1 & 1 \\ 0 & 1 & 2 & 3 \end{bmatrix} \begin{bmatrix} c_0 \\ c_1 \\ c_2 \\ c_3 \end{bmatrix} \tag{16}$$

According to formula 6, we can get:

$$\begin{bmatrix} c_0 \\ c_1 \\ c_2 \\ c_3 \end{bmatrix} = \begin{bmatrix} 1 & 0 & 0 & 0 \\ 0 & 1 & 0 & 0 \\ -3 & -2 & 3 & -1 \\ 2 & 1 & -2 & 1 \end{bmatrix} \begin{bmatrix} x_1 \\ \dot{x}_1 \\ x_2 \\ \dot{x}_2 \end{bmatrix} \tag{17}$$

Solve the system of linear equations above, we can get the formula at the midpoint $s = 1/2$ of the subinterval:

$$x_{ci} = \frac{x_i + x_{i+1}}{2} + \frac{h_i}{8}(f_i - f_{i+1}) \tag{18}$$

$$\dot{x}_{ci} = -\frac{3(x_i - x_{i+1})}{2h_i} - \frac{f_i + f_{i+1}}{4} \tag{19}$$

Where f_i, f_{i+1} denote the value of the nodes in subinterval i of state function $f(x, u, t)$, namely:

$$f_i = f(x_i, u_i, t_i) \tag{20}$$

$$f_{i+1} = f(x_{i+1}, u_{i+1}, t_{i+1}) \tag{21}$$

We define the midpoint of the subinterval as the collocation point. The state and control variable of every node and the control variable of the collocation point are regarded as the optimal decision variables in third order Simpson method.

$$Z = [x_0^T, u_{c0}^T, u_0^T, \ldots, x_N^T, u_{cN}^T, u_N^T] \tag{22}$$

There are $(n + m) \times (N + 1) + m \times N$ decision variables in the optimal problem in this study, n is the dimension of state variable and m is dimension of control variable. According to third order Simpson method, the calculated $\dot{x}_{ci}$ should be equal to the derivation of the state variable of collocation point in order to best fit the changing curve of optimal states. Simpson integral formula should be used to complete the integration of state formula in an short interval according to the calculation of the function $f(x_{bi}, u_{ci}, t_{ci})$ in the state formula of collocation point.

$$x_{i+1} = x_i + \frac{h_i}{6}[f(x_i, u_i, t_i) + 4f(x_{bi}, u_{ci}, t_{ci}) + f(x_{i+1}, u_{i+1}, t_{i+1})] \tag{23}$$

Then the vector of Hermite-Simpson Defect is calculated:

$$\Delta_i = x_i + \frac{h_i}{6}[f(x_i, u_i, t_i) + 4f(x_{bi}, u_{ci}, t_{ci}) + f(x_{i+1}, u_{i+1}, t_{i+1})] - x_{i+1} \tag{24}$$

The vector above composes the nonlinear constraints of system and then the optimum control problem is transformed into NLP problem. Use nonlinear programming algorithm to find the optimal value of decision variable which can force the Defect vector to zero. Finally, we can get the optimal control variables with the satisfaction of system's inequality constraints and terminal constraints. The value of optimal control in next time area is calculated by the same method as we mentioned above, and a group of state feedback can be acquired by cyclic operation.

5 The Approximation of Index Function with Neural Network Method

According to the method mentioned above, we can find that collocation points and index functions needs to be calculated during the process of acquiring the optimal control value. So we put forward a neural networked approximation method to get close to the value of index function in order to improve the online calculating performance of the algorithm. Then map the state variables and the optimal control values to neural net to start real-time online calculation after a lot of trainings.

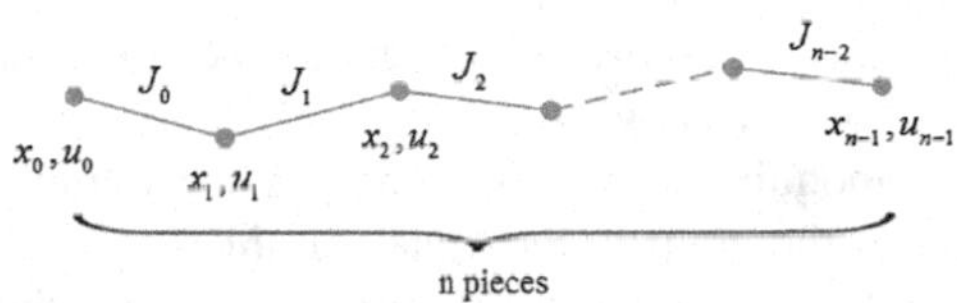

Fig. 3. Trajectory division by direct collocation method

In RHC method, the period $[t_d, t_d + T]$ is given as Fig. 3 describes, the trajectory is divided into N parts according to direct collocation method, trajectory i needs time τ_i, and the control variable of trajectory i needs linear interpolation by the starting control variable $\boldsymbol{u}_{i0}$ and the terminal control variable $\boldsymbol{u}_{i1}$.

$$\boldsymbol{u}_i(t) = \boldsymbol{u}_{i0} + (\boldsymbol{u}_{i1} - \boldsymbol{u}_{i0})\frac{t - t_{i0}}{t_{i1} - t_{i0}} \tag{25}$$

The index function of trajectory i can be described as below:

$$J_i = \int_{t_i}^{t_i+\tau_i} J_i(\boldsymbol{x}_i, \boldsymbol{u}_i)dt \tag{26}$$

From the formula, we can see that the index function is affected by current state, control input and next control input, we can use following formula to approximate the index function.

$$J_i = Y_J(\boldsymbol{x}_i, \boldsymbol{u}_i, \boldsymbol{u}_{i+1}) \tag{27}$$

Y_J is neural network mapping method, all the target functions in programming time-domain can be set up by making the terminal state of the latest trajectory part as the initial state in current time. Then it is easy to find that the whole index function depends on the initial state of the first node and the control variables of each node.

$$J = \sum_{i=0}^{N-2} J_i = \sum_{i=0}^{N-2} Y_J(\boldsymbol{x}_i, \boldsymbol{u}_i, \boldsymbol{u}_{i+1}) \tag{28}$$

According to the six index function mentioned above, a three-layer-BP network is presented, $[\boldsymbol{x}_0, \boldsymbol{u}_0, \boldsymbol{u}_1]^T$ is the input, and there are 14 nodes in the input layer and 15 nodes in the hidden layer, the output is an scalar value approximating the objective function. The *transig* transfer function is used in input layer and hidden layer while the *logsig* transfer function is used in output layer. The concrete net training method is given as follow:

(1) Create the initial position of dual fighters and the target randomly within a given area,
(2) The enemy's maneuver can be set freely,
(3) Our fighter can get optimal control value and collocation point by using the RHC method mentioned above at any time,
(4) Calculate the corresponding index functions as the training example of neural network according to the control and state variables,
(5) The system stops when reaching the terminal condition, then the whole system starts from the beginning again. All the training data will be acquired after one thousand cycles,
(6) Choose eighty percent of the training examples randomly to start network training and the other twenty percent are used as testing examples.

6 Simulation Analysis

The tactical scenario is given as below:

The dual fighters are flying towards the target, while the target makes uniform linear motion towards the dual fighters at the height of 5 km. We apply the method mentioned above to dual fighters to acquire the optimal control policy. The weight factors of index (w_i^1) are 0.3, 0.4 and 0.3, w_i^2 are 0.5, 0.3 and 0.2, w_i^3 are 0.6, 0.3 and 0.1, the simulation step is 0.02 s. The traditional scenario of tactical maneuver tracks are described in Fig. 4.

From Fig. 4, we can see that during the process I, general performance index such as time index plays an important role, and dual fighters maneuver towards the target. During the process II and III, the specific index should be put into the first place in cooperation between dual fighters, and the general performance index such as the stealth performance should also be considered. After 18651 steps, the main approach situation is composed, and simulation ends. The inner optimization toolbox of Matlab

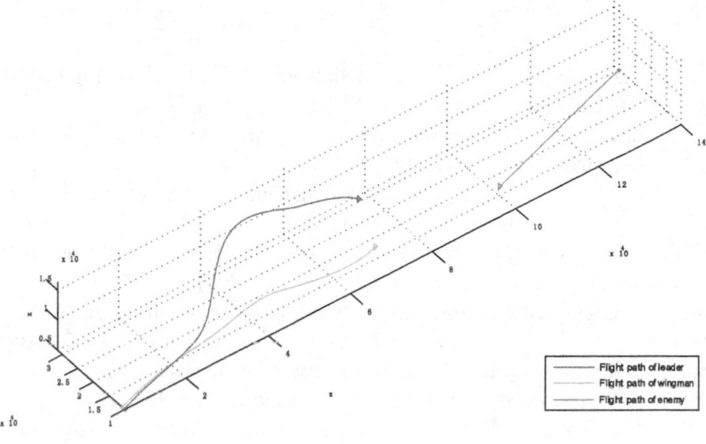

Fig. 4. Tracks of cooperative tactical maneuver

is adopted to solve the nonlinear programming problem, it cost about eight minutes to complete the whole simulation (CPU: P4 2.8 GHz, RAM: 1.5G, Matlab 2009a).

In the simulation, process I costs 4043 steps, process II costs 8322 steps, and process III costs 6286 steps. The collocation point in every simulation step and the value of index function are regarded as the training examples to serve for the neural network to approximate the index function. Several groups of experiments collect all the training examples to train the neural network in order to improve the real-time calculating performance.

7 Conclusion

The trajectory optimization problem during the cooperative tactical maneuver process is studied in this research. The optimum control model of cooperative tactical maneuver process is set up based on the analysis of cooperative vertical spread method and the index function approximation method based on BP neural network is given. Finally, the feasibility of the method we mentioned towards cooperative tactical maneuver decision problem is verified by mathematical simulation.

It is not easy to give the terminal condition of cooperative tactical maneuver, we usually need some experts' knowledge and academic information, and it needs to be studied deeply. The efficiency of solving the nonlinear problem and the method of acquiring the neural network training examples should also be studied deeply in the future.

References

1. Virtanen, K., Karelahti, J., Raivio, T.: Modeling air combat by a moving horizon influence diagram game. J. Guidance Control Dyn. **29**(5), 1080–1091 (2006)
2. Yong, E., Chen, L., Tang, G.: A survey of numerical methods for trajectory optimization of spacecraft. J. Astronaut. **29**(3), 397–406 (2008)
3. Mukai, H., Tanikawa, A., Schatler, H.: Sequential linear-quadratic method for differential games with air combat applications. Comput. Optim. Appl. **25**, 193–222 (2003)
4. Wan, W., Jiang, C., Wu, Q.: Application of one-step prediction influence diagram in air combat maneuvering decision. Electron. Opt. Control **16**(7), 13–17 (2009)
5. Air combat strategy using approximate dynamic programming. In: AIAA Guidance, Navigation and Control Conference and Exhibit, Honolulu, Hawaii, 18–21 August 2008
6. Li, F., Sun, L., Tong, M.: A tactical decision support system for BVR air combat based on neural network. J. Northwest. Polytechnical Univ. **19**(2), 317–322 (2001)
7. Li, M., Jiang, C., Yang, C.: A fuzzy-neural network method of occupying attack seat in air combat of attacker. Fire Control Command Control **27**(3), 18–20 (2002)
8. Yuan, F.: The application of optimum process theory in calculation of optimal trajectory for fighter. Flight Dyn. **18**(3), 50–53 (2000)
9. Liu, J., Tao, G., Xu, G.: Cooperative team tactics decision-making based on petri network in air combat. Fire Control Command Control **35**(10), 70–73 (2010)
10. Yuan, Z., Luo, J., Yu, L.: Tactical-guiding computation of "vertical dipersing" based on virtual tracking. Electron. Opt. Control **18**(6), 13–17 (2011)
11. Karelahti, J., Virtanen, K., Raivio, T.: Near-optimal missile avoidance trajectories via receding horizon control. J. Guidance Control Dyn. **30**(5), 1287–1298 (2007)
12. Kuwata, Y., Richards, A., Schouwenaars, T.: Distributed robust receding horizon control for multivehicle guidance. IEEE Trans. Control Syst. Technol. **15**(4), 627–631 (2007)
13. Tu, L., Yuan, J., Yue, X., Luo, J.: Improving design of reentry vehicle trajectory optimization using direct collocation method. J. Northwest. Polytechnical Univ. **24**(5), 653–657 (2006)

Graph-Based Semi-Supervised Learning on Evolutionary Data

Yanglei Song[1], Yifei Yang[2], Weibei Dou[1], and Changshui Zhang[2](✉)

[1] Department of Electronic Engineering, Tsinghua University, Beijing 100084, China
[2] Department of Automation, Tsinghua University, Beijing 100084, China
zcs@mail.tsinghua.edu.cn

Abstract. This paper presents a graph based semi-supervised learning algorithm on evolutionary data. By applying evolutionary smoothness assumption and incorporating it to the general framework of graph-based semi-supervised learning, we got a new algorithm called GSSLE. Empirical evaluations show that our method outperforms other state-of-the-art methods in terms of stability. It is able to deal with dynamic feature space tasks and proves efficient even if we do not have many unlabeled samples in the semi-supervised procedure.

Keywords: Machine learning · Semi-supervised learning · GSSLE

1 Introduction

Traditional machine learning algorithms assume that all the data are independently and identically distributed (i.i.d) and not time-varying. However, in many cases, such as online text analysis, the distribution of data evolves as time goes by. This kind of data is called evolutionary data. Evolutionary data is a challenge to traditional machine learning theories, as evolutionary data means changing distributions and will certainly not satisfy the i.i.d assumption. Actually, traditional ways learn on static data set are not able to reflect the changing discipline of data changing.

Evolutionary clustering is widely used in online text analysis. Clustering is suitable for tasks to understand data set more than to classifying it. Chakrabati and his partners [1] firstly came up with the concept of evolutionary data and proposed an algorithm to do clustering on it. They hope to maximize the snapshot quality and minimize historical cost at the same time. Many later efforts benefit from their work much. There were also some trials in supervised learning ways to solve the problem. Classification on evolutionary data is also called concept drift problems. In these works, the emphasis is on how to adjust classifiers to satisfy changing data.

Semi-supervised learning on evolutionary data use both labeled data and unlabeled data to train and improve a series of classifiers. In this paper, we pay most attention to graph-based and manifold-based semi-supervised learning algorithms. Jia and his partners [2] came up with the idea to do semi-supervised

X. He et al. (Eds.): IScIDE 2015, Part II, LNCS 9243, pp. 467–476, 2015.
DOI: 10.1007/978-3-319-23862-3_46

learning on evolutionary data at first, which is an extension of manifold based semi-supervised learning on evolutionary data.

In this paper, we propose a new graph-based semi-supervised learning algorithm. It does not suffer the risk of ill-posed problem faced by Jia [2] and performs better than existing algorithms in experiment. What's more, it can be easily generated to deal with dynamic feature space problems. The rest of this paper is organized as follows. Section 2 introduces our algorithm called GSSLE in detail. Section 3 is experiments on two different data sets. Section 4 concludes our contributions.

2 GSSLE for Evolutionary Data Learning

In this section we will introduce a new graph-based method for semi-supervised learning(SSL). We call the algorithm GSSLE(Graph-based Semi-Supervised Learning on Evolutionary data) for short.

2.1 Basic Settings

We are given a data set of $\{(x_i^{(t)}, y_i^{(t)})\}_{i=1}^{l_t}$ and $\{x_j^{(t)}\}_{j=l_t+1}^{l_t+u_t}$ at time t. Let $n_t = l_t + u_t$, so that there are n_t data points at time t. Also at time t we have already labeled all the points of time $t-1$, noted as $\{(x_i^{(t-1)}, y_i^{(t-1)})\}_{i=1}^{n_{t-1}}$. The task can be described as follow: using the labeled information $\{(x_i^{(t)}, y_i^{(t)})\}_{i=1}^{l_t}$ at time t and history information $\{(x_i^{(t-1)}, y_i^{(t-1)})\}_{i=1}^{n_{t-1}}$ at time $t-1$ to label the unlabeled data points $\{x_j^{(t)}\}_{j=1}^{n_t}$ at time t.

2.2 Graph Building and Evolutionary Smoothness Assumption

Graph-based learning demands we form graph $G < V, E >$ from original data. The vertex set $V = \{v_1, \ldots, v_{l+u}\}$ represents the points, and the weight is the similarity between two points. We use $W = [w_{ij}]_{(l+u)\times(l+u)}$ to record the weight of edge set E and we call $D = diag(d_{ii})$ the Degree Matrix, where $d_{ii} = \sum_{j=1}^{n_t} w_{ij}$. $L = D - W$ is the graph based Laplacian Matrix. There are two normalized forms of L:$L_{rw} = D^{-1}L$, $L_{sym} = D^{-1/2}LD^{-1/2}$, which is called the Normalized Laplacian Matrix [3].

Similar to traditional graph-based semi-supervised learning algorithms, we will construct graph at first. Here two graphs are needed: graph of points at time t and graph describing data "evolving" of points from time $t-1$ to t. We will explain that in detail.

On the one hand, we will construct graph of time t. There are several ways to build graph using known data points and here we adopt the widely used KNN graph.

$$w_{ij} = \begin{cases} \exp \dfrac{-d(x_i, x_j)^2}{2\sigma^2}, & \text{if } x_j \text{ is knn of } x_i \\ 0 & \text{otherwise} \end{cases} \tag{1}$$

To get a symmetric matrix W, we let $w_{ij} = w_{ji} = max(w_{ij}, w_{ji})$.

On the other hand, we need a graph to describe the data evolving from time $t-1$ to time t. Take the example of the k-nearest neighbors algorithm, we form the evolving graph as follow. Firstly, there are only edges between $x_j^{(t-1)}$ and $x_j^{(t)}$, which implies that we do not care about the relation between points generated at the same time. Secondly, if $x_j^{(t-1)}$ is one of k-nearest neighbors of $x_i^{(t)}$ at time $t-1$, we calculate the similarity h_{ij}^t between $x_j^{(t-1)}$ and $x_i^{(t)}$. Thirdly, we demand that h_{ij}^t should be normalized. As a result, we have the formula as follows: (we call this evolving graph $H^{(t)}$).

$$h_{ij}^t = \begin{cases} C \exp \dfrac{-d(x_j^{(t-1)}, x_i^{(t)})^2}{2\sigma^2}, & \text{if}\, x_j^{(t-1)} \text{is knn of}\, x_i^{(t)}, \sum_{j=1}^{n_{t-1}} h_{ij}^t = 1 \\ 0 \quad \text{otherwise} \end{cases} \tag{2}$$

To make full use of history information, we used the Evolutionary Smoothness Assumption propose in [2]:

Theorem 1 (Evolutionary Smoothness Assumption). *Evolving graph should be smooth, which argues that if point $x_i^{(t)}$ is close to $x_i^{(t-1)}$ in the feature space, the corresponding label $f_i^{(t)}$ should be close to $f_i^{(t-1)}$.*

We will explain in detail how to apply this assumption to our algorithm.

2.3 The Objective Function

We have already built two graphs: the time-t graph $W^{(t)}$ and the evolving graph $H^{(t)}$. Now we introduce a new mark f_i. For time t data, $f_i^{(t)} \in [-1, 1]$ represents the prediction value of point $x_i^{(t)}$. What we want is to introduce one way to label all data points at time t with $f_i^{(t)}$ and then get their labels.

In [4] O. Chapelle proposed a basic framework of semi-supervised learning as follow:

$$f = argmin \left(\sum_{i=1}^{l} (f_i - y_i)^2 + \lambda_1 f^T L f + \lambda_2 \varepsilon \sum_{i=l+1}^{l+u} f_i^2 \right) \tag{3}$$

By applying assumption (1) to the this framework, we get the following objective function with $\varepsilon = 0$ and $L = L_{sym}$:

$$\begin{aligned} Q(f) = \frac{1}{l_t} \sum_{i=1}^{l_t} (f_i^{(t)} - y_i^{(t)})^2 + \lambda_1 \sum_{i,j}^{n_t} w_{ij}^{(t)} \Big(\frac{1}{\sqrt{d_{ii}^{(t)}}} f_i^{(t)} - \frac{1}{\sqrt{d_{jj}^{(t)}}} f_j^{(t)}\Big)^2 \\ + \lambda_2 \sum_{i}^{n_t} \sum_{j}^{n_{t-1}} h_{ij}^{(t)} (f_i^{(t)} - f_j^{(t-1)})^2 \end{aligned} \tag{4}$$

In this formula, the first part means that for labeled points, the predicted value $f_i^{(t)}$ should be close to its corresponding label $y_i^{(t)}$. The second part of (4) is a trivial extension of classical SSL, which represents data smoothness at time t. The third part, states the smoothness of the evolving graph, or assumption (1). In fact, the larger $h_{ij}^{(t)}$ is, the closer $x_i^{(t)}, x_j^{(t-1)}$ is, and $f_i^{(t)}, f_j^{(t-1)}$ has to be closer to get minimum of the objective function.

2.4 Algorithm Analysis

Analytic Solution. In fact, formula (4) is a quadratic function of f, so we can get the minimum value using calculus method.

With some derivation, we can have the analytic solution to (4):

$$f^{(t)} = \begin{cases} [I-(1-\lambda)S^{(t)}]^{-1}y^{(t)}, & t=1 \\ [I+\alpha C^{(t)}-(1-\beta)S^{(t)}]^{-1}(\alpha C^{(t)}y^{(t)}+\beta H f^{(t-1)}), & t>1 \end{cases} \tag{5}$$

And then we can predict the label of $x_i^{(t)}$ using

$$y_i(t) = sign(f_i^{(t)}) = \begin{cases} 1, & f_i^{(t)} > 0 \\ -1, & f_i^{(t)} < 0 \end{cases} \tag{6}$$

Noticed that this solution is involved in matrix inversion, we will soon prove the existence of $[I+\alpha C^{(t)}-(1-\beta)S^{(t)}]^{-1}$.

2.5 Existence of Solution

The solution of (4) is given by (5). At time $t=1$, it degrades into a classical SSL algorithm and is perfectly solved in [5]. Here we'll demonstrate the existence of the solution at time $t>1$.

Lemma 1. *Let W be the weight matrix of graph G meeting $W=W^T, w_{ij}\geq 0$. Let $S=D^{-1/2}WD^{-1/2}$, the normalized Laplacian Matrix $L_{sym}=I-S$, thus L_{sym} is a positive semi-definite matrix.*

Lemma 2. *Suppose A is a positive semi-definite matrix, C is a $n\times n$ matrix and satisfies*

$$c_{ij} = \begin{cases} \lambda_i, & i=j \\ 0, & otherwise \end{cases} \quad \lambda_i \geq 0 \quad \textit{and at least one of} \quad \{\lambda_i\}_{i=1}^n \quad \textit{is positive.} \tag{7}$$

Then we have the conclusion: $\forall\alpha>0, A+\alpha C$ is positive definite.

With the two lemmas above, we can prove the existence of our solution as follow:

Proof. Proof to Existence of Analytic Solution In (5) $A=I+\alpha C^{(t)}-(1-\beta)S^{(t)}=(1-\beta)L_{sym}+(\beta I+\alpha C^{(t)})$, in which $\alpha>0, 0<\beta<1$. According to Lemma 1, $(1-\beta)L_{sym})$ is positive semi-definite. Furthermore, $(\beta I+\alpha C^{(t)})$ satisfies the definition of matrix C in Lemma 2, thus A is positive definite, and which is invertible.

2.6 Generalization to Dynamic Feature Space

To generalize our algorithm GSSLE to dynamic feature space, which can not be dealt with in SSLE_Jia, we have a natural strategy as follow. GSSLE constructs graph at first step and learns on it later. The learning procedure depends on the graph W, H. At time t, all data points have the same dimension and only the evolutionary graph H is affected by the changing of feature space. In (2), ignoring the constant factor

$$h_{ij}^{(t)} = \exp \frac{-d(x_j^{(t-1)}, x_i^{(t)})^2}{2\sigma^2} \tag{8}$$

We find that as long as we can reasonably define the distance $d(x_j^{(t-1)}, x_i^{(t)})$, we will not meet any problem in constructing H. In the following experiments we'll show an example of this generalization.

3 Experiment

3.1 Introduction to Dataset and Feature Extraction

NSF Dataset. NSF is short for National Science Foundation Research Awards Abstracts 1990–2003[1] which consists of abstracts issued by the National Science Foundation during 1990–2003. [6] has researched evolutionary clustering algorithms on this dataset. In our experiment we just use the data from 1990 to 2002 due to lack of information in 2003. Based on the value "NSF program" of every abstract, we generated a few articles from NSF in 3 categories: biology(bio), mathematics(math) and sociology(soc).

We used the TMG software [7] to extract features. In a collection of articles, the collection of all words occur is noted as $W_D = \{w_1, w_2, \ldots, w_N\}$, or a dictionary. We removed the stopping words and remained just the parent word of a series of derivatives. The collection of articles is recorded as $D = \{d_1, d_2, \ldots, d_M\}$ with feature set $F = \{f_1, f_2, \ldots, f_N\}$.

As long as we get the dictionary W_D, every articles can be expressed as an N-dimension vector. In the simplest case, we have $f_i(j) = 1$ if the word w_j occurs in article d_i, otherwise $f_i(j)$ will be set zero. In our experiment we adopted the TF-IDF treatment.

Theorem 2 (TF(Term Frequence)). *$TF_i(j) = \frac{n_{ij}}{\sum_{i=1}^{N} n_{ik}}$, here n_{ik} is the number of word $w_k (1 \leq k \leq N)$ in article d_i.*

Theorem 3 (TF-IDF(Term Frequence- inverse document frequency)).

$$\begin{aligned} IDF_j &= \log_2 \frac{M}{|\{d_i | w_j \in d_i\}|} \\ TF\text{-}IDF_i(j) &= TF_i(j) * IDF_j \end{aligned} \tag{9}$$

[1] http://kdd.ics.uci.edu/databases/nsfabs/nsfawards.html.

Mailing-List Dataset. This dataset has been used in [2], which has the same researching background as ours. In [2] there is detailed introduction to the dataset and we just use three classes in it to do our experiment: sqlite-users(sql), tutor-python(tut), and wine-devel(win). We adopt the same duration of 2005. 1 to 2008. 8, data generated from 44 months for experiment. The original data is formed in TF-IDF features and we do not need to do more processing.

3.2 Settings

Contrast Algorithms. Other than our GSSLE algorithm, we implemented 3 more algorithms: SSLE_Jia [2], LLGC [1,2,6] and MR [8]. Among them GSSLE and SSLE_Jia are evolutionary semi-supervised learning algorithms, while LLGC and MR are classical semi-supervised algorithms working just on the time-t set. For the classical algorithms, we just used time-t data to design the time-t classifier.

Distance Measurement. We adopt K-nearest strategy to form the graph and the distance is measured as follow

$$d(x_i, x_j) = \arccos < \frac{x_i}{|x_i|}, \frac{x_j}{|x_j|} > \tag{10}$$

Settings. Similar to [2], the graph W is constructed in a 10-nearest neighbor way. σ is not set in advance and we use median of $\{d(x_i, x_j)\}_{i,j=1}^{n}$ instead. Other parameters are determined by grid search to get highest average accuracy on bio/math and all experiments share the same parameters once determined (Table 1).

3.3 Results

On both NSF and Mailing-list dataset we did pairwise classification among three categories. At each certain time t, we randomly labeled the label of $N(N = 5, 10, 15, 20, 25, 30)$ samples(at least one point is labeled in each class). We denote that we have T time slices, for the case of NSF, T is 13 and for Mailing-list T is 44.

For GSSLE and SSLE_Jia, we do not have history information at time t=1, thus we collect results of time $t = 2 \sim T$ to calculate the average accuracy and standard deviation.

Table 1. Settings

algorithm	parameters
GSSLE	5-nearest graph H, $\lambda = 0.075, \alpha = 300, \beta = 0.2$
SSLE_Jia [2]	Linear kernel, $k(x_i, x_j) = x_i^T x_j$ similar to [2], $\lambda_A = 0.01, \lambda_I = 500$
LLGC [1,2,6]	$\lambda = 0.075$
MR [8]	Linear kernel $k(x_i, x_j) = x_i^T x_j$, $\lambda_A = 0.005, \lambda_I = 1000$

Data Filtering Strategy. In (4) it is easy to find that we have to make full use of history labels. At time $t-1$, all the previous points have been labeled, but there may be mistakes. We want to retain the points that we have more confidence in. Graph based SSL gives us a natural strategy.

x_i is thought more likely to be in class 1 if f_i corresponding to x_i is close to 1, and in the opposite situation, f_i close to -1 implicates x_i more probable to be in the other class. The lager the absolute value of f_i is, the more confident we are in predicting the label of x_i. For example, in Fig. 3 we predict the labels of time $t-1$ and the left half represents points with true label 1, and 94. 57 % of them is assigned correct f.

To retain the same number of points in each class, we hold the $n = 0.6 \times \min(n1, n2)$ points with largest absolute value each. In Fig. (1(a)) we mean the points marked red. In this way we got less history information but get higher accuracy of 99. 78 % at time $t-1$. Note that for data points at time $t-1$, what we retain is the label

$$f_i^{(t-1)} = \begin{cases} 1, f_i^{(t-1)} > 0 \\ -1, f_i^{(t-1)} < 0 \end{cases} \tag{11}$$

Rather than the prediction function value on the region $[-1, 1]$. To make the distinction more clear, in (4) we use $y_i^{(t-1)}$ instead of $f_i^{(t-1)}$ to emphasis that we've already known the label of time-t points and we have much confidence in it though some of the labels came from our previous prediction.

Applying the data filtering strategy to history data can help us be more confident in history information.

For labeled points, we mark them as $y_i = 1$ or $y_i = -1, i \leq l$, thus the absolute value of f_i reflects our confidence in x_i 's label. The following experiment proved the effectiveness of this strategy.

For $N = 5$, we calculated the accuracy and variance of the retained data. Also we list the original accuracy and variance of the untreated data at each corresponding time (Table 2).

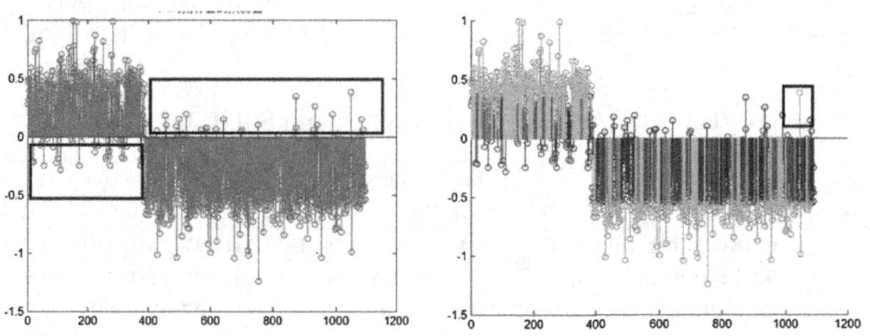

(a) predicted value f of time t-1, mistakes in black rectangle

(b) data filtered and predicted value f mistakes in black rectangle.

Fig. 1. Data filtering strategy

Table 2. Acc(Var) at each time

	bio vs math	bio vs soc	math vs soc	sql vs tut	sql vs win	tut vs win
Acc(Var) of Retaining data	99.67(0.30)	99.74(0.29)	99.96(0.09)	99.57(0.47)	99.52(0.1)	99.70(0.32)
Acc(Var) of untreated data	97.18(0.89)	94.65(1.01)	97.75(0.81)	93.14(1.24)	93.38(1.76)	94.80(1.16)

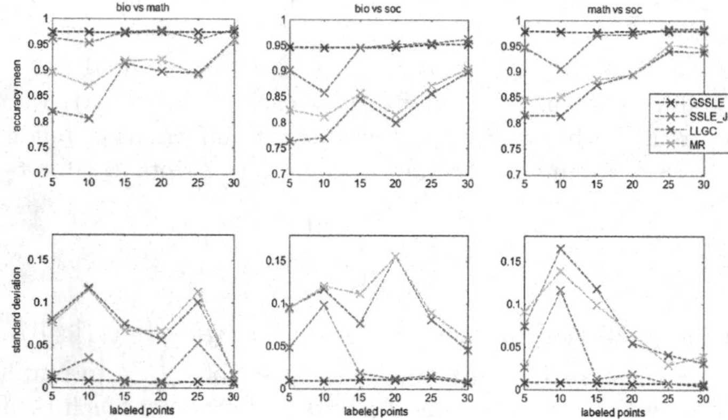

Fig. 2. Performance of 4 algorithms on NSF data

Obviously as we see, this filtering strategy effectively made our data more reliable (Fig. 2).

Result on NSF. Obviously, GSSLE and SSLE_Jia perform better than the other two algorithms, so we will focus on these two later (Table 3).

Result on Mailing-List. Now that the other two algorithms perform not so good as GSSLE and SSLE_Jia, we will just focus on these two algorithms (Table 4).

Table 3. Acc and Var of GSSLE and SSLE_Jia

Number of	bio vs math		bio vs soc		math vs soc	
labeled points	GSSLE	SSLE_Jia	GSSLE	SSLE_Jia	GSSLE	SSLE_Jia
5	**97.34(1.00)**	96.22(1.53)	**94.72(0.99)**	90.14(4.84)	**97.78(0.78)**	94.74(2.57)
10	**97.35(0.81)**	95.30(3.48)	**94.51(0.92)**	85.78(9.80)	**97.70(0.79)**	90.61(11.7)
15	97.28(0.84)	**97.51(0.67)**	**94.53(1.01)**	94.53(1.84)	**97.59(0.80)**	97.08(1.25)
20	97.48(0.58)	**97.73(0.71)**	**94.72(1.02)**	95.16(1.19)	**97.88(0.64)**	97.25(1.82)
25	**97.38(0.76)**	95.93(5.25)	95.20(1.23)	95.47(1.59)	97.82(0.70)	98.23(0.76)
30	97.45(0.77)	**98.06(0.52)**	95.26(0.74)	96.27(0.90)	97.98(0.44)	98.30(0.67)

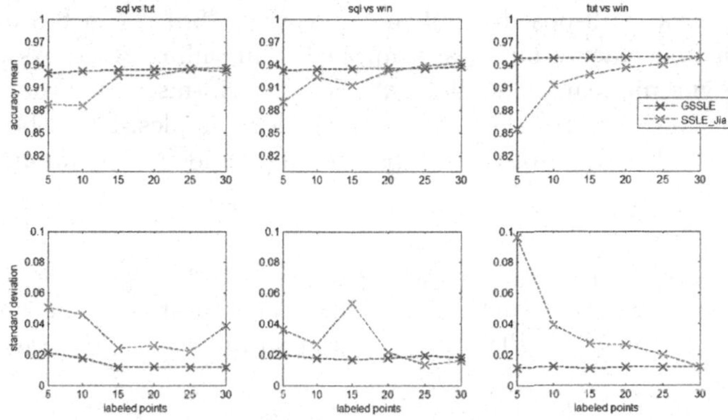

Fig. 3. Performance of GSSLE and SSLE_Jia on Mailing-list data

Table 4. GSSLE and SSLE_Jia on Mailing-list

Number of	sql vs tut		sql vs win		tut vs win	
labeled points	GSSLE	SSLE_Jia	GSSLE	SSLE_Jia	GSSLE	SSLE_Jia
5	**92.88(2.11)**	(88.78)(5.06)	**93.23(1.98)**	89.14(3.63)	**94.79(1.14)**	85.52(9.60)
10	**93.13(1.76)**	88.57(4.59)	**93.41(1.79)**	92.32(2.68)	**94.80(1.20)**	91.33(3.92)
15	**93.30(1.19)**	92.63(2.43)	**93.44(1.70)**	91.27(5.31)	**94.90(1.07)**	92.66(2.72)
20	**93.38(1.21)**	93.37(2.57)	**93.56(1.78)**	93.09(2.14)	**94.96(1.18)**	93.53(2.62)
25	**93.42(1.18)**	93.37(2.20)	93.43(1.96)	**93.78(1.35)**	**94.97(1.16)**	94.04(1.97)
30	**93.46(1.17)**	93.05(3.87)	93.73(1.82)	**94.22(1.60)**	**94.99(1.16)**	94.97(1.18)

Analysis. We have the conclusions as follow:

Firstly, using historical information can observably improve the performance of classifiers. In fact, the author of [2] holds the similar opinion. Secondly, GSSLE does better when we have fewer labeled samples ($N <= 10$). As the number of labeled samples grew larger, two algorithms reached similar performance. Thirdly, the performance of GSSLE does not depend much on the amount of labeled samples. We remain lots of points of time $t-1$ and use them to help classify points at time t. Moreover, the retained points have very high accuracy and supply much information.

4 Conclusion

Semi-supervised learning is a challenge in machine learning research. In this paper we propose a new algorithm GSSLE based on graph evolutionary smoothness. We have confidence in that history information indeed helped improve the performance due to the contrast experiments on MR and LLGC. Further more, experiments on NSF and Mailing-list dataset showed that compared to SSLE_Jia our GSSLE algorithm reached higher accuracy and smaller variance when there is fewer labeled points and performed similarly when we got more labeled information. Our analysis also pointed out the limit SSLE_Jia faced and GSSLE do not suffer the risk of ill-posed problem. Moreover, GSSLE can be easily generalized to dynamic feature space and applied it in online cases. As a result of introducing dynamic feature similarity, GSSLE performs better than SSLE_Jia if our dataset does not provide so many unlabeled samples. It can be concluded that GSSLE reaches state-of-the-art performance and has advantage over less unlabeled samples.

Acknowledgments. This research was supported by 973 Program(2013CB329503), NSFC (Grant No. 91120301) and Beijing Municipal Education Commission Science and Technology Development Plan key project under grant KZ201210005007.

References

1. Chakrabarti, D., Kumar, R., Tomkins, A.: Evolutionary clustering. In: Proceedings of the 12th ACM SIGKDD International Conference on Knowledge Discovery and Data Mining, pp. 554–560. ACM (2006)
2. Jia, Y., Yan, S., Zhang, C., et al.: Semi-supervised classification on evolutionary data. In: IJCAI, pp. 1083–1088 (2009)
3. Von Luxburg, U.: A tutorial on spectral clustering. Stat. Comput. **17**(4), 395–416 (2007)
4. Chapelle, O., Schölkopf, B., Zien, A., et al.: Semi-supervised learning, vol. 2. MIT press, Cambridge (2006)
5. Zhou, D., Bousquet, O., Lal, T.N., Weston, J., Schölkopf, B.: Learning with local and global consistency. Adv. Neural Inf. Process. Syst. **16**(16), 321–328 (2004)
6. Zhang, J., Song, Y., Chen, G., Zhang, C.: On-line evolutionary exponential family mixture. In: IJCAI, pp. 1610–1615 (2009)
7. Zeimpekis, D., Gallopoulos, E.: TMG: A matlab toolbox for generating term-document matrices from text collections. In: Kogan, J., Nicholas, C., Teboulle, M. (eds.) Grouping Multidimensional Data, pp. 187–210. Springer, Heidelberg (2006)
8. Belkin, M., Niyogi, P., Sindhwani, V.: Manifold regularization: A geometric framework for learning from labeled and unlabeled examples. J. Mach. Learn. Res. **7**, 2399–2434 (2006)

Predicting Drug-Target Interactions Between New Drugs and New Targets via Pairwise K-nearest Neighbor and Automatic Similarity Selection

Jian-Yu Shi[1(✉)], Jia-Xin Li[1], Hui-Meng Lu[1], and Yong Zhang[2,3]

[1] School of Life Sciences, Northwestern Polytechnical University, Xi'an, China
{jianyushi, luhuimeng}@nwpu.edu.cn,
lijiaxin0932@mail.nwpu.edu.cn

[2] Chinese Academy of Sciences, Shenzhen Institutes of Advanced Technology, Shenzhen, China
yzhang@cs.hku.hk

[3] Department of Computer Science, The University of Hong Kong, Hong Kong, China

Abstract. Predicting drug-target interaction (DTI) by computational methods has gained more and more concerns in both drug discovery and repositioning. However, several inherent difficulties in DTI data have not yet been addressed appropriately, including the powerless prediction of interactions for new drugs and/or new targets, the biased predicting model derived from imbalanced samples and the inadequate solution to missing interactions and multiple similarities. Moreover, assessed on inappropriate scenarios, existing methods may generate over-optimistic predictions. In this paper, we predict the potential interactions between new drugs and new targets based on pairwise K-nearest neighbor. With lower computational complexity, the proposed approach is able to obtain the less biased prediction and to relax the difficulty caused by missing interactions. Moreover, we develop a strategy to automatically select the best among multiple similarities to train classifiers. Based on four benchmark datasets, the effectiveness of our approach is demonstrated by an appropriate cross validation.

Keywords: Drug-target interaction · K nearest neighbor · Similarity selection · Missing data

1 Introduction

In pharmacology and drug discovery, drug candidates are designed to target the proteins related to diseases. Identifying new drug-target interactions (DTI) is crucial [1, 2]. However, testing candidates against each possible drug/target in wet lab would require a huge amount of money and may take a very long time.

Computational methods have shown their power when predicting novel drug–target interactions for drug discovery and repositioning [3]. Network-based algorithms (e.g. [3]) try to predict new interactions based on the topology of drug-target graph, in which

X. He et al. (Eds.): IScIDE 2015, Part II, LNCS 9243, pp. 477–486, 2015.
DOI: 10.1007/978-3-319-23862-3_47

drugs and targets are nodes and the interactions are undirected edges between these nodes. However, these algorithms cannot work well in the case involving a new drug and/or a new target which have/has no connection to the rest of the graph. Matrix factorization have been recently applied to DTI prediction [5], under the assumption that a drug and a target may interact with each other if they share similar features in a common latent feature space. However, like a cold-start problem, it is difficult to predict interaction for new drugs or new targets. Another popular approach to predict DTIs is to build a trained model based on known DTIs and similarities between drugs and targets (supervised classification). Holding an advantage of handling different scenarios, supervised classification models have gained a lot of concerns [6–10]. However, they suffer from the imbalance between few known interactions and many unknown drug-target pairs.

The assessment of existing methods are mainly based on cross validation. However, the current assessment are usually biased because of the mixture of different scenarios [11]. For example, [7] obtained the over-optimistic prediction of the interactions between new drugs and new targets. By contrast, [9, 10] predicted DTI interactions for new drugs under a more clean scenario. Recently, Pahikkala et al. [11] first presented four distinct scenarios and obtained more realistic assessments by designing scenario-based cross-validations (CV). The four scenarios are briefly depicted as follows. Given a set of approved (known) DTI interactions, one can predict new DTI: (S1) between known drugs and known targets; (S2) between new drugs and known targets; (S3) between known drugs and new targets; and (S4) between new drugs and new targets. Like the "cold start" problem in recommendation system [12], S2, S3 and S4 are still challenging to existing approaches.

Based on our previous work [9] which focuses on scenarios S2 and S3, we are currently concerned about the hardiest scenario **S4** in this paper. DTI prediction was modelled as a problem of supervised classification. We applied pairwise K nearest neighbor classifier to cope with the training bias caused by few positive samples and propose a "super" operation to relax the difficulty caused by missing interactions. In addition, we developed a strategy to automatically select the best similarity among similarities in hand. Finally, four benchmark datasets were used to validate the effectiveness of our approach [11].

2 Method

2.1 K Nearest Neighbor Algorithm

The interactions between m known drugs and n known targets could be represented by an interaction matrix $A_{m \times n}$, in which $a(i,j) = 1$ when there is a known interaction between drug d_i and target t_j, and $a(i,j) = 0$ otherwise. In addition, both the pairwise similarities between drugs and the pairwise similarities between targets would be used for DTI prediction.

Given a new drug d_x, one aims to predict the interaction between it and a known target t_j (referred to as S2). Inspired by the multi-label K nearest neighbor algorithm (ML-KNN) which is able to relax the bias caused by few positive samples [13], we treated the DTI prediction between d_x and t_j as probabilistic events [9]. The confidence score of the potential interaction between d_x and t_j is treated as the probability $\Pr[e^{d_x,t_j}]$ which can be calculated as:

$$\begin{aligned}\Pr[e^{d_x,t_j}] &= \Pr[a(x,j) = 1 | n(x,j,K) = c] \\ &= \frac{Pr[a(x,j) = 1] \cdot Pr[n(x,j,K) = c | a(x,j) = 1]}{\sum_{b=\{0,1\}} Pr[a(x,j) = b] \cdot Pr[n(x,j,K) = c | a(x,j) = b]},\end{aligned} \tag{1}$$

where $n(x,j,K) = \sum_{d_i \in N(x,K)} a(i,j) = c$ be the number of d_x's neighbors that interact with target t_j in known drugs, $c = 0, 1, \ldots, K$, $Pr[a(x,j) = b]$ is the prior probability reflecting the event that a drug interacts with t_j or not, and $Pr[n(x,j,K) = c | a(x,j) = b]$ is the conditional probability reflecting the event that the number of a drug's K-nearest neighbors that interact with target t_j when it interacts with t_j or not. They can be estimated as follows

$$Pr[a(x,j) = 1] \approx \frac{1 + \sum_{i=1}^{m} a(i,j)}{m+2}, Pr[a(x,j) = 0] = 1 - Pr[a(x,j) = 1], \tag{2}$$

$$Pr[n(x,j,K) = c | a(x,j) = b] \approx \frac{1 + \sum_{i} S[a(i,j) = b \& n(i,j,K) = c]}{(K+1) + \sum_{c'=0}^{K} \sum_{i} S[a(i,j) = b \& n(i,j,K) = c']}, \tag{3}$$

where $S[p]$ is the binary indication of whether the statement p is correct or not.

2.2 "Super" Operation

The above KNN has no consideration of the dependency or the correlation among targets. Besides, known interactions are probably missing in a given dataset. Thus, we clustered all targets which are similar to t_j into one group, named Super-Target (more details can be found in [9]). If d_x interacts with a Super-Target st_q, it would very likely interact with t_j in st_q as well. In this point of view, let e^{d_x,st_q} be the event that d_x interacts with st_q, and $e^{d_x,t_j \in st_q}$ be the event that d_x interacts with t_j when d_x interacts with st_q. Thus, the confidence score of d_x interacting with t_j can be defined by $\Pr[e^{d_x,t_j}] = \Pr[e^{d_x,st_q}] \cdot \Pr[e^{d_x,t_j \in st_q}]$.

Consequently, we first calculated the confidence score $\Pr[e^{d_x,st_q}]$ between d_x and the super-target st_q where $t_j \in st_q$ by the aforementioned procedure. Then we calculated the confidence score $\Pr[e^{d_x,t_j \in st_q}]$ between d_x and t_j within super-target st_q. The final confidence score between d_x and t_j is defined as $C_d(x,j) = \Pr[e^{d_x,st_q}] \cdot \Pr[e^{d_x,t_j \in st_q}]$, which will be directly used to calculate the area under curve (AUC) or area under precision-recall (AUPR) curve when assessing the performance of our proposed method (Sect. 2.5).

The above approach was designed for S2. Because of the technical symmetry between S2 and S3, it can predict interactions for a new target (S3) as well.

Accordingly, we may group similar drugs into a Super-Drug in S3. Likewise, the confidence score of the potential interaction between a known drug d_i and a new target t_y, can be defined as $C_t(i,y) = \Pr[e^{sd_p,t_y}] \cdot \Pr[e^{d_i \in sd_p,t_y}]$, where sd_p is the super-drug where $d_i \in sd_p$.

2.3 Predicting DTI by Pairwise KNN

The above approach has advantages derived from KNN and the "Super" operation. KNN has an advantage of reducing the bias when training a classifier with few positive samples [13]. The "Super" operation can relax the difficulty of missing interactions by considering the dependency between known targets in S2 or between known drugs in S3 [9]. In addition, "Super" is conceptually consistent with docking or inverse docking [14], because it can indicate whether a drug possibly interacts with a group of similar targets or whether a target possibly interacts with a group of similar chemical compounds (drug candidates). These advantages will remain after extending the above approach to preform DTI prediction in S4.

"Super" plays an important role in S4 because it can indicate how likely a *new* target interacts with a Super-Drug containing a *new* drug, as well as how likely a *new* drug interacts with a Super-Target containing a *new* target. For the pair of a new drug and a new target, we applied pairwise KNN to calculate the confidence score of being a potential interaction from two sides of S2 and S3 respectively.

In scenario S4, given a new drug d_x and a new target t_y, we intend to predict the interaction between them (Fig. 1). However, because of no interaction (no positive sample) between known drugs $\{d_i\}$ and t_y, we can only obtain a random value of $\Pr[e^{d_x,t_y \in st_q}]$. But, after clustering known targets and t_y, we're still able to calculate the probability $\Pr[e^{d_x,st_q}]$ where the super-targets st_q contains t_y. In addition, using the prediction of S2, we can calculate the confidence scores $\Pr[e^{d_x,t_j \in st_q}]$ of potential interactions between d_x and the known target $t_j \in st_q$(see also Sect. 2.1). As an alternate of $\Pr[e^{d_x,t_y \in st_q}]$, we calculate its estimation by

$$\Pr_d[e^{d_x,t_y \in st_q}] = \sum_{j=1}^{|st_q|} \left(\Pr[e^{d_x,t_j \in st_q}] \times Sim_t(y,j)\right) \Big/ \sum_{j=1}^{|st_q|} Sim_t(y,j), \tag{4}$$

where $Sim_t(y,j)$ is the similarity between t_y and t_j. Therefore, the confidence score of between d_x and t_y can be defined as $C_d(x,y) = \Pr[e^{d_x,st_q}] \cdot \Pr_d[e^{d_x,t_y \in st_q}]$.

Likewise, using the prediction of S3, we can calculate the confidence scores $\Pr[e^{d_i \in sd_p,t_y}]$ of interactions between $d_i \in sd_p$ and t_y. Besides, after clustering known drugs and d_x, we can obtain the probability $\Pr[e^{sd_p,t_y}]$ where the super-targets sd_p contains d_x. Therefore, the confidence score of between d_x and t_y can be defined as $C_t(x,y) = \Pr[e^{sd_p,t_y}] \cdot \Pr_t[e^{d_x \in sd_p,t_y}]$, where

$$\Pr_t\left[e^{d_x \in sd_p, t_y}\right] = \sum_{i=1}^{|sd_p|} \left(\Pr\left[e^{d_i \in sd_p, t_y}\right] \times Sim_d(x, i)\right) \Big/ \sum_{i=1}^{|sd_p|} Sim_d(x, i). \tag{5}$$

Therefore, the final confidence score of between d_x and t_y can calculated by aggregating two scores generated from pairwise KNN:

$$C\left(d_i, t_j\right) = \left(\Pr\left[e^{d_x, st_q}\right] \cdot \Pr_d\left[e^{d_x, t_y \in st_q}\right] + \Pr\left[e^{sd_p, t_y}\right] \cdot \Pr_t\left[e^{d_x \in sd_p, t_y}\right]\right) \Big/ 2. \tag{6}$$

	t_1			...		t_n	t_y
d_1					1		?
	1				1		?
...		1				1	?
		1	1				?
d_m				1		1	?
d_x	?	?	?	?	?	?	?

Fig. 1. The illustration of S4. The cell marked by "?" and filled by colors denotes the pair of the new drug d_x and the new target t_y to be predicted. The given DTI matrix is filled by gray. The cells corresponding to known DTIs in the matrix are marked by "1".

2.4 Similarities and Automatic Selection Among Similarities

The drug similarities and target similarities widely used in former publications are chemical-structure-based and sequence-based respectively [6–11]. The pairwise similarity between drugs (S_{chem}) is measured by aligning their chemical structures [15]. The pairwise similarity between targets (S_{seq}) is derived from by Smith-Waterman alignment [16]. Their detailed definitions can be found in [17].

To obtain better predictions, we also presented Anatomical Therapeutic Chemical (ATC)-based semantic similarity for drug (S_{ATC}) and functional category (FC)-based semantic similarity for targets (S_{FC}) and linearly incorporated them with S_{chem} and S_{seq} respectively [9].

To investigate the better combination among similarities when predicting DTI, we introduced three ways to combine different similarities: *Similarity Fusion* where the average of different similarities are used as the input similarity to calculate the confidence score for each drug-target pair; *Confidence Fusion* where the confidence scores are generated by individual similarities respectively and future aggregated into the final scores; and *Similarity Selection* where an adaptive selection among similarities are designed to automatically choose an appropriate similarity for each target or each drug. The adaptive selection is depicted as follows.

(1) For each drug interacting with target t_j, we first count the occurring number n_u^j of its K nearest neighbors among other drugs interacting with t_j.
(2) We then repeat the counting and calculate the average occurring number by $\tilde{n}_j = \sum_{u=1}^{U} n_u^j / U$, where U is the number of drugs interacting with t_j.

(3) Given L drug similarities, we use a set of $\left\{\tilde{n}_j^l, l = 1, \ldots, L\right\}$ to determine the best drug similarity l^* for target t_j by $l^* = \max\left(\left\{\tilde{n}_j^l\right\}\right)$.
(4) We select the best similarity for both t_j and the Super-Target target st_q where $t_j \in st_q$, and use it to build a drug-based classifiers for target t_j or st_q as described in Sects. 2.1 and 2.2.

For drug d_i, in a similar way, we select the best target similarity for d_i or the Super-Drug target sd_p where $d_i \in sd_p$ to train the target-based classifier.

2.5 Assessment

The cross validation we adopted was designed for S4 with the careful consideration that both the drug and the target of the testing pair should remain unknown in the training data [11]. Thus, neither drug nor target are shared between the testing pairs (cells marked with "?" and filled with red in Fig. 1) and the training pairs (cells filled with gray in Fig. 1). Besides, there are a portion of pairs (cells marked with "?" and filled with white in Fig. 1) which should not be used in either the training data or the testing data in each round of CV.

In practice, all known drugs and all known target are randomly partitioned into five non-overlapping subsets respectively. In each round of CV, each subset of drugs is removed as the testing drugs Tst_d, each subset of target is removed as the testing targets Tst_t and the remaining drugs and target are severally referred to the training drugs Trn_d and the training targets Trn_t. The drug-target pairs between Tst_d and Trn_t, and those between Trn_d and Trn_t are removed as unknown pairs which don't attend the assessment. However they would play the role of middleware in our approach even though they are unknown in the test (see also Sect. 2.3). Only the drug-target pairs consisting of Tst_d and Tst_t are taken as the testing samples and the drug-target pairs consisting of Trn_d and Trn_t are taken as the training samples in S4. In total, five by five-fold CV are used when assessing the performance of prediction.

Like what existing approaches did, we also adopted Receiver Operator Characteristic (ROC) curve and Precision-Recall (PR) curve to measure the performance by calculating the areas under curves (AUC and AUPR). The value of AUC depends on the average ranks of all true drug-target interactions. The value of AUPR depends on both the average ranks of true drug-target interactions as well as the correctness of top ranking predictions. AUPR can punish incorrect top ranking predictions more than AUC when the number of negative samples is much larger than the number of positive samples [18]. The bigger the value of AUC and the value of AUPR are, the better performance the method has. In practice, the averages of AUC and AUPR are recorded to assess the performance of approaches after running cross validation respectively.

Table 1. The comparison with the state-of-the-art method.

	EN(AUC|AUPR)	IC(AUC|AUPR)	GPCR(AUC|AUPR)	NR(AUC|AUPR)
[11]	0.764|0.250	0.678|0.189	0.786|0.175	0.677|0.193
Ours	**0.904|0.301**	**0.865|0.261**	**0.848|0.197**	**0.685|0.258**

3 Experiments

3.1 Benchmark Datasets

The adopted datasets in this paper, were originally from [17] and further used in subsequent works [6–11] as the benchmark. All of drug-target interactions in the original datasets were collected from KEGG database and grouped into four datasets in terms of the type of protein targets. For short, the four DTI datasets are denoted as EN, IC, GPCR and NR respectively. More details can be found in [17].

3.2 Comparison with the State-of-the-Art Work

Our approach was compared with [11] which firstly categorized DTI predictions into four scenarios and proposed the more appropriate cross validations to obtain more realistic assessments. In the experiments, the same similarities as those in [11] were used, including the drug similarity based on chemical structure alignment (S_{chem}) and the target similarity based on sequence alignment (S_{seq}). The performances of two approaches were measured by both AUC and AUPR (Table 1).

Since the number of known DTIs is much less than that of unknown DTIs, one should pay more attention to AUPR than AUC. Obviously, our results are better than those in [11] in terms of both AUC and AUPR. Though [11] didn't show the ROC curve, we show our ROC curves as follows (Fig. 2). In addition, [11] applied Kronecker Regularized Least Squared which has the computational complexity of $O(m^3 + n^3)$, whereas our method has the lower complexity of $O(kmn)$.

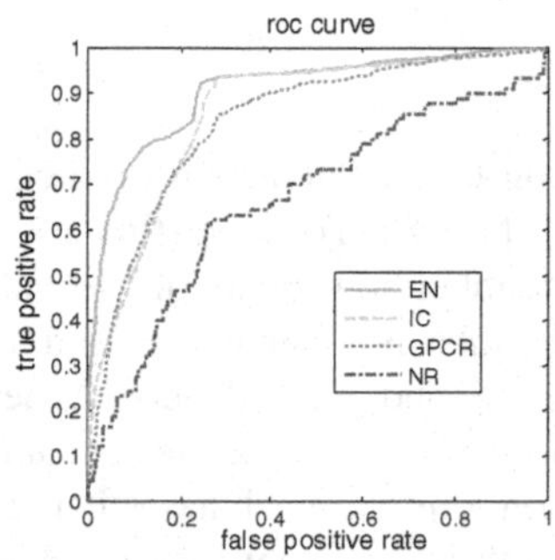

Fig. 2. The ROC curves of DTI Prediction on EN, IC, GPCR and NR datasets.

3.3 The Effective Combination of Different Similarities

In order to achieve better DTI prediction when given multiple similarities, we designed five strategies: (Set 1) using S_{chem} and S_{seq}; (Set 2) using S_{ATC} and S_{FC}; (Set 3) using $\bar{S}_d = (S_{chem} + S_{ATC})/2$ and $\bar{S}_t = \left(S_{seq} + S_{FC}\right)/2$ as drug similarity and target similarity in pairwise KNN respectively; (Set 4) aggregating the confidence scores generated individually by each of $\{S_{chem}, S_{ATC}, \bar{S}_d\}$ and each of $\left\{S_{seq}, S_{FC}, \bar{S}_t\right\}$; and (Set 5) selecting the best among $\{S_{chem}, S_{ATC}, \bar{S}_d\}$ and the best among $\left\{S_{seq}, S_{FC}, \bar{S}_t\right\}$. We measured the predicting performance by five strategies (Table 2).

Compared with Set 2, Set 1 shows that the structure-based drug similarity S_{chem} and the sequence-based target similarity S_{seq} are the similarities majorly considered during drug discovery. Set 3 (*Similarity Fusion*) shows that the average drug/target similarity works usually better than any of individual similarities, except for the dataset EN. The possible reason is that ATC may not measure drug similarity well in EN. The results on EN by using Set 2 also shows the similar results (the worse performance). Set 4 (*Confidence Fusion*) shows the slightly worse performance than Set 3. Set 5 (*Similarity Selection*) achieves the best results in terms of both AUC and AUPR, except for NR (only slightly worse than Set 3). Since NR has fewer drugs and fewer targets than other datasets, the average occurring number $\tilde{n}_j$ calculated by few samples may bias to the real one when selecting the best similarity during each round of CV. In general, the proposed *Similarity Selection* is the best strategy when given multiple similarities to predict DTIs.

Table 2. The comparison of strategies of multiple similarities.

	EN(AUC|AUPR)	IC(AUC|AUPR)	GPCR(AUC|AUPR)	NR(AUC|AUPR)
Set 1	0.904|0.301	0.865|0.261	0.848|0.197	0.685|0.258
Set 2	0.825|0.147	0.833|0.218	0.811|0.194	0.587|0.241
Set 3	0.902|0.254	0.890|0.274	0.872|0.246	**0.713|0.355**
Set 4	0.896|0.275	0.862|0.284	0.843|0.250	0.618|0.333
Set 5	**0.909|0.314**	**0.890|0.301**	**0.880|0.268**	0.703|0.353

4 Conclusion

The most difficult issue in drug-target interaction prediction is to predict interactions between new drugs and new targets. To solve this issue, we have introduced the pairwise KNN to cope with the inherent sparsity of known DTIs, the Super-Target and the Super-Drug to deal with the missing interactions, and integrated them to predict the interactions between new drugs and new targets. When multiple similarities are available, we have also proposed an adaptive strategy that can select the best similarity to train the classifier. In addition, our approach has other advantages, such as the lower computational complexity and the significant biological meaning for prediction. The experimental results on benchmark datasets under more realistic assessments have demonstrated the effectiveness of our approach.

However, predicting interactions not only between new drugs/targets and known targets/drugs but also between new drugs and new targets are still challenging so far, just like the "cold start" problems in recommender system. For example, the predicting performance in the case that a drug only interacts with a target is always very bad. We suggest that more non-structure properties of drugs, non-global-sequence properties of targets or other heterogeneous information should be incorporated together. For instance, one may use drug-disease associations (network) to turn new drugs in DTI network into known drugs.

Acknowledgments. This work was supported by the Fundamental Research Funds for the Central Universities (Grant No. 3102015ZY081), China Postdoctoral Science Foundation (Grant No. 2012M521803), and partially supported by NSFC (Grant No. 31402019), the Natural Science Foundation of Shaanxi Province, China (Grant No. 2014JQ4140), Shenzhen basic research grant (JCYJ20120615140531560) and Chinese Academy of Sciences research grant (No. KGZD-EW-103-5(9)).

References

1. Keiser, M.J., Setola, V., Irwin, J.J., Laggner, C., Abbas, A.I., Hufeisen, S.J., Jensen, N.H., Kuijer, M.B., Matos, R.C., Tran, T.B., Whaley, R., Glennon, R.A., Hert, J., Thomas, K.L., Edwards, D.D., Shoichet, B.K., Roth, B.L.: Predicting new molecular targets for known drugs. Nature **462**, 175–181 (2009)
2. Li, Y.Y., An, J., Jones, S.J.: A computational approach to finding novel targets for existing drugs. PLoS Comput. Biol. **7**, e1002139 (2011)
3. Cheng, F., Liu, C., Jiang, J., Lu, W., Li, W., Liu, G., Zhou, W., Huang, J., Tang, Y.: Prediction of drug-target interactions and drug repositioning via network-based inference. PLoS Comput. Biol. **8**, e1002503 (2012)
4. Yıldırım, M.A., Goh, K.I., Cusick, M.E., Barabasi, A.L., Vidal, M.: Drug-target network. Nat. Biotechnol. **25**, 1119–1126 (2007)
5. Gönen, M.: Predicting drug-target interactions from chemical and genomic kernels using Bayesian matrix factorization. Bioinformatics **28**, 2304–2310 (2012)
6. Bleakley, K., Yamanishi, Y.: Supervised prediction of drug-target interactions using bipartite local models. Bioinformatics **25**, 2397–2403 (2009)
7. Mei, J.P., Kwoh, C.K., Yang, P., Li, X.L., Zheng, J.: Drug-target interaction prediction by learning from local information and neighbors. Bioinformatics **29**, 238–245 (2013)
8. van Laarhoven, T., Nabuurs, S.B., Marchiori, E.: Gaussian interaction profile kernels for predicting drug-target interaction. Bioinformatics **27**, 3036–3043 (2011)
9. Shi, J.Y., Yiu, S.M., Li, Y.M., Leung, H.C.M., Chin, F.Y.L.: Predicting drug-target interaction for new drugs using enhanced similarity measures and super-target clustering. In: Proceedings of the IEEE International Conference on Bioinformatics and Biomedicine, pp. 45–50. IEEE Press, New York (2014)
10. van Laarhoven, T., Marchiori, E.: Predicting drug-target interactions for new drug compounds using a weighted nearest neighbor profile. PLoS ONE **8**, e66952 (2013)
11. Pahikkala, T., Airola, A., Pietila, S., Shakyawar, S., Szwajda, A., Tang, J., Aittokallio, T.: Toward more realistic drug-target interaction predictions. Brief Bioinform. **16**, 325–337 (2015)

12. Lam, X.N., Vu, T., Le, T.D., Duong, A.D.: Addressing cold-start problem in recommendation systems. In: Proceedings of the 2nd International Conference on Ubiquitous Information Management and Communication, pp. 208–211. New York (2008)
13. Zhang, M.L., Zhou, Z.H.: ML-KNN: a lazy learning approach to multi-label learning. Pattern Recogn. **40**, 2038–2048 (2007)
14. Vasseur, R., Baud, S., Steffenel, L.A., Vigouroux, X., Martiny, L., Krajecki, M., Dauchez, M.: Inverse docking method for new proteins targets identification: a parallel approach. Parallel Comput. **42**, 48–59 (2014)
15. Hattori, M., Okuno, Y., Goto, S., Kanehisa, M.: Development of a chemical structure comparison method for integrated analysis of chemical and genomic information in the metabolic pathways. J. Am. Chem. Soc. **125**, 11853–11865 (2003)
16. Smith, T.F., Waterman, M.S.: Identification of common molecular subsequences. J. Mol. Biol. **147**, 195–197 (1981)
17. Yamanishi, Y., Araki, M., Gutteridge, A., Honda, W., Kanehisa, M.: Prediction of drug-target interaction networks from the integration of chemical and genomic spaces. Bioinformatics **24**, i232–i240 (2008)
18. Davis, J., Goadrich, M.: The relationship between precision-recall and ROC curves. In: Proceedings of the 23rd International Conference on Machine Learning, pp. 233–240. ACM, New York (2006)

A Novel Complex-Events Analytical System Using Episode Pattern Mining Techniques

Jerry C.C. Tseng[1], Jia-Yuan Gu[1], P.F. Wang[2], Ching-Yu Chen[2], and Vincent S. Tseng[1(✉)]

[1] Department of Computer Science and Information Engineering, National Cheng Kung University, Tainan 701, Taiwan
{jerry.cc.tseng, lester13.gu, tsengsm}@gmail.com

[2] Institute for Information Industry, Taipei 106, Taiwan
{pfwang, chingyuchen}@iii.org.tw

Abstract. Along with the rapid development of *IoT* (Internet of Things), there comes the 'Big Data' era with the fast growth of digital data and the requirements rise for gaining useful knowledge by analyzing the rich data of complex types. How to effectively and efficiently apply data mining techniques to analyze the big data plays a crucial role in real-world use cases. In this paper, we propose a novel complex-events analytical system based on episode pattern mining techniques. The proposed system consists of four major components, including data preprocessing, pattern mining, rules management and prediction modules. For the core mining process, we proposed a new algorithm named *EM-CES* (*Episode Mining over Complex Event Sequences*) based on the sliding window approach. We also make the proposed system integrable with other application platform for complex event analysis, such that users can easily and quickly make use of it to gain the valuable information from complex data. Finally, excellent experimental results on a real-life dataset for electric power consumption monitoring validate the efficiency and effectiveness of the proposed system.

Keywords: Data mining · Complex event analytics · Episode pattern mining · Multivariate sequence mining

1 Introduction

In this 'Big Data' age, more and more digital data is collected around us and people hope to dig informative gold from the data mines, tons of data with complex types. We have learned that data mining is a simple-to-understand, but difficult-to-execute process. One of the main difficulties is the lack of an effective and efficient data mining system for mining complex-events data with real-life applications, such as network

This study is conducted under the "Advanced Sensing Platform and Green Energy Application Technology Project" of the Institute for Information Industry which is subsidized by the Ministry of Economy Affairs of the Republic of China.

X. He et al. (Eds.): IScIDE 2015, Part II, LNCS 9243, pp. 487–498, 2015.
DOI: 10.1007/978-3-319-23862-3_48

attack detection, weather forecast [1], and stock market analysis [5]. As to the sustainability of life, those mining techniques can be applied in energy management, not only to cut down the electricity consumption but also to reduce the carbon emission [8].

To the best of our study, there are still very limited research works focused on the topic of complex-events patterns mining though it is so useful for many real-life applications. Some relevant works like [6, 14] proposed good mining methods on this topic; however, the full data mining process was not incorporated to help users conduct the mining task efficiently.

The abovementioned situation motivated us to develop a system to provide people an effective and efficient way for conducting such data mining jobs. In this work, we aim to facilitate users to easily perform the major tasks on mining complex events data, such as data preprocessing, pattern mining, rule management, and prediction. To this end for developing such a system with general applications, a number of challenging issues need to be taken into account and overcome.

The first issue we took into consideration is the diverse type of input data, in particular that most of them are large-scale numeric data collected by various kinds of sensors. Besides, it's very common that different sources of data are collected simultaneous as multiple time series events, for such sequences we call complex event sequences. In order to feed the data into mining stage, we need to preprocess the data at first with key tasks like data transformation. *SAX* [4] is the most widely used symbolic representation for time series event, so we apply this technique to transform the numeric data into symbols for the further mining process. Furthermore, the episode mining techniques [2, 7, 9–11] and rule-based mining techniques [3, 13] can be used to discover useful patterns from the past events such as to foresee the possible events that will happen in the future. Now the mining techniques are applied in many real-world cases, such as the network error or attack detection, weather forecast [1], and investment in the stock market [5]. However, most of the existing studies on episode mining focused on mining a single event sequence and were not applicable to mining complex sequences. We developed in this work a novel algorithm named *EM-CES* (*Episode Mining over Complex Event Sequences*), which was extended from the sliding window-based algorithms *WINEPI* [9] to make it work well for the multivariate complex event sequence.

In this paper, we aim to develop an effective system to solve the underlying challenges as mentioned above. Our goals are as follows: (1) To design an approach for data transformation, which is one of the most important jobs in the data preprocessing stage; (2) To develop an effective episode pattern mining algorithm, especially in considering the multivariate features of complex event sequences; (3) To develop the software modules of the above techniques and make them integrable with other application platforms through flexible interfaces like JSON files.

The rest of the paper is organized as follows. We describe the problem definition and review the related works in Sect. 2. The design of the proposed system is developed in Sect. 3. Section 4 presents the experimental design and results. The paper is concluded in Sect. 5.

2 Problem Definitions and Related Work

2.1 Definitions

In this subsection, we adopt the notations used in [10] to introduce the definitions related to episode mining based on sliding widows.

Definition 1. ***Complex event sequence.*** A *complex event sequence* is a sequence of events, where each event has an associated time of occurrence. Given a set of *event* types E, an *event* is a pair (A, t) where $A \in E$ is an event type and t is an integer, the occurrence time of the event. A complex event sequence S on E is a triple (s, T_s, T_e), where $s = \langle (A_1, t_1), (A_2, t_2), \ldots, (A_n, t_n) \rangle$ is an ordered sequence of events such that $A_i \in E$ for all $i = 1, \ldots, n$, and $t_i \leq t_j$ for all $1 \leq i \leq j \leq n$. Furthermore, T_s and T_e are the starting time and the ending time of S respectively, and $T_s \leq t_i \leq T_e$ for all $i = 1, \ldots, n$. The size of a complex event sequence is equal to the number of time points in complex event sequence, and it is defined as $T_e - T_s + 1$. Given two sequences s and s', s' is a subsequence of s, if s' can be obtained by removing some events in s.

Definition 2. ***Episode.*** An episode f is an ordered tuple of simultaneous event set with the form $\langle SE_1, SE_2, \ldots, SE_k \rangle$, SE_i appears before SE_j for all i, j $(1 \leq i < j \leq k)$. Base on the definition, there are two different relationships in the episode, which included simultaneous and serial relationships. The length of α is denoted by $|f|$ and is equal to the number of events of f.

Definition 3. ***Prefix of episode.*** Given an episode $f = \langle SE_1, SE_2, \ldots, SE_k \rangle$, its sub-episode $\langle SE_1, SE_2, \ldots, SE_m -1, SE'_m \rangle$ $(m \leq k)$ is called *the prefix* of f iff SE'_m is sub-simultaneous event set of SE_m, and the events in $(SE_m - SE'_m)$ are alphabetically after those in SE'_m.

Definition 4. ***Support of an episode.*** Given an event sequence S, *the support of an episode* f is denoted by $sup(f)$, and it is formally defined as $sup(f) = |sc(f)|/N$, where N is the number of sliding windows of S, and $sc(f)$ is the support count of f.

Definition 5. ***Frequent episode.*** Given a user-specified *minimum support threshold minsup*, an episode f is said to be frequent iff $sup(f)$ is no less than *minsup*. *Minimum support count* is given as the product of *minsup* and N, where N is the size of S.

Example 1. Figure 1 shows an example of a complex event sequence, the complex event sequence $S = (s, 1, 16)$, where $s = \langle$(A, 1), (E, 1), (D, 2), …, (C, 16)$\rangle$. Given an episode $f = \langle$(B, F), (A)$\rangle$, a sliding window of length 4, and there are 4 windows ([1, 4], [5, 8], [9, 12] and [13, 16]). The episode f occurs in window [1, 4] and [5, 8], so $sc(f)$ is 2 and $sup(f)$ is 2/4 = 50 %. If *minsup* = 30 %, f is a frequent episode because $sup(f) \geq minsup$.

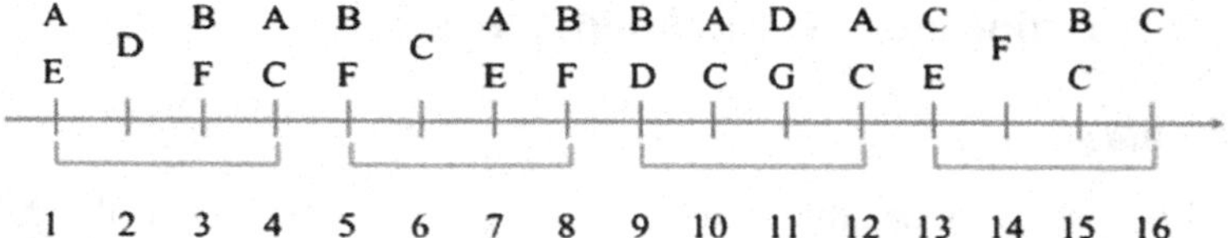

Fig. 1. An illustration of a complex event sequence

2.2 Related Work

Mannila et al. [9] introduced frequent episode first, where data comes in a single time series, and such data can be viewed as a sequence of events. Each event was associated with the time that it occurred, and we can use a pair (*e*, *t*) to represent, where means that event *e* occurred in time *t*. For example, Fig. 2 shows an event sequence. In Fig. 2, A, B, C, D, E and F are event types, and they respectively occur in a specific time. In this kind of data, they want to find some patterns, known as episodes that are partially ordered sets of events.

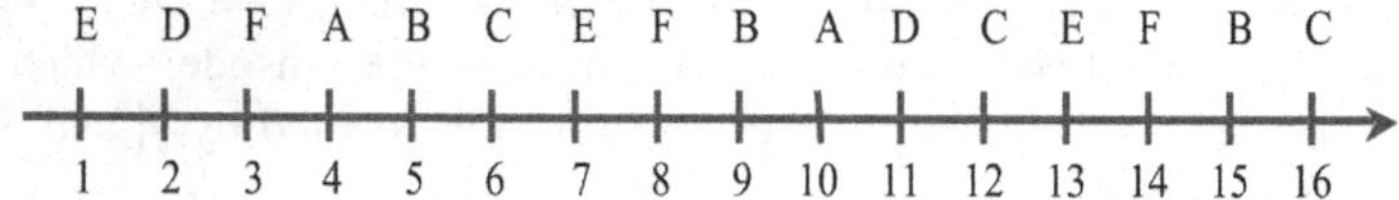

Fig. 2. An example of a simple event sequence

In [10], episodes were categorized into three types, shown in Fig. 3, according to the relationship among them: (1) *serial episodes*, only consider the serial relationship between the events, any event in the episode exists an order between each other event; (2) *parallel episodes*, only consider the parallel relationship between the events, there are no constraints on the relative order between any two events; (3) *composite episodes*, it's the mixture of serial and parallel episode.

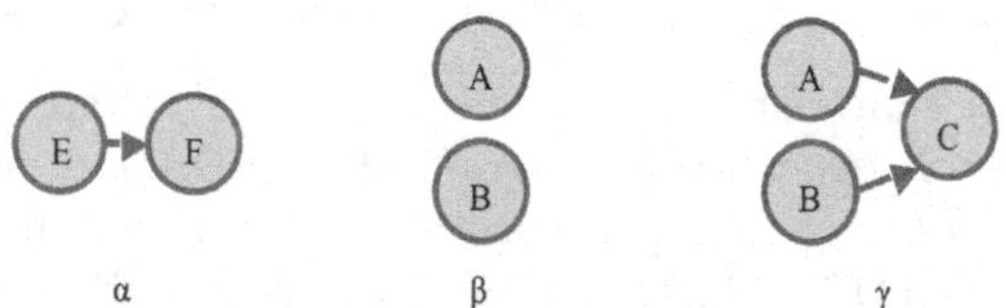

Fig. 3. Different types of episode

There are two ways to count the support of an episode: (1). the number of sliding windows [9]; and (2). the number of minimal occurrences [7, 11]. We can generate rules from the frequent episodes, and use it to help users make decisions. Episode mining has been of great interest in many applications, including network attack detection, biomedical data analysis, stock trend prediction and drought risk management in climatology data sets. Therefore, the frequent episode mining algorithms, as a tool for discovering frequent episodes, shows a practical purpose.

Based on sliding windows, to be considered interesting, the events of an episode must occur close enough in time. The user defines how close is close enough by giving the width of the window within which the episode must occur. A window is a slice of an event sequence, it can be denoted by $[t_s, t_e]$, which starting time is t_s, ending time is t_e, and the time span $t_e - t_s + 1$ is the width of the window. By the definition, the first window contains only the first time point of the sequence, and the last window contains only the last time point. And we define the support count of an episode as the number of windows in which the episode occurs. For example, consider the event sequence in Fig. 4, and set the width of window 5, then it shows two windows, the window starting at time 1 is drawn in solid line, and the immediately following window, starting at time 2 is drawn in dashed line. A serial episode f = <A, B> is supported by 4 sliding windows, which are ([1, 5], [2, 6], [3, 7], and [4, 8]). If the minimum support count equals to 3, episode α is a frequent episode, because that the support count of α is equal to 4 and it is no less than minimum support count.

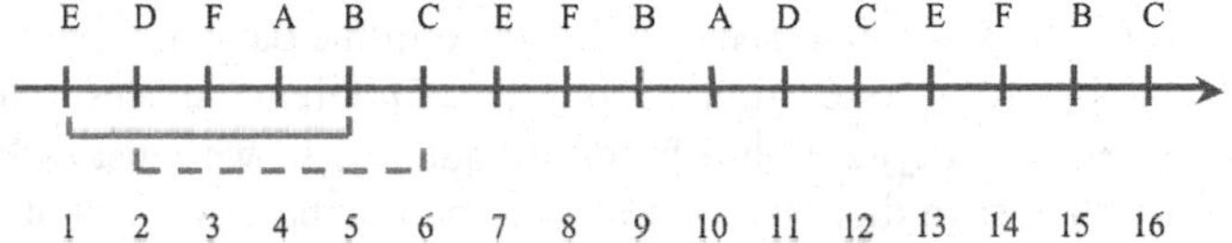

Fig. 4. Illustrations of sliding windows

For some applications, they are interested in a special type of episodes where the last event of the episode is the target event type, the episode is called target episode [6].

3 The Proposed System

Figure 5 shows the framework of our proposed system, namely, *CEAS* (*Complex Event Analytical System*). The input data here are streaming complex event sequences, and most of them are continuous data from various kinds of data sources, e.g., digital sensors or intelligent meters. Besides, the *CEAS* system can be integrated with other application platforms using JSON as the data exchange interface.

In the proposed system *CEAS*, there are four major components, namely, data preprocessing, pattern mining, rule management and prediction modules. The details are described below.

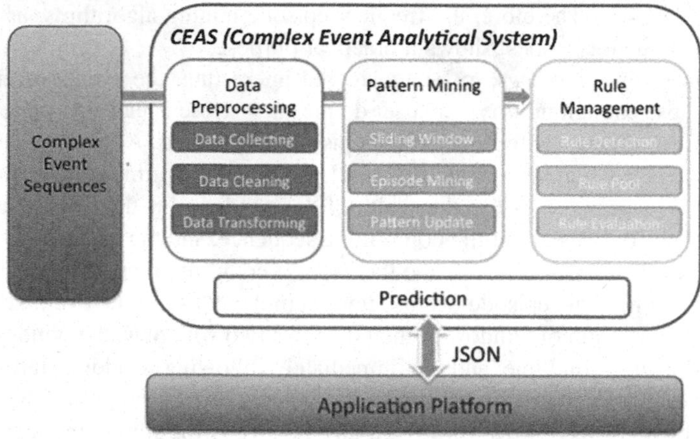

Fig. 5. System framework of CEAS

3.1 Data Preprocessing

The primary goal in this stage is to remove the noise in the data, and converts the data into a format suitable for further mining process. Different devices often generate various data in diverse data type with different frequency, so we must well preprocess the input data, and then save them as the transformed database for next step of pattern mining process. As to the numeric data, they need to be transformed to a symbolic presentation, or we can't pass the data to the mining phase.

Lin et al. [4] proposed *Symbolic Aggregate approXimation* (*SAX*), which is a symbolic representation of time series data. *SAX* allows a time series of arbitrary length n to be reduced to a string of arbitrary length w, ($w < n$, typically $w << n$). The alphabet size is also an arbitrary integer a, where $a > 2$.

According to our best study, there are two key features of *SAX*: dimensionality reduction and lower bounding. These two features are important advantages for the general real-world use cases, so we considered that *SAX* is much better than other presentation methods. We applied *SAX* first to transform the data into the *Piecewise Aggregate Approximation* (*PAA*) representation, and then symbolize the *PAA* representation into a discrete string.

3.2 Pattern Mining

In our system development, we found that finding the recurrent combinations of events, known as episodes, within the complex event sequence is very critical to most cases. People can use those episode patterns to learn the relationship among events, and to predict what may happen in the near future.

To determine whether an episode is a pattern, there are two ways to count the support of an episode: (1) the number of sliding windows which contains the episode, and (2) the number of minimal occurrences. To our study, sliding is more useful and basic, because for most real-world applications, since users would like to know 'what's going on' within a time period. Most of the existing episode mining algorithms with sliding window adopt an *Apriori-like* approach that need to generate candidates and then calculates the support count. It costs too much time and is not fitting to the streaming environment and multivariate complex event sequences. Therefore, we proposed a new method, named *Episode Mining over Complex Event Sequence* (*EM-CES*).

Since the *Aprori-like* approach does not meet our requirement due to the low efficiency and re-scanning, we referred to the *PrefixSpan* [12] approach and extend the approach to make it work well for episode pattern mining in complex event sequences. The basic steps are as follows:

1. Get the segment set from the complex event sequence by sliding window;
2. Set $k = 1$ and *pref* = < >;
3. Find the episode patterns that length equal to k from the segment set;
4. Generate the projected segment set for each pattern found in the previous step;
5. Set $k = k + 1$ and go to step 2 until no more patterns found;

First, we crop sort copies of the complex event sequence by sliding window to get the set of complex event segments, i.e., the subsequences of the complex event sequence which are contained in the sliding windows. Before starting mining the patterns, we initialize the length parameter k as 1 and the prefix episode *pref* as an empty sequence.

Then we count the support of each event to find the patterns by extending *pref* with each frequent event from the segment set, and generate the projected segment set for each pattern with length k we found. Then, the same process with $k = k + 1$ repeat again. If there are no episode patterns found in this step, we choose another prefix episode and mining with its projected segment set.

For example, the complex event sequence is illustrated in Fig. 1, and the sliding window is set with size 4 and sliding distance 4. Without loss of generality, we assume that the events appeared within the same time stamp are sorted in an alphabetical order. First, we get four segments by the sliding window, which are shown in Table 1. If *minsup* is 30 %, {<A>, <B>, <C>, <D>, <E>, <F>} are episode patterns with $k = 1$. After generating their projected segment sets, we choose <A> as the next prefix episode, and its projected segment set is shown in Table 2. The event denoted as an underline is the first end of *pref* in that segment. After counting the support, we can find {<B>, <C>, <D>, <F>, <_C>, <_E>} are frequent episodes in <A>'s projected segment set, and represent the episode patterns as {<A, B>, <A, C>, <A, D>, <A, F>, <(A, C)>, <(A, E)>}. Then we generate their projected segment sets from <A>'s projected segment set, choose <A, B> as the new prefix episode and go on with $k = 3$. After all episode patterns with <A> as the first event are found, the process choose a new prefix episode from the remaining to mine the patterns, until all patterns are found.

Table 1. Segment set of the complex event sequence

SID	Segment
1	<(A, 1), (E, 1), (D, 2), (B, 3), (F, 3), (A, 4), (C, 4)>
2	<(B, 5), (F, 5), (C, 6), (A, 7), (E, 7), (B, 8), (F, 8)>
3	<(B, 9), (D, 9), (A, 10), (C, 10), (D, 11), (G, 11), (A, 12), (C, 12)>
4	<(C, 13), (E, 13), (F, 14), (B, 15), (C, 15), (C, 16)>

Table 2. Projected set with *pref* = < A>

SID	Segment
1	<(_, 1), (E, 1), (D, 2), (B, 3), (F, 3), (A, 4), (C, 4)>
2	<(_, 7), (E, 7), (B, 8), (F, 8)>
3	<(_, 10), (C, 10), (D, 11), (A, 12), (C, 12)>

3.3 Rule Management

After the frequent episode patterns are mined from the complex event sequence, we would generate rules from the patterns with at least two complex events to make predictions with new coming streaming data. A rule can be denoted as LHS → RHS, which means if a complex event RHS may happen after a sequence of events LHS. A rule can be evaluated with its confidence, which is defined as *conf*(LHS → RHS) = *sup*(<LHS, RHS>)/*sup*(LHS), where <LHS, RHS> is a sequence that contains only LHS and RHS, and LHS happens before RHS. For example, the confidence of rule <A> → <B> is equal to *sup*(<A, B>)/*sup*(<A>). With a user-specified *minimum confidence*, *minconf*, we can only keep the rules whose confidence are no less than *minconf* in the *rule pool*.

To make predictions, we have to detect whether the LHS part of a rule happens in the new coming sequence. Furthermore, if there are multiple rules whose LHS part occurs, we need to make a final decision. To address these issues, there are two approaches: *straight forward* approach and *tree-based* approach. These approaches both have their advantages and drawbacks, according to their data structures, and are describe in the following.

Straight Forward Approach: We sort the rules with their confidence as the primary key, support and total length as the secondary keys. For each new coming segment, i.e., subsequence contained in a window, find the first rule whose LHS part is contained in the segment and take the RHS part as the prediction.

Tree-based Approach: We build a rule tree which contains all the rules whose confidence is no less than *minconf*. Each node except the root represents an event. If a node is the end of LHS, the RHS part is denoted on it as the predicted result. Whenever a new event in a segment comes, it will check all the child nodes of any visited node to find out the child nodes with the same event, and add them into the visited node list. Finally, collect all the predicted result of the visited nodes and make the final decision with voting.

Comparing with the *tree-based* approach, the *straight forward* approach is simple, easy to be updated, and is not necessary to load all the rule set into main memory. But it may loss some accuracy within its greedy manner. For the *tree-based* approach, the whole rule tree have to be loaded into main memory, and have to store a list of visited node. Because we make the final decision with several matched rules, it has a better performance on accuracy.

For example, there are three rules in the rule pool, which are shown in Table 3. The *straight forward* approach loads these rules with the order (2, 1, 3), sorted by the keys,

Table 3. An example of the rule pool

No.	LHS	RHS	Conf.	Sup.
1	<A>	<B>	66.7 %	3
2	<(A, E)>	<F>	100 %	2
3	<C>	<B>	66.7 %	3

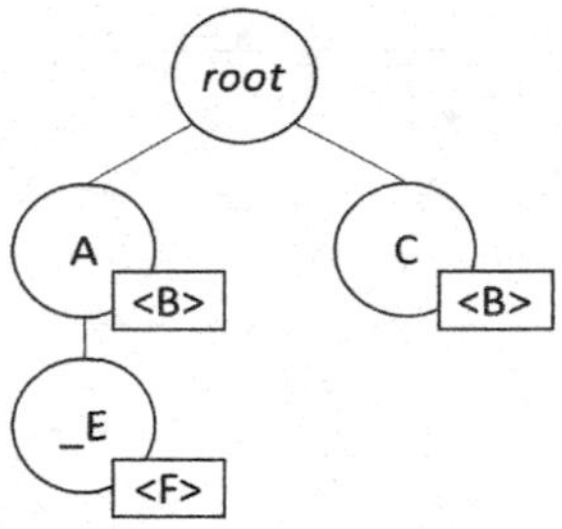

Fig. 6. An example of the rule tree

and the *tree-based* approach builds the rule tree, shown in Fig. 6. Then a new segment <(A, 11), (E, 11), (C, 13)> is coming. The *straight forward* approach finds the first matched rule, which is the rule No. 2, and makes the prediction as <F>. And the *tree-based* approach reaches the end nodes of all the three rules, and make the prediction as <B> after voting.

4 Experimental Evaluation

In this section, we evaluate the performance of a real-world case, an electric power consumption data of a store. The application scenario is that the power consumption is monitored by the power supply company under a contract of power consumption capacity to see whether it exceeds the contract capacity within every 15 min. The excessive electricity usage will result in some higher charge rate, even a fine as a penalty. Finding the episode patterns of the excessive electricity usage is very beneficial to electric power consumers because they can use that information for power management and avoidance of unexpected excessive usage of electricity.

4.1 Dataset

The dataset consists of the data collected by twenty-nine intelligent power meters within five months and the frequency of data collection is one record per minute. The total number of records is almost 6.4 millions, and each record consists of 6 attributes of recording_time, voltage (V), current (A), power (P), power factor (PF), Kilowatt-hour (kWh). Except for the recording_time, all the other attributes are of numeric data type. It's a typical case of multivariate complex event sequence, and we hope to mine the useful episode patterns for the excessive electricity usage.

4.2 Experiment Results

We first evaluated the execution time of the proposed algorithm *EM-CES* for mining frequent episodes under varied minimum support thresholds. Execution time of the algorithm and the number of episode patterns are respectively given in Fig. 7. We can

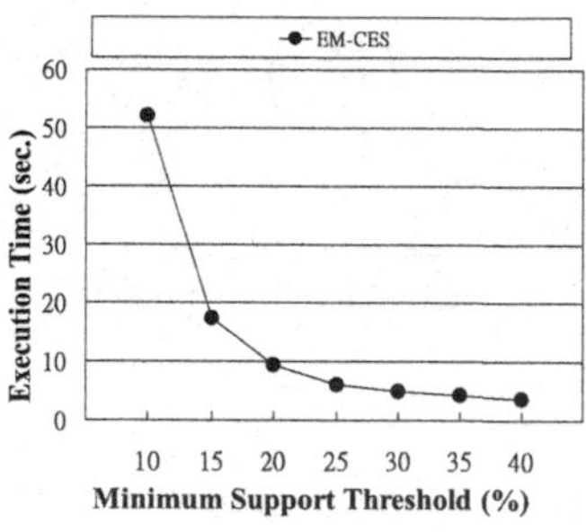

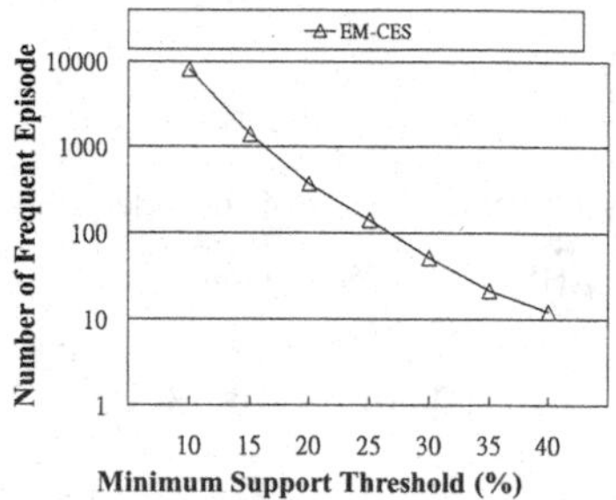

Fig. 7. Performance of EM-CES for mining episode patterns

learn the execution time of 10 % was less than 1 min with about 6.4 M records. It's an amazing performance while there are more than 8 K patterns found.

Then we evaluated the accuracy of the proposed algorithm for rules generating under varied minimum support thresholds. We choose the *straight forward* approach to make predictions due to the testing environment and set *minconf* = 70 %. Accuracy and the number of rules are respectively given in Fig. 8. In Fig. 8(a), we found that the accuracy of testing data is about 75 % when the minimum support is 10 %. But we also found the accuracy drops rapidly when minimum support is no less than 25 % since the number of rules is less than 10, which is shown in Fig. 8(b). To address this issue, we found that when the number of rules is too few, the predictions are usually made by the default rule, which is the based on the average electricity usage in the training data.

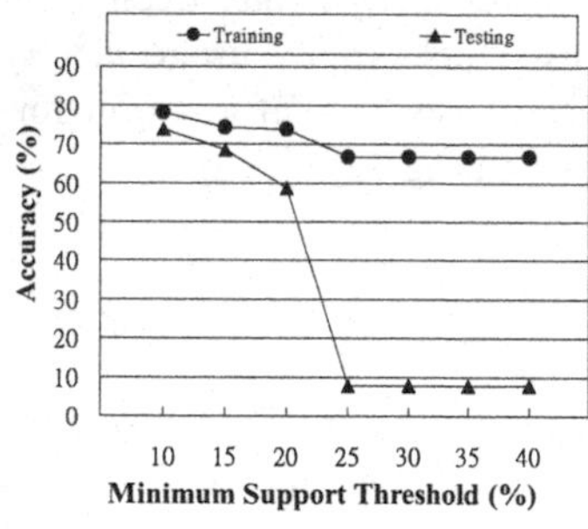

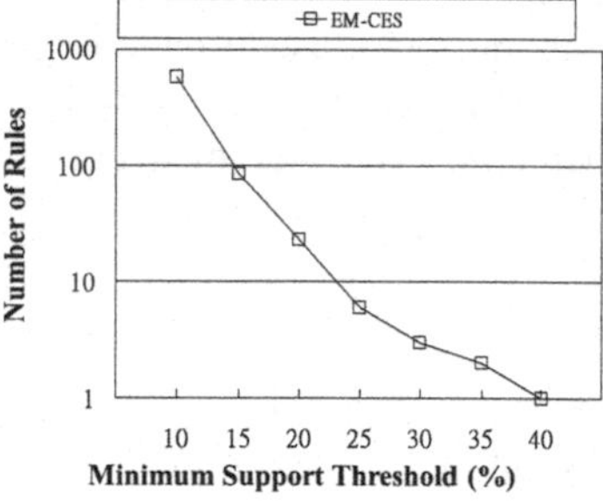

Fig. 8. Performance of EM-CES for generating episode rules

5 Conclusion and Future Works

In this paper, we have proposed an effective and efficient complex-events analytical system using episode pattern mining techniques. Four major components, namely, data preprocessing, pattern mining and rule management and prediction modules, are developed in the proposed system. We successfully applied and extended several well-known data mining techniques to make them work well for complex events analysis. By the experimental evaluation results, the proposed system can help users conduct the data mining jobs easily and effectively, and the proposed algorithm *EM-CES* shows excellent performance for the analytical applications on real-world complex events.

This is a beginning in developing a complex event analytical system, and there exist lots of enhancements that we could explore in the future: (1). Other than episode mining, we hope to incorporate more useful mining algorithms into this system to make it fit for more applications; (2). We plan to design an interface for users to make use of this system more easily; (3). We will keep improving the performance of the system to make it more efficient for the mining jobs.

References

1. Aboalsamh, H.A., Hafez, A.M., Assassa, G.M.R.: An efficient stream mining technique. WSEAS Trans. Inf. Sci. Appl. **5**(7), 1272–1281 (2008)
2. Casas-Garriga, G.: Discovering unbounded episodes in sequential data. In: Lavrač, N., Gamberger, D., Todorovski, L., Blockeel, H. (eds.) PKDD 2003. LNCS (LNAI), vol. 2838, pp. 83–94. Springer, Heidelberg (2003)
3. Fournier-Viger, P., Nkambou, R., Tseng, V.S.: RuleGrowth: mining sequential rules common to several sequences by pattern-growth. In: ACM Symposium on Applied Computing, pp. 956–961, Taiwan (2011)
4. Lin, J., Keogh, E., Lonardi, S., Chiu, B.: A symbolic representation of time series, with implications for streaming algorithms. In: Proceedings of ACM SIGMOD Workshop on Research Issues in Data Mining and Knowledge Discovery, pp. 2–11 (2003)
5. Lin, Y., Huang, C., Tseng, V.S.: A novel mining methodology for stock investment. J. Inf. Sci. Eng. **30**, 571–585 (2014)
6. Lin, Y., Jiang, P., Tseng, V.S.: Efficient mining of frequent target episodes from complex event sequences. In: Proceedings of International Computer Symposium (2014)
7. Ma, X., Pang, H., Tan, K.L.: Finding constrained frequent episodes using minimal occurrences. In: Proceedings of IEEE International Conference on Data Mining, pp. 471 474 (2004)
8. Mallik, R., Kargupta, H.: A sustainable approach for demand prediction in smart grids using a distributed local asynchronous algorithm. In: Proceedings of CIDU Conference of Intelligent Data Understanding, pp. 01–15 (2011)
9. Mannila, H., Toivonen, H., Verkamo, A.I.: Discovering frequent episodes in sequences. In: Proceedings of ACM SIGKDD International Conference on Knowledge Discovery and Data Mining, pp. 210–215 (1995)
10. Mannila, H., Toivonen, H., Verkamo, A.I.: Discovering frequent episodes in sequences. Data Min. Knowl. Disc. **1**(3), 259–289 (1997)
11. Mannila, H., Toivonen, H.: Discovering generalized episodes using minimal occurrences. In: Proceedings of ACM SIGKDD International Conference on Knowledge Discovery and Data Mining, pp. 146–151 (1996)
12. Pei, J., Han, J., Mortazavi-Asl, B., Pinto, H., Chen, Q., Dayal, U., Hsu, M.: PrefixSpan: mining sequential patterns efficiently by prefix-projected pattern growth. In: Proceedings of International Conference on Data Engineering, pp. 215–226 (2001)

13. Tseng, V.S., Lee, C.-H.: Effective temporal data classification by integrating sequential pattern mining and probabilistic induction. Expert Syst. Appl. (ESWA) **36**(5), 9524–9532 (2009)
14. Wu, C.-W., Lin, Y.-F., Yu, P.S., Tseng, V.S.: Mining high utility episodes in complex event sequences. In: Proceedings of ACM SIGKDD International Conference on Knowledge Discovery and Data Mining, pp. 536–544 (2013)

A New Method to Finding All Nash Equilibria

Zhengtian Wu[1,2](✉), Chuangyin Dang[2], Fuyuan Hu[1], and Baochuan Fu[1]

[1] School of Electronic and Information Engineering, Suzhou University of Science and Technology, Suzhou, People's Republic of China
wzht8@mail.ustc.edu.cn

[2] Department of Systems Engineering and Engineering Management, City University of Hong Kong, Kowloon, Hong Kong
mecdang@cityu.edu.hk

Abstract. It is a main concern in applications of game theory to effectively select a Nash equilibrium. All Nash equilibria is often required to be computed for this selection process. However, it is well known that the problem of finding only one mixed-strategy Nash equilibrium is a PPAD-complete process. Therefore, it is very hard to find all Nash equilibrium for a certain problem by traditional methods. By exploiting the properties of multilinear terms in the payoff functions, this paper presents a good approximation of the multilinear terms and develops a mixed-integer linear programming for finding all mixed-strategy Nash equilibria. An example of this method will be given too.

Keywords: Game theory · Nash equilibrium · PPAD-complete · Multilinear terms · Mixed-integer linear program

1 Introduction

Nash equilibrium, which is defined in [1], plays a fundamental role in the development of game theory and its diverse applications in such areas as economics, computer science, management, biology and social sciences. In these applications, a main concern is how to effectively select a Nash equilibrium, especially a mixed-strategy Nash equilibrium for a finite n-person game in normal form. To tackle this problem, many contributions have been made in the following literature.

In terms of computational complexity, it was shown in [2,3] that the problem of finding only one mixed-strategy Nash equilibrium is PPAD-complete. The first approach for finding Nash equilibrium of a two-person game was presented in [4] and was extended to finding Nash equilibria of n-person games respectively in [5,6]. The existence of Nash equilibrium was developed from an application of Brouwer and Kakutani fixed-point theorems in [7]. To approximate fixed points of continuous mappings, simplicial methods were originated in [8] and substantially developed in the literature such as [9,10]. Simplicial methods were adapted to compute Nash equilibria in [11]. Several more advanced simplicial methods for computing Nash equilibria were developed in [12,13]. By the linear tracing

X. He et al. (Eds.): IScIDE 2015, Part II, LNCS 9243, pp. 499–507, 2015.
DOI: 10.1007/978-3-319-23862-3_49

procedure in [14], a pivoting procedure was proposed in [15] for computing the Nash equilibria of two-person games selected. This procedure was extended to finding n-person games in [16]. A survey of the literature on the computation of Nash equilibria before 1996 can be found in [17] and on computing equilibria for two-person games in [18].

All the methods mentioned above can only find a sample Nash equilibrium. Unfortunately, in practices, a game can have many Nash equilibria and how to effectively select a suitable equilibrium to play is a very challenging issue. This equilibrium selection process often requires to compute all Nash equilibria of the game. To address this issue, several methods have been developed in the literature. A path-following algorithm was presented in [19] to compute all Nash equilibria of a bimatrix game. A homotopy algorithm was developed in [20] to compute all Nash equilibria of an n-person game. Both of these algorithms are based on the systems of polynomial equations. A survey on polyhedral homotopy continuation methods for all Nash equilibria can be found in [21]. In [22], a mixed-integer program (MIP) was formulated to find Nash equilibria of a bimatrix game, which has been employed to compute all Nash equilibria of a bimatrix game. Recently, by enumerating vertices of polyhedrons, two approaches were developed in [23] for computing all Nash equilibria of a bimatrix game.

This paper studies the computation of all mixed-strategy Nash equilibria of a finite n-person game in normal form. To solve this problem, we presents a good approximation of the multilinear terms and formulates a mixed-integer linear program by exploiting the properties of multilinear terms in the payoff functions. An enumeration of all the feasible solutions of the mixed-integer linear program yields all mixed-strategy Nash equilibria. Besides, as an advantage of the formulation, one can apply distributed computation to dramatically speed up the computation.

The rest of the paper is organized as follows. Section 2 formulates a new formulation for all Nash equilibria of a finite game. In Sect. 3, the approximation of the new formulation will be presented. There are two parts in this section. The first part introduces a good approximation of $y = x_1 x_2$. A mixed-integer linear programming formulation for finding all Nash equilibria of a finite game is built in the second part. Section 4 will conclude the paper with some remarks and present some future work.

2 A New Formulation for All Nash Equilibria of a Finite Game

Let $N = \{1, 2, \ldots, n\}$ be the set of players. The pure strategy set of player $i \in N$ is denoted by $S^i = \{s^i_j \mid j \in M_i\}$ with $M_i = \{1, 2, \ldots, m_i\}$. Given S^i, $i \in N$, the set of all pure strategy profiles is $S = \prod_{i=1}^n S^i$. We denote the payoff function of player $i \in N$ by $u^i : S \to R$. For $i \in N$, let $S^{-i} = \prod_{k \in N \setminus \{i\}} S^k$. Then, $s = (s^1_{j_1}, s^2_{j_2}, \ldots, s^n_{j_n}) \in S$ can be rewritten as $s = (s^i_{j_i}, s^{-i})$ with $s^{-i} = (s^1_{j_1}, \ldots, s^{i-1}_{j_{i-1}}, s^{i+1}_{j_{i+1}}, \ldots, s^n_{j_n}) \in S^{-i}$. A mixed strategy of player i is a probability distribution on S^i denoted by $x^i = (x^i_1, x^i_2, \ldots, x^i_{m_i})$. Let X^i be the set of all mixed strategies of player i. Then, $X^i = \{x^i = (x^i_1, x^i_2, \ldots, x^i_{m_i}) \in R^{m_i}_+ \mid \sum_{j=1}^{m_i} x^i_j = 1\}$. Thus, for $x^i \in X^i$, the probability assigned to pure strategy $s^i_j \in S^i$ is equal to x^i_j. Given X^i, $i \in N$, the set of all mixed strategy profiles is $X = \prod_{i=1}^n X^i$. For $i \in N$, let $X^{-i} = \prod_{k \in N \setminus \{i\}} X^k$. Then, $x = (x^1, x^2, \ldots, x^n) \in X$ can be rewritten as $x = (x^i, x^{-i})$ with $x^{-i} = (x^1, \ldots, x^{i-1}, x^{i+1}, \ldots, x^n) \in X^{-i}$. If $x \in X$ is played, then the probability that a pure strategy profile $s = (s^1_{j_1}, s^2_{j_2}, \ldots, s^n_{j_n}) \in S$ occurs is $\prod_{i=1}^n x^i_{j_i}$. Therefore, for $x \in X$, the expected payoff of player i is given by $u^i(x) = \sum_{s \in S} u^i(s) \prod_{i=1}^n x^i_{j_i}$. With these notations, a finite n-person game in normal form can be represented as $\Gamma = \langle N, S, \{u^i\}_{i \in N} \rangle$ or $\Gamma = \langle N, X, \{u^i\}_{i \in N} \rangle$.

Definition 1 (Nash (1951)). *A mixed strategy profile $x^* \in X$ is a Nash equilibrium of game Γ if $u^i(x^*) \geq u^i(x^i, x^{*-i})$ for all $i \in N$ and $x^i \in X^i$.*

With this definition, an application of the optimality condition leads to that x^* is a Nash equilibrium if and only if there are λ^* and μ^* together with x^* satisfying the system of

$$\begin{array}{l} u^i(s^i_j, x^{-i}) + \lambda^i_j - \mu_i = 0, \\ e^{i\top} x^i - 1 = 0, \\ x^i_j \lambda^i_j = 0, \\ x^i_j \geq 0, \ \lambda^i_j \geq 0, \\ j = 1, 2, \ldots, m_i, \ i = 1, 2, \ldots, n, \end{array} \tag{1}$$

where $e^i = (1, 1, \ldots, 1) \in R^{m_i}$.

By introducing a positive number β and 0-1 variables v^i_j, Wu and Dang in [24] have proved the following lemma 1 which can linearize the nonlinear term $x^i_j \lambda^i_j = 0$.

Lemma 1. *Let β be a given positive number such that*

$$\beta \geq \max_{i \in N} \{\max_{s \in S} u^i(s) - \min_{s \in S} u^i(s)\}.$$

Then, (1) is equivalent to

$$\begin{array}{l} u^i(s^i_j, x^{-i}) + \lambda^i_j - \mu_i = 0, \\ e^{i\top} x^i - 1 = 0, \\ x^i_j \leq v^i_j, \\ \lambda^i_j \leq \beta(1 - v^i_j), \\ v^i_j \in \{0, 1\}, \\ x^i_j \geq 0, \ \lambda^i_j \geq 0, \\ j = 1, 2, \ldots, m_i, \ i = 1, 2, \ldots, n. \end{array} \tag{2}$$

This lemma implies that finding a mixed-strategy Nash equilibrium is equivalent to finding a solution of the system 2. However, in the system 2 there are still multilinear terms $u^i(s^i_j, x^{-i})$ which make this system hard to solve by traditional methods. In the following of this paper, a good approximation of the multilinear terms will be presented.

For any $s^i_j \in S^i$, one can obtain from $u^i(x)$ that

$$u^i(s^i_j, x^{-i}) = \sum_{s^{-i}=(s^1_{j_1},\ldots,s^{i-1}_{j_{i-1}},s^{i+1}_{j_{i+1}},\ldots,s^n_{j_n})\in S^{-i}} u^i(s^i_j, s^{-i})\prod_{k\neq i} x^k_{j_k}.$$

Let $q(s^{-i}) = \prod_{k\neq i} x^k_{j_k}$ for any $s^{-i} = (s^1_{j_1},\ldots,s^{i-1}_{j_{i-1}},s^{i+1}_{j_{i+1}},\ldots,s^n_{j_n}) \in S^{-i}$. Then,

$$u^i(s^i_j, x^{-i}) = \sum_{s^{-i}=(s^1_{j_1},\ldots,s^{i-1}_{j_{i-1}},s^{i+1}_{j_{i+1}},\ldots,s^n_{j_n})\in S^{-i}} u^i(s^i_j, s^{-i})q(s^{-i}).$$

Substituting this into (2) yields

$$\begin{array}{l}
\sum_{s^{-i}\in S^{-i}} u^i(s^i_j, s^{-i})q(s^{-i}) + \lambda^i_j - \mu_i = 0,\\
e^{i\top}x^i - 1 = 0,\\
x^i_j \leq v^i_j,\\
\lambda^i_j \leq \beta(1 - v^i_j),\\
v^i_j \in \{0, 1\},\\
x^i_j \geq 0,\ \lambda^i_j \geq 0,\\
j = 1, 2, \ldots, m_i,\ i = 1, 2, \ldots, n,\\
q(s^{-i}) = \prod_{k\neq i} x^k_{j_k},\ s^{-i} \in S^{-i},\ i = 1, 2, \ldots, n.
\end{array} \tag{3}$$

For $i \in N$ and $s^{-i} = (s^1_{j_1},\ldots,s^{i-1}_{j_{i-1}},s^{i+1}_{j_{i+1}},\ldots,s^n_{j_n}) \in S^{-i}$, let

$$\begin{array}{l}
y_0(s^{-i}) = 1,\\
y_k(s^{-i}) = x^k_{j_k} y_{k-1}(s^{-i}),\ k = 1, 2, \ldots, i-1,\\
y_{i+1}(s^{-i}) = x^{i+1}_{j_{i+1}} y_{i-1}(s^{-i}),\\
y_k(s^{-i}) = x^k_{j_k} y_{k-1}(s^{-i}),\ k = i+2, i+3, \ldots, n.
\end{array} \tag{4}$$

Then,

$$q(s^{-i}) = \begin{cases} y_n(s^{-i}) & \text{if } i < n,\\ y_{n-1}(s^{-i}) & \text{if } i = n. \end{cases}$$

Therefore, one can solve the system 3 to find all Nash equilibrium of the system 1.

3 The Approximation of the New Formulation

3.1 A Good Approximation of $y = x_1x_2$

[25] studies approaches for obtaining convex relaxation of global optimization problems containing multilinear functions. Consider $y = x_1x_2$ with $x_1 \in [l_1, u_1]$ and $x_2 \in [l_2, u_2]$, where $u_1 > l_1 \geq 0$ and $u_2 > l_2 \geq 0$.

Lemma 2. *Let*

$$D = \left\{ (x_1, x_2, y)^\top \;\middle|\; \begin{array}{l} y \geq l_2 x_1 + l_1 x_2 - l_1 l_2, \\ y \geq u_2 x_1 + u_1 x_2 - u_1 u_2, \\ y \leq l_2 x_1 + u_1 x_2 - l_2 u_1, \\ y \leq u_2 x_1 + l_1 x_2 - l_1 u_2 \end{array} \right\}.$$

Then, D is the convex hull of

$$\{(x_1, x_2, y)^\top \mid y = x_1 x_2,\ l_1 \leq x_1 \leq u_1,\ l_2 \leq x_2 \leq u_2\}.$$

The proof of the lemma 2 can be found in [26].

Let d be any given positive integer. For $h = 1, 2, \ldots, d$, let $z_h \in \{0, 1\}$ such that

$$(1 - z_h) l_2 + \{l_2 + \frac{h-1}{d}(u_2 - l_2)\} z_h \leq x_2 \leq z_h \{l_2 + \frac{h}{d}(u_2 - l_2)\} + (1 - z_h) u_2.$$

Then, for any $x_1 \in [l_1, u_1]$ and $x_2 \in [l_2, u_2]$, y satisfies

$$\begin{array}{l} y \geq \{l_2 + \frac{h-1}{d}(u_2 - l_2)\} x_1 + l_1 x_2 - l_1 \{l_2 + \frac{h-1}{d}(u_2 - l_2)\} - (1 - z_h)(u_1 u_2 - l_1 l_2), \\ y \geq \{l_2 + \frac{h}{d}(u_2 - l_2)\} x_1 + u_1 x_2 - u_1 \{l_2 + \frac{h}{d}(u_2 - l_2)\} - (1 - z_h)(u_1 u_2 - l_1 l_2), \\ y < \{l_2 + \frac{h-1}{d}(u_2 - l_2)\} x_1 + u_1 x_2 - \{l_2 + \frac{h-1}{d}(u_2 - l_2)\} u_1 + (1 - z_h)(u_1 u_2 - l_1 l_2), \\ y \leq \{l_2 + \frac{h}{d}(u_2 - l_2)\} x_1 + l_1 x_2 - l_1 \{l_2 + \frac{h}{d}(u_2 - l_2)\} + (1 - z_h)(u_1 u_2 - l_1 l_2), \\ (1 - z_h) l_2 + \{l_2 + \frac{h-1}{d}(u_2 - l_2)\} z_h \leq x_2 \leq z_h \{l_2 + \frac{h}{d}(u_2 - l_2)\} + (1 - z_h) u_2, \\ \\ z_h \in \{0, 1\},\ h = 1, 2, \ldots, d, \\ \sum_{j=1}^{d} z_j = 1, \end{array}$$

yields a good approximation of $x_1 x_2$.

3.2 A Mixed-Integer Linear Programming Formulation of All Nash Equilibria of a Finite Game

In this subsection, a mixed-integer linear programming will be formulated based on the approximation method in above subsection. All Nash equilibrium in system 1 can be found though the solving this mixed-integer linear programming.

For $h = 1, 2, \ldots, d$, and $x_{j_k}^k \in [\frac{h-1}{d}, \frac{h}{d}]$, it is easy to see that

$$0 \leq y_k(s^{-i}) = y_{k-1}(s^{-i}) x_{j_k}^k \leq \frac{h}{d}.$$

For $h = 1, 2, \ldots, d$, let $z_h^{kj_k} \in \{0, 1\}$ such that

$$\frac{h-1}{d} z_h^{kj_k} \leq x_{j_k}^k \leq z_h^{kj_k} \frac{h}{d} + 1 - z_h^{kj_k}.$$

Then, any point in

$$\left\{ (y_{k-1}(s^{-i}), x^k_{j_k}, y_k(s^{-i}))^\top \;\middle|\; \begin{array}{l} y_k(s^{-i}) \geq \frac{h-1}{d} y_{k-1}(s^{-i}) - (1 - z_h^{kj_k}), \\ y_k(s^{-i}) \geq \frac{h}{d} y_{k-1}(s^{-i}) + x^k_{j_k} - \frac{h}{d} - (1 - z_h^{kj_k}), \\ y_k(s^{-i}) \leq \frac{h-1}{d} y_{k-1}(s^{-i}) + x^k_{j_k} - \frac{h-1}{d} + (1 - z_h^{kj_k}), \\ y_k(s^{-i}) \leq \frac{h}{d} y_{k-1}(s^{-i}) + (1 - z_h^{kj_k}), \\ \frac{h-1}{d} z_h^{kj_k} \leq x^k_{j_k} \leq z_h^{kj_k} \frac{h}{d} + 1 - z_h^{kj_k}, \\ z_h^{kj_k} \in \{0,1\}, \\ h = 1, 2, \ldots, d, \\ \sum_{h=1}^d z_h^{kj_k} = 1 \end{array} \right\}$$

yields a good approximation of $(y_{k-1}(s^{-i}), x^k_{j_k}, y_k(s^{-i}))^\top$ with $y_k(s^{-i}) = y_{k-1}(s^{-i}) x^k_{j_k}$, $y_{k-1}(s^{-i}) \in [0,1]$ and $x^k_{j_k} \in [0,1]$. This with 4 leads to

$$\begin{array}{l} y_0(s^{-i}) = 1, \\ y_k(s^{-i}) \geq \frac{h-1}{d} y_{k-1}(s^{-i}) - (1 - z_h^{kj_k}), \\ y_k(s^{-i}) \geq \frac{h}{d} y_{k-1}(s^{-i}) + x^k_{j_k} - \frac{h}{d} - (1 - z_h^{kj_k}), \\ y_k(s^{-i}) \leq \frac{h-1}{d} y_{k-1}(s^{-i}) + x^k_{j_k} - \frac{h-1}{d} + (1 - z_h^{kj_k}), \\ y_k(s^{-i}) \leq \frac{h}{d} y_{k-1}(s^{-i}) + (1 - z_h^{kj_k}), \\ \frac{h-1}{d} z_h^{kj_k} \leq x^k_{j_k} \leq z_h^{kj_k} \frac{h}{d} + 1 - z_h^{kj_k}, \\ z_h^{kj_k} \in \{0,1\}, \\ h = 1, 2, \ldots, d, \\ \sum_{h=1}^d z_h^{kj_k} = 1 \\ k = 1, 2, \ldots, i-1, \end{array} \tag{5}$$

$$\begin{array}{l} y_{i+1}(s^{-i}) \geq \frac{h-1}{d} y_{i-1}(s^{-i}) - (1 - z_h^{i+1,j_{i+1}}), \\ y_{i+1}(s^{-i}) \geq \frac{h}{d} y_{i-1}(s^{-i}) + x^{i+1}_{j_{i+1}} - \frac{h}{d} - (1 - z_h^{i+1,j_{i+1}}), \\ y_{i+1}(s^{-i}) \leq \frac{h-1}{d} y_{i-1}(s^{-i}) + x^{i+1}_{j_{i+1}} - \frac{h-1}{d} + (1 - z_h^{i+1,j_{i+1}}), \\ y_{i+1}(s^{-i}) \leq \frac{h}{d} y_{i-1}(s^{-i}) + (1 - z_h^{i+1,j_{i+1}}), \\ \frac{h-1}{d} z_h^{i+1,j_{i+1}} \leq x^{i+1}_{j_{i+1}} \leq z_h^{i+1,j_{i+1}} \frac{h}{d} + 1 - z_h^{i+1,j_{i+1}}, \\ z_h^{i+1,j_{i+1}} \in \{0,1\}, \\ h = 1, 2, \ldots, d, \\ \sum_{h=1}^d z_h^{i+1,j_{i+1}} = 1 \\ y_k(s^{-i}) \geq \frac{h-1}{d} y_{k-1}(s^{-i}) - (1 - z_h^{kj_k}), \\ y_k(s^{-i}) \geq \frac{h}{d} y_{k-1}(s^{-i}) + x^k_{j_k} - \frac{h}{d} - (1 - z_h^{kj_k}), \\ y_k(s^{-i}) \leq \frac{h-1}{d} y_{k-1}(s^{-i}) + x^k_{j_k} - \frac{h-1}{d} + (1 - z_h^{kj_k}), \\ y_k(s^{-i}) \leq \frac{h}{d} y_{k-1}(s^{-i}) + (1 - z_h^{kj_k}), \\ \frac{h-1}{d} z_h^{kj_k} \leq x^k_{j_k} \leq z_h^{kj_k} \frac{h}{d} + 1 - z_h^{kj_k}, \\ z_h^{kj_k} \in \{0,1\}, \\ h = 1, 2, \ldots, d, \\ \sum_{h=1}^d z_h^{kj_k} = 1 \\ k = i+2, i+3, \ldots, n, \end{array} \tag{6}$$

Replacing $q(s^{-i})$ of the system (3) with the systems (5) and (6), we obtain a mixed-integer linear programming as follows(system 7).

$$
\begin{array}{l}
\sum_{s^{-i}\in S^{-i}} u^i(s^i_j, s^{-i})q(s^{-i}) + \lambda^i_j - \mu_i = 0,\\
e^{i\top}x^i - 1 = 0,\\
x^i_j \le v^i_j,\\
\lambda^i_j \le \beta(1 - v^i_j),
\end{array}
$$

$$
\begin{array}{l}
v^i_j \in \{0,1\}, x^i_j \ge 0,\ \lambda^i_j \ge 0, j = 1,2,\dots,m_i,\ i = 1,2,\dots,n,\\
y_0(s^{-i}) = 1,\\
y_k(s^{-i}) \ge \frac{h-1}{d} y_{k-1}(s^{-i}) - (1 - z^{kj_k}_h),\\
y_k(s^{-i}) \ge \frac{h}{d} y_{k-1}(s^{-i}) + x^k_{j_k} - \frac{h}{d} - (1 - z^{kj_k}_h),\\
y_k(s^{-i}) \le \frac{h-1}{d} y_{k-1}(s^{-i}) + x^k_{j_k} - \frac{h-1}{d} + (1 - z^{kj_k}_h),\\
y_k(s^{-i}) \le \frac{h}{d} y_{k-1}(s^{-i}) + (1 - z^{kj_k}_h),\\
\frac{h-1}{d} z^{kj_k}_h \le x^k_{j_k} \le z^{kj_k}_h \frac{h}{d} + 1 - z^{kj_k}_h,\\
z^{kj_k}_h \in \{0,1\}, h = 1,2,\dots,d,\\
\sum_{h=1}^d z^{kj_k}_h = 1, k = 1,2,\dots,i-1,\\
y_{i+1}(s^{-i}) \ge \frac{h-1}{d} y_{i-1}(s^{-i}) - (1 - z^{i+1,j_{i+1}}_h),\\
y_{i+1}(s^{-i}) > \frac{h}{d} y_{i-1}(s^{-i}) + x^{i+1}_{j_{i+1}} - \frac{h}{d} - (1 - z^{i+1,j_{i+1}}_h),\\
y_{i+1}(s^{-i}) \le \frac{h-1}{d} y_{i-1}(s^{-i}) + x^{i+1}_{j_{i+1}} - \frac{h-1}{d} + (1 - z^{i+1,j_{i+1}}_h),\\
y_{i+1}(s^{-i}) \le \frac{h}{d} y_{i-1}(s^{-i}) + (1 - z^{i+1,j_{i+1}}_h),\\
\frac{h-1}{d} z^{i+1,j_{i+1}}_h \le x^{i+1}_{j_{i+1}} \le z^{i+1,j_{i+1}}_h \frac{h}{d} + 1 - z^{i+1,j_{i+1}}_h,\\
z^{i+1,j_{i+1}}_h \in \{0,1\}, h = 1,2,\dots,d,\\
\sum_{h=1}^d z^{i+1,j_{i+1}}_h = 1\\
y_k(s^{-i}) \ge \frac{h-1}{d} y_{k-1}(s^{-i}) - (1 - z^{kj_k}_h),\\
y_k(s^{-i}) \ge \frac{h}{d} y_{k-1}(s^{-i}) + x^k_{j_k} - \frac{h}{d} - (1 - z^{kj_k}_h),\\
y_k(s^{-i}) \le \frac{h-1}{d} y_{k-1}(s^{-i}) + x^k_{j_k} - \frac{h-1}{d} + (1 - z^{kj_k}_h),\\
y_k(s^{-i}) \le \frac{h}{d} y_{k-1}(s^{-i}) + (1 - z^{kj_k}_h),\\
\frac{h-1}{d} z^{kj_k}_h \le x^k_{j_k} \le z^{kj_k}_h \frac{h}{d} + 1 - z^{kj_k}_h,\\
z^{kj_k}_h \in \{0,1\}, h = 1,2,\dots,d,\\
\sum_{h=1}^d z^{kj_k}_h = 1, k = i+2, i+3, \dots, n,\\
q(s^{-i}) = \begin{cases} y_n(s^{-i}) & \text{if } i < n,\\ y_{n-1}(s^{-i}) & \text{if } i = n. \end{cases}, s^{-i} \in S^{-i},\ i = 1,2,\dots,n.
\end{array}
\qquad (7)
$$

Therefore, we can solve this a mixed-integer linear programming system 7 to find all mixed-strategy Nash equilibrium in normal form as described in system 1.

4 Conclusions and Future Work

In this paper, a new mixed-integer linear programming approach has been developed to find all mixed-strategy Nash equilibria of a finite n-person game in normal form. This method is based on the properties of multilinear terms in the payoff functions. An example has been given too. Some future work can be concluded as follows: Firstly, some numerical results of this method will be exploited in the next step. Secondly, this method will be extended to other similar problems. Lastly, as a feature of mixed-integer programming, this method will be implemented in a distributed way.

Acknowledgement. This work was partially supported by GRF(CityU 112910 of Hong Kong SAR Government), ARG(CityU 9667080 of Hong Kong SAR Government), the National Natural Science Foundation of China under Grant No.61472267 and Nature Foundation of Jiangsu Province under Grant No.BK2012166.

References

1. Nash, J.F.: Non-cooperative games. Ann. Math. **54**(2), 286–295 (1951)
2. Daskalakis, C., Goldberg, P.W., Papadimitriou, C.H.: The complexity of computing a nash equilibrium. In: Proceedings of the 38th Annual ACM Symposium on Theory of Computing, pp. 71–78 (2006)
3. Chen, X., Deng, X.: Settling the complexity of two-player nash equilibrium. In: Proceedings of the 47th Annual Symposium on Foundations of Computer Science (FOCS), pp. 261–272 (2006)
4. Lemke, C.E., Howson, J.T.: Equilibrium points of bimatrix games. J. Soc. Ind. Appl. Math. **12**(2), 413–423 (1964)
5. Wilson, R.: Computing equilibria of n-person games. SIAM J. Appl. Math. **21**(1), 80–87 (1971)
6. Rosenmüller, J.: On a generalization of the lemke-howson algorithm to noncooperative n-person games. SIAM J. Appl. Math. **21**(1), 73–79 (1971)
7. Nash, J.F.: Equilibrium points in n-person games. Proc. Nat. Acad. Sci. **36**(1), 48–49 (1950)
8. Scarf, H.E.: The approximation of fixed points of a continuous mapping. SIAM J. Appl. Math. **15**(5), 1328–1343 (1967)
9. Dang, C.: The D1-triangulation of Rn for simplicial algorithms for computing solutions of nonlinear equations. Math. Oper. Res. **16**(1), 148–161 (1991)
10. Allgower, E.L., Georg, K.: Piecewise linear methods for nonlinear equations and optimization. J. Comput. Appl. Math. **124**(1), 245–261 (2000)
11. Garcia, C.B., Lemke, C.E., Luethi, H.: Simplicial approximation of an equilibrium point of noncooperative n-person games. Math. Program. **4**, 227–260 (1973)
12. Van der Laan, G., Talman, A.J.J., Van der Heyden, L.: Simplicial variable dimension algorithms for solving the nonlinear complementarity problem on a product of unit simplices using a general labelling. Math. Oper. Res. **12**(3), 377–397 (1987)
13. Doup, T.M., Talman, A.J.J.: A new simplicial variable dimension algorithm to find equilibria on the product space of unit simplices. Math. Program. **37**(3), 319–355 (1987)
14. Harsanyi, J.C.: The tracing procedure: a bayesian approach to defining a solution for n-person noncooperative games. Int. J. Game Theor. **4**(2), 61–94 (1975)
15. Van den Elzen, A.H., Talman, A.J.J.: A procedure for finding nash equilibria in bi-matrix games. Math. Methods Oper. Res. **35**(1), 27–43 (1991)
16. Herings, P.J.J., Van den Elzen, A.: Computation of the nash equilibrium selected by the tracing procedure in n-person games. Game Econ. Behav. **38**(1), 89–117 (2002)

17. McKelvey, R.D., McLennan, A.: Computation of equilibria in finite games. Handb. Comput. Econ. **1**, 87–142 (1996)
18. Von Stengel, B.: Computing equilibria for two-person games. Handb. game Theor. Econ. Appl. **3**, 1723–1759 (2002)
19. Kosrnnva, M.M., Kinard, L.A.: A differentiable homotopy approach for solving polynomial optimization problems and noncooperative games. Comput. Math. Appl. **21**(6–7), 135–143 (1991)
20. Herings, P.J.J., Peeters, R.J.A.P.: A globally convergent algorithm to compute all nash equilibria for n-person games. Ann. Oper. Res. **137**(1), 349–368 (2005)
21. Datta, R.S.: Finding all nash equilibria of a finite game using polynomial algebra. Econ. Theor. **42**(1), 55–96 (2010)
22. Sandholm, T., Gilpin, A., Conitzer, V.: Mixed-integer programming methods for finding nash equilibria. In: Proceedings of the National Conference on Artificial Intelligence, pp. 495–501 (2005)
23. Avis, D., Rosenberg, G.D., Savani, R., Von Stengel, B.: Enumeration of nash equilibria for two-player games. Econ. Theor. **42**(1), 9–37 (2010)
24. Wu, Z., Dang, C., Karimi, H.R., Zhu, C., Gao, Q.: A mixed 0-1 linear programming approach to the computation of all pure-strategy nash equilibria of a finite n-person game in normal form. Mathematical Problems in Engineering, vol. 2014, p. 8 (2014)
25. Luedtke, J., Namazifar, M., Linderoth, J.: Some results on the strength of relaxations of multilinear functions. Math. Program. **136**(2), 325–351 (2012)
26. Al-Khayyal, F.A., Falk, J.E.: Jointly constrained biconvex programming. Math. Oper. Res. **8**(2), 273–286 (1983)

A Set of Metrics for Measuring Interestingness of Theorems in Automated Theorem Finding by Forward Reasoning: A Case Study in NBG Set Theory

Hongbiao Gao, Yuichi Goto, and Jingde Cheng(✉)

Department of Information and Computer Sciences,
Saitama University, Saitama 338-8570, Japan
{gaohongbiao,gotoh,cheng}@aise.ics.saitama-u.ac.jp

Abstract. The problem of automated theorem finding is one of 33 basic research problems in automated reasoning which was originally proposed by Wos in 1988, and it is still an open problem. The problem implicitly requires some metrics to be used for measuring interestingness of found theorems. However, no one addresses that requirement until now. This paper proposes the first set of metrics for measuring interestingness of theorems. The paper also presents a case study in NBG set theory, in which we use the proposed metrics to measure the interestingness of the theorems of NBG set theory obtained by using forward reasoning approach and confirms the effectiveness of the metrics.

Keywords: Metric · Automated theorem finding · Forward reasoning · Strong relevant logic · NBG set theory

1 Introduction

The problem of automated theorem finding (ATF for short) is one of the 33 basic research problems in automated reasoning which was originally proposed by Wos in 1988 [15,16], and it is still an open problem until now [8]. The problem of ATF is "What properties can be identified to permit an automated reasoning program to find new and interesting theorems, as opposed to proving conjectured theorems?" [15,16]. The most important and difficult requirement of the problem is that, in contrast to proving conjectured theorems supplied by the user, it asks for the criteria that an automated reasoning program can use to find some theorems in a field that must be evaluated by theorists of the field as new and interesting theorems. The significance of solving the problem is obvious because an automated reasoning program satisfying the requirement can provide great assistance for scientists in various fields [1–3].

To solve the ATF problem, a systematic methodology for ATF by using forward reasoning approach based on strong relevant logics was proposed [1–3,8]. The systematic methodology uses a filtering method to filter explicitly uninteresting theorems and presents the rest of theorems as the candidates

X. He et al. (Eds.): IScIDE 2015, Part II, LNCS 9243, pp. 508–517, 2015.
DOI: 10.1007/978-3-319-23862-3_50

of interesting theorems. From the viewpoint of the most important requirement of the ATF problem, the filtering method is not enough and some metrics to estimate the interestingness of reasoned out theorems are necessary. On the other hand, a few works aimed to automated theorem discovery (ATD) and automated theorem generation (ATG) have been done and some metrics have been proposed in those works [5–7, 9–12, 14]. However, the problem of ATF is different from the ATD and ATG such that their metrics are not suitable to be used in ATF. In fact, Wos's problem can be regarded as an attempt to find a systematic methodology in automated reasoning area, but the works on ATD and ATG almost aim to one certain mathematical field and their metrics are aimed to one certain mathematical field. Besides, the works of ATD and ATG rely on the approach of automated theorem proving such that their metrics are not suitable to measure interestingness of theorems found by forward reasoning approach.

This paper proposes the first set of metrics for measuring interestingness of theorems found in ATF. The paper also presents a case study in NBG set theory [13], in which we use the proposed metrics to measure the theorems of NBG set theory obtained by using forward reasoning approach and confirms the effectiveness of the metrics. The rest of the paper is organized as follows: Sect. 2 explains the basic notions and notations used in the paper. Section 3 shows factors related to interestingness of theorems. Section 4 shows the metrics for measuring interestingness of theorems found in ATF. Section 5 presents the case study in NBG set theory. Finally, concluding remarks are given in Sect. 6.

2 Basic Notions and Notations

A formal logic system L is an ordered pair $(F(L), \vdash_L)$ where $F(L)$ is the set of well formed formulas of L, and $\vdash_L$ is the consequence relation of L such that for a set P of formulas and a formula C, $P \vdash_L C$ means that within the framework of L taking P as premises we can obtain C as a valid conclusion. $Th(L)$ is the set of logical theorems of L such that $\phi \vdash_L T$ holds for any $T \in Th(L)$. According to the representation of the consequence relation of a logic, the logic can be represented as a Hilbert style system, a Gentzen sequent calculus system, a Gentzen natural deduction system, and so on [3].

Let $(F(L), \vdash_L)$ be a formal logic system and $P \subseteq F(L)$ be a non-empty set of sentences. A formal theory with premises P based on L, called a L-theory with premises P and denoted by $T_L(P)$, is defined as $T_L(P) =_{df} Th(L) \cup Th^e_L(P)$ where $Th^e_L(P) =_{df} \{A|P \vdash_L A \textit{ and } A \notin Th(L)\}$, $Th(L)$ and $Th^e_L(P)$ are called the logical part and the empirical part of the formal theory, respectively, and any element of $Th^e_L(P)$ is called an empirical theorem of the formal theory [3].

Based on the definition above, the problem of ATF can be said as "for any given premises P, how to construct a meaningful formal theory $T_L(P)$ and then find new and interesting theorems in $Th^e_L(P)$ automatically?" [3].

The notion of the degree [3] of a connective is defined as follows: Let θ be an arbitrary n-ary $(1 \leq n)$ connective of logic L and A be a formula of L, the degree of θ in A, denoted by $D_\theta(A)$, is defined as follows: (1) $D_\theta(A) = 0$ if and only if there is no occurrence of θ in A, (2) if A is in the form $\theta(a_1, a_2, ..., a_n)$ where $a_1, a_2, ..., a_n$ are formulas, then $D_\theta(A) = max\{D_\theta(a_1), D_\theta(a_2), ..., D_\theta(a_n)\} + 1$, (3) if A is in the form $\sigma(a_1, a_2, ..., a_n)$ where σ is a connective different from θ and $a_1, a_2, ..., a_n$ are formulas, then $D_\theta(A) = max\{D_\theta(a_1), D_\theta(a_2), ..., D_\theta(a_n)\}$, and (4) if A is in the form QB where B is a formula and Q is the quantifier prefix of B, then $D_\theta(A) = D_\theta(B)$.

The notion of predicate abstract level [8] is defined as follows: (1) Let $pal(X) = k$ denote that an abstract level of a predicate X is k where k is a natural number, (2) $pal(X) = 1$ if X is the most primitive predicate in the target field, (3) $pal(X) = max(pal(Y_1), pal(Y_2), ..., pal(Y_n)) + 1$ if a predicate X is defined by other predicates $Y_1, Y_2, ..., Y_n$ in the target field where n is a natural number. A predicate X is called k-level predicate, if $pal(X) = k$. If $pal(X) < pal(Y)$, we call the abstract level of predicate X is lower than Y, and Y is higher than X.

The notion of function abstract level [8] is defined as follows: (1) Let $fal(f) = k$ denote that an abstract level of a function f is k where k is a natural number, (2) $fal(f) = 1$ if f is the most primitive function in the target field, (3) $fal(f) = max(fal(g_1), fal(g_2), ..., fal(g_n)) + 1$ if a function f is defined by other functions $g_1, g_2, ..., g_n$ in the target field where n is a natural number. A function f is k-level function, if $fal(f) = k$. If $fal(f) < fal(g)$, we call the abstract level of function f is lower than g, and g is higher than f.

The notion of abstract level [8] of a formula is defined as follows: (1) $lfal(A) = (k, m)$ denote that an abstract level of a formula A where $k = pal(A)$ and $m = fal(A)$, (2) $pal(A) = max(pal(Q_1), pal(Q_2), ..., pal(Q_n))$ where Q_i is a predicate and occurs in A $(1 \leq i \leq n)$, or $pal(A) = 0$, if there is not any predicate in A, (3) $fal(A) = max(fal(g_1), fal(g_2), ..., fal(g_n))$ where g_i is a function and occurs in A $(1 \leq i \leq n)$, or $fal(A) = 0$, if there is not any function in A. A formula A is (k, m)-level formula, if $lfal(A) = (k, m)$.

The deduction distance by using Modus Ponens is defined as below: (1) $Dist(A) = 0$, if A an axiom; (2) $Dist(A) = max(Dist(\alpha), Dist(\beta)) + 1$, if A is deduced from two empirical theorems α and β by using Modus Ponens; (3) $Dist(A) = Dist(\alpha) + 1$, if A is deduced from an empirical theorem α and a logical theorem β by using Modus Ponens; (4) $Dist(A) = Dist(\alpha)$, if A is abstracted from α.

The propositional schema of a first-order logical formula can be obtained by replacing all of the atomic formulas of a first-order logical formula with propositional atomic formulas. For example, $((x = y) \Rightarrow (y = x)) \Rightarrow (x \subseteq y)$ is translated into $(A \Rightarrow B) \Rightarrow C$.

3 Factors Related to Interestingness of Theorems

Interestingness of theorems is related to plural factors. Here, we discuss the degree of logical connectives in empirical theorems, propositional schema of empirical theorems, abstract level of empirical theorems, and deduction distance of empirical theorems.

Degree of Logical Connectives

The first factor related to interestingness of theorems is the degree of logical connectives in empirical theorems. We have analyzed more than 400 known theorems of NBG set theory in Quaife's book [13] about the degree of logical connectives, and our analysis results are shown in Table 1. We found the degrees of the logical connectives of those known theorems are almost lower than 2. Therefore, the degree of logical connectives is related to the interestingness of empirical theorems, and interesting theorems always hold lower degree of logical connectives. The reason is that those theorems holding high degree of logical connectives are hard to be understood and mathematicians always introduce new predicates to abstract the formula holding higher degree of logical connectives. Cheng conjectured that almost all new theorems and questions of a formal theory can be deduced from the premises of that theory by finite inference steps concerned with finite number of low degree entailments [3].

Table 1. Degree of logical connectives of collected known theorems

⇒,0	242	56 %	∧,0	356	83 %	¬,0	404	94 %
⇒,1	187	44 %	∧,1	64	15 %	¬,1	25	6 %
⇒,2	0	0 %	∧,2	7	<2 %	¬,2	0	0 %
⇒,3	0	0 %	∧,3	2	<1 %	¬,3	0	0 %
⇒,4	0	0 %	∧,4	0	0 %	¬,4	0	0 %

Propositional Schema of Formula

The second factor is propositional schema of formula. We consider that the interesting theorems hold some frequent propositional schemata, after we investigated the propositional schemata of more than 400 known theorems. The most frequent propositional schemata of known theorems is A. A theorem is always interesting if the theorem does not contain any logical connective, because it holds clear and concise semantics. The second frequent propositional schema is $A \Rightarrow B$. We think the reason is that "if A then B"is a very frequent conditional propositional schema in any fields. Other frequent propositional schemata have been also shown in Table 2. The analysis results show that known theorems always hold some frequent propositional schemata. We can see known theorems as found interesting theorems, so we consider that the new and interesting theorems may also holds those frequent propositional schemata.

Abstract Level

The third factor is the abstract level of predicates and functions in one theorem. In the mathematical fields, mathematicians always make definition from simple to complex. For example, the predicate "$\in$" is the most basic predicate in the set theory. Then the mathematicians define the predicate "$\subseteq$"which is a higher level predicate than "$\in$", and abstracts from "$\in$"by the definition of "$\subseteq$": $\forall x \forall y (\forall u ((u \in x) \Rightarrow (u \in y)) \Leftrightarrow (x \subseteq y))$. Then the mathematicians define the

Table 2. Frequent propositional schemata of collected known theorems

Propositional schema	Appeared time	Appeared rate
A	186	43 %
$\neg A$	14	3 %
$A \Rightarrow B$	108	25 %
$(A \wedge B) \Rightarrow C$	54	13 %
$\neg(A \wedge B)$	10	2 %
$\neg(A \wedge B \wedge C)$	1	1 %
$A \vee B$	26	6 %
$A \vee B \vee C$	5	1 %
$(A \wedge B \wedge C) \Rightarrow D$	6	1 %
$(A \wedge B \wedge C \wedge D) \Rightarrow E$	2	<1 %
$A \Rightarrow (B \vee C)$	17	4 %

predicate "="which is a higher level predicate than "$\subseteq$", and abstracts from "$\subseteq$" by the axiom: $\forall x \forall y(((x \subseteq y) \wedge (y \subseteq x)) \Leftrightarrow (x = y))$. Based on the fact, we can consider that a theorem holds higher abstract level predicates and functions, the theorem is a more interesting theorem from the viewpoint of semantics.

Deduction Distance
The fourth factor is deduction distance. If a theorem can be reasoned out by several steps, the theorem is easy to be found and is too obvious to be understood by observing used premises. The interesting theorems are those theorems which are difficult to be reasoned out from premises. Therefore, if the deduction distance of a theorem is longer, the possibility to be an interesting theorem is higher.

4 A Set of Metrics for Measuring Interestingness of Theorems

Our metrics are to measure the interestingness for each empirical theorem by using the degree of logical connectives, propositional schema of formula, abstract level and deduction distance of empirical theorems and we use four variables Vd, Vp, Va, Ve to represent four values of interestingness respectively. In detail, the value of interestingness of the degree $Vd = Value_{\Rightarrow} * Value_{\wedge} * Value_{\neg}$. We showed the value of interestingness of the degree in Table 3. Second, we presented the value of interestingness of the propositional schemata of formula in Table 4. We assign the value 0 for empirical theorems containing a tautology part, because if one theorem contains a tautology part, this empirical theorem must not be an interesting empirical theorem. Third, if the abstract level of one

Table 3. Interesting values of degree of connectives

Degree	$Value_{\Rightarrow}$	Degree	$Value_{\wedge}$	Degree	$Value_{\neg}$
$\Rightarrow$,0	1	$\wedge$,0	1	$\neg$,0	1
$\Rightarrow$,1	1	$\wedge$,1	1	$\neg$,1	1
$\Rightarrow$,2	1/2	$\wedge$,2	1/2	$\neg$,2	1/2
$\Rightarrow$,3	1/3	$\wedge$,3	1/3	$\neg$,3	1/3
$\Rightarrow$,n	1/n	$\wedge$,n	1/n	$\neg$,n	1/n

Table 4. Interesting values of propositional schemata of formula

Propositional schema	Value
A	3
$A \Rightarrow B$	3
$\neg A$	2
$(A \wedge B) \Rightarrow C$	2
$(A \wedge B \wedge C) \Rightarrow D$	2
$A \Rightarrow (B \vee C)$	2
$\neg(A \wedge B)$	2
$\neg(A \wedge B \wedge C)$	2
$A \vee B$	2
$A \vee B \vee C$	2
Infrequent propositional schema	1
Propositional schema containing tautology	0

empirical theorem is (k, m), then the value of interestingness of abstract level of one theorem is counted by the formula $Va = k + m$. Fourth, if the deduction distance of one empirical theorem is $Dist(A)$, then the value of interestingness of deduction distance is $Ve = Dist(A)$. Then, we present a set of metrics to measure the value of interestingness of empirical theorem found in ATF: Vd, Vp, Va, Ve, $Vd{*}Vp$, $Vd{*}Va$, $Vd{*}Ve$, $Vp{*}Va$, $Vp{*}Ve$, $Va{*}Ve$, $Vd{*}Vp{*}Va$, $Vd * Vp * Ve$, $Vp * Va * Ve$, $Vd * Vp * Va * Ve$. The value is bigger, theorem is more interesting.

5 Case Study in NBG Set Theory

The purpose of the case study was to confirm the effectiveness of the proposed metrics. To confirm the effectiveness of those metrics, we used them to measure interestingness of empirical theorems of NBG set theory obtained by forward reasoning approach and compare the evaluated results. In detail, Quaife recorded the axioms and definitions of NBG set theory in his book [13]. We inputted all of axioms and definitions of NBG set theory in Quaife's book and use FreeEnCal [4] to reason out 149 empirical theorems. Then, we used the proposed metrics to evaluate the value of interestingness for each empirical theorem. In detail, to measure Vd and Vp, we have analyzed them by Tables 3 and 4 as shown in Sect. 4. To measure Va, we summarized the abstract levels of the predicates appeared in Quaife's book as shown in Table 5. Then, we also summarized the abstract levels of the functions appeared in Quaife's book as shown in Table 6. We also recorded Ve for each empirical theorem based on the information provided by FreeEnCal.

Table 5. Predicate abstract level in NBG set theory

Predicate	Abstract from	Level
$\in$	none	1
$\subseteq$	$\in$	2
$=$	$\subseteq$	3
INDUCTIVE	$\in$, $\subseteq$	3
SINGVAL	$\subseteq$	3
FUNCTION	$\subseteq$, SINGVAL	4
ONEONE	FUNCTION	5
OPERATION	FUNCTION, $=$, $\subseteq$	5
COMPATIBLE	FUNCTION, $=$, $\subseteq$	5
HOM	OPERATION, COMPATIBLE, $=$, $\in$	6

We recorded range of value, average value and deviation by each metric in Table 7. The combination $Vp{*}Va{*}Ve$ and $Vd{*}Vp{*}Va{*}Ve$ are the good choices, because range of values is wide and deviation is obvious such that we can easily distinguish the weight of interestingness for empirical theorems. We consider the $Vd{*}Vp{*}Va{*}Ve$ is the best choice, because the Vd is useful when we reason out the high degree empirical theorems. Therefore, we also investigated how many empirical theorems on each value of the combination $Vd{*}Vp{*}Va{*}Ve$ and showed the results in Fig. 1. The result shows that the metric measures interestingness of empirical theorems well. First, the value of uninteresting theorems are all shown by 0, which are those theorems containing tautology. We can easily distinguish the uninteresting theorems from all of the reasoned out empirical theorems. Second, in the rest of empirical theorems, the theorems holding higher value are fewer. The result satisfies our expectations, because the metric is helpful to distinguish the interesting theorems from large amount of reasoned out empirical theorems. Therefore the combination $Vd * Vp * Va * Ve$ is hopeful to measure interestingness of empirical theorems in ATF.

Table 6. Function abstract level in NBG set theory

Function	Abstract from	Level
{, }	none	1
∩	none	1
∼	none	1
{ }	{, }	2
∪	∼, ∩	2
+	∼, ∩	2
regular	∩	2
<, >	{, }, { }	3
succ	∪, { }	3
×	<, >	4
restrict	∩, ×	5
rotate	<, >, ×	5
flip	<, >, ×	5
D	restrict, { }	6
inverse	D, flip, ×	7
diag	∼, D, ∩	7
U	D, restrict	7
R	D, inverse	8
“	R, restrict	9
P	“,∼	10
∘	“, { }, ×, <, >	10
‘	U, “, { }	10
cantor	D, diag, inverse, ∘, ∩	11

Table 7. The result of the case study

	Range of value	Average value	Deviation
Vd	0.5–1	1.0	0.5
Vp	0–3	2.4	2.4
Va	2–14	6.2	7.8
Ve	1–6	2.0	4.0
$Vd * Vp$	0–3	2.4	2.4
$Vd * Va$	1–14	6.2	7.8
$Vd * Ve$	0.5–6	1.9	4.1
$Vp * Va$	0–42	16.3	25.7
$Vp * Ve$	0–18	5.0	13.0
$Va * Ve$	2–70	12.5	57.5
$Vd * Vp * Va$	0–42	16.2	25.8
$Vd * Vp * Ve$	0–18	4.9	13.1
$Vp * Va * Ve$	0–210	34.0	176.0
$Vd * Vp * Va * Ve$	0–210	33.8	176.2

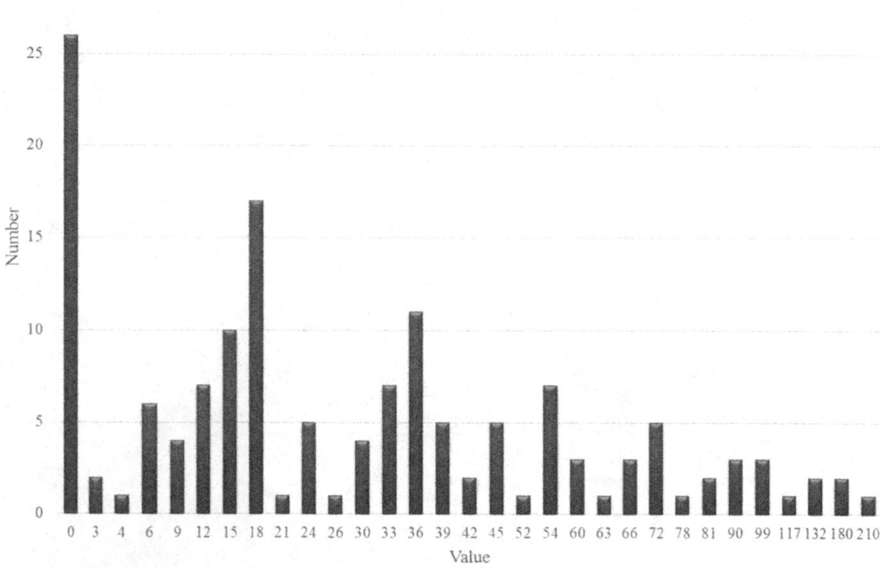

Fig. 1. The number of empirical theorems on each interesting value

6 Concluding Remarks

We have proposed a set of metrics for measuring interestingness of theorems found in ATF. We also presented a case study in NBG set theory, in which we used the proposed metrics to measure the interestingness of empirical theorems of NBG set theory reasoned out by forward reasoning approach. The results of the case study showed that our metrics are hopeful for satisfying the requirement of ATF.

There are many interesting and challenging research problems in our future works. First, we will used the proposed metrics to measure the interestingness of known theorems of NBG set theory appeared in some mathematical books and sort the order from low value to high value of interestingness, then we compare the sorted order with the appearing order of those known theorems in mathematical books. We expect two orders are almost same, because known theorems in mathematical books are always recorded from simple to complex. Second, we will do case studies of ATF in other fields to confirm the generality of the metrics, such as number theory, graph theory, and lattice theory.

References

1. Cheng, J.: A relevant logic approach to automated theorem finding. In: The Workshop on Automated Theorem Proving attached to International Symposium on Fifth Generation Computer Systems, pp. 8–15 (1994)

2. Cheng, J.: Entailment calculus as the logical basis of automated theorem finding in scientific discovery. In: Systematic Methods of Scientific Discovery: Papers from the 1995 Spring Symposium, AAAI Press - American Association for Artificial Intelligence, pp. 105–110 (1995)
3. Cheng, J.: A strong relevant logic model of epistemic processes in scientific discovery. In: Information Modelling and Knowledge Bases XI, Frontiers in Artificial Intelligence and Applications, vol. 61, pp. 136–159. IOS Press (2000)
4. Cheng, J., Nara, S., Goto, Y.: FreeEnCal: a forward reasoning engine with general-purpose. In: Apolloni, B., Howlett, R.J., Jain, L. (eds.) KES 2007, Part II. LNCS (LNAI), vol. 4693, pp. 444–452. Springer, Heidelberg (2007)
5. Colton, S.: Automated theorem discovery: a future direction for theorem provers. In: Proceedings of 1st Automated Reasoning: International Joint Conference, Workshop on Future Directions in Automated Reasoning, pp. 38–47 (2001)
6. Colton, S., Meier, A., Sorge, V., McCasland, R.: Automatic generation of classification theorems for finite algebras. In: Basin, D., Rusinowitch, M. (eds.) IJCAR 2004. LNCS (LNAI), vol. 3097, pp. 400–414. Springer, Heidelberg (2004)
7. Dalzotto, G., Recio, T.: On protocols for the automated discovery of theorems in elementary geometry. J. Autom. Reasoning **43**(2), 203–236 (2009)
8. Gao, H., Goto, Y., Cheng, J.: A systematic methodology for automated theorem finding. Theoret. Comput. Sci. **554**, 2–21 (2014). Elsevier
9. Gao, H., Goto, Y., Cheng, J.: Research on automated theorem finding: current state and future directions. In: Park, J.J., Pan, Y., Kim, C.-S., Yang, Y. (eds.) Future Information Technology. LNEE, vol. 309, pp. 105–110. Springer, Heidelberg (2014)
10. McCasland, R., Bundy, A., Autexier, S.: Automated discovery of inductive theorems. J. Stud. Log. Gramm. Rhetor. **10**(23), 135–149 (2007)
11. Montes, A., Recio, T.: Automatic discovery of geometry theorems using minimal canonical comprehensive gröbner systems. In: Botana, F., Recio, T. (eds.) ADG 2006. LNCS (LNAI), vol. 4869, pp. 113–138. Springer, Heidelberg (2007)
12. Puzis, Y., Gao, Y., Sutcliffe, G.: Automated generation of interesting theorems. In: Proceedings of 19th International Florida Artificial Intelligence Research Society Conference, AAAI press-The Association for the Advancement of Artificial Intelligence, pp. 49–54 (2006)
13. Quaife, A.: Automated Development of Fundamental Mathematical Theories. Kluwer Academic, Dordrecht (1992)
14. Recio, T., Velez, M.Z.: Automatic discovery of theorems in elementary geometry. J. Autom. Reasoning **23**(1), 63–82 (1999)
15. Wos, L.: Automated Reasoning: 33 Basic Research Problem. Prentice-Hall, Upper Saddle River (1988)
16. Wos, L.: The problem of automated theorem finding. J. Autom. Reasoning **10**(1), 137–138 (1993)

Multiview Correlation Feature Learning with Multiple Kernels

Yun-Hao Yuan[1,2(✉)], Xiao-Bo Shen[3,4], Zhi-Yong Xiao[1], Jin-Long Yang[1], Hong-Wei Ge[1], and Quan-Sen Sun[3]

[1] Department of Computer Science and Technology, Jiangnan University, Wuxi, China
yhyuan@jiangnan.edu.cn

[2] Key Laboratory of Advanced Process Control for Light Industry of Ministry of Education, School of IoT Engineering, Jiangnan University, Wuxi, China

[3] School of Computer Science and Engineering, Nanjing University of Science and Technology, Nanjing, China

[4] School of Information Technology and Electrical Engineering, The University of Queensland, Brisbane, Australia

Abstract. Recent researches have shown the necessity to consider multiple kernels rather than a single fixed kernel in real-world applications. The learning performance can be significantly improved if multiple kernel functions or kernel matrices are considered. Motivated by the recent progress, in this paper we present a multiple kernel multiview correlation feature learning method for multiview dimensionality reduction. In our proposed method, the input data of each view are mapped into multiple higher dimensional feature spaces by implicitly nonlinear mappings. Three experiments on face and handwritten digit recognition have demonstrated the effectiveness of the proposed method.

Keywords: Image recognition · Multiple kernels · Canonical correlation analysis · Multiset canonical correlations · Multiview feature learning

1 Introduction

Multiset canonical correlation analysis (MCCA) [1, 2] is a powerful technique for finding the linear correlations among multiple (more than two) random vectors. The multiple random vectors can be associated with three or more different views of the same objects. MCCA seeks one set of linear transformations for multiple multidimensional variables such that the projected variables in the low-dimensional space are maximally correlated. It can subsume a number of representative techniques of multivariate data analysis as special cases, for example, principal component analysis (PCA) [3] and canonical correlation analysis (CCA) [4]. At present, MCCA has been applied to various real-world applications such as blind source separation [5], functional magnetic resonance imaging (fMRI) analysis [6], remote sensing image analysis [7], and target recognition [8].

Recently, the extension of MCCA has attracted increasing attention and some impressive results have been obtained based on different motivations. Takane et al. [9]

X. He et al. (Eds.): IScIDE 2015, Part II, LNCS 9243, pp. 518–528, 2015.
DOI: 10.1007/978-3-319-23862-3_51

proposed a regularized MCCA approach by using a ridge type of regularization technique. In contrast with MCCA, the regularized MCCA can prevent the overfitting and avoid the singularity of within-set covariance matrices. Since MCCA is essentially an unsupervised learning method, it can not effectively reveal discriminant information in multiple canonical subspaces. To solve this issue, Su et al. [10] presented a multiset discriminant canonical correlation method, called multiple principal angle (MPA), where within-class subspaces possess the minimal principal angles and between-class subspaces have the maximal ones. Many experimental results show that MPA is very effective for visual recognition tasks.

On the other hand, the sample covariance matrices in MCCA usually deviate from the true ones due to noise and the limited number of training samples. To reduce the negative effect, Yuan and Sun [11] proposed a fractional-order embedding multiset canonical correlations (FEMCC) method based on fractional-order scatter matrices for multiple feature fusion. The fused features are more discriminative for recognition tasks. From the nonlinear viewpoint, Rupnik and Shawe-Taylor [12] proposed a kernel MCCA (KMCCA) by using implicitly nonlinear mappings for cross-lingual information retrieval tasks. Moreover, some other methods [13–15] have also been proposed.

Although KMCCA can discover the nonlinear correlation information among multiview high-dimensional data of the same objects, it is essentially a single-kernel-learning method, i.e., each view with only a kernel function. Recent researches [16, 17] have shown the necessity to consider multiple kernels rather than a single fixed kernel in practical applications. The learning performance can be significantly improved if multiple kernel functions or kernel matrices are considered. Motivated by recent progress in MCCA and multiple kernel learning, we present in this paper a multiple kernels based multiview correlation feature learning method. In the proposed method, the input data of each view are mapped into multiple higher dimensional feature spaces by implicitly nonlinear mappings. Finally, our proposed method is applied to face and handwritten digit recognition and yield encouraging results.

2 Review on Kernel MCCA

Assume m sets of vectors (views) from the same n objects are given as $\{X^{(i)} \in \Re^{p_i \times n}\}_{i=1}^{m}$, where $X^{(i)} = (x_1^{(i)}, x_2^{(i)}, \cdots, x_n^{(i)})$ describes a data matrix of the ith view containing p_i-dimensional observation vectors in its columns. For each view $X^{(i)}$, assume there is a nonlinear mapping $\phi_i : x^{(i)} \mapsto \phi_i(x^{(i)})$, which implicitly projects the original data into a higher-dimensional *feature space* $\mathcal{F}_i$. Let $\phi_i(X^{(i)}) = (\phi_i(x_1^{(i)}), \phi_i(x_2^{(i)}), \cdots, \phi_i(x_n^{(i)}))$ denote the transformed data in $\mathcal{F}_i$. KMCCA computes one set of projection directions $\{\alpha^{(i)} \in \mathcal{F}_i\}_{i=1}^{m}$ by the following optimization problem [12]:

$$\max_{\alpha} \rho_K(\alpha) = \sum_{i=1}^{m} \sum_{j=i+1}^{m} \alpha^{(i)T} \phi_i(X^{(i)}) \phi_j(X^{(j)})^T \alpha^{(j)}$$
$$s.t. \alpha^{(i)T} \phi_i(X^{(i)}) \phi_i(X^{(i)})^T \alpha^{(i)} = 1, \ i = 1, 2, \cdots, m \tag{1}$$

where $\alpha^T = (\alpha^{(1)T}, \alpha^{(2)T}, \cdots, \alpha^{(m)T})$. Note that we assume that $\phi_i(X^{(i)})$ has been centered in the problem (1), i.e., $\sum_{j=1}^{n} \phi_i(x_j^{(i)}) = 0$, $i = 1, 2, \cdots, m$. The details about the centering process can be found in [18].

Using the equalities $\alpha^{(i)T} \phi_i(X^{(i)}) \phi_j(X^{(j)})^T \alpha^{(j)} = \alpha^{(j)T} \phi_j(X^{(j)}) \phi_i(X^{(i)})^T \alpha^{(i)}$ and $\alpha^{(i)T} \phi_i(X^{(i)}) \phi_i(X^{(i)})^T \alpha^{(i)} = 1$, we can equivalently transform the optimization problem (1) into the following problem (2):

$$\max_{\alpha} \rho_K(\alpha) = \sum_{i=1}^{m} \sum_{j=1}^{m} \alpha^{(i)T} \phi_i(X^{(i)}) \phi_j(X^{(j)})^T \alpha^{(j)}$$
$$s.t. \quad \alpha^{(i)T} \phi_i(X^{(i)}) \phi_i(X^{(i)})^T \alpha^{(i)} = 1, \ i = 1, 2, \cdots, m. \tag{2}$$

Let $\alpha^{(i)} = \phi_i(X^{(i)}) \beta^{(i)}$ with $\beta^{(i)} \in \Re^n$. By using the kernel trick [18] and Lagrange multiplier technique, we can obtain a multivariate eigenvalue problem (MEP) [19]:

$$K\beta = \Lambda K_D \beta \tag{3}$$

where $\beta^T = (\beta^{(1)T}, \beta^{(2)T}, \cdots, \beta^{(m)T})$, $K \in \Re^{mn \times mn}$ is a block matrix with (i, j)th block element as $K_i K_j$, $K_D = diag(K_1^2, K_2^2, \cdots, K_m^2) \in \Re^{mn \times mn}$, $K_i = \phi_i(X^{(i)})^T \phi_i(X^{(i)})$, and $\Lambda = diag(\lambda_1 I_n, \lambda_2 I_n, \cdots, \lambda_m I_n)$ with $\{\lambda_i \in \Re\}_{i=1}^{m}$ as multivariate eigenvalues and $I_n \in \Re^{n \times n}$ as the identity matrix, $i, j = 1, 2, \cdots, m$. Since the MEP has no closed-form solutions, some iterative algorithms [12, 19] have been designed for its solutions.

3 Multiple Kernel Multiview Correlation Feature Learning

In this section, we consider multiple kernel functions for each original view and build a multiple kernel multiview correlation feature learning method for multiview dimensionality reduction.

3.1 Formulation with Multiple Kernels

Concretely, for any view $X^{(i)} = (x_1^{(i)}, x_2^{(i)}, \cdots, x_n^{(i)}) \in \Re^{p_i \times n}$ with $i \in \{1, 2, \cdots, m\}$, assume there are $n_i \geq 2$ nonlinear mappings: $\{\phi_j^{(i)} : x^{(i)} \mapsto \phi_j^{(i)}(x^{(i)})\}_{j=1}^{n_i}$, which implicitly map the original view $X^{(i)}$ into n_i different higher-dimensional feature spaces, respectively. Let us denote

$$\phi_i^f(X^{(i)}) = (\phi_i^f(x_1^{(i)}), \phi_i^f(x_2^{(i)}), \cdots, \phi_i^f(x_n^{(i)})) \tag{4}$$

with

$$\phi_i^f(x_k^{(i)}) = f_i\left(\phi_1^{(i)}(x_k^{(i)}), \phi_2^{(i)}(x_k^{(i)}), \cdots, \phi_{n_i}^{(i)}(x_k^{(i)})\right) \tag{5}$$

where $f_i(\cdot)$ is an ensemble mapping function, $i = 1, 2, \cdots, m$ and $k = 1, 2, \cdots, n$. Let $\alpha^{(i)}$ be the projection axis of $\phi_i^f(X^{(i)})$ in the feature space. Then, the proposed method can be formulated as

$$\begin{aligned} \max_{\alpha} J(\alpha) &= \sum_{i=1}^{m}\sum_{j=1}^{m} \alpha^{(i)T}\phi_i^f(X^{(i)})\phi_j^f(X^{(j)})^T\alpha^{(j)} \\ s.t.\quad & \alpha^{(i)T}\phi_i^f(X^{(i)})\phi_i^f(X^{(i)})^T\alpha^{(i)} = 1\,,\ i = 1, 2, \cdots, m. \end{aligned} \tag{6}$$

where $\alpha^T = (\alpha^{(1)T}, \alpha^{(2)T}, \cdots, \alpha^{(m)T})$. Note that, (6) shows that different ensemble mapping functions, i.e., $\{f_i\}_{i=1}^m$, will lead to different multiple kernel multiview correlation methods. In this paper, we define the m ensemble mapping functions as

$$f_i\left(\phi_1^{(i)}(x^{(i)}), \phi_2^{(i)}(x^{(i)}), \cdots, \phi_{n_i}^{(i)}(x^{(i)})\right) = \left(\phi_1^{(i)}(x^{(i)})^T, \phi_2^{(i)}(x^{(i)})^T, \cdots, \phi_{n_i}^{(i)}(x^{(i)})^T\right)^T \tag{7}$$

where $i = 1, 2, \cdots, m$.

Let $\alpha^{(i)} = \sum_{k=1}^{n}\phi_i^f(x_k^{(i)})\beta_k^{(i)} = \phi_i^f(X^{(i)})\beta^{(i)}$ with $\beta^{(i)T} = (\beta_1^{(i)}, \beta_2^{(i)}, \cdots \beta_n^{(i)}) \in \Re^n$. The problem (6) can be then reformulated as

$$\begin{aligned} \max_{\beta} J(\beta) &= \sum_{i=1}^{m}\sum_{j=1}^{m} \beta^{(i)T}\left[\phi_i^f(X^{(i)})^T\phi_i^f(X^{(i)})\right]\left[\phi_j^f(X^{(j)})^T\phi_j^f(X^{(j)})\right]\beta^{(j)} \\ s.t.\quad & \beta^{(i)T}\left[\phi_i^f(X^{(i)})^T\phi_i^f(X^{(i)})\right]\left[\phi_i^f(X^{(i)})^T\phi_i^f(X^{(i)})\right]\beta^{(i)} = 1,\ i = 1, 2, \cdots, m \end{aligned} \tag{8}$$

where $\beta^T = (\beta^{(1)T}, \beta^{(2)T}, \cdots, \beta^{(m)T}) \in \Re^{mn}$. According to (7), the problem (8) can be further converted as

$$\begin{aligned} \max_{\beta} J(\beta) &= \sum_{i=1}^{m}\sum_{j=1}^{m} \beta^{(i)T}\left[\sum_{k=1}^{n_i}\phi_k^{(i)}(X^{(i)})^T\phi_k^{(i)}(X^{(i)})\right]\left[\sum_{t=1}^{n_j}\phi_t^{(j)}(X^{(j)})^T\phi_t^{(j)}(X^{(j)})\right]\beta^{(j)} \\ s.t.\quad & \beta^{(i)T}\left[\sum_{k=1}^{n_i}\phi_k^{(i)}(X^{(i)})^T\phi_k^{(i)}(X^{(i)})\right]\left[\sum_{t=1}^{n_i}\phi_t^{(i)}(X^{(i)})^T\phi_t^{(i)}(X^{(i)})\right]\beta^{(i)} = 1,\ i = 1, 2, \cdots, m. \end{aligned} \tag{9}$$

Solving the problem (9), we can implicitly obtain all projection directions of the proposed method.

3.2 Solution

To solve the problem (9), we define $K_k^{(i)} = \phi_k^{(i)}(X^{(i)})^T \phi_k^{(i)}(X^{(i)}) \in \Re^{n\times n}$ using the kernel trick [18], where $K_k^{(i)}$ denotes the kernel matrix corresponding to the kth nonlinear mapping in the ith view, and $k = 1, 2, \cdots, n_i$. Now, the problem (9) can be formulated equivalently as

$$\begin{aligned} &\max_{\beta} J(\beta) = \sum_{i=1}^{m}\sum_{j=1}^{m} \beta^{(i)T}\left(\sum_{k=1}^{n_i} K_k^{(i)} \sum_{t=1}^{n_j} K_t^{(j)}\right)\beta^{(j)} \\ &s.t. \quad \beta^{(i)T}\left(\sum_{k=1}^{n_i} K_k^{(i)} \sum_{t=1}^{n_i} K_t^{(i)}\right)\beta^{(i)} = 1,\ i = 1, 2, \cdots, m \end{aligned} \tag{10}$$

Let us set

$$K^{(ij)} = \sum_{k=1}^{n_i} K_k^{(i)} \sum_{t=1}^{n_j} K_t^{(j)} = \sum_{k=1}^{n_i}\sum_{t=1}^{n_j} K_k^{(i)} K_t^{(j)}, \tag{11}$$

which is called *cross-composite kernel matrix*. With (11), the problem (10) can be rewritten concisely as

$$\begin{aligned} &\max_{\beta} J(\beta) = \sum_{i=1}^{m}\sum_{j=1}^{m} \beta^{(i)T} K^{(ij)} \beta^{(j)} \\ &s.t. \quad \beta^{(i)T} K^{(ii)} \beta^{(i)} = 1,\ i = 1, 2, \cdots, m. \end{aligned} \tag{12}$$

Similar to KMCCA, we can obtain a multivariate eigenvalue problem of the problem (12) using the Lagrange multiplier technique for solution. Existing study [19] has shown that MEP is very difficult and has no analytical solutions (i.e., exact solutions) in $m > 2$ case. Thus, in this paper, we provide an alternative way to solve the optimization problem (12). That is, we couple the constraints of the problem (12) to get a relaxed version with a single constraint as

$$\begin{aligned} &\max_{\beta} J(\beta) = \sum_{i=1}^{m}\sum_{j=1}^{m} \beta^{(i)T} K^{(ij)} \beta^{(j)} \\ &s.t. \quad \sum_{i=1}^{m} \beta^{(i)T} K^{(ii)} \beta^{(i)} = 1\,. \end{aligned} \tag{13}$$

Using the Lagrange multiplier technique, we can obtain the following generalized eigenvalue problem:

$$S_K \beta = \lambda S_{KD} \beta, \tag{14}$$

where $S_{KD} = diag(K^{(11)}, K^{(22)}, \cdots, K^{(mm)})$, $S_K \in \Re^{mn\times mn}$ with (i, j)th block element as $K^{(ij)}$, and $i, j = 1, 2, \cdots, m$.

It shows that the objective function in (13) can be maximized by calculating the eigenvectors of the generalized eigen-equation in (14). Thus, we choose a set of eigenvectors $\{\beta_k^T = (\beta_k^{(1)T}, \beta_k^{(2)T}, \cdots, \beta_k^{(m)T})\}_{k=1}^d$ corresponding to the first d largest eigenvalues as the dual solution vectors of the proposed method. Once dual solution vectors are obtained, the nonlinear feature extraction of multiview high-dimensional data can be carried out for subsequent classification tasks by

$$\begin{aligned} Y^{(i)} = P^{(i)T}\phi_i^f(X^{(i)}) &= \tilde{P}^{(i)T}\phi_i^f(X^{(i)})^T\phi_i^f(X^{(i)}) \\ &= \tilde{P}^{(i)T}\sum\nolimits_{j=1}^{n_i}\phi_j^{(i)}(X^{(i)})^T\phi_j^{(i)}(X^{(i)}) \\ &= \tilde{P}^{(i)T}\sum\nolimits_{j=1}^{n_i}K_j^{(i)}, \end{aligned} \tag{15}$$

where $P^{(i)} = (\alpha_1^{(i)}, \alpha_2^{(i)}, \cdots, \alpha_d^{(i)})$ and $\tilde{P}^{(i)} = (\beta_1^{(i)}, \beta_2^{(i)}, \cdots, \beta_d^{(i)})$, $i = 1, 2, \cdots, m$.

4 Experimental Results

To examine the performance of our proposed method, three experiments have been performed in face and handwritten digit image recognition. In addition, we compare it with kernel PCA (KPCA) and KMCCA for revealing the effectiveness. In all the experiments, the nearest neighbor (NN) classifier is used for recognition tasks.

4.1 Candidate Kernels

In our experiments, we adopt three views in total from the same objects and we use three kinds of kernel functions for the ith view in the proposed algorithm, i.e., linear kernel $k(x_j^{(i)}, x_t^{(i)}) = x_j^{(i)T}x_t^{(i)}$, RBF kernel $k(x_j^{(i)}, x_t^{(i)}) = \exp(-\left\|x_j^{(i)} - x_t^{(i)}\right\|_2^2 / 2\sigma_i^2)$, and polynomial kernel $k(x_j^{(i)}, x_t^{(i)}) = (x_j^{(i)T}x_t^{(i)} + 1)^{d_i}$, where σ_i in RBF kernel is set as the average value of all the l_2-norm distances $\left\|x_j^{(i)} - x_t^{(i)}\right\|_2$ as used in [20], and d_i in polynomial kernel is set as $i + 1, i = 1, 2, 3$ and $j, t = 1, 2, \cdots, n$. In KMCCA, we use the above three kinds of kernels with the same parameters, i.e., linear kernel for the first view, RBF kernel for the second, and polynomial kernel for the last. Moreover, for a fair comparison with our proposed method and KMCCA, we perform KPCA by first stacking three views together into a new single view and then using one of the above-described kernels.

4.2 Experiment Using the AT&T Database

The AT&T database[1] contains 400 face images from 40 persons. There are 10 grayscale images per person with a resolution of 92 × 112. In some persons, the images are taken at

[1] http://www.cl.cam.ac.uk/research/dtg/attarchive/facedatabase.html.

different times. The lighting, facial expressions and facial details are also varied. The images are taken with a tolerance for some tilting and rotation of the face up to 20°, and have some variation in the scale up to about 10 %. Ten images of one person are shown in Fig. 1.

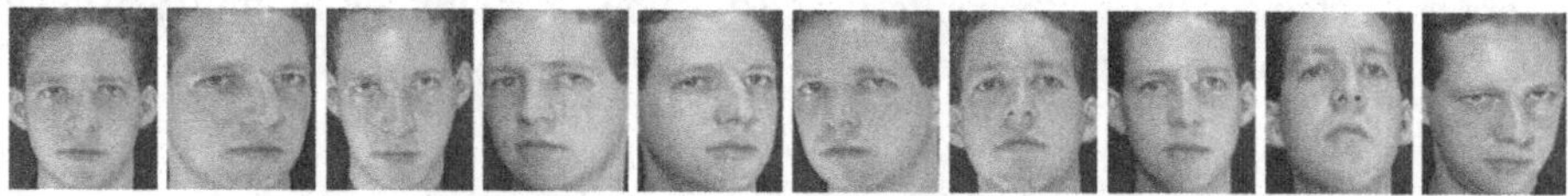

Fig. 1. Ten face images of one person in the AT&T database.

In this experiment, we employ the same preprocessing method as used in [11, 21] to obtain three-view data. That is, we first perform Coiflets, Daubechies, and Symlets orthonormal wavelet transforms to get three-set low-frequency sub-images (i.e., three views) from original images, respectively. Then, the K-L transform is employed to reduce the dimensionality of each view to 150. The final formed three views, each with 150 dimensions, are used in our experiment.

On this database, N images ($N = 4$, 5, 6, and 7) per person are randomly chosen for training, while the remaining $10 - N$ images are used for testing. For each N, we perform 10 independent recognition tests to evaluate the performances of KPCA, KMCCA, and our proposed method. Table 1 shows the average recognition rates of each method under NN classifier and their corresponding standard deviations.

From Table 1, we can see that our proposed method outperforms KMCCA and the baseline algorithm KPCA, no matter how many training samples per person are used. Particularly when the number of training samples is less, our method improves more compared with other methods. On the whole, KMCCA achieves better recognition results than KPCA. Moreover, KPCA with RBF kernel performs better than with linear and polynomial kernels.

Table 1. Ten-run average recognition rates (%) of KPCA, KMCCA, and our proposed method on the AT&T database and their corresponding standard deviations.

Method	4 Train.	5 Train.	6 Train.	7 Train.
KPCA_Lin	85.54±3.37	88.95±3.07	92.94±1.98	94.58±1.63
KPCA_RBF	87.96±2.72	90.20±3.08	93.38±1.70	95.25±2.04
KPCA_Pol2	85.58±3.21	88.75±2.77	92.88±1.94	94.58±1.81
KPCA_Pol3	85.63±3.29	88.55±3.02	92.56±1.94	94.42±1.84
KPCA_Pol4	85.17±3.47	87.95±3.00	92.31±1.77	94.25±1.49
KMCCA_Pol2	87.17±3.43	90.15±2.94	94.69±1.05	95.42±2.19
KMCCA_Pol3	87.33±3.79	89.95±2.96	92.69±2.97	96.17±2.52
KMCCA_Pol4	87.25±3.79	90.35±1.84	93.44±3.03	94.25±2.30
Ours	**91.54±2.45**	**93.60±1.66**	**96.31±1.23**	**96.67±1.71**

Note: In KPCA, Lin denotes linear kernel, RBF denotes RBF kernel, PolA denotes polynomial kernel with order A. KMCCA_PolA denotes that one of three views uses polynomial kernel with order A in KMCCA. All such abbreviations have the same meanings in the following tables unless otherwise stated.

4.3 Experiment Using the Yale Database

The Yale database contains 165 grayscale images of 15 persons. Each person has 11 images with different facial expressions and lighting conditions, i.e., center-light, with glasses, happy, left-light, without glasses, normal, right-light, sad, sleepy, surprised, and wink. Each image is cropped and resized to 100 × 80 pixels. Figure 2 shows eleven images of one person.

Fig. 2. Eleven images of one person in the Yale face database.

In this experiment, the Coiflets, Daubechies and Symlets wavelet transforms are again performed on original images to form three-view data. Also, their dimensions are, respectively, reduced to 75, 75, and 75 by K-L transform. For each person, five images are randomly selected for training, and the remaining six images for testing. Thus, the total number of training samples and testing samples is, respectively, 75 and 90. Ten-run tests are performed to examine the recognition performances of KPCA, KMCCA, and our proposed method. Table 2 summarizes the average recognition rates of each method under NN classifier and their corresponding standard deviations.

As can be seen, our proposed method is superior to KPCA and KMCCA. Its recognition rate exceeds KPCA 8.34 % and KMCCA 1.67 %. KPCA performs the worst and the RBF kernel is still more effective than other kernels in KPCA. These conclusions are overall consistent with those drawn from the experiment in Sect. 4.2.

Table 2. Ten-run average recognition rates (%) of KPCA, KMCCA, and our proposed method on the Yale database and their corresponding standard deviations (Std).

Method	Accuracy	Std
KPCA_Lin	81.00	2.25
KPCA_RBF	83.22	1.85
KPCA_Pol2	81.22	2.25
KPCA_Pol3	81.11	2.40
KPCA_Pol4	80.11	2.25
KMCCA_Pol2	88.89	3.35
KMCCA_Pol3	88.56	3.59
KMCCA_Pol4	89.89	2.83
Ours	**91.56**	**2.35**

4.4 Experiment Using Multiple Feature Dataset

The multiple feature dataset (MFD) from UCI contains 10 classes of handwritten numbers from 0 to 9. It includes six feature sets: 76-dimensional Fourier coefficients (Fou), 64-dimensional Karhunen-Loève coefficients (Kar), 47-dimensional Zernike moments (Zer), 216-dimensional profile correlations (Fac), 240-dimensional pixel averages (Pix), and 6-dimensional morphological features (Mor). Each class includes 200 samples and the total sample size is 2000. For computational efficiency, we adopt three feature sets Fou, Kar, and Zer and take the first 50 samples from each class in every feature set as our dataset, denoted as subMFD. Obviously, the total sample size is 500 in subMFD and each sample is represented by three views with 187 dimensions in total.

In this experiment, N samples (N = 10, 15, 20, and 25) per class are randomly chosen for training, while the remaining 50-N samples are used for testing. For given N, Ten independent tests are carried out for the performance evaluation of KPCA, KMCCA, and our proposed method. Table 3 lists the average recognition results of each method under the NN classifier and their corresponding standard deviations.

As we can see, Table 3 demonstrates again that our method outperforms KPCA and KMCCA on all cases. On this dataset, KPCA performs the second best and we can also see an inconsistent point in contrast with the previous results in Sects. 5.2 and 5.3, that is, KPCA performs better than KMCCA. Moreover, KPCA with RBF kernel achieves better performance than with other kernels. Together with the conclusions drawn from the foregoing two experiments, we can conclude that RBF kernel should be the first choice of KPCA for feature extraction and classification.

Table 3. Ten-run average recognition rates (%) of KPCA, KMCCA, and our proposed method on subMFD and their corresponding standard deviations.

Method	10 Train.	15 Train.	20 Train.	25 Train.
KPCA_Lin	90.40±1.66	93.80±0.92	93.83±0.67	93.80±1.12
KPCA_RBF	91.98±1.09	94.17±0.76	94.43±0.86	94.88±0.96
KPCA_Pol2	90.68±1.52	93.94±0.60	93.87±0.63	94.04±0.99
KPCA_Pol3	91.00±1.39	93.83±0.49	94.07±0.90	94.44±1.01
KPCA_Pol4	91.40±1.46	93.83±0.85	94.07±0.80	94.48±1.14
KMCCA_Pol2	83.80±3.11	91.89±4.22	93.63±2.84	94.76±4.05
KMCCA_Pol3	88.75±2.55	91.31±3.58	94.30±2.49	91.64±3.45
KMCCA_Pol4	89.05±1.50	89.37±1.30	92.30±2.61	85.28±2.88
Ours	**92.25±1.05**	**94.86±1.00**	**95.37±1.01**	**96.16±1.24**

5 Conclusions

In this paper, we have proposed a multiple kernel multiview correlation feature learning method for multiview dimensionality reduction and recognition tasks. The central idea of the proposed method is to map any view in multiple views to multiple higher dimensional feature spaces by multiple nonlinear mappings. This consideration makes our proposed method can discover multiple kinds of useful information of each original view in the feature spaces. Three experiments on face and handwritten digit recognition demonstrate the effectiveness of our method.

Acknowledgments. This work is supported by the National Science Foundation of China under Grant Nos. 61402203, 61273251, and 61305017, and the Fundamental Research Funds for the Central Universities under Grant No. JUSRP11458.

References

1. Horst, P.: Relations among m sets of measures. Psychometrika **26**, 129–149 (1961)
2. Kettenring, J.R.: Canonical analysis of several sets of variables. Biometrika **58**, 433–451 (1971)
3. Jolliffe, I.T.: Principal Component Analysis, 2nd edn. Springer, New York (2002)
4. Hotelling, H.: Relations between two sets of variates. Biometrika **28**, 321–377 (1936)
5. Li, Y.O., Adali, T., Wang, W., Calhoun, V.D.: Joint blind source separation by multiset canonical correlation analysis. IEEE Trans. Sig. Process. **57**, 3918–3929 (2009)
6. Correa, N.M., Eichele, T., Adali, T., Li, Y.O., Calhoun, V.D.: Multi-set canonical correlation analysis for the fusion of concurrent single trial ERP and functional MRI. NeuroImage **50**, 1438–1445 (2010)
7. Nielsen, A.A.: Multiset canonical correlations analysis and multispectral, truly multitemporal remote sensing data. IEEE Trans. Image Process. **11**, 293–305 (2002)
8. Thompson, B., Cartmill, J., Azimi-Sadjadi, M.R., Schock, S.G.: A multichannel canonical correlation analysis feature extraction with application to buried underwater target classification. In: Proceedings of International Joint Conference on Neural Networks, pp. 4413–4420 (2006)
9. Takane, Y., Hwang, H.: Regularized multiple-set canonical correlation analysis. Psychometrika **73**, 753–775 (2008)
10. Su, Y., Fu, Y., Gao, X., Tian, Q.: Discriminant learning through multiple principal angles for visual recognition. IEEE Trans. Image Process. **21**, 1381–1390 (2012)
11. Yuan, Y.-H., Sun, Q.-S.: Fractional-order embedding multiset canonical correlations with applications to multi-feature fusion and recognition. Neurocomputing **122**, 229–238 (2013)
12. Rupnik, J., Shawe-Taylor, J.: Multi-view canonical correlation analysis. In: SiKDD (2010). http://ailab.ijs.si/dunja/SiKDD2010/Papers/Rupnik_Final.pdf
13. Yuan, Y.-H., Sun, Q.-S., Zhou, Q., Xia, D.-S.: A novel multiset integrated canonical correlation analysis framework and its application in feature fusion. Pattern Recogn. **44**, 1031–1040 (2011)
14. Jing, X., Li, S., Lan, C., Zhang, D., Yang, J., Liu, Q.: Color image canonical correlation analysis for face feature extraction and recognition. Sig. Process. **91**, 2132–2140 (2011)
15. Shen, X.B., Sun, Q.S., Yuan, Y.H.: A unified multiset canonical correlation analysis framework based on graph embedding for multiple feature extraction. Neurocomputing **148**, 397–408 (2015)
16. Yan, F., Kittler, J., Mikolajczyk, K., Tahir, A.: Non-sparse multiple kernel fisher discriminant analysis. J. Mach. Learn. Res. **13**, 607–642 (2012)
17. Lin, Y.-Y., Liu, T.-L., Fuh, C.-S.: Multiple kernel learning for dimensionality reduction. IEEE Trans. Pattern Anal. Mach. Intell. **33**, 1147–1160 (2011)

18. Schölkopf, B., Smola, A., Müller, K.-R.: Nonlinear component analysis as a kernel eigenvalue problem. Neural Comput. **10**, 1299–1319 (1998)
19. Chu, M.T., Watterson, J.L.: On a multivariate eigenvalue problem: i. algebraic theory and power method. SIAM J. Sci. Comput. **14**, 1089–1106 (1993)
20. Wang, Z., Chen, S., Sun, T.: MultiK-MHKS: a novel multiple kernel learning algorithm. IEEE Trans. Pattern Anal. Mach. Intell. **30**, 348–353 (2008)
21. Yuan, Y.-H., Sun, Q.-S.: Graph regularized multiset canonical correlations with applications to joint feature extraction. Pattern Recogn. **47**, 3907–3919 (2014)

CRF-TM: A Conditional Random Field Method for Predicting Transmembrane Topology

Weizhong Lu[1], Baochuan Fu[1], Hongjie Wu[1,3(✉)], Qiang Lü[2,3], Kun Wang[1], and Min Jiang[4]

[1] School of Electronic and Information Engineering, Suzhou University of Science and Technology, Suzhou 215009, China
hongjiewu@mail.usts.edu.cn

[2] School of Computer Science and Technology, Soochow University, Suzhou 215006, China

[3] Jiangsu Provincial Key Lab for Information Processing Technologies, Suzhou 215006, China

[4] The First Affiliated Hospital of Soochow University, Suzhou 215006, China

Abstract. Transmembrane proteins are important for cell transport biology and in the treatment of disease. Understanding the helix count and locations in transmembrane proteins is a key problem for structural and functional analyses. But there is a lack of high resolution three-dimensional structures. In this study, we propose a method based on conditional random fields for predicting the helix count and locations, CRF-TM, which reflects long-range correlations in the full-length sequence as joint probabilities. Two datasets are employed in the performance validation. Our results show that CRF-TM can rank the first group better compared with other widely used TM predictors. The results obtained by CRF-TM are also used to predict the three-dimensional structures of GPCRs, which is crucial drug targets and also a subclass of transmembrane with seven spanning α-helices.

Keywords: Conditional random fields · Helix · Transmembrane

1 Introduction

Transmembrane proteins (TMPs) play important roles in cell transport biology and in the treatment of disease. They are responsible for some of the most important functions of cells such as signaling, transport, and catalyzing reactions [1]. However, the available high resolution membrane protein structures cannot satisfy current demands in terms of understanding the functions of TMPs, disease research, and the rapidly growing pharmaceutical industry. Indeed, TMPs only comprise about 1.5 % of the total

This paper is supported by grants no. 61170125, 61202290 under the National Natural Science Foundation of China (http://www.nsfc.gov.cn) and grants no. BK20131154 under Natural Science Foundation of Jiangsu Province.

X. He et al. (Eds.): IScIDE 2015, Part II, LNCS 9243, pp. 529–537, 2015.
DOI: 10.1007/978-3-319-23862-3_52

determined structures in the Protein Data Bank (PDB), although 20 %–30 % of all genes in genomes encode TMPs [2]. This is due to various difficult problems, such as insufficient expression, purification of the protein in a stable form, and the challenges of data collection from microcrystals [3]. Although there is a lack of high resolution three-dimensional structures of TMPs, it is still essential to understanding their topology, particularly the total number of TM helices, their boundaries, and their interior/exterior orientations relative to the membrane, to facilitate structural and functional analyses, which may direct further experimental work. Furthermore, topology can provide useful information when modeling three-dimensional structures of TMPs.

Many machine learning or stochastic methods have been developed for predicting the topology of TMPs. Yu et al. [4] proposed an approach based on parallel fusion of multi-view features to predict membrane protein subcellular localization. In particular, hidden Markov models (HMMs) were the earliest models employed in topology prediction. TMHMM [5] can be used to build an architecture that corresponds closely to the biological system. The TMP model is cyclic with seven types of states: helix core, helix caps on either side, a loop on the cytoplasmic side, two loops on the non-cytoplasmic side, and a globular domain state in the middle of each loop. HMM-TOP [6] is based on the hypothesis that the localizations of the transmembrane segments and their topology are determined by differences in the amino acid distributions in various structural parts of TMPs, rather than by the specific amino acid compositions of these parts. Thus, a HMM with a special architecture was developed to search for the transmembrane topology based on the maximum likelihood among all possible topologies of a given protein. Later, neural networks (NNs) and support vector machines (SVMs) were employed for predicting the topology of TMPs. PHDhtm [7] used evolutionary information as the input for an NN system, which significantly improved the results obtained by previous NN prediction methods based on single sequence information. Nugent and Jones [8] proposed an SVM-based TMP topology predictor that integrates both signal peptide and re-entrant helix prediction, which was benchmarked by a full cross-validation using a novel dataset of 131 sequences with known crystal structures. In addition, hybrid methods are also used frequently. For example, OCTOPUS [9] combined HMMs and artificial NNs to predict the correct topology for 94 % of TMP sequences.

One problem that must be addressed by these topology predictors is how to represent long-range dependencies in sequences because the inference problem is intractable for such models. It is well known that HMMs and their extended versions are generative models based on a joint probability mapping of paired observations and labeled sequences. However, the sequence diversity and remote homology in TMPs is sufficiently high such that HMM-based methods have problems modeling the correlations between long-range distance residues [10]. By contrast, conditional random fields (CRFs) [11] can specify the probabilities of possible labeled sequences given an observed sequence and they can represent attributes at different levels without assuming the independence of the attributes. CRF methods have been used successfully to resolve problems in the field of computational linguistics. Recently, CRF has also been applied in bioinformatics. For example, Li et al. [12] applied CRFs to the prediction of protein–protein interaction sites, and Wang and Sauer [13] proposed

OnD-CRF for predicting order and disorder in proteins. However, the use of CRFs for predicting transmembrane topologies has not been explored previously.

We consider that in addition to obtaining high accuracy predictions, it is very important that the model corresponds well to the biological reality. Thus, a CRF-based model can be used to represent long-range correlations in a full-length sequence. In this study, we propose a CRF-based model called CRF-TM for predicting transmembrane helix count, helix location.

2 Materials and Methods

2.1 Data Sets

A wide-used data set and a newest data set are employed to validate the performance of CRF-TM. The wide-used data set, we named as TMPDB106 selected from TMPDB, was created by Ikeda et al. [14] at 2003, then it was used in ConPredII [15], Phobius [16] and MEMSAT-SVM [17]. TMPDB is a database of experimentally-characterized transmembrane topology, which were determined experimentally by means of X-ray crystallography, NMR, gene fusion technique, substituted cysteine accessibility method, N-linked glycosylation experiment and other biochemical methods. TMPDB106 includes 106 sequences after removing the undetermined and redundant sequences. Another data set is PDBTM [18] which we named as PDBTM 472. PDBTM was created at 2004 and has been continuously updated every week by the TMDET algorithm that is able to distinguish between transmembrane and non-transmembrane proteins using their 3D atomic coordinates. The data set, we downloaded, is version 2014-11-28 containing 2340 transmembrane chains. Amone them 2033 chains are alpha-helix transmembrane and 299 chains are beta transmembane. And we removed three undetermined sequences (2AXT_C,1C51_B,3J5P_B) from 475 non-redundant alpha-helix chains. The summary of two data sets was listed in Table 1. The chain averagely contains 221.2 and 295.7 residues in TMPDB and PDBTM data sets respectively.

Table 1. Summary of two data sets

Data set	Chains	TMs	Res.	Avg length
TMPDB106	106	389	23475	221.5
PDBTM472	472	2473	139561	295.7

2.2 Conditional Random Fields

The task of predicting membrane topology is a classical a sequence labeling problem. The continued residues can be seen as a sequence and the aim is to determine whether the residue is embedded into the membrane or not. We expected to label the residues using CRFs to reflect the long-distance correlation. Conditional random fields (CRFs) were proposed by Lafferty et al. for labeling sequence data. Given a sequence of observations X = (x_1, x_2, …, x_n), we want to get the most probable label sequence Y = (y_1, y_2, …, y_n), i.e. $Y^* = argmax_Y P(Y|X)$. CRFs are undirected graphical models and the conditional probability P(Y|X) is computed directly. Both CRFs and HMMs suit to label sequence, differing from the probability solution formulation. HMMs obtain the target label sequence Y by maximizing the joint probability of X and Y, but HMM cannot use long distance features, which limits the broad application of this method. CRFs are exponential or log-linear models that can use any kind of features. By the fundamental theorem of random field, the joint distribution over label sequence Y given X can be given by the following conditional probability:

$$\mathrm{P(Y|X)} = \frac{1}{Z(X)}\exp\left(\sum_i \sum_j \lambda_j t_j(y_{i-1}, y_i, x, i) + \sum_i \sum_j \mu_j s_j(y_i, x, i)\right) \quad (1)$$

where, $t_j(y_{i-1}, y_i, x, i)$ is a transition feature function of the entire observation sequence and the labels at position i and i − 1 in the label sequence; $s_j(y_i, x, i)$ is a state feature function of the label at position i and the observation sequence. Z(X) is a normalization factor. More details about CRFs can be referred from [11].

2.3 Predicting TM Topology Based on CRF

To represent the predicted TMP topology as a sequence labeling problem, we label the set of residues with three labels as L = {I, O, M} or with two labels as L = {N, M}, where "I" indicates membrane inside, "O" indicates membrane outside, "M" indicates membrane helix, and "N" indicates non-membrane. The features employed in CRF can be divided to residue-related and sequence-related. The residue-related features are comprised by physical features, including hydrophobicity, hydrophilicity, flexibility, free energy transfer, and polarity. The sequence-related features are comprised by the sequence evolution profile which is calculated by PSI-BLAST based on the NCBI nonredundant database. CRF can model long-range correlations in a sequence, but nearby residues still affect the attributes of the focal residue. Therefore, we employ a window to extract physical and profile features from nearby residues. Different window sizes (20, 25, 30, and 35) are tested to explore their effects on the accuracy of transmembrane topology prediction. The dimension of a feature is: (five physical features + 20 profile features) * W, where W is the window size. Next, we apply CRF++, which is a widely used CRF toolkit, to train the model where parameter cis = 3 and we reference the label. Figure 1 illustrates the flow of the process employed for predicting TMP topology using CRF.

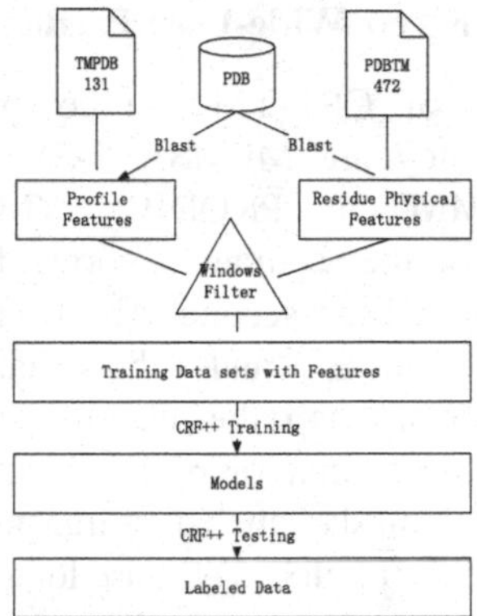

Fig. 1. The flow chart of CRF-TM

3 Results and Discussion

The performance upon each data set is measured using 10-fold cross-validation.

3.1 The Size of Window Affects the Results Less

Different window sizes (20, 25, 30, 35) were tried on both data sets and the results are listed in Table 2. Correct helix count and locations are the two important aspects of transmembrane topology. Correct helix count represents the fraction of the number of predicted correct proteins and total proteins. Correct helix location represents the fraction of the number of predicted correct helixes and total helixes. The deviation of the accuracies of correct helix locations upon different window sizes is very low in both data sets. It shows that CRF-TM can work robustly for different data sets or different window sizes.

Table 2. Accuracy of different window sizes

Data set	Size	Correct helix count	Correct helix location	Avg of correct helix locations	Dev of correct helix locations	Pearson correlation
TMPDB 106	20	81/106 = 76 %	370/389 = 95 %	95 %	0.0044	0.51
	25	83/106 = 78 %	366/389 = 94 %			0.51
	30	83/106 = 78 %	368/389 = 94 %			0.54
	35	80/106 = 75 %	367/389 = 94 %			0.53
PDBTM 472	20	275/472 = 58 %	2274/2473 = 92 %	91 %	0.0070	0.42
	25	279/472 = 59 %	2271/2473 = 92 %			0.44
	30	281/472 = 59 %	2264/2473 = 91 %			0.47
	35	253/472 = 53 %	2236/2473 = 90 %			0.45

3.2 Comparison with Other Ten Wide-Used Predictors on TMPDB106

To validate the performance of CRF-TM, we compared the accurcy with ten well-known TM predictors. They are MEMSATS-SVM, OCTOPUS, MEMSAT3, ENSEMBLE, PHOBIUS, HMMTOP, PRODIV, SVMTOP, TMHMM, PHDhtm. CRF-TM ranked the top 95 % on the accuracy of correct helix location, improved 4 % to MEMSATS-SVM which ranked the second. And CRF-TM ranks the sixth on the accuracy of correct helix count. Tables 1 and 2 show that CRF-TM always get higher rank on the accuracy of helix locations rather than the accuracy of on helix count. It could be caused by CRF-TM over estimation of the transmembrane helix, so that some nontransmembrane residues are missly labeled as transmembrane. Another reason is residue mis-prediction on the mid of helix may cause long helix is considered as two or more short helixes. While on this situation the helix locations can still be correctly predicted, because even if errors occur on 1–2 residues, most of helix residues are still covered by prediction.

Table 3. Comparison with other ten wide-used predictors on TMPDB106

Method	Algotithm	Correct helix count	Correct helix locations
CRF-TM	CRF	77 %	95 %
MEMSATS-SVM	SVM	95 %	91 %
OCTOPUS	NN + HMM	86 %	83 %
MEMSAT3	NN	84 %	76 %
ENSEMBLE	NN + HMM	77 %	76 %
PHOBIUS	HMM	75 %	76 %
HMMTOP	HMM	77 %	76 %
PRODIV	HMM	79 %	64 %
SVMTOP	SVM	66 %	64 %
TMHMM	HMM	75 %	68 %
PHDhtm	NN	75 %	54 %

3.3 The Performance on PDBTM472

Because there is not membrane-inside and membrane-outside information in the raw data set of PDBTM and the nonmembrane residue cannot be distinguished from membrane-inside or membrane-outside, we transferred the raw data set to a two label (Nonmembrane, Mmembrane) training set and test set. After we explored the average accuracy 0.91 % of helix location shown in Table 1, other statistical analysis, TP (True Positive), TN (True Negative), FP (False Positive), FN(False Negative), ACC(Accuracy), MCC (Matthews correlation coefficient), Sensitivity, Specificity, were listed in Table 4.

Table 4. Statistical analysis on residues on different window sizes

Window size	TP	TN	FN	FP	MCC	ACC	Sensitivity	Specificity
20	37751	81336	10855	9619	0.675	0.853	0.777	0.894
25	37605	81351	11001	9604	0.672	0.852	0.774	0.894
30	37429	81161	11177	9794	0.667	0.850	0.770	0.892
35	37241	80853	11365	10102	0.659	0.846	0.766	0.889

3.4 Cross Validation on Different Data Sets

In each of the previous experiments, training set and testing set are extracted from the same data set. For exploring the real ability of CRF-TM, cross validate experiments were launched. TMPDB106 was used as training set, and PDBTM472 was used as testing set. Then, we exchanged the training set and testing set. The results showed the accuracy of correct helix locations can keep a robust performance on different data set as Table 5. Accuracy of correct helix locations is only 1.6 % lower than 95 % in Table 3.

Table 5. The result of cross validation

Training set	Testing set	Correct helix count	Correct helix locations
TMPDB106	PDBTM472	266/472 = 56.3 %	2283/2473 = 92.3 %
PDBTM472	TMPDB106	86/106 = 81.1 %	368/389 = 94.6 %

3.5 Case Study on Predicting 3D Structures Using CRF-TM Results

Accurate helix count and helix locations can help predicting high-resolution 3D structure of membrane. G-protein-coupled receptors (GPCRs) is a special kind of transmembrane protein with seven membrane-spanning α-helices (TM-1 to TM-7) connected by three intracellular (IL-1 to IL-3) and three extracellular loops (EL-1 to EL-3), and finally an intracellular C-terminus. We use CRF-TM to test the helix locations of CXCR4 and A2A. The results show in column 2 of Table 6. Then patGPCR [19, 20] employed to modeling the 3D structures. The root-mean-square deviations (RMSDs) between predicted conformations and natives are listed in column 3 of Table 6. A direct visual comparison with native is depicted in Fig. 2. In Fig. 2, most of the boundaries of TM regions are coherent with the natives in three dimensions. The inconsistent regions are occurred by helix movement and loop modeling in red circles.

Table 6. Helix locations and RMSD of 3D structure of CXCR4

Name	Locations of 7 membrane-spanning regions	RMSD
A2A	3–37,41–66,78–101,117–138,170–208,219–225,268–306	2.987
CXCR4	5–28,45–67,72–87, 110–135,163–190,206–230,247–263	3.570

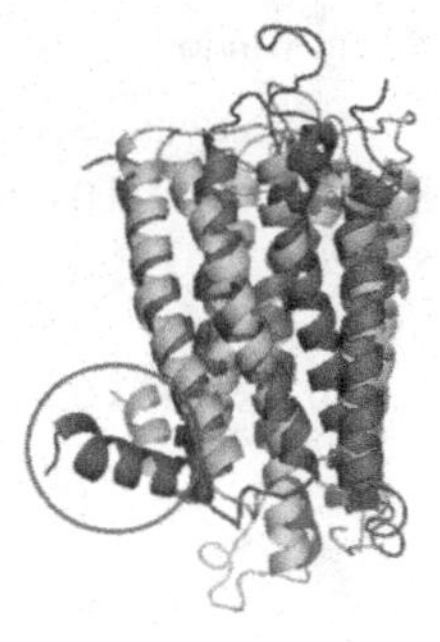

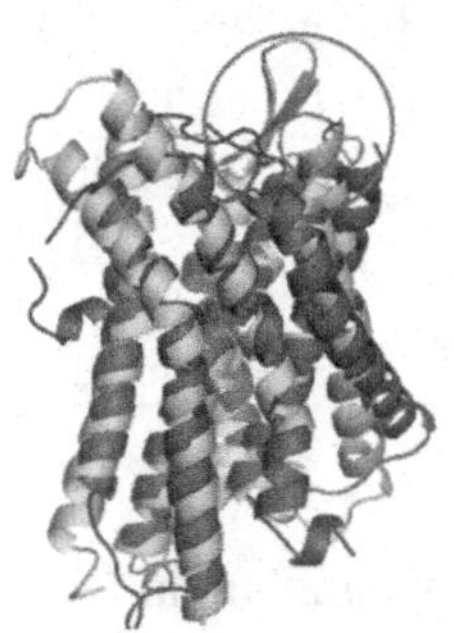

Fig. 2. A visual comparison between predicted 3D structures and natives. Native structures are colored red and predicted structures are colored green. A2A is depicted on left and CXCR4 is depicted on right (Color figure online).

4 Conclusions

In this paper, we presented a novel approach based on conditional random fields based, CRF-TM, to predicting the helix count and helix locations of the transmembrane proteins. CRF-TM reflects the long-range correlation under the full-length sequence as joint probability. Two datasets were employed in the performance validations. The results showed that the CRF-TM could rank in the first group comparing with recent wide-used TM predictors.

But the ability of predicting helix count is still week. The reason may be the window based feature extraction cannot filter the most valuable features from high dimension space. Manifold Learning can distinguish the most informative features and records by incorporating into the kernel space by using graph Laplacian [21]. So in the future works, I expect to improve the CRF-TM to reflect the variant of different residues.

Acknowledgments. This paper is supported by grants no. 61170125, 61202290 under the National Natural Science Foundation of China (http://www.nsfc.gov.cn) and grants no. BK20131154 under Natural Science Foundation of Jiangsu Province. The funders had no role in study design, data collection and analysis, decision to publish, or preparation of the paper. The authors thank Jin Wang and Shimin Chen for helping with the analysis of the experiment.

References

1. Hediger, M.A., Clémençon, B., Burrier, R.E., et al.: The ABCs of membrane transporters in health and disease (SLC series): introduction. Mol. Aspects Med. **34**(2), 95–107 (2013)
2. Coskun, Ü., Simons, K.: Cell membranes: the lipid perspective. Structure **19**(11), 1543–1548 (2011)
3. Bill, R.M., Henderson, P.J.F., Iwata, S., et al.: Overcoming barriers to membrane protein structure determination. Nat. Biotechnol. **29**(4), 335–340 (2011)

4. Yu, D., Wu, X., Shen, H., Yang, J.: Enhancing membrane protein subcellular localization prediction by parallel fusion of multi-view features. IEEE Trans. Nanobiosci. **11**(4), 375–385 (2012)
5. Sonnhammer, E.L.L., Von Heijne, G., Krogh, A.: A hidden Markov model for predicting transmembrane helices in protein sequences. Ismb **6**, 175–182 (1998)
6. Tusnady, G.E., Simon, I.: Principles governing amino acid composition of integral membrane proteins: application to topology prediction. J. Mol. Biol. **283**(2), 489–506 (1998)
7. Rost, B., Casadio, R., Fariselli, P., et al.: Transmembrane helices predicted at 95 % accuracy. Protein Sci. Publ. Protein Soc. **4**(3), 521 (1995)
8. Nugent, T., Jones, D.T.: Transmembrane protein topology prediction using support vector machines. BMC Bioinform. **10**(1), 159 (2009)
9. Viklund, H., Elofsson, A.: OCTOPUS: improving topology prediction by two-track ANN-based preference scores and an extended topological grammar. Bioinformatics **24**(15), 1662–1668 (2008)
10. Hopf, T., Colwell, L., Sheridan, R., et al.: Three-dimensional structures of membrane proteins from genomic sequencing. Cell **149**(7), 1607–1621 (2012)
11. Lafferty J., McCallum A., Pereira F.C.N.: Conditional random fields: probabilistic models for segmenting and labeling sequence data. In: Proceedings of the Eighteenth International Conference on Machine Learning, ICML 2001 (2001)
12. Li, M., Lin, L., Wang, X., et al.: Protein–protein interaction site prediction based on conditional random fields. Bioinformatics **23**(5), 597–604 (2007)
13. Wang, L., Sauer, U.H.: OnD-CRF: predicting order and disorder in proteins conditional random fields. Bioinformatics **24**(11), 1401–1402 (2008)
14. Ikeda, M., Arai, M., Okuno, M., Shimizu, T.: Toshio: TMPDB: a database of experimentally-characterized transmembrane topologies. Nucleic Acids Res. **31**(1), 406–409 (2003)
15. Arai, M., Mitsuke, H., Ikeda, M., et al.: ConPred II: a consensus prediction method for obtaining transmembrane topology models with high reliability. Nucleic Acids Res. **32**, W390–W393 (2004)
16. Lukas, K., Anders, K., Erik, L.L.S.: A combined transmembrane topology and signal peptide prediction method. J. Mol. Biol. **338**, 1027–1036 (2004)
17. Nugent, T., Jones, D.T.: Transmembrane protein topology prediction using support vector machines. BMC Bioinform. **10**, 159 (2009)
18. Kozma, D., Simon, I., Tusnady, G.E.: PDBTM: protein data bank of transmembrane proteins after 8 years. Nucleic Acids Res. 1–6 (2012)
19. Wu, H., Lü, Q., Quan, L., Qian, P.: PatGPCR: a multitemplate approach for improving 3D structure prediction of transmembrane helices of G-protein-coupled receptors. Article ID 486125 (2013). doi:10.1155/2013/486125
20. Wu, H., Lü, Q., Quan, L., et al.: Modeling the structural topology and predicting the three-dimensional structure for transmembrane helixes of GPCR. Chin. J. Comput. **10**, 2168–2178 (2013)
21. Cai, D., He, X.F.: Manifold adaptive experimental design for text categorization. IEEE Trans. Knowl. Data Eng. **24**(4), 707–719 (2012)

Bootstrapped Integrative Hypothesis Test, COPD-Lung Cancer Differentiation, and Joint miRNAs Biomarkers

Kai-Ming Jiang[1,2], Bao-Liang Lu[1,2], and Lei Xu[1,2,3(✉)]

[1] Department of Computer Science and Engineering, Shanghai Jiao Tong University, 800 Dong Chuan Road, Shanghai 200240, China
lxu@cse.cuhk.edu.hk

[2] The Key Laboratory of Shanghai Education Commission for Intelligent Interaction and Cognitive Engineering, Shanghai Jiao Tong University, 800 Dong Chuan Road, Shanghai 200240, China

[3] Department of Computer Science and Engineering, The Chinese University of Hong Kong, Shatin, NT, Hong Kong, China

Abstract. Integrative Hypothesis Test (IHT) has been recently proposed for an integrated study of hypothesis test, classification analysis and feature selection. This paper not only applies IHT to identifying miRNAs biomarkers for the differentiation of lung cancer and Chronic Obstructive Pulmonary Disease (COPD), but also proposes a bootstrapping method to enhance the reliability of IHT ranking on samples with a small size and missing values. On the GEO data set GSE24709, the previously reported fourteen differentially expressed miRNAs have been re-confirmed via one by one enumeration of their IHT ranking, with two doubtful miRNAs identified. Moreover, every pair of miRNAs is also exhaustively enumerated to examine the pairwise effect via the p-value, misclassification, and correlation, further identifying those that take core roles in coordinated effects. Furthermore, linked cliques are found featured with joint differentiation performances, which motivates us to identify such clique patterns as joint miRNAs biomarkers.

Keywords: IHT · Bootstrapping · Differential gene expression

1 Introduction

Based on the differential expression of miRNAs in tumors, efforts have been made on finding miRNA expression signatures of lung cancer and subtypes via not only tumor cells [1, 2] but also sera and peripheral blood cells from cancer patients [3–6]. Lung cancer closely relates to Chronic Obstructive Pulmonary Disease (COPD), a common pulmonary affliction encompassing chronic obstructive bronchitis and lung emphysema [7]. COPD is a global burden affecting 10–15 % of adults older than 40 years [8] and precedes lung cancer in 50–90 % of cases [9]. This paper also works such a topic via performing differentiation analyses on the expression of 863 miRNAs in blood cells of lung cancer patients and patients suffering from COPD, with data available in the Gene Expression Omnibus (GEO, http://www.ncbi.nlm.nih.gov/geo/, GSE24709) [9, 10].

X. He et al. (Eds.): IScIDE 2015, Part II, LNCS 9243, pp. 538–547, 2015.
DOI: 10.1007/978-3-319-23862-3_53

Differentiation analyses on miRNA expression and gene expression are made in one of two typical methods that are generally used in various tasks of case-control studies or binary classification. Under the name of *two sample test* or *model comparison* in general, the first method evaluates the overall difference between two populations of samples with each population described by a parametric model, usually a normal distribution. One widely used example on gene expression differentiation is t-test and Welch test. Under the name of *classification or model prediction*, the second method evaluates the performance of discriminative boundary that classifies each sample into its corresponding population. Each of the two methods has been extensively studied individually.

Though there are also some studies that separately use two methods on one experiment and report the performances obtained by both the methods, there lacks effort on systematically integrating the performances of two methods. Integrative Hypothesis Test (IHT) has been recently proposed towards this purpose [11, 12]. This paper applies IHT to identifying miRNAs biomarkers for the differentiation of lung cancer and COPD. Moreover, a bootstrapping method is proposed to enhance the reliability of IHT ranking on a small size of samples and many missing values among the samples.

First, we adopt typical practice of evaluating each miRNA one by one by using this bootstrapped based IHT ranking, resulting in a list of miRNAs with the top 15 IHT ranks (See Table 2) that covers all the 14 differentially expressed ones identified in [9, 10] but in a different order of reliability. Also, we found that among the 14 ones, hsa-miR-513b and hsa-miR-93* are really doubtful in their reliability. We checked the Human microRNA Disease Database (HMDD) version two [13], and find no report of these two miRNAs related to any kind of cancer.

Moreover, we examine paired miRNAs not just following typical practice via Pearson correlation, but also exhaustively evaluating every pair of miRNAs by the integrated performances of the p-value and classification accuracy with help of this bootstrapped based IHT ranking, resulting in a list of top-20 pairs of miRNAs (See Table 4). Interestingly, each of 19 pairs contains one miRNA that locates within top-10 in Table 2. Especially hsa-miR-675 and hsa-miR-92a each in 6 pairs while hsa-miR-369-5p in 5 pairs, hsa-miR-641 and hsa-miR-662 each in 2 pairs seemly take important roles in the differentiation of lung cancer and COPD, featured with classification accuracy (90 % – 94.9 %) and the p-value (in around 10^{-7}– 10^{-10}).

In most of pairs, two miRNAs in a pair are not correlated or weakly correlated. There are also cliques that demonstrate joint miRNAs activities. One is featured by hsa-miR-369-5p pairing each of five miRNAs, with considerable negative or positive correlations. The other is featured by hsa-miR-675 pairing each of six miRNAs, all with negative correlations, and another is featured by hsa-miR-92a pairing each of six miRNAs, with a half in weak correlations and the other half in considerable correlations. Each of these pairs also reaches high differentiation performances with classification accuracy (89 % – 95 %) and the p-value (around 10^{-7}– 10^{-10}). Therefore, we may identify such clique patterns as joint miRNAs biomarkers.

2 Methods

2.1 Integrative Hypothesis Tests

The name of integrative hypothesis tests (IHT) was previously advocated in Reference [11] for an integrative study of case-control problems from not only a model based perspective such as two-sample test or model comparison to evaluate an overall difference but also a boundary based perspective such as classification or model prediction about boundary distinguishability and prediction performance. There are two basic tasks for each of the two perspectives, as outlined in Reference [12] by its Table 1.

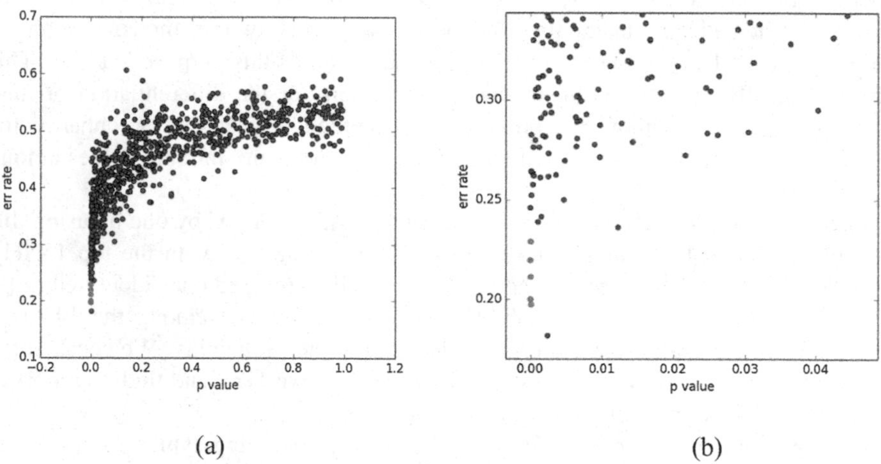

Fig. 1. 2D scatter plot (a) overall view, (b) zoom in view.

Given a set $X\omega = \{\mathbf{x}_{t,\omega}, t = 1, \cdots, N\omega\}$ of samples from two populations $\omega = 0,1$ (i.e., $\omega = 0$ for COPD and $\omega = 1$ for lung cancer), where each $\mathbf{x}_{t,\omega}$ is a vector that consists of all the features (i.e., miRNAs) and N_ω is the number of patients for the ωth population. A model based perspective study involves Task A and Task B, namely getting each population of samples to be described by a parametric model $q(x|\theta_\omega)$ (e.g., usually a normal distribution) and then compares the resulted models to examine the overall difference between two populations of samples. On the other hand, a boundary based study involves Task C and Task D, namely, classifying each sample to either $\omega = 1$ or 0 and examining whether a reliable separating boundary exists between the two populations of samples. Moreover, all the four tasks are associated with another problem called feature selection (i.e., identification of miRNAs).

Each of four tasks has been studied individually in the existing efforts, with each having its strength and limited coverage. However, performances of these tasks are coupled, and thus the best set of features for one task may not be necessarily the best for the others. Naturally, it was motivated to consider whether the performances of all the four tasks or at least more than one tasks can be jointly optimized. The necessarily and feasibility have been addressed in Reference [12] and also empirically justified via a so called 2D scattering plots (see Fig. 4 in [12]).

In this paper, we adopt a simplified IHT implementation that only considers Task B and Task C. For Task B, we consider $q(x|\theta_{\omega})$ to be either a univariate normal distribution when we evaluate each miRNA one by one exhaustively by the Welch's t-test or a two-variate normal distribution when we examine every pair of miRNAs exhaustively by the Hotelling T-squared test. The performance of Task B is the resulted p-value. For Task C, in the current preliminary study we simply use the linear discriminating analysis, with its performance measured by the misclassification rate. How comparative studies by using Bayes classifier and support vector machine (SVM) will be further conducted.

Also, we adopt the above mentioned 2D scattering plots to help us interactively to observe the joint performances of p-value and misclassification rate. As illustrated in Fig. 1, a small p-value indicates big difference of the two distributions and a small misclassification rate indicates a well classification of samples by a separating boundary. We are interested in those candidate points that are nearest to the origin of the coordinate space.

Though such 2D or 3D plots provide a possible joint evaluation, how to appropriately scaling each measure is a challenging issue. As addressed in Reference [12], we need to integrate multiple measures into a scalar index based on which the joint performance can be evaluated.

Table 1. A list of top-15 obtained by IHT

Rank	Gene id	p value	accuracy
1	hsa-miR-369-5p	1.24E-05	81.32 %
2	hsa-miR-675	2E-05	77.78 %
3	hsa-miR-662	9.25E-06	76.88 %
4	hsa-miR-641	4.86E-05	76.77 %
5	hsa-miR-767-3p	4.55E-06	76.46 %
6	hsa-miR-888*	0.00076	78.47 %
7	hsa-miR-26a	1.55E-06	75.59 %
8	hsa-miR-1299	0.000567	75.45 %
9	hsa-miR-95	7.46E-05	74.20 %
10	hsa-miR-636	0.000192	73.68 %
11	hsa-miR-1308	0.000196	72.95 %
12	hsa-miR-513b	0.000366	72.15 %
13	hsa-miR-668	0.000934	72.92 %
14	hsa-miR-130b	0.000349	71.18 %
15	hsa-miR-875-3p	0.001364	74.41 %

2.2 Rank Bootstrapping

We may turn the scattering plots in Fig. 1 into a rank list from increasingly sorting the distance to the origin of the coordinate space. Given in Table 1 is such a list obtained from evaluating each miRNA one by one exhaustively, with the p value obtained by the

Welch's t-test and the accuracy is equal to one minus the misclassification rate resulted from a linear discriminating analysis. Only the top 15 miRNAs are listed. However, the result maybe not reliable enough because there are only $N_1 = 28$ lung cancer patients and $N_0 = 24$ COPD samples. That is, we encounter a typical small sample size problem, which is actually widely encountered in the case-control studies.

Such a small sample size problem makes the resulted p value and the accuracy become random variables and thus each point in Fig. 1 may randomly move, resulting in its distance to the origin varies too, namely, the resulted rank in Table 1 is unreliable.

Bootstrapping is a widely used practice of estimating properties of an estimator when the sample size is insufficient, by measuring those properties via sampling from an approximating distribution. Also, this technique allows estimation of the sampling distribution of almost any statistic. Moreover, bootstrapping provides a way to account for the distortions caused by the specific sample that may not be fully representative of the population.

Table 2. A list of top-15 obtained with Rank Bootstrapping

Rank	Gene ID	Avg Rank	Std Rank
1	hsa-miR-662	2.8	1.30384
2	hsa-miR-636	3	1.224745
3	hsa-miR-675	3.4	1.516575
4	hsa-miR-369-5p	4.2	4.494441
5	hsa-miR-940	7.6	3.361547
6	hsa-miR-92a	8.4	5.029911
7	hsa-miR-1224-3p	8.6	4.505552
8	hsa-miR-26a	10.6	4.615192
9	hsa-miR-328	11.2	5.80517
10	hsa-miR-641	14.2	3.701351
11	hsa-miR-383	17	5.43139
12	hsa-let-7d*	21.2	4.086563
13	hsa-miR-93*	24	10.90871
14	hsa-miR-323-3p	24.8	6.220932
15	hsa-miR-513b	26.6	4.27785
	Excluded genes due to large rank std		
	hsa-miR-875-3p	20.2	17.32628
	hsa-miR-30e*	22	13.32291
	hsa-miR-139-5p	22.6	13.01153
	hsa-miR-1911	23.8	12.43785
	hsa-miR-130b	24	15.23155

Therefore, the bootstrapping provides a useful tool for improving the unreliable rank in Table 1, not just for a small size of samples but also for the problem of missing data, which also happens seriously in miRNAs expression profile. The bootstrapping will help to estimate the p-value and the misclassification rate, not for estimating their means and variances, but for estimating the mean and variance of the ranks of the corresponding miRNA, in the list obtained from increasingly sorting the distance to the origin of the coordinate space, in a way similar to Table 1 but not just for the top 15 but all the miRNAs in consideration.

Instead of sampling from an approximating distribution, resampling is implemented by constructing a number of resamples with replacement from the miRNAs expression data set. In each bootstrapping implementation, we will get a ranking list. The rank of each miRNA is a statistic in our consideration, the smaller the statistic is, the more we are interested in considering it as a candidate. After a large enough number of bootstrapping implementations, we may get the mean ranking and standard deviation ranking of each statistics, by which we may further generate a new ranking such that each miRNA is sorted increasingly according its mean ranking, subject to that its standard deviation is less than a threshold.

Though all the ranking lists addressed in this subsection consider each miRNA one by one, i.e., each row in Table 1 and Table 2 represents one miRNA, extension can be easily made to examine every pair of miRNAs simply with each row replaced by a pair of two miRNAs.

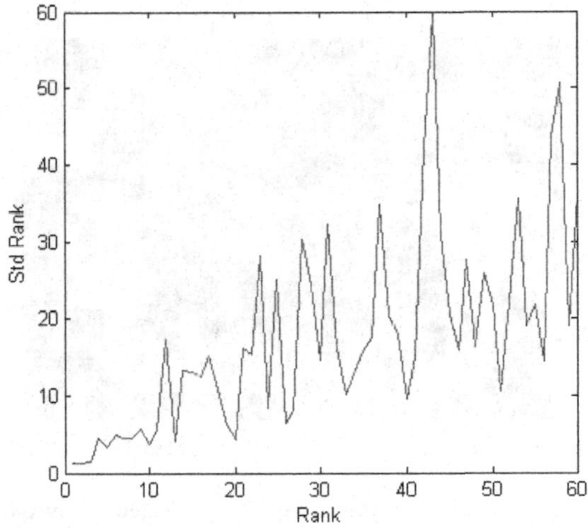

Fig. 2. Selection of a threshold for standard deviation.

3 Empirical Results

The Microarray data used in this paper are downloaded from the Gene Expression Omnibus (GEO, http://www.ncbi.nlm.nih.gov/geo/, GSE24709), consisting of the expression of the 863 human miRNAs annotated in miRBase version 12.0. In total, the data includes the miRNA expression in blood cells from 71 different individuals, including 28 lung cancer patients, 24 COPD patients, and 19 healthy controls. In our study we only concern about the COPD patients and lung cancer patients.

Evaluating each miRNA one by one exhaustively, the counterpart of Table 1 is shown in Table 2, obtained by rank bootstrapping. In each bootstrapping implementation, a size 45 of resampling samples are obtained. After 100 bootstrapping implementations, we get the mean ranking and standard deviation ranking for each statistics, by which we list the top-15 in Table 2 with miRNAs sorted increasingly according its mean ranking, subject to that its standard deviation is less than a threshold 12. As illustrated in Fig. 2, this threshold is obtained by sorting the mean ranks along the horizontal axis with the vertical axis for standard deviation, with the threshold found at the point that the standard deviation suddenly changes. The excluded genes with relatively small mean ranking but large standard deviation is also shown in Table 2.

We observe the 15 miRNAs covers all the 14 differentially expressed ones given in Table 3 but in a different order. The last three ones in Table 2, including hsa-miR-513b and hsa-miR-93* in Table 3, are really doubtful because of their reliability, we checked the Human microRNA Disease Database (HMDD) version two [13], and find no report of these two miRNAs related to any kind of cancer, while most of the other 12 genes appear in a few reports.

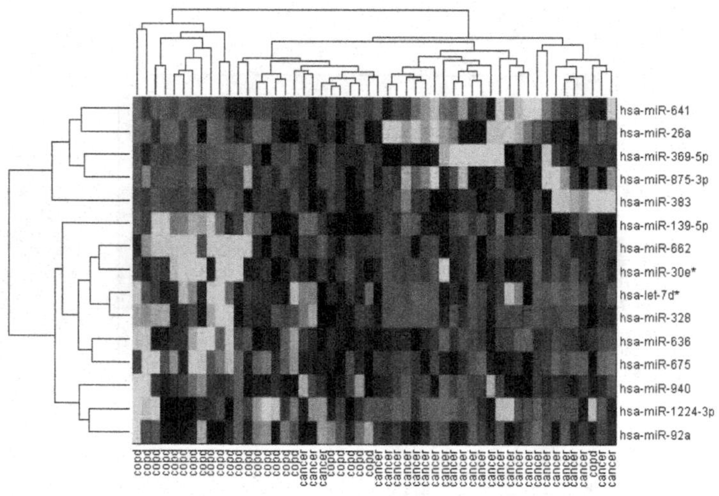

Fig. 3. Hierarchical clustering result of selected 15 miRNAs.

We apply hierarchical clustering to the miRNA expression data of those selected miRNAs and the clustering is visualized as heatmap in Fig. 3. As can be seen, obvious patterns of miRNAs can be found for COPD and lung cancer and the selected miRNAs can be divided into 2 clusters which one up-regulates with the disease and the other down-regulates.

Moreover, we examine paired miRNAs not just following the typical practice via Pearson correlation, but also exhaustively evaluate every pair of miRNAs by the integrated performances of the p-value and classification accuracy with help of this bootstrapped based IHT ranking. Table 4 gives a list of top-20 pairs of miRNAs. Being different from Table 2, each row represents a pair of miRNAs. The Pearson correlation between the two miRNAs is listed in the last column but did not join the integrative sorting. As expected, two individually best miRNAs does not necessarily form the best pair miRNAs.

Table 3. Significant markers identified in [9, 10] for differentiation of lung cancer versus COPD (p-value 0.01)

miRNA	Control	COPD	Lung cancer	Control vs. COPD	Control vs. Lung cancer	Lung cancer vs. COPD
hsa-miR-641	76.68	143.15	59.58	0.00013	0.90088	0.00075
hsa-miR-662	90.65	23.1	95.46	0.0003	0.5175	0.0001
hsa-miR-369-5p	33.46	97.1	33.25	0.00041	0.60298	0.0001
hsa-miR-383	74.96	142.06	73.83	0.00122	0.87052	0.00316
hsa-miR-636	246.59	106.39	222.87	0.00186	0.72712	0.00016
hsa-miR-940	225.92	152.89	247.83	0.00583	0.94678	0.00683
hsa-miR-26a	7269.84	7975.44	5568.45	0.00931	0.21746	0.00047
hsa-miR-92a	13651.44	9554.17	13651.44	0.00957	0.80809	0.00156
hsa-miR-328	59.92	76.93	208.31	0.96379	0.00428	0.00126
hsa-let-7d*	70.76	102.75	250.42	0.05763	0.00006	0.00278
hsa-miR-1224-3p	137.63	109.61	233.37	0.08731	0.86406	0.00316
hsa-miR-513b	66.76	80.41	39.04	0.03264	0.12765	0.00411
hsa-miR-93*	893.5	1303.7	2321.35	0.99299	0.01562	0.0068
hsa-miR-675	254.2	149.11	287.83	0.04421	0.04842	0.00156

Interestingly, each of 19 pairs contains one miRNA that locates within top-10 in Table 3. Especially hsa-miR-675 and hsa-miR-92a each in 6 pairs while hsa-miR-369-5p in 5 pairs, hsa-miR-641 and hsa-miR-662 each in 2 pairs seemly take important roles in the differentiation of lung cancer and COPD, featured with the classification accuracy (90 % – 94.9 %) and the p-value (in around 10^{-7}– 10^{-10}). Moreover, hsa-miR-26a, hsa-let-7d*, hsa-miR-636, hsa-miR-93* are also appear in both Tables 2 and 4. Thus, a total of 9 of miRNAs in Table 2 have appeared in Table 4 and draw our particular attention for further investigation.

In most of pairs, two miRNAs in a pair are not correlated or weakly correlated. There are also cliques that demonstrate joint miRNAs activities. One consists of (has-miR-369-5p, hsa-miR-92a), (has-miR-369-5p, hsa-miR-675), (has-miR-369-5p, hsa-miR-26a), (has-miR-369-5p, hsa-miR-183), (has-miR-369-5p, hsa-miR-940), with each pair in a considerable negative or positive correlation. The other consists of (hsa-miR-675, *hsa-miR-1271*), (hsa-miR-675, hsa-miR-641), (hsa-miR-675, hsa-miR-1299), (hsa-miR-675, hsa-miR-627), (hsa-miR-675, hsa-miR-489), all with negative correlations. The another one consists of (hsa-miR-92a, has-miR-369-5p), (hsa-miR-92a, hsa-miR-1204), (hsa-miR-92a, hsa-miR-376a*), (hsa-miR-92a, hsa-miR-767-3p), (hsa-miR-92a, hsa-miR-93*), with a half in weak correlations and the other half in

Table 4. A list of top-20 pairs obtained by IHT (Pearson Correlation did not join the sorting)

Gene ID1	Gene ID2	P value	Accuracy	Pearson
hsa-miR-1271	hsa-miR-675	5.74E-10	0.948958	-0.19585
hsa-miR-369-5p	hsa-miR-92a	3.19E-09	0.930903	-0.42127
hsa-miR-641	hsa-miR-675	8.1E-09	0.929514	-0.21791
hsa-miR-1204	hsa-miR-92a	8.74E-08	0.91875	0.013134
hsa-miR-1302	hsa-miR-662	4E-08	0.917014	-0.00473
hsa-miR-369-5p	hsa-miR-675	2.05E-10	0.915972	-0.30058
hsa-miR-1299	hsa-miR-675	2.17E-06	0.9125	-0.03089
hsa-miR-627	hsa-miR-675	4.21E-08	0.909375	-0.02967
hsa-miR-610	hsa-miR-662	5.5E-11	0.907639	0.114439
hsa-miR-26a	hsa-miR-369-5p	4.11E-09	0.905208	0.524418
hsa-miR-376a*	hsa-miR-92a	3.5E-08	0.904167	-0.11793
hsa-miR-767-3p	hsa-miR-92a	1.23E-07	0.900694	-0.16173
hsa-miR-183	hsa-miR-369-5p	7.83E-10	0.899306	0.211596
hsa-let-7d*	hsa-miR-1226*	9.54E-09	0.898958	-0.27561
hsa-miR-636	hsa-miR-888*	2.33E-07	0.897222	-0.28287
hsa-miR-92a	hsa-miR-93*	2.51E-07	0.896181	0.212414
hsa-miR-489	hsa-miR-675	1.99E-08	0.895139	-0.29558
hsa-miR-1248	hsa-miR-641	9.83E-08	0.893403	-0.07723
hsa-miR-875-5p	hsa-miR-92a	7.4E-07	0.893403	0.003946
hsa-miR-369-5p	hsa-miR-940	4.88E-09	0.892014	-0.35694

considerable correlations. The three cliques are also linked via (hsa-miR-675 and hsa-miR-92a. Each pair in these cliques reaches high differentiation performances with classification accuracy (89 % – 95 %) and the p-value (around 10^{-7}– 10^{-10}). We are thus motivated to identify such clique patterns as joint miRNAs biomarkers.

4 Conclusion

This paper applies Integrative Hypothesis Test (IHT) to identifying miRNAs biomarkers for the differentiation of lung cancer and COPD via an integrative perspective of both hypothesis test and linear classification. A bootstrapping method is proposed to enhance the reliability of IHT ranking on samples with a small size and missing values. Empirical study has been made on the GEO data set GSE24709. First, we exhaustively evaluate miRNAs one by one by the bootstrapped based IHT ranking, re-discover the 14 differentially expressed ones identified in [9, 10] but two of them regarded as really doubtful in their reliability. We checked the Human microRNA Disease Database (HMDD) version two [13], and find no report about the two miRNAs are related to any kind of cancer. Second, we also exhaustively evaluate every pair of miRNAs by this IHT ranking and found that the majority of those found one by one take the roles in a list of top miRNA pairs. Furthermore, we found three mutually linked cliques that demonstrate joint miRNAs activities featured with differentiation performances in the classification accuracy (89 % – 95 %) and the p-value (around 10^{-7}– 10^{-10}, which motivates us to identify such clique patterns as joint miRNAs biomarkers.

Acknowledgment. This work was partially supported by the National Natural Science Foundation of China (Grant No. 61272248), the National Basic Research Program of China (Grant No. 2013CB329401, the Science and Technology Commission of Shanghai Municipality (Grant No.13511500200), and Shanghai Jiao Tong University fund for Zhiyuan Chair Professorship.

References

1. Lu, J., et al.: MicroRNA expression profiles classify human cancers. Nature **435**(7043), 834–838 (2005)
2. Barshack, I., et al.: MicroRNA expression differentiates between primary lung tumors and metastases to the lung. Pathol. Res. Pract. **206**(8), 578–584 (2010)
3. Cortez, M.A., Calin, G.A.: MicroRNA identification in plasma and serum: a new tool to diagnose and monitor diseases (2009)
4. Gilad, S., et al.: Serum microRNAs are promising novel biomarkers. PLoS ONE **3**(9), e3148 (2008)
5. Wang, J., et al.: MicroRNAs in plasma of pancreatic ductal adenocarcinoma patients as novel blood-based biomarkers of disease. Cancer Prev. Res. **2**(9), 807–813 (2009)
6. Tammemagi, C.M., et al.: Impact of comorbidity on lung cancer survival. Int. J. Cancer **103** (6), 792–802 (2003)
7. van Gestel, Y.R., et al.: COPD and cancer mortality: the influence of statins. Thorax **64**(11), 963–967 (2009)
8. Young, R.P., et al.: COPD prevalence is increased in lung cancer, independent of age, sex and smoking history. Eur. Respir. J. **34**(2), 380–386 (2009)
9. Keller, A., Leidinger, P.: Peripheral profiles from patients with cancerous and non cancerous lung diseases, Gene Expression Omnibus (GEO, GSE24709) (2011). http://www.ncbi.nlm.nih.gov/geo/
10. Leidinger, P., et al.: Specific peripheral miRNA profiles for distinguishing lung cancer from COPD. Lung Cancer **74**(1), 41–47 (2011)
11. Xu, L.: Integrative hypothesis test and A5 formulation: sample pairing delta, case control study, and boundary based statistics. In: Sun, C., Fang, F., Zhou, Z.-H., Yang, W., Liu, Z.-Y. (eds.) IScIDE 2013. LNCS, vol. 8261, pp. 887–902. Springer, Heidelberg (2013)
12. Xu, L.: Bi-linear Matrix-variate Analyses, Integrative Hypothesis Tests, and Case-control Studies. To appear on Springer OA J. Appl. Inform., 1(1) (2015)
13. Human microRNA Disease Database. http://202.38.126.151/hmdd/tools/hmdd2.html

Research of Traffic Flow Forecasting Based on the Information Fusion of BP Network Sequence

Wei Zhang[1,2(✉)], Ridong Xiao[2], and Jing Deng[2]

[1] Chang an University, Xi'An 710064, China
zhwyl@163.com
[2] China Highway Engineering Consulting Corporation, Beijing 100097, China

Abstract. Traffic flow forecasting is an important aspect of the ITS as accurate traffic predication can alleviate congestion, save traveling time and reduce economical loses. The forecasting process may rely on historical data, current data, or both, to forecast the traffic volume in the future. In this paper, we compare three different approaches in traffic forecasting, study the input data and output data for these approaches, as well as some general insights, and also propose BP neural network to estimate accurate traffic flow for a roadway section. By means of three layers-BP neutral network model, in which mechanism algorithm are used to preprocess the multi-source data, error data is eliminated, multi-source data fusion is realized and accurate traffic forecasting is achieved.

Keywords: Information fusion · BP neural network · Traffic volume forecasting

1 Introduction

In recent years, traffic forecasting has become a crucial task in the area of intelligent transportation system (ITS), playing a fundamental role in the planning and development of traffic management and control systems. The goal is to predict traffic conditions in a transportation network based on its past behavior. Improving predictive accuracy within this context would be of extreme importance, not only to inform travelers about traffic conditions but also to better design infrastructures and mobility services, as well as schedule interventions. For this reason, in recent years, there has been great effort in developing sensor instruments supporting traffic control systems.

With the need of traffic control system, several forecasting models were applied in traffic flow prediction, such as Autoregressive model (AR), moving average model (MA), autoregressive moving average model (ARMA), the historical average model (HA) [1] and so on. Parameters of these model are generally estimated by the least squares method (LS). The model factors are relatively simple, so these models have the advantages of simple calculation, easy to update in real-time, convenient in widely application [2]. However, the actual traffic flow is uncertain and nonlinear, and these models cannot overcome random interference factors (season, climate, traffic accident, the driver mental state, etc.) [3]. Aiming at this situation, people begin to explore other

X. He et al. (Eds.): IScIDE 2015, Part II, LNCS 9243, pp. 548–558, 2015.
DOI: 10.1007/978-3-319-23862-3_54

advanced intelligent model for traffic volume forecasting. In this paper we use the information fusion model based on artificial neural network to forecast the traffic flow.

Artificial neural network, based on human understanding of brain neural network, is actually information processing system consisting of a large number of interconnected simple elements [4]. As a result, the artificial neural network has many biological advantages: (1) high parallelism; (2) high non-linear global scope; (3) excellent error-tolerance and associative memory function; (4) strong adaptive and self-learning ability. The application of artificial neural network model in traffic control system can solve the problem of uncertainty and non-linear in predicting traffic flow, which gets rid of the difficulties to establish accurate mathematical model in early time, and proposes a new thought for the traffic control system research.

Neural network model is a potential model in traffic flow forecasting [5–7]. However, the real-time traffic data in monitoring center is frequently received from traffic parameter detectors distributing in various circuits, such as coil detector, ultrasonic detector, infrared detector, microwave detector, video detector and so on. The performance parameters of various detectors are different, so each of them has its own advantages and disadvantages. In addition, the detection precision is further limited by the limitation of detection principles [8, 9]. Therefore, first we make data preprocessing and normalization in order to reduce the performance differences among different sensors at different sites. Then, we process data by BP neural network model to get the ideal traffic volume forecasting results.

2 Information Fusion and Artificial Neural Network

Information fusion, or data fusion, has many similarities with neural network in characteristic and structure since information fusion is based on intelligent thoughts, whose important prototype is human mind [10]. The similarity of neural network and information fusion can solve a portion of information fusion problem using neural network, providing a useful method for traffic information processing and treatment [11] (Fig. 1).

Artificial neuron model structure has n input $x_i(i = 1, 2, \cdots, n)$, w_i is the corresponding connecting weight. θ is the threshold of neurons, we use $x_0 = -1$ as fixed bias, make its corresponding connecting weight $w_0 = \theta$, the usual choice is:

$$u = f(\sum_{i=1}^{n} (w_i x_i) - \theta) = f(\sum_{i=0}^{n} w_i x_i) = f(W^T X). \tag{1}$$

$f(\cdot)$ is transfer function (activation function), whose function is to transform infinite domain to one specified limited scope output, it simulates the nonlinear processing ability of biological neurons. Meanwhile $f(\cdot)$ is the first element in the design of artificial neural network. W and X are the column vector of w_i and $x_i(i = 1, 2, \cdots, n)$ respectively, W^T is transposition of W.

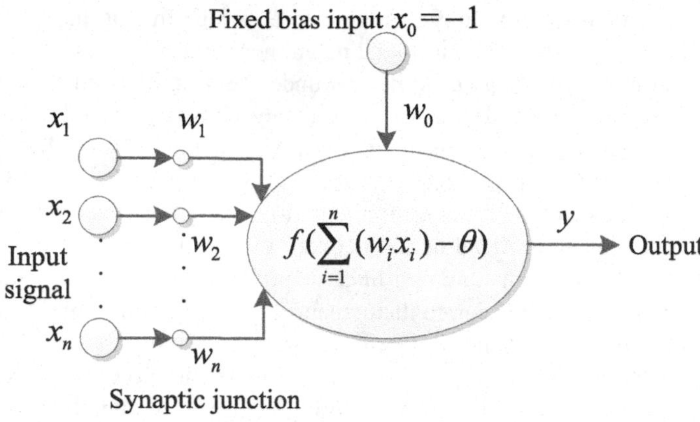

Fig. 1. Artificial neuron model

3 Information Fusion Design Based on Neural Network

Road traffic parameters (volume, occupancy, speed and travel time, etc.) and historical traffic data are interconnected; in addition, as a part of the road network, the traffic condition of the road section is affected by its upstream and downstream section of road. As a result, the traffic flow parameters between the experimental section and its related road sections are inevitably internally linked [12, 13]. It would therefore be feasible to predict the traffic parameters in future based on the data sequence consisted of historical traffic parameters. However, this method requires highly reliable real-time test method to achieve accurate traffic flow data. This paper is mainly aimed at the traffic volume and travel time of history collection in upstream and downstream sections, predicting future traffic volume with data fusion of BP neural network.

3.1 Data Preprocessing

The survey equipment gathering traffic volume based on different design philosophy have their own advantages and disadvantages [14], which may lead to different deviations. To improve the accuracy of traffic volume forecast by the BP neural network information fusion model, the input and output data should be preprocessed. With the traffic volume mechanism, this research tests various sources of data and rules out error data in the data pool.

The algorithm of traffic volume mechanism is to study the consistency of various parameters simultaneously through the correlations among traffic volume parameters [15]. The algorithm of traffic volume mechanism mainly refers to the traffic volume rules algorithm or traffic volume area algorithm. The algorithm of traffic volume rules: a few rules are determined based on the traffic volume mechanism, if the collected data satisfies one or more rules, these data are wrong. For example, the rules could be: the average occupancy is 0, and the traffic is not 0; Traffic volume is 0, and the average occupancy is not 0, obviously the above two rules cannot be satisfied at the same time, so if there are data which can satisfy these two rules, the data must be wrong. Based on the traffic volume mechanism, a relationship model in regards of the two parameters can be built, such as traffic volume and occupancy, traffic volume and travel speed, travel time and crowded length, etc.

(1) Traffic volume and occupancy model

$$a \cdot O_d^2 + b \cdot O_d - k_s \cdot \sigma_s \le q_d \le a \cdot O_d^2 + b \cdot O_d + k_s \cdot \sigma_s. \tag{2}$$

Type: a, b are the model parameters, which can be obtained through the historical data regression analysis; σ_s is the standard deviation for traffic volume; k_s is the correction coefficient of standard deviation.

(2) Traffic volume and speed model

$$\frac{1}{a(1-\frac{q_d}{C})} + \frac{f \cdot b}{1-\frac{q_d}{\lambda S}} - k_v \sigma_v \le \frac{1}{v_d} \le \frac{1}{a(1-\frac{q_d}{C})} + \frac{f \cdot b}{1-\frac{q_d}{\lambda S}} + k_v \sigma_v. \tag{3}$$

Type: a, b are the model parameters, f is signal intersections per kilometer, λ is green ratio, $\lambda = g/c$, S is saturation volume rate, σ_v is standard deviation of speed, k_v is the correction coefficient of standard deviation.

(3) Travel time and Jam length model

$$\frac{l}{a_1 \frac{N_l \cdot l_c}{C} + a_2 \cdot \frac{l - l_c}{v_m} a_3} - k_a \cdot \sigma_a \le \frac{1}{t_p} \le \frac{l}{a_1 \cdot \frac{N_l \cdot l_c}{C} + a_2 \cdot \frac{l - l_c}{v_m} + a_3} + k_a \cdot \sigma_a \tag{4}$$

Type: a_1, a_2, a_3 are the model parameters, σ_a is the standard deviation of the number of floating cars in main road, k_a is the correction coefficient of standard deviation, N_l is lane number of main road, C, v_m, l as above.

3.2 BP Network Structure and Network Training

Information fusion of the traffic volume forecasting model is composed of data processing unit and the BP neural network. Traffic parameters are processed by data processing units before turning into sample data; BP neural network consists of 3 layers: input layer, middle layer and output layer, the node numbers of input and output layer are determined by the sample dimension of data structure [16]. Input samples of the BP neural network forecasting model are two dimensions: travel time and traffic volume; Output samples, by contrast, are one dimension, traffic volume. As for the middle layer node number, it is determined by the nodes number of input layer together with output layer, which could be estimated by the following formula: $p = mq(n + p)$ and $p = \sqrt{np}$. In this type: p is the node number of middle layer. n, q are nodes number of input and output layer, m is the sample volume. Then within the scope of estimate, after the training of MATLAB, the best node number of middle layer can be obtained. The model structure is shown as Fig. 2.

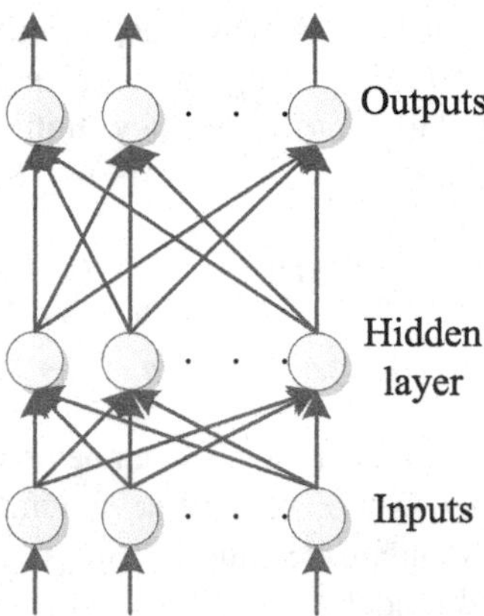

Fig. 2. Neural network structure

The process of BP neural network training is optimized step by step: in the forward process, input information transfers to the output layer through the hidden layer units step by step; If the output layer cannot get the desired output, input information moves back to propagation process. The error between the actual value and network output value therefore returns to the original connecting route, and can be reduced by modifying the connection weights of each neuron. After that, it turns back to positive propagation, and the whole calculation process will be repeated until the error is less than the set value [17].

The nonlinear mapping relation of the collected road traffic volume samples in this process could be effectively extracted by the trained neural network, and be saved as the form of weights and thresholds. After being trained with forecast data based on historical traffic volume, the network model has reached the preset accuracy requirement.

In operation process, when the neural network input non-sample, it could accomplish the arbitrary nonlinear mapping from n dimensional space of input to m dimensional space of output. Here nonlinear mapping from the input of historical traffic data in two dimension spaces to the output of traffic volume forecasting data in one dimension space is accompanied. Therefore, the neural network could get close to the desired nonlinear mapping by learning. The traffic volume regularity, unable to be described by traditional mathematical relationships, in this way can be interpreted relative accurately, and the road traffic volume for the future can be forecast.

The transfer function of basic neurons in BP network must be differentiable, so the S shaped logarithmic or tangent transfer function is often used. For example,

$$f(x) = \frac{1}{1 + e^{-(x)}}. \tag{5}$$

3.3 Traffic Parameter Estimation

Set $u_i(\tau)$ is vector in τ moment of i road section,$u_i(\tau-1)$ is the road traffic volume vector in a period of time before τ moment of i road section. Order $U(\tau) = [u_1(\tau), u_2(\tau), \cdots, u_d(\tau)]$, d is the total number of considered roads, if take solely the road traffic volume into consideration, d=1. Set $t_i(\tau)$ as vector in τ moment of i road section, $t_i(\tau-1)$ is the road traffic volume vector in a period of time before τ moment of i road section. Order $T(\tau) = [t_1(\tau), t_2(\tau), \cdots, t_d(\tau)]$.

Based on the characteristics of road length and traffic volume, we predict traffic volume of future period by using the traffic volume in current period and travel time in last s periods. So, we could take $U(\tau), U(\tau-1), \cdots, U(\tau-s)$ and $T(\tau), T(\tau-1), \cdots, T(\tau-s)$ as the input sample of No. τ, $U(\tau+1)$ as the output value of No.τ sample.

From the response function $f(x)$ characteristics of the BP network, we know the interval of node output value is (0, 1). As a result, the training samples should be numerically processed. In this paper, the geometric transformation method will be adopted: dividing the various indexes by a constant value, the input and output values of BP network is limited within the interval (0, 1); training samples in this way are numerically disposed before entering into the network. Specific algorithm is as follows:

(1) Input data columns $X = [T(\tau), U(\tau)|\tau = 1, 2, \cdots]$;
(2) Deal with data column, make it importable in BP network;
(3) Initialize the weights and threshold for each layer of BP network, take a random number of every layer as the initial weight and threshold value, $w_{ij} = random(\cdot)$, $v_{jt} = random(\cdot)$; Each model has the following cycle;
(4) Calculate b_j and C_t as follows,

$$b_j = 1/[1 + \exp(-\sum_{i=1}^{n} w_{ij}x_i + \theta_j)], \quad (6)$$

$$C_t = 1/[1 + \exp(-\sum_{j=1}^{n} v_{jt}b_j + \gamma_t)]. \quad (7)$$

Calculate error of each layer

$$d_t^k = (y_j^k - c_j^k)c_t^k(1 - c_t^k), \quad (8)$$

$$e_j^k = (\sum_{t=1}^{q} d_t^k \cdot v_{jt}) \cdot b_j \cdot (1 - b_j). \quad (9)$$

(5) Determine whether to circulate the total number of sample set, otherwise, return to step (4);

(6) Calculate the total error E, E is sum of sample errors; And determine whether E satisfy the accuracy requirements, if $E < \varepsilon$, stop learning, otherwise continue;
(7) Modify weights and thresholds following the formula, and return to step (4),

$$\Delta w_{ij}(l+1) = \beta \cdot e_j^k \cdot x_i + \eta \cdot \Delta w_{ij}(l), \tag{10}$$

$$\Delta v_{ij}(l+1) = \alpha \cdot d_t^k \cdot b_j + \eta \cdot \Delta v_{jt}(l), \tag{11}$$

$$\Delta \theta_j(l+1) = \beta \cdot e_j^k + \eta \cdot \Delta \theta_j(l), \tag{12}$$

$$\Delta \gamma_t(l+1) = \alpha \cdot d_t^k + \eta \cdot \Delta \gamma_t(l). \tag{13}$$

(8) Save w_{ij}, v_{jt} in case of calculation of predicted value;
(9) Calculate forecasting of traffic volume:
After the network training, forecast of traffic volume can be obtained in this type:

$$U_t = 1/[1 + \exp(-\sum_{j=1}^{n} v_{jt}/(1 + \exp(-\sum_{i=1}^{n} w_{ij}x_i + \theta_j)) + \gamma_t)]. \tag{14}$$

(10) Do the data processing of U_t value, which is the forecasting result of traffic volume.

4 The Experimental Results and Analysis

4.1 Results Analysis of BP Neural Network

Data is collected from a road section in ShunYi district of Beijing city on September 1st to 15th, 2014 (8:00 a.m. to 6:00 p.m.). After data pretreatment, traffic volume parameters are accurately predicted based on the BP neural network.

In the experiment, data collected from 1st to 12th is selected to be sample for training in the model, data from 13th to 15th is collected to test, forecast is considered to be correct if the predicted values from 13th to 15th and the actual value satisfy type (15), the experiment result is shown in Fig. 3.

$$\frac{|Actual\ value - Predicted\ value|}{Actual\ value} \prec 10\%. \tag{15}$$

In this paper, we adopt S shaped activation function in the hidden layer of network, and linear activation function in the output layer. The number of hidden layer nodes is 11, the maximum training number is 10000, and the initial learning rate is 0.21. First, process the experimental data to make the training data mapping in different order magnitude be inside the interval [−1, 1]. Then BP network is adopted for model training. In the process, the error squares gradually decrease as the number of training; the training won't stop until the desired error is achieved.

From Fig. 3, the predicted output curve of trained network model is very close to the original curve, the forecast result meets the requirements in terms of precision, which is highly consistent with the actual situation, and also the conclusion is perceivable.

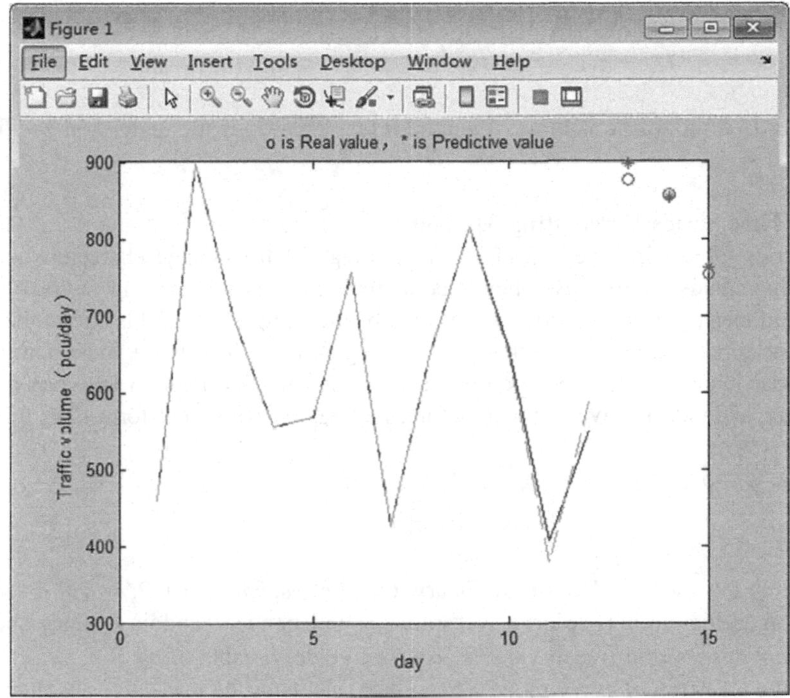

Fig. 3. The comparison figure of predicted value and actual value

4.2 Comparative Analysis of Several Forecasting Methods

In order to further study the forecast result of road traffic volume, we adopt the same data samples to predict based on the linear system theory, and compares with neural network model forecast.

Forecast method of linear system theory mainly includes the historical trend method, multivariate linear statistical regression, time series forecast method and so on.

4.2.1 Historical Trend Method

The premise of historical trend method is that traffic volume always follows certain rules. It usually takes simply the average traffic volume of historical time t as the predicted value of the current time t, which is a static method. So the historical trend method could not make an accurate forecast for actual traffic volume because of uncertain and nonlinear characteristics, this paper would not describe this in detail.

4.2.2 Multivariate Linear Statistical Regression

Multivariate linear regression model targets at finding the mathematical function relationship between the independent variable and dependent variable by statistical analysis on historical data, and then the predicted value can be output after taking the measured independent variables values. The model is widely applied in traffic volume forecasting of large-scale road network.

The multiple linear regression model of predicting traffic volume and historical traffic volume is:

$$U_t = a_1 U_1 + a_2 U_2 + a_3 U_3 + \ldots + a_{t-1} U_{t-1} + \delta. \tag{16}$$

Type: U is the traffic volume, δ is random variable, a_i is the regression coefficient of U_i.

4.2.3 Time Series Forecasting Method

Time series forecasting method targets at finding out the change characteristics and developing trends of forecast phenomenon through the analysis and study of time series, and then the extrapolation forecast can be made. Linear model is one of the most basic time series model, by contrast, the autoregressive model is the most commonly used linear model, which could describe linear dependence relation between data at some time and data at several times before. Autoregressive model form is as follows:

$$\left.\begin{array}{c} U_i = \mu + \varphi_{i-1}(U_{i-1} - \mu) + \varphi_{i-2}(U_{i-2} - \mu) + \cdots + \varphi_p(U_{p-1} - \mu) + \varepsilon_i \\ \varepsilon_i \sim N(0, \sigma_\delta^2) \end{array}\right\}. \tag{17}$$

Type: μ is the mean value of stationary time series; $\varphi_k (k = 1, 2, \cdots, p)$ is model regression coefficient: p is order; ε_i is residual, an independent random variable obeying the normal distribution (mean value 1, contrast grade variable of σ_δ^2).

For the precision of every model, the author calculates the respective absolute error and makes the statistical figure (Fig. 4).

From Fig. 4, the absolute error is small and relatively stable in the neural network for road traffic volume forecast. Apart from a few big fluctuations in a couple of special values, there are very small deviations. By contrast, the autoregressive models of multiple linear regression and time series forecast method shows relatively poor stability and significant deviations. Overall, the neural network forccast result is more effective and accurate than other methods.

To overcome some shortcomings like unscientific, unreliable and inconvincible in model performance evaluation with a single error indicator, the author adopts the MSE (mean square error), MSPE (mean square percentage error), MAE (mean absolute error) and MAPE (mean absolute relative error), 4 indicators altogether in evaluation of these traffic volume forecast models. The calculation result is as follows (Table 1).

From the calculation result of various indicators, the deviation of neural network is smaller than the other two models. The average deviation of neural network is 6.65 %, the average deviation of multivariate linear regression model is 14.30 %, and the average deviation of regression model is 12.96 %. It is shown that the neural network

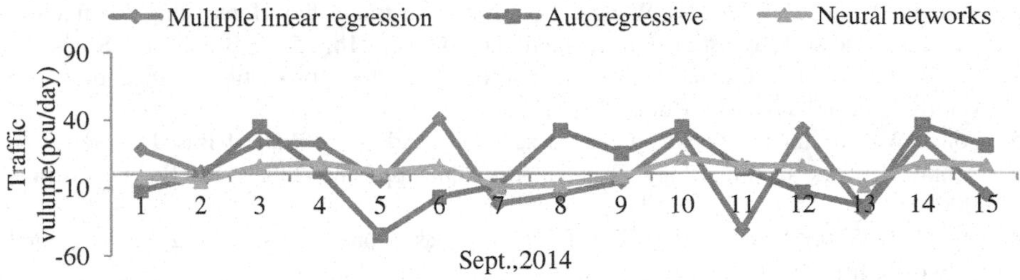

Fig. 4. The absolute error of three models forecasting

Table 1. Forecasting performance of three models

Model	MSE	MSPE	MAE	MAPE
Multiple linear regression	2.50258	0.015564	21.65485	0.14304
Autoregressive model	2.355996	0.014822	20.02648	0.129604
Neural network	1.063862	0.007646	9.69082	0.066582

of traffic volume forecast has significant advantage in accuracy and stability over multiple linear regression model and autoregressive model.

5 Conclusion

Information fusion is an emerging technology field, which provides a good method for traffic information processing. The biggest advantage of information fusion technology is to reasonably coordinate multi-source data, fully integrate useful information and improve the ability of correct decision making in changeable environments, which can be precisely and widely employed in the intelligent transportation field. Transportation system is a complex, nonlinear and time-varying system where people, vehicles and road are interacted. Therefore a high degree of uncertainty is an outstanding feature. However, information fusion based on the artificial neural network can make reasonable forecasts of traffic volume. Accurate and reliable traffic volume forecast information is the premise and foundation to achieve traffic guidance, and plays a positive role in broadening the forecasting algorithm design of traffic volume.

References

1. Weng, X.X., Du, G.I.: Hybrid Elman Neural Network Model for Short-Term Traffic Prediction, pp. 235–280. IASTED International Conference Press, Anaheim (2006)
2. Research Institute of Highway Ministry of Transport. Research Report of China Intelligent Transportation System Framework. Research Institute of Highway Ministry of Transport Press, Beijing (2010)

3. Yua, R., Laoa, Y., Maa, X., Wanga, Y.: Short-term traffic flow forecasting for freeway incident-induced delay estimation. J. Intell. Transp. Syst. **18**, 254–263 (2015). USA
4. Ke L., Jianjun,,T. Yingyuan, L: A short-term traffic flow forecasting method based on MapReduce, pp. 174–179. China (2015)
5. Cong, W.: An Internet Traffic Forecasting Model Adopting Radical Based on Function Neural Network Optimized by Genetic Algorithm, pp. 385–388. IEEE Press, Australia (2008)
6. Liang, Z.: Short-Term Traffic Flow Prediction Based on Interval Type-2 Fuzzy Neural Networks. IEEE Press, China (2010)
7. Chrobok, R., Kaumann, O., Wahle, J.: Different methods of traffic forecast based on real data. J. Eur. J. Oper. Res. **155**, 558–568 (2004)
8. Cetin, M., Comert, G.: Short-term traffic flow prediction with regime-switching models. J. Transp. Res. Rec. **1965**, 23–31 (2006)
9. Eleni, I.V., Matthew, G.K., John, C.G.: Optimized and meta-optimized neural networks for short-term traffic flow prediction: a genetic approach. J. Transp. Res. Part C Emerg. Technol. **13**, 211–234 (2005)
10. Chang, G., Tongming, G.: Comparison of Missing Data Imputation Methods for Traffic Flow, pp. 639–642. IEEE Press, USA (2011)
11. An, Y., Song, Q.: Short-term traffic flow forecasting via echo state neural networks, pp. 844–847. In: Ding, Y., Wang, H., Xiong, N., Hao, K., Wang, L. (Eds.) Proceedings of 7th ICNC, Press, China (2011)
12. Tao, Z., Lifang, H., Zhixin, L., Yuejie, Z.: Nonparametric Regression for the Short-Term Traffic Flow Forecasting. MACE Press, USA (2010)
13. Kamga, C.N., Mouskos, K.C., Paaswell, R.E.: A methodology to estimate travel time using dynamic traffic assignment (DTA) under incident conditions. J. Transp. Res. Part C Emer. Technol. **19**, 1215–1224 (2011)
14. Cheevarunothai, P., Zhang, G., Wang, Y.: Using precise time offset to improve freeway vehicle delay estimates. J. Intell. Transp. Syst. Technol. Plan. Oper. **16**, 82–93 (2012)
15. Chung, Y.: Quantification of non-recurrent congestion delay caused by freeway accidents and analysis of causal factors. Transp. Res. Rec. J. Transp. Res. Board **2229**, 8–18 (2011)
16. Horvitz, E., Apacible, J., Sarin, R., Liao, L..: Prediction, expectation, and surprise: methods, designs, and study of a deployed traffic forecasting service. CoRR, vol. abs/1207.1352, USA (2012)
17. Lippi, M., Bertini, M., Frasconi, P.: Short-Term Traffic Flow Forecasting: An Experimental Comparison of Time-Series Analysis and Supervised Learning. IEEE Press, New York (2013)

A Consistent Hashing Based Data Redistribution Algorithm

Xiang Fu(✉), Can Peng, and Weihong Han

College of Computer, National University of Defense Technology, Changsha, Hunan, China
xjtusefox@163.com, {kenneth_peng,hangweihong}@139.com

Abstract. Data partitioning is the basis of massive data distributed storage management and consistent hashing algorithm is one of the most popular data partitioning methods. We analyzes consistent hashing algorithm and find it can not guarantee the dynamic load balancing of the nodes in the system during the running time. To solve this problem,we propose a consistent hashing based data redistribution algorithm(CHRA). And the comparison of our simulation experiments verifies that the CHRA can balance the load of nodes in the system.

Keywords: Data partitioning · Consistent hashing · Load balancing · Data redistribution

1 Introduction

With the development of information technology, the data of peoples daily life is exploding. Applications have to handle vast amounts of data every day. Only distributed storage systems can meet the demand for high-efficiency storage and management. The distribution of the data will directly affect the speed of data access, the cost of system expansion and so on. Therefore, a good data distribution method means a high performance distributed storage system. Consistent hashing algorithm [1] is currently a very popular data partitioning [2,3] method which has been successfully applied in a variety of mainstream distributed storage systems such as Cassandra [8], Dynamo [9], Voldemort [10] and etc. However, the consistent hashing algorithm has an obvious weak point which is that it can not guarantee the load balancing [4,7] of the nodes consisting in the system. But data redistribution [5,6] is born to solve this problem. Aiming at finding an efficient and practical data redistribution algorithm, we have done a lot of work-related.

2 Consistent Hashing Algorithm

Consistent hashing algorithm is a special hash algorithm. The greatest advantage of the algorithm is that when nodes change (add nodes or remove nodes) in the

X. He et al. (Eds.): IScIDE 2015, Part II, LNCS 9243, pp. 559–566, 2015.
DOI: 10.1007/978-3-319-23862-3_55

system, most data will be mapped to the same node before the change. When adding a node into the system, it will only get the data from the adjacent node without affecting the rest of the non-adjacent nodes. When a node is removed from the system, the data in this node will only be reassigned to the adjacent node. The works of consistent hashing algorithm can be divided into the following three steps:

1, The Hash Circle. In general situation, traditional hash algorithm will map the values to the value space from 0 to $2^{32}-1$. We can image this space as a head-tail circle, then the values will be located on the circle, as shown in Fig. 1(a).

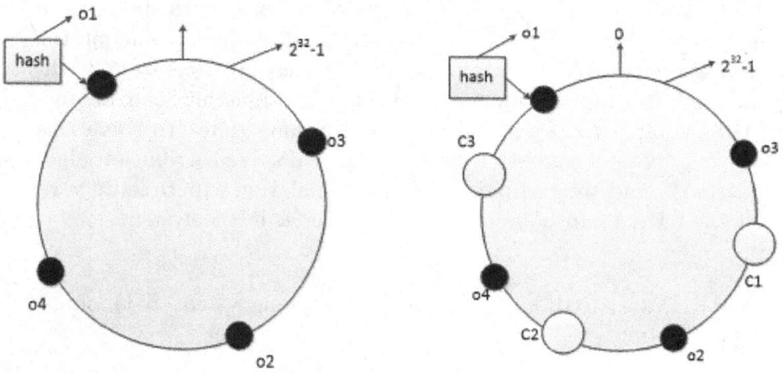

(a) Values mapped on the hash circle (b) Nodes and objects on the hash circle

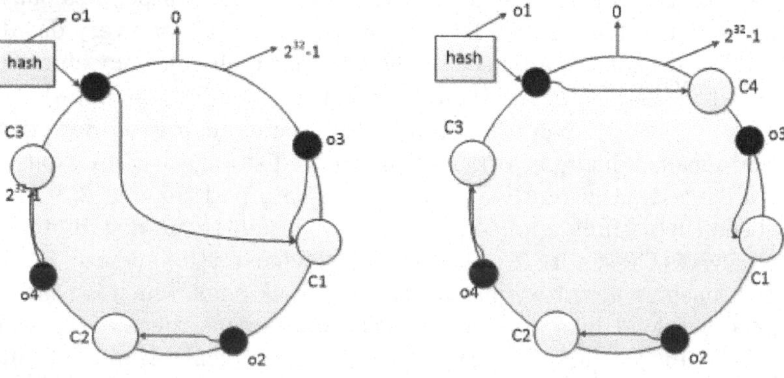

(c) Mapping result of 3 nodes and 4 objects

(d) Mapping result after adding C4

Fig. 1. The consistent hashing algorithm

2, Map Objects and Nodes onto the Hash Circle. The basic idea of consistent hashing algorithm is using the same hash algorithm to map nodes and objects into the same hash value space. Suppose there are three nodes marked as C1, C2, C3 and four objects marked as o1, o2, o3 and o4, then the mapping result will be shown on in Fig. 1(b).

3, Map Objects to Nodes. Each object is mapped to the first node it encounters in clockwise direction. According to the rule, Object o1 and o3 will be stored on C1; Object o2 will be stored on C2; Object o4 will be stored on C3. The mapping result is shown in Fig. 1(c).

Now we consider the impact of nodes change to the mapping result. The biggest flaw of traditional hash algorithm is that adding or removing nodes will cause a great change of the mapping result. What about the consistent hashing algorithm? Assume adding node C4 into the system, only Object o1 will be migrated to C4 according to the above mapping rule. As shown in Fig. 1 (d).

Assume removing node C2 from the system, similarly, only those nodes between node C2 and C1 will be affected. So object o2 will be re-mapped to C3.

Now consider this situation, in Fig. 1(c), if we remove C3 from the system, according to the mapping rule, node C2 will only store o2 while node C1 will store o1, o3 and o4 which means the system is quite imbalanced. Once the hash function is set, the mapping result can not be changed. So this algorithm can not guarantee the dynamic load balancing. Data need to be redistributed for high performance of the system.

3 Data Redistribution

3.1 Definition of Terms

In order to describe the CHRA comfortably, the terms that are used in this paper will be introduced below.

Node Load Capability. There are many factors infect the load of nodes in practical applications, such as CPU, memory (RAM), storage capacity, network bandwidth and ect.. In this research, we just focus on the top three factors mentioned above. We will not concern the specific calculations of load in this study, we just use a CPU, memory and storage capacity-related function to represent node load capability. In the function, ω_{cpu} represents the CPU of node, ω_{RAM} represents the memory and ω_c represents the storage, as shown in (1)

$$l = f\left(\omega_{cpu}, \omega_{RAM}, \omega_c\right) \tag{1}$$

Load Forecasting. Different data size or access frequency causes different node lord. In our research, we use the data size, history access frequency and the time when the data was inserted to predict the load caused by the data. Here we use (2) to represent the load forecasting. In this function, ω_s represents the data size, ω_f represents the history access frequency and ω_t represents the time when the data was inserted.

$$\mathrm{r} = \mathrm{g}\left(\omega_s, \omega_f, \omega_t\right) \tag{2}$$

Balancing Standard Deviation. To balance the node load of the system, nodes with bigger load capability(calculated by (1)) should store the data with bigger load forecasting(calculated by (2)). The balancing standard deviation can tell us the balancing degree of the system. When should we do data redistribution? In our research, A data redistribution is needed when the balancing standard deviation of the system is greater than a given threshold. We use l_i and r_i to represent the load capability of node i and the laod forecasting of data i , so the balancing standard deviation σ is calculated as follows:

$$\mathrm{L} = \sum_{i=1}^{n} l_i \tag{3}$$

$$R = \sum_{i=1}^{m} r_i \tag{4}$$

$$l_i^* = l_i \cdot R/L \tag{5}$$

$$\sigma = \sqrt{\frac{1}{n} \cdot \sum_{i=1}^{n} \left(\frac{l_i - l_i^*}{l_i^*}\right)^2} \tag{6}$$

Table 1. The node position table

The ip od the node	Positon
192.168.1.1	10245
192.168.1.122	200836
192.168.1.105	435068
......	......
192.168.1.253	$2^{32}-1$

Node Position Table. Unlike consistent hashing algorithm, the node position on the hash circle in CHRA is not calculated by the hash function which is used to calculate the position of data. We specify the node position according to the distribution of the data. The system needs to maintain a node position table which contains a unique identifier for each node (e.g. the ip of the node) and its position ($0 \sim 2^{32}-1$) on the hash circle, as shown in Table 1.

3.2 Algorithms in CHRA

The CHRA can only run in the system that uses consistent hashing algorithm as the data partitioning method such as Cassandra, Dynamo, Voldemort and etc. The data redistribution algorithm and the data access algorithm are mian algorithms in CHRA.

In Algorithm 1, we use the mapping rule in consistent hashing algorithm to map the data and nodes. We can see that most of the data migration happens between the neighboring nodes and data migration can be completed without shutting dwon the system. To read or write data, we should firstly get the storage node of it. Algorithm 2 will show you how.

Algorithm 1. The data redistribution algorithm.

Input:
The hash circle.
Output:
The node position table.;

Set a balancing standard deviation threshold σ^*;
Calculate the balancing standard deviation σ of the system;
if $\sigma > \sigma^*$ **then**
  Set the position of node n as $2^{32}-1$ in the node position table;
  for each data i **do**
    if $\sum_1^i r_i > l_k^*$ **then**
      Set the position of node k as the position of data (i-1) in the node position table;
      k++;
    else
      Migrate data i to node k if data i not in node k; // k initialized as 1
    end if
  end for
  return The new node position table;
end if
return The old node position table;

Algorithm 2. The data access algorithm.

Input:
The key of the data needed
Output:
The ip of the storage node.;

1: Use the hash function to calculate the position of the data on the hash circle, P;
2: Search the node position table for those nodes whoes position is not smaller than P and sort these nodes based on position into an array Connodes[] ;
3: Send the operation to node Connodes[0] ;
4: **return** Connodes[0];

4 Simulation

Simulation distributed database system A, B and C all use the consistent hashing algorithm as the data partitioning method. The only difference is that system B and C use CHRA to balance the load. The balancing standard deviation threshold σ^* for system B is 0.4 and for system C is 0.5. There are 8 nodes and 1000 data records in these systems at the beginning. To simplify the calculations, we assume that each node has the same load capability and each record has the same load forecasting. Then we send 100 operations to all of the systems. The operations are divided into 10 groups and each group contains 9 operations on data and 1 operation on nodes. Table 2 describes theses two kinds of operations.

The balancing standard deviation are recorded after every operation. Figure 2 shows the result of the first comparison which is between system A and B. We can see that our CHRA can guarantee the dynamic load balancing in a certain extent.

The comparison between system B and C in Fig. 3 shows that different balancing standard deviation threshold result in different load balancing effect. During the 100-operation process, although system B is more lord-balanced than system C, system C only need 3 times of data migration while system B need 6.

In practical applications, we should set the balancing standard deviation threshold for CHRA according to the migration costs and system load balancing needs to guarantee the high performance of the system.

Table 2. Two kinds of the simulation operations

Operation on data	Randomly insert or delete 100–200 records
Operation on nodes	Randomly add or remove 1–2 nodes

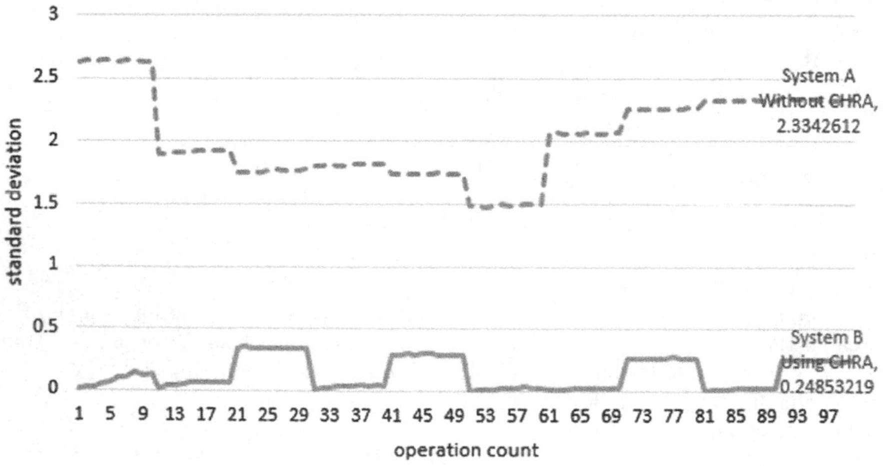

Fig. 2. the comparison between system A and B

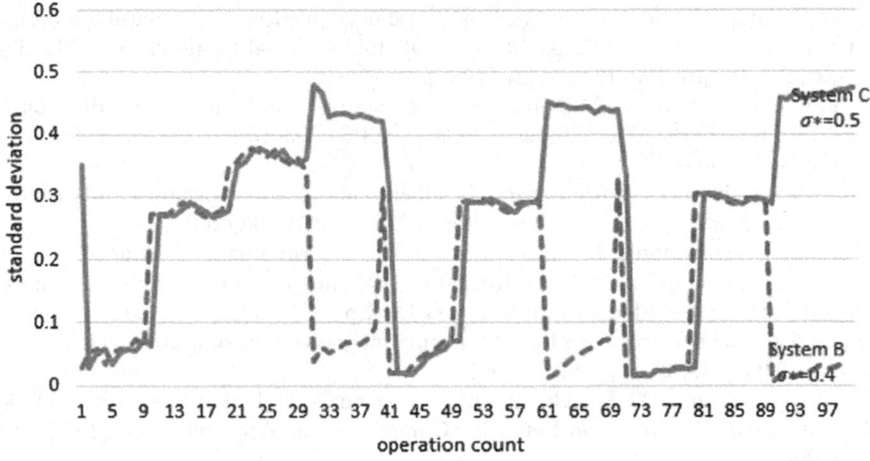

Fig. 3. the comparison between system B and C

5 Conclusion

In this paper, we carry out a detailed analysis of the consistent hashing algorithm which is currently very popular as a data partitioning method. But this algorithm cannot guarantee load balancing. We propose a consistent hashing based data redistribution algorithm(CHRA) to solve this problem. In CHRA, we follow the concept of hash circle and the mapping rule in consistent hashing algorithm. We propose a function to calculate the balancing standard deviation of a system. If the balancing standard deviation of the system is higher than the pre-set threshold, data redistribution is required. We use algorithm 2.1 to balance the node load. After the redistribution, the load forecasting of the data assigned to each node can match the node load capability which means that the system is well load-balanced.

Acknowledgment. This paper is supported by the national high technology research and developmentprogramme (863programme) (2012AA01A401), the national natural science foundation (60933005, 91124002) and the national information security program242 (2011A010).

References

1. Karger, D., Lehman, E., Leighton, T., et al.: Consistent hashing and random trees: distributed caching protocols for relieving hot spots on the World Wide Web. In: Proceedings of the Twenty-Ninth Annual ACM Symposium on Theory of Computing, pp. 654–663. ACM (1997)
2. Scheuermann, P., Weikum, G., Zabback, P.: Data partitioning and load balancing in parallel disk systems. VLDB J. **7**(1), 48–66 (1998)

3. Ceri, S., Negri, M., Pelagatti, G.: Horizontal data partitioning in database design. In: Proceedings of the 1982 ACM SIGMOD International Conference on Management of Data, pp. 128–136. ACM (1982)
4. Byers, J., Considine, J., Mitzenmacher, M.: Simple load balancing for distributed hash. In: Kaashoek, M.F., Stoica, I. (eds.) IPTPS 2003. LNCS, vol. 2735. Springer, Heidelberg (2003)
5. Guo, M., Nakata, I.: A framework for efficient data redistribution on distributed memory multicomputers. J. Supercomput. **20**(3), 243–265 (2001)
6. Herault, T., Herrmann, J., Marchal, L., et al.: Determining the optimal redistribution for a given data partition. In: 2014 IEEE 13th International Symposium on Parallel and Distributed Computing (ISPDC), pp. 95–102. IEEE (2014)
7. Yousefi'zadeh, H.: Database load balancing for multi-tier computer systems: U.S. Patent 6,950,848, 27 September 2005
8. Han, J., Haihong, E., Le, G., et al.: Survey on NoSQL database. In: 2011 6th International Conference on Pervasive Computing and Applications (ICPCA), pp. 363–366. IEEE (2011)
9. DeCandia, G., Hastorun, D., Jampani, M., et al.: Dynamo: amazon's highly available key-value store. ACM SIGOPS Oper. Syst. Rev. **41**(6), 205–220 (2007)
10. Feinberg, A.: Project voldemort: reliable distributed storage. In: Proceedings of the 10th IEEE International Conference on Data Engineering (2011)

Semantic Parsing Using Hierarchical Concept Base

Yi Gao, Caifu Hong, and Xihong Wu(✉)

Speech and Hearing Research Center, Key Laboratory of Machine Perception and Intelligence(Ministry of Education), School of Electronics Engineering and Computer Science, Peking University, Beijing, China
{gaoyi,hongcf,wxh}@cis.pku.edu.cn

Abstract. Compositional question answering first maps natural language sentences into meaning representations, then a meaning interpreter is used to evaluate the corresponding answers against a database. A novel approach is proposed in this paper which involves a concept base with rich hierarchical information. A new meaning representation form is introduced correspondingly to match the hierarchical concept base. A set of constructions which encode the correspondence of concept sequences and their meaning representations are used for parsing. The experimental results show that the proposed semantic parser performs favorably in terms of both accuracy and generalization performance compared to existing semantic parsers.

Keywords: Semantic parsing · Concept base · Hierarchical information

1 Introduction

Compositional question answering is important in natural language understanding. Previous works involve semantic parsing which maps a natural language(NL) sentence into its meaning representation(MR). Semantic parsers based on syntax first derive syntactic trees from the NL sentences, then the syntactic trees are converted to the corresponding MRs [1–3]. Semantic parsers based on machine translation technologies use synchronous grammars, which match NL string patterns and construct MRs synchronously [4–7]. Semantic parsers using dependency-based compositional semantics(DCS) derive all the possible MRs from NL sentences, then a probabilistic model is used to find the answers [8,9]. Recent years, semantic parsers using knowledge bases [10–13] such as Freebase [14], or using grounded information [15,16] are developed to handle domain-independent, large-scale corpus.

Overall, two kinds of information are used to improve the generalization performance of a semantic parser. One is the syntactic information. Words or phrases are generalized into syntactic non-terminals, which capture the unseen

Xihong Wu—IEEE Senior member.

X. He et al. (Eds.): IScIDE 2015, Part II, LNCS 9243, pp. 567–575, 2015.
DOI: 10.1007/978-3-319-23862-3_56

phrases in the training corpus. Another is the semantic information such as knowledge bases, which capture the unseen relations in the training corpus. However, traditional methods using syntactic information always tend to overgeneralize the NL words, and the methods using semantic information do not capture the hierarchical relations which would further improve the generalization of semantic parsers.

A novel approach for semantic parsing is proposed in this paper. A concept base is introduced which contains hierarchical relations between concepts. A new form of MR for the semantic parser is proposed to match the concept base. It shares the same structure with the concept base. Compared to the widely used Montague semantics based on lambda calculus, the proposed MR is easily to combine with a concept base. Compared to DCS which constrains the concept relations to several kinds, the proposed MR is free to contain all relations.

To integrate the concept base into the semantic parser, a set of constructions are introduced which map concept sequences into their MRs. The constructions have some resemblances to the rules in synchronous grammars. They both capture the syntactic information in a specific way, which avoids the overgeneralization of NL words. But constructions are based on the concept base. They can also captures the semantic information between concepts.

The experiments in GeoQuery [17], a benchmark dataset, have shown that the proposed system outperforms all existing systems both in accuracy and generalization performance.

2 Concept Base

The basic elements in the concept base are concepts. A string with the first letter capitalized denotes the corresponding concept if no ambiguities exist. The formal definition of the concept base is $K =< E, A, R, E_i, A_i, R_i, R_h >$, in which:

E represents entity, examples include $City$, $State$, $Person$, etc. E_i represents the instances of entities. The extension of an entity is the set of the instances of that entity. For example, $Austin$ is an instance of $City$, so it's an element in the extension of $City$.

A represents attribute, examples include $Height$, $Length$, $Area$, etc. A_i represents the instances of attributes. The extension of an attribute is the set of the instances of that attribute.

R represents relations. A relation relates a set of concepts which are relative elements of that relation. R_i are the instances of relations. The extension of a relation is the set of instances of that relation. For example, a relation $Loc(City, State)$ means a $City$ is located in a $State$, while $Loc(Austin, Texas)$ represents an instance of the relation $Loc(City, State)$, where $City$ and $State$ are instantiated as $Austin$ and $Texas$.

R_h represents the hierarchical relations. They are a kind of relations. If a concept C_1 is the hypernym of another concept C_2, then there exists a hierarchical relation between C_1 and C_2. Examples include the relation between $City$ and $Capital$, the relation between $Attribute$ and $Area$, etc.

The hypernym of a concept C is denoted as C_h^C. The extension of a concept C is denoted as $Ext(C)$. Note that words in NL are also regarded as concepts, they are instances of the entity $Word$.

Since there are no appropriate hierarchical concept bases for GeoQuery currently, a manually annotated one is used to conduct the experiments. Theoretically the hierarchical concept base is domain-independent, and can be used in any other systems.

3 Meaning Representation

3.1 Semantic Tree

The MRs of sentences also consist of concepts. A basic assumption is made that all the MRs are trees. A semantic tree is represented as $t =< Root\ (C_1)\ (C_2)\ \ldots\ (C_m) >$, where $Root$ is the root of the tree, and $C_1, C_2, \ldots, C_m$ are the child trees of $Root$. Figure 1 shows some examples. Each concept in the tree is a hyponym of some concept in the concept base, because they have different extensions. In Fig. 1(a), the extension of $State'$ should be the states bordering Texas, while its hypernym, $State$, has all the known states in its extension. This specific form of MR shares the same structure with the concept base, which allows convenient computation of the semantic tree.

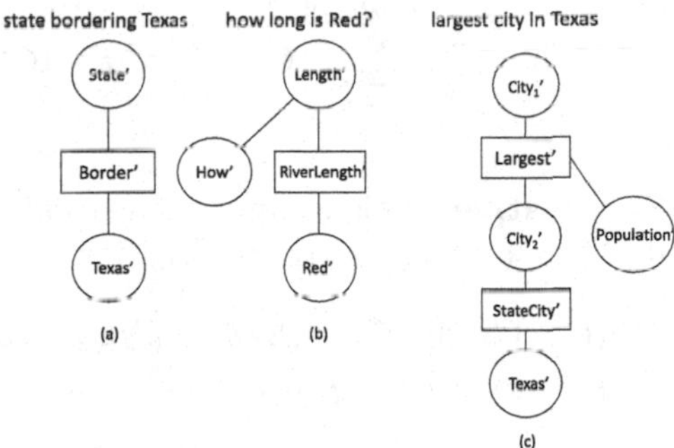

Fig. 1. Some examples of semantic trees in our system

3.2 Computation

The computation of a semantic tree is defined as the procedure of finding the extension of the focus concept in the semantic tree. Normally, the focus concept is the root of the tree. If there exists interrogatives in the tree, such as $When$, $Where$, $What$, How, etc., then focus concept is the concept connected to that interrogative. There are two typical cases:

(1) The focus concept is a relation R, and all the concepts connected to it are its relative elements, denoted as $C_1, C_2, \ldots, C_m$. The instances of the focus

concept should be included in the extension of its hypernym C_h^R. Denote an instance in the extension of R_h^R as R_i. Its relative elements are $C_{1i}, C_{2i}, \ldots, C_{mi}$, which are all instances. If for every $j \in 1, 2, \ldots, m$, C_{ji} is an instance of C_j, then R_i is an instance of R. Formally:

$$\begin{aligned} Ext(R) = \{&R_i(C_{1i}, C_{2i}, \ldots, C_{mi}) \in Ext(C_h^R)| \\ &C_{1i} \in Ext(C_1) \wedge C_{2i} \in Ext(C_2) \wedge \cdots \wedge C_{mi} \in Ext(C_m)\} \end{aligned} \tag{1}$$

This is called the "match" method. Since not all relation instances are known in advance, several computing methods are needed in GeoQuery:

Count. This computing method is used when counting the number of instances of a concept. Formally:

$$Ext(R) = \{R_i(C_{1i}, C_2) \in Ext(C_h^R)|C_{1i} \in Ext(C_1) \wedge C_{1i} = |Ext(C_2)|\} \tag{2}$$

Quantification. Examples of Quantification include *Total*, *Average*, etc. For example, the computing method of *Average* is:

$$\begin{aligned} Ext(R) = \{R_i(C_{1i}, C_2) \in Ext(C_h^R)|C_{1i} \in Ext(C_1)\wedge \\ C_{1i} = \frac{1}{N}\sum_{j=1}^{|Ext(C_2)|} C_{2j}, C_{2j} \in Ext(C_2)\} \end{aligned} \tag{3}$$

Comparative and Superlative. This computing method is used when comparing entities C_1 and C_2, on a specific attribute A_3. Formally:

$$\begin{aligned} Ext(R) = \{R_i(C_{1i}, C_2, A_3)|C_{1i} \in Ext(C_1), \forall C_{2i} \in Ext(C_2), \\ R_i(C_{1i}, C_{2i}, A_3) \in Ext(C_h^R)\} \end{aligned} \tag{4}$$

Negation. If the relation is negative, this computing method is used. Formally:

$$\begin{aligned} &Ext(R) = \{R_i(C_{1i}, C_{2i}, \ldots, C_{mi})|C_{1i} \in Ext(C_1) \wedge C_{2i} \in Ext(C_2)\wedge \\ &\cdots \wedge C_{mi} \in Ext(C_m) \wedge R_i(C_{1i}, C_{2i}, \ldots, C_{mi}) \notin Ext(C_h^R)\} \end{aligned} \tag{5}$$

(2) The concepts connected to the focus concept are relations, and each relation takes the focus concept as one of its relative elements. Denote the tree as $t =< C(R_1)(R_2)\ldots(R_m) >$. The extension of C is first set to be the extension of its hypernym C_h^C. Then the extensions of those relations are computed using the computing methods introduced above. For each instance C_i in the extension of C, if C_i is one relative element of some instance of R_1, as well as $R_2, \ldots, R_m$, then C_i is an instance of C. Usually, there exists a computing order. Absolute

relations, such as relations computed using "match", are computed before relative relations, such as ones computed using "comparative and superlative".

$$S = \bigcap_{j=1}^{m} \{x \in Ext(C_h^R) | \exists C_{1i}^j, C_{2i}^j, \dots C_{mi}^j, R_j(C_{1i}^j, C_{2i}^j, \dots, x, \dots, C_{mi}^j) \in Ext(R^j), R^j \textit{ is an absolute relation}\} \tag{6}$$

$$Ext(R) = \bigcap_{j=1}^{m} \{x \in S | \exists C_{1i}^j, C_{2i}^j, \dots C_{mi}^j, R_j(C_{1i}^j, C_{2i}^j, \dots, x, \dots, C_{mi}^j) \in Ext(R^j), R^j \textit{ is a relative relation}\} \tag{7}$$

For semantic trees containing the both cases, the result should be the intersection of the results obtained in them. Using the computing methods above mentioned, a semantic tree can be recursively computed.

4 Construction

Construction Representation. A construction encodes the correspondence of a concept sequence and its MR. Note that words in NL are regarded as concepts, so constructions can also be used to denote the correspondence of words and their MRs. A basic assumption for constructions is that except for the constructions of words, the MR of a construction should have one and only one relation concept, and all the other concepts in the MR are the relative elements of that relation. Formally, $cons =< P; F; T; H; R >$. Here P represents the concept sequence. F represents morphological and semantic features(MSFs), such as number for entities, participle for relations, affirmative or negative for relations, etc. They are used to restrict the concept usage to appropriate syntax and semantic context. They are universal and domain-independent. T is the corresponding semantic tree which consists of the concepts in P, and H is the root of the semantic tree. R represents the only relation in T.

Construction Annotation. The sentences in training corpus are manually annotated into constructions. For a sentence S, the procedure is as follows:

(1)For every word W in S, the corresponding MSF is extracted as F_w. Assume that W corresponds to only one MR in this context, denoted as C. The construction is annotated as: $< W; F_w; < C >; C; \epsilon >$.

(2)For every phrase $P_{nl} =< W_1, W_2, \dots, W_m >$ in S, first map every word in the phrase to corresponding MR, $P =< C_1, C_2, \dots, C_m >$, with the extracted MSF sequence as F. Then for every concepts in P, find its highest level of hypernym in the hierarchical concept base. This forms a new concept sequence, denoted as P_h. Annotate the corresponding semantic tree T with head H for P_h. Note that there should be only one relation R in T, otherwise the phrase should be split into smaller phrases. Thus the construction is $cons =< P_h; F; T; H; R >$.

Probabilistic Construction. In general, a word or a phrase may correspond to multiple MRs. These ambiguities are the main motivation for extending constructions into probabilistic constructions. Given a word or a phrase, the probability of a MR derived from this phrase is $p(T, H, R|P, F)$. It can be obtained from the corpus:

$$P(T, H, R|P, F) = \frac{count(< P; F; T; H; R >)}{\sum_{(T,H,R)} count(< P; F; T; H; R >)} \tag{8}$$

5 Semantic Parsing

A construction can be split into two parts, a production with H as its left hand side (LHS) and P as its right hand side (RHS), and a semantic tree T. Denote a random subsequence of the input concept sequence as $P_s = (C_{s1}, C_{s2}, \ldots, C_{sm})$, whose MSF sequence is F_s. For a construction $cons =< C_{c1}, C_{c2}, \ldots, C_{cm}; F_c; T; H; R >$, and for every $j \in 1, 2, \ldots, m$, check the following propositions: (1) C_{sj} is C_{cj}; (2) C_{sj} is one hyponym of C_{cj}; (3) C_{sj} is an instance of C_{cj}. If one of them is true, then replace C_{cj} in T as C_{sj}. This forms a new semantic tree T_s, which is part of the meaning representation of the input concept sequence. Denote the root of T_s as H_s. H_s can be further used to combine with other concepts in the input concept sequence. The parsing task has some resemblance with the probabilistic context-free grammar (PCFG) parsing, and the Earley's context-free parsing algorithm [18] is used. Unlike Earley's algorithm which is operated on words and syntactic non-terminals, here the parser operates on concepts. The probability of a semantic tree is obtained by multiplying the probabilities of all the constructions used in that semantic tree. Finally, the most probable one is selected from the semantic tree set T_{set}:

$$\begin{aligned} T_{best} &= argmax_{T \in T_{set}}(P(T, H, R|S, F) \\ &= argmax_{T \in T_{set}}(\prod_{T_j \in T} P(T_j, H_j, R_j|P_j, F_j))) \end{aligned} \tag{9}$$

6 Experiments and Results

The experiments on GeoQuery are conducted. The dataset contains about 800 facts asserting relational information about U.S. geography, and 880 questions annotated with the corresponding MRs. The average length of a sentence is 7.48 words. The proposed system is compared to: (1)WASP [4], which is based on machine translation techniques; (2) λ-WASP [5], an extension of WASP for handling MRs; (3) SYNSEM [1], which combines syntactic information and semantic information together, here we choose its result based on the gold-standard syntactic parses; (4) L2013 [9], which uses DCS as MRs; (5) W2014 [3], which performs best in current CCG-based parsers.

In the first experiment, the accuracy of the proposed system is tested. The corpus is split as 600 questions for training and 280 questions for testing.

Table 1. The Accuracy on GeoQuery

System	Accuracy(%)
WASP	74.8
λ-WASP	86.6
SYNSEM	88.2
L2013	91.4
W2014	90.4
New System	93.4

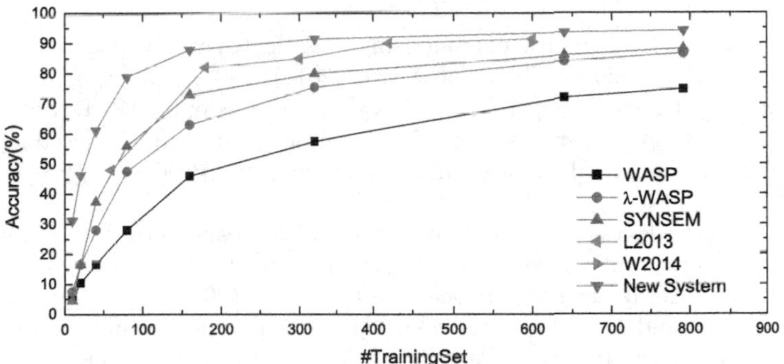

Fig. 2. Learning curves for various parsing algorithms on the GeoQuery corpus

Table 1 shows the results. A few observations can be made: (1)The new system outperforms all existing systems; (2)Though the new system needs more annotation, compared to SYNSEM which uses gold-standard syntactic parses, it still performs better.

The second experiment is about the generalization performance. The standard 10-fold cross validation is used. Figure 2 shows the learning curves of difference systems. It can be observed that, the new system outperforms other systems by wide margins, matching their best final accuracy with only 50 % of the total training examples. This can greatly alleviate the burden of annotation.

7 Conclusion

Hierarchical information can greatly improve the performance of semantic parsers. It includes a concept base which encodes the hierarchical relations between concepts, and a set of constructions which encodes the correspondences of concept sequences and their MRs. By using Earley's algorithm in the semantic parser, the accuracy and generalization performance on the standard semantic parsing dataset, GeoQuery, is clearly improved.

However, the constructions in the system were manually annotated. Learning these constructions automatically will be the future work. This could alleviate the burden of annotation, and also reduce the errors in the annotated corpus.

Acknowledgments. The research was partially supported by the National Basic Research Program of China (973 Program) under grant 2013CB329304, the Major Project of National Social Science Foundation of China under grant 12&ZD119, and the National Natural Science Foundation of China under grant 61121002.

References

1. Ge, R.: Learning for semantic parsing using statistical syntactic parsing techniques. Ph.D. thesis, University of Texas at Austin (2010)
2. Kwiatkowski, T., Zettlemoyer, L., Goldwater, S., Steedman, M.: Lexical generalization in CCG grammar induction for semantic parsing. In: Proceedings of the Conference on Empirical Methods in Natural Language Processing, pp. 1512–1523 (2011)
3. Wang, A., Kwiatkowski, T., Zettlemoyer, L.: Morpho-syntactic lexical generalization for CCG semantic parsing. In: Proceedings of the Conference on Empirical Methods in Natural Language Processing (EMNLP) (2014)
4. Wong, Y. W., and Mooney, R. J.: Learning for semantic parsing with statistical machine translation. In: Proceedings of HLT/NAACL 2006, pp. 439–446, New York City (2006)
5. Wong, Y.W., Mooney R.J.: Learning synchronous grammars for semantic parsing with lambda calculus. In: Association for Computational Linguistics (ACL), pp. 960–967, Prague, Czech Republic (2007)
6. Jones, B. K., Johnson, M., Goldwater, S.: Semantic parsing with bayesian tree transducers. In: Proceedings of the 50th Annual Meeting of the Association for Computational Linguistics, pp. 488–496 (2012)
7. Andreas, J., Vlachos, A., and Clark, S.: Semantic parsing as machine translation. In: Proceedings of the Conference of the Association for Computational Linguistics (ACL), pp. 47–52 (2013)
8. Liang, P., Jordan, M. I., and Klein, D.: Learning Dependency-Based Compositional Semantics. In: Association for Computational Linguistics (ACL), pp. 590–599, Portland (2011)
9. Liang, P., Jordan, M.I., Klein, D.: Learning dependency-based compositional semantics. Comput. Linguist. **39**(2), 389–446 (2013)
10. Krishnamurthy, J. and Mitchell, T.: Weakly Supervised Training of Semantic Parsers. In: Proceedings of the Joint Conference on Empirical Methods in Natural Language Processing and Computational Natural Language Learning, pp. 754–765 (2012)
11. Cai, Q. and Yates, A.: Semantic parsing freebase: towards open-domain semantic parsing. In: Proceedings of the Joint Conference on Lexical and Computational Semantics (2013)
12. Berant, J., Chou, A., Frostig, R., and Liang, P.: Semantic parsing on freebase from question-answer Pairs. In: Proceedings of the 2013 Conference on Empirical Methods in Natural Language Processing(EMNLP), pp. 1533–1544 (2013)

13. Kwiatkowski, T., Choi, E., Artzi, Y., and Zettlemoyer, L.: Scaling Semantic Parsers with On-the-Fly Ontology Matching. In: Proceedings of the 2013 Conference on Empirical Methods in Natural Language Processing. Seattle, Washington (2013)
14. Bollacker, K., Evans, C., Paritosh, P., Sturge, T., Taylor, J.: Freebase: a collaboratively created graph database for structuring human knowledge. In: Proceedings of the International Conference on Management of Data (SIGMOD), pp. 1247–1250 (2008)
15. Chen, D., Mooney, R.: Learning to interpret natural language navigation instructions from observations. In: Proceedings of the National Conference on Artificial Intelligence(AAAI), vol. 2, pp. 1–2 (2011)
16. Poon, H.: Grounded unsupervised semantic parsing. Assoc. Comput. Linguist. (ACL) **1**, 933–943 (2013)
17. Zelle, M., Mooney, R.J.: Learning to parse database queries using inductive logic proramming. In: Association for the Advancement of Artificial Intelligence (AAAI). MIT Press, Cambridge (1996)
18. Earley, J.: An efficient context-free parsing algorithm. Commun. Assoc. Comput. Mach. **6**(8), 451–455 (1970)

Semantic Parsing Using Construction Categorization

Yi Gao, Caifu Hong, and Xihong Wu(✉)

Speech and Hearing Research Center, Key Laboratory of Machine Perception and Intelligence(Ministry of Education), School of Electronics Engineering and Computer Science, Peking University, Beijing, China
{gaoyi,hongcf,wxh}@cis.pku.edu.cn

Abstract. Semantic parsing which maps a natural language sentence into its meaning representation is considered in this paper. A novel approach which involves construction categorization is proposed. Constructions encode the correspondences of concept sequences and their meaning representations. They are categorized using the syntactic relations included in a predefined hierarchical concept base. The semantic parser construct the meaning representations for the input sentences based on these constructions. Evaluations on a benchmark dataset, GeoQuery, demonstrate that the proposed semantic parser provides favorable accuracy, as well as the generalization performance.

Keywords: Semantic parsing · Hierarchical information · Construction categorization

1 Introduction

Semantic parsing is the task of mapping a natural language(NL) sentence into a completely formal meaning representation(MR). Previous works have proposed a number of systems for automatically learning semantic parsers. Semantic parsers based on combinatory categorial grammars(CCGs) map words in NL sentences into syntactic types, then derive syntactic trees based on a set of CCG rules, finally construct the corresponding MRs automatically according to the syntactic trees [1–4]. Semantic parsers based on semantically augmented parse trees(SAPTs) map words into internal nodes which include both syntactic and semantic labels, then derive SAPTs using existing syntactic parsers, finally translate the trees into corresponding MRs [5–8]. Semantic parsers based on machine translation technologies use synchronous grammars, which match NL string patterns and construct the corresponding MRs synchronously [9–11]. The string patterns here have some resemblances with the productions in syntactic parsing, but they are more specific.

Mapping words into non-terminals in the parse trees can improve the generalization of semantic parsers, but they bring ambiguities in at the same time.

X.Wu—IEEE Senior member.

X. He et al. (Eds.): IScIDE 2015, Part II, LNCS 9243, pp. 576–584, 2015.
DOI: 10.1007/978-3-319-23862-3_57

String patterns in synchronous grammars introduce less ambiguities, but they lack generalization. There exists a trade-off between the generalization and ambiguities, and a proper generalization method is needed for semantic parsers.

Recent years, a number of works involve knowledge bases or grounded information in their systems [1,3,12,13]. Knowledge bases can improve the generalization of semantic parsers because they can capture the unseen relations in the training corpus, especially when handling domain-independent, large-scale corpus.

Our previous work on semantic parsing involves a concept base which includes hierarchical relations between concepts. Constructions are used to map concept sequences into their MRs. The concepts in constructions are categorized using the hierarchical concept base, which improves the generalization of the semantic parser. A new method is proposed in this paper to further improve the generalization. Constructions are categorized using syntactic relations. Different constructions are allowed to share a same concept pattern, even though they have no semantic similarities. Compared to traditional methods using syntactic information, the categorized constructions improve the generalization of the semantic parser in a proper degree. The experiments in GeoQuery [14], a benchmark dataset, have shown that the proposed system outperforms all existing systems in accuracy, as well as generalization performance.

2 Background

This section gives a brief introduction to our previous work on semantic parsing, based on which the construction categorization method in this paper is proposed.

2.1 Hierarchical Concept Base

The concept base consists of concepts. In this paper, strings with the first letters capitalized are used to denote concepts. The formal definition is $K =< E, A, R, E_i, A_i, R_i, R_h >$. Here E represents entities, e.g., *City*. A represents attributes, e.g., *Height*. R represents relations, e.g., the relation that a city is located in a state. A relation relates several concepts which are relative elements of the relation. E_i, A_i and R_i are the instances of entities, attributes and relations respectively. R_h represents the hierarchical relations, which are also relations. The hypernym of a concept C is denoted as C_h^C. The extension of C is the set consisting of the instances of C, denoted as $Ext(C)$. Words in NL are also regarded as concepts, they are instances of the entity $Word$.

2.2 Meaning Representation

The MRs are trees consisting of concepts. The formal definition is $t =< Root\ (C_1)\ (C_2)\ \ldots\ (C_m) >$, where *Root* is the root of the tree, and $C_1, C_2, \ldots, C_m$ are the children of *Root*. Each concept in the tree is a hyponym

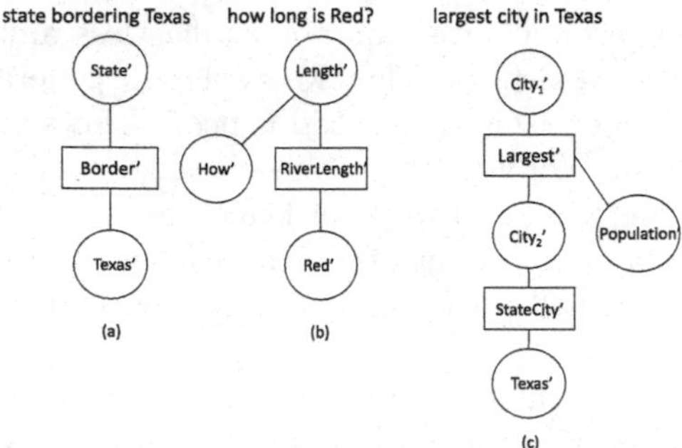

Fig. 1. Several examples of semantic trees in our system

of some concept in the concept base, because they have different extensions. Figure 1 shows some examples.

The computation of a semantic tree is defined as the procedure of computing the extension of the focus concept in the semantic tree. Normally, the focus concept is the root of the semantic tree. If the tree has interrogatives, such as $When$, $Where$, etc., the focus concept is the concept connected to the interrogative.

If the focus concept is a relation R, and all its children are its relative elements, denoted as $C_1, C_2, \ldots, C_m$, theoretically the extension of R can be obtained from the extensions of the relative elements of R. Formally:

$$\begin{aligned} Ext(R) = \{ & R_i(C_{1i}, C_{2i}, \ldots, C_{mi}) \in Ext(R_h^R) | \\ & C_{1i} \in Ext(C_1) \wedge C_{2i} \in Ext(C_2) \wedge \cdots \wedge C_{mi} \in Ext(C_m)\} \end{aligned} \tag{1}$$

If all the concepts connected to the focus concept C are relations, each of them takes C as one of its relative elements. Denote the relations as $R_1, R_2, \ldots, R_m$. The extension of C can be obtained from the extensions of these relations. Formally:

$$\begin{aligned} Ext(C) = \bigcap_{j=1}^{m} \{ & x \in Ext(C_h^C) | \exists C_{1i}^j, C_{2i}^j, \ldots C_{mi}^j, \\ & R_i^j(C_{1i}^j, C_{2i}^j, \ldots, x, \ldots, C_{mi}^j) \in Ext(R^j)\} \end{aligned} \tag{2}$$

For trees containing the both cases above, the result is the intersection of the results obtained in them.

2.3 Construction

Constructions map concept sequences into their semantic representations. The formal definition is $cons = < P; F; T; H; R >$, where P represents the concept sequence, F represents morphological and semantic features(MSFs) which are used to restrict the concept usage to appropriate syntax and semantic context,

T is the corresponding semantic tree with root H, and R is the only relation in T. All the other concepts in T are the relative elements of R.

Constructions are annotated from the training corpus. For every construction, the relative elements of the only relation are categorized using the hierarchical concept base. The semantic parser using these constructions capture the semantic relations between concepts, which can greatly improve the generalization performance.

Ambiguities exist when a concept sequence corresponds to multiple MRs. Probability is introduced to solve this problem. The probability of a MR derived from a concept sequence is $P(T, H, R|P, F)$. It can be obtained from the corpus:

$$P(T, H, R|P, F) = \frac{count(< P; F; T; H; R >)}{\sum_{(T,H,R)} count(< P; F; T; H; R >)} \tag{3}$$

2.4 Semantic Parsing

Earley's context-free parsing algorithm [15] is used for parsing. During parsing, the parser checks if a subsequence of the input concepts matches the concept sequence of a construction, i.e., every input concept is the corresponding concept or its hyponym or its instance in the construction, and they have the same MSFs. If so, the concepts in the semantic tree of the construction are replaced by the corresponding concepts in the input concept sequence. This forms a new semantic tree, whose head can be further combined with other concepts in the input sequence. Finally, the most probable semantic tree is selected from the tree set T_{set}:

$$\begin{aligned} T_{best} &= argmax_{T \in T_{set}} P(T, H, R|S, F) \\ &= argmax_{T \in T_{set}} \Big(\prod_{T_j \in T} P(T_j, H_j, R_j|P_j, F_j) \Big) \end{aligned} \tag{4}$$

3 Construction Categorization

In this section, the hierarchical concept base is augmented to include syntactic information, based on which the constructions are categorized. The MR form and the parsing method are the same as in Sect. 2.

3.1 Augmented Concept Base

The hierarchical concept base is augmented by introducing syntactic relations. The formal definition is $K =< E, A, R, E_i, A_i, R_i, R_h, R_s >$, where R_s represents the syntactic relations, and other symbols are the same as in Sect. 2.1. A syntactic relation relates several concepts syntactically. Examples include transitive verbs and prepositions. A transitive verb may relate two entities together, which can be denoted as $TransitiveVerb(Entity_1, Entity_2)$. Syntactic relations are regarded as concepts in the concept base. They can be related to other relations in the concept base through hierarchical relations. The concept *Border* is a hyponym of the concept $TransitiveVerb$. *Border* has the same relative elements as $TransitiveVerb$, denoted as $Border(Entity_1, Entity_2)$.

The hierarchical relations between syntactic relations and their hyponyms are like the correspondences of syntactic non-terminals and words in syntactic parsing. Syntactic trees mainly capture the syntactic relations between words, which leads to the generalization of syntactic parsers. By using the hierarchical concept base including syntactic information, the proposed semantic parser also captures these relations. However, syntactic trees do not consider the semantic relations between words, which leads to ambiguities of the words with the same non-terminal. While the proposed semantic parser uses semantic information included in the hierarchical concept base, which allows for proper semantic generalization and reduces the ambiguities between words.

Since currently there are no appropriate concept bases containing both the hierarchical information and the syntactic information for GeoQuery, a manually annotated hierarchical concept base including the syntactic relations is used to conduct experiments. Theoretically, the concept base is domain-independent, and can be used in any other systems.

3.2 Categorized Constructions

Using the syntactic relations in the augmented concept base, the constructions are categorized. Consider the two phrases: (1) *states bordering* $Texas$;(2) $rivers\ traversing\ Texas$. According to Sect. 2, the constructions corresponding to these phrases are:

$cons1 =< State_1, Border, State_2; plural, present\ \ participle, instance; < State_1(Border(State_2)) >; State_1; Border >$

$cons2 =< River, Traverse, State; plural, present\ \ participle, instance; < River(Traverse(State)) >; River; Traverse >$

These two constructions have some resemblances. First, both the concept sequences consist of an entity followed by a relation which is followed by another entity. Second, they have the same MSFs. Third, the semantic trees in the two constructions have the same structure, where the first entity is the root with the relation as its child, and the second entity is the child of that relation. A more general construction can be used to represent them ignoring the differences:

$cons3 =< Entity_1, TransitiveVerb, Entity_2; plural, presentparticiple, instance; < Entity_1\ (TransitiveVerb\ (Entity_2)) >; Entity_1; TransitiveVerb >$

The concept sequence in this construction are concepts in the syntactic relation $TansitiveVerb(Entity_1, Entity_2)$. Obviously this construction captures the syntactic information of the two phrases mentioned above. The two hyponyms of $TransitiveVerb$, $Border(State_1, State_2)$ and $Traverse(River, State)$, capture the semantic information. Construction has been further categorized using the syntactic information. Accurate parsing results can be obtained by the semantic parser using this construction and these relations.

For GeoQuery, the constructions are manually annotated and categorized. For every sentence S, the annotation procedure is as follows:

(1) For every word W with its MSF as F_w in S, assume that W corresponds to only one MR in this context, denoted as C. The construction is annotated as: $< W; F_w; < C >; C; \epsilon >$.
(2) For every phrase in S, first map every word in the phrase to corresponding MR, $P =< C_1, C_2, \ldots, C_m >$, with the extracted MSF sequence as F. For every concept in P, use its highest level hypernym in the concept base to replace it. This forms a new concept sequence P_h. Then annotate the corresponding semantic tree T with head H for P_h, and the only relation denoted as R. The construction is $cons =< P_h; F; T; H; R >$.
(3) For a construction $cons =< P; F; T; H; R >$, where R is not ϵ, and R has its syntactic hypernym R^R_{sh}. For every relative element of R in P, T and H, the corresponding relative element in R^R_{sh} is used to replace it, and change P, T, H to P_{sh}, T_{sh}, H_{sh} respectively. This forms a new construction, which can be denoted as $cons_h =< P_{sh}; F; T_{sh}; H_{sh}; R^R_{sh} >$.

If a construction has been categorized into another construction, the original one is discarded to reduce the time complexity of the semantic parser. The occurrence of the original construction is added to the occurrence of the categorized one, which is used to compute the probability of the categorized construction.

4 Experiments and Results

The experiments are conducted on GeoQuery. The dataset contains about 800 facts about U.S. geography, and 880 questions with the corresponding MRs in Prolog. The average length of a sentence is 7.48 words. The proposed system, the construction categorized version(CCV), is compared to: (1)WASP [9], which is based on machine translation techniques; (2) λ-WASP [10], an extension of WASP for handling MRs; (3) SYNSEM [8], which combines syntactic information and semantic information together, here we choose its result based on the gold-standard syntactic parses; (4) L2013 [16], which uses dependency-based compositional semantics(DCS) as the MR form; (5) W2014 [4], which performs best in current CCG-based parsers; (6)The basic version(BV) of the proposed system described in Sect. 2, which does not use construction categorization.

In the first experiment, the accuracy of the proposed system is tested. The corpus is split as 600 questions for training and 280 questions for testing. Table 1 shows the results. Several observations can be made. Both versions of the proposed system achieve higher accuracies on this corpus. Compared to the basic version, the construction categorized version also gains 0.5 % absolute improvement in accuracy. Though the proposed system needs more annotation, compared to SYNSEM which annotated the gold-standard parse trees, it also performs better.

Table 1. The Accuracy on GeoQuery

System	Accuracy(%)
WASP	74.8
λ-WASP	86.6
SYNSEM	88.2
L2013	91.4
W2014	90.4
New System(BV)	93.4
New System(CCV)	93.9

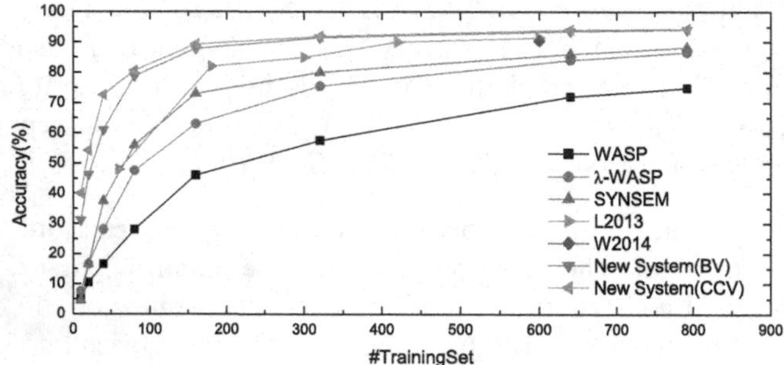

Fig. 2. Learning curves for various parsing algorithms on the GeoQuery corpus

The second experiment is conducted to obtain the learning curves. The standard 10-fold cross validation is used. Figure 2 shows the results. It can be observed that, the basic version and construction categorized version both outperform other systems by wide margins, and match the best final accuracy of other systems with only 50 % of the total training examples. Moreover, when the training data are extremely sparse, the construction categorized version shows superior to the basic version. When applied to a new domain, the construction categorized version can alleviate the burden of annotation greatly.

5 Conclusion

Using syntactic information for construction categorization can greatly improve the generalization performance of semantic parsers. Syntactic relations, as well as the hierarchical relations between these relations and other concepts, are introduced as part of the hierarchical concept base. Constructions are categorized using the syntactic information, which makes the proposed system gain great improvement in performance. Though the proposed system needs a little more annotation, its generalization is better than existing systems. Compared to the basic version of the semantic parser, the proposed system further alleviate the burden of annotation, especially on small training sets.

For now, the constructions in the proposed system were manually annotated and categorized. To automatically annotate and categorize the constructions using the hierarchical concept base will be the future work. This could further alleviate the burden of annotation, and reduce the errors in the annotated corpus.

Acknowledgments. The research was partially supported by the National Basic Research Program of China (973 Program) under grant 2013CB329304, the Major Project of National Social Science Foundation of China under grant 12&ZD119, and the National Natural Science Foundation of China under grant 61121002.

References

1. Cai, Q., Yates, A.: Semantic parsing freebase: towards open-domain semantic parsing. In: Proceedings of the Joint Conference on Lexical and Computational Semantics, vol. 30, pp. 328 (2013)
2. Kwiatkowski, T., Zettlemoyer, L., Goldwater, S., and Steedman, M.: Lexical generalization in CCG grammar induction for semantic parsing. In: Proceedings of the Conference on Empirical Methods in Natural Language Processing, pp. 1512–1523 (2011)
3. Kwiatkowski, T., Choi, E., Artzi, Y., and Zettlemoyer, L.: Scaling semantic parsers with on-the-fly ontology matching. In: Proceedings of the 2013 Conference on Empirical Methods in Natural Language Processing. Seattle, Washington (2013)
4. Wang, A., Kwiatkowski, T. and Zettlemoyer, L.: Morpho-syntactic lexical generalization for CCG semantic parsing. In: Proceedings of the Conference on Empirical Methods in Natural Language Processing (EMNLP) (2014)
5. Ge, R., Mooney, R.J.: A statistical semantic parser that integrates syntax and semantics. In: Proceedings of the Ninth Conference on Computational Natural Language Learning, Association for Computational Linguistics, pp. 9–16 (2005)
6. Ge, R., Mooney, R. J.: Discriminative reranking for semantic parsing. In: Proceedings of the COLING/ACL on Main Conference Poster Sessions, Association for Computational Linguistics, pp. 263–270 (2006)
7. Ge, R., Mooney, R.J.: Learning a compositional semantic parser using an existing syntactic parser. In: Proceedings of the Joint Conference of the 47th Annual Meeting of the ACL and the 4th International Joint Conference on Natural Language Processing. Association for Computational Linguistics, vol. 2, pp. 611–619 (2009)
8. Ge, R.: Learning for semantic parsing using statistical syntactic parsing techniques. Ph.D. thesis, University of Texas at Austin (2010)
9. Wong, Y.W., and Mooney, R.J.: Learning for semantic parsing with statistical machine translation. In: Proceedings of HLT/NAACL 2006, pp. 439–446, New York City (2006)
10. Wong, Y.W. and Mooney, R.J.: Learning synchronous grammars for semantic parsing with lambda calculus. In: Association for Computational Linguistics (ACL), pp. 960–967, Prague, Czech Republic (2007)
11. Andreas, J., Vlachos, A., and Clark, S.: Semantic parsing as machine translation. In: Proceedings of the Conference of the Association for Computational Linguistics (ACL), pp. 47–52 (2013)

12. Krishnamurthy, J., Mitchell, T.: Weakly supervised training of semantic parsers. In: Proceedings of the Joint Conference on Empirical Methods in Natural Language Processing and Computational Natural Language Learning, pp. 754–765 (2012)
13. Berant, J., Chou, A., Frostig, R., Liang, P.: Semantic parsing on freebase from question-answer pairs. In: Proceedings of the 2013 Conference on Empirical Methods in Natural Language Processing (EMNLP), pp. 1533–1544 (2013)
14. Zelle, M., Mooney, R.J.: Learning to parse database queries using inductive logic proramming. In: Association for the Advancement of Artificial Intelligence (AAAI). MIT Press, Cambridge (1996)
15. Earley, J.: An efficient context-free parsing algorithm. Commun. Assoc. Comput. Mach. **6**(8), 451–455 (1970)
16. Liang, P., Jordan, M.I., Klein, D.: Learning dependency-based compositional semantics. Comput. Linguist. **39**(2), 389–446 (2013)

Tunable Discounting Mechanisms for Language Modeling

Junfei Guo[1,2(✉)], Juan Liu[1(✉)], Xianlong Chen[3], Qi Han[4], and Kunxiao Zhou[3]

[1] School of Computer, Wuhan University, Wuhan, China
guojf@ims.uni-stuttgart.de, liujuan@whu.edu.cn
[2] Institute for Natural Language Processing, University of Stuttgart, Stuttgart, Germany
[3] City College of Dongguan University of Technology, Dongguan, China
[4] Institute for Visualization and Interactive Systems, University of Stuttgart, Stuttgart, Germany

Abstract. Language models are fundamental to many applications in natural language processing. Most language models are trained on training data that do not support discount parameters tuning. In this work, we present novel language models based on tunable discounting mechanisms. The language models are trained on a large training set, but their discount parameters can be tuned to a target set. We explore tunable discounting and polynomial discounting based on modified Kneser-Ney models. With the resulting implementation, our language models achieve perplexity improvements in in-domain and out-of-domain evaluation. The experimental results indicate that our new models significantly outperform the baseline model and are especially suited for domain adaptation.

Keywords: Language model · Tunable discounting · Polynomial discounting · Domain adaptation

1 Introduction

Language modeling is a well-studied topic in natural language processing (NLP) since language models play a role in many language technology tasks such as speech recognition [1], information retrieval [2] and machine translation [3]. A language model assigns probability to sequences of n words. One of the most popular language models is modified Kneser-Ney model [4], which have been implemented in language model toolkits such as SRILM [5] and KenLM [6].

The simplest method to compute the sequence probability is maximum likelihood estimate (MLE). Unfortunately, MLE overestimates the probability of

Junfei Guo acknowledges the support by Chinese Scholarship Council (CSC) during his PhD studies at the University of Stuttgart. This work was supported by the Young Teachers Development Fund of City College of Dongguan University of Technology and Natural Science Foundation of Guangdong Province, China (Grant No.2014A030310375).

X. He et al. (Eds.): IScIDE 2015, Part II, LNCS 9243, pp. 585–594, 2015.
DOI: 10.1007/978-3-319-23862-3_58

rare events and assigns zero probability to unseen word sequences. Smoothing techniques have been proposed to address the estimation challenge. Some of the previous language models [4,7] used absolute discounting or Kneser-Ney discounting to perform smoothing. They suggest to subtract fixed discounts through estimation on training data. Considering the mismatch between training and test data (possibly even from different domains), accurate estimation is difficult to achieve with only knowledge about training data.

We explore novel tunable discount mechanisms to tune language models to adjust to the target set. We train the language model on a training set as usual but tune the discounting parameters on the validation set, which we assume to be beneficial for domain adaptation. Compared the previous discounting methods, our tunable discount and polynomial discount techniques offer more flexible methods for adapting language models to a new domain. We tune the optimal discount parameters to minimize the validation set perplexity [8].

In our experiments, we use the well-know modified Kneser-Ney smoothing model as baseline. The performance of all the models is evaluated by using perplexity on in-domain and out-of-domain data. In-domain, our models have lower perplexity scores than the baseline model. Cross-domain, our language models achieve significantly better perplexity than the competitor. The experimental results demonstrate that our tunable discounting models outperform the modified Kneser-Ney (mKN) model. We expect the improvements of our models could potentially further optimize the related NLP application performance.

2 Language Models

In this section, we recall the commonly used modified Kneser-Ney [4] model, which is used in our contrastive model. We then present the novel variations of KN models with novel discounting techniques including the tunable discount and polynomial discounts. The notation we use throughout the paper is shown in Table 1.

2.1 Modified Kneser-Ney Model

Instead of absolute discounting by subtracting a fixed discount D, Kneser and Ney [7] developed the optimal value of their discount parameter D through

Table 1. Notation

symbol	denotation
w_i	A unigram word w_i
w_i^j	a segment from word w_i to word w_j
c	count of a n-gram
$c(w_i^j)$	count of w_i^j
$P(w_i\|w_{i-n+1}^{i-1})$	the of probability of w_i given a history n-gram

deleted estimation on training data. The modified Kneser-Ney model (mKN) [4] was proposed by Chen and Goodman, who derived the analogous values for the Kneser-Ney (KN) model [7]. Instead of a single discount parameter D, they proposed to use three discount parameters D_1, D_2, D_{3+}.

The discount function for modified Kneser-Ney model are

$$D(c) = \begin{cases} 0 & \text{if } c = 0 \\ D_1 = 1 - 2Y\frac{n_2}{n_1} & \text{if } c = 1 \\ D_2 = 2 - 3Y\frac{n_3}{n_2} & \text{if } c = 2 \\ D_{3+} = 3 - 4Y\frac{n_4}{n_3} & \text{if } c \geq 3 \end{cases}$$

where c represents the counts of the n-grams, n_i is the number of n-grams that appear exactly i times in the training data and

$$Y = \frac{n_1}{n_1 + 2n_2}.$$

They estimate the model parameters on the training set as follows.

$$P_{kn}(w_i|w_{i-n+1}^{i-1}) = \frac{c(w_{i-n+1}^{i}) - D(c(w_{i-n+1}^{i}))}{\sum_{w_i} c(w_{i-n+1}^{i})} + \gamma(w_{i-n+1}^{i-1})P_{kn}(w_i|w_{i-n+2}^{i-1}) \quad (1)$$

To make the distribution sum to 1, they take

$$\gamma(w_{i-n+1}^{i-1}) = \frac{D_1 N_1(w_{i-n+1}^{i-1}\cdot) + D_2 N_2(w_{i-n+1}^{i-1}\cdot) + D_{3+} N_{3+}(w_{i-n+1}^{i-1}\cdot)}{\sum_{w_i} c(w_{i-n+1}^{i})} \quad (2)$$

2.2 Tunable KN Model

We expect that modified KN smoothing achieves better performance with tuned discount parameters, especially on out-of-domain data. It is possible to add some tunable optimal parameters to the modified KN discounts. Therefore, we replace the discount function D by a tunable function T

$$T(c) = D(c) + \begin{cases} t_c & \text{if } c \leq 3 \\ t_3 & \text{otherwise} \end{cases}$$

t_c represents three constant parameters t_1, t_2 and t_3, respectively. The parameters added to the modified KN discount function $D(c)$ need to be optimized on the target data. We call this tunable discounting n-gram KN based model TKN and define it as follows.

$$P_{tkn}(w_i|w_{i-n+1}^{i-1}) = \begin{cases} \frac{c(w_{i-n+1}^i) - T(c(w_{i-n+1}^i))}{\sum_{w_i} c(w_{i-n+1}^i)} & \text{if } c(w_{i-n+1}^i) > 0 \\ \beta(w_{i-n+1}^i) P_{tkn}(w_i|w_{i-n+2}) & \text{if } c(w_{i-n+1}^i) = 0 \end{cases} \tag{3}$$

To make the distribution sum to 1, we take

$$\beta(w_{i-n+1}^i) = \frac{\sum_{w_i \text{ s.t. } c(w_{i-n+1}^i)>0} T(c(w_{i-n+1}^i))}{\sum_{w_i \text{ s.t. } c(w_{i-n+1}^i)=0} P_{tkn}(w_i|w_{i-1}^{i-n+2})} \cdot \frac{1}{\sum_{w_i} c(w_{i-n+1}^i)} \tag{4}$$

We use a simple back-off scheme in this model. The parameters are optimized on a validation set by tuning the additional discount parameters to minimize validation set perplexity. We use heuristic grid search to find the optimal parameters.

2.3 Domain Adaptation Model

Schütze presented a polynomial only discounting [9] which is a simpler way of discounting than Kneser-Ney discounting. Guo et al. applied the polynomial only discounting (POLO) to the n-gram based language model and proposed a domain adaptation (DA) model [10] replacing the modified KN discount by the POLO discount. The discount function E is defined for two parameters ρ and γ as $E(c) = \rho \cdot c^\gamma$. We would like to study this polynomial discounting mechanism and apply it to our tunable discounting parameters later.

2.4 Polynomial Discounting KN Model

Schütze also proposed the polynomial and KN discounting mechanism for POLKN [9] model which is interpolated model over classes. He considers the default discounting mechanism does not reallocate enough probability mass from high-frequency to low-frequency events in the class-based models. He replaces the discount function D by the function D'

$$D'(c) = D(c) + \begin{cases} 0 & \text{if } c \leq 3 \\ \rho \cdot c^\gamma & \text{otherwise} \end{cases}$$

The parameter γ controls the growth rate of the discount as a function of c, and the parameter ρ is a classical discount factor that can be scaled for optimal performance.

Compared with the DA model, we would like to retain the modified KN discounts for lower-frequency events ($c \leq 3$) and add a polynomial discount for higher frequency events ($c > 3$). Therefore, we present polynomial discounting modified KN-based language model (PKN) by using same approach as in modified KN models, but we replace the modified KN discount function $D(c)$ by the Polynomial KN Discount function $D'(c)$. We define this novel n-gram model PKN as follows.

$$P_{pkn}(w_i|w_{i-n+1}^{i-1}) = \begin{cases} \frac{c(w_{i-n+1}^i) - D'(c(w_{i-n+1}^i))}{\sum_{w_i} c(w_{i-n+1}^i)} & \text{if } c(w_{i-n+1}^i) > 0 \\ \beta(w_{i-n+1}^i) P_{pkn}(w_i|w_{i-n+2}) & \text{if } c(w_{i-n+1}^i) = 0 \end{cases} \tag{5}$$

Compared with the class-based trigram POLKN model [9], our language model is a simple n-gram model with polynomial KN discounts. We simply use n-grams which is cheaper and also makes it easier to generate the ARPA format language model. Since our model is a back-off model, the algorithm is valid for different n-gram orders including 5-grams. we also use heuristic grid search to find the optimal parameters.

2.5 Tunable and Polynomial Discounting KN Model

We note that the DA model only have one discount function with two parameters ρ and γ for unigrams, bigrams and any n-grams. Since the discount for unigrams, bigrams and other n-grams are different, more different functions for different counts could help the estimation. The original modified KN model has different discounts for different n-grams. In order to use the advantages of polynomial discount and the Kneser-Ney discount, we add tunable discounts and polynomial discount to Kneser-Ney discount for different count cases. From these, we introduce a tunable discounting and polynomial discounting mechanism. We replace the discount function D by the function E'

$$E'(c) = D(c) + \begin{cases} t_c & \text{if } c \leq 3 \\ \rho \cdot c^{\gamma} & \text{otherwise} \end{cases}$$

We have 5 parameters in this model, t_c represents three discount constant parameters t_1,t_2 and t_3 the same as in TKN model. When the n-grams are high-frequency events ($c > 3$) , we use the sum of modified KN discount $D(c)$ and the polynomial discount function with parameters ρ and γ. We call this tunable and polynomial discounting KN Model (TPKN) defined as:

$$P_{tpkn}(w_i|w_{i-n+1}^{i-1}) = \begin{cases} \frac{c(w_{i-n+1}^i) - E'(c(w_{i-n+1}^i))}{\sum_{w_i} c(w_{i-n+1}^i)} & \text{if } c(w_{i-n+1}^i) > 0 \\ \beta(w_{i-n+1}^i) P_{tpkn}(w_i|w_{i-n+2}) & \text{if } c(w_{i-n+1}^i) = 0 \end{cases} \tag{6}$$

The TPKN model setups are similar to the TKN and PKN models, only the discount function is different. We implemented all our models using the SRILM toolkit with back-off smoothing.

3 Experimental Setup

3.1 Corpora

We summarize the corpora used in Table 2. We run the in-domain experiments on the Wall Street Journal (WSJ). The training, validation (Val.) and test data are taken from WSJ, for the in-domain language models experiment. The training set contains more than 1.6 million sentences and both the validation and the test set have roughly 100,000 sentences.

For the out-of-domain experiments, we use a special release of the MultiUN corpus [11] as training data. It is a multilingual parallel corpus extracted from official documents published by the United Nations from 2000 to 2009. The English part was made available in August 2011 for IWSLT 2011. In this work, we only use the English part of all the data. The evaluation is based mostly on NIST data. We use NIST 2004 as validation data and NIST 2005 as test data. The NIST data consists of newswire documents, human transcriptions of broadcast news as well as web documents. Consequently, we consider the MultiUN and the NIST data as data from different domains. We perform our out-of-domain perplexity language model experiments on these data.

Table 2. Corpora used in our experiments

Corpus	Domain	Usage	Sentences
Wall Street Journal	News	Training	1,634,529
Wall Street Journal	News	Val. and Test	2,05,000
MultiUN	Official	Training	8,820,000
NIST 2004, -05	News	Val. and Test	10,716

3.2 Setup

For the language modeling part, we use SRILM to generate ARPA format language models on English training data. We have modified the SRILM implementation of discounting in order to implement our own model. We use heuristic grid search in order to determine the optimal discounting parameters. During the search, we explore the space with step size 0.01.

4 Experimental Results and Discussion

The outcomes of our experimental results together with the determined optimal parameters used are listed in Tables 3–4. They show the performance of the language models on both the in-domain and out-of-domain data. We will now discuss the individual experiments.

4.1 In-domain Language Model Perplexity

The first experiment investigates the performance of the different language models on in-domain data. In this case, the training, validation and test set are from the *Wall Street Journal*. All perplexity values are reported for the validation (Val) and the test (Tst) set. Table 3 shows the perplexity performance of all language models and their parameters. Our baseline is the modified KN model offered in SRILM. Additionally, we report results for the modified KN model including interpolation. As can be seen from the table, all the tunable language models have perplexities lower than the modified KN model (94.38 for Val. and 94.34 for Tst.). The results indicate that the tunable discounting techniques improve the language model performance.

We observe slight improvements for the backoff TKN models in perplexity (from 94.38 to 92.30) compared to our baseline on the validation set. Similarly, we observe a slight improvement in perplexity (from 94.34 to 92.25) on the test set. The tunable additional discounts $(-0.16 - 0.4and - 0.72)$ and the lower perplexities results suggest that the modified KN discounts are too large in the in-domain scenarios. The models with lower discount values work well in this case.

To explore the polynomial discounts, we also compare the polynomial only discounting DA model with the PKN and TPKN model. Comparing with DA model (92.32), the PKN model (93.22) does not obtain any improvement. However, the TPKN model (92.19) outperforms the DA model on the test set. We could find possible reasons when looking at the TKN parameters t_1, t_2 and t_3, we can see that the modified KN discounts not only the D_{3+}, but also D_1 and D_2 needed be tuned. So the PKN model does not perform well as DA and TPKN.

Table 3. Perplexities and parameters of the models on the in-domain data

Model	t_1	t_2	t_3	ρ	γ	Val	Tst
mKN						94.38	94.34
mKN+interp						92.48	92.42
TKN	−0.16	−0.40	−0.72			92.30	92.25
DA				0.61	0.03	92.37	92.32
PKN				−0.46	0.42	93.25	93.22
TPKN	−0.14	−0.36	−0.62	−0.40	0.34	**92.24**	**92.19**

When we compare the interpolated modified KN models (mKN+interp), we see that they achieve similar performance (92.42) as our models. This confirms the findings of [4], who also addressed that the perplexity improves when the various language model orders are interpolated.

Overall, it appears like modified KN discounts are too large for estimation in this in-domain case. The novel tunable TKN and TPKN models with subtracting optimal discounts to modified KN discounts improve the performance. The results thus far indicate good performance when compared with our baseline model.

4.2 Cross-Domain Language Model Perplexity

For the out-of-domain experiment we use the English part of the MultiUN corpus [11] as training corpus. As mentioned, we use NIST 2004 as validation set and NIST 2005 as the test set. The NIST documents include newswire, broadcast news, and web data, so they are quite different (in style and language) from the contract style documents of MultiUN.

Since we confirmed that interpolation improves performance in SRILM, we only use the interpolated variant of the modified KN model from now on. The obtained perplexities for the validation and the test set are shown in Table 4.

Comparing Tables 3 and 4, we immediately notice that the perplexities increase from approximately 90 by roughly 200 points, which confirms that the validation and test data are quite different from the training data. We cannot expect low perplexities performance on new domains as well as in-domain. However, the results in Table 4 show that our model can achieve considerable perplexity improvements for out-of-domain data. All the model achieve perplexities lower than the modified KN model.

The perplexity drops from 254.65 (mKN) to 237.76 (TPKN) on the validation data. The results are mirrored on the test set, where we observe an improvement from 312.30 (KN) to 284.99 (TPKN). The scores on the validation and test set are similar as was to be expected because the various NIST data sets are comparable. The parameters tuned on validation set were found to yield lower perplexities performance on the test data. Overall, our tunable models outperform the modified KN model in SRILM even though it uses interpolation.

The tunable discount KN-based TKN model shows better performance than modified KN model. The parameters $(-0.04, -0.1, 0.24)$ of the TKN model suggests us that modified KN discounts are not accurate enough in this out-of-domain scenario, the modified KN model keeps too much information from the old domain, which is not helpful for the out-of-domain data. Our model is actually beneficial in this out-of-domain scenario since we tuned our discounts by adding discounting parameters.

Table 4. Perplexity and parameters of the models on the out-of-domain data.

Model	t_1	t_2	t_3	ρ	γ	Val	Tst
mKN						254.65	312.30
TKN	−0.04	−0.10	0.24			241.42	291.30
DA				0.63	0.59	238.62	286.53
PKN				0.04	1.04	238.36	285.57
TPKN	−0.08	−0.12	−0.14	0.05	1.03	**237.76**	**284.99**

Comparing the DA model with PKN model and TPKN model on the test set, we can see that the perplexity of the PKN model (285.57) is lower than the DA models (286.53). From the TKN model parameters $(-0.04, -0.1, 0.24)$, we can see that we do not need to tune discounts for d_1 and d_2 too much, but d_{3+} seems not enough, we add 0.24 to d_{3+}. However, the DA model only use one discount function for the three discounts (d_1, d_2, and d_{3+}). Therefore, we guess this could be the explanation the PKN model and TPKN model can outperform DA model. In TPKN model, The t_c parameters of TKN $(-0.04, -0.1, 0.24)$ is similar as TPKN $(-0.08, -0.12, -0.14)$, especially the t_1 and t_2. Meanwhile, the ρ and γ parameters of PKN (0.04, 1.04) is very close to TPKN (0.05, 1.03). The parameters show that TPKN model retains the good performance of the TKN and PKN models by tuning all the discounts. Consequently, TPKN performance in terms of lowest perplexity (284.99).

The difference between the domains becomes apparent when comparing the best parameters settings in Tables 3 and 4. As expected, we observe a drastic change of parameter values comparing the in-domain scenario to the out-of-domain scenario. For the TKN model, parameters change from $(-0.16, -0.4, -0.72)$ to $(-0.04, -0.1, 0.24)$. In PKN and TPKN model, the parameters for the out-of-domain test are particularly large compared to the in-domain cases. PKN model parameters shift from $(-0.46, 0.42)$ to (0.04, 1.04). TPKN model $(-0.40, 0.34)$ to (0.05, 1.03). All the above parameters indicate that we need to subtract the tunable discounts to the modified KN discounts for in-domain scenarios, but more discounts help for the out-of-domain scenarios especially for the high-frequency events ($c > 3$). This means that moving to out-of-domain, the discounts are increased in comparison to the modified Kneser-Ney model, which yields that more probability mass is shifted from the observed n-grams to the unobserved n-grams. We assume that the difference between the training set and the validation set needs more discounts for more unseen n-grams. KN discounts is not sufficient enough for the out-of-domain scenario. The tuning of the discount parameters for our model on the domain validation set helps our model adapt to different data.

In summary, the perplexity experiments show that our models do not suffer worse performance than the competitors on in-domain data, but offer benefits on out-of-domain data. This suggests that we can safely use our model both on in-domain and out-of-domain data.

5 Conclusion

In this paper, we introduced novel discounting language models using tunable discounting and polynomial discounting. The discount parameters are tuned to the validation set to adjust the models to the target set. The experiments empirically determined that the tunable discounting language models perform as well as modified KN model on in-domain data. In the out-of-domain scenarios, we observed significant improvements.

In future work, we plan to improve the parameter optimization algorithm and implement interpolated models. We will apply our model to more target languages and other NLP tasks.

Acknowledgments. Junfei Guo acknowledges the support by Chinese Scholarship Council (CSC) during his PhD studies at the University of Stuttgart. All authors want to sincerely thank the colleagues and anonymous reviewers for their helpful comments, especially Jason Utt, Abhijeet Gupta and Florian Heimerl at the University of Stuttgart.

References

1. Rabiner, L., Juang, B.H.: Fundamentals of Speech Recognition. Prentice-Hall, Inc., Upper Saddle River (1993)
2. Manning, C.D., Raghavan, P., Schütze, H.: Introduction to Information Retrieval. Cambridge University Press, Cambridge (2008)
3. Koehn, P.: Statistical Machine Translation. Cambridge University Press, Cambridge (2010)
4. Chen, S.F., Goodman, J.: An empirical study of smoothing techniques for language modeling. In: Proceedings of ACL, pp. 310–318. ACL (1996)
5. Stolcke, A.: SRILM – an extensible language modeling toolkit. In: Proceedings of INTERSPEECH, ISCA, pp. 901–904 (2002)
6. Heafield, K., Pouzyrevsky, I., Clark, J.H., Koehn, P.: Scalable modified Kneser-Ney language model estimation. In: Proceedings of ACL, pp. 690–696. ACL (2013)
7. Kneser, R., Ney, H.: Improved backing-off for m-gram language modeling. In: Proceedings of ICASSP, Vol. 1, pp. 181–184. IEEE (1995)
8. Jelinek, F., Mercer, R.L., Bahl, L.R., Baker, J.K.: Perplexity-a measure of the difficulty of speech recognition tasks. J. Acoust. Soc. Am. **62**, S63 (1977)
9. Schütze, H.: Integrating history-length interpolation and classes in language modeling. In: Proceedings of ACL, pp. 1516–1525. ACL (2011)
10. Guo, J., Liu, J., Han, Q., Maletti, A.: A tunable language model for statistical machine translation. In: Proceeding of The Eleventh Biennial Conference of the Association for Machine Translation in the Americas (AMTA) (2014)
11. Eisele, A., Chen, Y.: MultiUN: a multilingual corpus from United Nation documents. In: Proceedings of LREC, ELRA, pp. 2868–2872 (2010)

Understanding Air Quality Challenges Through Simulation and Big Data Science for Low-Load Homes

Haiyan Xie[1(✉)], Tingting Liang[2], Hui Li[3], and Yao Shi[1]

[1] College of Applied Science and Technology, Illinois State University, Normal, USA
hxie@ilstu.edu

[2] College of Huaqing, Xian University of A&T, Xi'an, China
christinelove2011@163.com

[3] College of Civil Engineer, Changan University, Xi'an, China
hli123@ilstu.edu

Abstract. The goal of this research is to determine the prominent problems and challenges of the low-load homes in the aspects of high performance ventilation systems and indoor air quality strategies. The authors will first categorize the residential buildings according to their load capacities. The characteristics of the energy-consumption mode that residents value the most will also be investigated. Data will be gathered through accessing the database of building permits, approval, and commissioning. Data for space heating and cooling load information and designed occupancy can also be collected through sensors. Big data analysis tools will be used to examine the relationship between the construction technology selections and the importance of certain design decision factors. Building Information Modeling (BIM) technology will be implemented to simulate the alternative strategies to conventional central ducted space conditioning systems that will provide thermal comfort for the occupants.

Keywords: Big data science · Simulation · Low-Load homes · Air quality challenges

1 Introduction

To improve the performance of ventilation systems in the low-load homes will help with the reduction of energy consumption, making buildings more environmental friendly, and pursuing sustainable development. The effectiveness and efficiency of a ventilation system can affect the indoor air quality of a building. A heating, ventilation, and air-conditioning (HVAC) system adjusts temperature and humidity levels within a thermal enclosure that meet occupants' comfort expectations along with sufficient fresh air for good indoor air quality. There are three inter-related core technical challenges which are necessary to make buildings more energy efficient, productive, and affordable. The three challenges include: (A) building envelope assemblies and systems to achieve low heating and cooling loads; (B) comfort systems (HVAC and distribution) for low-load homes; and (C) ventilation systems and indoor air quality strategies for low-load homes

X. He et al. (Eds.): IScIDE 2015, Part II, LNCS 9243, pp. 595–602, 2015.
DOI: 10.1007/978-3-319-23862-3_59

(Communication from the Commission 2010). Low load is defined as a house with a thermal enclosure that yields a maximum space heating and cooling load of less than 10 Btu/h/ft^2 of conditioned floor area (31.5 W/m^2) (USDOE 2014). There are still many design and technology difficulties to satisfy that definition by using today's typical ducted forced-air systems. In July 2008, the Illinois Sustainable Technology Center (ISTC) was established in the University of Illinois, to study in the areas of pollution prevention; water and energy conservation; and materials recycling and beneficial reuse. While these studies provide valuable insight into the sustainable issues of infrastructure, they lack firm evaluation criteria to the low-load homes.

2 Literature Review

This study will examine the situations and problems in the low-load homes, such as freestanding houses, townhomes, or multifamily buildings, with the focus on indoor air ventilation and air quality. When the performance of the thermal enclosure is improved, such as in a "low-load" house, the requirements for the HVAC system change such that a traditional ducted forced-air system may no longer be capable of meeting those requirements. Noise and air pollution problems in a crowded city such as Chicago or New York may make the simple solution such as operable windows become less acceptable. In addition, operable windows cannot fulfill all the functions of noise reduction, lighting, natural ventilation, and temperature consistency. There are new design ideas and technology improvements for high performance ventilation systems and indoor air quality strategies. For example, Wang (2014) designed a ventilated system, which combined the multiple quarter-wave resonators (silencer) with the new wing wall. The ventilated system can balance between acoustic and ventilation performances. Wang (2014) claimed that the best ventilation performance of the wing wall is at an incident angle of 45°. With the ideal angle, the outlet air flow rate of Wang's design was doubled than that of the operable window (Wang 2014). Tantasavasdi et al. (2001) explored the potential of using natural ventilation as a passive cooling system in Thailand. They analyzed characteristics of past and present Thai houses in the areas of climate, culture, and technology. With the considerations of the thermal comfort requirements and the climate conditions in Bangkok, they found that natural ventilation could possibly create a thermally comfortable indoor environment during 20 % of the year. As many researchers have pointed, because of climate and geological differences, the air quality strategies vary at different places.

The current criteria of low-load homes are ambiguous. For example, Integrated Building and Construction Solutions (IBACOS) described the anticipated performance of a low-load house to achieve at least 50 % whole-house-source energy savings (USDOE 2012). At the Building America technical update meeting in July 2012, several definitions of a low-load house were articulated. Out of those articulations, the following description seems to be more practical in implementation: A load density per unit floor area (watt/square meter or Btu/(hour * square feet)), below which the infiltration and conduction through the thermal enclosure (i.e. walls) are no longer the only factors influencing the occupant comfort in the house (USDOE 2012). Bergey (2011) designed an experiment on a house with a radically simplified space conditioning system and interior partition doors open. His test results showed that the simplified house could attain sufficient temperature uniformity throughout the space; however, when those doors were closed, temperatures started fluctuation. But the real residential houses are much more complicated than that situation. People need to find a systematical approach to the sustainable growth.

Our focus on the indoor air ventilation and air quality of low-load homes motivates us to utilize the big data analysis tools and BIM simulation systems to seek possible solutions to the challenge. The proposed method will enable us to study the details of the identified possible solutions and their fullest resource efficiency without harming future generations' ability to enjoy an ecologically healthy and profitable future. The list of possible solutions in improving indoor air ventilation and air quality is growing longer. Cities and towns, energy producers, manufacturers, transporters, and consumers are increasingly looking for better ways to provide better ways to their living environments. To identify the possible solutions to the air quality challenge, Building Information Modeling technology will help the simulation and analysis. Currently, computer model integration helps people to reduce errors and increase performance of the design and development process (Merschbrock and Munkvold 2012). Software programs allow architects to hand-draw their ideas on paper and scan them into digital pictures start CAD drawings. With the increasing use of iPads and other tablet computers, designers can use the touch screen as paper and directly draw plans on the screens. Three dimensional modeling helped designers and architects to reduce time spend on sketching. Another significant benefit of 3D design is that it becomes easier to handle changes in design. There is no need to redraw all the drawings if a design is changed. With that aid, architect makes changes in one drawing and those changes automatically would be adjusted to all other drawings.

According to Kensek and Noble (2014), BIM has gained rapid acceptance in architecture and engineering schools, by building design and delivery professions, by the manufacturing and construction industries, and by building owners and managers. The main purpose of BIM is to integrate knowledge from various project participants that traditionally work in different phases of the building and maintaining processes. Sebastian (2010) discussed that, the decisions made during design phase affected, on average, 70 % of the life-cycle cost of a building. It is essential for collaborative design to rely on multidisciplinary knowledge for a building's life cycle. Using digital BIM platforms allows designers to associate data with geometry. Through that, designers can build parametric models for building design. BIM system developers designed the platforms so that architects, structural engineers, electrical engineers, plumbing and ventilation engineers, landscape architects, construction firms, and specialized subcontractors can be involved at design stage and provide benefits to projects with their knowledge (Merschbrock and Munkvold 2012). BIM developers pay close attentions to different energy simulation software. Kensek and Nobel (2014) argued that there were significant improvements in building performance simulations over the past two decades. As a result of those improvements current BIM platforms can conduct not only energy consumption tests which help people to make optimal lighting solutions but also many other different simulations (i.e., wind load simulation) that all together contribute to efficient decision-making process. Figure 1 shows an example of a BIM model built by using Autodesk Revit, which can later be analyzed in Green Building Studio system. The analysis result will be used in EnergyPlus tools.

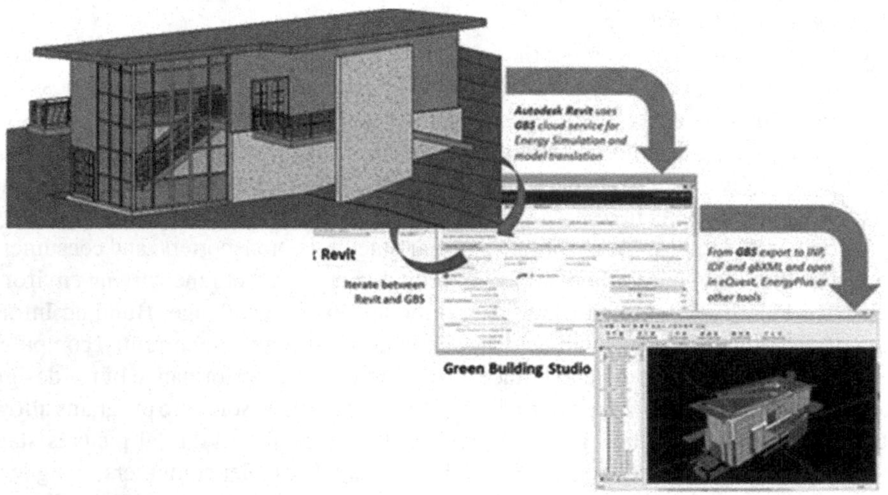

Fig. 1. Example of using BIM model for simulation and analysis

Thermal conductivity is an ability of materials to conduct heat. The faster heat flows through material the higher conductivity it has. Thermal resistance of a material is calculated as an R-value to show its ability to resist heat flow (Gooch 2010). It is measured in hours needed for 1 Btu to flow through 1 square feet of a given thickness of a material when the temperature difference is 1 °F (Gooch 2010). Thicker material has higher R-value than a thin one. U factor is the reciprocal of R-value and usually used for assemblies Autodesk provides basic information about material's thermal properties. Every material used in an envelope assembly has fundamental physical properties that determine their energy performance such as conductivity and resistance. In order to make efficient design decisions designers should be aware of these properties.

An underlying principle of low-load homes is that developed countries need to reduce the emissions from space conditioning their homes by at least 90 % on average. The purpose is to bring the emissions to be within sustainable levels. There are performance-based limits for maximum heating/cooling load, annual space conditioning demand, and total annual home energy consumption. There are several factors that can affect energy consumption, including: heating-degree days, appliance efficiency, fuel substitution for space and domestic water heating, windows, energy-efficient lighting and heating, ventilation, air conditioning (HVAC) systems (Shrestha and Kulkarni 2013), building envelope shape (Granadeiro et al. 2013), and building materials (thermal mass) (Andjelković et al. 2012). Andjelković et al. (2012) concluded that simulation results indicated that by adding thermal mass to building envelope and structure, the following improvements can be achieved: (1) 100 % of all simulated cases experienced reduced annual space heating energy requirements; (2) 67 % of all simulated cases experienced reduced annual space cooling energy requirements; (3) 83 % of all simulated cases experienced reduced peak space heating demand; and (4) 50 % of all simulated cases experienced reduced peak space cooling demand (Andjelković et al. 2012).

When evaluating new design and construction method, reduction of energy consumption is one of the main requirements (Bolotin et al. 2013). Because modeling for energy simulation is a time-consuming task, frequently this process was simply overlooked (Granadeiro et al. 2013). Nowadays developers now can receive benefits from all sorts of simulation software systems. Examples of simulation software include: Green Building Studio, BEopt, Building Energy Modeling and Simulation, etc. In this research, the authors will use Green Building Studio (GBS) to simulate energy consumption. GBS is an Autodesk product that allows architects and designers to perform an extended building energy and water consumption analysis, and helps to make optimal decisions regarding carbon-neutral building designs (Green Building Studio n.d.). The functions of GBS include; (1) it analyzes the entire energy-usage of the systems and provides energy cost projections; (2) it takes into consideration weather data based on the location of the project; (3) process is web based, therefore, it simulation process is rapid; (4) it is able to compare design alternative. The advantages of GBS web service are listed as follows: (1) interface of the software is very user-friendly; (2) it saves designers time and effort to calculate a significant amount of information; (3) all of the simulations are carried out on remote servers; (4) provided results are easy to understand and can be easily compare with results of different buildings design. In the proposed research, GBS's ability to provide results for design alternatives is crucial. Using the results, the authors can study how materials with different R-values can affect a building's annual energy consumption.

The simulation results will be analyzed using big data tools, such as Oracle NoSQL Database. Our strategic goal is to significantly improve the energy efficiency of buildings for the purpose to reduce national energy demand. The intended result will be greater energy independence and a cleaner environment.

3 Objectives

The objectives of this project are to:

- Determine alternative solutions that have reduced implementation costs with improved energy efficiency, while providing occupant requirements for comfort and fresh air. Strategies being explored include increased equipment modularity, decreased equipment size, etc.
- New knowledge and understanding will be achieved to help with the sustainability.
- Sustainability can be a part of the engine that drives economic growth. The research will illustrate how the implementation of new technology in this research can be applied to incorporate research and economic development.

4 Methodology

In this proposed project, the authors will first categorize the different types of residential buildings according to their load capacities. The characteristics of the energy-consumption mode that residents value the most will also be investigated. Data will be gathered through accessing the database of building permits, approval, and commissioning. The County Department of Building and Zoning will provide access to their current building database with space heating and cooling load information and designed occupancy. Data will also be collected via online survey on architects, builders, Heating/Ventilating/Air-Conditioning (HVAC) subcontractors, property managers, and owners. The questionnaire will include questions on the location and type of building; space heating and cooling load; conditioned floor area; factors influencing design decisions; and the characteristics of air ventilation and air quality control. Descriptive statistics and significance tests will be used to examine differences and similarities between small and large buildings. Big data analysis tools will be used to examine the relationships between the environmental factors. House owners and environmental agencies will benefit from this study. Findings could improve the ventilation and air quality products and services available to Illinois house owners and make the federal energy saving program more effective in enhancing houses' ability to control air quality.

Step 1. Identify residential building project parameters for thermal comfort.
The first task to study the low-load homes is to identify the alternative strategies to conventional central ducted space conditioning systems that will provide thermal comfort for the occupants according to ASHRAE 55-2010, ACCA, and others (air temperature, relative humidity, air speed, and mean radiant temperature). We plan to search for successful practices and examples through literature review.

Step 2. Pilot-test the survey in a sample project.
Step 2 can be concurrent with Step 1. In Step 2, a survey will be sent to architects, builders, Heating/Ventilating/Air-Conditioning (HVAC) subcontractors, property managers, and owners. The survey will provide feedback on the design idea for air ventilation and air quality control, as well as the result of the design in reality. We will ask further questions to the feedbacks that show great potentials of design or technology innovation. We will also ask the economic trade-offs involved with changing from a central ducted space conditioning system.

Step 3. Analyze data from simulations.
We will use BIM technology to build models for the terminal conditions and parameters needed for simplified space conditioning systems (e.g., face velocity, Btu/cfm, duration of run cycle) to help understand how to provide thermal comfort in new and existing homes in Illinois. After the completion of the 3D model of the residence with all areas and volumes being defined, the authors will perform energy consumption simulation.

Data will be examined using Green Building Studio, a parametric analysis technique that calculate energy consumption and air flow to aid decision making. GBS is a web-based application which can analyze gbXML type files that are exported form Autodesk Revit. All the building geometry comes from Revit model, including the number of rooms and their relationship to the exterior. User needs to provide some building information like building type and postal code. After all the information is collected, GBS will provide whole building energy analysis, carbon-emission estimates, water use, and cost estimates, Energy Star scoring, LEED daylight credit potential, natural ventilation, and thermal performance (Autodesk n.d.).

Interviews will be analyzed and additional survey items will be generated, if necessary. We will also study such issues as: what implementation issues—including technology gaps, code restriction, or installation issues—need to be resolved to enable simplified space conditioning strategies to succeed in providing good thermal comfort, humidity control, and ventilation air distribution?

5 Big Data Analysis

There are many sensors installed in a HVAC system. Use a public university as an example. It has 20,000 students and around 20 buildings. There are approximately 300,000 sensors installed to the entire HVAC systems on campus. In this study, sensors can be installed to multiple locations of a house to monitor its performance. Since the sensors detect surrounding environment without interruption, the amount of data generated will be in significant volume. In this research, the authors analyzed the complexity of the data and established interrelationships of the sensors data with energy-correlated factors. For example, exterior air temperature, room or building occupancy, level of operations, individual activity schedule or program, and specific building location. The authors also performed correlative analytics on those data and HVAC performance data, such as system operation time and energy consumption, maintenance histories and manufacturers' data.

6 Conclusion

In this research, the authors performed correlative analytics on the big data generated from sensors, HVAC systems, and environmental conditions. Air quality and HVAC system performance can be improved by adjusting the zones of building HVAC system. It is becoming critical to a company's success as HVAC system performance helps with occupant health, comfort, productivity and compliance. Building owners or occupants can change the HVAC system programs through their cell phones. When a house is un-occupied or in vacancy, building owner or tenant can turn off the chilling and heating functions of the HVAC system and just leave the airside function for ventilating purpose.

The future challenge of this research includes developing apps or flexible platforms to create visualization reports and dashboards on the fly to meet individual needs. People can use their smart phones of mobile devices to control the performance of HVAC systems.

References

Andjelković, B.V., Stojanović, B.V., Stojiljković, M.M., Janevskić, J.N., Stojanović, M.B.: Thermal mass impact on energy performance of a low, medium, and heavy mass building in Belgrade. Therm. Sci. **16**(2), S447 (2012)

Bolotin, S.A., Gurinov, A.I., Dadar, A.H., Oolakay, Z.H.: An energy efficiency evaluation of architectural and construction solutions of an initial design stage in autodesk REVIT architecture. Mag. Civ. Eng. **8**, 64–91 (2013). (English)

BTO: Building technologies office: energy efficiency starts here (2014). http://www1.eere.energy.gov/buildings/pdfs/bto_overview_risser_040213.pdf

Communication from the Commission: Europe 2020: A Strategy for Smart, Sustainable and Inclusive Growth. European Commission, Brussels (2010)

Gooch, J.W. (ed.): Encyclopedic Dictionary of Polymers [electronic resource]. Springer, London (2010)

Granadeiro, V., Duarte, J., Correia, J., Leal, V.: Building envelope shape design in early stages of the design process: Integrating architectural design systems and energy simulation. Autom. Constr. **32**, 196–209 (2013)

Green Building Studio: Cloud-based energy analysis software (n.d.). http://www.autodesk.com/products/green-building-studio/overview. Accessed 15 December 2014

Kensek, K.M., Noble, D. (eds.): Building Information Modeling: BIM in Current and Future Practice. Wiley, Hoboken (2014)

Merschbrock, C., Munkvold, B.: A research review on building information modeling in construction - an area ripe for is research. Commun. Assoc. Inf. Syst. **31**, 207–228 (2012)

Sebastian, R.: Integrated design and engineering using building information modelling: a pilot project of small-scale housing development in the Netherlands. Archit. Eng. Des. Manage. **6**(2), 103–110 (2010). doi:10.3763/aedm.2010.0116

Shrestha, P.P., Kulkarni, P.: Factors influencing energy consumption of energy star and non-energy star homes. J. Manage. Eng. **29**(3), 269–278 (2013)

Tantasavasdi, C., Srebric, J., Chen, Q.: Natural ventilation design for houses in Thailand. Energy Build. **33**(8), 815–824 (2001)

USDOE: Simplified space conditioning in low-load homes: results from the Fresno, California, retrofit unoccupied test house (2014). http://www.nrel.gov/docs/fy14osti/60712.pdf

USDOE: Expert meeting: simplified space conditioning systems for energy efficient homes (2012). http://www1.eere.energy.gov/buildings/residential/pdfs/ibacos_2012_simplified_space_conditioning_expert_meeting_invitation.pdf

Wang, Z.: The control of airflow and acoustic energy for ventilation system in sustainable building. Doctoral dissertation, The Hong Kong Polytechnic University. Appendix: Springer-Author Discount (2014)

Initial Seeds Selection in Dynamic Clustering Method Based on Data Depth

Caiya Zhang[1(✉)] and Ze Jin[2]

[1] School of Computer and Computing Science, Zhejiang University City College, Hangzhou, China
zhangcy@zucc.edu.cn
[2] Department of Statistics Science, Cornell University, Ithaca, USA
zj58@cornell.edu

Abstract. Resorting to the theory of atomic models and the tool of data depth, we propose a novel method for initial seeds selection in dynamic clustering method. We define the *cohesion* of a point in a given data set, which includes the information of the significance and locations of neighboring points together. Then, the dynamic clustering algorithm based on *cohesion* is proposed. Compared with the *density*-based dynamic clustering algorithm, the clustering results demonstrate that our proposed method is more effective and robust.

Keywords: Clustering · Initial seeds · Projection depth

1 Introduction

Cluster analysis seeks to divide a set of objects into relatively homogeneous groups, which has been applied widely in machine learning, data mining, pattern recognition and bioinformatics so on. There are two main types of methods: hierarchical clustering and dynamic clustering. In this paper we focus on the dynamic clustering, the main advantages of which are its simplicity and fast speed which allows it to run on data with large sample size. However, the clustering results by dynamic method strongly depend on initial seeds selection.

It is well known that *density*-method is a typical type of initial seeds selection. The density of a point in a data is estimated by the number of its neighboring points within the given distance d, the furthest distance between two data points in the same cluster. The main shortcoming of *density*-based initial seeds selection is that it does not concern about the information whether the points are at the center of the data set and how close they are to the initial seeds. Hence, *density*-based initial seeds selection is very sensitive to the two parameters: the shortest distance between two different clusters (denoted by D)

The work is supported by Natural Science Foundation of Zhejiang Province of China (LY14A010003) **AMS Subject Classification (2000):** Primary 62H30; Secondary 62-07.

X. He et al. (Eds.): IScIDE 2015, Part II, LNCS 9243, pp. 603–611, 2015.
DOI: 10.1007/978-3-319-23862-3_60

and the furthest distance between two data points in the same cluster, that is d. Then cluster boundaries may become obscure, which will directly influence the stability of the cluster results.

It is well known that K-means clustering is one of popular dynamic clustering algorithms. Redmond and Heneghan (2007) provided a brief summary of techniques in initial seeds selection and presented a method for initialising the K-means clustering algorithm by using a *kd*-tree to perform a density estimation of the data at various locations. Pavan et al. (2011) proposed a new seed selection algorithm, named SPSS, that produces single, optimal solution which is outlier insensitive. Recently, Mavroeidis and Marchiori (2014) proposed an algorithm based on a sparse PCA approach to discuss the feature selection for k-means clustering stability. However, before applying the K-means algorithm, except the two above parameters D and d need to be known in advance, the number k of initial clustering also needs to be given. Then the clustering seeds will be renewed with any one data point being assigned to the nearest cluster.

In our paper, we are interested in a type of dynamic clustering, which needs not the number of initial clustering to be given in advance and the clustering seeds will not be renewed until all the data points being assigned according to the nearest distance principle. Based on the theory of atomic model and the tool of data depth, we define the concept *cohesion* of a point in a given multivariate data and propose a new initial seeds selection method in Sect. 2. In Sect. 3, extensive experiments will be done to show that our proposed *cohesion*-based algorithm performs more robust and efficient than the traditional *density*-based algorithm.

2 Clustering Algorithm Based on Cohesion

To define the concept *cohesion* of a point in a multivariate data set, we need introduce some knowledge about atomic models and data depth.

2.1 Atomic Models and Data Depth

Atomic models have their beginnings in the early atomic theory. In atomic physics, the Bohr model (Bohr (1913a, 1913b)), depicts the atom as a small, positively charged nucleus surrounded by electrons that travel in circular orbits, which is similar in structure to the solar system, but with electrostatic forces providing attraction rather than gravity.

Let e and m_e be the elementary charge and the electron rest mass, respectively. And denote the radius and the atomic energy of the $i-$th steady path by r_i and E_i, respectively. According to the Bohr model, we have the following conclusion.

Conclusion A. *E_i is directly proportional to the electron rest mass m_e and inversely proportional to the radiu r_i, that is, $E_i = \frac{km_e}{r_i}$, where k is a known positive constant.*

Data depth is a measure of depth of a given point with respect to a multivariate data cloud or its underlying distribution. It provides a center-outward ordering of multivariate observations. Points deep inside a data cloud get high depth values and those on the outskirts get low depth values. During the past ten years, more and more researchers have adopted the tool of data depth for clustering and classification. For example, Ghosh & Chaudhuri (2005) developed the maximum depth classifers where half-space depth and simplicial depth were used for clustering. Liu et al. (2006) summarized the theory and applications of data depth in multivariate analysis. In our paper, we take projection depth (PD) into account since it is of good statistical properties (see Zuo & Serfling, 2000).

Definition A (Projection Depth) (see Zuo & Serfling, *2000*). *Suppose* $X' = \{X_1, \cdots, X_n\}$ *is a data set in* $p-$*dimensional space. The projection depth of the point* X_i *with respect to the data set* X *is defined by*

$$PD(X_i) = \frac{1}{1+O(X_i)}, \quad i = 1, \cdots, n,$$

where

$$O(X_i) = \sup_{\|u\|=1} \frac{u'X_i - med(u'X)}{mad(u'X)}, \quad i - 1, \cdots, n,$$

u is a p-dimensional column vector, the pair median (med) is the median of $u'X_1, \cdots, u'X_n$*, the median absolute deviation (mad) is the median of* $|u'X_1 - med(u'X)|, \cdots, |u'X_n - med(u'X)|$.

2.2 Cohesion

To reflect the importance of a point in a multi-dimensional data set, we can give a weight to every point. According to the properties of data depth, higher data depth leads to higher weight. Moreover, data weight in the form of convex function will be of good mathematical properties. Here, we define the data weight by exponential function and projection depth as follows.

Definition 2.1 (Exponential Weight). *Suppose* $X' = \{X_1, \cdots, X_n\}$ *is a data set in* $p-$*dimensional space. The exponential weight of the point* X_i *with respect to the data set* X *is defined by*

$$w(X_i) = exp\{PD(X_i)\}, \quad i = 1, \cdots, n,$$

where $PD(X_i)$ *is the projection depth of* X_i *and* $exp\{\cdot\}$ *is the exponential function.*

Then the concept of *Cohesion* will be defined as follows.

Definition 2.2 (Cohesion). *Suppose $X' = \{X_1, \cdots, X_n\}$ is a data set in $p-$dimensional space and d is a given positive constant. For any point X_i, denote the number of its neighboring data points within distance d by n_i (not including X_i itself) and denote these corresponding specific data points by $X_1^*, \cdots, X_{n_i}^*$. Let $PD(X_j^*)$ and $w(X_j^*)$ be the projection depth and the exponential weight of X_j^* with respect to the data set X, respectively, and $r(X_i, X_j^*)$ be the Euclidean distance between X_i and X_j^*.Then the cohesion of X_i is defined by*

$$Cs(X_i) = \sum_{j=1}^{n_i} \frac{w(X_j^*)}{r(X_i, X_j^*)} = \sum_{j=1}^{n_i} \frac{exp\{PD(X_j^*)\}}{r(X_i, X_j^*)}, \quad i = 1, \cdots, n,$$

From Definition 2.2, we can see that, the cohesion is very likely with the atomic model. If we view a data point as a nucleus, then its neighboring points can be viewed as the orbital electrons of the nucleus. So, the cohesion $Cs(X_i)$ is equivalent to the $i-$th path atomic energy E_i, the exponential weight $w(X_j^*)$ is equivalent to the electron rest mass m_e, and the distance $r(X_i, X_j^*)$ is equivalent with the radius r_i. Similarly to the atomic model, the cohesion $Cs(X_i)$ is directly proportional to the exponential weight $w(X_j^*)$, which reflect the importance degree of X_j^*, and inversely proportional to the distance $r(X_i, X_j^*)$.

To design the algorithm of dynamic clustering based on *cohesion*, we need to give the values of the two parameters D and d in advance. Then the algorithm can be concluded as follows.

Step 1: Initial seeds selection. Calculate the *cohesion* of each point under the given constant d, and select the point with the greatest *cohesion* as the first initial seed. Pick out the point with the second greatest *cohesion*, and select it as the second initial seed if the distance between it and the first initial seed is larger than D, otherwise omit it and turn to the next greatest one. In the same way, pick out each point in order of decreasing *cohesion* and select it as an initial seeds if the distance between it and any of the other previously chosen seeds is larger than D.

Step 2: Initial clustering. Assign each point to the cluster with its nearest initial seed, and calculate cluster centers.

Step 3: Iterative clustering. Replace former seeds with the cluster centers, assign each point to the cluster with its nearest seed. Recalculate the cluster centers and reassign each point. The iterative process will stop when all cluster centers coincide with the previous ones. If it doesn't happen, we can set a rule to stop the iterative process, such as limiting the most times of iteration.

Since the density of a data point is only related with the number of its neighboring points within the given distance d, while the cohesion is also related with the importance degree and the distance of its neighboring points, the *cohesion* includes more information than the *density*. It means that the algorithm of *cohesion*-based dynamic clustering (CBDC) will be less sensitive to the two constants D and d than the *density*-based dynamic clustering (DBDC). And so the clustering results by CBDC will be more effective and robust than those by DBDC. This point will be supported further more through the results on experiments in next section.

3 Results on Experiments

Assume that $X' = \{X_1, \cdots, X_n\}$ is a given data in R^p. Throughout the following experiments, the parameters D and d satisfy that $D = 2d$. R-Square statistic (RS) and pseudo F statistic (PFS) are used to determine the number of clusters and to measure the effect and rationality of clustering (See Milligan &Cooper, 1985). RS is defined as $RS = 1 - SSE/SST$ and PFS is defined as $PFS = [(SST - SSE)/(c-1)]/[SSE/(n-c)]$, where SST is the total sum of squares, SSE is the within sum of squares, c is the number of clusters, n is the number of observations. Large values of RS and PFS represent that the clusters obtained at a given step are significantly different among each other, which suggest high rationality in a given clustering.

3.1 Designed Data Sets

Firstly, we design three data sets with unknown clustering from one-dimensional space to three-dimensional space, both DBDC and CBDC are applied. Then we compare the performance of these two methods under some fixed parameters, and evaluate the clustering results by the statistics RS and PFS. Although the density and cohesion of a data point depend on the parameter d, the clustering results will be improved by iterations of the algorithms. Here, we synthesize the clustering analysis idea and pairwise distance to decide the value of d, and let $D = 2d$.

One-dimensional case $(p = 1, n = 9)$
Denote the designed data set by
$X' = (0.3000\ 0.6500\ 1.0000\ 1.7500\ 2.1500\ 2.5500\ 3.2500\ 3.6000\ 3.9500)$.

Two-dimensional case $(p = 2, n = 10)$
Denote the designed data set by

$$X' = \begin{pmatrix} 1.0873 & 0.6898 & 2.1226 & 1.5124 & 2.8655 & 3.0323 & 4.2627 & 4.7709 & 3.4115 & 4.7347 \\ 4.5019 & 3.2120 & 4.8551 & 3.8974 & 2.7705 & 2.4773 & 1.1360 & 1.8319 & 0.4699 & 0.5658 \end{pmatrix}.$$

Three-dimensional case $(p = 3, n = 8)$

Denote the designed data set by

$$X' = \begin{pmatrix} 0.0000 & 1.5191 & 0.9888 & 2.4216 & 2.6202 & 3.9325 & 3.4333 & 4.9996 \\ 0.5115 & 0.0000 & 1.0043 & 2.3669 & 2.6327 & 3.9510 & 5.0015 & 4.5448 \\ 0.5098 & 1.5344 & 1.0024 & 2.3566 & 2.5544 & 3.9515 & 3.5596 & 4.4447 \end{pmatrix}.$$

From Tables 1, 2, 3, no matter in one-dimensional case, two-dimensional case or in three-dimensional case, we can see that any two different points have different values of cohesion, which means that these points are distinguishable perfectly and can be ranked by cohesion. However, several different points holding the same value of density, which may lead to confusion. Moreover, the clustering results by CBDC are more reasonable than those by DBDC according to the values of statistics RS and PSF.

Table 1. Dynamic clustering on one-dimensional data under $D = 1.56, d = 0.78$.

	X_1	X_2	X_3	X_4	X_5	X_6	X_7	X_8	X_9
PD	0.3833	0.4340	0.5000	0.7419	1.0000	0.7419	0.5111	0.4423	0.3898
density	2	2	3	2	2	2	3	2	2
cohesion	6.7649	8.9025	9.3055	8.9940	10.5000	9.1773	9.5562	8.9825	6.8282

	initial seeds	initial clustering	final clustering	RS	PSF
DBDC	3	$\{1,2,3,4\}$	$\{1,2,3,4\}$	0.7582	21.9500
	7	$\{5,6,7,8,9\}$	$\{5,6,7,8,9\}$		
CBDC	1	$\{1,2,3\}$	$\{1,2,3\}$	0.9416	48.3519
	5	$\{4,5,6\}$	$\{4,5,6\}$		
	9	$\{7,8,9\}$	$\{7,8,9\}$		

Table 2. Dynamic clustering on two-dimensional data under $D = 2.28, d = 1.14$.

	X_1	X_2	X_3	X_4	X_5	X_6	X_7	X_8	X_9	X_{10}
PD	0.3961	0.0761	0.0846	0.4244	0.4670	0.7559	0.4544	0.1029	0.0792	0.3726
density	2	1	2	3	1	1	3	1	1	1
cohesion	3.0634	1.4277	2.7047	3.9771	6.3128	4.7287	4.2486	1.8280	1.4574	2.1281

	initial seeds	initial clustering	final clustering	RS	PSF
DBDC	4	$\{1,2,3,4,5\}$	$\{1,2,3,4,5\}$	0.7286	21.4728
	7	$\{6,7,8,9,10\}$	$\{6,7,8,9,10\}$		
CBDC	1	$\{1,2,3,4\}$	$\{1,2,3,4\}$	0.8776	25.0937
	5	$\{5,6\}$	$\{5,6\}$		
	9	$\{7,8,9,10\}$	$\{7,8,9,10\}$		

3.2 Real Sample

We choose the well-known IRIS data published by Fisher (1936) as a real sample. The IRIS data contains four quantitative variables measured on 150 specimens of iris plants. These include sepal length (SEPALLEN), sepal width (SEPALWID), petal length (PETALLEN), and petal width (PETALWID). The classification variable, SPECIES, represents the species of iris from which the measurements were taken. There are three species in the data: Iris setosa, Iris versicolor, and Iris virginica. Each specie includes 50 observations. For convenience, we choose two of variables: PETALLEN and PETALWID to analyze and denote the values of these three species by 1, 2 and 3, respectively. To evaluate the clustering results by DBDC and CBDC comprehensively, we carry out our experiments

Table 3. Dynamic clustering on three-dimensional data under $D = 2.80, d = 1.40$.

	X_1	X_2	X_3	X_4	X_5	X_6	X_7	X_8
PD	0.0745	0.0155	0.0921	0.3963	0.3474	0.0655	0.0168	0.0755
density	1	1	2	1	1	2	1	1
cohesion	0.9064	0.8743	1.7004	3.6641	3.8478	1.6474	0.8699	0.8107

	initial seeds	initial clustering	final clustering	RS	PSF
DBDC	3	$\{1,2,3,4\}$	$\{1,2,3,4\}$	0.7215	15.5423
	6	$\{5,6,7,8\}$	$\{5,6,7,8\}$		
CBDC	1	$\{1,2,3\}$	$\{1,2,3\}$	0.9219	29.4997
	5	$\{4,5\}$	$\{4,5\}$		
	8	$\{6,7,8\}$	$\{6,7,8\}$		

under varied values of the parameters d and D with the sample size from small to large.

Small sample case ($n = 15$)

Here, we pick up five data points randomly in each specie of IRIS data.

$$X' = \begin{pmatrix} 1 & 1 & 1 & 1 & 1 & 2 & 2 & 2 & 2 & 2 & 3 & 3 & 3 & 3 & 3 \\ 1.3 & 1.5 & 1.4 & 1.7 & 1.4 & 4.7 & 4.5 & 4.9 & 4.0 & 4.6 & 5.2 & 5.0 & 5.2 & 5.4 & 5.1 \\ 0.2 & 0.2 & 0.2 & 0.4 & 0.3 & 1.4 & 1.5 & 1.5 & 1.3 & 1.5 & 2.3 & 1.9 & 2.0 & 2.3 & 1.8 \end{pmatrix} \begin{matrix} SPECIE \\ PETALLEN \\ PETALWID \end{matrix}.$$

Now, we apply the algorithms of DBDC and CBDC on these 15 chosen observations under the parameter d from 0.35 to 0.5 by step size 0.05. The clustering results are listed on Tables 4, 5, 6.

From Tables 4, 5, 6, we can see that with the increase of the parameter d, the clustering results by DBDC vary greatly, from 4 groups to 2 groups. However, the clustering results by CBDC keep identical completely to original clustering

Table 4. Dynamic clustering on Iris data ($n = 15$) under $D = 0.70, d = 0.35$.

	Initial seeds	Initial clustering	Final clustering
DBDC	2	$\{1,2,3,4,5\}$	$\{1,2,3,4,5\}$
	6	$\{6,7,8,10\}$	$\{6,7,8,10\}$
	9	$\{9\}$	$\{9\}$
	13	$\{11,12,13,14,15\}$	$\{11,12,13,14,15\}$
CBDC	3	$\{1,2,3,4,5\}$	$\{1,2,3,4,5\}$
	10	$\{6,7,8,9,10\}$	$\{6,7,8,9,10\}$
	11	$\{11,12,13,14,15\}$	$\{11,12,13,14,15\}$

Table 5. Dynamic clustering on Iris data ($n = 15$) under $D = 0.80, d = 0.40$.

	Initial seeds	Initial clustering	Final clustering
DBDC	2	$\{1, 2, 3, 4, 5\}$	$\{1, 2, 3, 4, 5\}$
	7	$\{6, 7, 8, 9, 10\}$	$\{6, 7, 8, 9, 10\}$
	13	$\{11, 12, 13, 14, 15\}$	$\{11, 12, 13, 14, 15\}$
CBDC	3	$\{1, 2, 3, 4, 5\}$	$\{1, 2, 3, 4, 5\}$
	10	$\{6, 7, 8, 9, 10\}$	$\{6, 7, 8, 9, 10\}$
	11	$\{11, 12, 13, 14, 15\}$	$\{11, 12, 13, 14, 15\}$

Table 6. Dynamic clustering on Iris data ($n = 15$) under $D = 1.00, d = 0.50$.

	Initial seeds	Initial clustering	Final clustering
DBDC	1	$\{1, 2, 3, 4, 5\}$	$\{1, 2, 3, 4, 5\}$
	8	$\{6, 7, 8, 9, 10, 11, 12, 13, 14, 15\}$	$\{6, 7, 8, 9, 10, 11, 12, 13, 14, 15\}$
CBDC	3	$\{1, 2, 3, 4, 5\}$	$\{1, 2, 3, 4, 5\}$
	10	$\{6, 7, 8, 9, 10\}$	$\{6, 7, 8, 9, 10\}$
	11	$\{11, 12, 13, 14, 15\}$	$\{11, 12, 13, 14, 15\}$

Table 7. Dynamic clustering on Iris data ($n = 150$) under $D = 1.70, d = 0.85$.

	Number of clusters	Number of points	RS	PFS
DBDC	4	$N_1 = 50$ $N_2 = 29$ $N_3 = 40$ $N_4 = 31$	0.964503	1322.345
CBDC	3	$N_1 = 50$ $N_2 = 54$ $N_3 = 46$	0.942923	1214.226

Table 8. Dynamic clustering on Iris data ($n = 150$) under $D = 1.90, d = 0.95$.

	Number of clusters	Number of points	RS	PFS
DBDC	3	$N_1 = 50$ $N_2 = 54$ $N_3 = 46$	0.942923	1214.226
CBDC	3	$N_1 = 50$ $N_2 = 54$ $N_3 = 46$	0.942923	1214.226

all long, which provides further evidence that CBDC are more reasonable and robust than DBDC

Large sample case ($n = 150$)

Next, we apply the algorithms of DBDC and CBDC on the all observations of Iris data under the parameter d from 0.85 to 1.10. The number of observations in each group and the values of statistics RS and PFS are listed on Tables 7, 8, 9.

From Tables 7, 8, 9, when $D = 1.70, d = 0.85$, although the values of RS and PFS in DBDC are larger than those in CBDC, the clustering results by DBDC differ remarkably with the original clustering where each group has 50

Table 9. Dynamic clustering on Iris data ($n = 150$) under $D = 2.20, d = 1.10$.

	Number of clusters	Number of points	RS	PFS
DBDC	2	$N_1 = 51$ $N_2 = 99$	0.843085	795.189
CBDC	3	$N_1 = 50$ $N_2 = 54$ $N_3 = 46$	0.942923	1214.226

observations. When $D = 1.90, d = 0.95$, the clustering results by DBDC are the same as by CBDC. When $D = 2.20, d = 1.10$, the performance of DBDC is obviously inferior to CBDC.

4 Conclusion

From what has been discussed above, we can draw some conclusions as follows. The cohesion of a data point provides more information than the density. Then the cohesion-based initial seeds selection seems not so sensitive to the two parameters D and d, which represent the shortest distance between two different clusters and the farthest distance any two points in the same cluster, respectively. Therefore, the clustering results by CBDC are more robust and rational than by DBDC. In some other words, CBDC is more competitive than DBDC.

References

Bohr, N.: On the constitution of atoms and molecules, part I binding of electrons by positive nuclei. Phil. Mag. **26**, 1–24 (1913a)

Bohr, N.: On the constitution of atoms and molecules, part II systems containing only a single nucleus. Phil. Mag. **26**, 476–502 (1913b)

Fisher, R.A.: The use of multiple measurements in taxonomic problems. Ann. Eugenics **7**, 179–188 (1936)

Ghosh, A.K., Chaudhuri, P.: On maximum depth and related classifiers. Scand. J. Stat. **32**, 328–350 (2005)

Liu, R., Serfling, R., Souvaine, D.: Depth functions in nonparametric multivariate inference. DIMACS Ser. Discrete Math. Theoret. Comput. Sci. **72**, 1–16 (2006)

Mavroeidis, D., Marchiori, E.: Feature selection for Dynamic clustering stability: theoretical analysis and analgorithm. Data Min. Knowl. Disc. **28**, 918–960 (2014)

Pavav, K.K., Rao, A.A., Rao, A.V.D., Sridhar, G.R.: Robust seed selection algorithm for dynamic type algorithms. Int. J. Comput. Sci. & Inf. Technol. **3**, 147–163 (2011)

Redmond, S.J., Heneghan, C.: A method for initialising the dynamic clustering algorithm using kd-trees, Pattern. Recogn. Lett. **28**, 965–973 (2007)

Zuo, Y., Serfling, R.: General notions of statistical depth function. Ann. Stat. **28**, 461–482 (2000)

The Data Quality Evaluation that Under the Background of Wisdom City

FengJing Li(✉) and YongLi Wang

College of Computer Science and Engineering,
Nanjing University of Science and Technology, Nanjing, China
fengjingli_fj@163.com

Abstract. Current sensor-based monitoring systems use multiple sensors in order to identify high-level information based on the events that take place in the monitored environment. This information is obtained through low-level processing of sensory media streams, which are usually noisy and imprecise, leading to many undesired consequences such as incorrect data or incomplete data, inconsistent data. Therefore, we need a mechanism to compute the quality of sensor-driven information that would help a user or a system in making an informed decision and improve the automated monitoring process. In this article, with wisdom city management as the application background, the inclinometer data as the research object, researching a kind of efficient data quality evaluation method based on sensor observations. And we propose a model to characterize such quality of information in a multisensory multimedia monitoring system in terms of certainty, accuracy/confidence and timeliness.

Keywords: Sensors · Data quality · The quality evaluation

1 Introduction

The data quality issues of each link, including data cleaning, data integration, similarity record detection, data quality evaluation, data quality process control and management, etc., it has made a lot of academic research and practical exploration. In these links, data quality assessment is foundation and essential prerequisite of improving the quality of the data, it can gives a reasonable assessment about the quality of the whole or part of the application system data situation, which can help users understand the application system of data quality well, and then take the corresponding treatment process to improve data quality.

1.1 Research Background

Wisdom urban management system is a typical Internet application, one of the most important reason that Internet of things technology hard to practical is that the data quality problem. Data quality is mainly refers to the extent that an information system has realized the pattern and the consistency of the data instance, meaning mode and how the data instance implemented on the correctness, consistency, completeness, real-time and the minimality. The Iot mixed with non-deterministic, high and strong correlation characteristics poses challenges for the data quality assurance.

X. He et al. (Eds.): IScIDE 2015, Part II, LNCS 9243, pp. 612–621, 2015.
DOI: 10.1007/978-3-319-23862-3_61

1.2 Research Significance

This paper presents a QoI's properties calculation algorithms under the background of multi-sensor multimedia monitoring system. Achievement of this paper are: (1) describing a clear calculation model researched the uncertainty, real-time and accuracy/confidence level; (2) using multimodal integration schemes for certainty and accuracy; (3) considering environment such as sensor environment to select appropriate sensors used for QoI calculation; (4) researching the QoI characteristics information-level, quality attributes-level and system-level.

2 Model and Definition

2.1 Data Quality Measurement Model

2.1.1 Problem Formalization

That S as a multi-sensor system developed in the intelligent monitoring environment, used to identify a set of information items $\prod_r = \{I_1, I_2, \ldots .I_r\}$, r is the total number of information items $r = |\Pi_r|$ some examples of information item are human body detection and recognition.

- let the system S use a collection $M_n = \{M_1, M_2, \ldots .M_n\}, n \geq 1$, the number of media stream from already bought sensors, to detect all kinds of information.
- $q_{b,j} \in [0, 1], (1 \leq b \leq k, 1 \leq j \leq r)$ is the bth quality attributes, as the jth information items (I_j) in a special environment, k is the total number of quality attributes, in our case, k = 3.
- $w_{I_j}(1 \leq j \leq r)$ and $w_{q_b}(1 \leq b < k)$ is the weight assigned to r information items and k quality attributes. The weight will be based on user requirements and the environment.

(1) For each quality attribute based on multi-sensor observation provides a dynamic calculation model, the model will be used to calculate the bth information quality attribute values in a single item $q_{b,j} \in [0, 1]$.
(2) Measuring the quality of the individual information item $QoI_j, 1 \leq j \leq r$, accurately, this process can be expressed as:

$$QoI_j = f(q_{b,j}, w_{qb}) \quad (1)$$

(3) To measure system quality attributes based on all individual information items quality attributes, can be expressed as:

$$QoI_{qb} = g(q_{b,j}, w_{I_j}) \quad (2)$$

(4) To measure the information quality of the whole system, paying attention to, f and g are two aggregation function, described in subsequent sections.

2.1.2 The Model of Quality Attributes Assessment

I. Simulating quality attributes
In this section, we describe three quality attributes of a multi-sensor device (associated with an information item) $q_{b,j}(1 \le b \le k, k = 3, 1 \le j \le r)$.

a. Certainty
All sorts of feeling media flows in the multi-sensor system provide different probability grades for different mission task. Assumes that the media stream used to observe an event alone, but when providing data, the single data stream probability score can fuse in together, in order to obtain the overall probability of an event. Therefore, the system use a media stream collection $M_n = \{M_1, M_2, \ldots, M_n\}$, $n \ge 2$, that the media stream is a specific observation task of n different sensors at a specific moment. $p(I_j|M_1)$ (indicated for p_1^j) as a probability score, it based on local decisions of the media stream M_1, jth information item I_j. In fact, for an information item, multiple sensors local decisions may be similar or contradictory. In order to consider this similarity and contradiction between the local decision, They can be divided into two subsets ϕ1 and ϕ2, set 0.5 as a group critical value, before subsets represent events happen and after represent events did not happen. Intuitive said, this means that in a probabilistic framework, when the sensor observations scores higher than 0.5, represent the hypothesis is correct.
Local decisions of each group are fused respectively to determine which group has the high total score. In this paper, we use Bayes mechanism as the fusion process. Note that this fusion process using the sensors confidence and consistent/divisions of the past. In this method, every two representative information item score fusing use the following equation:

$$pi, m = \frac{(p_i^j)^{f_i'i} * (p_m^j)^{f_m'} * e^{r_{i,m}^j}}{(p_i^j)^{fi} * (p_m^j)^{f_m} * e^{r_{i,m}^j} + (p_i^j)^{fi} * (1 - p_m^j)^{f_m} * e^{-r_{i,m}^j}} \tag{3}$$

$p_{i,m}$ is the total probability score, based sensor media flows M_i and M_m on the time t. Note that the time parameter t is related to the right and left sides of the Eq. (3), but it does not clearly described in the equation. The denominator in the Eq. (3) is used as a standardized factor, limiting values between [0, 1]. Term $f_i' = f_i(f_i + f_m)$, $f_m' = f_m(f_m + f_i)$ are two weighted factor, they calculate the past confidence that two sensors participated in. Here, f_i and f_m is the sensor initial confidence or is M_i and M_m media stream confidence at time t−1. This article will demonstrate confidence is how to model in Sect. 3.2.3.
In Eq. (3), $\gamma_{i,m}^j \in [-1, 1]$ refers to the agreement/disagreement between two sensors (aka consistent coefficient). When fused probability given certain weights to observe in sensor, it was as a growth factor. The $\gamma_{i,m}^j$ value of two sensors in time t can calculate the agreement/disagreement in current with time t−1:

$$r^j_{i,m} = \beta\left[1 - 2 * \left|p^j_i(t) - p^j_m(t)\right|\right] + (1 - \beta)\left[r^j_{i,m}(t-1)\right] \tag{4}$$

$1 - 2 \times \left|P^j_i(t) - P^j_m(t)\right|$ is the current agreement/disagreement between two sensors. Note that when two sensors is exactly the value is 1, and when two sensors are divided the value is 1. Weighted factor β and 1−β was assigned to the current and past consistent coefficient respectively. When calculating consistent coefficient between a set of sensor and a sensor, the two-way value of consistent coefficient be average used in average chain group. For example, consistent coefficient calculation formula of a set of sensors $(M_{\rm i}, M_m)$ and a sensor M_s is: $\gamma^j_{i,m,s} = (\gamma^j_{i,s}, \gamma^j_{m,s})/2$. Eqs. (3) and (4) are repeatedly used in combined with the current observation points based on the media stream φ_1, to get $p^j_{\varphi 1}(t)$ cumulative score. Similarly, for not received events get group φ_2 score $p^j_{\varphi 2}(t)$, in the end, if $P^j_{\varphi 1}(t) \geq P^j_{\varphi 2}(t)$, based on group φ_1 decision is right, it represents the jth information item happened, on the other hand, the decision based on group φ_2 was wrong, because it represents the events did not happen.

Sensor only if all observed, however, belong to a group, then the fusion of grouping score will be as certainty value of the item information. As a result, the certainty value calculated using the following form:

$$q_{1,j}(t) = \{\frac{P^j_{\varphi 1}(t), P^j_{\varphi 1}(t) \geq P^j_{\varphi 2}(t)}{P^j_{\varphi 2}(t), otherwise} \tag{5}$$

$q_{1,j}(t)$ is the certainty quality attribute of jth information item at time t.

b. Accuracy

The accuracy of a sensor media M_i can be calculated using four possible parameter when detecting the information item I_j, including: true positive TP^j_i, error certainly FP^j_i, false negative FN^j_i and correct negative TN^j_i, as follows:

$$Accuracy^j_i(t) = \frac{TP^j_i(t) + TN^j_i(t)}{TP^j_i(t) + TN^j_i(t) + FN^i_i(t) + TN^j_i(t)}, 1 \leq j \leq r \tag{6}$$

We simulated the confidence level $f^j_i(t)$ of a single sensor at time t, which is based on the fused probability score $p^j_{\varphi 1}(t)$ and $\overline{p^j_{\varphi 2}}(t)$, the probability score is gained by repeated using the Eq. (3). As mentioned earlier, if $p^j_{\varphi 1} \geq \overline{p^j_{\varphi 2}}(t)$, we think events based on group $\varphi 1$ happen, otherwise think events based on group $\varphi 2$ happen. Intuitively, therefore, we use the rewards and punishment mechanism, and if the final decision is based on the results of ϕ1, the corresponding increase the confidence level of grouping $\varphi 1$ sensor, at the same time, reduce the confidence value of group $\varphi 2$, vice versa. It is said the general model below:

$$f^j_i(t) = f^j_i(\mathrm{t} - 1) \pm \mathrm{pt}_i(\mathrm{t}) \tag{7}$$

$f_i^j(\mathrm{t}-1)$ is the confidence of ith sensor for the jth information items in the past. The punishment $\mathrm{pt}_i(\mathrm{t})$ will be assigned to a failure group.

Once you have the confidence level of a single sensor, summarizing the value of the sensor in successful group, to get the confidence level of I_j in system, therefore, the $q_{2,j}(\mathrm{t})$ accurate calculation of jth information item in time t is:

$$q_{2,j}(t) = \frac{1}{p}\sum_{y=1}^{p} f_y^j(t) \tag{8}$$

p is the number of sensors.

c. Real-time

To simulate real-time, we argue that a system is detected the jth information item (event) in a unit time T_j, here, an information item value T_j is determined by the experiment, have to note, however, due to the change of the operating environment and other problems involving multiple media, per unit time of actual detection target event is $T_j \pm \delta_j$, δ_j is jitter error related to task I_j. In general, jitter is associated with task based on different types of achievement, a mission on hardware equipment compared to perform tasks on software equipment with low jitter. Based on the above illustration, this paper can define a current real-time of a information item as follows:

$$q_{3,j}(t) = \{^{1,\delta_j \le Th_j}_{0,otherwise} \tag{9}$$

$Th_j \pm \delta_j$ is the amount of time actual information item I_j, and if δ_j is less than or equal to the critical value Th_j, we assume that the system is real-time, this means that Th_j is the allowed range, the response of the system can be different from the previous calculation that regarded as the real-time T_j in the meantime. On the other hand, we will consider if δ_j beyond the permissible range, the system will not be able to access to real-time information item I_j in a specific instance. The value of Th_j in real-time software system is more common than hardware real-time systems, including $Th_j \to 0$.

II. Aggregation of information quality

(1) The aggregation of information item level: in this method (see Fig. 2(a)), the quality attribute of the information item I_j aggregation use the linear weighted sum fusion scheme to obtain the quality of information item level QoI_j, using Eq. (11) for all items, further aggregating for the whole system QoI.

$$QoI_j = \sum_{b=1}^{k} q_{b,j} \times w_{qb}, 1 \le j \le r \tag{10}$$

$$QoI = \sum_{j=1}^{r} QoI_j \times w_{I_j} \tag{11}$$

(2) The aggregation of quality attribute level : in this method (see Fig. 2 (b)), using the Eq. (12), aggregate a particular quality attribute of all items, obtain the attribute quality QoI_{q_b}, using Eq. (13), further aggregating attribute quality will be get the overall quality of the system.

$$QoI_{q_b} = \sum_{j=1}^{r} q_{b,j} \times w_{I_j}, 1 \leq b \leq k \tag{12}$$

$$QoI = \sum_{b=1}^{k} QoI_{q_b} \times w_{q_b} \tag{13}$$

The letter b, j, r and k semantics described earlier. q_b refer to all information of a single attribute value. K = 3 and r is the number of information items. $w_{I_j} \in [0,1], \sum_{j=1}^{r} w_{I_j} = 1$ is the bth quality attribute weight, the choice of weight w_{q_b} based on the demand of the user or application (Fig. 1).

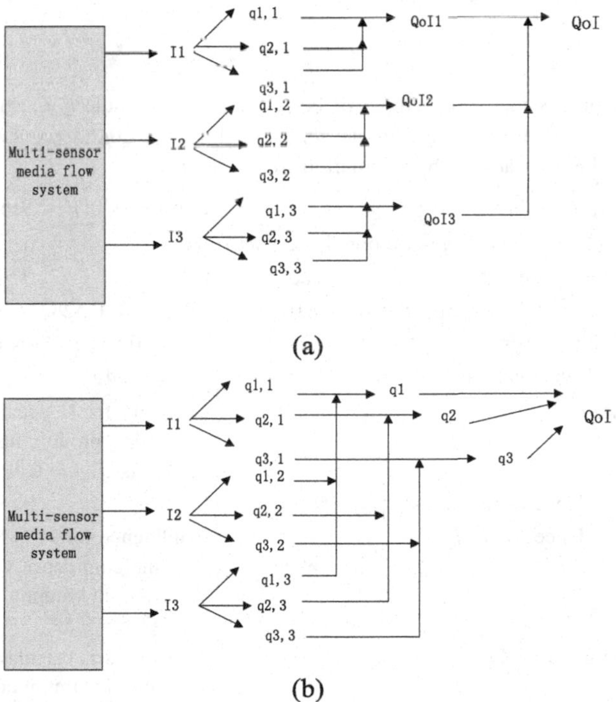

Fig. 1. Two kinds of aggregation methods of quality attribute, in (a) method, aggregating all quality attribute QoI_j of a information item, then aggregating all information items obtain the system-level QoI;in (b) method, aggregating all information item of each quality attribute QoI_{q_b}, then aggregate all quality attribute obtain system level QoI.

3 Experimental Analysis

3.1 QoI Assessment

The observation points from real-time media processing can be divided into two subsets (such as φ_1 and φ_2), as described in Sect. 2.2.2. For example (see Table 1 line 3), we consider {0.49, 0.55} as observation points, and φ_2 = {1-0.49}.Iterated using the Eq. (3) for fused scores of each group, obtain two groups' total score $p_{\varphi1}$ and $p_{\varphi2}$ (see Table 1 line 6). Note that the fusion process considers the consistent/different between sensors (different tasks) and previous confidence level of target information item. Using Eq. (4) calculate the agreement/disagreement between any two observation tasks sensor (see Table 1 line 4). For the current agreement/disambiguation, we put β = 0.6 as an example of weight distribution. Initial agreement/disambiguation between the two sensors or corresponding is 0.0, it changing over time.

Initial confidence at the beginning of the test phase is previous accuracy of sensors' related tasks, and the initial value of two sensors is {0.5, 0.5} (see Table 1 row 2), it will change according to the confidence's computation with time, and it described in the next section. In the instance, $p_{\varphi1}$ and $p_{\varphi2}$ is 0.49 and 0.35. Now according to the two total score, we determined when $p_{\varphi1} \geq p_{\varphi2}$ the information items I_1 happened, otherwise we determine it is not happened.

Table 1. A QoI processing steps sample of computing QoI attribute $q_{(b,j)}$, given the current observation points, and for information item I_1, initialing the confidence level of a single processing tasks (the accuracy of the calculation in advance).

Instance	QoI processing steps	SensorId (T1)	SensorId (T2)
t = 0	Initial confidence (pre-computed accuracy)	0.5	0.5
t = 1	Current observation (probability score)	0.49	0.55
	Agreement/disagreement (pair-wise)	(T1,T2) = 0.528	
	Construct group	φ_1 = {0.55}	φ_2 = {1-0.49}
	Fused observations	$p_{\varphi1}$ = {0.49}	$p_{\varphi2}$ = {0.35}
	Certainty (q1,1)	Certainty for I1 equals the winning group's fused score. That is q1,1 = 0.49	
	Confidence in individual tasks (f1,1)	0.51	0.49
	Confidence (q2,1)	Confidence for I1 is the average of the confidence values of sensors in winning group, that is q2,1 = 0.53	
	timeliness(q3,1)	Timeliness is calculated based on the maximum acceptable time taken in obtaining the decision about I1. In this instance, timeliness is found as q3,1 = 1.0	

3.2 Result Analysis

In this section, we describes how to evaluate QoI of system, the system is based on the processing tasks that different sensory media stream performed. we sampling every 1 min, and deal with captured media of different sensors to identify these items.

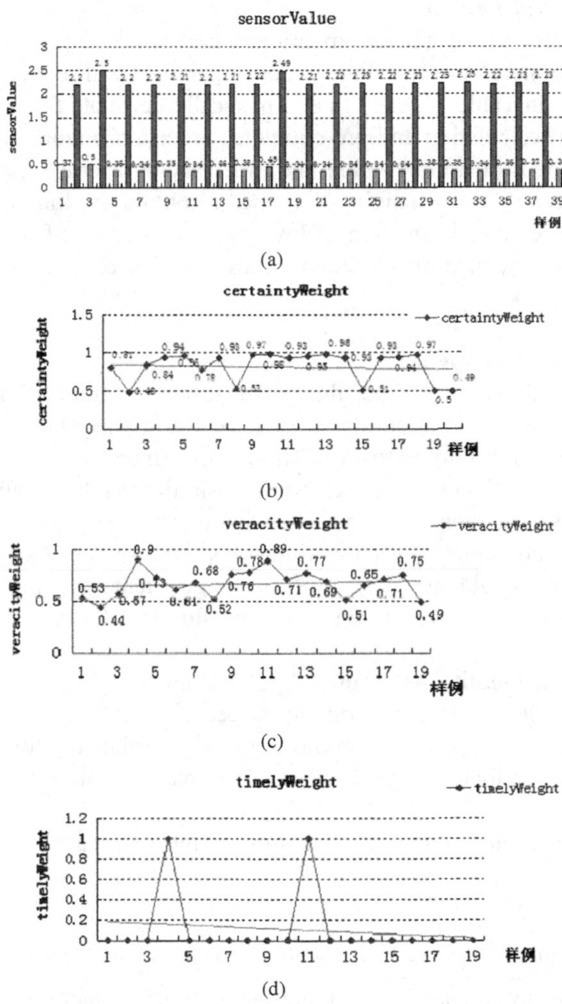

Fig. 2. quality attributes of 40 examples

3.2.1 Certainty Evaluation

Figure 2(b) shows the instantaneous certainty value of I_1 40 samples instances (blue lines) and its time-varying average (green lines). Note that in the diagram, the X axis represents the 40 samples in the test phase. When there are events occur in the test sequence, those samples was selected as a discrete event simulation, evaluating the performance of this system. When there is no event occurs, therefore, we ignore other instances. In Fig. 2(b), there are some fluctuations in the certainty value, it can be expected, because the certainty refers to the probability of the current monitoring, and it is obtained by processing two sensor flow and the fusion of subsequent observations. In contrast, average certainty value is almost no fluctuation.

3.2.2 Accuracy Assessment

Accuracy is calculated by Eq. (10).As mentioned earlier, the sensors' confidence of the winning group will increase, and failure group confidence will reduce, it was reflected in Table 1 (line 8). In Table 1, in 9th line dynamically calculate the accuracy value on the basis of current calculation and pre-computed precision. Figure 2(c) shows the all instances' instantaneous certainty value and time-varying average associated with I1. Attention, as sensors or way changing that used to obtain the final decision of some event, the confidence value is changing. However, like accuracy, from Fig. 9(a and b), observed that the confidence of multimedia sensor will also base on the current confidence value of sensors.

3.2.3 Real-Time Assessment

We observed that, about most time, the system can determine an information item within 10 s in the pre-computation phase. In some cases, when the system is takes up more than T_j value 5 s, we argue that the value is maximum acceptable jitter value δ_j. Now, using the Eq. (12) found that the real-time is similar to other quality properties. In Fig. 2(d), we can see moments of time value and mean time change about I_1. Differs from certainty and credibility, instantly time value has a huge fluctuations, this is a result of the system could not determine the target item in a timely, as a result, according to the Eq. (12),we can achieve a zero value in a flash.

3.2.4 Information Quality Attributes Aggregation

Now we introduce how to use method introduced in Sect. 3.2.2 aggregate different information quality attributes. Based on the weight of distributed information items and information quality attributes to evaluate the information item and quality attributes aggregation.

In Table 2, through an instance (I_1) data to show the aggregation methods of quality attributes.

Table 2. Aggregating a quality attribute information item

Information items	Quality attributes			Information item level quality Wq1 = 0.3, Wq2 = 0.5, Wq3 = 0.2
I_1	$q_{1,1}$	$q_{2,1}$	$q_{3,1}$	QoI_1
	0.49	0.53	1	0.63

Setting Wq1 = 0.3, wq2 = 0.5, wq3 = 0.2. QoI_1 = 0.63 using the formula (14) to aggregate.

4 Conclusion and Future Work

In order to demonstrate the applicability of the proposed method, we put some sensors in the key position in the laboratory environment, do experiment of billboard monitoring and identification. As a future work, some problems can be exploring, such as:

(1) Researching dynamic *QoI* calculation on intelligent environment, choosing and arranging different time for users service;
(2) analyzing environmental background how to affect the calculation of *QoI* in a distributed sensing environment;
(3) Researching how apply the algorithm to the actual sensor real-time work environment.

References

1. Chen, D., Yang, J., et al.: Detecting social interactions of the elderly in a nursing home environment. ACM Trans. Multimedia. Commun. **4**, 1–6 (2007)
2. Wald, L.: Some terms of reference in data fusion. IEEE Trans. Geosci. Remote Sens. **37**(3), 1190–1193 (1999)
3. Mariano, V., Min, et al.: Performance evaluation of object detection algorithms. In: Proceedings of the 16th International Conference on Pattern Recognition (ICPR), vol. 3, pp. 965–969 (2002)
4. MulledSchneiders, S., Jager, et al.: Performance evaluation of a real time video surveillance systems. In: Proceedings of the Joint IEEE International Workshop on Visual Surveillance and Performance Evaluation of Tracking and Surveilance (VS-PETS), pp. 137–143 (2005)
5. Nascimento, J.C., Marques, J.S.: Performance evaluation of object detection algorithms for video surveillance. IEEE Trans. Multimedia **8**(4), 761–774 (2006)
6. Ziliani, F., Velastin, et al.: Performance evaluation of event detection solutions: the CREDS experience. In: Proceedings of the IEEE Conference on Advanced Video and Signal-Based Surveillance, pp.201–206 (2005)
7. Klein, A., Do, et al.: Representing data quality for streaming and static data. In: Proceedings of the IEEE ICDE Workshop on Ambient Intelligence, Media, and Sensing(AIMSA), pp. 3–10 (2007)
8. Yates, D.J., Nahum, E.M., et al.: Data quality and query cost in pervasive sensing systems. Mobile Comput. **4**(6), 851–870 (2008)
9. Bisdikian, C.: On sensor sampling and quality of information: a starting point. In: Proceedings of the Workshop on Pervasive Communications, pp. 279–284 (2007)
10. Han, Q., Venkatasubramanian, N.: Timeliness-accuracy balanced collection of dynamic context data. IEEE Trans. Para. Distrib. Syst **18**(2), 158–171 (2007)

4 Conclusion and Future Work

Author Index

Printed in the United States
By Bookmasters